Cyber-Physical-Human Systems

Cyber–Physical–Human Systems

Fundamentals and Applications

Edited by

Anuradha M. Annaswamy
Massachusetts Institute of Technology
Cambridge, MA
USA

Pramod P. Khargonekar
University of California
Irvine, CA
USA

Françoise Lamnabhi-Lagarrigue
CNRS, CentraleSupelec, University of Paris-Saclay
Gif-sur-Yvette
France

Sarah K. Spurgeon
University College London
London
UK

IEEE Press Series on Technology Management, Innovation, and Leadership

IEEE**TEMS**
Technology & Engineering
Management Society

IEEE PRESS
WILEY

Published by John Wiley & Sons, Inc., Hoboken, New Jersey.
Published simultaneously in Canada.

For general information on our other products and services or for technical support, please contact our Customer Care Department within the United States at (800) 762-2974, outside the United States at (317) 572-3993 or fax (317) 572-4002.

Wiley also publishes its books in a variety of electronic formats. Some content that appears in print may not be available in electronic formats. For more information about Wiley products, visit our web site at www.wiley.com.

Library of Congress Cataloging-in-Publication Data:

Names: Annaswamy, Anuradha M., 1956- contributor. | Khargonekar, P. (Pramod), contributor. | Lamnabhi-Lagarrigue, F. (Françoise), 1953- contributor. | Spurgeon, Sarah K, contributor.
Title: Cyber–physical–human systems : fundamentals and applications / Anuradha M. Annaswamy, Massachusetts Institute of Technology, Cambridge, MA, USA, Pramod P. Khargonekar, University of California, Irvine, CA, USA, Françoise Lamnabhi-Lagarrigue, CNRS, CentraleSupelec, Gif-sur-Yvette, France, Sarah K. Spurgeon, University College, London, London, UK.
Description: First edition. | Hoboken, New Jersey : Wiley, [2023] | Includes bibliographical references and index.
Identifiers: LCCN 2022058865 (print) | LCCN 2022058866 (ebook) | ISBN 9781119857402 (hardback) | ISBN 9781119857419 (adobe pdf) | ISBN 9781119857426 (epub)
Subjects: LCSH: Human-machine systems. | Sociotechnical systems. | Cooperating objects (Computer systems)
Classification: LCC TA167 .C885 2023 (print) | LCC TA167 (ebook) | DDC 620.8/2–dc23/eng/20230111
LC record available at https://lccn.loc.gov/2022058865
LC ebook record available at https://lccn.loc.gov/2022058866

Cover Design: Wiley
Cover Image: © Phonlamai Photo/Shutterstock

Set in 9.5/12.5pt STIXTwoText by Straive, Chennai, India

Contents

A Note from the Series Editor

Welcome to the Wiley–IEEE Press Series on Technology Management, Innovation, and Leadership! The IEEE Press imprint of John Wiley & Sons is well known for its books on technical and engineering topics. This new series extends the reach of the imprint, from engineering and scientific developments to innovation and business models, policy and regulation, and ultimately to societal impact. For those who are seeking to make a positive difference for themselves, their organization, and the world, technology management, innovation, and leadership are essential skills to home.

The world today is increasingly technological in many ways. Yet, while scientific and technical breakthroughs remain important, it is connecting the dots from invention to innovation to the betterment of humanity and our ecosphere that has become increasingly critical. Whether it is climate change or water management or space exploration or global healthcare, a technological breakthrough is just the first step. Further requirements can include prototyping and validation, system or ecosystem integration, intellectual property protection, supply/value chain set-up, manufacturing capacity, regulatory and certification compliance, market studies, distribution channels, cost estimation and revenue projection, environmental sustainability assessment, and more. The time, effort, and funding required for realizing real-world impact dwarf what was expended on the invention. There are no generic answers to the big-picture questions either, the considerations vary by industry sector, technology area, geography, and other factors.

Volumes in the series will address related topics both in general – e.g. frameworks that can be applied across many industry sectors – and in the context of one or more application domains. Examples of the latter include transportation and energy, smart cities and infrastructure, and biomedicine and healthcare. The series scope also covers the role of government and policy, particularly in an international technological context.

With 30 years of corporate experience behind me and about five years now in the role of leading a Management of Technology program at a university, I see a broad-based need for this series that extends across industry, academia, government, and nongovernmental organization. We expect to produce titles that are relevant for researchers, practitioners, educators, and others.

I am honored to be leading this important and timely publication venture.

Tariq Samad
Senior Fellow and Honeywell/W.R. Sweatt Chair in Technology Management
Director of Graduate Studies, M.S. Management of Technology
Technological Leadership Institute | University of Minnesota
samad@iccc.org

About the Editors

Dr. Anuradha M. Annaswamy is founder and director of the Active-Adaptive Control Laboratory in the Department of Mechanical Engineering at MIT. Her research interests span adaptive control theory and its applications to aerospace, automotive, propulsion, energy systems, smart grids, and smart cities. She has received best paper awards (Axelby; CSM), as well as Distinguished Member and Distinguished Lecturer awards from the IEEE Control Systems Society (CSS) and a Presidential Young Investigator award from NSF. She is a Fellow of IEEE and IFAC. She is the recipient of the Distinguished Alumni award from Indian Institute of Science for 2021. She is the author of a graduate textbook on adaptive control, coeditor of two vision documents on smart grids and two editions of the Impact of Control Technology report, and a coauthor of two National Academy of Sciences, Engineering, and Medicine Committee reports related to electricity grids. She served as the President of CSS in 2020.

Pramod P. Khargonekar is vice chancellor for Research and Distinguished Professor of Electrical Engineering and Computer Science at the University of California, Irvine. He was chairman of the Department of Electrical Engineering and Computer Science at the University of Michigan, dean of the College of Engineering at the University of Florida, and was assistant director for the Directorate of Engineering at the National Science Foundation He has received numerous honors and awards including IEEE Control Systems Award, IEEE Baker Prize, IEEE Control Systems Society Bode Lecture Prize, IEEE Control Systems Axelby Award, NSF Presidential Young Investigator Award, AACC Eckman Award, and is a Fellow of IEEE, IFAC, and AIAA.

Françoise Lamnabhi-Lagarrigue, IFAC Fellow, is CNRS Emeritus Distinguished Research Fellow, CentraleSupelec, Paris-Saclay University. She obtained the Habilitation Doctorate degree in 1985. Her main recent research interests include observer design, performance, and robustness issues in control systems. She has supervised 26 PhD theses. She founded and chaired the EECI International Graduate School on Control. She is the editor-in-chief of *Annual Reviews in Control*. She is the prizewinner of the 2008 French Academy of Science Michel Monpetit prize and the 2019 Irène Joliot-Curie prize, Woman Scientist of the Year. She is knight of the Legion of Honor and officer of the National Order of Merit.

Sarah K. Spurgeon is head of the Department of Electronic and Electrical Engineering and professor of Control Engineering at UCL. She is currently vice-president (publications) for the International Federation of Automatic Control and editor-in-chief of IEEE Press. She was awarded the Honeywell International Medal for "distinguished contribution as a control and measurement technologist to developing the theory of control" in 2010 and an IEEE Millennium Medal in 2000. Within the United Kingdom, she is a fellow of the Royal Academy of Engineering (2008) and was awarded an OBE for services to engineering in 2015.

List of Contributors

Kumar Akash
Honda Research Institute USA, Inc.
San Jose
CA
USA

Anuradha M. Annaswamy
Department of Mechanical Engineering
Massachusetts Institute of Technology
Cambridge
MA
USA

Noreen Anwar
The State Key Laboratory for Management and
Control of Complex Systems
Institute of Automation
Chinese Academy of Sciences
Beijing
China

Michel Audiffren
Centre de Recherches sur la Cognition et
l'Apprentissage, UMR CNRS 7295
Université de Poitiers
Poitiers
France

Balázs Benyó
Department of Control Engineering and
Information Technology
Budapest University of Technology and
Economics
Budapest
Hungary

Bruno Berberian
Information Processing and Systems
Department
ONERA
Salon-de-Provence
France

Ming Cao
Faculty of Science and Engineering
University of Groningen
Groningen
The Netherlands

Maria Castaldo
Université Grenoble Alpes
CNRS, Inria, GrenobleINP
GIPSA-lab
Grenoble
France

J. Geoffrey Chase
Department of Mechanical Engineering
Centre for Bio-Engineering
University of Canterbury
Christchurch
New Zealand

Xiaoyu Chen
The State Key Laboratory for Management and
Control of Complex Systems
Institute of Automation
Chinese Academy of Sciences
Beijing
China

and

School of Artificial Intelligence
University of Chinese Academy of Sciences
Beijing
China

Yeong S. Chiew
Department of Mechanical Engineering
School of Engineering
Monash University Malaysia
Selangor
Malaysia

Murat Cubuktepe
Dematic Corp.
Austin
TX
USA

Thomas Desaive
GIGA In Silico Medicine
Liege University
Liege
Belgium

Aleksandr V. Efremov
Department of Aeronautical Engineering
Moscow Aviation Institute
National Research University
Moscow
Russian Federation

Emre Eraslan
Department of Mechanical Science &
Engineering
University of Illinois at Urbana-Champaign
Urbana-Champaign
IL
USA

Eduard Fosch-Villaronga
eLaw Center for Law and Digital Technologies
Leiden University
Leiden
The Netherlands

Paolo Frasca
University GrenobleAlpes
CNRS, Inria, GrenobleINP
GIPSA-lab
Grenoble
France

Masayuki Fujita
Department of Information Physics and
Computing
The University of Tokyo
Tokyo
Japan

Sebin Gracy
Department of Electrical and Computer
Engineering
Rice University
Houston
TX
USA

Hiroyuki Handa
Tsukuba Research Laboratory
YASKAWA Electric Corporation
Tsukuba
Ibaraki
Japan

Takeshi Hatanaka
Department of Systems and Control
Engineering
School of Engineering
Tokyo Institute of Technology
Tokyo
Japan

Jacob Hunter
School of Mechanical Engineering
Purdue University
West Lafayette
IN
USA

Neera Jain
School of Mechanical Engineering
Purdue University
West Lafayette
IN
USA

Nils Jansen
Department of Software Science
Institute for Computing and Information
Science
Radboud University Nijmegen
Nijmegen
The Netherlands

Qing-Shan Jia
Department of Automation
Center for Intelligent and Networked Systems
(CFINS)
Beijing National Research Center for
Information Science and Technology (BNRist)
Tsinghua University
Beijing
China

Frank J. Jiang
Division of Decision and Control Systems,
Department of Intelligent Systems, EECS
KTH Royal Institute of Technology
Stockholm
Sweden

Victor Díaz Benito Jiménez
University of Alcalá
University Campus - Calle 19
Madrid
Spain

Karl H. Johansson
Division of Decision and Control Systems,
Department of Intelligent Systems, EECS
KTH Royal Institute of Technology
Stockholm
Sweden

Pramod P. Khargonekar
Department of Electrical Engineering and
Computer Science
University of California
Irvine
CA
USA

Jennifer L. Knopp
Department of Mechanical Engineering
Centre for Bio-Engineering
University of Canterbury
Christchurch
New Zealand

Bernard Lambermont
Department of Intensive Care
CHU de Liege
Liege
Belgium

Françoise Lamnabhi-Lagarrigue
CNRS
CentraleSupelec
University of Paris-Saclay
Gif-sur-Yvette
France

Xiaoshuang Li
The State Key Laboratory for Management and
Control of Complex Systems
Institute of Automation
Chinese Academy of Sciences
Beijing
China

and

The School of Artificial Intelligence
University of Chinese Academy of Sciences
Beijing
China

Ryan W. Liu
Department of Navigation Engineering
School of Navigation, School of Computer
Science and Artificial Intelligence
Wuhan University of Technology
Wuhan
China

Teng Long
Department of Automation
Center for Intelligent and Networked Systems
(CFINS)
Beijing National Research Center for
Information Science and Technology (BNRist)
Tsinghua University
Beijing
China

Yisheng Lv
The State Key Laboratory for Management and
Control of Complex Systems
Institute of Automation
Chinese Academy of Sciences
Beijing
China

Sanna Malinen
Deparament of Management, Marketing, and
Entrepreneurship
University of Canterbury
Christchurch
New Zealand

Jonas Mårtensson
Division of Decision and Control Systems,
Department of Intelligent Systems, EECS
KTH Royal Institute of Technology
Stockholm
Sweden

Knut Moeller
Department of Biomedical Engineering
Institute of Technical Medicine
Furtwangen University
Villingen-Schwenningen
Germany

Juan C. Moreno
Neural Rehabilitation Group
Translational Neuroscience Department
Cajal Institute
Spanish National Research Council
Madrid
Spain

Vineet Jagadeesan Nair
Department of Mechanical Engineering
Massachusetts Institute of Technology
Cambridge
MA
USA

Katharina Naswall
School of Psychology, Speech and Hearing
University of Canterbury
Christchurch
New Zealand

Philip E. Paré
Elmore Family School of Electrical and
Computer Engineering
Purdue University
West Lafayette
IN
USA

S. M. Mizanoor Rahman
Department of Mechanical Engineering
Pennsylvania State University
Dunmore
PA
USA

Tahira Reid
Mechanical Engineering and Engineering
Design
The Pennsylvania State University
State College
PA
USA

Behzad Sadrfaridpour
Department of Mechanical Engineering
Clemson University
Clemson
SC
USA

Mike Salomone
Laboratoire de Psychologie et NeuroCognition
Univ. Grenoble Alpes
Univ. Savoie Mont Blanc, CNRS
Grenoble
France

Tariq Samad
Technological Leadership Institute
University of Minnesota
Minneapolis
MN
USA

Henrik Sandberg
Division of Decision and Control Systems,
Department of Intelligent Systems, EECS
KTH Royal Institute of Technology
Stockholm
Sweden

Thomas Schauer
Department of Electrical Engineering and
Computer Science
Control Systems Group
Technische Universität Berlin
Berlin
Germany

Geoffrey M. Shaw
Department of Intensive Care
Christchurch Hospital
Christchurch
New Zealand

Baike She
Department of Mechanical and Aerospace
Engineering
University of Florida
Gainesville
FL
USA

Sarah K. Spurgeon
Department of Electronic and Electrical
Engineering
University College London
London
UK

Shreyas Sundaram
Elmore Family School of Electrical and
Computer Engineering
Purdue University
West Lafayette
IN
USA

Ufuk Topcu
Department of Aerospace Engineering and
Engineering Mechanics
Oden Institute for Computational Engineering
and Science
The University of Texas at Austin
Austin
TX
USA

Frédéric Vanderhaegen
Department on Automatic Control and
Human-Machine Systems
Université Polytechnique Hauts-de-France,
LAMIH lab
UMR CNRS 8201
Valenciennes
France

and

INSA Hauts-de-France
Valenciennes
France

Tommaso Venturini
Medialab
Université de Genève
Geneva
Switzerland

and

Centre Internet et Societé
CNRS
Paris
France

Ian D. Walker
Department of Electrical and Computer
Engineering
Clemson University
Clemson
SC
USA

Fei-Yue Wang
The State Key Laboratory for Management and
Control of Complex Systems
Institute of Automation
Chinese Academy of Sciences
Beijing
China

Xiao Wang
The State Key Laboratory for Management and
Control of Complex Systems
Institute of Automation
Chinese Academy of Sciences
Beijing
China

Yue Wang
Department of Mechanical Engineering
Clemson University
Clemson
SC
USA

Jennifer H. K. Wong
School of Psychology, Speech and Hearing
University of Canterbury
Christchurch
New Zealand

Gang Xiong
The Beijing Engineering Research Center of
Intelligent Systems and Technology
Institute of Automation
Chinese Academy of Sciences
Beijing
China

and

The Guangdong Engineering Research Center
of 3D Printing and Intelligent Manufacturing
The Cloud Computing Center
Chinese Academy of Sciences
Beijing
China

Junya Yamauchi
Graduate School of Information Science and
Technology
Department of Information Physics and
Computing
The University of Tokyo
Tokyo
Japan

Jing Yang
The State Key Laboratory for Management and
Control of Complex Systems
Institute of Automation
Chinese Academy of Sciences
Beijing
China

and

The School of Artificial Intelligence
University of Chinese Academy of Sciences
Beijing
China

Peijun Ye
The State Key Laboratory for Management and
Control of Complex Systems
Institute of Automation
Chinese Academy of Sciences
Beijing
China

Yildiray Yildiz
Department of Mechanical Engineering
Bilkent University
Ankara
Turkey

Madeleine Yuh
School of Mechanical Engineering
Purdue University
West Lafayette
IN
USA

Hongxin Zhang
The State Key Laboratory of CAD & CG
Zhejiang University
Hangzhou
China

Hongxia Zhao
The State Key Laboratory for Management and
Control of Complex Systems
Institute of Automation
Chinese Academy of Sciences
Beijing
China

Cong Zhou
Department of Mechanical Engineering
Centre for Bio-Engineering
University of Canterbury
Christchurch
New Zealand

and

School of Civil Aviation
Northwestern Polytechnical University
Taicang
China

Xu Zhou
Computer Network Information Center
Chinese Academy of Sciences
Beijing
China

Fenghua Zhu
The Beijing Engineering Research Center of
Intelligent Systems and Technology
Institute of Automation
Chinese Academy of Sciences
Beijing
China

and

The Guangdong Engineering Research Center
of 3D Printing and Intelligent Manufacturing
The Cloud Computing Center
Chinese Academy of Sciences
Beijing
China

Lorenzo Zino
Faculty of Science and Engineering
University of Groningen
Groningen
The Netherlands

Introduction

Anuradha M. Annaswamy[1], Pramod P. Khargonekar[2], Françoise Lamnabhi-Lagarrigue[3], and Sarah K. Spurgeon[4]

[1]*Department of Mechanical Engineering, Massachusetts Institute of Technology, Cambridge, MA, USA*
[2]*Department of Electrical Engineering and Computer Science, University of California, Irvine, CA, USA*
[3]*CNRS, CentraleSupelec, University of Paris-Saclay, Gif-sur-Yvette, France*
[4]*Department of Electronic and Electrical Engineering, University College London, London, UK*

Cyber–Physical–Human Systems (CPHS) are defined as interconnected systems that include physical systems, computing and communication systems, and humans, and allow these entities to communicate with each other and make decisions across space and time (Sowe et al. 2016). The concept is a natural extension of the notion of Cyber–Physical Systems (CPS) which, for more than 20 years, has successfully enabled multidisciplinary research involving control systems, communications, networking, sensing, and computing to develop new theoretical foundations and addressed several major technological applications. The more recent area of CPHS focuses on the increasing interactions between CPS and humans. CPHS are no different than other engineering systems in terms of functionality: they are collections of various interacting components put together to achieve a specified objective. What distinguishes these systems is their central ingredient: the human. These interactions occur at various levels, and are leading to novel hybrid intelligent systems in almost all sectors in society. They range from physiological signal interaction, to individual cognitive and behavioral interactions with engineered systems in various sectors, to social networks with interactions at various scales between individuals and populations. These systems are being designed to address major technological applications that contribute to human welfare in a wide range of domains, including transportation, aerospace, health and medicine, robotics, manufacturing, energy, and the environment. This perspective is profoundly different from the conventional understanding where humans are treated as isolated elements who operate or benefit from the system. Humans are no longer passive consumers or actors. They are empowered decision-makers and drive the evolution of the technology. For example, an individual might interact with a robot at home or an autonomous vehicle on the road; a manufacturing firm may integrate robots and intelligent operators to maximize its productivity; and a city or state may leverage smart grid technologies and renewable energy together with control room operators to reach its clean energy goals.

In many of these domains, the interaction in a CPHS between the human and the system requires an in-depth understanding for the development of novel control technologies. More precisely, in order to succeed in delivering what is required for these applications – i.e. in designing these individual smart systems and coordinating them in a stable, optimal, and economically efficient fashion – CPHS necessitates new concepts, methods, and tools. New problems result from emerging interactions between cyber–physical systems and humans. Categories that have been identified include: (i) human–machine symbiosis (e.g. smart prosthetics, exoskeletons); (ii) humans as supervisors/operators of complex engineering systems (e.g. aircraft pilots, car drivers, process plant operators, robotic surgery operators); (iii) humans as control agents in multiagent systems (e.g. road automation, traffic management, electric grid); (iv) humans as elements in controlled systems (e.g. home comfort control, home security systems); (v) humans working in parallel with digital entities (e.g. digital twins) in both physical and virtual spaces, i.e. parallel intelligence. CPHS also raise a variety of specific technical challenges, including modeling human behavior across a range of levels and control architectures, determining the cognitive science principles needed for the design of autonomous or semiautonomous cyber–physical systems, and identifying key factors that enable cyber–physical systems to augment human performance across a range of interactions. This understanding will enable the best design of the overall system and will achieve useful outcomes for individuals, organizations, and society.

The emerging area of CPHS started around 10 years ago. In particular, a series of successful CPHS workshops (H-CPS-I 2014, CPHS 2016, CPHS 2018, CPHS 2020, CPHS 2022), technically co-sponsored by IEEE CSS and IFAC, have been organized biennially since 2014, stimulated by the CPHS Steering Committee (*). Our vision for this first CPHS edited book is grounded in the idea that CPHS is at an embryonic stage as a possible new discipline which is emerging at the intersection between engineering and social-behavioral sciences.

Seven chapters describe **CPHS concepts, methods, tools, and techniques** by focusing on paradigms that explore the integration of engineering and social-behavioral sciences. Each chapter develops some future research challenges and provides a vision that will be very useful in particular for researchers relatively new to the field. More precisely, in Chapter 1, *Human-in-the-Loop Control and Cyber–Physical–Human Systems: Applications and Categorization*, four different architectural patterns are differentiated and discussed: human-in-the-plant, human-in-the-controller, human–machine control symbiosis, and humans-in-control-loops; in Chapter 2, *Human Behavioral Models Using Utility Theory and Prospect Theory*, models of human behavior for characterizing human decisions in the presence of stochastic uncertainties and risks are outlined. In Chapter 3, *Social Diffusion Dynamics in Cyber–Physical–Human Systems*, a model of multilayer complex networks that captures social diffusion dynamics is presented and possible control actions to accelerate or decelerate the diffusion processes are introduced. Chapter 4, *Opportunities and Threats of Interactions Between Humans and Cyber–Physical Systems – Integration and Inclusion Approaches for CPHS*, develops a new concept on the inclusion of human systems based on that of the management of dissonance, when discrepancies occur between groups of people or between human and autonomous systems. A new framework where closed-loop interactions between humans and autonomous systems based on calibration of human cognitive states is also introduced in Chapter 5, *Enabling Human-Aware Autonomy through Cognitive Modeling and Feedback Control*. From the basis of a new model of the interaction between the human and the robot, an accurate characterization of human behavior is obtained in Chapter 6, *Shared Control with Human Trust and Workload Models*; in Chapter 7, *Parallel Intelligence for CPHS: An ACP (Artificial societies, Computational experiments, Parallel execution) Approach*, is devoted to the presentation of an approach to Parallel Intelligence in order to ensure lifelong developmental AI and ongoing learning through smart infrastructures constructed by CPHS.

In the following parts, key goals and drivers for different important application contexts will be described with five chapters on Transportation (Part 2), three chapters on Robotics (Part 3), three chapters on Healthcare (Part 4), and two chapters on Sociotechnical systems (Part 5). For each of these chapters, the specific CPHS concepts, methods, tools, and techniques that are key to advancing from a technical perspective are delineated, and recent research advances are articulated alongside future research challenges. Visionary perspectives for these application areas are also highlighted.

Part 2 on **Transportation** starts with Chapter 8, *Regularities of Human Operator Behavior and its Modeling*, which introduces a highly augmented control system with a new generation of interfaces (screens and inceptors) in order to ensure the level of safety required for the use of the vehicle. In Chapter 9, *Safe Shared Control Between Pilots and Autopilots in the Face of Anomalies*, a taxonomy in flight control gathering various types of interactions between human operators and autonomy are laid out, and different types of responsibility sharing between human operators and autonomy are proposed. Chapter 10, *Safe Teleoperation of Connected and Automated Vehicles*, surveys the current teleoperation systems that allow remote human operators to effectively supervise a connected vehicle and ensure that the vehicle and its environment remain safe, despite possible issues in the wireless network or human error. Chapter 11, *Charging Behavior of Electric Vehicles*, reviews the existing paradigm on the charging behavior of electric vehicles as a demand that is necessary to satisfy. It then focuses on the potential opportunity to use electric vehicles as mobile storage and concludes with a study of the large-scale coordination problem for the charging behaviors of Electric Vehicles.

The next part is dedicated to **Robotics**. Chapter 12, *Trust-Triggered Robot-Human Handovers Using Kinematic Redundancy for Collaborative Assembly in Flexible Manufacturing*, proposes computational models of robot trust in humans for human–robot collaborative assembly tasks and develops real-time measurement methods of trust. In Chapter 13, *Fusing Neuro-Prostheses and Wearable Robots with Humans to Restore and Enhance Mobility*, the basics of neuro-prostheses, neuro-modulation, and wearable robotics are introduced, and recent technological developments with application examples as well as open challenges will be highlighted. The third chapter of this part, Chapter 14, *Contemporary Issues and Advances in Human-Robot Collaborations*, gives a comprehensive overview of contemporary issues and recent advances in human–robot collaboration.

Another important application domain of interest in the CPHS framework, **Healthcare,** is developed in Part 4. The first contribution, Chapter 15, *Overview and Perspectives on the Assessment and Mitigation of Cognitive Fatigue in Operational Settings*, addresses cognitive fatigue from a design point of view, in particular by introducing the way in which the system could consider the state of operator fatigue in order to adapt and improve cooperation. On a different topic, Chapter 16, *Epidemics Spread Over Networks: Influence of Infrastructure and Opinions*, provides background on promising modeling, analysis, and applications of networked epidemic spreading models. The last chapter of this part, Chapter 17, *Digital Twins and Automation of Care in the Intensive Care Unit*, covers firstly the growing development of digital twins for medicine to improve care and productivity to meet growing demand, and second, explores the integrated role played by social sciences and other human factors in the development, translation, and adoption of urgently needed innovations in intensive care.

Part 5 is devoted to some recent developments around **Socio-technical Systems**. In Chapter 18, *Online Attention Dynamics in Social Media*, insights to dynamical evolution of online social networks, their reactions to external stimuli, their potential pathologies and degenerations, and their sensitivity to manipulations are presented together with a focus on attention dynamics in social media, from the perspective of encouraging contributions from the control systems community. Chapter 19, *Cyber-Physical-Social Systems for Smart City*, discusses how CPHS have the potential to adaptively optimize operation toward intelligent transportation systems with continuous real-time monitoring of its state, environment, and related behaviors, while providing real-time recommendations.

The last Chapter 20, *Conclusion and Perspectives*, emphasizes in particular the open directions provided by this new discipline. The exciting opportunities for the control community are emphasized. The study of CPHS will have profound, wide-ranging, and irreversible implications for the future of humanity and the biosphere in general. While these technologies have the potential to bring about benefits to humankind, they also give rise to profound ethical and social concerns. As such, the design, development, and application of these technologies require careful reflection, debate, and deliberation. This important facet of CPHS is also discussed in this last part.

Reference

Sowe, S.K., Simmon, E., Zettsu, K. et al. (2016). *Cyber-physical-human systems: putting people in the loop. IT Professional* 18 (1): 10–13.

(*) CPHS Steering Committee

Aaron D. Ames <ames@caltech.edu>,
Saurabh Amin <amins@mit.edu>,
Anuradha M. Annaswamy<aanna@mit.edu>,
John Baras <baras@isr.umd.edu>,
Masayuki Fujita <masayuki_fujita@ipc.i.u-tokyo.ac.jp>,
Takanori Ida <ida@econ.kyoto-u.ac.jp>,
Karl H. Johansson <kallej@kth.se>,
Pramod P. Khargonekar <pramod.khargonekar@uci.edu>,
(Chair) Francoise Lamnabhi-Lagarrigue <francoise.lamnabhi-lagarrigue@centralesupelec.fr>,
Mariana Netto <mariana.netto@univ-eiffel.fr>,
Tariq Samad <tsamad@umn.edu>,
Sarah K. Spurgeon <s.spurgeon@ucl.ac.uk>,
Dawn Tilbury <tilbury@umich.edu>.

Part I

Fundamental Concepts and Methods

1

Human-in-the-Loop Control and Cyber–Physical–Human Systems: Applications and Categorization

Tariq Samad

Technological Leadership Institute, University of Minnesota, Minneapolis, MN, USA

1.1   Introduction

Designers, developers, managers, and researchers in engineering tend to focus exclusively on the technological aspects of their products, whether tangible or intellectual. In the case of automation and control systems, topics of modeling, optimization, estimation, feedback algorithm design, sensor fusion, and numerous others are pursued. The goal, depending on the nature of the application, may be to design a sophisticated advanced device or service, or, in the case of research, to uncover some mathematical truth.

What's missing in the picture, often literally, is the human. Sensors, actuators, processing platforms, data stores, and communication networks are widely depicted. Human interfaces are often shown too, but the humans themselves rarely are.

Yet, people are an inherent part of many engineered systems, especially in the realm of automation and control. Control systems keep people warm in winter and cool in summer, in homes and buildings; we drive cars to get from points A to B; we rely on control systems to sustain our lives and to improve our health if we're ill, and these systems can be implanted within us too; power grid operators help keep the lights on and industry running. Speaking of industry, the performance and safety of process plants and factories is in the hands of personnel, who rely extensively on technology. With digital transformation a pervasive trend, the boundaries of engineered systems are being expanded too – encompassing sectors such as healthcare and infrastructure.

Cyber–physical–human systems (CPHS) is a relatively recent term that has emerged to encompass human connections with cyberphysical systems (CPS). CPHS is a term of recent coinage; the first use of it appears to have been about a decade ago (Smirnov et al. 2013). A closely related term – it might even be considered synonymous with CPHS – is human-in-the-loop systems. In the control context, the human-in-the-loop has been explicitly discussed since at least the 1950s – Fontaine (1954) refers to prior work that "developed a transfer function for the human in the closed-loop system" (p. 35).

As the examples above (and many others can be cited) suggest, people are involved in automation and control systems in multifarious ways. But is there a better way to characterize the CPHS or human-in-the-loop space than by a long list of examples? What are the categories that can capture the roles and functions of people in control systems? My intent in this chapter is to address these and associated questions.

Cyber–Physical–Human Systems: Fundamentals and Applications, First Edition.
Edited by Anuradha M. Annaswamy, Pramod P. Khargonekar, Françoise Lamnabhi-Lagarrigue, and Sarah K. Spurgeon.
© 2023 The Institute of Electrical and Electronics Engineers, Inc. Published 2023 by John Wiley & Sons, Inc.

Before we dive into details, some limitations of scope should be noted:

- People are involved in numerous ways in control systems. The focus here is on operational roles, as distinct from, say, design and commissioning.
- Control is pervasive not only in the human-engineered world but also in nature and biology as well, and human-in-the-loop examples can be adduced in this realm too…but I will limit my examples to engineered systems, albeit broadly writ.
- The history of control engineering goes back millennia, to the water clocks of antiquity; these applications correspond to the "prehistory of automatic control" (Lewis, 1992). We do not discuss them here.

In the next section I analyze the cyber–physical–human terminology, teasing apart subsets of the tripartite term. Four categories of human-in-the-loop control systems are then introduced and discussed, with extensive examples: human-in-the-plant, human-in-the-controller, human–machine control symbiosis, and humans-in-multiagent-loops. A timeline that indicates milestones in the advancement of human-in-the-loop-control applications for the four categories is presented. Deeper discussion of two future application scenarios, related to (semi-)autonomous driving in urban environments and smart grids for mitigating climate change, are also included. Broader concerns and considerations are then discussed, related to other ways of classifying CPHS, modeling human decision-making, and the increasing need for ethical awareness.

An earlier version of this chapter appeared as (Samad, 2020).

1.2 Cyber + Physical + Human

The trigger event for the increasing interest in CPHS was a workshop in 2014 in Paris, titled "Human Cyber-Physical-System Interaction: Control for Human Welfare." This workshop was led by M. Netto, S. Spurgeon, and F. Lamnabhi-Lagarrigue and initiated a sequence of biannual CPHS workshops – the "1st IFAC Conference on Cyber-Physical & Human Systems" in Florianopolis, Brazil, in 2016, led by M. Netto and S. Spurgeon; the "2nd IFAC Conference on Cyber-Physical and Human Systems" in Miami, 2018, led by A. Annaswamy and D. Tilbury; and the "3rd IFAC Workshop on Cyber-Physical & Human Systems" in Beijing (and virtually) in 2020, chaired by F.-Y. Wang and T. Samad. Most recently, CPHS 2022 was held in Houston, USA, in December 2022, chaired by M. Oishi. Proceedings of the past IFAC-sponsored CPHS events are available (Becker 2016; Mettler 2019; Namerikawa 2020; Vinod 2022).

As the term suggests, CPHS are systems that integrate three elements: the cyber, the physical, and the human. Figure 1.1 shows a Venn diagram that separately and conjointly covers the different domains involved in the term. It is instructive to note the various spaces in the diagram, beginning with the singleton disciplines:

- **Cyber**: The word today refers to digital computation and its attendant elements and can perhaps be considered synonymous with information and communication technologies (ICT). However, its origin can be traced to the rise of cybernetics (Wiener 1948), predating the digital era, as pointed out by Coe (2015): "Before there was cyber-anything, there was the field of cybernetics. Pioneered in the late 1940s by a group of specialists in fields ranging from biology to engineering to social sciences, cybernetics was concerned with the study of communication and control systems in living beings and machines. The interest in how systems work is reflected in the

Figure 1.1 Cyber (C), Physical (P), and Human (H) spaces comprising cyberphysical systems.

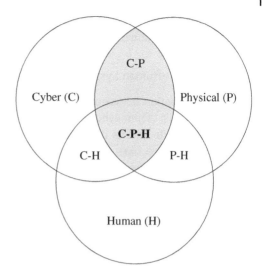

etymology of cybernetic, which comes from the Greek word kubernētēs (κυβερνᾶν), 'steersman', from kubernan 'to steer.'"

Thus, signaling and control via analog mechanisms, particularly analog electronics, are very much part of "cyber." Whether pneumatics, hydraulics, and other nonelectronic mechanisms for signaling and simple computation can be considered "cyber" can be debated, but automation systems relying solely on these would certainly be excluded from what today are called CPHS.

- **Physical**: By this term we mean "real," i.e. physical, systems and phenomena, as distinct from the digital realm, involving disciplines, with their attendant laws, of physics, chemistry, biology, and other fields of science. Not all science falls under this term, especially in the CPHS context. The social and economic sciences are excluded and arguably computer sciences as well – the control of a purely digital system, not meaningfully affecting the physical world (with apologies for the circular argument), would generally not be seen as CPHS.

- **Human**: Various human properties and functions are relevant in our context. These include sensing and perception, actuation, ergonomics, and cognition, all of which are involved when humans interact with automation and control systems. Social and team interactions will also be increasingly implicated as visions of CPHS progress.

Each pairwise combination of these disciplines is worth discussing briefly as well. I note that, as with all post-hoc categorizations, the dividing lines are not clean and gray areas are unavoidable.

1.2.1 Cyberphysical Systems

CPSs, as distinct from CPHSs, involve an integration of computation and communication with physical systems. Today's technologically intensive world is replete with examples, such as antilock braking or fuel injection control in automotive vehicles and control systems in appliances such as washing machines and refrigerators.

Even though CPSs are often deemed synonymous with "embedded" systems – integrations of sensing, computation, and actuation that are automatic and do not involve people for their operation – if we trace the operation and impact of a CPS we will, almost inevitably, see connections with humans. Consider the antilock braking system (ABS). It is the driver that initiates it, the effect on the driver is noticeable, and there is evidence that the braking behavior of drivers has changed as a result of ABS adoption (Sagberg et al. 1997).

1.2.2 Physical–Human Systems

Not every interaction between people and artifacts is mediated by electronic computation and control, but the list of non-cyber physical–human systems is steadily getting shorter. Many decades ago, an automobile and its driver would have been considered a physical–human system, but today's cars can have 100+ microprocessors, responsible for functions as varied as fuel injection, safety systems, cabin environment, and infotainment.

Nondigital control systems with human interaction fall into this region as well. The 4–20 mA standard is still widely used – e.g. for thermostat/furnace signaling in homes and for process control applications. Pneumatic and hydraulic systems are also extant, including in process plants, transportation, construction, and machine tools. These nondigital technologies are not only for communication. Calculations, including those involved in proportional-integral-derivative (PID) control, can be done. Indeed, the first PID controller predates the computer age (Astrom and Kumar 2014).

The first controllable prosthetics predate engineered computation too, with reports of prosthetic arms that enabled the wearer to perform dexterous actions going back to the sixteenth century (Romm 1989).

1.2.3 Cyber–Human Systems

Most of the control system examples we will be considering in this chapter relate to plants that are physical systems. Thus, sensors and actuators are physical devices, for example to measure temperatures and pressures and to effect flows of materials and movements of components. But control systems can also operate entirely in the cyber realm, with human interfaces. A close-at-hand example as I write is the operating system of a computer, which allocates computing and memory resources to different processes based on measurements of usage and other factors. Physical considerations can intrude, however, as in when the heating of components results in cyber (reducing CPU activity) or physical (turning on a fan inside the computer) actions being taken.

Other examples of (nonphysical) cyber–human systems include financial trading platforms and online computer games. In the latter case, the online game may include a simulation of a putatively real world, but the distinction between simulation and reality is all-important.

1.3 Categorizing Human-in-the-Loop Control Systems

I will use the prototypical closed-loop control architecture of Figure 1.2 as a frame for human-in-the-loop categorization. This diagram elides any number of complexities and complications. But it is familiar to students, researchers, and practitioners in control and thus helpful for the purposes of the categorization below.

The figure distinguishes between the "Controller" and the "Plant," the two major elements of the control loop. The latter is the system under control – e.g. a home or building, a car or an airplane,

Figure 1.2 Simple control system architecture.

a robot or even a part of a human or animal body. Measurements from the plant are taken using sensors and communicated to the "Controller," which affects the behavior of the plant by sending commands to actuators, thereby closing the loop.

The controller determines what changes should be made to the plant so as to ensure it functions in desirable ways. These changes could be to increase or decrease input flows, move equipment and subsystems, provide less or more energy, stop or start pumps or motors, and effect other actions, all with objectives such as ensuring safety and reliability, maximizing response time, and reducing operational costs. The changes to be made are calculated by algorithms and, in advanced control, these algorithms rely crucially on models of the plant that are incorporated within the controller. The incorporation can be explicit or implicit. Model-predictive control is an example of the former, in that the controller, for its real-time determination of the next signal to actuators, relies on a mathematical model of the plant. But any controller, if it is capable of any level of performance, incorporates some information about the plant. A PID controller, although not conventionally considered model-based (indeed, usually considered as the representative of nonmodel-based control), still incorporates some knowledge about the plant. For tuning a PID controller – identifying appropriate values of its parameters – at a minimum, the sign of the gain and some rough approximation of the time constant of the plant must be known.

There is much more to a control application than shown in Figure 1.2. Information about control objectives, such as the "set-point," is another input to the Controller. Sensors and actuators, the interfacial elements between the Controller and Plant, are also elided in the figure. Other algorithms, such as for estimation, are often included. The environment within which the closed-loop system sits is an important consideration as well, and becomes increasingly critical for larger-scale and more autonomous control systems. In control systems, environmental influences are usually considered as disturbances, and illustrated as such, that the controller has to reject.

With this as background, we now discuss architectural templates for human-in-the-loop control. Four categories are depicted in Figure 1.3 and elaborated on below. (With a topic as

Figure 1.3 Categories of human-in-the-loop control.

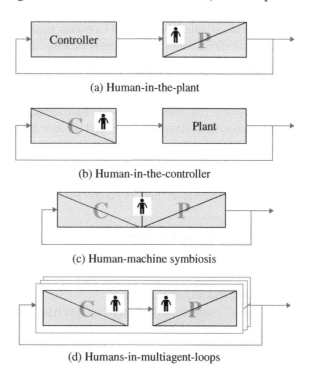

(a) Human-in-the-plant

(b) Human-in-the-controller

(c) Human-machine symbiosis

(d) Humans-in-multiagent-loops

cross-disciplinary as CPHS, there's more than one way to effect a categorization. For an alternative view, see Hirche et al. (in press).)

1.3.1 Human-in-the-Plant

This category represents cases where a person is part of the "plant" and affected by controller actions. For example, the occupant in an office building is a human in a plant. Modern building management systems regulate temperature, humidity levels, lighting, and possibly other parameters with complex, distributed network architectures (Braun 2015). Rate structures from utilities may be incorporated as well.

Most occupants in commercial buildings cannot readily control how comfortable their spaces are, and these can be too warm or cool for their preferences. Occupant comfort is one of many factors that a building management system takes into account. The control system for a building will also seek to minimize energy cost and wear and tear on equipment. In specialized facilities, other criteria apply as well – pollutant levels are tightly controlled in clean rooms and laboratories.

The comfort of the human-in-the-plant is also important in aircraft cabins. The first fully pressurized aircraft to enter commercial service was the Boeing 307, launched by TWA in 1940 (Anon. 2013). Most recently, cabin control has become a topic of increased attention in the Covid-19 pandemic. As discussed in Moody (n.d.), state-of-the-art environmental control systems for aircraft cabins control temperature, humidity, and air quality. Most airflows are from the ceiling to the floor, minimizing lateral airflow among the passengers. Outside air is mixed with HEPA-filtered inside air in about a 50–50 ratio, with air exchanged 6–20 times per hour. Humidity is added, especially in parts of the cabin with lower passenger density (premium class) – humidity is naturally at a higher level where more passengers are present. Volatile organic compounds (VOCs) and ozone are catalytically removed.

The examples above are primarily related to the comfort of the human in the plant, but control systems also ensure the safety and well-being of people. Sometimes safety and comfort go hand-in-hand, as when cabin control systems regulate cabin pressure. As a prospective application, surgical robots (of the autonomous variety) can be envisioned. The importance of modeling the plant for advanced control has been noted above and is equally relevant when the plant includes or is a human. Digital twin technologies are being explored toward this end for healthcare (see the chapter in this volume titled "Digital Twins and Automation of Care in the Intensive Care Unit" by J.G. Chase et al.). Less futuristic systems have already been prototyped, such as for the closed-loop control of anesthesia (Dumont 2012) and autocompression clothing to treat autism, hypotension, and other afflictions (Duvall et al. 2016).

This category also covers applications where the system under control is "part" of – including implanted within – a person. Today's pacemakers and insulin pumps integrate sensors, control algorithms, and actuators that dose medication or transmit electrical signals.

1.3.2 Human-in-the-Controller

In this category, the human is the controlling entity or part of it. Examples abound. To note a few: operators of building management systems and manufacturing plants, drivers of cars and trucks, aircraft pilots and drone operators.

The "cyber" element in these applications typically relates to the automation technology that assists humans, or collaborates with them, in control tasks. In modern complex control systems, human operators are not performing the control function manually; a host of sophisticated technologies is involved. Technological advances have enabled human controllers to enhance their awareness of the state of the plant, to predict outcomes of potential actions they may take, to communicate with other people with a stake in the operation of the closed-loop system, to simplify or facilitate the actions they implement, to reduce their physical and cognitive workload, and much else besides.

Early work in human-in-the-loop control targeted aerospace and defense applications: "Specific knowledge of the characteristics of the human while in a closed-loop control situation would be of considerable value in the design of [machines such as airplanes, certain radars, and gun fire-control systems]" (Fontaine 1954, p. 1). This domain has continued to influence the field, with control-theoretic models of pilots being developed in detail (Hess 2021).

Commercial aviation furnishes an illuminating example of how automation has changed the nature of control tasks that operators perform. In the earliest days of heavier-than-air powered aircraft (beginning in 1903), automation was nonexistent: the control surfaces of the Wright Flyer were manually manipulated by the Wright brothers. The first automated functionality was the autopilot, demonstrated in 1914 (Scheck 2004). These autopilots relied on the new Sperry gyroscope, another application of which was a fully autonomous flying bomb, but the "Sperry Gyroscope fared better with a control system that worked in conjunction with a human operator" (Mindell 2002, p. 76). Autopilots debuted in commercial aviation in 1933, with the Boeing 247 (Boeing, n.d.). The next step in the evolution was the "handling qualities controller," deployed in the Boeing 707 and the McDonnell-Douglas DC-8. Today, flight management systems allow aircraft pilots to input a flight route by specifying a series of space–time waypoints; under normal conditions, the airplane will then fly itself (Krisch 2014). Figure 1.4 illustrates this architecture; the successive outer loops mirror its historical evolution.

Catastrophic consequences can result if the interaction between the human and automated elements of a controller go awry, and several accidents in aviation and industrial plants bear witness to this observation. These include the Chernobyl nuclear plant explosion (Stein 2003); the crash of American Airlines Flight 965 in Cali, Colombia, in 1995 (Rodrigo et al. 1996); and the crashes of two Boeing 737 MAX aircraft in Asia and Africa in 2018 and 2019 (KNKT 2019; Aircraft Accident Investigation Bureau 2019). It was the combination of the operators and the automation that resulted in these accidents, not either factor by itself.

For example, AA 965 was attempting a direct route to the runway, instead of a longer instrument landing system (ILS) precision approach, in good weather, relying on the FMS for guiding the aircraft to the destination. The pilots deployed the airbrakes to slow the aircraft and expedite the descent. Because of an ambiguity in the FMS waypoint labels, a waypoint near Bogota was mistakenly selected instead of the waypoint in the approach path. This resulted in the autopilot changing the course of the plane automatically toward Bogota, involving an attempted wide U-turn. The airplane was in a valley. The ground proximity warning system detected an imminent collision

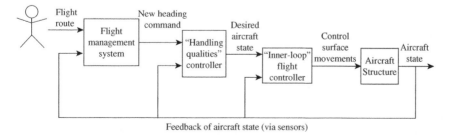

Figure 1.4 Today's flight control system architecture. The role of the human pilot has steadily been "upgraded" since the Wright Flyer. Source: Samad and Cofer (2001); © Elsevier B.V.

with the terrain. The pilots turned off the autopilot and selected take-off power in an attempt to clear the mountain, but without retracting the speed brakes.

A number of safety recommendations were issued as a result of the analysis of the accident (Skybrary n.d.), many of which related to the human–machine aspects of flying FMS-equipped aircraft. These included automatic retraction of speed brakes under maximum thrust commands, presenting pilots with angle-of-attack information, retention of certain waypoints in the FMS, and presentation of accurate terrain information in flight displays.

It is common for the same system to have humans as part of both plant and controller. For example, commercial passenger aircraft have pilots ("in the controller") and passengers ("in the plant"). The presence of passengers affects the work of the pilot – considerations apply that would not in the case of a cargo-only flight. Just as multiple people can be in the same plant at the same time, similarly multiple people can be involved in control functions simultaneously. The pilot/copilot tandem in the aircraft cockpit is an example. Analogously, a refinery control room may have a number of operators and engineers who as a group are responsible for the operation of the plant.

In many human-in-the-controller systems, the human is in the driver's seat, so to speak; the automation is there to facilitate the human's task by highlighting relevant information, enabling situational awareness, effecting actions reliably, and the like. But this "human-in-control" approach is not the only model. Automation systems can be better at certain control tasks than human operators and can override human operators as well. Emergency braking systems in modern cars are one example – even if the human doesn't apply brakes when a collision is all-but-inevitable, the automated car will. More fluid leader–follower roles are also possible. In domains such as robotics and automobile driving, haptic shared control can allow control authority to be smoothly transferred between the operator and the automation as needed for better performance (Abbink et al. 2012; Rahal et al. 2020).

The human-in-the-controller category is well represented in this volume. A.V. Efremov, in the chapter "Regularities of Human Operator Behavior and Its Modeling," focuses on modeling operator behavior specifically in the aerospace context. Control-relevant human modeling studies have been done for other domains as well, such as automobile drivers (see the chapter "Driver Modeling for the Design of Advanced Driver Assistance Systems: an Interdisciplinary Approach" by F. Mars, S. Mammar, and P. Chevrel). The multivehicle context is the focus of "Safe Teleoperation of Connected and Automated Vehicles" by F.J. Jiang, J. Märtensson, and K.H. Johansson. Safety is a paramount consideration in CPHS research, and in "Safe Shared Control Between Pilots and Autopilots in the Face of Anomalies," E. Eraslan, Y. Yildiz, and A.M. Annaswamy focus on collaborative human/machine control for safety-critical systems. A specific aspect of human behavior in CPHS, cognitive fatigue, is examined in "Overview and Perspectives on the Assessment and Mitigation of Cognitive Fatigue in Operational Settings," by M. Salomone, M. Audiffren, and B. Berberian. A broad review of human–robot interaction, encompassing multiple of the CPHS categories identified in this chapter, can be found in "Contemporary Issues and Advances in Human-Robot Collaborations," by T. Hatanaka et al.

1.3.3 Human–Machine Control Symbiosis

Often, the same person can be in both "in-the-plant" and "in-the-controller" roles simultaneously: a homeowner can adjust a thermostat for their own comfort, and the driver of an automobile is also a passenger. The two roles for the human in these cases can be considered separately; they are not inherently indivisible or symbiotic.

However, in some human-in-the-loop systems, categorizing the human's role as either plant or controller is misleading – a person is both, and indivisibly. Powered exoskeletons, which can augment human physical effort for applications in healthcare, defense, and package delivery, illustrate the symbiotic possibilities (Kazerooni 2008). The control of exoskeletons integrates human actions with computerized reactions and the very action of the human affects their abilities in real time. Just as it would be nonsensical to distinguish the two roles in the case of a nonbionically assisted individual who is able to pick up an object, the roles cannot be distinguished in the cyberphysically assisted instance. Powered as well as passive exoskeletons have seen substantial progress over the past decade and more, with products now commercially available. The company Ekso Bionics (http://eksobionics.com), for example, has products that can help patients with brain or spinal cord injuries gain mobility and workers in manufacturing and construction lift larger weights or use tools for longer durations. Soft-robotic wearable clothing can also provide support for people, to improve physical performance and treat medical problems (Granberry et al. 2021). Powered and actuated wingsuits with active control are also under development (D'Andrea et al. 2014).

Another example at the technological forefront is brain implants for prosthetics and wearable systems. In prosthetic control systems, the person is involved in the control decisions and is also being actuated. These systems rely on brain–computer interfaces (Lebedev and Nicolelis 2006; Vidal et al. 2016) through which artificial sensory feedback can be communicated to the brain and neural electrode measurement and processing effects prosthetic movement. Other examples with wearable robots for healthcare can be found in the chapter "Fusing Electrical Stimulation and Wearable Robots with Humans to Restore and Enhance Mobility" by T. Schauer, E. Fosch-Villaronga, and J.C. Moreno.

Closed-loop automated pacemakers were mentioned as a human-in-the-plant example above. But biomedical devices can also be patient-controlled – thus, again, the human is involved in the control and is also subject to it. Such devices are not entirely a recent innovation, although they have become considerably more sophisticated over the decades; the first patient-controlled artificial pacemaker was implanted in 1960 (University of Birmingham 2011).

Further progress in these application areas will, we expect, not only lead to dramatic enhancements – but also, as with many developments in human-in-the-loop control, raise complex and challenging ethical and moral issues. I discuss this issue further later.

1.3.4 Humans-in-Multiagent-Loops

Finally, in many cases, human-in-the-loop systems inherently involve multiple people, in conjunction with artificial agents. The humans-in-the-multiagent-loops category covers systems that cannot be analyzed as simple combinations of single-person systems. Thus, in this category we do not include buildings with many occupants or commercial aircraft with hundreds of passengers – or cruise ships with thousands. In such cases, the individuals are not meaningfully interacting among themselves for purposes related to the operation of the plant; they can be seen as "copies" of the "human-in-the-plant" category.

Complications arise when multiple people jointly and actively engage in the loop, especially in the "human-in-the-controller" role. For training purposes, dual controls may be used – for example in aircraft and cars. Trainer aircraft with dual controls appeared first, by 1912 (Cameron 1999, p. 37). Dual controls for cars appear to have come more than a decade later (Veigel 2023). In these cases, the plant is a single physical (or cyberphysical) system, the vehicle.

In today's aircraft cockpits dual controls also exist, but, more often than not, they are for redundancy and safety. Either the pilot or copilot can operate the aircraft alone, aided by avionics.

However, when the two individuals need to cooperate for some operational reason – e.g. in cases of faults and incidents – the multiagent classification is appropriate.

Building management systems were placed in the humans-in-the-plant category earlier. But in some environments, multiple inhabitants are inseparably both controllers and controlled. Astronauts in a manned spacecraft are one example. Another is the Biosphere 2 experimental facility, in which several people lived within an enclosed, isolated structure for two years, managing, with automation, their environment and their lives within it (Marino and Odum 1999).

In most complex engineered systems, control architectures are structured, often hierarchically. This is especially true in this human-in-the-loop category. An example is shown in Figure 1.5.

Hierarchical command-and-control or leader–follower structures imply that the same person can be in the "human-in-the-plant" role and "human-in-the controller" role. An example of this is air traffic control (ATC), where control connections exist among air traffic controllers, aircraft pilots, and passengers. From the perspective of air traffic controllers, the plants they need to control are the aircraft in the airspace. ATC personnel do not have direct control of aircraft; instead, they communicate with the cockpit crew. The latter are thus "in the plant" for the air traffic controllers. But the pilots are controllers themselves too, of their individual aircraft. Sophisticated automation equipment is used at both levels, and the human–automation interactions are intricate and crucial for performance and safety.

The true multiagent nature of air traffic control is evident, too, from the nature of the controllers' work. They are not controlling each aircraft in their airspace as separate entities; they must consider the airspace and the aircraft within it as a whole. Commands to different aircraft must be coordinated, especially in times of congestion or safety concerns. Multiple air traffic controllers will themselves need to collaborate.

With further development of unmanned aerial vehicles, a new kind of agent will be introduced into this scenario. Now air traffic controllers, and possibly other aircraft, will need to interact with fully autonomous systems (or close-to-fully autonomous systems – it is hard to imagine that ATC will not be able to override a UAV's automation, although there will be limitations to the extent of control that could be exercised).

Analogously, in a process plant such as an oil refinery, engineers and supervisors may be directing operations and controlling equipment from a central control room, but field personnel are also deployed in the plant and can be directed to check out units and equipment, manually operate valves and pumps, and take other control actions that cannot, or in a particular context should not, be taken remotely. Centralized control rooms first appeared in the late 1930s, after the development of the PID controller (Beniger 1986, p. 31), which automated low-level loop control. Process automation architectures have evolved considerably since (Samad et al. 2007). Today, some industrial facilities (e.g. wastewater treatment plants) and some other facilities (e.g. data centers) are remotely operated and monitored, going without onsite staff for extended durations. For more

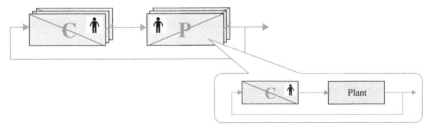

Figure 1.5 An example of a hierarchical CPHS architecture, combining the humans-in-multiagent-loops and human-in-the-controller categories.

complex industrial plants – e.g. chemicals and refineries – "lights out" operation is a topic for research (Olivier 2017).

As with the human-in-the-controller category, several chapters in this volume can be classified under humans-in-multiagent-loops. A couple of these relate to CPHS applications in social media and online networks: "Social Diffusion Dynamics in Cyber-Physical-Human Systems" by L. Zino and M. Cao and "On Online Attention Dynamics in Social Media" by M. Castaldo, P. Frasca, and T. Venturini. In a related vein, a framework for "cyberphysical social systems" is presented in "Parallel Intelligence for CPHS: An ACP Approach" by X. Wang et al.

The modeling and control of pandemics and epidemics can also benefit from a multiagent CPHS perspective, as discussed in the chapter "Epidemics Spread Over Networks: Influence of Infrastructure and Opinions" by B. She et al. Additional articles on this timely, indeed urgent, topic can be found in (Hernandez-Vargas et al. 2021).

1.4 A Roadmap for Human-in-the-Loop Control

Figure 1.6 provides a visual reference to the evolution of human-in-the-loop control, classified by the categories presented in this chapter. Examples from application domains covering healthcare, buildings, industry, automotive, and aerospace are included. In general, the dates, approximate as they are, refer to operational and/or commercial applications, not research or prototype developments. The timeline extends over about a century and a half, which means that the first half of it, approximately, is predigital computation. Some of the projected developments on the far right may take longer than the horizon depicted to mature.

The late initiation of human–machine control symbiosis is evident from the timeline, but it is multiagent CPHS that is the most futuristic category, in its ultimate visions. The two applications listed in that part of the roadmap are discussed further below.

1.4.1 Self- and Human-Driven Cars on Urban Roads

A few years ago, many automotive companies, and some companies that are not automobile manufacturers, were stating that, by 2019 or 2020, they would deploy self-driving cars on roads, including

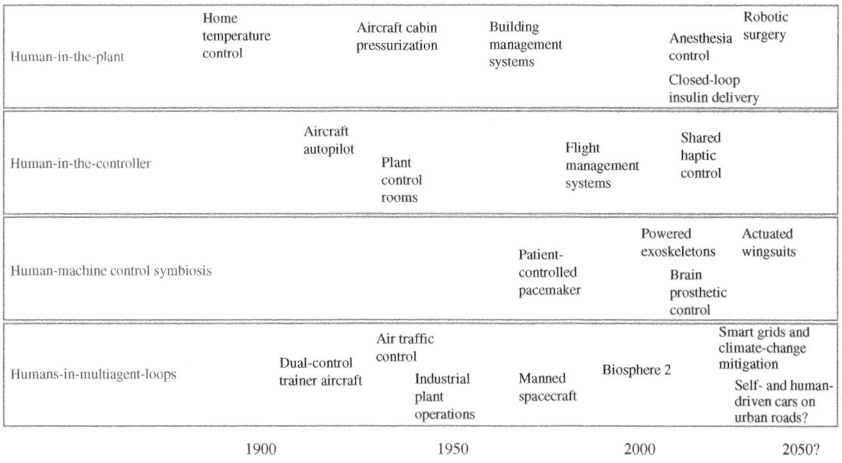

Human-in-the-plant		Home temperature control	Aircraft cabin pressurization	Building management systems	Anesthesia control	Robotic surgery
					Closed-loop insulin delivery	
Human-in-the-controller		Aircraft autopilot	Plant control rooms		Flight management systems	Shared haptic control
Human-machine control symbiosis				Patient-controlled pacemaker	Powered exoskeletons / Brain prosthetic control	Actuated wingsuits
Humans-in-multiagent-loops		Dual-control trainer aircraft	Air traffic control / Industrial plant operations	Manned spacecraft	Biosphere 2	Smart grids and climate-change mitigation / Self- and human-driven cars on urban roads?
	1900		1950		2000	2050?

Figure 1.6 An approximate timeline/roadmap for human-in-the-loop control developments, by category.

in urban environments. When the target dates approached, it was universally acknowledged that the problem was harder than anticipated, and that the challenge was less with the technology than with the human element. It is because of the pervasive human element that skepticism about autonomous cars has increased (Madrigal 2018).

In this hypothetical future urban road traffic scenario, human-driven and autonomous cars will share the road. The humans are both controllers (drivers) and the controlled (passengers). Passengers can be riding in either human-driven or autonomous cars. Pedestrians, bicyclists, and public transportation will be part of the system too, and a source of continuous disturbances. In this environment, people and automation will need to interoperate with each other dynamically.

The humans-in-multiagent-loops complexities involved are evident from examples such as the following:

- At a multiway stop sign, human drivers will resort to various signals and maneuvers to determine the order in which the vehicles should proceed. Some of the signals may be physical – a nodding of the head or wave of a hand or stop-and-start progress.
- Human drivers may rely on observations of others to determine safety conditions, such as seeing how other vehicles react to a wet surface to determine safe speeds while being cognizant of technological differences with the other vehicles (e.g. the presence or absence of traction control, the type and state of tires).

With a mix of autonomous and human-operated cars, similar multiagent behaviors will be needed. Research is required before this can be achieved, and it is again the human element that is the crux of the issue.

For complex multiagent CPHS such as hybrid urban traffic systems, the models required extend beyond the vehicle systems, the built infrastructure, and traffic regulations. Models of humans, covering their behavior, perception, and decision-making, are also needed. Consider an example that's often brought up (Nooteboom 2017): A ball rolls across the road in front of a car without impeding its motion – the ball has crossed the road by the time the car intersects its path, so there is no immediate danger to the car or its occupants. A human driver will know to slow down, perhaps even stop, because a child may be running after the ball. How will an autonomous car know to do the same? What models must be incorporated into the "intelligence" behind the autonomy to ensure safety?

Other examples can be imagined too. Each may be considered an unlikely "corner case," but there are innumerable corner cases. Autonomous systems that are part of complex CPHS, especially those operating in peopled environments, must integrate open-ended knowledge that includes knowledge of people that is currently unavailable, or at least unformalized.

1.4.2 Climate Change Mitigation and Smart Grids

The mitigation of climate change is another "grand challenge" for CPHS. That this is a CPHS problem might not even be obvious at first glance. Research into solutions to combat climate change is largely pursued by physicists, chemists, and engineers from various specializations (including control science and engineering). Yet, here too, people are at the heart of the problem.

One important area in which the controls community is extensively involved is in smart grids (Annaswamy and Amin 2013). As dispatchable fossil-fuel sources are phased out, intermittent and uncertain renewable generation becomes the predominant electricity source, and affordable storage at scale remains work in progress, coordination of generation and consumption across the grid and across time scales will be required.

Automation is being deployed for demand-side management to help with this balancing problem (Samad et al. 2016). Initial solutions often relied on direct load control, with the utility automatically managing "behind-the-meter" devices (e.g. automatic air-conditioner-cycling agreements with residential customers for hot summer days). In general, though, these have had limited appeal – in most cases, customers are not willing to abdicate control authority to the utility. More recently, solutions that leave the consumer in control have been developed and matured to the point of global standards being established (Bienert and Samad 2021). These allow the preferences of people and businesses (commercial buildings and industry facilities) to be incorporated. But deeper understanding of consumers' habits and decision-making is required for demand-side management to be widely and successfully deployed. Some compromise in manufacturing production, office productivity, and home comfort may be necessary, but the customer cannot be left "out of the loop."

Thus, a combination of agents, both autonomous artificial ones and automation-assisted human ones, will be involved, and at various levels and functions of the future grid. These include not only the utility and the consumer, but also market operators and other third-party providers. It is this ecosystem that constitutes the smart grid CPHS. Similar considerations apply for other smart grid and related areas, such as the charging of electric vehicles (see the chapter "Charging Behavior of Electric Vehicles" by Q.-S. Jia and T. Long) and smart cities (see "Cyber-Physical-Social Systems for Smart City" by G. Xiong et al.), and in many other areas of climate-change mitigation as well. In each case, the ecosystem that must be engaged involves automation and human elements.

1.5 Discussion

1.5.1 Other Ways of Classifying Human-in-the-Loop Control

In addition to distinguishing between "in-the-controller" and "in-the-plant" roles, distinctions can usefully be drawn based on the complexity of the interactions between human and machine elements. Two aspects of the complexity are the bandwidth and scope of the interaction. In the evolution of flight control associated with Figure 1.4, the frequency of pilot engagement with the aircraft has steadily decreased. Orville Wright needed to control the Wright Flyer in real time, making actuator moves at a subsecond frequency to ensure stable flight. With inner-loop autopilots, pilots could be hands-off as long as the airplane was in straight-and-level flight. Handling-qualities controllers allowed pilots to specify new headings and altitudes with automation taking care of the trajectory change. With flight-management systems, pilots can be hands-off through the entire airspace part of the flight. The level of abstraction of pilot commands has increased as well, from actuator changes to trajectory changes to heading changes to complete route plans.

Levels of complexity also arise for humans who are "under control." For indoor environmental control, the temperature of the occupied space is being controlled. In commercial buildings, humidity may also be controlled. In an airplane, cabin pressure and fresh air must also be regulated. These controls are usually setpoint-driven. A more elaborate control strategy would be to optimize the "comfort" of an occupant, a multivariable proposition. Thermal comfort models have been developed and standardized to this end, notably PMV/PPD (Predicted Mean Vote/Predicted Percentage Dissatisfied) (Fanger 1972; ISO 2005).

The human/machine balance of concern or responsibility is another dimension to consider. What aspects of the control task are the human or humans responsible for, versus the machine, and how important for the system under control are the human versus the nonhuman elements?

Although a wealth of literature exists on human-in-the-loop control, from both theoretical and practical perspectives, taxonomic reviews of the topic are lacking and this lacuna has

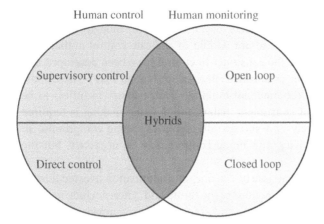

Figure 1.7 Taxonomy of human-in-the-loop applications. Source: Reproduced with permission from Nunes et al. (2015)/IEEE.

therefore been the principal motivation for this chapter (but see Hirche et al. (in press) for a recent and notable exception). An early categorization that does not adopt a comprehensive systems-and-controls perspective is Nunes, Zhang, and Silva (2015), building on prior work by Munir et al. (2013). See Figure 1.7 for the taxonomy of "human-in-the-loop cyber-physical systems" (HiTLCPS) developed therein. On the left side of the figure, higher-level supervisory control (e.g. setting set-points) is distinguished from the human directly controlling the system. On the monitoring side of the taxonomy humans are passively sensed, with the resulting data used for reporting (open loop) or for changing control actions (closed loop). Hybrid systems, in the center of the figure, integrate people-centric sensing and human inputs. Schirner et al. (2013) also review "HiLCPS," focusing on scenarios where the human is part of the control system and separate from the physical environment to be controlled.

1.5.2 Modeling Human Understanding and Decision-Making

Even if the term itself is of recent coinage, CPHS have been a crucial part of our technology-intensive world for decades. Aircraft with automation-assisted pilots, or cars with various driver-assistance features, or chemical plants with carefully designed control rooms, all have benefited from work that can legitimately be called CPHS research. Much of this research has been in human factors and ergonomics. Visual displays and aural signals, haptic and force feedback, safety limits related to movement and sensory modalities, pain and convenience thresholds, today's systems incorporate the state-of-the-art understanding of such phenomena.

The quality of performance obtained in any control system is ineluctably tied to the quality of the models used in the design and operation of the system. Methodologies such as system identi-fication, supported by sophisticated tools, are used for developing dynamic models. Data collected through designed experiments (e.g. wind-tunnel tests for aircraft, step tests for process plant units) are often crucial in this process. For people engaged in sensorimotor tasks, dynamic modeling of relevant human functions is required, and approaches have been successfully developed. Some of these are reviewed in Hess (1996), with a focus on the visual-input-to-physical-action transfer function.

As we take a more comprehensive view of human-in-the-loop control systems, the need for models that go beyond ergonomics and sensorimotor phenomena becomes evident. Cognition, comprehension, and decision-making need to be modeled as well. Models of the fidelity we expect in the engineering realm are hard to attain here. Indeed, "irrational" and "illogical" mental processes in people will need to be captured, both to represent humans as parts of controlled systems and as parts of controllers.

The seminal work of Kahneman and Tversky in what is now referred to as prospect theory has illuminated a number of biases and illusions that affect human decision-making. These are unconscious failings of our cognition, although Nesbitt (2015) suggests that, at least in some cases, awareness and training can help people overcome them. A large number of biases and illusions have been identified. A few especially relevant to CPHS are noted below.

- **Loss aversion**: People are asymmetrically loss-averse. This means that the absolute value of the (negative) psychological value of a loss of a certain amount is much higher than the psychological value of a gain of the same amount. Based on Kahneman and Tversky's experiments, it will take a gain of over US$200 to offset the loss of US$100 in our minds – people will tend to not take a fair 50–50 bet if the potential loss is US$100, and the potential gain is US$200!

- **Framing**: Would you have a more favorable view of a food product if it was advertised as "90% fat-free" or as "10% fat"? If so, you're not alone. Advertisers understand the importance of framing, and designers of CPHS need to, as well. Saad et al. (2016) suggest how framing along with loss aversion can be coopted to help consumers prefer smart grid solutions.

- **Sunk-cost fallacy**: In the words of Kahneman (2011): "The sunk-cost fallacy keeps people for too long in poor jobs, unhappy marriages, and unpromising research projects." The bias applies to decision-making in cyberphysical systems as well. Once an investment has been made in a certain operational strategy for a complex engineering system, a disincentive for change is immediately established. Even with evidence that the current approach is not working or if a rational analysis indicates that changing directions would be the better decision, human decision-makers will be biased towards staying on the path.

- **Endowment effect**: The ownership of a physical product or an idea results in its overvaluation. People are more hesitant to change direction than to adopt the direction in the first place.

- **Certainty and possibility effects**: As another example of the irrationality of human decision-making, Figure 1.8 illustrates our nonlinear interpretation of probability. We are overly sensitive to small differences in probability at the extremes of the range, and we "underweight"

Figure 1.8 Decision weight versus probability. Source: Based on data in Kahneman (2011).

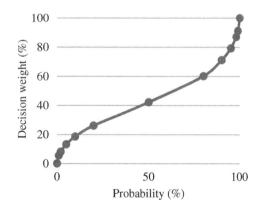

probability differences in the middle of the range. A question that arises is whether, in order to help people make decisions that are more rational, probability numbers should be mapped through an inverse of the curve before presenting them to human decision-makers!

It is such behavioral anomalies (except that they are the norm rather than the exception when humans are involved) that future CPHS will need to incorporate. Explorations of prospect theory for CPHS are happening – see Chapter 2 in this volume, "Human Behavioral Models Using Utility Theory and Prospect Theory," by A.M. Annaswamy and V.J. Nair, for more discussion.

Other areas of human modeling remain important as well. For some related research see the chapter "Enabling Human-Aware Autonomy Through Cognitive Modeling and Feedback Control" by N. Jain et al.

1.5.3 Ethics and CPHS

The products, services, and solutions developed by automation and control engineers impact people and communities in multifarious ways. These go beyond enhancing comfort, providing safe transportation, or providing reliable and low-cost goods and services. The scope of CPHS systems is steadily expanding, with greater effect on human lives and the environment. This progression is evident from the socially responsible automation pyramid shown in Figure 1.9 (Sampath and Khargonekar 2018). Automation is no longer employed in the service solely of cost reduction and performance improvements; people are at the center of it, and society as a whole is enmeshed too.

There are many aspects of ethics that impinge on the work of CPHS researchers and practitioners. The safety, health, and security of the operators and the users of products, services, and solutions need to be considered. The nature of an application will often imply environmental impacts as well. With emerging technologies such as machine learning and artificial intelligence, issues of bias, fairness, and transparency arise. There is typically a desired net societal benefit of such offerings, but tradeoffs and compromises have to be made. We must remember too that some CPHS products are, in circumscribed ways, intended to cause harm – usually in the interests of a greater good, in

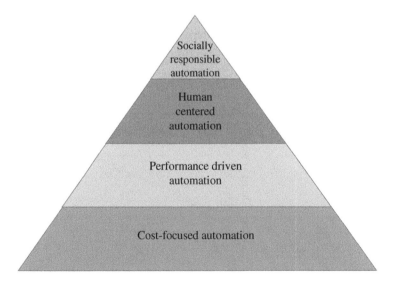

Figure 1.9 The "socially responsible automation" pyramid. Source: Adapted from Sampath and Khargonekar (2018).

the eyes of their developers and sponsors – and it would be nothing short of naive to ignore such work: "Thou shalt do no harm" is not a tenable ethical precept in today's or any feasible future world, at least not without considerable qualification.

Thus, ethics is a complex topic. It is not one that technologists are equipped to deal with, or have ever been equipped to deal with, by the education and training they receive. Perhaps in the past they could perform their work assuming that other stakeholders would assume responsibilities related to human, societal, and environmental impacts – they themselves could follow specifications they were given, standards and methodologies that had been established, etc. – but today they have to shoulder part of the burden of ethical decision-making themselves. A technology-centered mindset itself breeds a certain hubris and an attitude of technological determinism. It is all too easy, immersed in the possibilities offered by new research developments, to be enamored by the "cool" innovations one is working on, and to forget that the work will effect numerous people and the planet over the course of years and often decades. "Responsible innovation" is required (Lamnabhi-Lagarrigue and Samad 2023).

There is much more to be said about the ethics of CPHS than can be said here. Related chapters in this volume cover trust, threat, and inclusion; see "Trust-Triggered Robot-Human Handovers Using Kinematic Redundancy for Collaborative Assembly in Flexible Manufacturing" by S.M.M. Rahman et al.; "Opportunities and Threats of Interactions Between Humans and Cyber-Physical Systems – Integration and Inclusion Approaches for CPHS" by F. Vanderhaegen and V.D. Benito Jiménez; and "Shared Control with Human Trust and Workload Models" by M. Cubuktepe, N. Jansen, and U. Topcu.

1.6 Conclusions

Humanity's engagement with technology continues on a path of greater immersion, with the human element becoming increasingly intertwined with cyberphysical systems. At the same time, the researchers, developers, and practitioners of technological solutions often think of their work and its deliverables as a shifting of responsibilities from people to engineered products and services – attention focuses on the latter at the expense of the former. The recognition of CPHS as a discipline attests to the belated realization that greater, not less, attention needs to be devoted to people and society throughout the value stream and life cycle of automation systems.

Human-in-the-loop control systems – which, if they are not considered synonymous with CPHS, are subsumed by them – can be of various types. Regardless of the nature of the human–automation interplay – in the classification presented in this chapter, whether it is human-in-the-plant, human-in-the-controller, human–machine symbiosis, or humans-in-multiagent-loops – the costs and benefits as well as the risks and rewards of increasing technology immersion will continue to escalate. Impacts cannot be bounded; directly or through domino effects our society and ecosystem are at stake. CPHS may be an emerging discipline, but it is one that all science, engineering, and technology communities need to embrace.

Acknowledgments

A version of the four-level categorization of CPHS presented above was first developed for the inaugural CPHS workshop, held in Paris in September, 2014, under the leadership of Françoise Lamnabhi-Lagarrigue.

References

Abbink, D.A., Mulder, M., and Boer, E.R. (2012). Haptic shared control: smoothly shifting control authority? *Cognition, Technology & Work* 14 (1): 19–28.

Aircraft Accident Investigation Bureau (2019). Aircraft Accident Investigation Preliminary Report: Ethiopian Airlines Group. B737–8 (MAX) Registered ET-AVJ, 28 NM South East of Addis Ababa, Bole International Airport, March 10, 2019, Ministry of Transport, Federal Democratic Republic of Ethiopia. www.ecaa.gov.et/Home/wp-content/uploads/2019/07/Preliminary-Report-B737-800MAX-ET-AVJ.pdf (accessed 23 February 2023).

Annaswamy, A.M. and Amin, M. (2013). IEEE vision for smart grid controls: 2030 and beyond. *IEEE vision for smart grid controls: 2030 and beyond*, June 20, https://doi.org/10.1109/IEEESTD.2013.6577608.

Anon (2013). Boeing 307 stratoliner. Aviation History Online Museum. http://www.aviation-history.com/boeing/307.html (accessed 23 February 2023).

Astrom, K. and Kumar, P.R. (2014). Control: a perspective. *Automatica* 50: 3–43.

Becker, L.B. (ed.) (2016). 1st IFAC Conference on Cyber-Physical & Human-Systems CPHS 2016. *IFAC PapersOnLine* 49 (32): 1–252. https://www.sciencedirect.com/journal/ifac-papersonline/vol/49/issue/32.

Beniger, J.R. (1986). *The Control Revolution: Technological and Economic Origins of the Information Society*. Harvard University Press.

Bienert, R. and Samad, T. (2021). Standardizing demand-side management: OpenADR and its impact. *IEEE Electrification Magazine* 9 (3): 83–91.

Boeing (n.d.). Model 247/C-73 transport. https://www.boeing.com/history/products/model-247-c-73.page (accessed 25 June 2020).

Braun, J.E. (2015). Building control systems. In: *Encyclopedia of Systems and Control* (ed. J. Baillieul and T. Samad), 98–104. Springer.

Cameron, R.H. (1999). Training to fly: military flight training 1907–1945. Air Force History and Museums Program, https://media.defense.gov/2010/Dec/02/2001329902/-1/-1/0/training_to_fly-2.pdf (accessed 23 February 2023).

Coe, T. (2015). Where does the word *cyber* come from?. OUPblog, Oxford University Press, March 28. https://blog.oup.com/2015/03/cyber-word-origins (accessed 15 February 2022).

D'Andrea, R., Wyss, M., and Waibel, M. (2014). Actuated wingsuit for controlled, self-propelled flight. In: *The Impact of Control Technology*, 2e (ed. T. Samad and A.M. Annaswamy). IEEE Control Systems Society, available at www.ieeecss.org.

Dumont, G. (2012). Closed-loop control of anesthesia – a review. *IFAC Proceedings Volumes* 45 (18): 373–378. https://doi.org/10.3182/20120829-3-HU-2029.00102.

Duvall, J.C., Dunne, L.E., Schlief, N., and Holschuh, B. (2016). Active 'Hugging' vest for deep touch pressure therapy. *UbiComp '16: Proceedings of the 2016 ACM International Joint Conference on Pervasive and Ubiquitous Computing: Adjunct*, pp. 458–463, https://doi.org/10.1145/2968219.2971344.

Fanger, P.O. (1972). *Thermal Comfort: Analysis and Applications in Environmental Engineering*. McGraw Hill Book Company.

Fontaine, A.B. (1954). A model for human tracking behavior in a closed-loop control system. Ph.D. Dissertation. The Ohio State University. Columbus, Ohio, U.S.A.

Granberry, R.C., Compton, C., Woelfle, H. et al. (2021). Enhancing performance and reducing wearing variability for wearable technology system–body interfaces using shape memory materials. *Flexible and Printed Electronics* 6 (2): https://iopscience.iop.org/article/10.1088/2058-8585/abf848/meta (accessed 19 April 2022).

Hernandez-Vargas, E.A., Giordano, G., Sontag, E. et al. (ed.) (2021). First special section on systems and control research efforts against COVID-19 and future pandemics. *Annual Reviews in Control* 50: 343–344. https://doi.org/10.1016/j.arcontrol.2020.10.007.

Hess, R.A. (1996). Human-in-the-loop control. In: *The Control Handbook* (ed. W.S. Levine), 1497–1505. CRC Press.

Hess, R.A. (2021). Adaptive human pilot models for aircraft flight control. In: *Encyclopedia of Systems and Control*, 2e (ed. J. Baillieul and T. Samad), 40–45. Springer.

Hirche, S., Ames, A., Samad, T., Fontan, A., and Lamnabhi-Lagarrigue, F. (in press). Cyber-physical human systems. In: *Control for Societal-Scale Challenges Roadmap 2030* (ed. A. Annaswamy, K.H. Johansson, and G. Pappas). IEEE Control Systems Society.

ISO (2005). Ergonomics of the thermal environment – Analytical determination and interpretation of thermal comfort using calculation of the PMV and PPD indices and local thermal comfort criteria. ISO 7730:2005. https://www.iso.org/standard/39155.html (accessed 26 June 2020).

Kahneman, D. (2011). *Thinking Fast and Slow*. Farrar, Straus and Giroux.

Kazerooni, H. (2008). Exoskeletons for human performance augmentation. In: *Springer Handbook of Robotics* (ed. B. Sciliano and O. Khatib), 773–793. Springer.

KNKT (2019). Aircraft accident investigation report: PT. Lion Mentari Airlines, Boeing 737-8 (MAX): PK-LQP, Tanjung Karawang, West Java, Republic of Indonesia, 29 October 2018, Komite Nasional Keselamatan Transportasi, Jakarta, Indonesia. http://knkt.dephub.go.id/knkt/ntsc_aviation/baru/2018%20-%20035%20-%20PK-LQP%20Final%20Report.pdf (accessed 23 June 2020).

Krisch, J.A. (2014). What is the flight management system? A pilot explains. *Popular Mechanics*, 18 March. https://www.popularmechanics.com/flight/a10234/what-is-the-flight-management-system-a-pilot-explains-16606556 (accessed 23 June 2020).

Lamnabhi-Lagarrigue, F. and Samad, T. (2023). Social, organizational, and individual impacts of automation. In: *Handbook of Automation* (ed. S.Y. Nof). Springer (in press).

Lebedev, M.A. and Nicolelis, M.A. (2006). Brain-machine interfaces: past, present and future. *Trends Neuroscience* 29 (9, September): 536–546. https://doi.org/10.1016/j.tins.2006.07.004.

Lewis, F. (1992). *Applied Optimal Control and Estimation*. Prentice-Hall.

Madrigal, A.C. (2018). 7 arguments against the autonomous-vehicle utopia. *The Atlantic*, December 20. https://www.theatlantic.com/technology/archive/2018/12/7-arguments-against-the-autonomous-vehicle-utopia/578638 (accessed 26 June 2020).

Marino, B.D.V. and Odum, H.T. (ed.) (1999). *Biosphere 2: Research Past and Present*. Elsevier Science.

Mettler, B. (2019). 2nd IFAC Conference on Cyber-Physical and Human Systems CPHS 2018. *IFAC PapersOnLine* 51 (34): 1–410. https://www.sciencedirect.com/journal/ifac-papersonline/vol/51/issue/34.

Mindell, D. (2002). *Between Human and Machine: Feedback, Control, and Computing before Cybernetics*. The Johns Hopkins University Press.

Moody, M. (n.d.). A breath of fresh air: how Covid is making cabin environmental controls more important than ever. Aviation Business News. https://www.aviationbusinessnews.com/cabin/breath-of-fresh-air-cabin-covid (accessed 23 February 2023).

Munir, S., Stankovic, J.A., Liang, C.-J.M. et al. (2013). Cyber physical system challenges for human-in-the-loop control. *8th International Workshop on Feedback Computing*, San Jose, CA, USA, June. https://www.usenix.org/node/174694 (accessed 23 February 2023).

Namerikawa, T. (ed.) (2020). 3rd IFAC Workshop on Cyber-Physical & Human Systems CPHS 2020. *IFAC PapersOnLine* 53 (5): 1–880. https://www.sciencedirect.com/journal/ifac-papersonline/vol/53/issue/5.

Nisbett, R.E. (2015). *Mindware: Tools for Smart Thinking*. Farrar, Straus and Giroux.

Nooteboom, L. (2017). Child-friendly autonomous vehicles: designing autonomy with all road users in mind. Humanizing Autonomy, November 13. https://humanisingautonomy.medium.com/child-friendly-autonomous-vehicles-2880ca74165f (accessed 14 February 2022).

Nunes, D.S., Zhang, P., and Silva, J.S. (2015). A survey on human-in-the-loop applications towards an internet of all. *IEEE Communication Surveys & Tutorials* 17 (2): 944–965.

Olivier, L.E. (2017). On lights-out process control in the minerals processing industry. Ph.D. Thesis, Univ. of Pretoria, South Africa.

Rahal, R., Matarese, G., Gabiccini, M. et al. (2020). Caring about the human operator: haptic shared control for enhanced user comfort in robotic Telemanipulation. *IEEE Transactions on Haptics* 13 (1): 197–203. https://doi.org/10.1109/TOH.2020.2969662.

Rodrigo, C.C., Orlando, J.R., and Saul, P.G. (1996). Aircraft accident report: controlled flight into terrain. American Airlines Flight 965, Boeing 757-223, N651AA, Near Cali, Colombia, December 20, 1995, Aeronautica Civil of the Republic of Colombia. https://reports.aviation-safety.net/1995/19951220-1_B752_N651AA.pdf (accessed 23 June 2020).

Romm, S. (1989). Arms by design: from antiquity to the renaissance. *Plastic and Reconstructive Surgery* 84 (1): 158–163.

Saad, W., Glass, A.L., Mandayam, N.B. et al. (2016). Toward a consumer-centric grid: a behavioral perspective. *Proceedings of the IEEE* 104 (4): 865–882.

Sagberg, F., Fosser, S., and Sætermo, I.-A.F. (1997). An investigation of behavioural adaptation to airbags and antilock brakes among taxi drivers. *Accident Analysis & Prevention.* 29 (3): 293–302. https://doi.org/10.1016/S0001-4575(96)00083-8. PMID 9183467.

Samad, T. (2020). Human-in-the-loop control: applications and categorization. *Proceedings of the 3rd IFAC Workshop*, Beijing, IFAC-PapersOnLine 53 (5): 311–317.

Samad, T. and Cofer, D. (2001). Autonomy in automation: trends, technologies, tools. *Computer Aided Chemical Engineering* 9: 1–13. https://doi.org/10.1016/S1570-7946(01)80002-X.

Samad, T., Koch, E., and Stluka, P. (2016). Automated demand response for smart buildings and microgrids: the state of the practice and research challenges. *Proceedings of the IEEE* 104 (4): 726–744.

Samad, T., McLaughlin, P., and Lu, J. (2007). System architecture for process automation: review and trends. *Journal of Process Control* 17 (3): 191–201.

Sampath, M. and Khargonekar, P. (2018). Socially responsible automation. *Bridge* 48 (4): 45–72.

Scheck, W (2004). Lawrence sperry: genius on autopilot. *Aviation History*, November. https://www.historynet.com/lawrence-sperry-autopilot-inventor-and-aviation-innovator.htm (accessed 23 June 2020).

Schirner, G., Erdogmus, D., Chowdhury, K., and Padi, T. (2013). The future of human-in-the-loop cyber-physical systems. *Computer* 46 (1): 36–45.

Skybrary (n.d.). B752, evicinity Cali Colombia, 1995. https://skybrary.aero/accidents-and-incidents/b752-vicinity-cali-colombia-1995 (accessed 11 February 2022).

Smirnov, A., Kashevnik, A., Shilov, N. et al. (2013). Context-aware service composition in cyber physical human system for transportation safety. In: *Proceedings of 13th International Conference on ITS Telecommunications (ITST)*, Tampere, Finland (05–07 November 2013), 139–144. IEEE.

Stein, G. (2003). Respect the unstable. *IEEE Control Systems Magazine* 23 (4): 12–25.

University of Birmingham (2011). Blue plaque guide. www.birmingham.ac.uk/documents/culture/bookletfinalpdf.pdf (accessed 27 October 2020).

Veigel (2023). Dual controls. Veigel. https://www.veigel-automotive.de/en/driving-school-systems/dual-controls/ (accessed 23 February 2023).

Vidal, G.W.V., Rynes, M.L., Kelliher, Z. et al. (2016). Review of brain-machine interfaces used in neural prosthetics with new perspective on somatosensory feedback through method of signal breakdown. *Scientifica (Cairo)* 8956432. https://doi.org/10.1155/2016/8956432.

Vinod, A.P. (ed.). (2022). 4th IFAC Workshop on Cyber-Physical and Human Systems CPHS 2022. *IFAC-PapersOnLine* 55 (41): 1–192. https://www.sciencedirect.com/journal/ifac-papersonline/vol/55/issue/41.

Wiener, N. (1948). *Cybernetics; or control and communication in the animal and the machine.* Cambridge, Mass: *Technology Press, MIT*.

2

Human Behavioral Models Using Utility Theory and Prospect Theory

Anuradha M. Annaswamy and Vineet Jagadeesan Nair

Department of Mechanical Engineering, Massachusetts Institute of Technology, Cambridge, MA, USA

2.1 Introduction

Analysis and synthesis of large-scale systems require the understanding of cyber physical human systems. Interactions between humans and cyber-components that interact with the physical system are varied and depend on a variety of factors and the goals of the large-scale systems. If the problem at hand concerns the behavior of the cyber–physical human systems (CPHS) under emergency conditions, the interactions between humans and automation need to focus on a shared control architecture (Erslan et al. 2023) with appropriate granularity of task allocation and timeline. Typically, the human in this context is an expert and when anomalies occur, either takes over control from automation or provides close supervision to the automation to ensure an overall safe CPHS. Under normal circumstances, the interactions may include other architectures. The role of the human is not necessarily that of an operator or an expert, but a user. The human may be a component in the loop, responding to outputs from the automation, and making decisions that serve in turn as inputs or reference signals r to the physical system (see Figure 2.1).

Typical examples of such interactions have begun to occur both in power grids and transportation and can be grouped under the rubric of transactive control. The transactive control concept (Chassin et al. 2004; Bejestani et al. 2014; Annaswamy and Nudell 2015) consists of a feedback loop resulting from incentives provided to consumers. Introduced in the context of smart grids, a typical transactive controller consists of an incentive signal sent to the consumer from the infrastructure and a feedback signal received from the consumer, and together the goal is to ensure that the underlying resources are optimally utilized. This introduces a feedback loop, where empowered consumers serve as actuators into an infrastructure, and transactive control represents a feedback control design that ensures that the goals of the infrastructure are realized, very similar to Figure 2.1. The use of transactive control in smart grids can be traced to homeostatic control proposed in Schweppe (1978) and Schweppe et al. (1980), which suggested that demand-side assets can be engaged using economic signals. Transactive control in this context has come to denote market-based control mechanisms that incentivize responsive loads and engage them in providing services to the grid where and when of great need (as described in Katipamula et al. (2006), Li et al. (2015), Hao et al. (2016), Somasundaram et al. (2014), Hammerstrom et al. (2008), Widergren et al. (2014), Melton (2015), Kok (2013), and Bernards et al. (2016)), and has the ability to close the loop by integrating customers via the right incentives to meet their local objectives such as lowering their electric bill as well as meeting global system objectives such as voltage and frequency regulation.

Cyber–Physical–Human Systems: Fundamentals and Applications, First Edition.
Edited by Anuradha M. Annaswamy, Pramod P. Khargonekar, Françoise Lamnabhi-Lagarrigue, and Sarah K. Spurgeon.
© 2023 The Institute of Electrical and Electronics Engineers, Inc. Published 2023 by John Wiley & Sons, Inc.

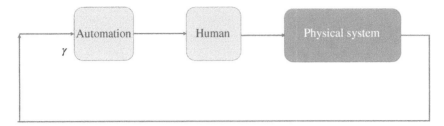

Figure 2.1 An Example of a CPHS with human-in-the-loop.

Another example of transactive control is in the context of congestion control in transportation (Phan et al. 2016; Annaswamy et al. 2018) – where dynamic toll pricing, determined by the controller, is used as an incentive signal to the drivers, who then decide whether or not to enter a tolled segment, thereby regulating traffic density and possibly alleviating congestion. In both examples, it is clear that an understanding of the human behavior is important, and the modeling of the overall socio-technical system that includes the human behavior and their interaction with the physical system is an important first step in designing the feedback controller.

This chapter focuses on two different tools that have been proposed for deriving behavioral models of human as a consumer. These tools include Utility Theory and Prospect Theory, and are described in Sections 2.2 and 2.3. Prospect Theory is a framework introduced by Nobel prize-winning behavioral economists and psychologists that has been extensively shown to better represent decision-making under uncertainty. It builds upon Utility Theory models by introducing additional nonlinear transformations on agents' objective utilities to model their irrational and subjective behaviors. Cumulative Prospect Theory (CPT) is an extension of Prospect Theory that also considers distortion of subjective probabilities through a weighting function. Examples are drawn from transportation to elucidate the impact of these tools, particularly in terms of predicting mode choice probabilities of passengers.

2.2 Utility Theory

The starting point for Utility Theory (Ben-Akiva et al. 1985) is the assignment of a value to an outcome in the form of a utility function. If in general there are N different possible choices, then U_i is utility of outcome $i, i = 1, \ldots, N$. The benefit of the utility function is that it provides a substrate for modeling a human's decision when faced with these choices. In particular, the probability that the human will select choice ℓ is determined using the utility function as:

$$p^\ell = \frac{e^{U^\ell}}{\sum_{j=1}^{N} e^{U^\ell}} \quad \ell \in \{1, \ldots, N\} \tag{2.1}$$

Equation (2.1) then serves as a simple behavioral model of the human which can be appropriately utilized in the overall problem of interest. For a simple problem with only two choices A and B, the probability p^A is given by

$$p^A = \frac{1}{1 + e^{-\Delta U}} \tag{2.2}$$

where $\Delta U = U^A - U^B$. Modeling the behavior of the human in the problem, whether as a consumer, an operator, a user, or an expert, the Utility Theory-based approach entails the characterization of all possible outcomes and the determination of the utility of each of the outcome i as U^i. A variety of factors contributes to the utility and hence determining the behavioral model in Equation (2.1) is nontrivial. Quantitative aspects related to economy, qualitative aspects such as comfort, and hard-to-identify aspects such as strategic behavior, negative externalities, global, and network-based expectations, are all factors that may need to be simultaneously accounted for. Nevertheless, Equation (2.1) serves as a cornerstone for many problems, where CPHS models have to be derived.

2.2.1 An Example

Suppose we consider a transportation example where a passenger intends to travel from points A to B and has two choices for transport: a shared ride service (SRS) and public transit. The utility function for this trip can be determined as:

$$u = \mathbf{a}^\mathsf{T} \mathbf{t} + b\gamma + c \tag{2.3}$$

where the components of $\mathbf{t} = [t_{walk}, t_{wait}, t_{ride}]^\mathsf{T}$ denote the walking, waiting, and riding times, respectively, γ denotes the ride tariff, $\mathbf{a} = [a_{walk}, a_{wait}, a_{ride}]^\mathsf{T}$ are suitable weights, and c denotes all other externalities that do not depend on either travel time or ride tariff. Both the weights $\mathbf{a}$ and tariff coefficient b are assumed to be negative since these represent disutilities to the rider arising from either longer travel times or higher prices, while c can be either positive or negative depending on the characteristics of the given travel option. All of the parameters $\mathbf{a}, b, c$ determine the behavior of a rider and could vary with time, the environment, or other factors. Equation (2.3) indicates that the choice of the rider of the SRS over other options like public transit is determined by whether $u(SRS) > u$(public transit) for a given ride.

A behavioral model of a driver as above was applied to the congestion control problem on a highway segment to determine a dynamic tolling price strategy (Ben-Akiva et al. 1985; Phan et al. 2016). It was shown that with the alternative u_0 corresponding to travel on a no-toll road, the toll price can be determined using a nonlinear proportional-integral (PI) controller using a socio-technical model that was a cascaded system with the behavioral model as in Equations (2.2) and (2.3) and an accumulator model of the traffic flow. Using actual data from a highway segment in the US city of Minneapolis, it was shown that such a model-based dynamic toll price leads to a much more efficient congestion control (Annaswamy et al. 2018). More recently, this approach has been extended to a more complex highway section with multiple merges and splits, and applied to data obtained from a highway section near Lisbon, Portugal (Lombardi et al. 2022). Here too, a behavioral model of the driver similar to Equation (2.3) was employed. The results obtained displayed a significant improvement compared to existing traffic flow conditions, with minimal changes to the toll price.

2.3 Prospect Theory

The behavioral model based on Utility Theory has two deficiencies. The first is that the utility function is embedded in a stochastic environment, causing the utility function model to be more complex than that considered in Equation (2.3). The second is that the model as considered in Equation (2.2) may not be adequate in capturing all aspects of decision-making of a human. Strategic decision-making, adjustments based upon the framing effect, loss aversion, and probability distortion are several key features related to subjective decision-making of individuals when facing uncertainty. It is in this context that *Prospect Theory* (Kahneman and Tversky 2012) in general, and *Cumulative Prospect Theory* (CPT) (Tversky and Kahneman 1992), in particular, provide an alternate tool that may be more appropriate. CPT builds upon Prospect Theory by using a probability weighting function to represent the agent's distortion of perceived probabilities of outcomes and uses these probability weights to compute the subjective utilities from subjective values. We briefly describe this tool below.

We first introduce a stochastic component into the problem and utilize the transportation example as the starting point. As travel times are subject to stochasticity, u becomes a random process. For simplicity, suppose we assume that there are only two possible travel time outcomes, $\bar{t}$ and $\underline{t}$ ($\underline{t} \leq \bar{t}$) having corresponding utilities $\underline{u}$ and $\bar{u}$ ($\underline{u} \leq \bar{u}$), occurring with probabilities of $p \in [0, 1]$ and $1 - p$, respectively. It follows that the utility function for the SRS is given by

$$\underline{u} = \mathbf{a}_{sm}^{\mathsf{T}} \bar{t} + b_{sm}\gamma_{sm} + c_{sm}$$
$$\bar{u} = \mathbf{a}_{sm}^{\mathsf{T}} \underline{t} + b_{sm}\gamma_{sm} + c_{sm} \tag{2.4}$$

If these outcomes follow a Bernoulli distribution, its cumulative distribution function (CDF) is defined on the support $[\underline{u}, \bar{u}]$:

$$F_U(u) = \begin{cases} 0 & \text{if } u < \underline{u} \\ p & \text{if } \underline{u} \leq u < \bar{u} \\ 1 & \text{if } u \geq \bar{u} \end{cases} \tag{2.5}$$

Suppose the alternative choice has a utility function u_o. If $u_o \leq \underline{u}$, the customer would always choose the SRS since it offers strictly better outcomes and conversely if $u_o \geq \bar{u}$. For all other cases, the underlying model becomes a combination of Equations (2.2) and (2.5).

We now address the second deficiency in Utility Theory. Conventional Utility Theory postulates that consumers choose among travel options based on their respective expected utilities (Fishburn 1988; Von Neumann and Morgenstern 2007). Alternatively, random utility models are another framework within Utility Theory that predict choice probabilities based on the utilities of different alternatives (computed using logit models) without accounting for risk, by assuming certain distributions for unobserved factors and error terms. However, both of these are inadequate when there is significant uncertainty involved. Prospect Theory (PT) is an alternative to Utility Theory that better describes subjective human decision-making in the presence of uncertainty and risk (Tversky and Kahneman 1992; Kahneman and Tversky 2012), and CPT is a variant of PT that weighs different outcomes using distorted subjective probabilities as perceived by passengers. This is needed since individuals have been shown to consistently underestimate the likelihood of high probability outcomes while overestimating the likelihood of less likely events (Tversky and Kahneman 1992). To describe CPT, we introduce a value function $V(\cdot)$ and a probability distortion $\pi(\cdot)$ given by Guan et al. (2019a) and Prelec (1998), with $\pi(0) = 0$ and $\pi(1) = 1$ by definition. These nonlinearities map the objective utilities (u) and probabilities (p) of each possible outcome to subjective values, as perceived by the passengers. Note here that the probability weighting function $\pi(\cdot)$ as described in Equation (2.7) is unique to CPT. The graphs in Figure 2.2 show examples of how the value and probability weighting functions may vary according to the objective utility u and actual probability p, respectively.

$$V(u) = \begin{cases} (u - R)^{\beta^+} & \text{if } u \geq R \\ -\lambda(R - u)^{\beta^-} & \text{if } u < R \end{cases} \tag{2.6}$$

$$\pi(p) = e^{-(-\ln(p))^{\alpha}} \tag{2.7}$$

The CPT parameters here describe loss aversion (λ), diminishing sensitivity in gains (β^+) and losses (β^-), and probability distortion (α). The reference R is the baseline against which users compare uncertain prospects. These can vary across individuals and also depending on the particular

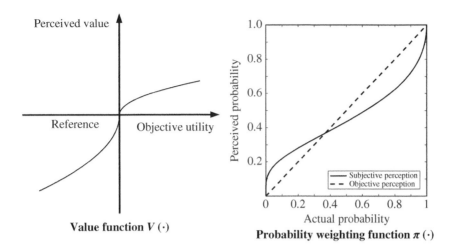

Figure 2.2 Illustrations of the CPT value function and probability weighting functions. Source: Adapted from (Guan et al. 2019a).

set of alternatives the customer is facing. Further details on the different possible types of references are provided in Section 2.3.2.

With the above distortions in the value function, the overall utility function for a given stochastic outcome gets modified. Note that the utility function U is now a random variable of the form

$$U = X + b\gamma = \mathbf{a}^\mathsf{T}\,\mathbf{T} + b\gamma \tag{2.8}$$

where the $b\gamma$ term is deterministic based on the SRS ride offer tariff and the random variable for the disutility (due to travel times) $X = \mathbf{a}^\mathsf{T}\mathbf{T}$ is stochastic and represents the uncertainty in travel times (walking, waiting, riding) $\mathbf{T} = [\underline{\mathbf{T}}, \overline{\mathbf{T}}]$. Similar to Equation (2.4), we can derive $X = [\underline{x}, \overline{x}]$ with CDF $F_X(x) = F_U(x + b\gamma)$, where $\underline{x}$ and $\overline{x}$ correspond to the shortest and longest travel times. These correspond to the best- and worst-case utilities $\overline{u}$ and $\underline{u}$, respectively, with $U = [\underline{u}, \overline{u}]$.

The Utility Theory-based derivation of the objective utility is as follows: if U takes on discrete values $u_i \in \mathbb{R}, \forall i \in \{1, \ldots, n\}$ and the outcomes are in ascending order, i.e. $u_1 < \cdots < u_n$, where $n \in \mathbb{Z}_{>0}$ is the number of possible outcomes, one can determine the objective utility U^o as the expectation of U according to Utility Theory as (Von Neumann and Morgenstern 2007), i.e.

$$U^o = \sum_{i=1}^{n} p_i u_i \tag{2.9}$$

where $p_i \in (0, 1)$ is the probability of outcome u_i, and $\sum_{i=1}^{n} p_i = 1$. In contrast, a Prospect Theory-based derivation of utility function includes the value function V rather than u_i and the probability distortion $\pi(.)$ is applied to the probabilities p_i. Suppose we define the corresponding utility function as U_R^s, which is the subjective utility perceived by the passenger according to Cumulative Prospect Theory, then

$$U_R^s = \sum_{i=1}^{n} w_i V(u_i) \tag{2.10}$$

where R denotes the reference corresponding to the framing effect mentioned above, and w_i denotes the weighting that represents the subjective perception of p_i. Suppose that k out of the n outcomes are losses, $0 \le k \le n, k \in \mathbb{Z}_{\ge 0}$, and the rest are nonlosses, i.e. $u_i < R$ if $1 \le i \le k$ and $u_i \ge R$

if $k < i \leq n$. We can then derive the subjective probability weights assigned by the decision-maker to each of these discrete outcomes from the cumulative distribution function of U given by $F_U(u)$, as follows:

$$w_i = \begin{cases} \pi\left[F_U(u_i)\right] - \pi\left[F_U(u_{i-1})\right], & \text{if } i \in [1,k] \text{ (losses)} \\ \pi\left[1 - F_U(u_{i-1})\right] - \pi\left[1 - F_U(u_i)\right], & \text{otherwise (non-losses)} \end{cases} \tag{2.11}$$

where we have assumed $F_U(u_0) = 0$ for ease of notation.

It is clear that in contrast to U^o, U_R^s is centered on R, loss aversion is captured by choosing $\lambda > 1$, and diminishing sensitivity by choosing $0 < \beta^+, \beta^- < 1$. The probability distortion is quantified by choosing $0 < \alpha < 1$. The extension from Equation (2.10) to the continuous case of U_R^s is

$$U_R^s = \int_{-\infty}^{R} V(u)\frac{d}{du}\left\{\pi\left[F_U(u)\right]\right\} du + \int_{R}^{\infty} V(u)\frac{d}{du}\left\{-\pi\left[1 - F_U(u)\right]\right\} du \tag{2.12}$$

With the above objective evaluation of a utility function as in Equation (2.9) and subjective evaluation as in Equation (2.10), one can now determine the probability of acceptance of an outcome as follows. As has been shown in Equation (2.1), the evaluation of the probability of acceptance of an outcome ℓ requires the utility of all alternate outcomes. Without loss of generality, suppose there are only two alternatives, with the objective and subjective utility of option $i \in \{1, 2\}$, given by U_i^0 and $U_{i_R}^s$, respectively. Then the objective probability of acceptance of choice 1 using Utility Theory is given by

$$p_1^0 = \frac{e^{U_1^0}}{e^{U_1^0} + e^{U_2^0}} \tag{2.13}$$

while the subjective probability of acceptance of option 1 using CPT is given by

$$p_{1_R}^s = \frac{e^{U_{1_R}^s}}{e^{U_{1_R}^s} + e^{U_{2_R}^s}} \tag{2.14}$$

2.3.1 An Example: CPT Modeling for SRS

We illustrate the Prospect Theory model using the transportation example considered above, extended to the case where a passenger now has three choices $i \in \{1, 2, 3\}$, (i) public transit like buses or the subway, (ii) using a shared ride pooling service (SRS), and (iii) another which may be an exclusive ride hailing service (such as UberX). The SRS has greater uncertainty in pick up, drop off, and travel times when compared to the UberX and transit alternatives due to the possibility of more passengers being added en route. Thus, both UberX and transit can be treated as certain prospects when compared to the SRS option, which has uncertain outcomes. CPT can then be used to model the passenger's risk preferences to predict their decision-making under such uncertainty. The discussions above show that a number of parameters related to the CPT framework have to be determined. These include $\alpha, \beta^+, \beta^-, \lambda$ defined in $V(\cdot)$ and $\pi(\cdot)$, which are in addition to the parameters $\mathbf{a}, b, c$ defined in (Equation (2.3)) associated with the travel times $t_{\text{walk}}, t_{\text{wait}}, t_{\text{ride}}$ and tariff coefficients, for all three travel modes. In order to estimate these parameters, we designed and conducted a comprehensive survey eliciting travel and passenger risk preferences from $N = 955$ respondents in the greater Boston metropolitan area (Jagadeesan Nair 2021). Note that the constant terms in the utility function, c_{UberX} and c_{SRS} were measured relative to public transit as a baseline, i.e. $c_{transit} = 0$.

Table 2.1 summarizes the mean values and standard deviations of the parameters that we estimated for the discrete mode choice model using maximum simulated likelihood estimation, along with their standard errors. We can also use the estimated travel time and price coefficients

Table 2.1 Parameters describing the discrete choice logit models for SRS and UberX.

Parameter	Mean	SE	SD	SE
a_{walk} [min^{-1}]	−0.0586	0.0053	0.1412	0.0079
a_{wait} [min^{-1}]	−0.0113	0.0182	0.1491	0.0356
$a_{ride,\ transit}$ [min^{-1}]	−0.0105	0.0013	0.0284	0.0017
$a_{ride,\ UberX}$ [min^{-1}]	−0.0086	0.0014	0.0058	0.0010
$a_{ride,\ SRS}$ [min^{-1}]	−0.0186	0.0013	0.0095	0.0007
b [\$$^{-1}$]	−0.0518	0.0050	0.0597	0.0042
c_{UberX}	−2.5926	0.1800	2.3034	0.1558
c_{SRS}	−2.2230	0.1497	1.8175	0.1530

Table 2.2 Value of time spent on different modes, obtained from the random parameters logit model.

Trip leg or mode	VOT (in \$/h)
Walking	67.8702
Waiting	13.1480
Transit ride	12.1703
Exclusive ride hailing	9.9466
Pooled ride sharing	21.5549

to determine the passengers' value of time (VOT) spent on different modes. The value of time is defined as the extra tariff that a person would be willing to pay or cost incurred to save an additional unit of time, i.e. it measures the willingness to pay (WTP) for extra time savings. In absolute terms, the VOT spent on mode i can be calculated as the ratio between the marginal utilities of travel time and trip cost:

$$VOT_i = \frac{\frac{\partial U_i}{\partial t}}{\frac{\partial U_i}{\partial \gamma}} = \frac{a_i}{b_i} \tag{2.15}$$

The VOT parameters estimated from our survey are shown in Table 2.2. For the CPT model, estimated, we slightly modified the model to allow for different probability distortion effects in the gain and loss regimes. Thus, the modified weighting function is given by

$$\pi_{\pm}(p) = e^{-(-ln(p))^{\alpha_{\pm}}} \tag{2.16}$$

The CPT risk parameters were estimated using the method of certainty equivalents (Rieger et al. 2017; Wang and Zhao 2019), by presenting surveyed passengers with a series of chance scenarios asking them to choose a travel mode by comparing the uncertain or risky SRS versus the certain UberX and transit options. Nonlinear least squares curve fitting was then used to estimate the CPT parameter values. A schematic for the overall estimation process is shown in Figure 2.3. More details on the survey design and estimation procedures can be found in Jagadeesan Nair (2021).

Having estimated these parameters, we can compute the subjective or perceived value of the SRS outcomes, and the passenger's subjective probability of accepting the shared ride service offer using

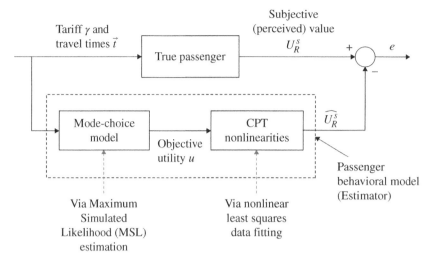

Figure 2.3 Estimation of mode choice models and CPT parameters from passenger survey data.

Equations (2.14)–(2.6). Since the CPT model better accounts for passengers' irrational preferences when faced with uncertainty and risk, this subjective acceptance probability is more accurate than what would be predicted using conventional Utility Theory alone.

2.3.1.1 Detection of CPT Effects via Lotteries
In addition to estimating the numerical values parametrizing the subjective value and probability weighting functions, we also used the survey responses to detect the key CPT effects, which are:

- **Framing effect**: Individuals value prospects with respect to a reference point instead of an absolute value, and perceive gains and losses differently.
- **Diminishing sensitivity**: In both gain and loss regimes, sensitivity diminishes when the prospect gets farther from the reference. Therefore, the perceived value is concave in the gain regime and convex for losses, implying that people are generally risk-averse in gains and risk-seeking in the loss regime.
- **Probability distortion**: Individuals overweight small probability events and underweight large probability events.
- **Loss aversion**: Individuals are affected much more by losses than gains.

In order to estimate these effects, we considered simplified choice scenarios involving monetary lotteries. Survey respondents were also asked a series of hypothetical lottery questions after completing the SRS travel choice scenarios. We can then test for the existence of CPT-like behaviors based on their lottery responses, as described in Rieger et al. (2017). More details on the survey and lottery scenarios can be found in Jagadeesan Nair (2021).

From Table 2.4, we see that the valid responses clearly display CPT effects. The reflection or framing effect is shown by nearly all the valid respondents, indicating that our proposed value function is likely an accurate descriptor of how the passengers perceive their gains and losses. The probability weighting effect is not as dominant, but it is still quite significant. We find that majority of them (>72%) show at least some overweighting of probabilities, and it is also most common in the lower probability ranges (between 10 and 60%). This agrees with CPT theory since it postulates that people tend to overestimate the likelihood of rare events. The relatively large value of the mean gain/loss ratio (>1) in the mixed lotteries indicates a significant degree of loss aversion

Table 2.3 Summary of CPT parameter estimates for SRS travel preferences.

	α^+	α^-	β^+	β^-	λ
Mean	0.4456	0.1315	0.2166	0.3550	20.0494
Median	0.4124	0.1320	0.2188	0.3649	11.8715
SD	0.1828	0.0448	0.0985	0.1906	25.8554

Table 2.4 Summary of key CPT effects observed from lotteries.

CPT effect tested	% of valid responses
Reflection effect (framing & diminishing sensitivity)	95.03
Probability overweighting between	
10% and 60% probability	62.56 %
60% and 90% probability	40.51 %
10% and 90% probability	51.05 %
Any probability weighting	72.44 %
Loss aversion	
Mean gain/loss ratio for mixed outcome lotteries	3.7254
Median gain/loss ratio for mixed outcome lotteries	1.0250

among the surveyed passengers. However, the median value is quite close to 1 indicating that loss aversion may not be as prevalent for a sizeable portion of the passengers sampled in this study. These results from Tables 2.3 and 2.4 demonstrate that we can use survey responses and data on passenger choices to (i) validate CPT behavioral effects and also (ii) estimate mathematical models and parameters to describe their risk preferences. Using real-time data from sources like ridesharing apps, these models can be continuously updated and improved in order to more accurately predict passenger choice and behavior.

2.3.2 Theoretical Implications of CPT

Before discussing the theoretical implications of CPT for SRS, we briefly outline the reference R which is one of the most important parameters for CPT modeling There are mainly two different types of references Wang and Zhao (2018):

1. **Static reference**: These are fixed values independent of the SRS offer. For example, this could be the objective utility of the alternative option to the SRS, i.e. $R = U_A^0 = A^0$.
2. **Dynamic reference**: These depend directly on the uncertain prospect (SRS offer) itself. For example, the reference could be $R = \tilde{x} + b\gamma$, where the travel time disutility $\tilde{x}$ may be set as the best case ($\underline{x}$), worst case ($\overline{x}$), or some value in between $\tilde{x} \sim F_X(x)$.

In Section 2.3.1, we determined mode choice and CPT behavioral models for passengers for a large population and over many different possible travel scenarios. In the following, we consider the case of a single trip (ride request and offer), in order to draw a few key insights about passenger risk preferences and travel behavior using computational experiments. Here, the passenger makes a choice between the SRS option (uncertain prospect) and an exclusive ride-hailing service like UberX (certain prospect). This section is adapted from our earlier work applying CPT-based dynamic pricing for SRS (see Guan et al. (2019a) for more details).

A dynamic routing problem of 16 passengers using real SRS request data from San Francisco was considered (Annaswamy et al. 2018), and the request from one of these passengers was utilized to run the computational experiments shown here. An alternating minimization-based algorithm developed in Guan et al. (2019b) was applied to determine the optimal routes and corresponding travel times of the SRS. The possible delays in travel time were constrained to be at most four minutes of extra wait and ride time, respectively. For each ride request, travel times and prices of the SRS option (UberX) were retrieved from Uber.[1]

Using the travel times and prices, along with utility coefficients from Table 2.1, the objective utility A^o of the alternative (UberX) and $\underline{x}$, $\overline{x}$ of the SRS are calculated, using Equation (2.4). Note that $A^o, \underline{x}, \overline{x}$ are negative as they represent disutilities incurred by the passenger due to travel times and tariffs. Using this numerical setup, we explore the three implications via simulations: (i) fourfold pattern of risk attitudes, (ii) strong aversion of mixed prospects, and (iii) self-reference. We use the following key properties of CPT-based behavioral models for our analysis; more details and derivations of these can be found in Guan et al. (2019a):

Property 1 is related to static and dynamic reference points as described above. These help decide the dynamic tariff γ that drives the subjective acceptance probability p_R^s to approach the desired probability of acceptance p^*. Defining the expected values as $\bar{U} = \mathbb{E}_{f_U}(u)$ and $\overline{X} = \mathbb{E}_{f_x}(x)$, Properties 2 and 3 are related to the subjective utility and subjective acceptance probability when using $R = \bar{U}$, i.e. $U_{\bar{U}}^s$ and $p_{\bar{U}}^s$, respectively, Guan et al. (2019a).

1. Given any static reference point $R \in \mathbb{R}$ or dynamic reference point of the form of $R = \tilde{x} + b\gamma$, $\tilde{x} \in \mathbb{R}$, p_R^s strictly decreases with the tariff γ.
2. Given any uncertain prospect, there exists a λ^*, such that $\forall \lambda > \lambda^*$, $U_{\bar{U}}^s < 0$.
3. For any uncertain prospect, given that λ is sufficiently large such that $U_{\bar{U}}^s < 0$, and within the price range $\gamma \in [\underline{\gamma}, \overline{\gamma})$, where $\underline{\gamma}$ satisfies $\overline{X} + b\underline{\gamma} = A^o$, and $\overline{\gamma}$ satisfies $\left[A^o - (\overline{X} + b\overline{\gamma})\right]^{\beta^+} - U_{\bar{U}}^s = A^o - (\overline{X} + b\overline{\gamma})$, $p_{\bar{U}}^s < p^o$. In the SRS context, this implies that under the stated conditions, the uncertain SRS is relatively less attractive to the passenger than the alternative A in terms of their probability of accepting the ride offer.

All of the above properties do agree with our intuition, but they allow us to formally and precisely express these ideas to provide a rigorous mathematical framework. In Sections 2.3.2.1–2.3.2.3, we use these along with the previously defined CPT-based behavioral model to quantify the theoretical consequences of CPT for SRS applications.

2.3.2.1 Implication I: Fourfold Pattern of Risk Attitudes

The fourfold pattern of risk attitudes is regarded as "the most distinctive implication of Prospect Theory" (Tversky and Kahneman 1992), which states that while facing an uncertain prospect, individual risk attitudes can be classified into four categories:

1. Risk averse over high probability gains.
2. Risk seeking over high probability losses.
3. Risk seeking over low probability gains.
4. Risk averse over low probability losses.

---

1 https://www.uber.com/.

These risk attitudes are often used to justify the subjective decision-making of individuals for application such as SRS as well as other problems such as playing lotteries (as described in Section 2.3.1.1) and getting insurance coverage, among others.

We now illustrate the fourfold pattern in the SRS context using the following scenario, which uses the classic setup for the analysis of the fourfold pattern (Tversky and Kahneman 1992): individuals decide between two options, a certain prospect and an uncertain prospect with two outcomes. The uncertain prospect is the SRS (i.e. UberX), which we assume obeys a truncated Poisson distribution with $K = 1$, i.e. the passenger is subjected to at most one delay. Therefore, the two possible SRS outcomes are $(\underline{x} + b\gamma)$ and $(\overline{x} + b\gamma)$, with the following corresponding probabilities derived from the Poisson probability mass function (PMF) $f_X^P(x)$:

$$f_X^P(x) = \begin{cases} \frac{1}{Z^P} \frac{(\lambda^P)^k e^{-\lambda^P}}{k!}, & \text{if } x = \overline{x} - k\frac{\overline{x}-\underline{x}}{K} \\ 0, & \text{otherwise} \end{cases} \tag{2.17}$$

$$f_X^P(\underline{x}) = \frac{\lambda^P}{\lambda^P + 1}, \quad f_X^P(\overline{x}) = \frac{1}{\lambda^P + 1} \tag{2.18}$$

The four scenarios above are realized through suitable choices of R and λ^P as follows: depending on the choice of the dynamic reference point, the SRS is a gain if $R = \underline{x} + b\gamma$ and a loss if $R = \overline{x} + b\gamma$. The SRS is considered high or low probability when the outcome that is not the reference can be realized with a probability of p_{NR} or $(1 - p_{NR})$, respectively, where p_{NR} is close to 1. In the computational experiments presented in Figure 2.4, $p_{NR} = 0.95$. Moreover, the range of the tariff is chosen as follows:

$$\begin{cases} \underline{x} + b\gamma < A^o & \text{if } R = \underline{x} + b\gamma \\ \overline{x} + b\gamma > A^o & \text{if } R = \overline{x} + b\gamma \end{cases} \tag{2.19}$$

such that the objective utility of the certain prospect, A^o, lies in the same gain or loss regime as the SRS and therefore represents a reasonable alternative to the SRS.

With the uncertain and the certain prospect defined in the SRS context above, we illustrate the fourfold pattern in Figure 2.4 using four quadrants. According to the fourfold pattern (a)–(d) in

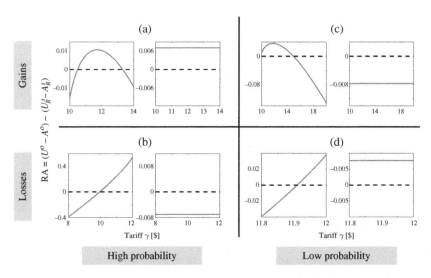

Figure 2.4 Illustration of the fourfold pattern of risk attitudes in the SRS context. (a) $R = \underline{x} + b\gamma$ and $f_X^P(\underline{x}) = 0.05$, (b) $R = \underline{x} + b\gamma$ and $f_X^P(\underline{x}) = 0.95$, (c) $R = \overline{x} + b\gamma$ and $f_X^P(\underline{x}) = 0.95$, and (d) $R = \overline{x} + b\gamma$ and $f_X^P(\underline{x}) = 0.05$.

Figure 2.4, the diagonal quadrants should correspond to risk-averse behavior while the off-diagonal ones are risk-seeking. In each quadrant, we plot a metric defined as RA $= (U^o - A^o) - (U_R^s - A_R^s)$ with respect to the tariff γ. This metric captures the Relative Attractiveness (RA) that the uncertain prospect has over the certain prospect for rational individuals (who can be modeled by conventional Utility Theory) versus individuals modeled with CPT. This follows from Equations (2.13) and (2.14), since RA $> 0 \implies p^o > p_R^s$. In Figure 2.4, we note that RA > 0 corresponds to all regions where the light gray curve is above zero and indicates risk-averse attitudes, as rational individuals have higher probability to accept the uncertain prospect than irrational ones. Similarly, RA < 0 corresponds to the light gray line being below zero and denotes risk-seeking attitudes. In each quadrant, two subplots are provided. The subplot on the right corresponds to a specific set of parameters $\beta^+ = \beta^- = \lambda = 1$ which completely removes the role of $V(\cdot)$ and hence removes the effects of framing, diminishing sensitivity and loss aversion (as specified in Section 2.3.1.1). The subplot on the left corresponds to standard CPT parameters chosen in the range $0 < \beta^+, \beta^- < 1$, $\lambda > 1$, thus producing a general CPT model. As explained before, each quadrant corresponds to a specific choice of R and λ^P, which together determine if an outcome is a gain or loss, and whether with high or low probability.

The most important observation from Figure 2.4 comes from the differences between the left and right subplots in each of the four quadrants. For example, from Figure 2.4a which is the case of high probability gains, all risk attitudes in the right subplot correspond to RA > 0 and therefore risk averse, while those on the left are only risk averse for a certain price range. That is, the fourfold pattern is violated in the left subplot. The same trend is exhibited in all four quadrants. This is because the fourfold pattern is due to the interplay between $\pi(\cdot)$ and $V(\cdot)$ and is valid only when the magnitude of $\pi(\cdot)$ is sufficiently large relative to that of $V(\cdot)$, such that probability distortion dominates (Harbaugh et al. 2009). This corresponds to the right subplots[2] as well as the left subplots within certain price ranges.

The insights drawn from this analysis of the fourfold risk attitudes pattern is that these four categories can suitably inform the dynamic pricing strategy for the SRS tariff design, through the left subplots in each quadrant of Figure 2.4. That is, it helps quantify two key qualitative statements: (i) presence of risk-seeking passengers allows for flexibility in charging higher tariffs, and (ii) presence of risk-averse passengers highlights the need for some additional constraints or limits on reasonable tariffs that can be imposed.

2.3.2.2 Implication II: Strong Risk Aversion Over Mixed Prospects
A mixed prospect is defined as an uncertain prospect whose portfolio of possible outcomes involves both gains and losses (Kahneman and Tversky 2012; Abdellaoui et al. 2008). Clearly, an uncertain prospect is always mixed when the reference point R is chosen to be its own expectation (i.e. expected value of its outcomes). The strong risk aversion of mixed prospects in the CPT framework stems from mainly loss aversion, as the impact of the loss often dominates its gain counterpart. This implication will be illustrated below in our SRS context using two different interpretations.

The first interpretation follows from property 2, which states that when $R = \bar{U}$, the subjective utility is strictly negative for a sufficiently large λ. Therefore, with $R = \bar{U}$ and such a large enough λ, the uncertain prospect is always subjectively perceived as a strict loss. This has been verified numerically with $\lambda > 1$. Since the objective utility determined by Utility Theory relative to the expectation is neutral (since $\bar{U} = U^0 = \mathbb{E}_{f_U}(u)$), strong aversion is exhibited.

The second interpretation follows from Property 3, which implies that when Property 2 holds, within the tariff range $[\underline{\lambda}, \bar{\lambda})$, the uncertain prospect is less likely to be accepted by the CPT-inclined passengers compared to the rational ones, as $p_{\bar{U}}^s < p^o$.

---

2 The right subplot in each quadrant corresponds to the case where individuals are risk neutral in the gain or loss regimes separately, and loss neutral, then $\pi(\cdot)$ alone is sufficient to generate the fourfold pattern.

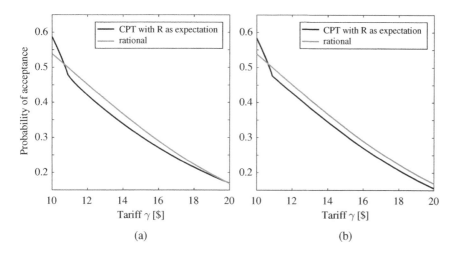

Figure 2.5 Comparison of $p_{\bar{U}}^s$ and p^o. For fair comparison, the tariff range of $\gamma \geq \frac{A^o - \bar{X}}{b}$ is plotted, where the alternative is nonloss. (a) CPT with $\beta^+ < 1$ and (b) CPT with $\beta^+ = 1$.

Figure 2.5 illustrates Property 3 with $f_X(x)$ obeying a Normal distribution, with the tariff range $\underline{\gamma}$ and $\overline{\gamma}$ approximated using the numerical setup. It is clear from the left subplot that within this price range, passengers exhibit strong risk aversion over the SRS, as the light gray curve is strictly above the dark gray one. It is interesting to note that when $\beta^+ = 1$, which corresponds to the scenario where passengers are risk neutral in the gain regime, the maximum possible tariff $\overline{\gamma} \to \infty$ (see Figure 2.5b).

The implication regarding strong risk aversion over mixed prospects is as follows: as the SRS has significant uncertainty, for passengers who regard the expected service quality as the reference, and when the alternative is relatively a nonloss prospect, strong risk aversion is exhibited. Hence, the SRS is strictly less attractive to these passengers when compared to rational ones. Therefore, the dynamic tariffs may need to be suitably designed by the SRS operator or server so as to compensate for these perceived losses. These could also take the form of special rebates, subsidies, or deals for SRS passengers.

2.3.2.3 Implication III: Effects of Self-Reference

In this section, we compare $p_{\bar{U}}^s$ with $p_{A^o}^s$, i.e. the subjective acceptance probability of the uncertain SRS while using (i) the expected SRS outcome as the dynamic reference versus, (ii) the objective utility of the certain alternative (e.g. UberX) as a static reference, respectively. Four different probability distributions are considered. In each case, how these two probabilities vary with the tariff γ was evaluated. The results are shown in Figure 2.6.

Figure 2.6 illustrates that for all four distributions, $p_{\bar{U}}^s \geq p_{A^o}^s, \forall \gamma$, which implies that the SRS is always more attractive when the reference is the expectation of itself, rather than the alternative. Note that $p_{\bar{U}}^s = p_{A^o}^s$ when $\gamma = \frac{A^o - \bar{X}}{b} \implies \bar{U} = A^o$ and hence the two reference points coincide.

The following summarizes the third implication inferred from Figure 2.6: $\bar{U}$ is essentially the rational counterpart of the uncertain prospect. Therefore, it could be argued that, when deciding between two prospects, the chance to accept one prospect is always higher if this prospect itself is regarded as the reference, compared with the case where the alternative is considered as the reference. This arises from loss aversion, i.e. $\lambda > 1$, and can be explained as follows: when one prospect is regarded as the reference, by definition, it would never be perceived as a loss and therefore not experience the magnified perception out of losses, whereas the alternative may be subject

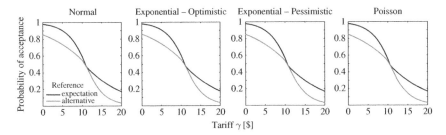

Figure 2.6 Comparison of $p_{\bar{U}}^s$ with $p_{A^\circ}^s$ using four different PDFs $f_X(x)$. In the truncated Poisson distribution, the parameters are set as $\lambda^P = 4$ and $K = 5$.

to being regarded as a loss and therefore can experience this skewed perception. In contrast, if the alternative is chosen as the reference, the roles are reversed.[3] Moreover, the statement is in fact intuitive as those passengers who regard the expectation as the reference have in some sense already subscribed to the SRS, and hence are naturally inclined to exhibit higher acceptance probabilities and therefore have higher willingness to pay. This partially explains the reason why converting customers to more uncertain SRS options from competitors and more traditional, predictable transportation modes (like driving, public transit, and exclusive ride-hailing) is typically more difficult than maintaining the current customer base. The last observation from Figure 2.6 is the invariance of the comparison with respect to the underlying probability distributions, which implies that the above results on self-reference are fairly general.

2.4 Summary and Conclusions

Several examples of CPHS include real-time decisions from humans as a necessary building block for the successful performance of the overall system. Many of these problems require behavioral models of humans that lead to these decisions. In this chapter, we describe two different tools that may be suitable for determining these behavioral models, which include Utility Theory and Prospect Theory. Tools from Utility Theory have been used successfully in several problems in transportation such as resource allocation and balance of supply and demand. This theory is described in Section 2.2 and illustrated using a transportation example that consists of a shared mobility problem where human riders are presented with the choice of different travel modes. We then show how these models can be used to address and mitigate traffic congestion. Cumulative Prospect Theory, an extension of Prospect Theory, is a modeling tool widely used in behavioral economics and cognitive psychology that captures subjective decision-making of individuals under risk or uncertainty. CPT is described in Section 2.3, with the same transportation example used to illustrate its application potential. Results from a survey conducted with about 1000 respondents are used to derive a CPT model and estimate its parameters. Lottery questions were also included in the survey to illustrate CPT effects, the results of which are described in this section as well. Finally, a few theoretical implications of CPT are presented that provide an overall quantitative structure to the qualitative behavior of humans in overall decision-making scenarios.

Human-in-the-loop behavioral models can be applied to several other applications beyond ridesharing, both within the transportation and mobility sector, and other domains that involve

---

3 Other effects of CPT due to $\alpha, \beta^+, \beta^- < 1$ may result in complicated nonlinearities which might alleviate loss aversion. Therefore, this statement is valid when λ is sufficiently larger than 1, such that loss aversion dominates.

humans interacting with cyber–physical systems. One such example includes demand response in power grids. Future research involves the development of more accurate models of human operators for which more quantitative data are required with human decision-makers as integral components of the overall infrastructure. For instance, these could include more sophisticated utility functions for different modes that also take into account factors other than price and travel time which could affect passengers' choices. In addition, more realistic data sets on ridesharing, mode choice, and dynamic pricing could be obtained either by conducting larger, representative surveys or through pilot studies and field trials performed in conjunction with ridesharing companies or transit authorities.

Acknowledgments

This work was supported by the Ford-MIT Alliance.

References

Abdellaoui, M., Bleichrodt, H., and l'Haridon, O. (2008). A tractable method to measure utility and loss aversion under prospect theory. *Journal of Risk and Uncertainty* 36 (3): 245.

Annaswamy, A. and Nudell, T. (2015). Transactive control–what's in a name. *IEEE Smart Grid Newsletter*

Annaswamy, A.M., Guan, Y., Tseng, H.E. et al. (2018). Transactive control in smart cities. *Proceedings of the IEEE* 106 (4): 518–537.

Bejestani, A.K., Annaswamy, A., and Samad, T. (2014). A hierarchical transactive control architecture for renewables integration in smart grids: analytical modeling and stability. *IEEE Transactions on Smart Grid* 5 (4): 2054–2065.

Ben-Akiva, M.E., Lerman, S.R., Lerman, S.R. (1985). *Discrete Choice Analysis: Theory and Application to Travel Demand*, vol. 9. MIT Press.

Bernards, R., Reinders, J., Klaassen, E. et al. (2016). Meta-analysis of the results of European smart grid projects to quantify residential flexibility. *CIRED Workshop 2016*, 1–4. IET.

Chassin, D.P., Malard, J.M., Posse, C. et al. (2004). Modeling Power Systems as Complex Adaptive Systems. *Technical Report No. PNNL-14987*. Richland, WA (United States): Pacific Northwest National Lab. (PNNL).

Erslan, E., Yildiz, Y., and Annaswamy, A.M. (2023). *Safe Shared Control in the Face of Anomalies*. IEEE-Press.

Fishburn, P.C. (1988). *Nonlinear Preference and Utility Theory*, No. 5. Baltimore, MD: Johns Hopkins University Press.

Guan, Y., Annaswamy, A.M., and Tseng, H.E. (2019a). Cumulative prospect theory based dynamic pricing for shared mobility on demand services. *2019 IEEE 58th Conference on Decision and Control (CDC)*, 2239–2244. IEEE.

Guan, Y., Annaswamy, A.M., and Tseng, H.E. (2019b). A dynamic routing framework for shared mobility services. *ACM Transactions on Cyber-Physical Systems* 4 (1): 6:1–6:28. https://doi.org/10.1145/3300181.

Hammerstrom, D.J., Ambrosio, R., Carlon, T.A. et al. (2008). Pacific Northwest GridWise™ testbed Demonstration Projects; Part I. Olympic Peninsula Project. *Technical Report No. PNNL-17167*. Richland, WA (United States): Pacific Northwest National Lab. (PNNL).

Hao, H., Corbin, C.D., Kalsi, K., and Pratt, R.G. (2016). Transactive control of commercial buildings for demand response. *IEEE Transactions on Power Systems* 32 (1): 774–783.

Harbaugh, W.T., Krause, K., and Vesterlund, L. (2009). The fourfold pattern of risk attitudes in choice and pricing tasks. *The Economic Journal* 120 (545): 595–611.

Jagadeesan Nair, V. (2021). Estimation of cumulative prospect theory-based passenger behavioral models for dynamic pricing & transactive control of shared mobility on demand. Master's thesis. Massachusetts Institute of Technology.

Kahneman, D. and Tversky, A. (2012). Prospect theory: an analysis of decision under risk. In: (ed. L.C. MacLean and W.T. Ziemba) *Handbook of the Fundamentals of Financial Decision Making, World Scientific Handbook in Financial Economics Series*, vol. 4, 99–127. World Scientific. ISBN 978-981-4417-34-1. https://doi.org/10.1142/9789814417358_0006. https://www.worldscientific.com/doi/abs/10.1142/9789814417358_0006.

Katipamula, S., Chassin, D.P., Hatley, D.D. et al. (2006). Transactive Controls: A Market-Based GridWise™ Controls for Building Systems. *Technical Report No. PNNL-15921*. Richland, WA (United States: Pacific Northwest National Lab. (PNNL).

Kok, K. (2013). *The PowerMatcher: Smart Coordination for the Smart Electricity Grid*, 241–250. The Netherlands: TNO.

Li, S., Zhang, W., Lian, J., and Kalsi, K. (2015). Market-based coordination of thermostatically controlled loads–Part I: a mechanism design formulation. *IEEE Transactions on Power Systems* 31 (2): 1170–1178.

Lombardi, C., Annaswamy, A., and Santos, L.P. (2022). Model-based dynamic toll pricing scheme for a congested suburban freeway with multiple access locations. *Journal of Intelligent Transportation Systems (provisionally accepted)*, 1–28.

Melton, R. (2015). Pacific Northwest Smart Grid Demonstration Project Technology Performance Report Volume 1: Technology Performance. *Technical Report PNW-SGDP-TPR-Vol.1-Rev.1.0; PNWD-4438, Volume 1*. Richland, WA (United States): Pacific Northwest National Lab. (PNNL).

Phan, T., Annaswamy, A.M., Yanakiev, D., and Tseng, E. (2016). A model-based dynamic toll pricing strategy for controlling highway traffic. *2016 American Control Conference (ACC)*, 6245–6252. IEEE.

Prelec, D. (1998). The probability weighting function. *Econometrica* 66 (3): 497–527. https://doi.org/10.2307/2998573.

Rieger, M.O., Wang, M., and Hens, T. (2017). Estimating cumulative prospect theory parameters from an international survey. *Theory and Decision* 82 (4): 567–596. https://doi.org/10.1007/s11238-016-9582-8.

Schweppe, F.C. (1978). Power systems2000': hierarchical control strategies. *IEEE Spectrum* 15 (7): 42–47.

Schweppe, F.C., Tabors, R.D., Kirtley, J.L. et al. (1980). Homeostatic utility control. *IEEE Transactions on Power Apparatus and Systems* (3): 1151–1163.

Somasundaram, S., Pratt, R.G., Akyol, B. et al. (2014). Reference Guide for a Transaction-Based Building Controls Framework. Pacific Northwest National Laboratory.

Tversky, A. and Kahneman, D. (1992). Advances in prospect theory: cumulative representation of uncertainty. *Journal of Risk and Uncertainty* 5 (4): 297–323. https://doi.org/10.1007/BF00122574.

Von Neumann, J. and Morgenstern, O. (2007). *Theory of Games and Economic Behavior (Commemorative Edition)*. Princeton University Press.

Wang, S. and Zhao, J. (2018). How Risk Preferences Influence the Usage of Autonomous Vehicles. *Technical Report number: 18-00785*. Washington DC, United States: Transportation Research Board 97th Annual Meeting.

Wang, S. and Zhao, J. (2019). Risk preference and adoption of autonomous vehicles. *Transportation Research Part A: Policy and Practice* 126: 215–229.

Widergren, S.E., Subbarao, K., Fuller, J.C. et al. (2014). AEP Ohio gridSMART Demonstration Project Real-Time Pricing Demonstration Analysis. *Technical Report No. PNNL-23192*. Richland, WA (United States): Pacific Northwest National Lab. (PNNL).

3

Social Diffusion Dynamics in Cyber–Physical–Human Systems

Lorenzo Zino and Ming Cao

Faculty of Science and Engineering, University of Groningen, Groningen, The Netherlands

3.1 Introduction

Social diffusion is a fundamental process in social communities, as it is key to the well-functioning and the evolution of human societies. Through social diffusion, new conventions, behaviors, ideas, and technological advances are introduced in a population, and the members of the population collectively adopt them, some of which change the existing status quo. To better understand the importance of such a process, we underscore that social diffusion has led to many important changes and advances in our society. Classical examples, having been extensively studied in the literature, include the abandoning of painful and dangerous cultural practices such as footbinding in rural China at the beginning of the twentieth century (Mackie 1996; Brown and Satterthwaite-Phillips 2018), the fast adoption of hybrid seed corn by farmers for improved crop yield in the 1930s (Ryan and Gross 1943, 1950), the usage of new medicines by medical professionals (Valente 1996), and the continuous evolution of linguistic conventions in spoken and written languages (Lieberman et al. 2007; Amato et al. 2018). Remarkably, the importance of social diffusion is being extensively witnessed during the COVID-19 pandemic, which is still ongoing at the moment of writing this chapter. In fact, the prompt and collective adoption of new practices and social norms, such as the implementation of working from home, the use of face masks, and the adoption of elbow-bumping as a greeting convention instead of shaking hands, are key for the success of nonpharmaceutical intervention policies in mitigating the spread of the disease and avoiding the collapse of the health-care system. See, for instance, West et al. (2020), Betsch et al. (2020), Prosser et al. (2020), and Martínez et al. (2021).

These examples have highlighted the importance of social diffusion and its pervasive presence in many real-world humans-in-the-loops systems, especially for the category of humans-in-the-multiagent-loops, discussed in Samad (2023); Chapter 1 in this book. Hence, it is no surprise that social psychologists have extensively studied such a process by means of experimental studies (see, for instance, Centola and Baronchelli (2015), Centola et al. (2018), and Andreoni et al. (2021)), which has been key for the development of sociological and psychological theories on social diffusion and social change, as in Lewis (2002), Rogers (2003), Bicchieri (2005), and Marmor (2009). In the last decade, the increasingly growing amount of data available and the technological advances that have allowed to process and analyze such data have paved the way for the foundation and the development of a new research field: computational social science. See Lazer et al. (2009). The use of tools and techniques from computational social science has been

Cyber–Physical–Human Systems: Fundamentals and Applications, First Edition.
Edited by Anuradha M. Annaswamy, Pramod P. Khargonekar, Françoise Lamnabhi-Lagarrigue, and Sarah K. Spurgeon.
© 2023 The Institute of Electrical and Electronics Engineers, Inc. Published 2023 by John Wiley & Sons, Inc.

fundamental in the understanding of many social diffusion processes, including the evolution of written languages (Amato et al. 2018), information diffusion during natural disasters (Dong et al. 2018; Cinelli et al. 2020), diffusion of online innovation (Karsai et al. 2014), and online attention dynamics, discussed in Castaldo et al. (2023); Chapter 18 in this book. More details on the recent advances of computational social science can be found in Edelmann et al. (2020).

Aside from these works, which focus on laboratory experiments and analysis of empirical data, mathematical models have emerged as powerful frameworks for studying social diffusion, beyond the practical limitations of experimental settings, utilizing tools from dynamical systems theory and control theory. The first class of models proposed in the 1960s are population models, which include the well-known Bass model introduced by the marketing scientist Frank Bass and its variants. See, Bass (1969), Mahajan et al. (1990), and Rogers (2003). This class of models captures the diffusion process at the macroscopic level, using only a few parameters to reproduce the diffusion process. However, despite the fact that population models successfully capture some important aspects of social diffusion processes, they are limited in their ability to explore the role of individual-level mechanisms and of the communication network layer on the emergent behavior of the system. To bridge this gap, agent-based models (ABMs) have been proposed and developed as valuable tools to capture dynamics involving humans-in-the-multiagent-loops. ABMs have been used to predict social diffusion processes, from individual-level details to the emergent behavior of the system, and to possibly design intervention policies to influence the final outcome of the dynamics, thereby favoring (or hindering) social diffusion. Relevant references are (Granovetter 1978; Goldenberg et al. 2001; Kempe et al. 2003; Bettencourt et al. 2006). In the past few decades, the study of social diffusion processes through mathematical models and, in particular, through ABMs has become very popular and has attracted the attention of scholars from many different fields and research communities, spanning from systems and control engineers, to physicists and network scientists, to marketing scientists and economists, to mathematical sociologists (Rogers 2003; Young 2009; Jusup et al. 2022). Hence, one of the purposes of this chapter is to present a general overview of all these important efforts from a dynamical systems and control perspective.

In ABMs, a population of individuals has specified some individual-level dynamics. These dynamics are typically assumed to be simple (often inspired by epidemic processes, complex contagion mechanisms, or game-theoretic interactions) and are designed to capture important social, psychological, and behavioral aspects of human decision-making processes, as detailed in Bonabeau (2002). As we shall illustrate in what follows, the tendency of being influenced by or of imitating others in the network, the sensitivity to emerging trends, the limited span of attention, the tendency to be consistent with previous choices, the bounded level of rationality in human decision-making, and many other features have all been successfully encapsulated within the ABM framework. While the individual-level dynamics are typically clean, complexity arises enabled by the communication network through which individuals interact and exchange information. The growing theory of complex networks (Newman 2010) has allowed for the representation of real-world cyber and physical constraints within the underlying individual-to-individual communication by including multiple layers of network that account for diverse interaction channels between the individuals in an ABM and, possibly, varieties of dynamics that co-evolve in an intertwined fashion. These recent advances, which have mostly been developed in the last decade, have empowered an increasingly faithful representation of real-world social diffusion processes in cyber–physical–human systems (CPHS) toward producing accurate predictions of their long-term behavior. Such an increased understanding of real-world social diffusion processes and the development of new tools to study them have enabled the systems and control community to study the complex problem of closed-loop control system architectures with humans-in-multiagent-loops (see Samad (2023); Chapter 1 in this book for more details), with the ultimate goal of proposing control-theoretic strategies to influence the final outcome of a social diffusion process in a CPHS.

The rest of this chapter is divided into four main parts. In Section 3.2, we briefly introduce a general formalism for social diffusion in CPHS and provide an overview of the main theoretical approaches that have been developed to study them. In Section 3.3, we provide an overview of the classical decision-making mechanisms that can be integrated within the general formalism for social diffusion and discuss the recent efforts and advances proposed toward realistic modeling of individual decision-making. In Section 3.4, we present some recent developments in the analysis of social diffusion in multiplex networks and on co-evolutionary dynamics, which are characteristics of real-world CPHS. In Section 3.5, we outline the main directions for the future research in this field, with a specific focus on the efforts toward controlling the emergent behavior of social diffusion processes.

3.2 General Formalism for Social Diffusion in CPHS

In this section, we discuss a general formalism for social diffusion in CPHS, which accounts for the presence of humans-in-multiagent-loops. First, we introduce the network-theoretic concept of multiplex networks, which can be used to describe the different cyber and physical communication channels that characterize CPHS. Second, we describe a general mathematical framework to study social diffusion processes occurring in these CPHS.

3.2.1 Complex and Multiplex Networks

In the last few decades, the developments in network theory have provided powerful tools to represent and study complex systems, including CPHS. More details on the recent advances of complex networks theory can be found in Newman (2010). In particular, the development of multiplex networks, that is, networks in which a set of nodes is connected through different types of edges, which may account for different communication channels, and may have different properties and features. Some surveys on multiplex networks (and, more in general, of multilayer networks) have appeared in Kivelä et al. (2014) and Boccaletti et al. (2014). In the following, we propose a simple formulation of multiplex networks. Such a formulation will be used in the rest of this chapter to represent CPHS.

A *multiplex network* is represented by a quadruple $\mathcal{G} = (\mathcal{V}, \mathcal{L}, \mathcal{E}, \mathcal{A})$, where $\mathcal{V} := \{1, \ldots, n\}$ is the *node set*, which represents the $n \in \mathbb{Z}_+$ individuals of the population. In general multiplex networks, the set of nodes may vary across the layers (see (Kivelä et al. 2014)). However, in the context of modeling CPHS, the simpler multiplex formulation with the same node set $\mathcal{V}$ across the layers suffices, and we make this assumption for the multiplex network to be discussed. The *layer set* $\mathcal{L} := \{1, \ldots, \ell\}$ denotes all the $\ell \in \mathbb{Z}_+$ different communication channels between the individuals of the population. Each layer is characterized by an *edge set*. Specifically, for each layer $k \in \mathcal{L}$, we define the set $\mathcal{E}_k \subseteq \mathcal{V} \times \mathcal{V}$, where the edge $(i, j) \in \mathcal{E}_k$ if and only if node i can communicate with node j on the kth layer. The edge sets are gathered in the set $\mathcal{E} := \{\mathcal{E}_k\}_{k \in \mathcal{L}}$. Each edge on each layer is associated with a positive weight, where $a_{ij}^{(k)} > 0$ is the *weight* associated with edge (i, j) on layer k and measures how much i interacts with j on the kth layer. If $(i, j) \notin \mathcal{E}_k$, then, by convention, $a_{ij}^{(k)} = 0$. Weights can be gathered in the (weighted) adjacency matrices $A^{(1)}, \ldots, A^{(k)}$. (Weighted) adjacency matrices are gathered in the set $\mathcal{A} := \{A^{(k)}\}_{k \in \mathcal{L}}$. The structure of a multiplex network is represented in Figure 3.1. In the special case in which $\ell = 1$, a multiplex network reduces to a standard network, also called *monoplex*. In this case, we simplify the notation by dropping the

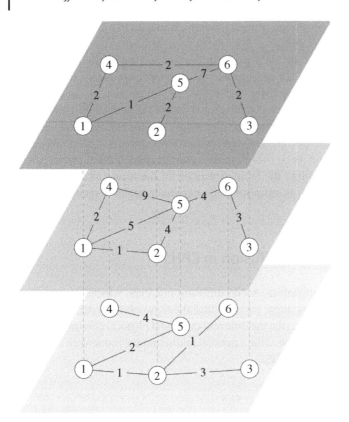

Figure 3.1 A multiplex network, with $n = 6$ nodes $\mathcal{V} = \{1, \ldots, n\}$ and three layers $\mathcal{L} = \{1, 2, 3\}$, is denoted in different shades of gray. The weights of the edges on each layer (denoted in the figure) are gathered in the weighted adjacency matrices $A^{(1)}$, $A^{(2)}$, and $A^{(3)}$.

index associated with the layer in the edge set and in the weighted adjacency matrix, denoting the (monoplex) network as a triple $\mathcal{G} = (\mathcal{V}, \mathcal{E}, A)$, and we will refer to the network instead of the layer when discussing its structure.

3.2.2 General Framework for Social Diffusion

A general framework for modeling and studying social diffusion in CPHS is defined in the following. We consider a scenario in which an innovation spreads in a population toward potentially replacing a status quo. We take innovation as a general concept, which may represent, for instance, a novel idea, product, technology, or convention, using the terminology proposed in Rogers (2003).

The population is formed by n individuals, which are connected through a (possibly multiplex) network $\mathcal{G} = (\mathcal{V}, \mathcal{L}, \mathcal{E}, A)$. Each individual in the population has to decide whether to adopt the innovation and may revise their decision during the diffusion process. The decision that individual $i \in \mathcal{V}$ makes at time t (which can be discrete, $t \in \mathbb{Z}_+$, or continuous, $t \in \mathbb{R}_+$, depending on the specific implementation of the model) is denoted by the state variable $x_i(t) \in \mathcal{S}$, where $\mathcal{S}$ is a finite set of states. In its simplest implementation, the set of states comprises two possible states, $\mathcal{S} = \{0, 1\}$, representing whether an individual adopts the status quo or the innovation. However, more states could be included to account for further refining of the categories of adopters and non-adopters. For instance, as we shall see in Section 3.3.1, the famous Daley–Kendall model, proposed in Daley and Kendall (1965), considers two classes of adopters, depending on whether they actively spread the innovation or not. The state of all nodes is gathered in the vector $x(t) \in S^n$, which is the state variable of the system.

Each individual revises their state according to some predetermined dynamics, which define the individuals' decision-making mechanisms. A high-level implementation can be formulated as follows: first, for each individual $i \in \mathcal{V}$ and pair of states $a \neq b \in S$, we define a decision-making function $f_{ab}^{(i)}(x(t), \mathcal{A}) \in \mathbb{R}_+$. This function captures the rate at which individual i changes their state from a to b, depending on the state of the system at time t ($x(t)$) and on the structure of the network underlying the CPHS, represented by the set of (weighted) adjacency matrices $\mathcal{A}$. Note that the network can pose constraints on the information that the individual gathers through the communication network. If the decision-making functions are the same for all the individuals, that is, $f_{ab}^{(i)}(x(t), \mathcal{A}) = f_{ab}(x(t), \mathcal{A})$ for all $i \in \mathcal{V}$, we say that the population is *homogeneous*; otherwise, we refer to it as an *heterogeneous* population. For the sake of simplicity, we will consider homogeneous populations, unless differently specified. The exact meaning of such a decision-making function depends on the formulation of the model.

In a discrete-time formulation, we write the rule that determines the evolution of the state of individual $i \in \mathcal{V}$ as follows:

$$\mathbb{P}[x_i(t+1) = b \mid x_i(t) = a] = f_{ab}(x(t), \mathcal{A}) \tag{3.1}$$

for all $a \neq b \in S$, and

$$\mathbb{P}[x_i(t+1) = a \mid x_i(t) = a] = 1 - \sum_{b \in S \setminus \{a\}} \mathbb{P}[x_i(t+1) = b \mid x_i(t) = a] \tag{3.2}$$

as illustrated in Figure 3.2. Here, the decision-making functions are interpreted as probabilities. Hence, in this scenario, we enforce $f_{ab}^{(i)}(x(t), \mathcal{A}) \in [0, 1]$ and $\sum_{b \in S \setminus \{a\}} \mathbb{P}[x_i(t+1) = b \mid x_i(t) = a] \in [0, 1]$, for any individual and state. The dynamics in Equation (3.1) induces a Markov chain on the state space S^n. For more details on Markov chains, we refer to Levin et al. (2006). Note that, in the specific scenario in which the decision-making functions assume only binary values, that is, $f_{ab}(x(t), \mathcal{A}) \in \{0, 1\}$ for all $a, b \in S$ and $i \in \mathcal{V}$, it follows that all the decisions are always taken in a deterministic fashion, and Equation (3.1) reduces to a deterministic recurrence equation.

In a similar fashion, a continuous time implementation of the revision rule can be formalized as follows, utilizing continuous-time Markov jump processes:

$$\lim_{\Delta t \to 0} \frac{1}{\Delta t} \mathbb{P}[x_i(t + \Delta t) = b \mid x_i(t) = a] = f_{ab}(x(t), \mathcal{A}) \tag{3.3}$$

for any $a \neq b \in S$. Note that, in the continuous-time formulation, the probability that an individual does not change state is defined implicitly, since it must be equal to the complementary of the probabilities that the individual changes state to any other possible state. For more details on continuous-time Markov jump processes, we refer to Levin et al. (2006).

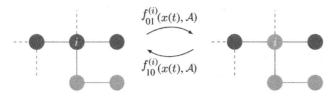

Figure 3.2 Schematic of the decision-making function in a scenario with $S = \{0, 1\}$. Individuals $j \in \mathcal{V}$ that are adopting the innovation ($x_j(t) = 1$) are denoted in light gray; those that are not adopting it ($x_j(t) = 0$) are denoted in dark gray.

Given this general formalization of social diffusion in CPHS, different models can be derived, depending on the explicit functional form of the decision-making functions. In Section 3.2.3, we will present and discuss some of the key approaches that have been proposed in the literature.

3.2.3 Main Theoretical Approaches

Before discussing in details the modeling approaches that have been proposed to capture decision-making mechanisms during social diffusion, we briefly illustrate some of the main mathematical methods and theoretical procedures that have been proposed and used to study dynamical processes in multiplex networks, especially when the complexity of the network structure and of the dynamics involved challenge the direct tractability of the dynamical system. These methods were originally proposed for the analysis of dynamics on monoplex networks, and they have been successively re-formulated in order to deal with more complex and realistic structures of communication networks.

For the study of many dynamics, *branching processes* have emerged as a simple but powerful framework to perform analytical studies. In plain words, a branching process is a stochastic process that starts with a population formed by a single individual and, at each discrete time step, each individual present in the population generates a random number of offspring that follows a predetermined distribution, and then it is removed. Utilizing the theory of branching processes and, in particular, the generating function approach, one can determine whether the process will eventually die, or if it will continue indefinitely. More details on the mathematical formulation of branching processes and on their analysis can be found in Athreya and Ney (1972). Classical epidemic models have been formulated as or bounded by branching processes, enabling their analysis on monoplex models. See, for instance, Becker (1977). Recently, analogies with branching processes have been successfully formulated for more general diffusion models and for dynamics in multiplex networks, see, for instance, Buono et al. (2014).

One of the main difficulties in the study of diffusion processes in networks lies in the discrete nature of the state space, where the state of each individual is selected by a set of states S. Consequently, the state space has a dimension that grows exponentially in the population size, practically hindering the direct analytical treatment of the dynamics. One of the approaches that have been proposed to address this issue is the so-called *microscopic Markov chain approximation* (MMCA). In the MMCA, instead of studying the state of each individual (i.e. whether the individual adopts or not the innovation), it studies the temporal evolution of the probability that the individual adopts the innovation, which is a continuous variable instead of a discrete one. Under some mild conditions, the evolution of such probabilities can be approximated using a deterministic system of equations. This would allow to study a dynamical system made by a number of equation that grows linearly in the number of individuals, instead of a stochastic process on a state space with an exponentially growing size. The MMCA, initially proposed in a discrete-time formulation in Chakrabarti et al. (2008) and in a continuous-time formulation in Van Mieghem et al. (2009) for monoplex networks, has become very popular in the last decade in the physics and complex systems communities, and has been successfully generalized to multiplex networks. See, for instance, Vida et al. (2015) and Yin et al. (2020).

In scenarios in which the high dimension of the network and the complexity of the dynamics prevent approaching the system even via an MMCA, *mean-field* theory can provide some tools to analyze the system and uncover important properties of the more complex original system. Mean-field theory exces in capturing the emergent behavior of a system at the population level by averaging the effect of all microscopic individual-level interactions and mechanisms through some macroscopic quantities. This procedure allows to further reduce the complexity of the diffusion process with respect to the MMCA by reducing the number of variables of the system to a few macroscopic ones. Hence, one can analyze them using differential equations and standard system-theoretic techniques. Also these methods have been successfully adopted to the analysis of diffusion dynamics in multiplex networks, as illustrated in Sahneh et al. (2013).

3.3 Modeling Decision-Making

In this section, we present the main modeling paradigms that have been developed toward a realistic representation of human understanding and decision-making mechanisms during social diffusion in networks. We discuss the key strong points and the main limitations of the different approaches that have been proposed, and we provide an overview of the most recent advances, which have allowed to incorporate many realistic features within these mathematical models.

The section is divided into three main parts. In the first one, we discuss the branch of mathematical models for social diffusion that are driven by pairwise interactions, inspired by the spread of epidemic diseases. In the second part, we present the modeling paradigms based on threshold mechanisms, which are able to capture the nonlinearity that characterizes many social contagion processes. In the third part, we introduce game-theoretic models, whose flexible formulation allows for the inclusion of many real-world features of social diffusion.

3.3.1 Pairwise Interaction Models

Important analogies between the spread of epidemic diseases and social diffusion processes were already observed and discussed more than 50 years ago. We mention the seminal work in Goffman and Newill (1964) and the other in Daley and Kendall (1964). These two papers, which both appeared in *Nature* in the second part of 1964, suggest that dynamical system theory can be leveraged in order to study the diffusion of a new idea or rumor in a population. In particular, both works suggest that the famous susceptible–infected–susceptible (SIS) and susceptible–infected–removed (SIR) epidemic model, originally proposed in Kermack and McKendrick (1927) for the spread of epidemic diseases, could be generalized in order to obtain a model for social diffusion processes. In an epidemic model, each individual of the population is characterized by a health state, which is selected from a finite set of compartments and can change in time according to some specified dynamics. The key dynamics are the "contagion process" and the "removal process." By means of the first, individuals that are susceptible to the disease become infected after a pairwise interaction with an infected individual; through the second, infected individuals spontaneously change their health state by recovering or dying.

In the general framework for social diffusion that we have proposed in Section 3.2.2, pairwise interaction models (sometimes referred to as epidemic-like models) can be easily formulated by observing that, whether all state changing spontaneous or triggered by pairwise interactions, the decision-making functions for a generic individual $i \in \mathcal{V}$ can be written as a linear combination of impact that each neighbor of i poses on it. Specifically, we can write

$$f_{ab}(\boldsymbol{x}(t), A) = \sum_{j \in \mathcal{V}} A_{ij} \phi_{ab}(x_j(t)) \tag{3.4}$$

where the sum in Equation (3.4) should be considered over all the layers of the network, in case of multiplex networks. For the sake of simplicity, we present the decision-making models in the scenario of a monoplex network. Here, $\phi_{ab}(c)$ can be interpreted as the probability (or rate) with which an individual changes their state from a to b after an interaction with an individual with state c, with

$a, b, c \in \mathcal{A}$. Note that spontaneous transitions can be modeled within this framework by simply including them in all the interaction kernels, independent of the state of the individual with whom i interacts. A pairwise interaction model can thus be described through the functions $\phi_{ab}(\cdot)$, which can be gathered in a set of matrices $\Phi(\cdot)$ and are called (pairwise) *interaction kernels*.

Pairwise interaction social diffusion models rely on the assumption that an innovation spreads in a population by means of pairwise interactions between the individuals, similar to what happens in the contagion mechanisms for epidemic diseases. Among the first and most known models developed within this paradigm, we mention the ignorant-spreader-stifler model. This model was originally proposed and formalized in Daley and Kendall (1965) as a stochastic model for the spread of a novel rumor in a population and is often referred to as the Daley–Kendall (DK) model. A refined and slightly simplified version of the ignorant-spreader-stifler model was proposed in Maki and Thompson (1973) and goes under the name of Maki–Thompson (MT) model.

Formally, both the classical implementations of the ignorant–spreader–stifler model, consider a continuous-time stochastic model for the diffusion of the innovation in a homogeneous population. In these models, the state of each individual is chosen among three possible categories. In particular, besides the category of nonadopters (denoted ignorants in the context of rumor spreading, 0), the category of adopters is split into two different subcategories: spreaders (S), and stiflers (R), that is, $x_i(t) \in \{0, S, R\}$. Ignorant individuals are not aware of the innovation that is spreading across the population; spreaders are aware of it and actively spread it; and stiflers are aware of it, but are no longer interested in spreading it. The innovation is propagated through pairwise interactions between spreaders and others. In particular, spreaders act similarly as infected individuals in a classical epidemic model (for more details, we refer to She et al. (2023); Chapter 16 in this book, and Zino and Cao (2021)): when an ignorant individual interacts with a spreader, the ignorant becomes aware of the innovation and becomes a spreader with a certain probability $\lambda > 0$. The key difference between epidemic models and these pairwise interaction social diffusion models lies in the "removal" process: while removal in epidemic models is assumed to be a spontaneous process through which infected individual eventually recover or die, in the ignorant–spreader–stifler models, spreaders become stiflers triggered by an interaction. In particular, in the MT model, it is assumed that a spreader that interacts with another spreader or with a stifler realizes that the innovation, that is diffusing in the population is not "novel" anymore, and may lose interest in spreading it. Hence, with a certain probability equal to $\mu > 0$, the spreader stops spreading it, becoming a stifler. Hence, the interaction kernels of the model, following Equation (3.4) are defined as

$$\Phi(0) = \begin{bmatrix} \cdot & 0 & 0 \\ 0 & \cdot & 0 \\ 0 & 0 & \cdot \end{bmatrix} \quad \Phi(S) = \begin{bmatrix} \cdot & \lambda & 0 \\ 0 & \cdot & \mu \\ 0 & 0 & \cdot \end{bmatrix} \quad \Phi(R) = \begin{bmatrix} \cdot & 0 & 0 \\ 0 & \cdot & \mu \\ 0 & 0 & \cdot \end{bmatrix} \tag{3.5}$$

where the first, second, and third rows and columns of the matrices correspond to the states ignorant, spreader, and stifler, respectively. The state transitions of this model are illustrated in Figure 3.3a. Note that the only difference in Equation (3.5) with respect to the standard SIR epidemic model from Kermack and McKendrick (1927) is in the entry $\phi_{SR}(0)$, which is equal to 0 for the ignorant–spreader–stifler model, while it is equal to μ in the classical SIR model.

Originally, the ignorant–spreader–stifler model was studied assuming no communication constraints between the individuals. Under this assumption, the n-dimensional Markov process could be reduced, by leveraging the symmetry in the population, to a lower-dimensional system, which can be studied analytically and approximated arbitrarily well by a system of ordinary differential equations (see Kurtz (1981) for more details on the approximation). More recently, the ignorant–spreader–stifler model was embedded and studied in network structures, following the seminal work in Lajmanovich and Yorke (1976) in which the standard SIS epidemic model

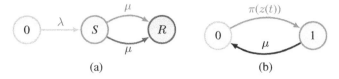

Figure 3.3 Representation of the transitions governed by the pairwise interaction kernels for (a) the ignorant–spreader–stifler model proposed in Maki and Thompson (1973) and (b) the model for diffusion of innovation proposed in Fagnani and Zino (2017). The difference in shades of the arrow is associated with the state of the individual with which an individual interacts; black arrows denote spontaneous transitions. Source: Adapted from Fagnani and Zino (2017).

was studied in networks. In Moreno et al. (2004), the authors studied the model on complex networks by means of a set of mean-field equations that approximate the stochastic system for large-scale populations. Similarly, other classical population models for social diffusion were embedded and studied on complex networks by utilizing the paradigm of pairwise interaction models. In Zhang and Moura (2014), a scaled SIS model in networks was proposed to capture social diffusion, and the authors provided a characterization of the equilibrium distribution of the process, depending on the model parameters. In Rizzo and Porfiri (2016), the authors have implemented and analyzed the classical Bass model on a time-varying network through a mechanism driven by pairwise interactions.

One key limitation of the use of classical pairwise interaction models for social diffusion lies in the fact that, different from epidemic diseases, which are transmitted through simple contagion mechanisms (i.e. typically with a constant infection probability through an interaction between a susceptible individual and an infected one), social diffusion is governed by nontrivial decision-making mechanisms through which individuals choose whether to adopt the novelty and spread it in the population. An extensive literature on the difference between the contagion mechanisms for epidemics and social contagion has been developed, focusing on the notion of complex contagion. Further details can be found in Centola and Macy (2007) and Centola (2018). A key characteristic of complex contagion is that the probability for an individual to "be infected" by the social diffusion process is typically varying and may depend (often in a nonlinear fashion) on the state of the rest of the system and, possibly, on many other endogenous and exogenous factors.

To partially address this issue, a novel model was proposed in Fagnani and Zino (2017). In this model, individuals have a binary state $S = \{0, 1\}$ to denote whether they adopt or not the innovation. The main difference with respect to a classical epidemic model is that the "contagion process" accounts for the state of the rest of the system. Specifically, when an individual, that is adopting the status quo interacts with another that adopts the innovation, then the first one becomes aware of the existence of the innovation and adopts it with a certain probability $\pi(\boldsymbol{x}(t))$, which, in general, may depend on the entire state of the system at time t. In particular, it is assumed that the function $\pi(\boldsymbol{x}(t))$ depends on the total fraction of adopters of the innovation in the network $z(t) := \frac{1}{n} \sum x_i(t)$. The "recovery process," instead, is assumed to be spontaneous, with rate $\mu > 0$, similar to the one of classical epidemic models. For this model, we can write the interaction kernels as

$$\Phi(0) = \begin{bmatrix} \cdot & 0 \\ \mu & \cdot \end{bmatrix} \qquad \Phi(1) = \begin{bmatrix} \cdot & \pi(z(t)), \\ \mu & \cdot \end{bmatrix} \tag{3.6}$$

as illustrated in Figure 3.3b. Note that, when $\phi(z) = \lambda$ for all $z \in [0, 1]$, the model reduces to the SIS epidemic model in networks, proposed, and studied in Lajmanovich and Yorke (1976).

In Fagnani and Zino (2017), the author studied the possible long-term behaviors of the system depending on the properties of function $\phi(z)$ and on the network through which individuals interact. Specifically, if we assume that the function $\phi(z)$ is nondecreasing and convex down, then,

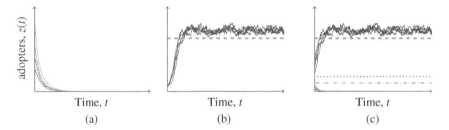

Figure 3.4 Some plots of social diffusion, generated with the model in Fagnani and Zino (2017). In (a), the innovation diffusion fails, for any fraction of initial adopters. In (b), social diffusion occurs for any fraction of initial adopters, guaranteeing a final fraction of adopters, that is above a threshold (dashed line). In (c), the fraction of initial adopters determines the outcome of the system: if above a threshold (dotted line), then social diffusion always occurs; if below a second threshold (dash-dotted line), then diffusion always fails. The two thresholds depend on the network structure. Source: Adapted from Fagnani and Zino (2017).

a rigorous analysis of the model is possible, which concludes that three scenarios could occur, depending on the model parameters and on the structure of the underlying communication network. In the first scenario, the diffusion of the innovation fails almost surely. In the second scenario, the innovation will spread almost surely reaching an equilibrium in which a fraction of the population adopts the innovation. Finally, in the third scenario, failure and success are both possible, and the outcome depends on the initial fraction of adopters in the population and their positioning in the network. Bounds on the initial fraction of adopters needed to guarantee success and failure depending on the network structure have been analytically identified. These three regimes are illustrated through some representative diffusion curves in Figure 3.4.

Even though this model and other extensions of classical pairwise interaction models are able to capture some of the characteristics of the complex contagion processes that occur during social diffusion, they are still not able to model the inherent nonlinearity that are often inherently present in decision-making processes. For this reason, further classes of models have been developed, in which the decision-making functions cannot be directly written as a linear combination of some pairwise interaction kernels.

3.3.2 Linear Threshold Models

In sociology, in particular in mathematical sociology literature, a very popular approach for modeling the adoption of innovation that started being developed in the second part of the last century are linear threshold (LT) models, which are extensively discussed in Granovetter (1978). In the LT model, individuals can decide from a binary set of states $S = \{0, 1\}$, which represent whether the individual adopts (1) or not adopts (0) the innovation. The main novelty of the LT model with respect to pairwise interaction models is that the impact of interactions with others is assumed to be inherently nonlinear. In particular, each individual $i \in \mathcal{V}$ is associated with a threshold value $\theta_i \in [0, 1]$ such that individual i will adopt the innovation only if a fraction θ_i of their neighbors adopt the innovation. In its simplest implementation, the LT model is a discrete-time deterministic process in which individuals that start adopting the innovation will never abandon it. Hence, the decision-making functions for the LT model are

$$f_{01}^{(i)}(\boldsymbol{x}(t), \mathcal{A}) = \begin{cases} 1 & \text{if } \frac{1}{d_i}\sum_{j \in \mathcal{V}} A_{ij}x_j(t) \geq \theta_i \\ 0 & \text{otherwise,} \end{cases} \tag{3.7}$$

and $f_{10}^{(i)}(\boldsymbol{x}(t), \mathcal{A}) = 0$, where $d_i = \sum_{j \in \mathcal{V}} A_{ij}$ is the degree of node i.

The model in Equation (3.7) has been extended along many different directions, and it has been studied on different network structures in order to disclose the role of the network on triggering diffusion cascades. An important, seminal contribution has been made in Watts (2002), in which the LT model is used to study the size of diffusion cascades on random networks. In this work, two different regimes with respect to the size of the diffusion cascade are identified depending on the properties of the network. In poorly connected networks, the size of a cascade follows a power-law distribution, whereby the majority of cascades fail to reach global diffusion. In highly connected networks, instead, the size distribution of cascades is proved to be bimodal, whereby, if a diffusion process is able to start the cascade process, then it will likely reach most of the network. In that study, the role of heterogeneity is also studied. In particular, it has been observed that heterogeneous thresholds favor diffusion, while heterogeneity in the degree distribution of the network hinders social diffusion. Recent works have followed this approach, with the aim of providing a further characterization of the role of the network on the outcome of cascading processes ruled by LT models. For instance, in Rossi et al. (2019), the asymptotic behavior of the diffusion process is studied on different configuration models, as a function of the distribution of degrees and of the (heterogeneous) thresholds. Among the extensions, we mention Grabisch and Li (2019), in which the long-term behavior of the system is studied in the presence of anti-conformist individuals, who adopt the innovation only when below a certain threshold.

Despite all these generalizations, the LT modeling approach has some limitations. A key limitation is that in order to capture real-world diffusion processes, LT models typically require heterogeneous thresholds. This translates, in general, into models with potentially as many parameters as the number of individuals, which are thus impossible to be calibrated in real-world scenarios, at least without the risk of overfitting the data.

3.3.3 Game-Theoretic Models

In the last two decades, game-theoretic models for decision-making have become increasingly popular. The reason for such a success lies in part in their flexibility. Within a game-theoretic framework, one can encapsulate many different real-world behavioral mechanisms that play a role in human decision-making processes utilizing a parsimonious model, using a few parameters. Therefore, in contrast to heterogeneous LT models, which require many parameters to be calibrated, game-theoretic models allow for a simple calibration of model parameters to real-world scenarios, as done, for instance, in Pagan and Dörfler (2019) and Ye et al. (2021a).

In the following, we formalize a game-theoretic decision-making mechanism in the simplest scenario of a binary state $S = \{0, 1\}$. First, for each individual $i \in \mathcal{V}$, we define two *payoff functions* $\pi_0^{(i)}(x(t), A)$ and $\pi_1^{(i)}(x(t), A)$, which model the payoff that individual i receives for choosing the status quo (state 0) or the innovation (state 1), respectively, and that may depend on the state of the others and on the network structure. Then, after introducing a rationality parameter $\beta \geq 0$, the decision-making functions are defined using the log-linear dynamics from Ellison (1993). Specifically, for each individual $i \in \mathcal{V}$, we set

$$
\begin{aligned}
f_{01}(x(t), A) &= \frac{e^{\beta \pi_1^{(i)}(x(t), A)}}{e^{\beta \pi_0^{(i)}(x(t), A)} + e^{\beta \pi_1^{(i)}(x(t), A)}} \\
f_{10}(x(t), A) &= \frac{e^{\beta \pi_0^{(i)}(x(t), A)}}{e^{\beta \pi_0^{(i)}(x(t), A)} + e^{\beta \pi_1^{(i)}(x(t), A)}}
\end{aligned}
\tag{3.8}
$$

Note that, $\beta \geq 0$ is a measure of the rationality of an individual in the decision-making: for $\beta = 0$, strategies are revised fully at random, for $\beta = \infty$, the individual will always choose the state that maximizes the payoff, yielding a deterministic model, which is called *(myopic) best-response dynamics*. Precisely, in the limit $\beta \to \infty$, the two equations in Equation (3.8) read

$$f_{01}(\boldsymbol{x}(t), A) = \begin{cases} 1 & \text{if } \pi_1^{(i)}(\boldsymbol{x}(t), A) > \pi_0^{(i)}(\boldsymbol{x}(t), A) \\ \frac{1}{2} & \text{if } \pi_1^{(i)}(\boldsymbol{x}(t), A) = \pi_0^{(i)}(\boldsymbol{x}(t), A) \\ 0 & \text{if } \pi_1^{(i)}(\boldsymbol{x}(t), A) < \pi_0^{(i)}(\boldsymbol{x}(t)A), \end{cases} \tag{3.9}$$

and

$$f_{10}(\boldsymbol{x}(t), A) = \begin{cases} 1 & \text{if } \pi_1^{(i)}(\boldsymbol{x}(t), A) < \pi_0^{(i)}(\boldsymbol{x}(t), A) \\ \frac{1}{2} & \text{if } \pi_1^{(i)}(\boldsymbol{x}(t), A) = \pi_0^{(i)}(\boldsymbol{x}(t), A) \\ 0 & \text{if } \pi_1^{(i)}(\boldsymbol{x}(t), A) > \pi_0^{(i)}(\boldsymbol{x}(t), A), \end{cases} \tag{3.10}$$

respectively. Note that other implementations of the deterministic (myopic) best-response may differ from Equations (3.9)–(3.10) for the value given to the decision-making function when $\pi_1^{(i)}(\boldsymbol{x}(t), A) = \pi_0^{(i)}(\boldsymbol{x}(t), A)$. For instance, in Zino et al. (2021), it is assumed to be equal to 0. More details on log-linear dynamics can be found in Ellison (1993).

Game-theoretic decision-making models are thus governed by the payoff functions. To model social diffusion processes in networks, a key mechanisms that has been typically incorporated into the payoff function is social influence, as suggested by the social psychology literature (Noelle-Neumann 1993; Cialdini and Goldstein 2004). Specifically, social influence is modeled through coordination games, as initially proposed in Morris (2000). In a network coordination game, it is assumed that each player plays a two-player coordination game with each of their neighbors on the network, receiving a unit payoff for coordinating on the status quo (state 0) and a payoff equal to $1 + \alpha$ for coordinating on the innovation (state 1), where the parameter $\alpha \in [(-1, \infty)$ represents a relative disadvantage (if negative) or advantage (if positive) of the innovation with respect to the status quo. Then, the overall payoff that an individual receives for choosing the status quo or the innovation is computed as a linear combination of the payoffs received in each game with each one of their neighbors, where the weights are determined by the adjacency matrix A. Hence, we obtain the following payoff functions:

$$\pi_0^{(i)}(\boldsymbol{x}(t), A) = \sum_{j \in \mathcal{V}} A_{ij}(1 - x_j(t)), \quad \pi_1^{(i)}(\boldsymbol{x}(t), A) = (1 + \alpha) \sum_{j \in \mathcal{V}} A_{ij} x_j(t) \tag{3.11}$$

as illustrated in Figure 3.5. Note that, in the limit case $\beta \to \infty$ in which Equation (3.8) reduces to the deterministic best-response dynamics in Equations (3.9)–(3.10), the coordination game in network induces a dynamics, that is very similar to the LT model, where the relative advantage α shapes the threshold θ. However, differently from the LT model, in the deterministic best-response dynamics, an individual who has adopted the innovation is allowed to switch back to the status quo, according to (3.10).

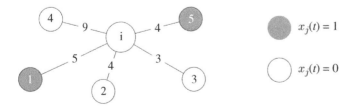

Figure 3.5 Example of a game-theoretic coordination game in which an individual receives a unit payoff for coordinating on the status quo (0, white) and a payoff equal to $1 + \alpha$ for coordinating on the innovation (1, gray). Overall, following Equation (3.11), player i receives $\pi_0^{(i)} = 16$ for choosing action 0 and $\pi_1^{(i)} = 9(1 + \alpha)$ for choosing action 1.

Coordination games in networks have been extensively studied in the literature. We mention the works in Montanari and Saberi (2010) and Young (2009, 2011), which have allowed us to shed light on the key role played by the network structure and the relative advantage in determining whether an innovation diffuses in a network, establishing also interesting results on the speed of the diffusion process. Specifically, it has been shown that an important characteristic for networks to favor social diffusion is the presence of clusters and cohesive sets, where the number of connections within the set is sufficiently large compared to the number of connections with individuals outside the set.

As already mentioned, one of the key advantages of the game theoretic setting lies in its flexibility. By changing the structure of the payoff function in Equation (3.11), one can incorporate further behavioral mechanisms that play a role in the decision-making process. For instance, based on experimental evidence, two additional terms were added to the payoff functions in Ye et al. (2021a): one term accounts for the individuals' tendency to be consistent with their past decisions (often referred to as inertia or status quo bias (Samuelson and Zeckhauser 1988) in the social psychology literature); the second one captures the sensitivity of the individuals to emerging trends in the population, which has been pervasively observed in the experimental psychology literature (Mortensen et al. 2019). However, the complexity of the proposed model hinders its analytical treatment, limiting the analysis to Monte Carlo numerical simulations. A simpler extension of the coordination game model, in which emerging trends are incorporated in a more simplistic fashion, is studied analytically in Zino et al. (2021).

The game-theoretic formalism presented in this section is amenable to several analytically tractable extensions. We mention the work in Ramazi et al. (2016) that studies the asymptotic behavior of network games on heterogeneous populations, in which some individuals perform an anti-coordination decision-making process (similar to the anti-conformist individuals in Grabisch and Li (2019)), followed by some more further theoretical developments in Vanelli et al. (2020).

In the above, we have illustrated how game-theoretic models offer a powerful tool to capture real-world decision-making processes in a parsimonious framework, addressing many of the limitations of the classical models based on pairwise interactions or on linear threshold mechanisms. However, their main drawback lies in their complexity, which often prevents a complete analysis of the system on complex networks, different from the simpler model presented in Sections 3.3.1 and 3.3.2. To sum up, to model decision-making during social diffusion, many different approaches have been proposed. All of these modeling approaches have indeed advantages and disadvantages. Therefore, in order to decide which modeling approach better addresses a certain problem that one wants to study, it is key to evaluate the trade-off between the ability of the model to capture real-world features of the decision-making process, and its amenability to analytical treatment.

3.4 Dynamics in CPHS

In Section 3.3, we have extensively discussed several different approaches that have been proposed in the literature to model human decision-making during social diffusion. In all these approaches, the individuals of the population make a decision on whether to adopt the innovation following some predetermined mechanisms using the information that the individuals gather from a communication network. As we have illustrated, the structure of the communication network can play an important role in shaping the emergent behavior of the system, determining the success or the failure of the social diffusion process.

A key characteristic of many real-world CPHS is the presence of multiple communications channels between its components. In other words, real-world human systems often integrate various interaction patterns due to features of cyber information flows and physical communication actions. Moreover, in many real-world scenarios, social diffusion processes co-evolve together with other dynamics on the (multiplex) network at comparable time-scales, mutually influencing one another. Hence, in order to study real-world social diffusion processes and understand collective human behavior, it is key to incorporate multiplex and co-evolving dynamics within the state-of-the-art frameworks for social diffusion described in the first part of this chapter. In the rest of this section, we provide an overview of recent developments in modeling and analyzing diffusion processes in multiplex networks and co-evolutionary processes.

3.4.1 Social Diffusion in Multiplex Networks

Considering how multiplex structures have recently emerged as powerful mathematical tools to represent real-world CPHS, it is not surprising that several important advances have been made in the last decade on modeling and analyzing social diffusion in multiplex networks. In the following, we provide an overview of the main results developed for the different decision-making mechanisms illustrated in Section 3.3.

The study of models for social diffusion based on pairwise interactions in multiplex networks can be traced back to the extension of classical network epidemic models to multiplex networks. See, for instance, Saumell-Mendiola et al. (2012). These studies, performed using the MMCA technique described in Section 3.2.3, suggest that the presence of multiple layers favor the spread of epidemic diseases. In particular, it has been observed that epidemics with parameters that do not lead to a successful spread in any of the layer are instead able to cause a spread in the multiplex structure. In Yagan et al. (2013), the authors propose the use of a multiplex network with two layers for the spread of information, with one layer that captures communication through physical interactions between individuals and a second layer that models communication flows through online social networks. Through a rigorous theoretical analysis of the dynamical system, the authors confirm the results of the MMCA analysis on the simpler epidemic model, proving that the simultaneous presence of multiple communication channels can dramatically impact the speed and scale of diffusion processes. Based on these seminal works, which have highlighted the important role that the presence of multiple layers can play in diffusion processes driven by pairwise-interaction decision-making mechanisms, several studies have been accomplished in the last few years, toward shedding lights on the role of directionality in the communication channels (see, for instance, Wang et al. (2019b)), or on the role of clusters on the different layers (Zhuang and Yağan 2016).

Also the LT model has been formalized in the context of multiplex networks, and several important studies have been performed. In Brummitt et al. (2012), the original model for innovation diffusion proposed in Watts (2002) has been extended by considering individuals interacting in multiplex networks, as illustrated in Figure 3.6. In particular, it is assumed that the condition in Equation (3.7) for the adoption of innovation is on the minimum over all the layers in which an individual has interactions. Namely, an individual adopts the innovation if a sufficiently large fraction of their neighbors in any layer do. Using the generating function to study the evolution of the stochastic cascading phenomenon, the authors have established that the presence of multiple layers favor diffusion, as already observed for models based on pairwise-interaction mechanisms. In fact, layers that are not susceptible to global cascades can instead yield them, if coupled. Similar conclusions have been established in Yağan and Gligor (2012), where a similar model is considered, in

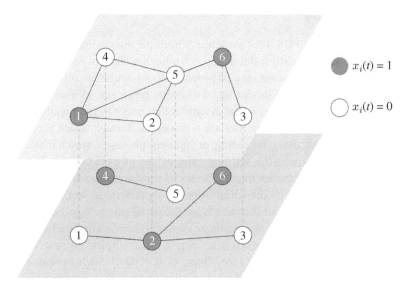

Figure 3.6 An LT model on a multiplex network with two layers $\mathcal{L} = \{1, 2\}$ denoted in dark gray and light gray, respectively. The dynamics depend on the implementation of the model in multiplex networks. For instance, if $\theta_5 = 0.5$, then node 5 will adopt the innovation according to Brummitt et al. (2012), since on layer 1, the individual has 2 of their 3 neighbors adopting the innovation. On the contrary, individual 5 may not adopt the innovation following Zhong et al. (2021), since only 1 of their 3 neighbors of the individual adopts the innovation on layer 2.

which an individual adopts the innovation if a weighted average of the fraction of adopters among their neighbors across the layers is greater than a threshold. These works have been extended in many directions. In Hu et al. (2014), correlation between the layers has been investigated using generating function techniques, showing that it is a key feature to ensure sustained diffusion cascades in the network. In Zhong et al. (2021), the framework is generalized by introducing a double threshold behavior in which an individual adopts the innovation if a sufficiently large amount of their neighbors has already adopted the innovation on a sufficiently large number of layers. This generalizes the original formulation in Brummitt et al. (2012), where it is sufficient that the threshold is reached on a single layer. Besides formally defining the model, the authors also provide theoretically grounded algorithms to estimate the fraction of adopters in the asymptotic state.

The success of game-theoretic models on monoplex networks in capturing complex human decision-making during social diffusion has motivated many researches to formulate generalizations of their formalism to more complex network structures, including multiplex networks. See Wang et al. (2015) for more details. In particular, concerning models for innovation diffusion, in Matamalas et al. (2015), different network games have been studied in multiplex networks, including a coordination game. For the coordination game, it has been observed that the presence of multiple layers of communication has a nontrivial effect on the diffusion processes, enabling the existence of stable configurations in which the population does not reach a consensus on whether to adopt the innovation or the status quo, with a nonnegligible part of the population that adopts the innovation, while the rest of the population maintains the status quo. In Ramezanian et al. (2015), the network coordination game in Equation (3.11) is generalized to multiplex networks and a lower bound for the success of an innovation is established. For more details and further discussion the extension of models for social diffusion in multiplex, we refer to Boccaletti et al. (2014), Salehi et al. (2015), and De Domenico and Stella (2021).

3.4.2 Co-Evolutionary Social Dynamics

One key characteristic of many real-world CPHS is the presence of multiple intertwined dynamics that co-evolve at comparable time scales, mutually influencing each other's evolution. For instance, in many real-world scenarios, multiple innovations may be proposed to a population at the same time, leading to potential competitions among different social diffusion processes. Moreover, the social diffusion process may be influenced by other social dynamics that occur in the population during the social diffusion. Examples include opinion formation processes, through which people may revise their opinions on the innovation, that is spreading, or epidemic processes, which impact individuals' decision on whether to adopt innovative self-protective behaviors or to take vaccines. It is important to notice that while these processes impact the social diffusion process, it is also often the case that their evolution is affected by the social diffusion dynamics. In fact, opinion formation processes are influenced by the behaviors observed (see the social psychology literature on intuitionist model in Haidt (2001) and norm interiorization processes in Gavrilets and Richerson (2017)), and human behavior can impact the course of an epidemic process, as empirically observed during the ongoing COVID-19 pandemic (see, for instance, Zhang et al. (2020)). Such a presence of mutual influences calls for the development of novel mathematical models that incorporate a co-evolution of the different dynamics involved in the social diffusion processes, toward deriving accurate predictions of the emergent behavior of the CPHS.

Formally, a co-evolutionary model can be formulated similar to a standard model for social diffusion, by simply expanding the state variable of each individual. In particular, in co-evolutionary models, the state variable of each individual $i \in \mathcal{V}$ becomes a vector $\boldsymbol{x}_i(t) = (x_i^{(1)}(t), \ldots, x_i^{(m)}(t)) \in S^{(1)} \times \cdots \times S^{(m)}$, where m is the number of dynamics involved in the process, $x_i^{(k)}(t)$ is the state of the kth dynamics for individual i, and $S^{(k)}$ are the possible states of the kth dynamics. Each entry of the state variable vector is revised according to the rules in Equation (3.1) or Equation (3.3) (depending on the continuous or discrete nature of the system), where the decision-making function, in general, can depend on the entire state variable vector and not just on the corresponding entry, capturing the interdependence between the dynamics.

The study of multiple diffusion processes co-evolving on the same network has witnessed a remarkable growth over the last decade. Inspired by the epidemic models with two or multiple co-evolving viruses (see, for instance, Liu et al. (2019)), several models have been proposed in which multiple ideas spread on a network, potentially competing one against the others. In Wang et al. (2012), a model for the diffusion of two ideas on a network is proposed and studied under two different assumptions on the level of competition between the ideas and on the role of neighbors' ideas. The first scenario is obtained by simply extending the formalism of pairwise-interaction mechanisms in 3.3.1 to multiple innovations. In a second scenario, instead, the presence of a neighbor that has adopted a certain idea reduces the probability that the individual adopts the opposite idea. Using a mean-field approach, the authors study the asymptotic behavior of the system in the two scenarios, showing how the first scenario may lead to stable configurations in which both ideas are present, while the second favors the establishment of a consensus). Similar models have been developed and analyzed via mean-field approaches on homogeneous networks in Wang et al. (2014), and expanded to more general network structures in Zhang and Zhu (2018). In Liu et al. (2018b), the authors consider a scenario in which the two diffusion processes occur sequentially, and whether an individual adopts the first innovation, impacts the probability of adopting the second one. The authors analytically characterize a phase transition between success and failure of the second diffusion process, which can be continuous or discontinuous depending on whether the first innovation has a synergistic or an inhibiting effect on the second. In Pathak et al. (2010), a generalized LT model with multiple cascades is proposed, in which individuals on a network can switch between the different diffusion cascades. The model is formalized as a Markov chain, and its asymptotic behavior is studied by characterizing its steady-state distribution. In Askarizadeh et al. (2019), a game-theoretic model is proposed, in which the diffusion of a rumor and an anti-rumor are studied on a network by means of numerical simulations, toward identifying under which conditions the diffusion of a suitable anti-rumor can be used as a strategy to control the spread of a rumor.

Many of these models have been generalized to more than two co-evolving diffusion processes (see Wang et al. (2019a) for a recent survey) and to multiplex networks, in which each social diffusion process occurs on one layer. In Liu et al. (2018a), two simple diffusion dynamics based on pairwise interactions are coupled on a two-layer multiplex network, where it is assumed that the adoption of an innovation on a layer enhances the probability of adopting the innovation on the other layer. Formally, it is imposed that the probability λ in Equation (3.5) is a function of the state of the individual on the other layer. Furthermore, researches have focused on unveiling the role of the network structure on the different layers. See, for instance, Chang and Fu (2019). While all these models assume that the diffusion processes on the two layers follow the same decision-making mechanisms, some efforts have been recently made to study the competition between diffusion processes driven by different mechanisms. An interesting example can be found in Czaplicka et al. (2016), where a diffusion driven by pairwise interactions and one with an LT decision-making model are coupled on a two-layer multiplex network, and different behaviors have been observed for the two diffusion dynamics: the former exhibits a discontinuous transition from maintenance of the status quo and social diffusion, while the latter exhibits a continuous phase transition. A similar approach is used in Yu et al. (2020), where the diffusion driven by pairwise interactions is interpreted as the diffusion of information on or awareness of an innovation, while the LT decision-making model represents the innovation adoption process.

All models presented and discussed so far consider two or multiple innovations that spread on the same networks. A second, important aspect of real-world CPHS is the presence of different dynamics that evolve together with the social diffusion process (or processes), such as opinion formation processes or epidemic spreading, potentially on different communication networks, as illustrated if Figure 3.7. In the rest of this section, we present and discuss some of the key recent advances in this research direction.

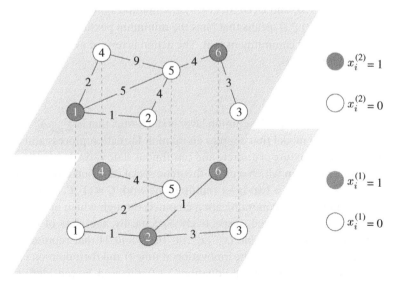

Figure 3.7 Structure of a coevolutionary model with two coevolving dynamics $\mathbf{x}_i(t) = (x_i^{(1)}(t), x_i^{(2)}(t))$ evolving on a two-layer multiplex, $\mathcal{L} = \{1, 2\}$.

The integration of social diffusion processes in epidemic models has been originally proposed in Granell et al. (2013), where an epidemic spreading is coupled with the diffusion of information awareness to prevent its infection. The two processes are both modeled as driven by pairwise-interaction mechanisms on a two-layer multiplex network, where epidemics spread on a "physical contact layer," while information diffuses on a "virtual contact layer." The two dynamics are coupled, as individuals that are aware of the disease decide to adopt preventive measures to reduce their infection probability, while it is also assumed that people become aware after contagion. This produces a coupled model where the "contagion" probability λ in Equation (3.5) on one layer depends on the state of the individual on the other layer: diffusion of awareness hinders the diffusion of the disease, while the diffusion of the disease favors the diffusion of awareness. The analysis of the model, performed via an MMCA approach, reveals that the structure of the two layers and the interactions between the two diffusion processes play a key role in determining whether any epidemics outbreak is quickly extinguished of if it becomes endemic, diffusing to the entire population. Recent efforts have been made to incorporate more complex decision-making models in epidemic spreading processes. In Ye et al. (2021b), a classical epidemic model is coupled with a game-theoretic decision-making mechanism that governs individuals' adoption of novel self-protective measures. The increased complexity of the framework leads to a deeper intertwining between the dynamics: similar to the model described above, the adoption of protective measures reduces the infection probability in Equation (3.5), while the payoff functions used for the decision-making process incorporate several realistic factors, including a risk perception that depends on the fraction of infected individuals in the entire population. Such a model, studied through a mean-field approach in Frieswijk et al. (2022), is able to reproduce a wide range of emergent phenomena, including successful collective adoptions of self-protective behaviors, periodic oscillations in both dynamics, and resurgent epidemic outbreaks.

The integration of opinion formation processes into social diffusion co-evolutionary models requires a slightly different formalization. In fact, mathematical models for opinion formation typically assume that the opinion of an individual is a continuously distributed variable, representing how much an individual is in favor of a certain innovation (Jia et al. 2015). In its simplest implementation, proposed by French (1956) and DeGroot (1974), the opinion $x_i^{(O)}(t)$ of individual $i \in \mathcal{V}$ belongs to the set $[0, 1]$, where $x_i^{(O)}(t) = 0$ means that i has the minimum possible support to the innovation and $x_i^{(O)}(t) = 1$ indicates full commitment to it. The action is revised as a weighted average of the opinion of the neighbors, that is,

$$x_i^{(O)}(t+1) = \sum_{j \in \mathcal{V}} A_{ij}^{(O)} x_j^{(O)}(t) \tag{3.12}$$

where it is assumed that the adjacency matrix on the opinion layer is stochastic, that is $\sum_{j \in \mathcal{V}} A_{ij}^{(O)} = 1$, for all $i \in \mathcal{V}$. Hence, a co-evolutionary model that couples an opinion formation process and a social diffusion one is characterized by a mixture of discrete and continuous state variables. In Zino et al. (2020b), the authors propose a model in which a game-theoretic social diffusion model and an opinion formation process are coupled on a two-layer multiplex network $\mathcal{G} = (\mathcal{V}, \mathcal{L}, \mathcal{E}, \mathcal{A})$, in which the two layers $\mathcal{L} = \{O, D\}$ represent the communication channels through which opinions are exchanged and others' behaviors are observed, respectively. Specifically, the state of each individual is captured by a two-dimensional variable $(x_i^{(D)}, x_i^{(O)}) \in \{0, 1\} \times [0, 1]$ that represents their behavior (i.e. whether individual i has adopted the innovation at time t) and their opinion on it, respectively. In this model, the opinion update rule in Equation (3.12) is revised, by considering a convex combination of the weighted average of the opinion of the neighbors and of their behavior as

$$x_i^{(O)}(t+1) = (1-\mu_i)\sum_{j\in\mathcal{V}}A_{ij}^{(O)}x_j^{(O)}(t) + \mu_i\sum_{j\in\mathcal{V}}A_{ij}^{(D)}x_j^{(D)}(t) \tag{3.13}$$

while the behavior is updated according to a game-theoretic decision-making rule in which a term that increases the payoff for behaving consistently with their own opinion in added the payoffs defined in Equation (3.11), thereby obtaining

$$\pi_0^{(i)}(\boldsymbol{x}(t),\mathcal{A}) = (1-\lambda_i)\frac{1}{d_i}\sum_{j\in\mathcal{V}}A_{ij}(1-x_j^{(D)}(t)) + \lambda_i(1-x_i^{(O)}(t)), \tag{3.14}$$

$$\pi_1^{(i)}(\boldsymbol{x}(t),\mathcal{A}) = (1-\lambda_i)\frac{1}{d_i}\sum_{j\in\mathcal{V}}A_{ij}(1+\alpha)x_j^{(D)}(t) + \lambda_i x_i^{(O)}(t). \tag{3.15}$$

Numerical and analytical studies of the model performed in Zino et al. (2020b) and Zino et al. (2020a), respectively, suggest that the characteristics of the interdependence between the two processes and the structure of the two layers can explain the occurrence of different emergent behaviors, from the establishment of robust unpopular norms to a collective paradigm shifts toward the innovation, as illustrated in Figure 3.8. A similar model, in which an opinion formation process is coupled with an information diffusion process, is presented in Soriano-Paños et al. (2019).

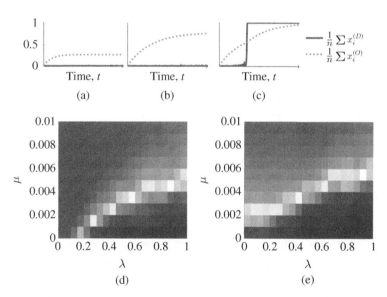

Figure 3.8 Different possible outcomes observed for the coevolutionary model proposed in Zino et al. (2020b), with different sets of parameters λ and μ (uniform across the population), starting from a status-quo majority. In (a), social diffusion fails, and people does not support the innovation (Regime I); in (b), a robust unpopular norm is established, in which the innovation is not adopted even though $\frac{1}{n}\sum x_i^{(O)}(t) > 0.5$, for t sufficiently large (Regime II); in (c), social diffusion and a paradigm shift toward the innovation is observed (Regime III). In (d) and (e), we illustrate the average fraction of final adopters, $\frac{1}{n}\sum x_i^{(D)}$, and the average final opinion, $\frac{1}{n}\sum x_i^{(O)}$, respectively, for different values of the parameters λ and μ, when both layers are Erdös–Rényi random graph. In the top-left corner values are negative, in the bottom-right corner values are positive; the color intensity denotes the magnitude, with corresponding to 0. The figures show that, for large values of μ, Regime I occurs, for small values of μ and λ, Regime II occurs, for small values of μ and λ sufficiently large, Regime III occurs. Source: Adapted from Zino et al. (2020b).

The asymptotic behavior of the system, explored via numerical simulations, exhibits a bi-stable scenario, with explosive transitions between the two stable configurations, which represent the consensus-informed population and disagreement-uninformed population, respectively.

3.5 Ongoing Efforts Toward Controlling Social Diffusion and Future Challenges

In this chapter, we have discussed the important problem of social diffusion, through which a novel product, idea, or behavior is collectively adopted by a population, and we have proposed a general formalism to model social diffusion in CPHS. Through detailed discussions on the main modeling approaches proposed to capture human decision-making concerning the adoption of an innovation, we have provided an overview of the broad and growing literature on social diffusion and on the ongoing efforts toward developing more realistic models, which are able to capture the complexity of CPHS with humans-in-multiagent-loops.

Besides gaining a better understanding of social diffusion as a collective phenomenon and unveiling the role of communication channels and constraints in determining the outcome of potential social diffusion processes, the development of accurate mathematical models for social diffusion in CPHS has another important objective: designing tools to control social diffusion processes toward reaching a desired outcome. In this section, we discuss the formalization of the problem of controlling social diffusion on complex networks and present some of the seminal works that have been proposed and the key recent advances toward controlling social diffusion. Such a problem can be classified under the humans-in-multiagent-loops category, introduced in Samad (2023); Chapter 1 in this book.

A popular approach to the control of social diffusion lies in identifying a small subset of initial adopters such that their total influence is sufficient to unlock social diffusion on a network, given a certain decision-making mechanism. Along this line, since the beginning of the 2000s, such an influence maximization problem has been attracting increasing interest in the scientific community. In Kempe et al. (2003), this problem has been formalized as a combinatorial optimization problem, and it has been proved to be NP-hard, paving the way for the design of algorithms to derive approximate solutions to the problem. See Li et al. (2018) for more details and a review of recent developments for monoplex networks.

Motivated by these advances in the control of social diffusion on monoplex networks, several researchers have recently expanded their studies to deal with multiplex networks. We mention the work in Kuhnle et al. (2018), where the authors leverage a submodularity property of the dynamics to approximate the influence maximization problem in multiplex networks for different diffusion dynamics (including some implementations of the LT models and of pairwise-interactions mechanisms in multiplex networks). In Yang et al. (2020), the authors investigate the opposite problem, that is, the influence minimization problem in multiplex networks for LT models by means of solving integer linear programs. Despite these recent advances, the design of algorithms to accurately approximate the solution of the influence maximization problem in multiplex networks is still an open problem for many decision-making mechanisms.

Other control approaches have been proposed in the literature. An increasingly popular research direction, in particular for models based on game-theoretic decision-making mechanisms, entails the design of interventions to appropriately change the payoff functions in individuals' decision-making processes and the devise of learning protocols, designed to favor the social diffusion process. This research line builds on the recent results on the control of games surveyed in Marden and Shamma (2015), Quijano et al. (2017), and Riehl et al. (2018).

Another approach, which is particularly promising for pairwise interaction models, entails the development of dynamical interventions, to be implemented during the social diffusion processes to boost it. Inspired by the growing literature on the dynamic control of epidemics (see Zino and Cao (2021) for a recent review), several works have been proposed for the dynamic control of social diffusion processes. See for instance, Zino et al. (2017), where a strategy to speed up a diffusion process is proposed by leveraging graph-theoretic tools. However, to the best of our knowledge, all the results proposed in the literature focus on monoplex networks, while the design of dynamical control policies for social diffusion in multiplex networks is still an open problem. Finally, some researchers have focused on the problem of finding the optimal network structure to favor social diffusion, given a specific type of decision-making mechanism. This problem is in general quite challenging like most inverse problems. For monoplex networks, some analytical insights for simple pairwise-interaction decision-making mechanisms have been provided in Pan et al. (2021), but further efforts are required to extend these findings to more complex decision-making mechanisms and more general network structures.

This chapter has highlighted how the mathematical modeling of social diffusion processes is a continuously evolving topic that attracts many researchers from different fields, helping to understand the very nature of many real-world social diffusion processes and the role of individual-level mechanisms and of the complexity of the communication network on shaping the emergent behavior of CPHS. Our overview has shown that despite we have a rather strong grip on the basic modeling of human decision-making and on the role of simple network structures, many challenges are yet to be addressed. For instance, the effects of distinct communication layers and of multiple coevolving dynamical processes are yet to be fully explored. Another important challenge, discussed in the above, concerns leveraging the network structure to design effective intervention policies to control social diffusion processes. By pursuing these important objectives, we wish to develop mathematically grounded toolboxes toward understanding, predicting, and, ultimately, controlling crucial CPHS.

Acknowledgments

The work was supported in part by the European Research Council (ERC-CoG-771687).

References

Amato, R., Lacasa, L., Díaz-Guilera, A., and Baronchelli, A. (2018). The dynamics of norm change in the cultural evolution of language. *Proceedings of the National Academy of Sciences of the United States of America* 115 (33): 8260–8265. https://doi.org/10.1073/pnas.1721059115.

Andreoni, J., Nikiforakis, N., and Siegenthaler, S. (2021). Predicting social tipping and norm change in controlled experiments. *Proceedings of the National Academy of Sciences of the United States of America* 118 (16): e2014893118. https://doi.org/10.1073/pnas.2014893118.

Askarizadeh, M., Ladani, B.T., and Manshaei, M.H. (2019). An evolutionary game model for analysis of rumor propagation and control in social networks. *Physica A: Statistical Mechanics and its Applications* 523: 21–39. https://doi.org/10.1016/j.physa.2019.01.147.

Athreya, K.B. and Ney, P.E. (1972). *Branching Processes*. Berlin, Heidelberg: Springer-Verlag. https://doi.org/10.1007/978-3-642-65371-1.

Bass, F.M. (1969). A new product growth for model consumer durables. *Management Science* 15 (5): 215–227.

Becker, N. (1977). Estimation for discrete time branching processes with application to epidemics. *Biometrics* 33 (3): 515. https://doi.org/10.2307/2529366.

Betsch, C., Korn, L., Sprengholz, P. et al. (2020). Social and behavioral consequences of mask policies during the COVID-19 pandemic. *Proceedings of the National Academy of Sciences of the United States of America* 117 (36): 21851–21853. https://doi.org/10.1073/pnas.2011674117.

Bettencourt, L.M.A., Cintrón-Arias, A., Kaiser, D.I., and Castillo-Chávez, C. (2006). The power of a good idea: quantitative modeling of the spread of ideas from epidemiological models. *Physica A: Statistical Mechanics and its Applications* 364: 513–536. https://doi.org/10.1016/j.physa.2005.08.083.

Bicchieri, C. (2005). *The Grammar of Society: The Nature and Dynamics of Social Norms*, 1e. Cambridge: Cambridge University Press.

Boccaletti, S., Bianconi, G., Criado, R. et al. (2014). The structure and dynamics of multilayer networks. *Physics Reports* 544 (1): 1–122. https://doi.org/10.1016/j.physrep.2014.07.001.

Bonabeau, E. (2002). Agent-based modeling: methods and techniques for simulating human systems. *Proceedings of the National Academy of Sciences of the United States of America* 99 (3): 7280–7287. https://doi.org/10.1073/pnas.082080899.

Brown, M.J. and Satterthwaite-Phillips, D. (2018). Economic correlates of footbinding: implications for the importance of Chinese daughters' labor. *PLoS One* 13 (9): e0201337. https://doi.org/10.1371/journal.pone.0201337.

Brummitt, C.D., Lee, K.-M., and Goh, K.-I. (2012). Multiplexity-facilitated cascades in networks. *Physical Review E* 85: 045102. https://doi.org/10.1103/PhysRevE.85.045102.

Buono, C., Alvarez-Zuzek, L.G., Macri, P.A., and Braunstein, L.A. (2014). Epidemics in partially overlapped multiplex networks. *PLoS One* 9 (3): e92200. https://doi.org/10.1371/journal.pone.0092200.

Castaldo, M., Frasca, P., and Venturini, T. (2023). On online attention dynamics in social media. In: *Cyber-Physical-Human Systems: Fundamentals and Applications* (ed. A. Annaswamy et al.). Hoboken, NJ: John Wiley & Sons, Inc.

Centola, D. (2018). *How Behavior Spreads: The Science of Complex Contagions*. Princeton NJ: Princeton University Press.

Centola, D. and Baronchelli, A. (2015). The spontaneous emergence of conventions: an experimental study of cultural evolution. *Proceedings of the National Academy of Sciences of the United States of America* 112 (7): 1989–1994. https://doi.org/10.1073/pnas.1418838112.

Centola, D. and Macy, M. (2007). Complex contagions and the weakness of long ties. *American Journal of Sociology* 113 (3): 702–734. https://doi.org/10.1086/521848.

Centola, D., Becker, J., Brackbill, D., and Baronchelli, A. (2018). Experimental evidence for tipping points in social convention. *Science* 360 (6393): 1116–1119. https://doi.org/10.1126/science.aas8827.

Chakrabarti, D., Wang, Y., Wang, C. et al. (2008). Epidemic thresholds in real networks. *ACM Transactions on Information and System Security* 10 (4): 1–26. https://doi.org/10.1145/1284680.1284681.

Chang, H.-C.H. and Fu, F. (2019). Co-contagion diffusion on multilayer networks. *Applied Network Science* 4 (1): 78. https://doi.org/10.1007/s41109-019-0176-6.

Cialdini, R.B. and Goldstein, N.J. (2004). Social influence: compliance and conformity. *Annual Reviews in Psychology* 55: 591–621. https://doi.org/10.1146/annurev.psych.55.090902.142015.

Cinelli, M., Quattrociocchi, W., Galeazzi, A. et al. (2020). The COVID-19 social media infodemic. *Scientific Reports* 10 (1). https://doi.org/10.1038/s41598-020-73510-5.

Czaplicka, A., Toral, R., and Miguel, M.S. (2016). Competition of simple and complex adoption on interdependent networks. *Physical Review E* 94: 062301. https://doi.org/10.1103/PhysRevE.94.062301.

Daley, D.J. and Kendall, D.G. (1964). Epidemics and rumours. *Nature* 204 (4963): 1118. https://doi.org/10.1038/2041118a0.

Daley, D.J. and Kendall, D.G. (1965). Stochastic rumours. *IMA Journal of Applied Mathematics* 1 (1): 42–55. https://doi.org/10.1093/imamat/1.1.42.

De Domenico, M. and Stella, M. (2021). *Dynamics on Multi-layer Networks*, 121–144. Cambridge University Press. https://doi.org/10.1017/9781108553711.012.

DeGroot, M.H. (1974). Reaching a consensus. *Journal of the American Statistical Association* 69 (345): 118–121. https://doi.org/10.1080/01621459.1974.10480137.

Dong, R., Li, L., Zhang, Q., and Cai, G. (2018). Information diffusion on social media during natural disasters. *IEEE Transactions on Computational Social Systems* 5 (1): 265–276. https://doi.org/10.1109/TCSS.2017.2786545.

Edelmann, A., Wolff, T., Montagne, D., and Bail, C.A. (2020). Computational social science and sociology. *Annual Review of Sociology* 46 (1): 61–81. https://doi.org/10.1146/annurev-soc-121919-054621.

Ellison, G. (1993). Learning, local interaction, and coordination. *Econometrica* 61 (5): 1047. https://doi.org/10.2307/2951493.

Fagnani, F. and Zino, L. (2017). Diffusion of innovation in large scale graphs. *IEEE Transactions on Network Science and Engineering* 4 (2): 100–111. https://doi.org/10.1109/TNSE.2017.2678202.

French, J.R.P. Jr. (1956). A formal theory of social power. *Psychological Review* 63 (3): 181–194. https://doi.org/10.1037/h0046123.

Frieswijk, K., Zino, L., Ye, M. et al. (2022). A mean-field analysis of a network behavioral-epidemic model. *IEEE Control Systems Letters* 6: 2533–2538. https://doi.org/10.1109/LCSYS.2022.3168260.

Gavrilets, S. and Richerson, P.J. (2017). Collective action and the evolution of social norm internalization. *Proceedings of the National Academy of Sciences of the United States of America* 114 (23): 6068–6073. https://doi.org/10.1073/pnas.1703857114.

Goffman, W. and Newill, V.A. (1964). Generalization of epidemic theory: an application to the transmission of ideas. *Nature* 204 (4955): 225–228. https://doi.org/10.1038/204225a0.

Goldenberg, J., Libai, B., and Muller, E. (2001). Talk of the network: a complex systems look at the underlying process of word-of-mouth. *Marketing Letters* 12 (3): 211–223. https://doi.org/10.1023/A:1011122126881.

Grabisch, M. and Li, F. (2019). Anti-conformism in the threshold model of collective behavior. *Dynamic Games and Applications* 10 (2): 444–477. https://doi.org/10.1007/s13235-019-00332-0.

Granell, C., Gómez, S., and Arenas, A. (2013). Dynamical interplay between awareness and epidemic spreading in multiplex networks. *Physical Review Letters* 111: 128701. https://doi.org/10.1103/PhysRevLett.111.128701.

Granovetter, M. (1978). Threshold models of collective behavior. *American Journal of Sociology* 83 (6): 1420–1443. https://doi.org/10.1086/226707.

Haidt, J. (2001). The emotional dog and its rational tail: a social intuitionist approach to moral judgment. *Psychological Review* 108: 814–834. https://doi.org/10.1037/0033-295x.108.4.814.

Hu, Y., Havlin, S., and Makse, H.A. (2014). Conditions for viral influence spreading through multiplex correlated social networks. *Physical Review X* 4: 021031. https://doi.org/10.1103/PhysRevX.4.021031.

Jia, P., MirTabatabaei, A., Friedkin, N.E., and Bullo, F. (2015). Opinion dynamics and the evolution of social power in influence networks. *SIAM Review* 57 (3): 367–397. https://doi.org/10.1137/130913250.

Jusup, M., Holme, P., Kanazawa, K. et al. (2022). Social physics. *Physics Reports* 948: 1–148. https://doi.org/10.1016/j.physrep.2021.10.005.

Karsai, M., Iñiguez, G., Kaski, K., and Kertész, J. (2014). Complex contagion process in spreading of online innovation. *Journal of The Royal Society Interface* 11 (101): 20140694. https://doi.org/10.1098/rsif.2014.0694.

Kempe, D., Kleinberg, J., and Tardos, É. (2003). Maximizing the spread of influence through a social network. *Proceedings of the 9th ACM SIGKDD International Conference on Knowledge Discovery and Data Mining*, 137–146.

Kermack, W.O. and McKendrick, A.G. (1927). A contribution to the mathematical theory of epidemics. *Proceedings of the Royal Society A* 115 (772): 700–721.

Kivelä, M., Arenas, A., Barthelemy, M. et al. (2014). Multilayer networks. *Journal of Complex Networks* 2 (3): 203–271. https://doi.org/10.1093/comnet/cnu016.

Kuhnle, A., Alim, Md.A., Li, X. et al. (2018). Multiplex influence maximization in online social networks with heterogeneous diffusion models. *IEEE Transactions on Computational Social Systems* 5 (2): 418–429. https://doi.org/10.1109/TCSS.2018.2813262.

Kurtz, T.G. (1981). *Approximation of Population Processes*, vol. 36. Philadelphia, PA: SIAM.

Lajmanovich, A. and Yorke, J.A. (1976). A deterministic model for gonorrhea in a nonhomogeneous population. *Mathematical Biosciences* 28 (3–4): 221–236. https://doi.org/10.1016/0025-5564(76)90125-5.

Lazer, D., Pentland, A., Adamic, L. et al. (2009). Computational social science. *Science* 323 (5915): 721–723. https://doi.org/10.1126/science.1167742.

Levin, D.A., Peres, Y., and Wilmer, E.L. (2006). *Markov Chains and Mixing Times*. Providence, RI: American Mathematical Society.

Lewis, D. (2002). *Convention: A Philosophical Study*, 1e. Hoboken, NJ: Wiley-Blackwell.

Li, Y., Fan, J., Wang, Y., and Tan, K.-L. (2018). Influence maximization on social graphs: a survey. *IEEE Transactions on Knowledge and Data Engineering* 30 (10): 1852–1872. https://doi.org/10.1109/TKDE.2018.2807843.

Lieberman, E., Michel, J.-B., Jackson, J. et al. (2007). Quantifying the evolutionary dynamics of language. *Nature* 449 (7163): 713–716. https://doi.org/10.1038/nature06137.

Liu, J., Paré, P.E., Nedić, A. et al. (2019). Analysis and control of a continuous-time bi-virus model. *IEEE Transactions on Automatic Control* 64 (12): 4891–4906. https://doi.org/10.1109/tac.2019.2898515.

Liu, Q.-H., Wang, W., Cai, S.-M. et al. (2018a). Synergistic interactions promote behavior spreading and alter phase transitions on multiplex networks. *Physical Review E* 97 (2): 022311. https://doi.org/10.1103/physreve.97.022311.

Liu, Q.-H., Zhong, L.-F., Wang, W. et al. (2018b). Interactive social contagions and co-infections on complex networks. *Chaos: An Interdisciplinary Journal of Nonlinear Science* 28 (1): 013120. https://doi.org/10.1063/1.5010002.

Mackie, G. (1996). Ending footbinding and infibulation: a convention account. *American Sociological Review* 61 (6): 999–1017. https://doi.org/10.2307/2096305.

Mahajan, V., Muller, E., and Bass, F.M. (1990). New product diffusion models in marketing: a review and directions for research. *Journal of Marketing* 54 (1): 1–26. https://doi.org/10.2307/1252170.

Maki, D.P. and Thompson, M. (1973). *Mathematical Models and Applications: With Emphasis on the Social, Life, and Management Sciences*. Hoboken, NJ: Prentice-Hall. ISBN 9780135616703.

Marden, J.R. and Shamma, J.S. (2015). *Game Theory and Distributed Control, Handbook of Game Theory with Economic Applications*, vol. 4, 861–899. Elsevier. https://doi.org/10.1016/B978-0-444-53766-9.00016-1.

Marmor, A. (2009). *Social Conventions: From Language to Law*, 1e. Princeton, NJ: Princeton University Press.

Martínez, D., Parilli, C., Scartascini, C., and Simpser, A. (2021). Let's (not) get together! The role of social norms on social distancing during COVID-19. *PLoS One* 16 (3): 1–14. https://doi.org/10.1371/journal.pone.0247454.

Matamalas, J.T., Poncela-Casasnovas, J., Gómez, S., and Arenas, A. (2015). Strategical incoherence regulates cooperation in social dilemmas on multiplex networks. *Scientific Reports* 5 (1): 9519. https://doi.org/10.1038/srep09519.

Montanari, A. and Saberi, A. (2010). The spread of innovations in social networks. *Proceedings of the National Academy of Sciences of the United States of America* 107 (47): 20196–20201. https://doi.org/10.1073/pnas.1004098107.

Moreno, Y., Nekovee, M., and Pacheco, A.F. (2004). Dynamics of rumor spreading in complex networks. *Physical Review E* 69: 066130. https://doi.org/10.1103/PhysRevE.69.066130.

Morris, S. (2000). Contagion. *The Review of Economic Studies* 67 (1): 57–78. https://doi.org/10.1111/1467-937X.00121.

Mortensen, C.R., Neel, R., Cialdini, R.B. et al. (2019). Trending norms: a lever for encouraging behaviors performed by the minority. *Social Psychological and Personality Science* 10 (2): 201–210. https://doi.org/10.1177/1948550617734615.

Newman, M. (2010). *Networks: An Introduction*. Oxford: Oxford University Press.

Noelle-Neumann, E. (1993). *The Spiral of Silence: Public Opinion, Our Social Skin*. Chicago, IL: University of Chicago Press.

Pagan, N. and Dörfler, F. (2019). Game theoretical inference of human behavior in social networks. *Nature Communications* 10 (1): 5507. https://doi.org/10.1038/s41467-019-13148-8.

Pan, L., Wang, W., Tian, L., and Lai, Y.-C. (2021). Optimal networks for dynamical spreading. *Physical Review E* 103: 012302. https://doi.org/10.1103/PhysRevE.103.012302.

Pathak, N., Banerjee, A., and Srivastava, J. (2010). A generalized linear threshold model for multiple cascades. *2010 IEEE International Conference on Data Mining*, 965–970. https://doi.org/10.1109/ICDM.2010.153.

Prosser, A.M.B., Judge, M., Bolderdijk, J.W. et al. (2020). 'Distancers' and 'non-distancers'? The potential social psychological impact of moralizing COVID-19 mitigating practices on sustained behaviour change. *British Journal of Social Psychology* 59 (3): 653–662. https://doi.org/10.1111/bjso.12399.

Quijano, N., Ocampo-Martinez, C., Barreiro-Gomez, J. et al. (2017). The role of population games and evolutionary dynamics in distributed control systems: the advantages of evolutionary game theory. *IEEE Control Systems Magazine* 37 (1): 70–97. https://doi.org/10.1109/MCS.2016.2621479.

Ramazi, P., Riehl, J., and Cao, M. (2016). Networks of conforming or nonconforming individuals tend to reach satisfactory decisions. *Proceedings of the National Academy of Sciences of the United States of America* 113 (46): 12985–12990. https://doi.org/10.1073/pnas.1610244113.

Ramezanian, R., Magnani, M., Salehi, M., and Montesi, D. (2015). Diffusion of innovations over multiplex social networks. *2015 The International Symposium on Artificial Intelligence and Signal Processing (AISP)*, 300–304. https://doi.org/10.1109/AISP.2015.7123501.

Riehl, J., Ramazi, P., and Cao, M. (2018). A survey on the analysis and control of evolutionary matrix games. *Annual Reviews in Control* 45: 87–106. https://doi.org/10.1016/j.arcontrol.2018.04.010.

Rizzo, A. and Porfiri, M. (2016). Innovation diffusion on time-varying activity driven networks. *The European Physical Journal B* 89 (1): 20. https://doi.org/10.1140/epjb/e2015-60933-3.

Rogers, E.M. (2003). *Diffusion of Innovations*, 5e. New York. ISBN 9780743258234.

Rossi, W.S., Como, G., and Fagnani, F. (2019). Threshold models of cascades in large-scale networks. *IEEE Transactions on Network Science and Engineering* 6 (2): 158–172. https://doi.org/10.1109/TNSE.2017.2777941.

Ryan, B. and Gross, N.C. (1943). The diffusion of hybrid seed corn in two Iowa communities. *Rural Sociology* 8 (1): 15.

Ryan, B. and Gross, N. (1950). Acceptance and diffusion of hybrid corn seed in two Iowa communities. *Research Bulletin* 29 (372): 663–708.

Sahneh, F.D., Scoglio, C., and Van Mieghem, P. (2013). Generalized epidemic mean-field model for spreading processes over multilayer complex networks. *IEEE/ACM Transactions on Networking* 21 (5): 1609–1620. https://doi.org/10.1109/TNET.2013.2239658.

Salehi, M., Sharma, R., Marzolla, M. et al. (2015). Spreading processes in multilayer networks. *IEEE Transactions on Network Science and Engineering* 2 (2): 65–83. https://doi.org/10.1109/TNSE.2015.2425961.

Samad, T. (2023). Human-in-the-loop control and cyber-physical-human systems: applications and categorization. In: *Cyber-Physical-Human Systems: Fundamentals and Applications* (ed. A. Annaswamy et al.). Hoboken, NJ: John Wiley & Sons, Inc.

Samuelson, W. and Zeckhauser, R. (1988). Status quo bias in decision making. *Journal of Risk and Uncertainty* 1 (1): 7–59. https://doi.org/10.1007/BF00055564.

Saumell-Mendiola, A., Ángeles Serrano, M., and Boguñá, M. (2012). Epidemic spreading on interconnected networks. *Physical Review E* 86: 026106. https://doi.org/10.1103/PhysRevE.86.026106.

She, B., Gracy, S., Sundaram, S. et al. (2023). Epidemics spread over networks: influence of infrastructure and opinions. In: *Cyber-Physical-Human Systems: Fundamentals and Applications* (ed. A. Annaswamy et al.). Hoboken, NJ: John Wiley & Sons, Inc.

Soriano-Paños, D., Guo, Q., Latora, V., and Gómez-Gardeñes, J. (2019). Explosive transitions induced by interdependent contagion-consensus dynamics in multiplex networks. *Physical Review E* 99: 062311. https://doi.org/10.1103/PhysRevE.99.062311.

Valente, T.W. (1996). Social network thresholds in the diffusion of innovations. *Social Networks* 18 (1): 69–89. https://doi.org/10.1016/0378-8733(95)00256-1.

Van Mieghem, P., Omic, J., and Kooij, R. (2009). Virus spread in networks. *IEEE/ACM Transactions on Networking* 17 (1): 1–14. https://doi.org/10.1109/tnet.2008.925623.

Vanelli, M., Arditti, L., Como, G., and Fagnani, F. (2020). On games with coordinating and anti-coordinating agents. *IFAC-PapersOnLine* 53 (2): 10975–10980. https://doi.org/10.1016/j.ifacol.2020.12.2848. 21st IFAC World Congress.

Vida, R., Galeano, J., and Cuenda, S. (2015). Vulnerability of state-interdependent networks under malware spreading. *Physica A: Statistical Mechanics and its Applications* 421: 134–140. https://doi.org/10.1016/j.physa.2014.11.029.

Wang, Y., Xiao, G., and Liu, J. (2012). Dynamics of competing ideas in complex social systems. *New Journal of Physics* 14 (1): 013015. https://doi.org/10.1088/1367-2630/14/1/013015.

Wang, J., Zhao, L., and Huang, R. (2014). 2Si2R rumor spreading model in homogeneous networks. *Physica A: Statistical Mechanics and its Applications* 413: 153–161. https://doi.org/10.1016/j.physa.2014.06.053.

Wang, Z., Wang, L., Szolnoki, A., and Perc, M. (2015). Evolutionary games on multilayer networks: a colloquium. *The European Physical Journal B* 88 (5): 124. https://doi.org/10.1140/epjb/e2015-60270-7.

Wang, W., Liu, Q.-H., Liang, J. et al. (2019a). Coevolution spreading in complex networks. *Physics Reports* 820: 1–51. https://doi.org/10.1016/j.physrep.2019.07.001. Coevolution spreading in complex networks.

Wang, X., Aleta, A., Lu, D., and Moreno, Y. (2019b). Directionality reduces the impact of epidemics in multilayer networks. *New Journal of Physics* 21 (9): 093026. https://doi.org/10.1088/1367-2630/ab3dd0.

Watts, D.J. (2002). A simple model of global cascades on random networks. *Proceedings of the National Academy of Sciences of the United States of America* 99 (9): 5766–5771. https://doi.org/10.1073/pnas.082090499.

West, R., Michie, S., Rubin, G.J., and Amlôt, R. (2020). Applying principles of behaviour change to reduce SARS-CoV-2 transmission. *Nature Human Behaviour* 4 (5): 451–459. https://doi.org/10.1038/s41562-020-0887-9.

Yagan, O., Qian, D., Zhang, J., and Cochran, D. (2013). Conjoining speeds up information diffusion in overlaying social-physical networks. *IEEE Journal on Selected Areas in Communications* 31 (6): 1038–1048. https://doi.org/10.1109/JSAC.2013.130606.

Yağan, O. and Gligor, V. (2012). Analysis of complex contagions in random multiplex networks. *Physical Review E* 86: 036103. https://doi.org/10.1103/PhysRevE.86.036103.

Yang, L., Yu, Z., El-Meligy, M.A. et al. (2020). On multiplexity-aware influence spread in social networks. *IEEE Access* 8: 106705–106713. https://doi.org/10.1109/ACCESS.2020.2999312.

Ye, M., Zino, L., Mlakar, v.Z. et al. (2021a). Collective patterns of social diffusion are shaped by individual inertia and trend-seeking. *Nature Communications* 12: 5698. https://doi.org/10.1038/s41467-021-25953-1.

Ye, M., Zino, L., Rizzo, A., and Cao, M. (2021b). Game-theoretic modeling of collective decision making during epidemics. *Physical Review E* 104: 024314. https://doi.org/10.1103/PhysRevE.104.024314.

Yin, H., Wang, Z., Gou, Y., and Xu, Z. (2020). Rumor diffusion and control based on double-layer dynamic evolution model. *IEEE Access* 8: 115273–115286. https://doi.org/10.1109/ACCESS.2020.3004455.

Young, H.P. (2009). Innovation diffusion in heterogeneous populations: contagion, social influence, and social learning. *American Economic Review* 99 (5): 1899–1924. https://doi.org/10.1257/aer.99.5.1899.

Young, H.P. (2011). The dynamics of social innovation. *Proceedings of the National Academy of Sciences of the United States of America* 108 (4): 21285–21291. https://doi.org/10.1073/pnas.1100973108.

Yu, Q., Yu, Z., and Ma, D. (2020). A multiplex network perspective of innovation diffusion: an information-behavior framework. *IEEE Access* 8: 36427–36440. https://doi.org/10.1109/ACCESS.2020.2975357.

Zhang, J. and Moura, J.M.F. (2014). Diffusion in social networks as sis epidemics: beyond full mixing and complete graphs. *IEEE Journal of Selected Topics in Signal Processing* 8 (4): 537–551. https://doi.org/10.1109/JSTSP.2014.2314858.

Zhang, Y. and Zhu, J. (2018). Stability analysis of I2S2R rumor spreading model in complex networks. *Physica A: Statistical Mechanics and its Applications* 503: 862–881. https://doi.org/10.1016/j.physa.2018.02.087.

Zhang, N., Jia, W., Lei, H. et al. (2020). Effects of human behavior changes during the coronavirus disease 2019 (COVID-19) pandemic on influenza spread in Hong Kong. *Clinical Infectious Diseases* 73 (5): e1142–e1150. https://doi.org/10.1093/cid/ciaa1818.

Zhong, Y.D., Srivastava, V., and Leonard, N.E. (2021). Influence spread in the heterogeneous multiplex linear threshold model. *IEEE Transactions on Control of Network Systems* 9 (3): 1080–1091. https://doi.org/10.1109/TCNS.2021.3088782.

Zhuang, Y. and Yağan, O. (2016). Information propagation in clustered multilayer networks. *IEEE Transactions on Network Science and Engineering* 3 (4): 211–224. https://doi.org/10.1109/TNSE.2016.2600059.

Zino, L. and Cao, M. (2021). Analysis, prediction, and control of epidemics: a survey from scalar to dynamic network models. *IEEE Circuits and Systems Magazine* 21 (4): 4–23. https://doi.org/10.1109/MCAS.2021.3118100.

Zino, L., Como, G., and Fagnani, F. (2017). Fast diffusion of a mutant in controlled evolutionary dynamics. *IFAC-PapersOnLine* 50 (1): 11908–11913. https://doi.org/10.1016/j.ifacol.2017.08.1429. 20th IFAC World Congress.

Zino, L., Ye, M., and Cao, M. (2020a). A coevolutionary model for actions and opinions in social networks. *2020 59th IEEE Conference on Decision and Control (CDC)*, 1110–1115. https://doi.org/10.1109/CDC42340.2020.9303954.

Zino, L., Ye, M., and Cao, M. (2020b). A two-layer model for coevolving opinion dynamics and collective decision-making in complex social systems. *Chaos* 30 (8): 083107. https://doi.org/10.1063/5.0004787.

Zino, L., Ye, M., and Cao, M. (2021). On modeling social diffusion under the impact of dynamic norms. *2021 60th IEEE Conference on Decision and Control (CDC)*, 4976–4981. https://doi.org/10.1109/CDC45484.2021.9682999.

4

Opportunities and Threats of Interactions Between Humans and Cyber–Physical Systems – Integration and Inclusion Approaches for CPHS

Frédéric Vanderhaegen[1,2] and Victor Díaz Benito Jiménez[3]

[1] *Université Polytechnique Hauts-de-France, LAMIH lab, UMR CNRS 8201, Valenciennes, France*
[2] *INSA Hauts-de-France, Valenciennes, France*
[3] *University of Alcalá, University Campus – Calle 19, Madrid, Spain*

The design of cyber–physical and human systems (CPHS) relates to user-centered automation and system engineering processes. The application of the "Tailor-made" metaphor is useful to develop CPHS that guarantees achievement of factors such as well-being, satisfaction, performance, safety, resilience, or sustainability. On the one hand, based on features of factories of the future with regard to Industry 4.0 or 5.0 concepts, cyber–physical systems (CPS) are more and more flexible because they become interoperable, connected, or digital, and their abilities increase to make them more autonomous and to transform them into smart, resilient, green, sustainable, safe, or reliable entities. On the other hand, the human-in-the-loop concept remains an important feature to be developed for CPHS. Such a CPHS involve different types of interactions between humans and CPS: workers on manufacturing systems, customers who will use manufactured products, staff who produces services, stakeholders who design, assess, analyze, or maintain the production systems, etc. Moreover, it implies shared control process in a workplace or between workplaces involving humans and CPS.

Among "tailor-made" integration methods, human-centric, design-based approaches have been in vogue for a long time. However, regarding the future evolution of industrial or transport systems involving CPS interacting with humans, this approach seems unsuitable or insufficient, mainly because it produces systems that may not be suitable for all users even though they were designed with the help of a low number of them. More recently, inclusion-driven approaches that attempt to satisfy a maximum number of users have been developed. They relate to the "all-in-one" metaphor that consists in designing CPHS for everybody, everywhere, and whatever social, economic, or cognitive conditions of stakeholders and by respecting cultural, environmental, and ecological constraints. Integration concept adapts people or situation to standards, while inclusion one adapts variability of people and of situation to standards.

This chapter discusses about these integration and inclusion approaches. It proposes an inclusive solution which is applicable for a large class of CPHS. It presents a theoretical framework for modeling CPHS based on the "all-inclusive" concept in order to design systems with abilities to cooperate, compete, learn, or teach by combining abilities to control any situations and to learn from them. Opportunities and threats of resulting CPHS are interpreted in terms of dissonances with positive or negative impacts, respectively. Dissonances are discrepancies about human, technical, or organizational factors. For instance, they are inconsistencies between perceived and measured factors, between groups of people, between human and CPHS or between workplaces. They are threats when discrepancies are uncomfortable or hazardous or are opportunities when they are pleasant or beneficial. They can be detected or hidden, and their detection can produce other dissonances due to for instance an increasing of stress or workload. The proposed framework is illustrated with practical examples to discuss different alternatives to the CPHS approach.

The first section of the chapter discusses on the principles about human-in-the-loop based CPHS. The second section details "tailor-made" approaches based on the human–systems integration concept. The third section presents "all-in-one" conceptual contributions about human–systems inclusion. A framework for human–CPS inclusion process based on dissonances between humans and CPS is developed in the fourth section, and the chapter finishes with results about practical examples to illustrate the feasibility of such a framework.

4.1 CPHS and Shared Control

A system is autonomous when it is capable to act alone without any human intervention. However, this autonomy can be temporary or permanently degraded or incomplete or insufficient to face unprecedented situations for instance.

Therefore, human autonomy, i.e. human ability to act without depending on other systems or humans, can be useful for recovering a lack of autonomy of CPS and vice versa: autonomy of CPS can increase by human autonomy. For being autonomous or for sharing control tasks, humans and CPS need three main prerequisites (Vanderhaegen 2016a, 2021a): (i) skills to be applied for achieving CPHS goals or for controlling any normal or abnormal situations; (ii) resource availability to be ready to act or interact; and (iii) devices for making actions and interactions possible. As a matter of fact, shared control process between human and CPS consists in sharing their skill, their availability, or their device to optimize risk prevention, recovery, or containment. It requires a switching system to share control tasks between human and CPS, and a communication system to act on the controlled process, to inform about task allocation and problem-solving processing or to exchange data on a workplace or between workplaces. Different modalities of allocation and communication systems are detailed on (Vanderhaegen 1993, 1999). For instance, task allocation can be managed manually or automatically, and the allocation of a task can be preemptive if it is interruptible and recoverable, or definitive if not. Communications can be done remotely or face to face and require dedicated human–machine interaction devices and protocols. Shared autonomy frameworks based on allocation and communication systems can be adapted to different domains of application such as manufacturing, robotics, or transport systems (Sheridan 1992; Vanderhaegen 1993, 2012; Zieba et al. 2010, 2011; Powell et al. 2016; Inagaki and Sheridan 2019; Pacaux-Lemoine and Flemisch 2019; Tiaglik et al. 2022).

Some studies on human-in-the-loop concept consider the development of advanced supports for improving human abilities or for sharing control tasks between humans and CPS. The "Tailor-made" metaphor has thus to be applied in order to make any situations under control, Figure 4.1. It consists in developing devices and communication protocols in order to make interaction and shared control feasible in a workplace or between workplaces. Devices for CPHS aim to exchange data or resources or to interact with the controlled process. Development of future CPHS is based on Operator 4.0 concepts related to cognitive, sensorial, or physical interaction devices (Romero et al. 2016; Segura et al. 2020). These devices apply technology such as virtual reality, augmented reality, mixed reality, assistive exoskeleton, cobot, or social network. Designing CPHS consists in considering humans as sensors or actuators and in assisting them by CPS (Ruppert et al. 2018). However, humans are also capable of controlling the sharing

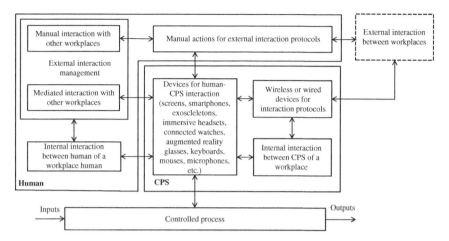

Figure 4.1 The "Tailor-made" metaphor for CPHS in a workplace and between workplaces. Source: Adapted from Vanderhaegen (1993, 1997).

process between CPS and themselves. When CPS control the sharing process, it is mandatory to develop adapted interaction devices in case of CPS failure to take back control manually in safe conditions. Indeed, shared control between human and CPS can lead to hazardous situations when human–CPS interaction devices are not adapted for understanding correctly situations by humans (Vanderhaegen 2021b). Development of CPHS abilities are then required to recover such brittleness to face surprising changes or unanticipated situations (Woods 2018; Woods and Shattuck 2000). Situation awareness is a key point for future CPHS to perceive and understand them, and to anticipate their possible evolution (Endsley 1995; Carsten and Vanderhaegen 2015; Dekker 2017). Distributed configurations of CPHS or cooperation activities between human and CPS are examples of solution to reduce such risks of loss or lack of situation awareness (Millot 2015; Salmon et al. 2015; Eraslan et al. 2022).

CPHS concerns different types of human actions for all life-cycle steps of a system or of a product. Human–systems integration approaches consider different stakeholders for a "tailor-made" system design process.

4.2 "Tailor-made" Principles for Human–CPS Integration

Approaches of human–systems integration can be gathered into seven families, Table 4.1. Worker centered CPHS results from anthropocentric factor integration by combining user-centered design and system engineering competences (Boy 2020; Rauch et al. 2020). It includes human–machine teaming or human–system symbiosis for which human and machine behave as a team and each one can recover error of the other. There is therefore a mutual empowerment between human and machine, but when symbiosis between human and machine is sustainable, a failure of a decision-maker can penalize the activities of the other (Vanderhaegen et al. 2019). Design process of end-user centered CPHS is an holistic integration of different stakeholders of a CPHS for optimizing facilities flexibility and by adapting human-in-the-loop concept with user-oriented automation processes (Umbrico et al. 2021; Boy and McGovern Narkeviciu 2013). Customer-centered CPHS focuses on customer's interests by replacing mass production goal by mass personalization or customization of product (Karaköse and Yetis 2017; Wang et al. 2017). Among customer-oriented CPHS approaches, there are services-oriented ones when a company provides its customers with

Table 4.1 Tailor-made metaphor applied to human–CPS integration.

Concepts	Principles	References
Worker-oriented CPHS	Antropocentric factors on workplace design	Vanderhaegen et al. (2019), Boy (2020), and Rauch et al. (2020)
End-user-oriented CPHS	Holistic integration of stakeholders	Umbrico et al. (2021) and Boy and McGovern Narkeviciu (2013)
Customization-oriented CPHS	Personnalization of products	Karaköse and Yetis (2017) and Wang et al. (2017)
Service-oriented CPHS	Servitization of companies	Wiesner et al. (2017) and Farsi and Erkoyuncu (2021)
Life-cycle-oriented CPHS	Product and service lifecycle consideration	Hien et al. (2021)
Daily-life-oriented CPHS	Social networking between people	Yin et al. (2018), Zeng et al. (2020), Kinsner (2021), and Yilma et al. (2021)

a turnkey solution, including new services as consulting activities (Wiesner et al. 2017; Farsi and Erkoyuncu 2021). Life-cycle centered CPHS relates to servitization of companies, but includes life-cycle steps of services. Daly-life centered CPHS uses social networking between people or meta-verses and proposes cyber-physical-social systems (Yin et al. 2018; Zeng et al. 2020; Kinsner 2021; Yilma et al. 2021).

Many of these human–systems integration processes are based on standards or norms. Therefore, they aim to design universal and compliant CPHS to socially shape human behaviors which must converge toward those foreseen during the design of the CPHS. Cooperation is usually applied for managing degrees of automation and replacing human actions. When CPHS is accepted by a majority of users, minorities of users for who it is not adapted are not immediately considered because it is easier to force them to adapt themselves to the systems via specific training programs for instance. The integration of stakeholders and of services for designing future CPHS is a challenging issue but any obstacle that was not anticipated can take time before being solved properly and this makes goals of tailor-made failed. Another complementary metaphor can recover these issues: the "all-in-one" metaphor that aims to design CPHS for any categories of users by implementing specific abilities of adaptation to CPHS.

4.3 "All-in-one" based Principles for Human–CPS Inclusion

The "all-in-one" metaphor is applied to human–systems inclusion principles that consist in designing CPHS by considering (i) variability of users in terms of expectations, autonomous abilities, or demands for instance, and (ii) a maximum number of criteria related to cognition, economy, ecology, environment, ethics, or society (see Table 4.2). This aims to empower people or society. To do so, inclusive design adapts CPHS for a maximum number of people. Due to ambitious goals of the inclusive design application, the concept of "Universal Design" or "Universal Usability" is invoked to respond to needs or expectations of persons with disabilities (Newell and Gregor 2000; Abascal and Nicolle 2005). This naming is therefore related to human–systems integration to design systems for as large a group as possible and consists of employing methods that are inclusive in nature. Inclusive manufacturing considers the possibility to organize production process by

Table 4.2 "All-in-one" metaphor applied to human–CPS inclusion.

Concept	Goal	Criteria	References
Inclusive design	Design for all	Ethics, society	Newell and Gregor (2000) and Abascal and Nicolle (2005)
Inclusive manufacturing	Production by all	Economy, ecology, and society	Garrido-Hidalgo et al. (2018) and Singh et al. (2019a,b)
Inclusive mobility	Mobility for all	Accessibility, human right	Gallez and Motte-Baumvol (2017), Fian and Hauger (2020), and Ranchordas (2020)
Inclusive education	Education for all	Upskill, sustainability, well-being, pedagogy	Florian and Linklater (2010), Haug (2017), and Ackah-Jnr (2020)
Inclusive robotics	Activity for all	Ethics, society, accessibility, well-being	Kremer et al. (2018), Stöhr et al. (2018), Kildal et al. (2019), Monasterio Astobiza et al. (2019), and Simões et al. (2022)
Human–systems inclusion	All-inclusive systems	Cooperation, competition, learning, education	Vanderhaegen (2021b) and Vanderhaegen et al. (2021)

everybody and everywhere regardless social, economic, physical, or cognitive obstacles of workers. This aims at empowering workers sustainably by transforming dependent societies into independent ones. Inclusive mobility is mobility for all and is an answer to international and national laws and rights for human accessibility to transport and for mobility for everybody and everywhere (Gallez and Motte-Baumvol 2017; Fian and Hauger 2020; Ranchordas 2020). Inclusive education seeks to achieve same goals than inclusive manufacturing or mobility, but focuses on education for all. With regard to inclusive robotics, activity for all is applied by designing assistive robots and by considering factors like ethics or well-being. Finally, human–systems inclusion paradigm propose to design CPHS by applying "all-inclusive" process where users are free to use interaction devices and autonomous CPS as they want to achieve their expected goals (Vanderhaegen 2021b; Vanderhaegen et al. 2021). Four main families of criteria are developed: cooperation, competition, learning, and education.

Difficulties can occur when designing CPHS that have to respect all these criteria for maximizing inclusion purposes. Social plasticity is the ability of a society to find sustainable equilibrium facing any disturbances (Little 2007). It is also ability of humans to change their relationships over the time and emergent feature to achieve this goal is the development of cooperative behaviors (Del Val et al. 2016). Structurally, it consists in stopping any failed relationships, researching beneficial ones and improving cooperative relationships to copy the best behaviors (Eguíluz et al. 2005; Del Val et al. 2013). The proposed inclusion framework extends this concept to CPHS plasticity by considering that emergent features of CPHS are not only cooperation and learning but also competition and pedagogy to optimize Human–CPS relationship and autonomy (Vanderhaegen et al. 2006; Vanderhaegen 2012, 2021b; Vanderhaegen and Carsten 2017). It is based on dissonances that are gaps between human, technical, or organizational factors. Their prospective, on-line, or retrospective study will increase identification or development of emergent plasticity features in term "CPHS" abilities and interactions for cooperation, competition, learning, and education, Figure 4.2. Dissonance-based Human-CPS inclusion adapts variability of people and of situation to multiple possible standards of CPHS uses. Development of task allocation and communication systems

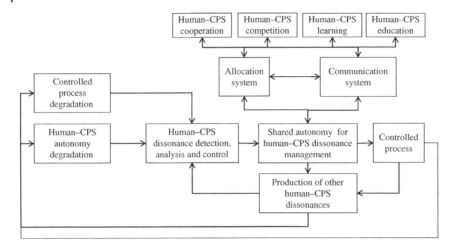

Figure 4.2 Human–system integration framework for "All-inclusive" CPHS design.

between human and CPS consists in proposing relevant devices to cooperate, compete, learn, and teach for improving autonomy characteristics for self-control or shared control activities.

Therefore, system plasticity is a prerequisite for CPHS resilience. Indeed, it relates to successful abilities to face and react to any changes of controlled process state or Human–CPS autonomy state in order to make CPHS resilient (Vanderhaegen 2017). Resilience process in human–machine system control engineering was mainly studied by applying cooperation abilities or learning ones (Zieba et al. 2010, 2011; Ouedraogo et al. 2013; Vanderhaegen and Zieba 2014; Enjalbert and Vanderhaegen 2017), but rarely via competition or education. Functional and structural features for CPHS plasticity achievement consider degradations of the controlled process or degradations of Human and CPS autonomy levels due to a lack of competence, of resource availability or of device to make possible actions on the controlled process or interactions between human and CPS. Detection, analysis, and control of Human–CPS dissonances aim to optimize shared control between human and CPS by activating allocation and communication systems and by managing activities based on cooperation, competition, learning, or education. The process is recursive because management of dissonances can generate other dissonances. For instance, as problem-solving of stressful or uncomfortable dissonance can be difficult or unprecedented, this can lead to a wrong dangerous action that may produce a new dissonance. Human–CPS Dissonance can be managed by behaviors like wait-and-see or trial-and-error based attitudes (Vanderhaegen, Caulier, 2011) or by dissonance discovery processes based on analysis of autonomy parameters (Vanderhaegen 2014, 2016b, 2021a, 2021c).

4.4 Dissonances, Opportunities, and Threats in a CPHS

The inclusion framework discussed in the previous section can be expanded upon to study emergent features in a CPHS based on the concept of dissonances defined as inconsistent differences between individual, collective, organizational, or technical factors. Some of these inconsistencies are usually named human errors and can be analyzed by applying many human reliability assessment methods (Reason 1990; Kiwan 1997; Vanderhaegen 2001, 2010; Hickling and Bowie 2013; Rangra et al. 2017). Human error is a gap between expected and actual uses in terms of human behaviors or results of these behaviors. The classical approach to its analysis assumes the existence of a prescription, a standard, or a frame of reference relating to expected uses of a CPHS.

It is possible to deal with human error when deviations in question exceeds a predefined threshold of acceptability. However, application of human error concept for CPHS is insufficient for several reasons. First, other configurations that are rarely considered by human error assessment methods must be treated: absence of a repository framework (i.e. there is no framework that defines behaviors or attitudes to be applied), presence of an incorrect repository framework (i.e. this framework exists but it is incorrect), or existence of multiple repository frameworks (i.e. there are several frameworks that guides behaviors or attitudes). Second, evolutions of acceptability threshold of human errors, human errors under this threshold, and cognitive effects of human errors detected by those who performed them are not appropriately addressed. Thus, what is interpreted as being erroneous by some people may not be by others due to the presence of different benchmarks. In addition, an action or a situation resulting from an action that is considered as erroneous initially, can in use become normality when the initial frame of reference absorbs this deviation that can become a new norm. Uses of CPHS are not static, but dynamic and must depend on the evolution of factors such as habits, motivation, or preferences. At each iteration of system use, differences between felt, expressed, or measured factors can be constant, irregular, variable, positive, or negative. When these differences are imperceptible or absorbed because they are interpreted as negligible for example, literature on weak signals can be useful for their analysis. In intelligent economics, a weak signal is a low frequency or low amplitude signal and is associated with two concepts (Ansoff 1975; Holopainen and Toivonen 2012; Garcia-Nunes and da Silva 2019): (i) positive impacts called opportunities, and (ii) negative impacts called threats. These are associated with warning signs of danger generated intentionally or unintentionally, with or without intention to harm. They are the subject of implausible, unthinkable, and even unpleasant extrapolations from imperceptible, incomplete, heterogeneous, and uncertain data. A small gap between what is expected and what is done can therefore be associated with a weak signal for which the opportunities and threats deserve to be analyzed. The stability or variability of this gap in the short, medium, and long terms must also be considered (Vanderhaegen, 2017; Jimenez and Vanderhaegen 2019). They must take into account not only causes of occurrence of gaps but also their consequences on cognitive factors when they are detected by the people who performed them or who are responsible to.

The concept of dissonance makes it possible to deal with possible diversity and evolution of frames of reference or of deviations in the course of CPHS uses. In musicology, dissonance is the simultaneous or sequential production of a set of notes that sound out of tune or are unpleasant. This concept has been adapted to systems engineering in the context of risk analyzes (Vanderhaegen 2014, 2016b). Cognitive dissonance is a conflict between cognitions, i.e. between knowledge or elements of knowledge such as attitudes, beliefs, skills, or intentions (Festiger 1957). A collective or organizational dissonance is an inconsistent divergence of viewpoints between several individuals, between different groups of people or between societies (Kervern 1994). This concept of dissonance is extended to the uses of CPHS and to treat Human–CPS dissonances. Uses are good or bad practices of operating CPHS at the individual, collective, or organizational level. A dissonance can generate situations of not only discomfort, embarrassment, overload, discontent, or stress but also satisfaction, joy, or well-being for example (Vanderhaegen, Carsten, 2017). These positive and negative consequences are interpreted in terms of opportunities or threats, respectively. Different strategies can be applied to accept or reject them by maintaining or modifying the inappropriate expectations, attitudes, believes, or behaviors that cause dissonances (Vanderhaegen 2021c):

- Indifference in the face of dissonances. Dissonances and their associated consequences go unrecognized, or it is a matter of turning a deaf ear, or pretending not to see any problem and considering the dissonance as nonexistent. While this may have negative impacts, it is not seen or ignored, and inappropriate behaviors that cause dissonance remain unaffected. In other words, even if threats really exist, they are not treated by considering actual behaviors as best opportunities to be applied.

- Valuing opportunities associated with dissonances in order to justify inappropriate behaviors that are still interpreted as right ones. Paying people to accept them, making the truth confusing, transforming a lie into benefits, doing good deeds to make people forget about inappropriate behaviors, pretending its because of others, or claiming that others are behaving similarly are examples of strategies that can increase opportunities valuation and encourage others to apply them (Barkan et al. 2015).
- Devaluation of threats is associated with dissonances to justify inappropriate behaviors that are still interpreted as right opportunities. Discrediting guilt with self-punishment and physical pain, seeking forgiveness while confessing, degrading threats by justifying opportunities positively in relation to the behavior of others, or hypocritically validating threats without changing viewpoint are examples of threats devaluation (Barkan et al. 2015). For instance, when people are conscious that their actions are producing serious threats, they feel the need to free themselves of their guilt. This is called the Macbeth effect (Gollwitzer and Melzer 2012).
- Recognize threats of dissonances by prohibiting inappropriate behaviors that are considered as wrong behavioral opportunities and interpreted in terms of right behavioral threats.

With regard to CPHS investigations based on dissonances, several opportunities and threats can be considered, Table 4.3. Main opportunities are adapted from hedonic, consequentialist, and relativist thoughts to maximize the effects of positive dissonances, to minimize the impacts of negative dissonances, and to consider any positive or negative behaviors as beneficial (Vanderhaegen et al. 2021). Posing the pros and cons of dissonances can be analyzed and assessed via the Benefit-Cost-Deficit (BCD) model to determine positive and negative effects (Vanderhaegen 2004; Vanderhaegen et al. 2011). In case of successful control processes, gains are expressed in terms of profits or benefits and acceptable losses in costs, in case of failed control process, and some losses become unacceptable because they are deficits or hazards. Gains are therefore opportunities and unacceptable losses are possible threats that have to be controlled. Plasticity features of the overall CPHS framework must respect also variability between human and CPS and mutual human–CPS engagement for Human–CPS dissonance identification and control, including weak-signal based

Table 4.3 Examples of opportunities and threats of human–CPS inclusion framework.

Opportunities	Threats
Benefits-costs-deficits based analysis	Hidden human–CPS dissonance occurrence or effect
High-level of variability control between people and CPS	Application of wrong beliefs about human–CPS dissonance
Mutual human–CPS engagement for human–CPS dissonance identification and control	Unethical dissonance occurrence
Human-CPS networking resilience	Planned obsolescence
Human and CPS upskill	High human dependency
Explanation of inclusion criteria	Contaminated plasticity features
Updating of multiple frames of reference	Time, space, and cost consuming of plasticity processes

Source: Adapted from Vanderhaegen and Jimenez (2018) and Vanderhaegen et al. (2021).

dissonances. For instance, mutual engagement process for controlling human cognitive state has to consider dissonances between human factors like those identified on (Vanderhaegen and Jimenez 2018; Vanderhaegen et al. 2020). Indeed, some factors can be used for assessing human workload or stress, but their interpretation can generate dissonances due to ambiguity of their relevancy depending on framing effect. Such framing effect can therefore be controlled by human–CPS networking and upskill, by developing pedagogical abilities to train human or CPS to inclusion criteria evolution, or by updating frames of references or standards.

Despite these opportunities, threats can still occur. Hidden and dangerous human-CPS dissonance can exist or wrong beliefs about dissonances cannot be detected immediately. Dissonances can therefore have latent consequences, and cooperating, competing, learning, or educating about them come up against the problem of levels of certainty about the repeatability of their consequences over time. Variability between people or society can produce ethical dissonance for some of them, but unethical for others. Serious disagreement between persons can therefore occur. Killer robots or love robots are examples of such ethical problems. Updating processes are important features for CPHS, but CPHS devices, software, or autonomy parameters can become obsolete and thus useless. Incompatibility between CPHS updates, increasing of social divide or undetectable wrong standards are examples of consequences. Even if it seems obvious that CPS can be helpful to reduce human workload or error, their long-term use can lead to degrading of human skill or health or make people highly dependent on technology, and transform people into "digital idiots" (Desmurget 2019). Contaminated plasticity features is another possible threat due to adversarial data, data poisoning, data leakage, erroneous data formatting, incomplete data, omitted data, or information withholding in the course of cooperation, competition, learning, or education activities. Finally updating of CPHS for plasticity management process can take time, space, or money.

4.5 Examples of Opportunities and Threats

Three examples are presented. The first example presents the interest of combining cooperation and competition activities for improving skills of decision support systems based on mirror effect learning process. The second example illustrates the risk of contaminated learning process when discovering content of interaction devices. The last one investigates human variability of interpretation in terms of risk perception, situation awareness, and inclusion features.

The first example studies human engagement by cooperation and competition to take benefits of human–CPS networking and improve CPHS skills in terms of electrical consumption. To do so, one of the communication system characteristics presented by Vanderhaegen (1993, 1999) may be applicable. It is based on serial networking between CPHS. On serial human–CPS networking, activities of a CPHS impact the activities of another, while CPHS of parallel networking are all connected (see Figure 4.3).

Based on train driving interface of the MissRail® simulator (Vanderhaegen and Richard 2014), an ecodriving-based experimental protocol was organized to implement three plasticity features on CPHS. Given a CPHS, human was free to compete or cooperate with the CPS which gave indications on the position of the driving manipulator in order to optimize consumption over a 4 km course while respecting speed limits and acceleration constraints. CPS learned about the users' activities by applying a mirror effect-based learning system that considers best parts of human activity in order to improve its knowledge and modify planned position indications (Vanderhaegen 2021b). CPS's updating with driver's activity of a CPHS will be used by a driver of another CPHS. From CPHS1 to CPHS6, CPS knowledge updating improves electrical train consumption, Figure 4.4.

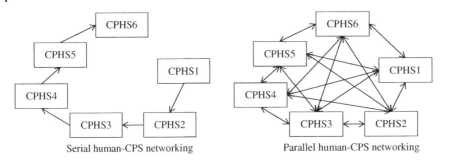

Figure 4.3 Serial and parallel Human-CPS networking.

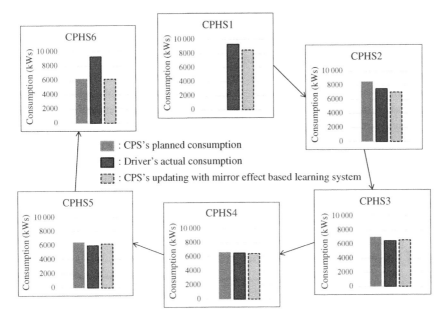

Figure 4.4 Result about consumption for each CPHS.

Drivers were able to cooperate with CPS by respecting the proposed positions of the train manipulator or to compete with it in order to try to perform best. A driver of CPHS6 failed when competing with CPS by degrading significantly the real consumption level. On CPHS1, there is no activated CPS because it begins to learn from the first driver's behavior. In Figure 4.4, consumption related to CPS knowledge updating for a given CPHS (right bar on all CPHS pictures) will be used at the following CPHS (left bar on all CPHS pictures except CPHS1), and consumption related to CPS advices were improved progressively. CPS knowledge base contained manipulator positions to be followed to obtain the expected consumption.

The second example is an experience about contaminated learning. It consists in presenting communication system content of three CPHS and looking for determining the meaning of indicator lights on or off, Figure 4.5.

Feedback of 62 persons was recorded by using the Reverse Comic Strip (RCS) method (Vanderhaegen 2013, 2021c). RCS aims to make relationships between felt emotions and human activities. Emotions are translated by selecting pictures' drawing faces with different expressions and by reporting what people said or thought. Participants must guess the correct rule to apply when the lights are on. After determining the rule for a given CPHS, the correct rule was

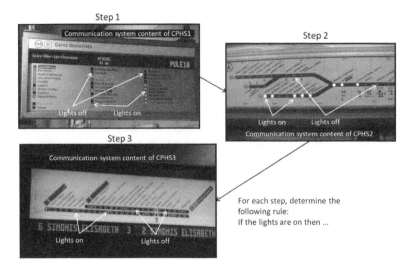

Figure 4.5 Human interpretation of communication system content of three CPHS.

communicated and the same exercise was done for the following CPHS. Rule for CPHS1 and CPHS2 is identical: if lights are on, then the arriving train will stop at the corresponding station. However, it differs for CPHS3: if lights are on then, there are trains at the indicating station. Most participants identified the correct rule for CPHS1 and CPHS2, but failed for guessing rule of CPHS3, Figure 4.6. Results about RCS give information about possible dissonance occurrence. Most selected pictures for CPHS1 and CPHS2 relate to positive emotion, while they are mainly negative for CPHS3 for which positive reporting of what people said or thought are less numerous. Contamination of learning process is due to similarity between communication system content for all CPHS3 and this make err participants for guessing the correct rule of the last CPHS.

	Construction of correct or wrong rule for each CPHS			
	Same rule as the previous CPHS	Wrong rule construction	No identified rule	Identification of correct rule
CPHS1	–	10	10	42
CPHS2	45	17	0	45
CPHS3	39	16	7	0

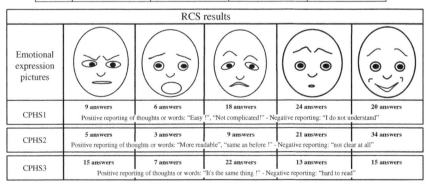

Figure 4.6 Results about rule construction and RCS.

The last example aims to study variability about situation awareness of allocation and communication systems of a CPHS and to identify plasticity features to improve it. On the studied CPHS, there are three flows of robot guided by a central rail: Line A, Line B, and Line C dedicated to three workshops A, B, and C, respectively. Robots from Lines A and B have two levels of autonomy:

– They are capable of stopping automatically when they detect a rear-end or lateral collision with another robot under 2 m. When the collision is over, they continue their mission by communicating their next intention. Robots from Line B have priority over the robots in Line A: when a collision between robots from Lines A and B will occur, robots of Line A stop and give priority to robots from Line B. When it is too late to give priority, robots of Line A continue their mission and the robots of Line B stop.
– They can recognize environmental conditions and priority constraints. When traffic light F1 is red, they stop and when it is green, they continue their mission.

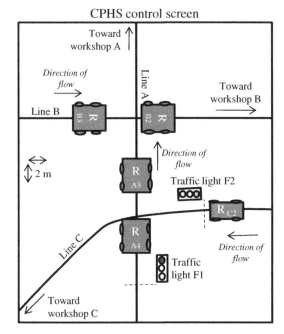

	Lieu de livraison	Robot intention communication	Confirmation of robot intention by human?
Robot R_{A3}	Atelier A	Move towards the Line A / Line B intersection	Yes ☐ No☐ No opinion ☐
Robot R_{A4}	Atelier A	Pass the Line A / Line C intersection	Yes ☐ No☐ No opinion ☐
Robot R_{B2}	Atelier B	Move to the exit via workshop B	Yes ☐ No☐ No opinion ☐
Robot R_{B3}	Atelier B	Move towards the Line A / Line B intersection	Yes ☐ No☐ No opinion ☐
Robot R_{C2}	Atelier C	Move towards the Line A / Line C intersection	Yes ☐ No☐ No opinion ☐

Figure 4.7 Allocation and communication systems of CPHS for studying variability about situation awareness and plasticity features.

Robots of Line C are only capable of recognizing their work environment, i.e. they are capable of detecting traffic light F2, to stop if it is red or to continue their mission if it is green. Robots communicate their intentions for allocation and communication system. Human supervises them by authorizing or blocking their intention achievement. This action is simulated by three possible human answers to validate robots' intentions: Yes, No, or No Opinion associated with 2, 1, or 0 score, respectively.

Eighteen future engineers were invited to assess several configurations, and Figure 4.7 is one of them. Danger of the situation simulated on Figure 4.7 occurs between robots from Line A and Line C. Traffic light F2 is green, robot R_{C2} can therefore move. Robot R_{A4} is already moving on Line A/Line C intersection due to previous situation for which traffic light F1 was green. Due to priority given by robots from Line A to robots from Line B, robots R_{A3} may stop and blocks robot R_{A4} that may stop at Line A/Line C intersection. A collision may occur between robots R_{C2} and R_{A4}. Therefore, the R_{C2} robot must be stopped by not authorizing its intention to move when the F2 light is green at this time. On Figure 4.7, the answer No is expected for robot R_{C2}. There is no risk of collision between robots from Lines A and B because robots are capable of managing alone possible rear-end or lateral collisions by stopping before the impact.

This analysis is supposed to be easy to do by using contents of control screen and allocation and communication systems. Tables 4.4 and 4.5 summarize the results of the experimental protocol.

Table 4.4 Results on variability about situation awareness.

Subjects	Intention of R_{C2} is authorized	The proposed CPHS facilitates situation awareness	Hazards are detected	Allocation and communication system is satisfactory	Control screen content is satisfactory
1	1	1	0	1	2
2	0	2	2	2	2
3	1	2	2	1	2
4	2	1	2	1	2
5	2	1	2	2	1
6	1	2	2	2	2
7	1	2	1	2	1
8	1	2	2	1	1
9	2	2	2	1	2
10	2	2	2	1	2
11	1	1	1	2	2
12	0	1	2	0	2
13	2	2	2	1	1
14	2	1	2	1	1
15	1	2	2	2	2
16	1	2	2	1	1
17	2	2	2	2	2
18	2	2	2	2	2

Note: 0: No opinion; 1: No; 2: Yes.

Table 4.5 Results on variability about plasticity features.

Subjects	The proposed CPHS is cooperative	The proposed CPHS needs competition abilities	The proposed CPHS needs learning abilities	The proposed CPHS needs education abilities
1	1	2	2	0
2	2	1	2	2
3	1	1	2	0
4	1	1	2	2
5	1	1	2	2
6	2	1	2	2
7	2	1	2	2
8	0	2	2	2
9	2	1	2	2
10	2	1	2	2
11	1	2	2	2
12	0	2	2	0
13	1	0	2	0
14	1	0	0	0
15	2	1	2	2
16	2	2	2	2
17	2	0	0	0
18	1	0	2	2

Note: 0: No opinion; 1: No; 2: Yes.

The first column of Table 4.4 corresponds to answers about authorization to achieve indicated intentions of robot R_{c2}. The rest of the columns concerns answers on different feeling about situation awareness characteristics: situation awareness facilitation, hazard detection, allocation and communication system efficiency, and control screen efficiency. Table 4.5 contains results about plasticity features in terms of cooperative abilities of the proposed CPHS and of the requirements for competition, learning, and education abilities to be developed. Education abilities are expressed in terms of pedagogical features. The lines that are shaded correspond to seven participants who identified wrong risk or who detected no risk at all. However, among them, two did not authorize the intention of R_{C2}, four said that the proposed CPHS facilitates situation awareness, five estimated that allocation and communication systems are satisfactory and six are satisfied by the screen control. Similarly, among the 11 people who correctly detected the risk of collision between R_{C2} and R_{A4}, eight authorized the intention of R_{C2}, three estimated that the proposed CPHS did not facilitate situation awareness, eight were disappointed by the allocation and communication system, and five considered the control screen as unsatisfactory.

Some human–robot dissonances occurred due to several reasons:

– People who think that the controlled situation is risky can detect wrong risks, but at the same time interrupt this dangerous situation without perceiving the right danger.

– People who detect correct risks can authorize hazardous intention achievement of CPS.
– Variability about relevancy of contents of control screen and allocation and communication system is high, and this is sometimes contradictory with the feeling about situation awareness facilitation by using the proposed CPHS.

Table 4.5 also displays a high variability for developing plasticity features such as competition, learning, and education. People who felt that the proposed CPHS is cooperative are not free from risk detection errors or from dissonance between right risk detection and incorrect authorization of intention achievement of robot R_{C2}. Learning and education ability development seems mandatory because a majority of participants mentioned it. However, even competition ability requirement is not relevant for a majority of people; it can be useful for a minority of them to reduce possible misunderstanding of a CPHS functioning and facilitating situation awareness.

4.6 Conclusions

This chapter discussed the approaches to CPHS based on human–CPS integration and human–CPS inclusion. It presented human–CPS integration frameworks via the "tailor-made" metaphor. State of the art identified several focuses for human–CPS integration: approaches oriented to worker, to end-user, to customization, to service, to life cycle, or to daily-life. Human–CPS integration frameworks rather adapt workers, end-users, customers, services, life cycles, or daily-lives to standards. Human–CPS inclusion ones consider variability of people, societies, or situations in order to adapt standards. Therefore, any behavior or experience of a CPHS can be beneficial to others and any CPHS might be designed by and for everybody and for achieving a maximum number of criteria as economic, social, or environmental ones. Human–CPS inclusion frameworks were discussed around the "all-in-one" metaphor by distinguishing the targets which the CPHS is designed to: CPHS designed for all, production of CPHS by all, CPHS dedicated to mobility for all, education-oriented CPHS for all, activity oriented CPHS for all, or all-inclusive centered CPHS.

With regard to a synthesis about human–CPS inclusion, the chapter proposed a new framework based on the management of human–CPS dissonances that are uncomfortable inconsistencies between human, technical, and organizational factors. These inconsistencies may be acceptable to some and unacceptable to others, or this acceptability may vary for an individual or a group of people. Acceptability of dissonances can therefore evolve and risks associated with a dissonance risk is identified in terms of opportunities or threats.

The proposed framework aims to control such opportunities and threats of dissonances and to improve emerging features for CPHS plasticity by developing abilities on cooperation, competition, learning, and education. Human–CPS dissonance management process aims to take benefit from opportunities such as mutual human–CPS work engagement, CPHS resilience, CPHS upskill, pedagogical abilities to explain CPHS results or behaviors, or updating of CPHS frames of reference. It also considers threats such as possible hidden dissonance, wrong beliefs, unethical dissonance, planned obsolescence, human dependency, contaminated plasticity feature, or time, space, or cost consuming of plasticity process control. Practical examples illustrated the interest of the proposed human–CPS integration framework and of the emerging plasticity features related to human and CPS abilities to cooperate, to compete, to learn, and to educate. Future research will extend the validation process of the human–CPS integration framework based on the all-inclusive paradigm and on emergent features for CPHS plasticity to adapt CPHS to everybody, everywhere, and whatever the economic, environmental, and social constraints by developing abilities for cooperation, competition, learning, and education.

Further studies about the management of dissonances between human and CPS are needed for a deeper analysis of the proposed human–CPS inclusion framework. Plasticity features related to cooperation, competition, learning, and education cannot be useful for everybody, but everybody may temporarily need them. Human–CPS inclusion considers such intermittent needs, and CPHS have to be designed by considering not only requirements from a majority of people but also from minority ones in order to maximize the inclusion principles, i.e. CPHS for everybody and everywhere, by respecting economic, ecological, social, or cognitive constraints of their users.

References

Abascal, J. and Nicolle, C. (2005). Moving towards inclusive design guidelines for socially and ethically aware HCI. *Interacting With Computers* 17: 485–505.

Ackah-Jnr, F.R. (2020). Inclusive education, a best practice, policy and provision in education systems and schools: the rationale ad critique. *European Journal of Education Studies* 6 (10): https://doi.org/10.5281/zenodo.3605128.

Ansoff, H.I. (1975). Managing strategic surprise by response to weak signals. *Californian Management Review* 18 (2): 21–33.

Barkan, R., Ayal, S., and Ariely, D. (2015). Ethical dissonance, justifications, and moral behavior. *Current Opinion in Psychology* 6: 157–161.

Boy, G. (2020). *Human-Systems Integration Design: From Virtual to Tangible*. Boca Raton, FL: CRC Press, Taylor & Francis Group.

Boy, G. and McGovern Narkeviciu, J. (2013). Unifying human centered design and systems engineering for human systems integration. In: *Complex Systems Design & Management 2013* (ed. M. Aiguier, F. Boulanger, D. Krob, and C. Marchal), 151–162. Cham, Switzerland: Springer International Publishing https://doi.org/10.1007/978-3-319-02812-5_12.

Carsten, O. and Vanderhaegen, F. (2015). Situation awareness: valid or fallacious? *Cognition Technology & Work* 17: 157–158.

Dekker, S.W.A. (2017). The danger of losing situation awareness. *Cognition Technology & Work* 17 (2): 159–161.

Del Val, E., Rebollo, M., and Botti, V. (2013). Promoting cooperation in service-oriented MAS through social plasticity and incentives. *The Journal of Systems and Software* 86: 520–537.

Del Val, E., Rebollo, M., and Botti, V. (2016). Self-organization in service discovery in presence of non-cooperative agents. *Neurocomputing* 176: 81–90.

Desmurget, M. (2019). *La Fabrique du crétin digital – Les dangers des écrans pour nos enfants (The factory of the digital moron – The dangers of screens for our children)*. Paris, France: Editions du Seuil.

Eguíluz, V.M., Zimmermann, M.G., Cela-Conde, C.J., and Miguel, M.S. (2005). Cooperation and emergence of role differentiation in the dynamics of social networks. *American Journal of Sociology* 110 (4): 977–1008.

Endsley, M.R. (1995). Toward a theory of situation awareness in dynamic systems. *Human Factors* 37 (1): 32–64.

Enjalbert, S. and Vanderhaegen, F. (2017). A hybrid reinforced learning system to estimate resilience indicators. *Engineering Applications of Artificial Intelligence* 64: 295–301.

Eraslan, E., Yildiz, Y., and Annaswamy, A. (2022). Safe shared control in the face of anomalies. In: *Cyber–Physical–Human Systems: Fundamentals and Applications (CPHS)* (ed. A. Annaswamy, P. Khargonekar, F. Lamnabhi-Lagarrigue, and S. Spurgeon). Wiley-IEEE Press.

Farsi, M. and Erkoyuncu, J. A. (2021). Industry 5.0 transition for an advanced service provision. *10th International Conference on Through-life Engineering Service* (16–17 November 2021), University of Twente, The Netherlands.

Festiger, L. (1957). *A Theory of Cognitive Dissonance*. Stanford, CA: Stanford University Press.

Fian, T. and Hauger, G. (2020). Composing a conceptual framework for an inclusive mobility System. *IOP Conference Series Materials Science and Engineering* 960 (3): https://doi.org/10.1088/1757-899X/960/3/032089.

Florian, L. and Linklater, H. (2010). Preparing teachers for inclusive education using inclusive pedagogy to enhance teaching and learning for all. *Cambridge Journal of Education* 40 (4): 369–386.

Gallez, C. and Motte-Baumvol, B. (2017). Inclusive mobility or inclusive accessibility? A European perspective. *Cuadernos Europeos de Deusto*, Governing Mobility in Europe: Interdisciplinary Perspectives 79–104.

Garcia-Nunes, P.I. and da Silva, A.E.A. (2019). Using a conceptual system for weak signals classification to detect threats and opportunities from web. *Futures* 107: 1–16.

Garrido-Hidalgo, C., Hortelano, D., Roda-Sanchez, L. et al. (2018). IoT heterogeneous mesh network deployment for human-in-the-loop challenges towards a social and sustainable Industry 4.0. *IEEE Access* 6: 28417–28437.

Gollwitzer, M. and Melzer, A. (2012). Macbeth and the Joystick: evidence for moral cleansing after playing a violent video game. *Journal of Experimental Social Psychology* 48: 1356–1360.

Haug, P. (2017). Understanding inclusive education: ideals and reality. *Scandinavian Journal of Disability Research* 19 (3): 206–217.

Hickling, E.M. and Bowie, J.E. (2013). Applicability of human reliability assessment methods to human–computer interfaces. *Cognition Technology & Work* 15 (1): 19–27.

Hien, N. N., Lasa, G., and Iriarte, I. (2021). Human-centred design in the context of servitization in industry 4.0: A collaborative approach. *30th RESER International Congress: Value Co-creation and Innovation in the New Service Economy*. http://hdl.handle.net/20.500.11984/5314.

Holopainen, M. and Toivonen, M. (2012). Weak signals: Ansoff today. *Futures* 44: 198–205.

Inagaki, T. and Sheridan, T.B. (2019). A critique of the SAE conditional driving automation definition, and analyses of options for improvement. *Cognition Technology & Work* 21: 569–578.

Jimenez, V. and Vanderhaegen, F. (2019). Dissonance oriented stability analysis of Cyber-Physical Systems. *IFAC-PapersOnLine* 51 (34): 230–235.

Karaköse, M. and Yetis, H. (2017). A Cyber-physical system based mass-customization approach with Integration of industry 4.0 and smart city. *Wireless Communications and Mobile Computing* 1058081.

Kervern, G.-Y. (1994). *Latest Advances in Cindynics*. Paris: Economica Editions.

Kildal, J., Martín, M., Ipiña, I., and Maurtua, I. (2019). Empowering assembly workers with cognitive disabilities by working with collaborative robots: a study to capture design requirements. *Procedia CIRP* 81: 797–802.

Kinsner, W. (2021). Digital twins for personalized education and lifelong learning. *IEEE Canadian Conference on Electrical and Computer Engineering (CCECE)* 1–6. https://doi.org/10.1109/CCECE53047.2021.9569178.

Kiwan, B. (1997). Validation of human reliability assessment techniques: Part 1 — validation issues. *Safety Science* 27 (1): 25–41.

Kremer, D., Hermann, S., Henkel, C., and Schneider, M. (2018). Inclusion through robotics: designing human-robot collaboration for handicapped workers. In: *25th International Conference on Transdisciplinary Engineering* (3–6 July 2018), Modena, Italy, 239–248. https://doi.org/10.3233/978-1-61499-898-3-239.

Little, D. (2007). *Plasticity of the Social*. University of Michigan-Dearborn, Social Science History Association, November, 2007.

Millot, P. (2015). Situation Awareness: Is the glass half empty or half full? *Cognition Technology & Work* 17: 169–177.

Monasterio Astobiza, A., Toboso, M., Aparicio, M. et al. (2019). Bringing inclusivity to robotics with INBOTS. *Nature Machine Intelligence* 1: 164. https://doi.org/10.1038/s42256-019-0040-5.

Newell, A.F. and Gregor, P. (2000). User sensitive inclusive design – in search of a new paradigm. In: *First ACM Conference on Universal Usability*, Arlington Virginia USA, (16–17 November), 39–44. https://dl.acm.org/doi/pdf/10.1145/355460.355470.

Ouedraogo, K., Enjalbert, S., and Vanderhaegen, F. (2013). How to learn from the resilience of Human–Machine Systems? *Engineering Applications of Artificial Intelligence* 26 (1): 24–34.

Pacaux-Lemoine, M.-P. and Flemisch, F. (2019). Layers of shared and cooperative control, assistance, and automation. *Cognition Technology & Work* 21: 579–591.

Powell, J.P., Fraszczyk, A., Cheong, C.N., and Yeung, H.K. (2016). Potential benefits and obstacles of implementing driverless train operation on the tyne and wear metro: a simulation exercise. *Urban Rail Transit* 2 (3-4): 114–127.

Ranchordas, S. (2020). Smart mobility, Transport poverty, and the right to inclusive mobility. University of Groningen Faculty of Law Research Paper Series, No. 13/2020, December 2020.

Rangra, S., Sallak, M., Schön, W., and Vanderhaegen, F. (2017). A graphical model based on performance shaping factors for assessing human reliability. *IEEE Transactions on Reliability* 66 (4): 1120–1143.

Rauch, E., Linder, C., and Dallaseg, P. (2020). Anthropocentric perspective of production before and within Industry 4.0. *Computers & Industrial Engineering* https://doi.org/10.1016/j.cie.2019.01.018.

Reason, J. (1990). *Human Error*. Cambridge, UK: Cambridge University Press.

Romero, D., Stahre, J., Wuest, T., Noran, O. (2016). Towards an Operator 4.0 typology: a human-centric perspective on the fourth industrial revolution technologies. *International Conference on Computers & Industrial Engineering (CIE46)* (29-31 October 2016), Tianjin, China.

Ruppert, T., Jaskó, S., Holczinger, T., and Abonyi, T. (2018). Enabling technologies for operator 4.0: a survey. *Applied Sciences* 8 (9): 1650. https://doi.org/10.3390/app8091650.

Salmon, P.M., Walker, G.H., and Stanton, N.A. (2015). Broken components versus broken systems: why it is systems not people that lose situation awareness. *Cognition, Technology & Work* 17 (2): 179–183.

Segura, A., Diez, H.V., Barandiaran, I. et al. (2020). Visual computing technologies to support the Operator 4.0. *Computers & Industrial Engineering* 139: 105550.

Sheridan, T.B. (1992). *Telerobotics, Automation, and Human Supervisory Control*. USA: MIT Press.

Simões, A.C., Pinto, A., Santos, J. et al. (2022). Designing human-robot collaboration (HRC) workspaces in industrial settings: a systematic literature review. *Journal of Manufacturing Systems* 62: 28–43.

Singh, S., Barde, A., Mahanty, B., and Tiwari, M.K. (2019a). Digital twin driven inclusive manufacturing using emerging technologies. *IFAC PapersOnLine* 52 (13): 2225–2230.

Singh, S., Mahanty, B., and Tiwari, M.K. (2019b). Framework and modelling of inclusive manufacturing system. *International Journal of Computer Integrated Manufacturing* 32 (2): 105–123.

Stöhr, M., Schneider, M., and Henkel, C. (2018). Adaptive work instructions for people with disabilities in the context of human-robot collaboration. In: *2018 IEEE 16th International Conference on Industrial Informatics (INDIN)*, 301–308. IEEE.

Tiaglik, M.S., Efremov, A.V., Irgaleev, I.K. et al. (2022). Significance of motion cues in research using flight simulators. *IFAC PapersOnLine* 55 (41): 131–135.

Umbrico, A., Orlandini, A., Cesta, A., et al. (2021). Towards user-awareness in human-robot collaboration for future cyber-physical systems. *26th IEEE International Conference on Emerging Technologies and Factory Automation (ETFA)* (September 7–10), Vasteras, Sweden.

Vanderhaegen, F. (1993). Multilevel human-machine cooperation between human operators and assistance tools – Application to Air Traffic Control. PhD dissertation, Université de Valenciennes et du Hainaut-Cambrésis, Valenciennes, France, December 21.

Vanderhaegen, F. (1997). Multilevel organization design: the case of the air traffic control. *Control Engineering Practice* 5: 391–399.

Vanderhaegen, F. (1999). Cooperative system organisation and task allocation: illustration of task allocation in air traffic control. *Le Travail Humain* 62: 197–222.

Vanderhaegen, F. (2001). A non-probabilistic prospective and retrospective human reliability analysis method – application to railway system. *Reliability Engineering and System Safety* 71: 1–13.

Vanderhaegen, F. (2004). The BCD model of human error analysis and control. *Proceedins of 9th IFAC/IFIP/IFORS/IEA Symposium on Analysis, Design and Evaluation of Man-Machine Systems*, Atlanta, USA, September.

Vanderhaegen, F. (2010). Human-error-based design of barriers and analysis of their uses. *Cognition Technology & Work* 12: 133–142.

Vanderhaegen, F. (2012). Cooperation and learning to increase the autonomy of ADAS. *Cognition Technology & Work* 14 (1): 61–69.

Vanderhaegen, F. (2013). Toward a reverse comic strip based approach to analyse human knowledge. *IFAC Proceedings Volumes* 46 (15): 304–309.

Vanderhaegen, F. (2014). Dissonance engineering: a new challenge to analyse risky knowledge when using a system. *International Journal of Computers Communications & Control* 9 (6): 750–759.

Vanderhaegen, F. (2016a). Toward a petri net based model to control conflicts of autonomy between cyber-physical and human-systems. *IFAC-PapersOnLine* 49 (2): 36–41.

Vanderhaegen, F. (2016b). A rule-based support system for dissonance discovery and control applied to car driving. *Expert Systems With Applications* 65: 361–371.

Vanderhaegen, F. (2017). Towards increased systems resilience: new challenges based on dissonance control for human reliability in Cyber-Physical&Human Systems. *Annual Reviews in Control* 44: 316–322.

Vanderhaegen, F. (2021a). Heuristic-based method for conflict discovery of shared control between humans and autonomous systems – a driving automation case study. *Robotics and Autonomous Systems* 146: 103867. https://doi.org/10.1016/j.robot.2021.103867.

Vanderhaegen, F. (2021b). Pedagogical learning supports based on human-systems inclusion applied to rail flow control. *Cognition Technology & Work* 23: 193–202.

Vanderhaegen, F. (2021c). Weak signal-oriented investigation of ethical dissonance applied to unsuccessful mobility experiences linked to human–machine interactions. *Science & Engineering Ethics* 27: 2. https://doi.org/10.1007/s11948-021-00284-y.

Vanderhaegen, F. and Carsten, O. (2017). Can dissonance engineering improve risk analysis of human–machine systems? *Cognition Technology & Work* 19 (1): 1–12.

Vanderhaegen, F. and Caulier, P. (2011). A multi-viewpoint system to support abductive reasoning. *Information Sciences* 181: 5349–5363.

Vanderhaegen, F. and Jimenez, V. (2018). The amazing human factors and their dissonances for autonomous Cyber-Physical & Human Systems. *IEEE Conference on Industrial Cyber-Physical Systems*, Saint-Petersburg, Russia, May.

Vanderhaegen, F. and Richard, P. (2014). MissRail: a platform dedicated to training and research in railway systems. In: *Proceedings of the international conference HCII*, 22–27 June 2014, Creta Maris, Heraklion, Crete, Greece, 544–549.

Vanderhaegen, F. and Zieba, S. (2014). Reinforced learning systems based on merged and cumulative knowledge to predict human actions. *Information Sciences* 276 (20): 146–159.

Vanderhaegen, F., Chalmé, S., Anceaux, F., and Millot, P. (2006). Principles of cooperation and competition – application to car driver behavior analysis. *Cognition, Technology, and Work* 8: 183–192.

Vanderhaegen, F., Zieba, S., Enjalbert, S., and Polet, P. (2011). A Benefit/Cost/Deficit (BCD) model for learning from human errors. *Reliability Engineering & System Safety* 96 (7): 757–766.

Vanderhaegen, F., Wolff, M., Ibarboure, S., and Mollard, R. (2019). Heart-computer synchronization interface to control human-machine symbiosis: a new human availability support for cooperative systems. *IFAC-PapersOnLine* 52 (19): 91–96.

Vanderhaegen, F., Wolff, M., and Mollard, R. (2020). Non-conscious errors in the control of dynamic events synchronized with heartbeats: a new challenge for human reliability study. *Safety Science* 129: https://doi.org/10.1016/j.ssci.2020.104814.

Vanderhaegen, F., Nelson, J., Wolff, M., and Mollard, R. (2021). From Human-Systems Integration to Human-Systems Inclusion for use-centred inclusive manufacturing control systems. *IFAC-PapersOnLine* 54 (1): 249–254.

Wang, Y., Ma, H.-S., Yang, J.-H., and Wang, K.-S. (2017). Industry 4.0: a way from mass customization to mass personalization production. *Advances in Manufacturing* 5: 311–320.

Wiesner, S., Marilungo, E., and Thoben, K.-D. (2017). Cyber-physical product-service systems – challenges for requirements engineering. *International Journal of Automation Technology* 11 (1): 17–28.

Woods, D.D. (2018). The theory of graceful extensibility: basic rules that govern adaptive systems. *Environment Systems and Decisions* 38: 433–457.

Woods, D.D. and Shattuck, L.G. (2000). Distant supervision-local action given the potential for surprise. *Cognition Technology & Work* 2: 242–245.

Yilma, B.A., Panetto, H., and Naudet, Y. (2021). Systemic formalisation of Cyber-Physical-Social System (CPSS): a systematic literature review. *Computers in Industry* 129: 103458.

Yin, D., Ming, X., and Zhang, X. (2018). Understanding data-driven cyber-physical-social system (D-CPSS) using a 7C framework in social manufacturing context. *Sensors* 20: https://doi.org/10.3390/s20185319.

Zeng, J., Yang, L.T., Lin, M. et al. (2020). A survey: cyber-physical-social systems and their system-level design methodology. *Future Generation Computer systems* 105: 1028–1042.

Zieba, S., Polet, P., Vanderhaegen, F., and Debernard, S. (2010). Principles of adjustable autonomy: a framework for resilient human machine cooperation. *Cognition, Technology & Work* 12 (3): 193–203.

Zieba, S., Polet, P., and Vanderhaegen, F. (2011). Using adjustable autonomy and human–machine cooperation to make a human–machine system resilient – Application to a ground robotic system. *Information Sciences* 181 (3): 379–397.

5

Enabling Human-Aware Autonomy Through Cognitive Modeling and Feedback Control

Neera Jain[1], Tahira Reid[3], Kumar Akash[2], Madeleine Yuh[1], and Jacob Hunter[1]

[1] *School of Mechanical Engineering, Purdue University, West Lafayette, IN, USA*
[2] *Honda Research Institute USA, Inc., San Jose, CA, USA*
[3] *Mechanical Engineering and Engineering Design, The Pennsylvania State University, State College, PA, USA*

5.1 Introduction

Effective human–automation interaction (HAI) in cyber–physical–human systems (CPHS) holds great promise for improved safety, performance, and efficiency across a variety of domains, ranging from transportation to healthcare to manufacturing. These settings have in common a vision of shared autonomy between human and automation. However, such a vision faces numerous engineering challenges. While the automotive industry has driven many recent research advances, highly publicized problems in HAI (Davies 2016; Board 2017) plague the industry as well as consumer confidence in such technologies (Lienert 2018). Additionally, the aviation industry has struggled for decades with the increased use of automation (Billings 1997; Abbot et al. 1996) and the unanticipated problems it created (such as mode confusion and inattentiveness). Similar stories exist in biomedical device design, power systems, and in other critical domains. Since human response to a machine is difficult to model mathematically, control theoretic approaches often make simplifying assumptions about human behavior that are unsupported by experimental evidence. Further, the need for transparency dictates that unlike a controller in a fully automated system, communication to the human about machine intent is imperative (Lee and Moray 1992; Hancock et al. 2011). Hence, for human–machine systems, feedback control is tightly coupled with communication to the human.

As aptly described in Chapter 2, humans can interact with both the cyber and physical worlds as described by the acronym CPHS, but similarly, a human's interactions with the cyber component (absent of the physical) is widespread. Whether considering a CPHS or a cyber–human system (CHS), many researchers have recognized that models that capture the dynamics of human cognition are necessary for improved HAI (Breazeal et al. 2016; Kress-Gazit et al. 2021). The incorporation of "Theory of Mind", which describes the ability to understand and predict human behavior and mental states (Byom and Mutlu 2013; Mou et al. 2021; Pynadath et al. 2013), could provide autonomous systems with cognitive models that estimate and predict human behaviors and subsequent decision-making, all the while accounting for human variability. To that end, researchers from multiple scientific disciplines have proposed conceptual frameworks of human

Cyber–Physical–Human Systems: Fundamentals and Applications, First Edition.
Edited by Anuradha M. Annaswamy, Pramod P. Khargonekar, Françoise Lamnabhi-Lagarrigue, and Sarah K. Spurgeon.
© 2023 The Institute of Electrical and Electronics Engineers, Inc. Published 2023 by John Wiley & Sons, Inc.

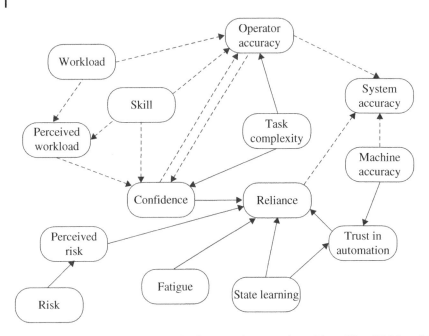

Figure 5.1 A conceptual framework of automation use adapted from Riley (1996) and Parasuraman and Riley (1997) showing the complex interactions between multiple human cognitive factors while interacting with automation and their effect on human reliance and performance. The solid arrows denote relationships supported by experimental data, and the dotted arrows denote hypothesized relationships or relationships that are dependent on the context of the system (Parasuraman and Riley 1997).

behavior and decision-making as it relates to autonomous systems. Figure 5.1 shows one example of a conceptual framework describing the complex relationship between cognitive factors and exogenous inputs on human behavior.

This particular framework highlights the dependency of the human's reliance on automation on cognitive factors such as their trust in the automation, their perception of risk, and their own confidence. It also describes how exogenous factors affect human cognitive factors. For example, the transparency of the system affects human cognitive factors of trust (Chen et al. 2014; Felfernig and Gula 2006; Tintarev and Masthoff 2007; Mercado et al. 2016; Akash et al. 2020b; Hosseini et al. 2018) and workload (Bohua et al. 2011; Helldin 2014; Alonso and de la Puente 2018), whereas the task complexity affects both the human's confidence and their performance (operator accuracy). Other frameworks corroborate similar findings, particularly as it relates to the dependency that human reliance on automation has on the human's trust in that automation.

Before defining the scope of this chapter, the importance of cognitive factors in HAI is first introduced, followed by a brief discussion of the combined roles of modeling, user study design, and controller design in successfully closing the loop between human and machine. The Introduction ends with a guide for reading this chapter.

5.1.1 Important Cognitive Factors in HAI

From these conceptual frameworks, several relevant cognitive factors that affect human behaviors and their decisions to rely on automation have been identified (Riley 1996; Lee and Moray 1994; Lee and See 2004; Gao and Lee 2006; Parasuraman et al. 2007; Hoff and Bashir 2015; Endsley 2017; Proctor and Van Zandt 2018). These include, but are not limited to, trust, self-confidence, workload, and perceived risk. According to Lee and See (2004), trust can be defined as "the attitude that an agent will help achieve an individual's goals in a situation characterized by uncertainty and vulnerability." From Briggs et al. (1998), self-confidence is often called self-efficacy, which can be best defined as "a belief in one's capability of performing a simple task" (Bandura 1986). Parasuraman et al. (2008) states that "workload can be described as the relation between the function relating the mental resources demanded by a task and those resources available to be supplied by the human operator." Workload has been characterized by the Yerkes–Dodson law which explains, and has empirically demonstrated, the relationship between human workload and performance (Yerkes and Dodson 1908). The human's workload, or perception of their workload, has long been considered critical to their performance. Therefore, the NASA Task Load Index (TLX) survey was developed to provide a multidimensional rating procedure for characterizing workload that considers several subfactors that affect it, including "mental demand, physical demand, temporal demand, performance, effort, and frustration" (Gore 2020; Hart and Staveland 1988). Perceived risk can be defined as the vulnerabilities of the system understood by a person in a given context or relationship (Lee and See 2004; Slovic et al. 2004). Other relevant factors include situational awareness (Endsley 2017; Parasuraman et al. 2008), fatigue (Hursh et al. 2004; Grech et al. 2009), and information retention (Anderson and Schooler 1991; Rubin et al. 1999).

From a modeler's perspective, an important feature of such frameworks is that they capture relevant input–state–output relationships. For example, machine accuracy is an input that affects human trust in automation, which is known to be dynamic. Human trust, in turn, affects the human's reliance on the automation. Such relationships can be used to structure dynamic modeling frameworks for cognitive factors, as is discussed more in Section 5.2.

5.1.2 Challenges with Existing CPHS Methods

Decades of research have established correlations between certain human behaviors and cognitive factors, as well as the fundamental notion that many cognitive factors are dynamic. However, building mathematical, control-oriented models of human cognition or decision-making remains a major challenge. What model structure is appropriate for capturing the relevant mathematical relationships? Are the dominant relationships linear or nonlinear? Probabilistic or deterministic? What level of modeling complexity is appropriate? In Section 5.2, the reader will first be introduced to an array of modeling architectures that can and have been used to capture human cognitive dynamics, and then be presented with a variety of computational models that are more suitable for algorithm design. These models can be used for both state estimation/prediction and model-based control design.

Of course, developing these types of cognitive models necessarily relies on human data, data that many engineers who conduct research in cyber–physical systems (CPS) are not familiar with collecting or analyzing. Furthermore, the design of human user studies for application to CPHS varies depending on the goal of the study. These goals may be more conventional, such as testing a hypothesis involving the correlation between two variables, or more algorithm-focused, such as developing a training dataset for said algorithm(s). Until now, many researchers interested in using black-box techniques, including machine learning, to identify models of complex systems (such as humans) have relied on naturalistic data. This is data "that make up records of human activities that are neither elicited by nor affected by the actions of social researchers" (Given 2008). While such data can be useful, particularly for online training or model adaptation, several complications may arise. First and foremost, such data are generally not labeled, making the interpretation of data-driven models difficult. Second, naturalistic data is often biased because it primarily represents recurring behavior. This leads to a well-known problem in machine learning in which models may be reliable "most of the time" but suffer in predicting less-expected or off-nominal outcomes.

This can be particularly dangerous in the context of models trained for improving human interactions with autonomous systems where misuse or disuse of the automation by the human, or a lack of transparency on the part of the automation, can have fatal consequences. An additional challenge arises in the heterogeneity of human data itself. As will be discussed more in Section 5.3, behavioral, psychophysiological, and self-report data must all be leveraged to achieve the broader goals of cyber–physical–human systems. However, the characteristics and availability of each type of data pose challenges for CPHS researchers that must be overcome.

With suitable models, it is possible to design and validate control algorithms that can enable autonomous systems to be responsive and adaptive to humans in real time. Concepts such as adaptive automation were first proposed by Rouse (1976). Since then, several researchers have explored the benefits and challenges of such concepts, mostly centered around systems that vary task allocation based on task demands, workload, and human performance (Dubois and Le Ny 2020; Parasuraman et al. 1999; Kaber and Riley 1999). From a control systems perspective, the human has often been modeled as a disturbance, with the control objective formulated as one of robustness to the human's input. However, the vision of human-aware CPHS is one in which humans actively interact and collaborate with the cyber–physical aspects of systems to jointly achieve shared goals. Cognitive feedback can enable these systems to *calibrate* the human's cognitive states so as to maximize the human's decision-making capability (Proctor and Van Zandt 2018; Chen et al. 2018; Akash et al. 2019a). Such systems would also be capable of adapting to system errors and external disruptions during operation by exploiting human skills in areas where the human is superior to the CPS.

A block diagram visualizing human-aware cognitive state feedback is shown in Figure 5.2. The scope of this chapter is on the interactions between human–machine dyads and considers the cognitive modeling of individual humans. Nevertheless, many of the fundamentals discussed in this chapter are likely of value to those interested in the interactions of more complex human-autonomy teaming scenarios. With respect to the taxonomy defined in Chapter 2, the methods described in this chapter are directly applicable to human-in-the-plant, human-in-the-controller, and human–machine symbiosis architectures, and may be extendable to humans-in-multiagent-loops.

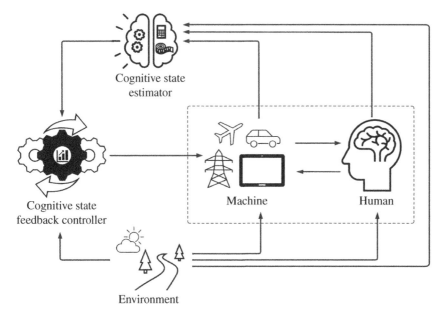

Figure 5.2 Block diagram depiction of human-aware cognitive state feedback control.

5.1.3   How to Read This Chapter

The field of cyber–physical–human systems (CPHS) is interdisciplinary. Advancing the state of the art relies on the expertise of researchers from a wide range of disciplines, including engineering, computer science, and the social sciences. This chapter is written to engage researchers across all of these disciplines; however, depending upon one's expertise, they may benefit from reading the sections in a particular order. Given the modeling and control focus of the chapter, it begins with an in-depth discussion of cognitive modeling. Specifically, Section 5.2 describes the computational modeling of human cognition that is applicable to algorithm design and real-time interactions between humans and autonomous systems. For researchers familiar with such models, this section is a suitable access point. Section 5.3 introduces the fundamentals of human subject study design within the CPHS context; therefore, researchers who are primarily familiar with the experimental study of human behavior may find it preferable to read Section 5.3 prior to Section 5.2. Finally, Section 5.4 provides considerations for feedback control and approaches to create human-aware systems that respond and adapt to humans in real time, thereby describing an important application of the models in Section 5.2.

Regardless of how the chapter is read, the different sections seek to inform both novice and expert CPHS researchers as they develop, train, and validate cognitive models, design human subject studies, and model-based control algorithms to calibrate human cognitive states. The chapter concludes in Section 5.5 with a discussion of opportunities for further investigation within the CPHS domain.

5.2   Cognitive Modeling

The cognitive modeling section is divided into three parts: modeling considerations, cognitive architectures, and computational cognitive models. Section 5.2.1 describes general factors that impact the training and interpretability of various cognitive models. Section 5.2.2 describes existing cognitive architectures that simulate human cognition using a holistic approach that does not depend on individual cognitive factors. Section 5.2.3 then discusses a broad range of computational modeling approaches that aim to represent specific cognitive factors.

5.2.1   Modeling Considerations

In this chapter, cognitive models are defined as those that capture human cognition or human cognitive factors explicitly, as shown in Figure 5.3. While there are several cognitive factors of interest in CPHS contexts, more recently, one of the most modeled factors is human trust in automation (Lee and See 2004; Chen et al. 2018; Xu and Dudek 2015; Mikulski et al. 2012; Juvina et al. 2015; Lee and Moray 1992; Fayazi et al. 2013; Sadrfaridpour and Wang 2018; Wagner et al. 2018; Freedy et al. 2007; Floyd et al. 2015; Xu and Dudek 2012; van Maanen et al. 2011; Azevedo-Sa et al. 2021; Lee and Moray 1994; Gao and Lee 2006; Saeidi and Wang 2019; Sadrfaridpour et al. 2016b; Akash et al. 2020b). Other cognitive models incorporate cognitive factors such as workload (Rouse et al. 1993; Akash et al. 2020b; Gil and Kaber 2012; Moore et al. 2014; Cao and Liu 2011; Gray et al. 2005; Wu and Liu 2007; Mitchell and Samms 2009; Riley et al. 1994; Wang et al. 2010; Aldrich and Szabo 1986), self-confidence (Lee and Moray 1994; Gao and Lee 2006; Saeidi and Wang 2019; Sadrfaridpour et al. 2016b; Tao et al. 2020), and perceived risk (Wagner et al. 2018; Freedy et al. 2007). Moreover, cognitive models are not necessarily limited to a single cognitive factor; there are existing cognitive frameworks that model multiple cognitive factors and how they interact with one another (Lee and Moray 1994; Gao and Lee 2006; Saeidi and Wang 2019; Akash et al. 2020b; Sadrfaridpour et al. 2016b). Nevertheless, these models typically consider at most two cognitive factors

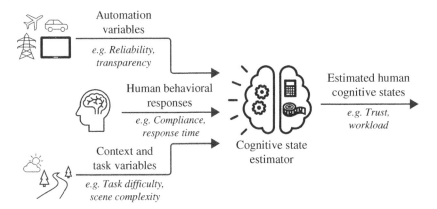

Figure 5.3 Diagram showing an example of a cognitive state estimator in the context of human–automation interaction.

in a single model. Given that the conceptual frameworks developed by human factors researchers show that multiple factors are indeed coupled in complex ways, models that capture the dynamic coupling between different cognitive factors continues to be an area in need of continued study and understanding. Note that in the context of computational modeling, cognitive factors are often referred to as cognitive states, as will be done throughout this chapter.

Before discussing different types of modeling approaches, it is helpful to underscore several considerations that one needs to make when modeling human cognitive dynamics, particularly for the purposes of state estimation, prediction, or calibration. The first affects the interpretability of the model. In general, data-based modeling can be classified as either gray-box or black-box. The former refers to the use of data for model parameterization and requires that the model structure is predefined (e.g. mathematical relationship between dynamic states, number of dynamic states). A black-box model, more common in modeling humans, can provide a very accurate mapping between the input and output data being used for system identification, but the mapping itself is typically not interpretable. This is not necessarily a problem if the goal of the model is to predict variables defined as outputs. However, cognitive states are generally not measurable and instead must be estimated based on other measurable quantities. In this case, a gray-box approach in which the states are defined can be leveraged to achieve interpretability. This issue will be further discussed in Section 5.2.3.

Another consideration when choosing the type of mathematical model to be identified or parameterized is whether it relies on continuous or discrete variables. While there is no clear consensus, research from cognitive psychology suggests that cognitive states, such as trust, do exist on a continuum (Muir 1994). Nonetheless, other researchers have demonstrated that a discrete definition may still be suitable for the purposes of cognitive state calibration (Akash et al. 2020a).

The final consideration is the number of parameters that must be estimated for a particular model structure, and whether that is appropriate given the volume of data available for model training and validation. While this consideration is important for any data-based modeling effort, it becomes more challenging to address in the context of cognitive modeling given the difficulties associated with collecting human data from large numbers of participants in controlled experimental settings. Moreover, data from several individuals often must be aggregated in order to train a single model as it is extremely rare to have enough data from a single individual to adequately train a model.

As will be evident from the models described in Section 5.2.3, these considerations can help guide the choice of model formulation. In the sections that follow, cognitive architectures will first be introduced to provide the reader with an important background on long-standing frameworks developed by the human factors community to model cognitive processes. This will be followed by a detailed discussion of various computational models that are more amenable to algorithm design and synthesis.

5.2.2 Cognitive Architectures

One approach for modeling human cognition is through a cognitive architecture. Cognitive architectures are theories for simulating and understanding the process of human cognition and are categorized as follows: connectionist models explain human behavior by utilizing a network of connected processing units, such as a neural network, to represent knowledge distribution. Symbolic models use a process that identifies human experience through the manipulation of symbolic representations. Connectionist and symbolic models can be combined to develop hybrid connectionist-symbolic models such as ACT-R (Adaptive Control of Thought-Rational) (Anderson 1996), Soar (Laird 2012), EPIC (Executive-Process/Interactive Control) (Kieras and Meyer 1997), and Connectionist Learning with Adaptive Rule Induction On-line (CLARION) (Sun 2006). ACT-R is made up of modules, buffers, and pattern-matching components, as shown in Figure 5.4. Perceptual-motor modules provide an interface or simulation of the real world. The declarative memory module represents factual knowledge, while the procedural memory module contains production rules that represent knowledge on how to do certain tasks. ACT-R accesses these modules through the use of buffers which serve as the interface to each module. The contents of the buffers represent the state of ACT-R at that time, and the pattern-matching searches for production rules that best fit the current buffer states (Ritter et al. 2019).

Next, the general theory of the Soar cognitive architecture "is based on goals, problem spaces, states, and operators" (Martin et al. 2021). The main processing cycle involves the interaction between procedural memory and working memory. Similarly to ACT-R, associative retrieval is used to access procedural knowledge. All long-term knowledge for controlling behavior is used

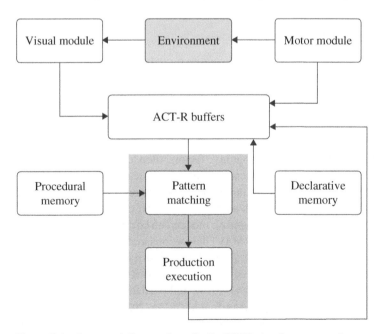

Figure 5.4 Recreated diagram from Budiu (2013) showing a general representation of the components involved in ACT-R. Source: Budiu (2013)/Carnegie Mellon University.

as production rules. Soar also implements reinforcement learning in addition to chunking in its procedural memory. While chunking learns new production rules, reinforcement learning adjusts the rules to create operator selection preferences. Additional modules for semantic and episodic memory are both accessed by the cues created in Soar's working memory. Through this process, Soar can be used to generate behavior in which the procedural knowledge provides the control, and the working memory acts as a global workspace (Laird 2012). EPIC, Executive-Process/Interactive Control, "is suited for modeling human multi-modal and multiple-task performance" (Martin et al. 2021). It not only captures cognitive processes, as is done in ACT-R and Soar, but also models sensory motor processes related to cognitive functions (Kieras and Meyer 1997). CLARION also aims to encompass a broader scope of human cognition by tackling meta-cognitive interaction and cognitive-motivational interaction in addition to implicit–explicit interaction (Sun 2006).

Cognitive architectures are intentionally defined to model a broad array of cognitive processes. As a result, they may not be able to model a context-specific, decision-making behavior such as a human's reliance on automation assistance (van Maanen et al. 2012). This leads to the usage of sequential sampling models, also known as accumulator models, which explain decision-making behavior as the accumulation of evidence when compared to a response threshold (Purcell and Palmeri 2017). Most "accumulator models are based on the drift diffusion model (DDM)" (Peters et al. 2015). Accumulator models have been used in combination with cognitive architectures. For example, the Retrieval by Accumulating Evidence in an Architecture (RACE/A) theory combines the level of detail from accumulator models with the cognitive architectures' task complexity to show how behavior is correlated to decision-making (van Maanen et al. 2012). Cognitive architectures serve to provide a holistic representation of human cognition. In order to further narrow down the scope to model specific aspects of cognition, computational cognitive models are used. For that reason, computational cognitive models are necessary for modeling specific behavioral dynamics and cognitive processes. A strength of these architectures is that they capture high-level interactions between different cognitive processes that a human undergoes when interacting with different kinds of systems, including autonomous ones.

5.2.3 Computational Cognitive Models

An intuitive approach toward developing computational cognitive models is to adapt the established cognitive architectures. For example, researchers have used factors based on ACT-R that explain a human's cognitive processes to collect data and develop a quantitative model of situational awareness (Kim and Myung 2016; Wiese et al. 2019). However, depending on the context, task complexity, and availability of data, the cognitive architectures can either become too limiting or too complex to be adapted quantitatively. If the purpose of a given cognitive model is the estimation, prediction, or calibration of one or more cognitive factors, then a model that explicitly describes the dynamics of the factor, or "state," is of increased interest. Researchers have predominantly focused on building models for predicting human trust in automation, while acknowledging the need for predicting other cognitive states, such as workload or self-confidence, to predict human reliance or decision-making. Computational cognitive models can be divided into four main categories: auto-regressive moving average vector (ARMAV) derivations and other linear models (Lee and Moray 1992; Hu et al. 2019; Azevedo-Sa et al. 2021; Hoogendoorn et al. 2013), Dynamic Bayesian Networks (Xu and Dudek 2015), decision analytical models based on decision or game theory (Freedy et al. 2007; Floyd et al. 2015; Wagner et al. 2018), and partially observable Markov decision process (POMDP) models (Chen et al. 2018; Akash et al. 2019a, 2020a). Albeit quantitative, the majority of these models are not amenable to control design, and none are complete in their characterization of all the cognitive states that are relevant for CPHS contexts.

For reference, examples of the four categorizations of models have been provided.

5.2.3.1 ARMAV and Deterministic Linear Models

An ARMAV model was originally presented by Lee and Moray (1992) to represent the relationship between trust and system performance. A regression model was used to identify factors which caused changes in trust. From this, the dynamics of trust was modeled using a trust transfer function. The model depicts trust as an output of a black-box model, using system specific measures (e.g. system fault occurrence within a pasteurization plant) as the inputs. As shown in Equation (5.1), trust T at time-step t was modeled as a function of system performance P, system fault occurrence F, and random noise perturbation a (Lee and Moray 1992; Sanders and Nam 2021).

$$T(t) = \phi_1 T(t-1) + A_1 P(t) + A_1 \phi_2 P(t-1)$$
$$+ A_2 F(t) + A_2 \phi_3 F(t-1) + a(t) \tag{5.1}$$

The variables A_1 and A_2 are the weightings associated with performance and fault occurrence, respectively. The variables ϕ_{1-3} are the auto-regressive parameters. Lee and Moray (1994) expand on the ARMAV model by incorporating self-confidence SC in addition to Trust T.

$$\%\text{Automatic}(t) = \phi_1(\%\text{Automatic}(t-1))$$
$$+ A_1((T - SC)(t)) + A_2(\text{Individual Bias}) + a(t) \tag{5.2}$$

From Equation (5.2), the usage of the automatic controller is dependent on the difference between the human's self-confidence and their trust in the automation, past use of the automatic controller ϕ_1, and individual bias operators had toward manual control. The parameters A_1 and A_2 represent the weights of the difference between trust and self-confidence and the individual bias, respectively. The variable $a(t)$ represents the normally distributed independent fluctuations of the ARMAV model (Lee and Moray 1994). Variations of the model originally published by Lee and Moray (1992) build on the ARMAV approach. For example, the trust function of the model developed by Sadrfaridpour et al. (2016a) accounts for both human performance P_H and robot performance P_R. These differences can be seen in Equation (5.3), in which trust T at time step $k+1$ is quantified as

$$T(k+1) = K_T T(k) + K_R P_R(k) + K_H P_H(k), \tag{5.3}$$

where K_T, K_R, and K_H are weightings applied to trust, robot performance, and human performance, respectively. Variations of the initial ARMAV model developed by Lee and Moray (1992) and other linear models are included in Table 5.1.

5.2.3.2 Dynamic Bayesian Models

Dynamic Bayesian networks are used to model stochastic processes and are capable of capturing how a process evolves over time (Murphy 2022; Sanders and Nam 2021). A highly referenced model that utilizes a dynamic Bayesian structure is OPTIMo or Online Probabilistic Trust Inference Model (Xu and Dudek 2015). OPTIMo quantifies the amount of trust a human supervisor has in an autonomous robot by defining the degree of human trust in automation at each time step as a random variable. Belief distributions are maintained through "performance-centric trust measures" that are deduced from various experiences and interactions (Xu and Dudek 2015). This gray-box approach effectively allows OPTIMo to consider "local relationships between trust and related factors" (e.g. robot task performance) and model trust evolution over time (Xu and Dudek 2015). It combines two approaches from other trust models: causal reasoning and evidential factors. Causal reasoning is applied to infer and update the robot's trustworthiness given the robot's task performance. Evidential factors are analyzed to quantify the human supervisor's degree of trust (Xu and Dudek 2015).

Table 5.1 Visual representation of different model types of existing computational models.

Model	Deterministic	Stochastic	T	W	SC	R	Automation
Lee and Moray (1992)	✓		✓				
Lee and Moray (1994)	✓		✓		✓		
Sadrfaridpour et al. (2016a)	✓		✓	✓			✓
Hu et al. (2019)		✓	✓				
Azevedo-Sa et al. (2021)	✓		✓				
Hoogendoorn et al. (2013)	✓		✓				
Xu and Dudek (2015)		✓	✓				
Soh et al. (2020)		✓	✓				
Freedy et al. (2007)	✓		✓				✓
Wagner et al. (2018)	✓		✓			✓	✓
Floyd et al. (2015)	✓		✓				✓
van Maanen et al. (2011)	✓		✓				✓
Gao and Lee (2006)		✓	✓		✓		
Akash et al. (2020a)		✓	✓	✓			✓
Chen et al. (2018)		✓	✓				
Rouse et al. (1993)	✓			✓			
Moore et al. (2014)	✓			✓			
Zhou et al. (2015)	✓			✓			✓
Gil and Kaber (2012)		✓		✓			
Hancock and Chignell (1988)	✓			✓			✓
Dubois and Le Ny (2020)	✓		✓	✓			✓

Trust, workload, self-confidence, and risk are denoted as T, W, SC, and R, respectively.

5.2.3.3 Decision Analytical Models

"Decision analytical" is a term used by Sanders and Nam (2021) to describe models that are grounded in game theory, decision field theory, economics, etc. For example, Freedy et al. (2007) establishes a score called the "relative expected loss," *REL*, by combining observed human task allocation decision behavior, risk associated with autonomous robot control, and observed robot performance. The *REL* score is calculated using the cost of robot failure C, the known probability of failure P, and the number of manual overrides K. In the model, this score is used to identify human operators with mis-calibrated trust for future training (Sanders and Nam 2021).

$$REL = \sum_{i=1}^{n} \frac{P_i C_i}{K} \tag{5.4}$$

Wagner et al. (2018) take a similar approach by relating trust to risk. Risk is calculated using the expected loss L from choosing event x when event y occurs and the probability P of event y occurring, as shown in Equation (5.5).

$$R(x, y) = \sum L(x, y) P(y) \tag{5.5}$$

In this model, the risk value is used in an economic game by a robot to infer if it is trusted (Wagner et al. 2018). Another decision analytical model proposed by van Maanen et al. (2011) models trust calibration. The appropriateness of trust α is calculated by evaluating the difference between the estimated descriptive trust τ_i^d and prescriptive trust τ_i^p at time t, as shown in Equation (5.6).

$$\alpha_i(t) = \tau_i^d(t) - \tau_i^p(t) \tag{5.6}$$

Descriptive trust is defined as the trust the human subject has in the different trustees, whereas prescriptive trust is defined as the trust a rational agent would have in the different trustees (van Maanen et al. 2011). Both descriptive trust and prescriptive trust range in value from 0 to 1. Positive values of appropriateness are correlated to over-trust, while negative values are correlated to under-trust. If the absolute value of the appropriateness value is below a threshold, it is assumed that trust is calibrated.

Xu and Dudek (2012) represents trust as a function based on "reputation." The model is used in the context of a supervisor–worker relationship, in which a robot issues internal commands via a planner to a behavior module that acts on these commands through the robot's actuators. At any time, a human supervisor can issue external commands to the behavior module, effectively overriding the robot's commands. In this model, supervisor commands $I_s(i)$ and worker commands $I_w(i)$ are defined as indicator functions within a discrete system, as shown in Equation (5.7).

$$I_s(i), I_w(i) \triangleq \begin{cases} 0, & \text{if command is issued during } (iQ - Q, iQ] \\ 1, & \text{otherwise} \end{cases} \tag{5.7}$$

The trust module then infers the worker's reputation $r(i)$ from the perspective of the supervisor depending on experiences and events during the interaction process. In Equation (5.8), reputation is lost if "the supervisor intervenes or if the robot's planner fails to generate a suitable command" (Xu and Dudek 2012). If the human does not override the robot's planner command, reputation is awarded (Xu and Dudek 2012).

$$r(i) = \begin{cases} r(i-1) - \Delta\rho, & \text{if } I_s(i) = 1 \text{ or } I_w(i) = 0 \\ r(i-1) + \Delta\rho, & \text{otherwise} \end{cases} \tag{5.8}$$

Xu and Dudek (2012) defines the supervisor's trust in the worker as "the utility of the robot worker's current reputation $U(r(i))$," as shown in Equation (5.9).

$$t(i) \triangleq U(r(i)) = \frac{\log(r(i) + 1)}{\log(\rho_{max} + 1)} \tag{5.9}$$

The utility function is a twice-differentiable function with a positive first derivative and negative second derivative which, according to utility theory from Norstad (1999), reflects that it is "much easier to lose trust than to gain trust" (Xu and Dudek 2012).

The final example of a decision analytical model is the extended decision field theory (EDFT) model proposed by Gao and Lee (2006). The EDFT model was used to represent the stochastic and dynamic decision process of choosing between manual and automatic control and, consequently, the evolution of operator preference when relying on automation. The preference P of automatic mode over manual model is the difference between trust T and self-confidence SC, as seen in Equations (5.12), (5.10), and (5.11), respectively.

$$T(n) = (1 - s) \times T(n - 1) + s \times B_{CA}(n) + \eta(n) \tag{5.10}$$

$$SC(n) = (1 - s) \times SC(n - 1) + s \times B_{CM}(n) + \eta(n) \tag{5.11}$$

$$P(n) = T(n) - SC(n) \tag{5.12}$$

B_{CA} and B_{CM} denote the input for the evolution of trust and self-confidence which represent the evolution of both the operator's trust and self-confidence. This is to represent that operators choose to rely or not rely on automation based on the capability of the manual and automatic control (Gao and Lee 2006). The variable η represents the uncertainty due to other factors and is a random variable with zero mean and variance.

5.2.3.4 POMDP Models

A partially observable Markov decision process, POMDP, is a "7-tuple $(S, A, O, T, E, R, \gamma)$ where S is a finite set of states, A is a finite set of actions, and O is a finite set of observations" (Akash 2020). The transition probability function, $T(s'|s, a)$, represents "the transition from the current state s to the next state s' given the action" and the emission probability function $E(o|s)$ represents "the likelihood of observing o, given that the subject is in state s" (Akash 2020). The reward function $R(s', s, a)$ and discount factor γ are used to find an optimal control policy using a completely trained model. POMDPs provide a framework "that accounts for partial observability through hidden states" (Akash 2020). This is useful when developing probabilistic dynamic models of human cognitive dynamics that are not able to be directly measured.

In the context of modeling human cognition, the cognitive factors are represented by cognitive states. This gray-box modeling framework approach can be utilized to estimate and predict human cognitive states that are parameterized using human subject data for increased model interpretability. Additionally, this allows for pre-existing relationships supported by literature between the cognitive states, actions, and observations to be explicitly defined within the framework. Akash et al. (2019a) use a POMDP to capture the dynamics of human cognitive factors of trust and workload in the context of a human being aided by a virtual robot in a reconnaissance mission. In the mission, a robot checks a target building for danger. It then reports back to the human, and the human chooses whether or not to wear protective gear before surveilling the building themselves. A finite set of states $S = [Trust, Workload]^T$ is defined in which either human trust or workload can be either low or high. "A finite set of actions is defined as $A = [Recommendation, Experience, Transparency]^T$" (Akash et al. 2019b). The recommendation action is the robot's report after checking the target building. The experience action is the reliability of the robot's last recommendation. The transparency action is an exogenous factor that refers to the amount of information provided to the human. This discrete action can take one of three values: low, medium, or high. The finite set of observations is defined as $O = [Compliance, Response\ time]^T$, in which the compliance observation is whether or not the human disagrees or agrees with the robot's current recommendation. The response time is categorized into three bins representing slow, medium, and fast response time. In a future iteration, response time was changed to a continuous observation. Each workload state had a characteristic response time distribution that was defined by an ex-Gaussian distribution (Akash et al. 2020a). Human subject data are used to estimate the transition probability function, observation probability functions, and previous probabilities of trust and workload states. It is important to note that the actions, states, and observations of a POMDP are context-specific. Akash et al. (2019a) use a POMDP to model the human cognitive states of trust and workload, in which the actions associated with robot and the observations are the human's reactions. Work by Chen et al. (2018) uses a POMDP in which the state is the fully observable world state and partially observable human trust. The actions include both robot and human actions.

To summarize, while ARMAV and other linear models are more suited for deterministic modeling of human behavior, dynamic Bayesian networks and POMDPs are used for dynamically modeling human behavior as a stochastic process. Models under the decision analytical umbrella term use game theory, economics, decision field theory, and more depending on the context of the task. These models can be specifically designed for deterministic or stochastic modeling of human behavior. It should be noted that there is no consensus on the "best" modeling framework for human cognitive dynamics. For different contexts, a given approach may work better than another, particularly when considering different types of cognitive factors or coupled factors, such as in the case of trust and self-confidence. As mentioned in the modeling considerations, the amount of data available and necessary for model training may affect modeling method options. For example, the number of actions, states, observations, and desired types of relationships may cause POMDPs to grow exponentially in the number of parameters to be estimated, causing model training to be computationally expensive. Additionally, the choice of black-box versus gray-box approaches inherently affects the choice of modeling method. ARMAV and other linear models are black-box approaches, whereas dynamic Bayesian networks and POMDPs are gray-box approaches. Context-specific factors such as the given task or goal, the environment in which the task will be completed, and the nature of the interaction or relationship between the human and CPS (e.g. supervisor–worker or human–robot teaming) will impact the modeling approach needed.

5.3 Study Design and Data Collection

As highlighted in the Introduction, a major challenge associated with modeling human cognitive dynamics is that the approach must be inherently data-driven. As shown in Figure 5.5, human data is typically collected to measure human response to changes in a set of independent variables. Engineers and algorithm designers typically have no qualms handling large data sets; however, designing an experiment to collect *human* data for model training or identification is an unfamiliar and even daunting task. In this section, the reader is introduced to the fundamentals of human subject study design within the cyber–physical–human systems (CPHS) context. Additionally, practical suggestions for CPHS researchers are shared in order to guide them in designing a study to meet their specific goals (e.g. collecting training data, validating an algorithmtc.). Furthermore, the reader is introduced to the types of human data that are available to collect as well as appropriate analysis methods to be employed.

The ultimate goal of any human study design, henceforth called "study design," is to answer research questions that address an important problem, expand the body of knowledge, and add value to society. Figure 5.6 is a flow diagram showing how the **research goal** determines the approach for conducting human subject studies in CPHS. At the inception of a project, the research team determines the nature of the research question. If the ultimate goal of the research is to make claims or draw conclusions about a population of people, then the hypothesis testing pathway should be followed. This pathway requires the use of statistical tests to make inferences about what

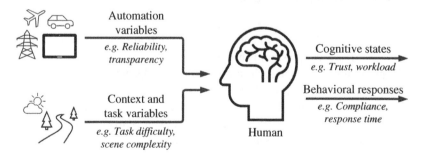

Figure 5.5 Diagram showing examples of independent variables (e.g. automation and context/task variables) and dependent variables (e.g. cognitive states, behavioral responses) considered in a human subject study in a CPHS context.

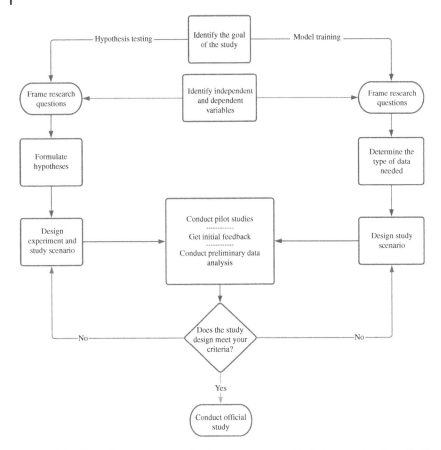

Figure 5.6 Flow diagram representing key processes and tasks to complete for effective study design.

is observed. Sample size and effect size are important parameters (see Sullivan and Feinn (2012) for more information). However, if the goal is to develop an algorithm or a mathematical model, then the pathway related to model training should be followed. In the field of CPHS, the priority is to design studies that stimulate a human response, capture the response, and then integrate that response into relevant models of interest.

5.3.1 Frame Research Questions and Identify Variables

Framing research questions is critical for conducting high quality and focused research. The absence of a research question can lead a team to conduct studies that are haphazard or inadvertently introduce bias. Once a research question is established, a team can determine what is already known about this question by reviewing relevant literature, analyzing preliminary data, and consulting subject matter experts (Banerjee et al. 2009).

Quantitative research questions are highly relevant for CPHS research as they generate numerical data that can be used for modeling, algorithm design, and making statistical inferences.[1] The wording of a quantitative research question should identify a clear independent and dependent

---

1 Where appropriate, the inclusion of qualitative data can add richness to quantitative results. Refer to Creswell and Plano Clark (2011) for more information.

variable of interest. Figure 5.5 shows the relationship between independent variables (inputs) and dependent variables (outputs). Specifically, the figure shows how one might manipulate independent variables related to the automation, the context, and the tasks. Cognitive states and behavioral responses are the dependent variables which capture how an individual is affected by changes to the independent variables. An example of a quantitative research question is: what is the effect of time pressure on braking behavior in a driving task? Time pressure is the independent variable and braking behavior is the dependent variable. Another example that includes a psychological dependent measure is: How does trust vary during take-over requests in a driving simulator? Here trust is the dependent variable while take-over request is the independent variable that a research team may systematically manipulate to assess a person's trust response. The research team should ensure that their research questions are aligned with the intended goal of the study as well as the time and resources available. This may require iteration. When it comes to fast-moving fields like CPHS, staying current with both the scientific literature as well as the popular press may be particularly important for determining relevant research questions.

5.3.2 Formulate Hypotheses or Determine the Data Needed

5.3.2.1 Hypothesis Testing Approach

Hypotheses should be formulated if the team intends to conduct statistical tests. They help researchers predict an outcome and provide a basis for statistical testing of the resultant data. Without a hypothesis, conducting statistical tests can lead to common errors (see (Banerjee et al. 2009)). Similar to research questions, hypotheses provide focus to a study. They help the researcher prioritize the aspects of a study that will best support and provide answers to the research question(s). Hypotheses should be grounded in the literature or preliminary research results. A hypothesis based on intuition should be a last resort. Banerjee et al. (2009) and Farrugia et al. (2010) provide relevant information on how to formulate hypotheses and test them. Examples of CPHS-related studies that include hypotheses are demonstrated in Huang et al. (2019) and Luster and Pitts (2021).

A hypothesis should make a prediction about the relationship between independent and dependent variables identified in the research question(s). For instance, the research question about time pressure and braking behavior may lead to the following null hypothesis: *There is no effect of time-pressure on braking behavior*. An alternative hypothesis that predicts a specific effect would also be valid: *As time pressure increases, braking behavior will increase*. Regardless of the type of hypothesis, statistical testing will prove whether that hypothesis can be accepted or rejected.

5.3.2.2 Model Training Approach

When a team's goal is to develop or train a model, the research question gives insight on the data to be obtained. For example, the goal of the paper by Akash et al. (2018) was to train a model that captured the relationship between trust level (the dependent variable) and participant experience with a certain sensor (the independent variable) in a driving context. The research question that stemmed from this goal was "How can trust be measured in real-time during an automation-assisted task?" This research question naturally led the team to the next step in Figure 5.6: "Determine the type of data needed." Since trust is intangible, the team brainstormed *physical* measurements that would provide data in order to model this interesting cognitive factor. They determined that psychophysiological measurements via electroencephalography (EEG) and galvanic skin response (GSR) would provide appropriate data for estimating trust.

There are three primary categories of dependent variables, or data, that can be collected in cyber–physical–human research: behavioral data, psychophysiological data, and self-report data.

To date, measurements of human subjects in CPHS contexts have largely focused on the use of behavioral data – any action that can be observed from the human, such as eye or body movements, facial expression, or speech production. More recently, physiological measures have also been used to infer human cognitive activity. These include involuntary (chemical or electrical) responses in bodily systems, such as heart rate, pupil dilation, or skin response (Gaffey and Wirth 2014). With measures such as the electroencephalogram (EEG), we also have access to changes in the human brain and central nervous system. When physiological signals are used to infer mental activity, they are referred to as psychophysiological signals. GSR "is a classical psychophysiological signal that captures arousal based upon the conductivity of the surface of the skin" (Akash et al. 2018) and has been used in measuring stress, trust, and cognitive load (Nikula 1991; Jacobs et al. 1994; Khawaji et al. 2015). Other common psychophysiological signals are the electrocardiogram (ECG) and functional near-infrared spectroscopy (fNIRS) (Fairclough and Gilleade 2014; Sibi et al. 2016; Verdière et al. 2018; Causse et al. 2017). Psychophysiological measures are continuous-time measurements and can provide insights that go beyond behavioral data or self-reports (Gaffey and Wirth 2014). The latter is important because people are sometimes not able to articulate their own higher-order mental processes in self-reports (Nisbett and Wilson 1977). However, given that the neurological processes that govern the human's psychophysiological response are complex, researchers typically rely on classification algorithms and other machine learning techniques, as well as simpler correlations, to map psychophysiological signals to useful interpretations such as mental workload (Lotte et al. 2007; Dussault et al. 2005; Hankins and Wilson 1998). Unlike psychophysiological data, behavioral data are often discrete measurements that may not be available at regular intervals. Another potential drawback of behavioral data is that they typically are defined relative to a specific context (e.g. letting go of a steering wheel in an autonomous vehicle), making them more difficult to generalize. Nevertheless, they contain rich information from which human trust, mental workload, and other cognitive states can be inferred. Table 5.2 summarizes the heterogeneity among behavioral, psychophysiological, and self-reported human data as well as their individual pros and cons for use in CPHS.

Table 5.2 The heterogeneity in behavioral, psychophysiological, and self-report human data.

	Behavioral	**Psychophysiological**	**Self-report**
Sampling frequency	Real time event-based (or continuous in some contexts)	Real time continuous	Real time[a]
Time scale	Highly variable and dependent on context)	On the order of seconds	On the order of minutes
Disadvantages	Dependency on context	Need for the sensors themselves; sensor faults	Human perception of one's own mental processes may not be accurate
Advantages	Easier to interpret as compared to psychophysiological data	Captures latent (involuntary) human response	Provides explicit quantification of the human's state of awareness

a) Self-reports could be solicited from a human every few minutes, but collection may need to be limited to avoid disrupting or aggravating the human.

5.3.3 Design Experiment and/or Study Scenario

Creating the experiment and study scenario requires careful consideration of the independent variables that will be manipulated, the dependent variables that will be measured, the study context, and the overall sequence of events.[2] It is recommended that decisions about the experiment design and study scenario be grounded in the literature or other reputable/valid sources. For instance, research teams who desire to conduct studies related to trust in autonomous systems must read recent and past work on the subject to learn how others assess trust. The objective is to discover if aspects of a published method, or even the entire method, can inspire their study design. Published psychometric scales can be used as is or with slight modifications so as not to lose the original intention of the scales. For example, many researchers use the popular trust survey developed by Jian et al. (2000) as a basis for acquiring participant feedback on their trust in automated systems. Additionally, in a study involving a reconnaissance mission (Akash et al. 2019b), aspects of the experiment were adapted from Wang et al. (2015). Although most CPHS studies may be novel, the extent to which aspects of the study design can be justified by previously published literature strengthens the quality of the work.

5.3.3.1 Hypothesis Testing Approach

After formulating a hypothesis, the next step is to determine the number of variations of each independent variable that should be studied. For instance, with the independent variable of "time pressure" discussed earlier, one may want to consider a discrete number – e.g. two to four – of different times for their study. For example, a simulator study conducted by Pawar and Velaga (2020) used "No time pressure," "Low time pressure," and "High time pressure" for a total of three variations. In human factors terms, independent variables are referred to as "factors" and the variations are known as "levels." If two or more independent variables are of interest, then a formal design of experiment (DOE) process should be considered (Montgomery 2012). Design of Experiments (DOEs) provide systematic ways to generate combinations of variables that constitute an experiment. For example, in a full-factorial experiment with two independent variables (factors) and three variations (levels), a total of $2^3 = 8$ possible experiment conditions can be tested. It is often best to start with two levels per factor leading to $2^2 = 4$ which reduces the number of conditions to be tested. There are a variety of DOEs that can be used. Extensive coverage of DOEs in psychological contexts is covered by Westfall et al. (2014) and Harris and Horst (2008).

Next, one needs to design the study scenario, context, and conditions in which to measure the effects of these independent variables. At this point, the research team has already determined research questions, variables, and hypotheses; thus, they probably have a strong inclination of what they want the study scenario to look like. For example, Pawar and Velaga (2020) conducted their studies in a driving simulator and used an urban road with stimuli representing "sudden events" (pedestrians crossing) and "noticeable events" (obstacle overtaking) to evaluate the effects of time pressure. Dependent measures included reaction time which incorporated actions that suggest braking where the accelerator pedal was released and pressure applied to the brake pedal. Additional details about the scenario are provided in the study.

5.3.3.2 Model Training Approach

Similar to the hypothesis testing approach, the model training approach requires a study scenario or context to be specified. For example, Akash et al. (2018) created a scenario where participants

---

2 See Hu et al. (2019) for an example of how sequence of events were organized within a trial (Figure 5.1) and between groups (Figure 5.3).

needed to respond to feedback provided by an image processing sensor during a simulated driving task. Details were given to the participants about the sensor's algorithm which detects obstacles in front of the car and asks them (the participants) to respond to the algorithm's report. The two main independent variables were obstacle presence – *obstacle detected* and *clear road* – and performance – *reliable* and *faulty*. The dependent measures were participants' reports on whether or not they *trusted* or *distrusted* the algorithm's report as well as their psychophysiological responses from the EEG and GSR measurements. The main takeaway when designing scenarios for the model training approach is to ensure that the scenarios elicit *measurable* human responses that provide data with the requisite granularity for a model to be trained.

5.3.4 Conduct Pilot Studies and Get Initial Feedback; Do Preliminary Analysis

Pilot studies are a set of preliminary experiments that enable researchers to iteratively examine the effectiveness of their study design. The research team should conduct pilot studies with the expectation that their experiment design may change. For example, even a conscientious researcher with a meticulously designed experiment may have failed to account for participant fatigue, issues with question wording, data collection channels interfering with one another, or necessary protocol steps. Oversights of this nature are almost always caught and resolved through pilot studies. Valuable feedback is obtained with enough time to perfect the experiment before expending too many resources prior to the study's official launch.

Conducting preliminary analyses gives the researcher early insight into the data before investing significant time and resources on what could be a flawed study. Note, a flawed study does not mean the results are contradictory; rather, a flawed study involves confounding variables, an unbalanced experiment, and so on. When it comes to the ***hypothesis testing approach***, the preliminary analysis procedure consists of verifying appropriate statistical methods that test the study's original hypotheses. Researchers use inferential statistics in hypothesis testing (Wetzels et al. 2011). The analysis depends on whether the dependent measures are numerical/continuous or categorical/discrete. For analysis of categorical variables, readers are referred to work by Agresti (2019). However, the more common inferential statistics are used to analyze continuous variables and include analysis of variance, regression analysis, two-sample t-tests and others. Correlation analysis is helpful for gaining quick insights on how dependent and independent variables relate. Three popular correlation analysis methods include Pearson correlation, Kendall rank correlation, and Spearman correlation. With several channels of data, it makes sense to understand which dependent variables are correlated so you can begin to understand which variables, and which interaction of variables, are affecting your participants.

For a study involving the ***model training approach***, the preliminary analysis consists of using pilot data to begin training a model or perhaps a pseudomodel. This is extremely important for studies that incorporate self-report and/or psychophysiological measurements in addition to a behavioral, or performance, measurement. An early start on model training allows one to verify that an experiment design is in fact eliciting sufficiently sensitive responses that meaningfully influence the model. For example, in the work by Akash et al. (2018), the experiment was first tested and refined by using self-report data collected from a large number of participants (>500) on Amazon Mechanical Turk (M-Turk[3]) (Akash et al. 2017). The authors established that the experiment was effective in eliciting trust dynamics in a systematic way and could therefore create a dynamic trust model. This gave the authors confidence to later incorporate

---

3 To learn more about M-Turk, see Paolacci et al. (2010).

psychophysiological measures, including an EEG and GSR, in the same experiment but with individuals in-person (Akash et al. 2018). Thus, researchers may often want to run an initial study as a pilot, even in a different context (e.g. online versus in-person), to understand if additional dependent variables are likely to capture a cognitive state and function as suitable model inputs. Then, if psychophysiological instruments are employed, an in-person pilot study with subsequent analyses should be conducted to again verify that the experiment design is eliciting responses as expected. To be clear, it is not the elicitation of the researchers' expected *results* that is important, but rather that the expected *responses* provide adequate sensitivity to function as notable model inputs.

5.3.5 A Note about Institutional Review Boards and Recruiting Participants

Since CPHS research involves humans, researchers must obtain approval from their organization's Institutional Review Board (IRB) before any data collection can begin. The IRB is a governing body that evaluates research involving human subjects. Their primary goal is to ensure that researchers follow proper ethics when engaging human participants and to ensure their rights are protected (Grady 2015). Once the IRB approves a study, recruitment can begin. Recruitment may involve e-mail, flyers, word-of-mouth, or other means to invite participants to an experiment. Several universities host a central page on which they advertise research studies from across campus who are looking for participants. Additionally, participants from the at-large community can be recruited through other online community pages and/or newsletters. The process may also require informed consent in which the individuals recruited are informed about the study goals, procedures, risks, benefits, compensation (if any), etc.

It is common for researchers in industry or academia to use convenience samples (Elfil and Negida 2017; Peterson and Merunka 2014). Students at a university are an example of a convenience sample, or a subset of a population in which researchers can easily draw from that does not necessarily represent for whom the research outcomes are intended (Peterson and Merunka 2014). Therefore, it is important to carefully craft a detailed plan to recruit a sample population representative of those who are expected to interact with the cyber–physical system(s) under consideration. Researchers in CPHS are highly encouraged to be intentional in recruiting diverse participants (Reid and Gibert 2022).

In summary, from obtaining IRB approval to experiment design, pilot testing, recruiting subjects, and analyzing data – study design efforts must be carefully thought out and executed. Researchers, especially those new to CPHS, should work with experts well-versed in these methods, such as those in human factors or psychology.

5.4 Cognitive Feedback Control

An interesting application of the models described in Section 5.2 is the design of human-aware algorithms that are responsive and adaptive to humans in real time, as shown in Figure 5.7. This can be realized by formulating a control problem with the human-automation dyad as the plant;[4] the automation interface/actions act as actuators that calibrate the human cognitive states. A block diagram visualizing such a feedback control system is shown in Figure 5.2, where the controller adapts the machine behavior based on estimates of the human cognitive states. This section presents the considerations essential to frame this control problem along with the fundamental approaches to

---

4 *plant* in control theory is a system to be controlled and is typically a combination of the process and actuator.

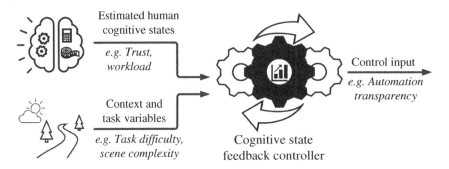

Figure 5.7 Diagram describing inputs and outputs of a cognitive state-based feedback controller.

design accompanying algorithms. Additionally, a case study is presented to demonstrate how each approach can be used in practice. Finally, evaluation methods to characterize the efficacy of these feedback systems are briefly discussed.

5.4.1 Considerations for Feedback Control

A primary consideration is identifying control variables in the given application-context that can be varied to perform cognitive state-based feedback control during human interactions with autonomous systems. To achieve this, a useful approach is to categorize the context based on the concept of "levels of automation" (Parasuraman et al. 2000). Adapting Sheridan's (Sheridan 1997) classification of tasks in ten functional levels, Parasuraman et al. (2000) proposed a simple four-stage model based on human information processing. Specifically, most systems comprise of four sequential tasks of information-processing: information-acquisition, information-analysis, decision-selection, and action-implementation. Thereby, each of these stages can be automated, resulting in four types of automation: acquisition, analysis, decision, and action automation, respectively. The degree of automation within each type of automation is called the level of automation (LOA) and can vary depending on the context. A schematic of the four-stage model and the corresponding automation types is shown in Figure 5.8. So depending on the extent of adaptation needed for an application, one could change the degree of automation within each automation stage, or vary the stage of automation, or a combination of both. For example, an autonomous vehicle could increase the driving functions it controls – i.e. change the level of action automation – if it detects low levels of the driver's attention. Or the autonomous vehicle could

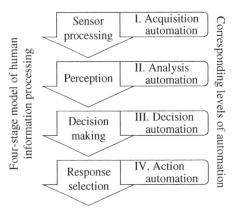

Figure 5.8 Simple four-stage model of human information processing and the corresponding types of automation (adapted from Parasuraman et al. (2000)).

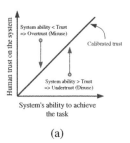 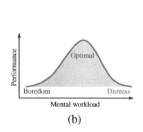 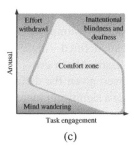

(a) (b) (c)

Figure 5.9 Optimal human cognitive states. (a) Calibrated trust, (b) optimal workload, and (c) optimal engagement. Figures adapted from Dehais et al. (2020).

instead give control to the driver and become a recommender system – i.e. reduce the automation stage from action automation to decision automation – if the driver's trust is too low.

A second key consideration is what cognitive states of the human should the CPHS detect and thereby respond to. In CPHS, the human can perform a variety of roles, ranging from supervisor to collaborator to observer. Depending on the human's role and the application context, some aspects of human cognition may affect system performance more than others. For example, in applications involving professionally trained users (e.g. military surveillance, aircraft pilots), cognitive states such as workload or situational awareness are especially critical; ill-calibrated cognitive states can potentially have fatal consequences due to the safety-critical nature of such contexts. Alternatively, in applications involving novice users (e.g. semiautonomous vehicles), cognitive states such as trust and self-confidence are also critical. Regardless, researchers agree that these cognitive states should be calibrated or maintained at an optimal level for improved CPHS performance. For example, trust should be calibrated to a system's reliability or trustworthiness to avoid disuse or misuse of automation (Lee and See 2004) (Figure 5.9a). Additionally, it has been argued that "workload should be maintained at an optimal level" to avoid both boredom due to low workload and distress due to high workload (Dehais et al. 2020) (Figure 5.9b). Finally, optimal performance is achieved in the region of arousal and task engagement labeled as "comfort zone" as shown in Figure 5.9c (Dehais et al. 2020).

5.4.2 Approaches

A human model can be implicitly or explicitly part of the overall system model to ensure that human cognition is accounted for in planning and control. Depending on the level of integration of the human model, the cognitive feedback control can be categorized as follows: heuristics-based planning, measurement-based feedback, or goal-oriented feedback.

5.4.2.1 Heuristics-Based Planning

A relatively simple and direct way to account for human cognition in planning and control is to modify system behavior based on salient and measurable variables such as environmental events, task characteristics, and/or human performance that impacts human cognition. Heuristics or other logic-based strategies that depend on these variables can be designed to ensure optimal human cognitive states based on prior knowledge. This includes defining strategies utilizing human performance models or the cognitive architecture-based models discussed in Section 5.2.2. Following this paradigm, adaptive task allocation based on human performance and task complexity has been proposed by various researchers (Parasuraman et al. 1992; Hockey 2003). For example, if the task load increases beyond a predefined threshold, the task is reallocated to the automation. Although studies have shown strong evidence for the effectiveness of such strategies, a primary limitation is the assumption that a given variation in the context/task variables – for which the thresholds are defined – always has the same impact on human cognitive states. This assumption may be reasonable in general, but may not apply across different humans or even strictly to a single person given the dynamic nature of most cognitive states. For example, one human may experience a significant increase in workload for a given task, whereas the same task may lessen the workload for another human. Therefore, paradigms that account for real-time measurements or estimates of the human cognitive states have the potential to further improve CPHS performance.

5.4.2.2 Measurement-Based Feedback

True implementation of cognitive feedback control requires quantitative measurements or estimates of the human cognitive states to inform the control algorithm. This can be achieved in real time using estimators based on the computational models discussed in Section 5.2.3. Additionally, recent studies have shown success in measuring human cognitive states using psychophysiological measurements (Akash et al. 2018; Berka et al. 2007), either alone or in combination with behavioral measurements. Using these measurements, heuristics or reference tracking control algorithms can be used to achieve optimal/calibrated levels of cognitive states. For example, Azevedo-Sa et al. (2020) used a rule set for the trust calibrator and trust estimates from a state-space model to vary the automation communication style. In Natarajan et al. (2022), trust measurements are used to vary the driving style of the automated vehicle. Similarly, studies have used measures derived from the EEG to optimize humans' task engagement and situational awareness (Prinzel et al. 2000; Pope et al. 1995; Bailey et al. 2003).

5.4.2.3 Goal-Oriented Feedback

A primary challenge of measurement-based feedback is identifying the optimal cognitive states. Although these have been theoretically defined, quantification of these values is not straightforward and can vary depending on the system or context. For example, the optimal workload value can depend on task as well as scene complexity, and the optimal trust value depends on the system's reliability. Moreover, these optimal values can be nonlinearly dependent on the context variables, making the problem even more complex. To address this issue, one can take a goal-oriented approach, where, instead of attempting to track the optimal values of the cognitive states, one optimizes the automation behavior toward a defined goal while considering the cognitive model. This requires human cognitive models that not only account for the effect of environment or system variables on the human cognitive states, but also capture the effect of human cognitive states on performance-related variables. For example, apart from modeling the effect of automation reliability on human trust, one must quantitatively model the effect of human trust on their reliance (or compliance) on the system. Using such models, we can formulate an optimal control problem with a cost function that optimizes the control objective while considering the human cognitive model. Such an approach not only optimizes the overall performance but the control policy can also provide insight into the optimal cognitive states.

5.4.2.4 Case Study

To illustrate these ideas, consider the scenario presented by Akash et al. (2020a). The participants were tasked with surveilling buildings in a reconnaissance mission while being assisted by a decision-aid robot. The goal of each mission was to search 15 buildings as fast as possible and mark them as safe or unsafe, based on the presence or absence of armed adversaries. The participants chose between using light armor or heavy armor while searching the building based on the recommendation from the robot and their past experience with the robot's recommendations. The least risky option was for the participant to search a building with heavy armor, expending seconds to do so. Alternatively, they could expend only three seconds by using light armor but risk a penalty of 20 seconds if there was danger present in the building. The robot could choose to present the recommendation at three information (or transparency) levels: low, medium, and high. As described in Section 5.2.3.4, the authors used a POMDP to model the dynamics of trust and workload in this context. The goal of the human-aware system in this scenario was to calibrate human trust and workload by varying the transparency of the robot's recommendation. The three control approaches described in Sections 5.4.2.1–5.4.2.3 can each be considered in the context of this scenario, as follows.

First, consider the heuristic-based planning approach. In the given scenario, higher levels of transparency can induce a higher workload but are necessary to empower the human to make the most informed decision and avoid errors. Therefore, a simple heuristic could be to use high transparency when the robot recommends light armor such that the human does not make the incorrect decision in the highest risk case. Otherwise, the robot can always provide its recommendation with low transparency. This heuristic can implicitly reduce workload by keeping the transparency low while ensuring calibrated trust in high-risk scenarios.

For a measurement-based feedback approach, the POMDP model developed by Akash et al. (2019a) could be used to estimate the belief state of trust and workload during the interaction. Assuming that the average reliability of the robot is known to be 70%, a set-point-based or reference tracking feedback controller, such as a proportional-integral-derivative (PID) controller, could be used to modify the transparency level such that trust is driven to 0.7 and workload is minimized to an arbitrary low value. The authors used a similar approach to design a state-based reward in Akash et al. (2019a). Note that this approach does not exploit the model dynamics as predicted by the POMDP model for control; instead, the POMDP model is used only for state estimation.

Finally, Akash et al. (2020a) used a goal-oriented feedback approach by minimizing the mission time in the task. Since the POMDP model captured the effect of the cognitive states on human compliance and response time, the model was then used to design a control policy that could change the transparency of the robot's recommendation in order to minimize the human's mission time. Specifically, the reward function $\mathcal{R} = \zeta \mathcal{R}_D + (1 - \zeta)\mathcal{R}_{RT}$ was designed to reduce the time spent by the human due to their decision $\mathcal{R}_D$ and response time $\mathcal{R}_{RT}$.

5.4.3 Evaluation Methods

The final step in designing any human-aware feedback algorithm is to evaluate or validate its effectiveness toward achieving specified performance objectives. A preliminary way to test the algorithm is through simulations using the models of each of the CPHS elements. However, given the lack of accuracy and variability associated with most of the human models, it is ideal to test such systems with actual human-in-the-loop testing. This requires another phase of user study design and data collection (see Section 5.3). The primary hypothesis in the evaluation phase is formulated toward testing if the designed feedback system performs better than a baseline (often open-loop) approach. The choice of baseline while evaluating such systems should not only include static system design but should also include previously designed adaptive systems.

5.5 Summary and Opportunities for Further Investigation

As cyber–physical systems become increasingly prevalent in the daily lives of humans, system designers must account explicitly for their interaction with humans. As such, the field of cyber–physical–human systems requires tools for modeling human cognition and designing the necessary algorithms to maximize system performance. In Section 5.2, important considerations for modeling human cognitive dynamics in CPHS contexts were described, along with a broad overview of relevant modeling frameworks and techniques. Next in Section 5.3, a framework for contextualizing the role and design of human subject studies for CPHS research was introduced and supported with detailed examples for the reader. Finally, in Section 5.4, the design of algorithms to govern closed-loop interactions based on the calibration of human cognitive states was discussed. Despite the progress that has been made, though, significant opportunity for further investigation awaits the interested researcher. Some open challenges are described below.

5.5.1 Model Generalizability and Adaptability

Control-oriented cognitive models and control policies should account for individual differences across a given population of humans. An open problem is how to utilize learning algorithms online for retraining or adapting models for improved accuracy. While building mathematical or algorithm-oriented models of human cognition is already difficult, the extendability of those models to individuals whose data was not in the original training dataset or to different HAI contexts remains an open challenge. This necessarily requires the synthesis of new data collected online for real-time adaptation of the models.

Another challenge is the absence of models that explicitly map controllable inputs to the human's cognitive states. The majority of models have been used for estimation or prediction of human cognitive states, but except for Akash et al. (2020a), have not been used for model-based control design. As highlighted in Section 5.3, in order to build input–output models suitable for control design, data that actually captures perturbations to controllable inputs is necessary. Only then can appropriate input–output models be identified or trained.

5.5.2 Measurement of Cognitive States

A major challenge is the lack of standardized methods for measuring cognitive states. Most cognitive states, if not all, have ambiguous interpretations in the literature and thereby lack objective measures. There have been several pioneering efforts to develop instruments to measure these states, mostly in the form of self-report scales. Scales such as the NASA TLX (Proctor and Van Zandt 2018), situation awareness global assessment technique (SAGAT) (Endsley 1988) and situational awareness rating technique (SART) (Taylor 2017) have been used across different contexts for measuring workload and situational awareness, respectively. However, context-independent scales for most other cognitive states are lacking. Recent trends show that greater consideration should be given to objective measures as opposed to self-report scales. These objective measures include behavioral or psychophysiological measures. Nevertheless, more research is needed to better characterize these different measures and develop rules or guidelines with regards to the contexts, conditions, or environments under which particular types of measurements are most suitable for cognitive state measurement or estimation. Techniques to leverage these heterogeneous measurements are also needed.

5.5.3 Human Subject Study Design

As mentioned in the Introduction, naturalistic data are often not suitable for human cognitive modeling. However, the process for acquiring the necessary human subject data through experiments also poses several challenges. First, there is the challenge of the ecological validity of laboratory-based experiments. While they provide an important foundation for understanding specific phenomena, it is understood that a number of factors that may be present in real-life scenarios may not be reproducible in a laboratory setting. This is particularly true when collecting data online. Platforms such as Amazon Mechanical Turk (Paolacci et al. 2010) have become useful for collecting data from a large number of participants in a relatively short amount of time; this becomes particularly important for model training when the number of model parameters is large. Such data, which is generally limited to behavioral or self-report (e.g. mouse-clicks and keyboard inputs (Cepeda et al. 2018)), can be valuable for addressing certain aspects of a research question, like experiment effectiveness (see for example Akash et al. (2018)). However, in online environments, researchers have limited control over the participant's environment which in turn can greatly affect the quality of the resulting data. While laboratory-based, in-person studies offer a more controlled environment than online studies, human subjects will still interact differently with *simulated* cyber–physical systems than those in real-life settings. Therefore, when data are collected in an online or laboratory setting, researchers must clearly identify and state potential weaknesses or limitations in the ecological validity of their research. Moreover, it is equally important for CPHS researchers to develop appropriate methods to safely translate both online and laboratory research to field settings.

It is also challenging for most researchers to recruit diverse participants in their studies (Reid and Gibert 2022), thereby compromising the quality of data-sets used for CPHS research. In fact, it is well established that data-sets lacking diversity lead to flawed algorithms and design outcomes (Howard and Borenstein 2018). Although some suggestions are provided by Reid and Gibert (2022), guidelines and best practices for how researchers can be more intentional in recruiting diverse subjects remains an open area of inquiry,

Developing cognitive models and control algorithms for CPHS inherently requires time and patience, in large part because these systems involve people—arguably one of the most complex systems on the planet! As humans continue interacting with CPS in increasingly diverse domains or environments, new factors will arise that influence human behavior and their cognitive states. Therefore, further research in CPHS will continue to require expertise from multiple disciplines. Collaboration between dynamical systems, controls, human factors, and psychology researchers, and other domain experts, is strongly encouraged in order to make meaningful advances toward a harmonious future between humans and machines.

References

Abbot, K., Slotte, S., and Stimson, D. (1996). The Interfaces Between Flightcrews and Modern Flight Deck Systems, June 1996. http://doi.apa.org/get-pe-doi.cfm?doi=10.1037/e664742007-001 (accessed 8 February 2023).

Agresti, A. (2019). *An Introduction to Categorical Data Analysis*, Wiley Series in Probability and Statistics, 3e. Hoboken, NJ: Wiley. ISBN 978-1-119-40526-9.

Akash, K. (2020). Reimagining human–machine interactions through trust-based feedback. Thesis. Purdue University Graduate School. https://hammer.purdue.edu/articles/thesis/Reimagining_Human-Machine_Interactions_through_Trust-Based_Feedback/12493007/1 (accessed 8 February 2023).

Akash, K., Hu, W.-L., Reid, T., and Jain, N. (2017). Dynamic modeling of trust in human–machine interactions. *2017 American Control Conference (ACC)*, 1542–1548, Seattle, WA, USA, May 2017. IEEE. ISBN 978-1-5090-5992-8. https://doi.org/10.23919/ACC.2017.7963172. https://ieeexplore.ieee.org/document/7963172/ (accessed 8 February 2023).

Akash, K., Hu, W.-L., Jain, N., and Reid, T. (2018). A classification model for sensing human trust in machines using EEG and GSR. *ACM Transactions on Interactive Intelligent Systems* 8 (4): 27:1–27:20. https://doi.org/10.1145/3132743.

Akash, K., Polson, K., Reid, T., and Jain, N. (2019a). Improving human–machine collaboration through transparency-based feedback – Part I: human trust and workload model. This material is based upon work supported by the National Science Foundation under Award No.1548616. Any opinions, findings, and conclusions or recommendations expressed in this material are those of the author(s) and do not necessarily reflect the views of the National Science Foundation. *IFAC-PapersOnLine* 51 (34): 315–321. https://doi.org/10.1016/j.ifacol.2019.01.028.

Akash, K., Reid, T., and Jain, N. (2019b). Improving human–machine collaboration through transparency-based feedback – Part II: Control design and synthesis. *IFAC-PapersOnLine* 51 (34): 322–328. https://doi.org/10.1016/j.ifacol.2019.01.026.

Akash, K., McMahon, G., Reid, T., and Jain, N. (2020a). Human trust-based feedback control: dynamically varying automation transparency to optimize human–machine interactions. *IEEE Control Systems Magazine* 40 (6): 98–116. https://doi.org/10.1109/MCS.2020.3019151.

Akash, K., Jain, N., and Misu, T. (2020b). Toward adaptive trust calibration for level 2 driving automation. Proceedings of the 2020 International Conference on Multimodal Interaction, ICMI '20, 538–547, New York, NY, USA, October 2020. Association for Computing Machinery. ISBN 978-1-4503-7581-8. https://doi.org/10.1145/3382507.3418885.

Aldrich, T.B. and Szabo, S.M. (1986). A methodology for predicting crew workload in new weapon systems. *Proceedings of the Human Factors Society Annual Meeting* 30 (7): 633–637. https://doi.org/10.1177/154193128603000705.

Alonso, V. and de la Puente, P. (2018). System transparency in shared autonomy: a mini review. *Frontiers in Neurorobotics* 12: 83. https://doi.org/10.3389/fnbot.2018.00083.

Anderson, J.R. (1996). ACT: a simple theory of complex cognition. *American Psychologist* 51 (4): 355–365. https://doi.org/10.1037/0003-066X.51.4.355.

Anderson, J.R. and Schooler, L.J. (1991). Reflections of the environment in memory. *Psychological Science* 2 (6): 396–408. https://doi.org/10.1111/j.1467-9280.1991.tb00174.x.

Azevedo-Sa, H., Jayaraman, S.K., Yang, X.J. et al. (2020). Context-adaptive management of drivers' trust in automated vehicles. *IEEE Robotics and Automation Letters* 5 (4): 6908–6915. https://doi.org/10.1109/LRA.2020.3025736.

Azevedo-Sa, H., Jayaraman, S.K., Esterwood, C.T. et al. (2021). Real-time estimation of drivers' trust in automated driving systems. *International Journal of Social Robotics* 13 (8): 1911–1927. https://doi.org/10.1007/s12369-020-00694-1.

Bailey, N.R., Scerbo, M.W., Freeman, F.G. et al. (2003). A brain-based adaptive automation system and situation awareness: the role of complacency potential. *Proceedings of the Human Factors and Ergonomics Society Annual Meeting* 47 (9): 1048–1052. https://doi.org/10.1177/154193120304700901.

Bandura, A. (1986). *Social Foundations of Thought and Action: A Social Cognitive Theory*. Englewood Cliffs, NJ: Prentice-Hall, Inc. ISBN 978-0-13-815614-5.

Banerjee, A., Chitnis, U.B., Jadhav, S.L. et al. (2009). Hypothesis testing, type I and type II errors. *Industrial Psychiatry Journal* 18 (2): 127–131. https://doi.org/10.4103/0972-6748.62274.

Berka, C., Levendowski, D.J., Lumicao, M.N. et al. (2007). EEG correlates of task engagement and mental workload in vigilance, learning, and memory tasks. *Aviation, Space, and Environmental Medicine* 78 (5): B231–B244.

Billings, C.E. (1997). *Aviation Automation: The Search for a Human-Centered Approach*. Mahwah, NJ: Erlbaum. ISBN 0-8058-2126-0.

National Transportation Safety Board (2017). Collision Between a Car Operating With Automated Vehicle Control Systems and a Tractor-Semitrailer Truck Near Williston, Florida, May 7, 2016.

Technical Report Highway Accident Report NTSB/HAR-17/02. Washington, DC: National Technical Information Service.

Bohua, L., Lishan, S., and Jian, R. (2011). Driver's visual cognition behaviors of traffic signs based on eye movement parameters. *Journal of Transportation Systems Engineering and Information Technology* 11 (4): 22–27.

Breazeal, C., Dautenhahn, K., and Kanda, T. (2016). Social robotics. In: *Springer Handbook of Robotics*, Springer Handbooks (ed. B. Siciliano and O. Khatib), 1935–1972. Cham: Springer International Publishing. ISBN 978-3-319-32552-1. https://doi.org/10.1007/978-3-319-32552-1_72.

Briggs, P., Burford, B., and Dracup, C. (1998). Modelling self-confidence in users of a computer-based system showing unrepresentative design. *International Journal of Human–Computer Studies* 49 (5): 717–742. https://doi.org/10.1006/ijhc.1998.0224.

Budiu, R. (2013). ACT-R about. http://act-r.psy.cmu.edu/about/ (accessed 8 February 2023).

Byom, L.J. and Mutlu, B. (2013). Theory of mind: mechanisms, methods, and new directions. *Frontiers in Human Neuroscience* 7: 413. https://doi.org/10.3389/fnhum.2013.00413.

Cao, S. and Liu, Y. (2011). Mental workload modeling in an integrated cognitive architecture. *Proceedings of the Human Factors and Ergonomics Society Annual Meeting* 55 (1): 2083–2087. https://doi.org/10.1177/1071181311551434.

Causse, M., Chua, Z., Peysakhovich, V. et al. (2017). Mental workload and neural efficiency quantified in the prefrontal cortex using fNIRS. *Scientific Reports* 7 (1): https://doi.org/10.1038/s41598-017-05378-x.

Cepeda, C., Rodrigues, J., Dias, M.C. et al. (2018). Mouse tracking measures and movement patterns with application for online surveys. In: *Machine Learning and Knowledge Extraction* (ed. A. Holzinger, P. Kieseberg, A.M. Tjoa, and E. Weippl), 28–42. Cham: Springer International Publishing. ISBN 978-3-319-99740-7. https://doi.org/10.1007/978-3-319-99740-7_3.

Chen, J.Y., Procci, K., Boyce, M. et al. (2014). Situation Awareness-Based Agent Transparency. *Technical Report*. Fort Belvoir, VA: Defense Technical Information Center. http://www.dtic.mil/docs/citations/ADA600351 (accessed 8 February 2023).

Chen, M., Nikolaidis, S., Soh, H. et al. (2018). Planning with trust for human–robot collaboration. *Proceedings of the 2018 ACM/IEEE International Conference on Human–Robot Interaction - HRI '18*, 307–315. Chicago, IL, USA: ACM Press. ISBN 978-1-4503-4953-6. https://doi.org/10.1145/3171221.3171264.

Creswell, J.W. and Plano Clark, V.L. (2011). *Designing and Conducting Mixed Methods Research*, 2e. Los Angeles, CA: SAGE Publications. ISBN 978-1-4129-7517-9.

Davies, A. (2016). Google's Self-Driving Car Caused Its First Crash. https://www.wired.com/2016/02/googles-self-driving-car-may-caused-first-crash/ (accessed 8 February 2023).

Dehais, F., Lafont, A., Roy, R.N., and Fairclough, S. (2020). A neuroergonomics approach to mental workload, engagement and human performance. *Frontiers in Neuroscience* 14: 1–17. https://doi.org/10.3389/fnins.2020.00268.

Dubois, C. and Ny, J.L. (2020). Adaptive task allocation in human–machine teams with trust and workload cognitive models. *2020 IEEE International Conference on Systems, Man, and Cybernetics (SMC)*, 3241–3246. https://doi.org/10.1109/SMC42975.2020.9283461.

Dussault, C., Jouanin, J.-C., Philippe, M., and Guezennec, C.-Y. (2005). EEG and ECG changes during simulator operation reflect mental workload and vigilance. *Aviation, Space, and Environmental Medicine* 76 (4): 344–351(8).

Elfil, M. and Negida, A. (2017). Sampling methods in clinical research; an educational review. *Emergency* 5 (1): e52. https://doi.org/10.22037/emergency.v5i1.15215.

Endsley, M.R. (1988). Situation awareness global assessment technique (SAGAT). *Proceedings of the IEEE 1988 National Aerospace and Electronics Conference*, 789–795. IEEE.

Endsley, M.R. (2017). From here to autonomy: lessons learned from human–automation research. *Human Factors: The Journal of the Human Factors and Ergonomics Society* 59 (1): 5–27. https://doi.org/10.1177/0018720816681350.

Fairclough, S. and Gilleade, K. (ed.) (2014). *Advances in Physiological Computing*, Human–Computer Interaction Series. London; New York Springer. ISBN 978-1-4471-6391-6.

Farrugia, P., Petrisor, B.A., Farrokhyar, F., and Bhandari, M. (2010). Practical tips for surgical research: research questions, hypotheses and objectives. *Canadian Journal of Surgery. Journal Canadien De Chirurgie* 53 (4): 278–281.

Fayazi, S.A., Wan, N., Lucich, S. et al. (2013). Optimal pacing in a cycling time-trial considering cyclist's fatigue dynamics. *2013 American Control Conference*, 6442–6447, Washington, DC, June 2013. IEEE. ISBN 978-1-4799-0178-4 978-1-4799-0177-7 978-1-4799-0175-3. https://doi.org/10.1109/ACC.2013.6580849.

Felfernig, A. and Gula, B. (2006). An empirical study on consumer behavior in the interaction with knowledge-based recommender applications. *The 8th IEEE International Conference on E-Commerce Technology and The 3rd IEEE International Conference on Enterprise Computing, E-Commerce, and E-Services (CEC/EEE'06)*, 37, June 2006. https://doi.org/10.1109/CEC-EEE.2006.14.

Floyd, M.W., Drinkwater, M., and Aha, D.W. (2015). Improving trust-guided behavior adaptation using operator feedback. In: *Case-Based Reasoning Research and Development*, Lecture Notes in Computer Science (ed. E. Hullermeier and M. Minor), 134–148. Cham: Springer International Publishing. ISBN 978-3-319-24586-7. https://doi.org/10.1007/978-3-319-24586-7_10.

Freedy, A., DeVisser, E., Weltman, G., and Coeyman, N. (2007). Measurement of trust in human–robot collaboration. *2007 International Symposium on Collaborative Technologies and Systems*, 106–114, May 2007. https://doi.org/10.1109/CTS.2007.4621745.

Gaffey, A.E. and Wirth, M.M. (2014). Psychophysiological measures. In: *Encyclopedia of Quality of Life and Well-Being Research* (ed. A.C. Michalos), 5181–5184. Dordrecht: Springer Netherlands. ISBN 978-94-007-0753-5. https://doi.org/10.1007/978-94-007-0753-5_2315.

Gao, J. and Lee, J.D. (2006). Extending the decision field theory to model operators' reliance on automation in supervisory control situations. *IEEE Transactions on Systems, Man, and Cybernetics - Part A: Systems and Humans* 36 (5): 943–959. https://doi.org/10.1109/TSMCA.2005.855783.

Gil, G.-H. and Kaber, D.B. (2012). An accessible cognitive modeling tool for evaluation of pilot–automation interaction. *The International Journal of Aviation Psychology* 22 (4): 319–342. https://doi.org/10.1080/10508414.2012.718236.

Given, L. (2008). *The SAGE Encyclopedia of Qualitative Research Methods*. Thousand Oaks, CA: SAGE Publications, Inc. ISBN 978-1-4129-4163-1 978-1-4129-6390-9. https://doi.org/10.4135/9781412963909. http://methods.sagepub.com/reference/sage-encyc-qualitative-research-methods.

Gore, B.F. (2020). TLX @ NASA Ames - Home. https://humansystems.arc.nasa.gov/groups/tlx/index.php (accessed 8 February 2023).

Grady, C. (2015). Institutional review boards: purpose and challenges. *Chest* 148 (5): 1148–1155. https://doi.org/10.1378/chest.15-0706.

Gray, W.D., Schoelles, M.J., and Sims, C. (2005). Cognitive metrics profiling. *Proceedings of the Human Factors and Ergonomics Society Annual Meeting* 49 (12): 1144–1148. https://doi.org/10.1177/154193120504901210.

Grech, M., Neal, A., Yeo, G. et al. (2009). An examination of the relationship between workload and fatigue within and across consecutive days of work: is the relationship static or dynamic? *Journal of Occupational Health Psychology* 14: 231–242. https://doi.org/10.1037/a0014952.

Hancock, P.A. and Chignell, M.H. (1988). Mental workload dynamics in adaptive interface design. *IEEE Transactions on Systems, Man, and Cybernetics* 18 (4): 647–658. https://doi.org/10.1109/21.17382.

Hancock, P.A., Billings, D.R., Schaefer, K.E. et al. (2011). A meta-analysis of factors affecting trust in human–robot interaction. *Human Factors* 53 (5): 517–527. https://doi.org/10.1177/0018720811417254.

Hankins, T.C. and Wilson, G.F. (1998). A comparison of heart rate, eye activity, EEG and subjective measures of pilot mental workload during flight. *Aviation, Space, and Environmental Medicine* 69 (4): 360–367.

Harris, P. and Horst, J.S. (2008). *Designing and Reporting Experiments in Psychology*. McGraw-Hill Education (UK). ISBN 978-0-335-22178-3.

Hart, S.G. and Staveland, L.E. (1988). Development of NASA-TLX (task load index): results of empirical and theoretical research. In: *Advances in Psychology*, Human Mental Workload, vol. 52 (ed. P.A. Hancock and N. Meshkati), 139–183. North-Holland. https://doi.org/10.1016/S0166-4115(08)62386-9. https://www.sciencedirect.com/science/article/pii/S0166411508623869.

Helldin, T. (2014). Transparency for future semi-automated systems effects of transparency on operator performance, workload and trust. PhD thesis. Örebro Universitet.

Robert, G. and Hockey, J. (2003). *Operator Functional State: The Assessment and Prediction of Human Performance Degradation in Complex Tasks*. IOS Press. ISBN 978-1-58603-362-0.

Hoff, K.A. and Bashir, M. (2015). Trust in automation: integrating empirical evidence on factors that influence trust. *Human Factors: The Journal of the Human Factors and Ergonomics Society* 57 (3): 407–434.

Hoogendoorn, M., Jaffry, S.W., Van Maanen, P.P., and Treur, J. (2013). Modelling biased human trust dynamics. *Web Intelligence and Agent Systems* 11 (1): 21–40.

Hosseini, M., Shahri, A., Phalp, K., and Ali, R. (2018). Four reference models for transparency requirements in information systems. *Requirements Engineering* 23 (2): 251–275. https://doi.org/10.1007/s00766-017-0265-y.

Howard, A. and Borenstein, J. (2018). The ugly truth about ourselves and our robot creations: the problem of bias and social inequity. *Science and Engineering Ethics* 24 (5): 1521–1536. https://doi.org/10.1007/s11948-017-9975-2.

Hu, W.-L., Akash, K., Reid, T., and Jain, N. (2019). Computational modeling of the dynamics of human trust during human–machine interactions. *IEEE Transactions on Human–Machine Systems* 49 (6): 485–497. https://doi.org/10.1109/THMS.2018.2874188.

Huang, G., Steele, C., Zhang, X., and Pitts, B.J. (2019). Multimodal cue combinations: a possible approach to designing in-vehicle takeover requests for semi-autonomous driving. *Proceedings of the Human Factors and Ergonomics Society Annual Meeting* 63 (1): 1739–1743. https://doi.org/10.1177/1071181319631053.

Hursh, S., Redmond, D., Johnson, M. et al. (2004). Fatigue models for applied research in warfighting. *Aviation, Space, and Environmental Medicine* 75: A44–A53; discussion A54.

Jacobs, S.C., Friedman, R., Parker, J.D. et al. (1994). Use of skin conductance changes during mental stress testing as an index of autonomic arousal in cardiovascular research. *American Heart Journal* 128 (6): 1170–1177.

Jian, J.-Y., Bisantz, A.M., and Drury, C.G. (2000). Foundations for an empirically determined scale of trust in automated systems. *International Journal of Cognitive Ergonomics* 4 (1): 53–71. https://doi.org/10.1207/S15327566IJCE0401_04.

Juvina, I., Lebiere, C., and Gonzalez, C. (2015). Modeling trust dynamics in strategic interaction. *Journal of Applied Research in Memory and Cognition* 4 (3): 197–211. https://doi.org/10.1016/j.jarmac.2014.09.004.

Kaber, D.B. and Riley, J.M. (1999). Adaptive automation of a dynamic control task based on secondary task workload measurement. *International Journal of Cognitive Ergonomics* 3 (3): 169–187. https://doi.org/10.1207/s15327566ijce0303_1.

Khawaji, A., Zhou, J., Chen, F., and Marcus, N. (2015). Using galvanic skin response (GSR) to measure trust and cognitive load in the text–chat environment. *Proceedings of the 33rd Annual ACM Conference Extended Abstracts on Human Factors in Computing Systems*, 1989–1994. ACM Press.

Kieras, D.E. and Meyer, D.E. (1997). An overview of the EPIC architecture for cognition and performance with application to human–computer interaction. *Human–Computer Interaction* 12 (4): 391–438. https://doi.org/10.1207/s15327051hci1204_4.

Kim, J. and Myung, R. (2016). A predictive model of situation awareness with ACT-R. *Journal of the Ergonomics Society of Korea* 35 (4): 225–235. https://doi.org/10.5143/JESK.2016.35.4.225.

Kress-Gazit, H., Eder, K., Hoffman, G. et al. (2021). Formalizing and guaranteeing human–robot interaction. *Communications of the ACM* 64 (9): 78–84. https://doi.org/10.1145/3433637.

Laird, J.E. (2012). *The Soar Cognitive Architecture*. Cambridge, MA: MIT Press. ISBN 978-0-262-12296-2.

Lee, J. and Moray, N. (1992). Trust, control strategies and allocation of function in human–machine systems. *Ergonomics* 35 (10): 1243–1270. https://doi.org/10.1080/00140139208967392.

Lee, J.D. and Moray, N. (1994). Trust, self-confidence, and operators' adaptation to automation. *International Journal of Human–Computer Studies* 40 (1): 153–184.

Lee, J.D. and See, K.A. (2004). Trust in automation: designing for appropriate reliance. *Human Factors: The Journal of the Human Factors and Ergonomics Society* 46 (1): 50–80. https://doi.org/10.1518/hfes .46.1.50_30392.

Lienert, P. (2018). Most Americans wary of self-driving cars: Reuters/Ipsos poll. https://www.reuters .com/article/us-autos-selfdriving-usa-poll-idUSKBN1FI034 (accessed 8 February 2023).

Lotte, F., Congedo, M., Lécuyer, A. et al. (2007). A review of classification algorithms for EEG-based brain–computer interfaces. *Journal of Neural Engineering* 4 (2): R1–R13. https://doi.org/10.1088/ 1741-2560/4/2/R01.

Luster, M.S. and Pitts, B.J. (2021). Trust in automation: the effects of system certainty on decision-making. *Proceedings of the Human Factors and Ergonomics Society Annual Meeting* 65 (1): 32–36. https://doi.org/10.1177/1071181321651079.

Martin, L., Jaime, K., Ramos, F., and Robles, F. (2021). Declarative working memory: a bio-inspired cognitive architecture proposal. *Cognitive Systems Research* 66: 30–45.

Mercado, J.E., Rupp, M.A., Chen, J.Y.C. et al. (2016). Intelligent agent transparency in human–agent teaming for multi-UxV management. *Human Factors* 58 (3): 401–415. https://doi.org/10.1177/ 0018720815621206.

Mikulski, D., Lewis, F., Gu, E., and Hudas, G. (2012). Trust method for multi-agent consensus. *Proceedings of SPIE - The International Society for Optical Engineering*, volume 8387, April 2012. https://doi.org/10.1117/12.918927.

Mitchell, D.K. and Samms, C. (2009). Workload warriors: lessons learned from a decade of mental workload prediction using human performance modeling. *Proceedings of the Human Factors and Ergonomics Society Annual Meeting* 53 (12): 819–823. https://doi.org/10.1177/154193120905301212.

Montgomery, D.C. (2012). *Design and Analysis of Experiments*, 8e. Hoboken, NJ: Wiley. ISBN 978-1-118-14692-7.

Moore, J.J., Ivie, R., Gledhill, T.J. et al. (2014). Modeling Human Workload in Unmanned Aerial Systems. *Formal Verification and Modeling in Human–Machine Systems: Papers from the AAAI Spring Symposium*, 6.

Mou, W., Ruocco, M., Zanatto, D., and Cangelosi, A. (2021). When Would You Trust a Robot? A Study on Trust and Theory of Mind in Human–Robot Interactions. *arXiv:2101.10819 [cs]*, January 2021. http://arxiv.org/abs/2101.10819.

Muir, B.M. (1994). Trust in automation: Part I. Theoretical issues in the study of trust and human intervention in automated systems. *Ergonomics* 37 (11): 1905–1922.

Murphy, K.P. (2022). Dynamic Bayesian networks: representation, inference and learning. University of California, Berkeley, 55.

Natarajan, M., Akash, K., and Misu, T. (2022). Toward adaptive driving styles for automated driving with users trust and preferences. *Proceedings of the 2022 ACM/IEEE International Conference on Human–Robot Interaction*, HRI '22. New York, NY, USA: Association for Computing Machinery.

Nikula, R. (1991). Psychological correlates of nonspecific skin conductance responses. *Psychophysiology* 28 (1): 86–90.

Nisbett, R.E. and Wilson, T.D. (1977). Telling more than we can know: verbal reports on mental processes. *Psychological Review* 84 (3): 231–259.

Norstad, J. (1999). An introduction to utility theory. http://homepage.mac.com/j.norstad/finance/util .pdf (accessed 8 February 2023).

Paolacci, G., Chandler, J., and Ipeirotis, P.G. (2010). Running Experiments on Amazon Mechanical Turk. SSRN Scholarly Paper ID 1626226. Rochester, NY: Social Science Research Network. https:// papers.ssrn.com/abstract=1626226 (accessed 8 February 2023).

Parasuraman, R. and Riley, V. (1997). Humans and automation: use, misuse, disuse, abuse. *Human factors* 39 (2): 230–253.

Parasuraman, R., Bahri, T., Deaton, J.E. et al. (1992). Theory and Design of Adaptive Automation in Aviation Systems. *Technical Report*. Catholic Univ of America Washington DC Cognitive Science Lab. https://apps.dtic.mil/sti/citations/ADA254595 (accessed 8 February 2023).

Parasuraman, R., Mouloua, M., and Hilburn, B. (1999). Adaptive aiding and adaptive task allocation enhance human–machine interaction. *Automation Technology and Human Performance: Current Research and Trends*, 119–123.

Parasuraman, R., Sheridan, T.B., and Wickens, C.D. (2000). A model for types and levels of human interaction with automation. *IEEE Transactions on Systems, Man, and Cybernetics - Part A: Systems and Humans* 30 (3): 286–297.

Parasuraman, R., Barnes, M., and Cosenzo, K.A. (2007). Adaptive Automation for Human–Robot Teaming in Future Command and Control Systems. 1 (2): 31.

Parasuraman, R., Sheridan, T.B., and Wickens, C.D. (2008). Situation awareness, mental workload, and trust in automation: viable, empirically supported cognitive engineering constructs. *Journal of Cognitive Engineering and Decision Making* 2 (2): 140–160.

Pawar, N.M. and Velaga, N.R. (2020). Modelling the influence of time pressure on reaction time of drivers. *Transportation Research Part F: Traffic Psychology and Behaviour* 72: 1–22. https://doi.org/10 .1016/j.trf.2020.04.017.

Peters, J.R., Srivastava, V., Taylor, G.S. et al. (2015). Human supervisory control of robotic teams: integrating cognitive modeling with engineering design. *IEEE Control Systems Magazine* 35 (6): 57–80. https://doi.org/10.1109/MCS.2015.2471056.

Peterson, R.A. and Merunka, D.R. (2014). Convenience samples of college students and research reproducibility. *Journal of Business Research* 67 (5): 1035–1041. https://doi.org/10.1016/j.jbusres .2013.08.010.

Pope, A.T., Bogart, E.H., and Bartolome, D.S. (1995). Biocybernetic system evaluates indices of operator engagement in automated task. *Biological Psychology* 40 (1): 187–195. https://doi.org/10.1016/0301-0511(95)05116-3.

Prinzel, L.J., Freeman, F.G., Scerbo, M.W. et al. (2000). A closed-loop system for examining psychophysiological measures for adaptive task allocation. *The International Journal of Aviation Psychology* 10 (4): 393–410. https://doi.org/10.1207/S15327108IJAP1004_6.

Proctor, R.W. and Van Zandt, T. (2018). *Human Factors in Simple and Complex Systems*, 3e. Boca Raton, FL: CRC Press.

Purcell, B.A. and Palmeri, T.J. (2017). Relating accumulator model parameters and neural dynamics. *Journal of Mathematical Psychology* 76 (B): 156–171. https://doi.org/10.1016/j.jmp.2016.07.001.

Pynadath, D.V., Si, M., and Marsella, S. (2013). Modeling theory of mind and cognitive appraisal with decision-theoretic agents. In: *Social Emotions in Nature and Artifact* (ed. J. Gratch and S. Marsella), 70–87. Oxford University Press. ISBN 978-0-19-538764-3. https://doi.org/10.1093/acprof:oso/9780195387643.003.0006. https://oxford.universitypressscholarship.com/view/10.1093/acprof:oso/9780195387643.001.0001/acprof-9780195387643-chapter-6 (accessed 8 February 2023).

Reid, T. and Gibert, J. (2022). Inclusion in human–machine interactions. *Science* 375 (6577): 149–150. https://doi.org/10.1126/science.abf2618.

Riley, V. (1996). Operator reliance on automation: theory and data. In: *Automation and Human Performance: Theory and Applications*, 1e (ed. R. Parasuraman and M. Mouloua), 19–35. Mahwah, NJ: CRC Press.

Riley, V., Lyall, E., and Wiener, E. (1994). Analytic workload models for flight deck design and evaluation. *Proceedings of the Human Factors and Ergonomics Society Annual Meeting* 38 (1): 81–84. https://doi.org/10.1177/154193129403800115.

Ritter, F.E., Tehranchi, F., and Oury, J.D. (2019). ACT-R: a cognitive architecture for modeling cognition. *WIREs Cognitive Science* 10 (3): e1488. https://doi.org/10.1002/wcs.1488.

Rouse, W.B. (1976). Adaptive allocation of decision making responsibility between supervisor and computer. In: *Monitoring Behavior and Supervisory Control*, NATO Conference Series (ed. T.B. Sheridan and G. Johannsen), 295–306. Boston, MA: Springer US. ISBN 978-1-4684-2523-9. https://doi.org/10.1007/978-1-4684-2523-9_24.

Rouse, W.B., Edwards, S.L., and Hammer, J.M. (1993). Modeling the dynamics of mental workload and human performance in complex systems. *IEEE Transactions on Systems, Man, and Cybernetics* 23 (6): 1662–1671. https://doi.org/10.1109/21.257761.

Rubin, D.C., Hinton, S., and Wenzel, A. (1999). The precise time course of retention. *Journal of Experimental Psychology: Learning, Memory, and Cognition* 25 (5): 1161–1176. https://doi.org/10.1037/0278-7393.25.5.1161.

Sadrfaridpour, B. and Wang, Y. (2018). Collaborative assembly in hybrid manufacturing cells: an integrated framework for human–robot interaction. *IEEE Transactions on Automation Science and Engineering* 15 (3): 1178–1192. https://doi.org/10.1109/TASE.2017.2748386.

Sadrfaridpour, B., Saeidi, H., Burke, J. et al. (2016a). Modeling and control of trust in human–robot collaborative manufacturing. In: *Robust Intelligence and Trust in Autonomous Systems* (ed. R. Mittu, D. Sofge, A. Wagner, and W.F. Lawless), 115–141. Boston, MA: Springer US. ISBN 978-1-4899-7668-0. https://doi.org/10.1007/978-1-4899-7668-0_7.

Sadrfaridpour, B., Saeidi, H., and Wang, Y. (2016b). An integrated framework for human–robot collaborative assembly in hybrid manufacturing cells. *2016 IEEE International Conference on Automation Science and Engineering (CASE)*, 462–467. https://doi.org/10.1109/COASE.2016.7743441.

Saeidi, H. and Wang, Y. (2019). Incorporating trust and self-confidence analysis in the guidance and control of (semi)autonomous mobile robotic systems. *IEEE Robotics and Automation Letters* 4 (2): 239–246. https://doi.org/10.1109/LRA.2018.2886406.

Sanders, N.E. and Nam, C.S. (2021). Applied quantitative models of trust in human–robot interaction. In: *Trust in Human–Robot Interaction*, 449–476. Elsevier. ISBN 978-0-12-819472-0. https://doi.org/10.1016/B978-0-12-819472-0.00019-8. https://linkinghub.elsevier.com/retrieve/pii/B9780128194720000198.

Sheridan, T.B. (1997). Task analysis, task allocation and supervisory control. In: *Handbook of Human–Computer Interaction*, 87–105. Elsevier.

Sibi, S., Ayaz, H., Kuhns, D.P. et al. (2016). Monitoring driver cognitive load using functional near infrared spectroscopy in partially autonomous cars. *2016 IEEE Intelligent Vehicles Symposium (IV)*, 419–425, Gotenburg, Sweden, June 2016. IEEE. ISBN 978-1-5090-1821-5. https://doi.org/10.1109/IVS.2016.7535420. http://ieeexplore.ieee.org/document/7535420/.

Slovic, P., Finucane, M.L., Peters, E., and MacGregor, D.G. (2004). Risk as analysis and risk as feelings: some thoughts about affect, reason, risk, and rationality. *Risk Analysis* 24 (2): 311–322. https://doi.org/10.1111/j.0272-4332.2004.00433.x.

Soh, H., Xie, Y., Chen, M., and Hsu, D. (2020). Multi-task trust transfer for human–robot interaction. *The International Journal of Robotics Research* 39 (2-3): 233–249. https://doi.org/10.1177/0278364919866905.

Sullivan, G.M. and Feinn, R. (2012). Using effect size–or why the *P* value is not enough. *Journal of Graduate Medical Education* 4 (3): 279–282. https://doi.org/10.4300/JGME-D-12-00156.1.

Sun, R. (2006). The CLARION cognitive architecture: extending cognitive modeling to social simulation. In: *Cognition and Multi-Agent Interaction: From Cognitive Modeling to Social Simulation*, 79–99. New York: Cambridge University Press. ISBN 978-0-521-83964-8.

Tao, Y., Coltey, E., Wang, T. et al. (2020). Confidence estimation using machine learning in immersive learning environments. *2020 IEEE Conference on Multimedia Information Processing and Retrieval (MIPR)*, 247–252, Shenzhen, Guangdong, China, August 2020. IEEE. ISBN 978-1-72814-272-2. https://doi.org/10.1109/MIPR49039.2020.00058. https://ieeexplore.ieee.org/document/9175522/.

Taylor, R.M. (2017). Situational awareness rating technique (SART): the development of a tool for aircrew systems design. In: *Situational Awareness*, 111–128. Routledge.

Tintarev, N. and Masthoff, J. (2007). A survey of explanations in recommender systems. *2007 IEEE 23rd International Conference on Data Engineering Workshop*, 801–810. https://doi.org/10.1109/ICDEW.2007.4401070.

van Maanen, P.-P., Wisse, F., van Diggelen, J., and Beun, R.-J. (2011). Effects of reliance support on team performance by advising and adaptive autonomy. *2011 IEEE/WIC/ACM International Conferences on Web Intelligence and Intelligent Agent Technology*, volume 2, 280–287, August 2011. https://doi.org/10.1109/WI-IAT.2011.117.

van Maanen, L., van Rijn, H., and Taatgen, N. (2012). RACE/A: an architectural account of the interactions between learning, task control, and retrieval dynamics. *Cognitive Science* 36 (1): 62–101. https://doi.org/10.1111/j.1551-6709.2011.01213.x.

Verdière, K.J., Roy, R.N., and Dehais, F. (2018). Detecting Pilot's engagement using fNIRS connectivity features in an automated vs. manual landing scenario. *Frontiers in Human Neuroscience* 12: 6. https://doi.org/10.3389/fnhum.2018.00006.

Wagner, A.R., Robinette, P., and Howard, A. (2018). Modeling the human–robot trust phenomenon: a conceptual framework based on risk. *ACM Transactions on Interactive Intelligent Systems* 8 (4): 26:1–26:24. https://doi.org/10.1145/3152890.

Wang, W., Cain, B., and Lu, X.L. (2010). Predicting operator mental workload using a time-based algorithm. *Proceedings of the Human Factors and Ergonomics Society Annual Meeting* 54 (13): 987–991. https://doi.org/10.1177/154193121005401314.

Wang, N., Pynadath, D.V., Unnikrishnan, K.V. et al. (2015). Intelligent agents for virtual simulation of human–robot interaction. In: *Virtual, Augmented and Mixed Reality*, Lecture Notes in Computer Science (ed. R. Shumaker and S. Lackey), 228–239. Cham: Springer International Publishing. ISBN 978-3-319-21067-4. https://doi.org/10.1007/978-3-319-21067-4_24.

Westfall, J., Kenny, D.A., and Judd, C.M. (2014). Statistical power and optimal design in experiments in which samples of participants respond to samples of stimuli. *Journal of Experimental Psychology: General* 143 (5): 2020–2045. https://doi.org/10.1037/xge0000014.

Wetzels, R., Matzke, D., Lee, M.D. et al. (2011). Statistical evidence in experimental psychology: an empirical comparison using 855 *t* tests. *Perspectives on Psychological Science* 6 (3): 291–298. https://doi.org/10.1177/1745691611406923.

Wiese, S., Lotz, A., and Russwinkel, N. (2019). SEEV-VM: ACT-R Visual Module based on SEEV theory, 301–307.

Wu, C. and Liu, Y. (2007). Queuing network modeling of driver workload and performance. *IEEE Transactions on Intelligent Transportation Systems* 8: 528–537. https://doi.org/10.1109/TITS.2007.903443.

Xu, A. and Dudek, G. (2012). Trust-driven interactive visual navigation for autonomous robots. *2012 IEEE International Conference on Robotics and Automation*, 3922–3929. https://doi.org/10.1109/ICRA.2012.6225171.

Xu, A. and Dudek, G. (2015). OPTIMo: online probabilistic trust inference model for asymmetric human–robot collaborations. *Proceedings of the 10th Annual ACM/IEEE International Conference on Human–Robot Interaction*, HRI '15, 221–228. New York, NY, USA: ACM. ISBN 978-1-4503-2883-8. https://doi.org/10.1145/2696454.2696492.

Yerkes, R.M. and Dodson, J.D. (1908). The relation of strength of stimulus to rapidity of habit-formation. *Journal of Comparative Neurology and Psychology* 18 (5): 459–482. https://doi.org/10.1002/cne.920180503.

Zhou, J., Jung, J.Y., and Chen, F. (2015). Dynamic workload adjustments in human–machine systems based on GSR features. In: *Human–Computer Interaction – INTERACT 2015*, vol. 9296 (ed. J. Abascal, S. Barbosa, M. Fetter et al.), 550–558. Cham: Springer International Publishing. ISBN 978-3-319-22700-9 978-3-319-22701-6. https://doi.org/10.1007/978-3-319-22701-6_40.

6

Shared Control with Human Trust and Workload Models

Murat Cubuktepe[1], Nils Jansen[2], and Ufuk Topcu[3]

[1]*Dematic Corp., Austin, TX, USA*
[2]*Department of Software Science, Institute for Computing and Information Science, Radboud University Nijmegen, Nijmegen, the Netherlands*
[3]*Department of Aerospace Engineering and Engineering Mechanics, Oden Institute for Computational Engineering and Science, The University of Texas at Austin, Austin, TX, USA*

6.1   Introduction

In shared control, a robot executes a task to accomplish the goals together with a human operator while adhering to additional safety and performance requirements. Applications of shared control in cyber–physical–human systems (CPHS) include remotely operated semiautonomous wheelchairs (Galán et al. 2016), robotic teleoperation (Javdani et al. 2015), and human-in-the-loop unmanned aerial vehicle mission planning (Feng et al. 2016). A human operator issues a command through an input interface, which maps the command directly to an action for the robot. The problem is that a sequence of such actions may fail to accomplish the task at hand, due to limitations of the interface or failure of the human operator to comprehend the complexity of the problem. Therefore, a so-called *autonomy protocol* provides assistance for the human in order to complete the task according to the given requirements.

At the heart of the shared control problem is the design of an autonomy protocol. In the literature, there are two main directions, based on either *switching* the control authority between human and autonomy protocol (Shen et al. 2004) or on *blending* their commands toward joined inputs for the robot (Dragan and Srinivasa 2013; Jansen et al. 2017).

One approach to switching the authority first determines the desired goal of the human operator with high confidence and then assists toward exactly this goal (Fagg et al. 2004; Kofman et al. 2005). In Fu and Topcu (2016), switching the control authority between the human and the autonomy protocol ensures the satisfaction of specifications that are formally expressed in temporal logic. In general, switching of authority may cause a decrease in human's satisfaction, who usually prefers to retain as much control as possible (Kim et al. 2012).

Blending incorporates providing an alternative command in addition to the command of the human operator. Both commands are then blended to form a joined input for the robot to introduce a more flexible trade-off between the human's control authority and the level of autonomous assistance. A *blending function* determines the emphasis that is put on the autonomy protocol in the blending, that is, regulating the amount of assistance provided to the human (Dragan and Srinivasa 2012, 2013; Leeper et al. 2012). Switching of authority can be seen as a special case of blending, as the blending function may assign full control to the autonomy protocol or the human. In general, putting more emphasis on the autonomy protocol in blending may lead to greater accuracy in

Cyber–Physical–Human Systems: Fundamentals and Applications, First Edition.
Edited by Anuradha M. Annaswamy, Pramod P. Khargonekar, Françoise Lamnabhi-Lagarrigue, and Sarah K. Spurgeon.

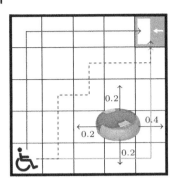

Figure 6.1 A wheelchair in a shared control setting with the human's perspective. Source: Tibor Antalóczy / Wikimedia Commons / CC BY-SA 3.0. (Readers are requested to refer to the online version for color representation.)

accomplishing the task. However, as humans prefer to retain control of the robot and may not approve if a robot issues a set of commands that are significantly different from the human's command (Javdani et al. 2015; Kim et al. 2012). However, none of the existing blending approaches provide *formal correctness* guarantees that go beyond statistical confidence bounds. Correctness here refers to ensuring safety and optimizing performance according to the given requirements. Our goal is to design an autonomy protocol that admits formal correctness while rendering the robot behavior as close to the human's commands as possible, which is shown to enhance the human experience.

A human may be uncertain about which command to issue in order to accomplish a task. More-over, a typical interface used to parse human's commands, such as a brain–computer interface, is inherently imperfect. To capture such uncertainties and imperfections in the human's decisions, we introduce *randomness* to the commands issued by humans. It may not be possible to blend two different deterministic commands. If the human's command is "up" and the autonomy proto-col's command is "right," we cannot blend these two commands to obtain another deterministic command. By introducing randomness to the human's commands and the autonomy protocol, we ensure that the blending is always well-defined.

Take as an example a scenario involving a semiautonomous wheelchair (Galán et al. 2016) whose navigation has to account for a randomly moving autonomous vacuum cleaner, see Figure 6.1. The wheelchair needs to navigate to the exit of a room, and the vacuum cleaner moves according to a probabilistic transition function. The task of the wheelchair is to reach the exit gate while not crashing into the vacuum cleaner. The human may not fully perceive the motion of the vacuum cleaner. Note that the human's commands, depicted with solid green line in Figure 6.1, may cause the wheelchair to crash into the vacuum cleaner. The autonomy protocol provides another set of commands, which is indicated by solid red line in Figure 6.1, to carry out the task safely. However, the autonomy protocol's commands deviate highly from the commands of the human. The two sets of commands are then blended into a new set of commands, depicted using the dashed red line in Figure 6.1. The blended commands perform the task safely while generating behavior as similar to the behavior induced by the human's commands as possible.

6.1.1 Review of Shared Control Methods

Most shared control methods in CPHS applications use the so-called *predict-then-act* approach. First, they determine the human's goal with high probability based on their inputs (predict), and then assist for that single goal (act). This approach has been used in Yu et al. (2005), Kragic et al. (2005), Kofman et al. (2005), Dragan and Srinivasa (2012, 2013), Hauser (2013), Muelling et al. (2015), and Javdani et al. (2015). While these approaches is shown to be effective in the domains that they have considered, they usually *do not assist the human* until they determine a sufficient prediction of the goal. Therefore, such procedures may lead to unsafe behavior during execution as the human's input may violate necessary specifications.

Iturrate et al. presented shared control using feedback based on electroencephalography (a method to record electrical activity of the brain) (Iturrate et al. 2013), where a robot is partly controlled via error signals from a brain–computer interface. In Trautman (2015), Trautman proposes to treat shared control broadly as a random process where different components are modeled by their joint probability distributions. As in our approach, randomness naturally prevents strange effects of blending: Consider actions "up" and "down" to be blended with equally distributed weight without having means to actually evaluating these weights. Finally, in Fu and Topcu (2016), a synthesis method switches authority between a human operator and the autonomy such that satisfaction of linear temporal logic constraints can be ensured. In this work, we present a general framework for the shared control that does not rely on predicting the goal of the human before providing assistance. Particularly, we develop a framework to assist the human while learning the human's strategy, and subsequently, the human's goal, to prevent unsafe behavior that might drive our system into an unsafe behavior during the whole execution.

In what follows, we call a formal interpretation of a sequence of the human's commands the *human strategy*, and the sequence of commands issued by the autonomy protocol the *autonomy strategy*. In Jansen et al. (2017), we formulated the problem of designing the autonomy protocol as a *nonlinear programming problem*. However, solving nonlinear programs is generally intractable (Bellare and Rogaway 1993). Therefore, we proposed a greedy algorithm that iteratively *repairs* the human strategy such that the specifications are satisfied without guaranteeing an optimal solution, based on Pathak et al. (2015). Here, we propose an alternative approach for the blending of the two strategies. We follow the approach of repairing the strategy of the human to compute an autonomy protocol. We ensure that the resulting robot behavior induced by the repaired strategy deviates minimally from the human strategy and satisfies safety and performance properties given in temporal logic specifications. We formally define the problem as a *nonlinear problem* and solve the resulting nonlinear problem by a sequential convex programming method for partially observable Markov decision process (POMDPs), similar to Cubuktepe et al. (2021b) and Suilen et al. (2020).

6.1.2 Contribution and Approach

In this chapter, we model the behavior of the robot as a POMDP (Puterman 2014), which captures the robot's actions inside a potentially stochastic environment, and also assumes partial observability of the state. A major difference between MDPs and POMDPs for policy synthesis is the intrinsic nonconvexity of policy synthesis in POMDPs, which yield a formulation of the shared control problem as a nonconvex optimization problem. On the other hand, shared control problem for MDPs can be formulated as a *quasiconvex optimization problem*, which can be solved in time polynomial in the size of an MDP (Cubuktepe et al. 2021a). It is known that this nonconvexity severely limits the scalability for synthesis in POMDPs (Vlassis et al. 2012). We develop an iterative algorithm that solves the resulting nonconvex problem in a scalable manner by adapting sequential convex programming (SCP) (Yuan 2015; Mao et al. 2018). In each iteration, it linearizes the underlying nonconvex problem around the solution from the previous iteration. The algorithm introduces several extensions to alleviate the errors resulting from the linearization. One of these extensions is a verification step not present in the existing SCP schemes.

Problem formulations with MDPs and POMDPs typically focus on maximizing an expected reward (or minimizing the expected cost). However, such formulations may not be sufficient to ensure safety or performance guarantees in a task that includes a human operator. Recently, it was shown that a reward structure is not sufficient to capture temporal logic constraints in general (Hahn et al. 2019). We design the autonomy protocol such that the resulting robot behavior satisfies probabilistic temporal logic specifications. Such verification problems have been extensively studied for MDPs (Baier and Katoen 2008), and mature tools exist for efficient verification (Kwiatkowska et al. 2011; Dehnert et al. 2017). On the other hand, the complexity of policy synthesis in POMDPs limit the scalability of existing POMDP solvers.

The question remains on how to obtain the human strategy in the first place. It may be unrealistic that a human can provide the strategy for an POMDP that models a realistic scenario. We use *inverse reinforcement learning* (IRL) to get a formal interpretation as a strategy based on human's inputs (Abbeel and Ng 2004; Ziebart et al. 2008). However, most existing work in IRL (Abbeel et al. 2010; Ziebart et al. 2008; Zhou et al. 2017; Ziebart 2010; Hadfield-Menell et al. 2016; Finn et al. 2016) has focused on Markov decision processes (MDPs), assuming that the learner can fully observe the state of the environment and expert demonstrations. However, the learner will not have access to full state observations in many applications. For example, a robot will never know everything about its environment (Ong et al. 2009; Bai et al. 2014; Zhang et al. 2017) and may not observe the internal states of a human with whom it works (Akash et al. 2019; Liu and Datta 2012), which is often the case in CPHS. Such information limitations violate the intrinsic assumptions made in most existing IRL techniques. In this work, we learn human strategies using our recent maximum-causal-entropy IRL method for POMDPs (Djeumou et al. 2022), which can deal with partial observability by incorporating side information into the learning process.

In summary, the main contribution of this chapter is to efficiently synthesize an autonomy protocol such that the resulting blended or repaired strategy meets all given specifications while only minimally deviating from the human strategy. We present a new technique based on sequential convex programming, which is done by solving a number of convex optimization problems (Boyd and Vandenberghe 2004).

6.1.3 Review of IRL Methods Under Partial Information

We also review the related work on IRL under partial information. The closest work to Djeumou et al. (2022) is by Choi and Kim (2011), where they extend classical maximum-margin-based IRL techniques for MDPs to POMDPs. However, even on MDPs, maximum-margin-based approaches cannot resolve the ambiguity caused by suboptimal demonstrations, and they work well when there is a single reward function that is clearly better than alternatives (Osa et al. 2018). In contrast, we adopt causal entropy that has been shown in Osa et al. (2018) and Ziebart (2010) to alleviate these limitations on MDPs. Besides, Choi and Kim (2011) rely on efficient off-the-shelf solvers to the forward problem. Instead, this chapter also describes an algorithm that outperforms off-the-shelf solvers and can scale to POMDPs that are orders of magnitude larger compared to the examples in Choi and Kim (2011). Further, Choi and Kim (2011) do not incorporate task specifications in their formulations.

IRL under partial information has been studied in prior work (Kitani et al. 2012; Boularias et al. 2012; Bogert and Doshi 2014, 2015; Bogert et al. 2016). Reference Boularias et al. (2012) considers the setting where the features of the reward function are partially specified as opposed to having partial information over the state of the environment. The work in Kitani et al. (2012) considers a special case of POMDPs. It only infers a distribution over the future trajectories of the expert given demonstrations as opposed to computing a policy that induces a similar behavior to the expert. The works in Bogert and Doshi (2014, 2015), and Bogert et al. (2016) assume that the states of the environment are either fully observable or fully hidden to the learning agent. Therefore, these approaches also consider a special case of POMDPs, like in Kitani et al. (2012). We also note that none of these methods incorporate side information into IRL and do not provide guarantees on the performance of the policy with respect to a task specification.

The idea of using side information expressed in temporal logic to guide and augment IRL has been explored in some previous works. In Papusha et al. (2018) and Wen et al. (2017), the authors incorporate side information as in temporal logic specification to learn policies that induce a behavior similar to the expert demonstrations and satisfies the specification. Reference Memarian et al. (2020) iteratively infers an underlying task specification that is consistent with the expert demonstrations and learns a policy and a reward function that satisfies the task specification. However, these methods also assume full information for both the expert and the agent.

6.1.3.1 Organization

We introduce all formal foundations that we need in Section 6.2. We provide an overview of the shared control problem in Chapter 6.3. We present the *shared control synthesis problem* to provide a solution based on sequential convex programming in Chapter 6.4. We indicate the applicability and scalability of our approach on experiments in Section 6.5 and draw a conclusion and critique of our approach in Chapter 6.6.

6.2 Preliminaries

In this section, we introduce the required formal models and specifications that we use to synthesize the autonomy protocol, and we give a short example illustrating the main concepts.

6.2.1 Markov Decision Processes

A *probability distribution* over a finite set X is a function $\mu : X \to [0,1] \subseteq \mathbb{R}$ with $\sum_{x \in X} \mu(x) = \mu(X) = 1$. The set $Distr(X)$ denotes all probability distributions over X.

Definition 6.1 (MDP): A *Markov decision process (MDP)* $\mathcal{M} = (S, s_I, Act, \mathcal{P})$ has a finite set S of states, an initial state $s_I \in S$, a finite set Act of actions, a transition probability function $\mathcal{P} : S \times Act \to Distr(S)$. The transition probability function $\mathcal{P}(s, \alpha, s')$ defines the probability of reaching a state s' in § given the current state $s \in S$ and action α in Act.

MDPs have *nondeterministic choices* of actions at states; the successors are determined *probabilistically* via associated probability distributions. A *cost function* $C : S \times Act \to \mathbb{R}_{\geq 0}$ associates cost with state-action pairs. If there is only a single action at each state, the MDP reduces to a *Markov chain (MC)*. We use *strategies* to resolve the choices of actions in order to define a probability and expected cost measure for MDPs.

Definition 6.2 (Strategy): A *memoryless and randomized strategy* for an MDP $\mathcal{M}$ is a function $\sigma : S \to Distr(Act)$. The set of all strategies over $\mathcal{M}$ is $Str^{\mathcal{M}}$.

Resolving all the nondeterminism for an MDP $\mathcal{M}$ with a strategy $\sigma \in Str^{\mathcal{M}}$ yields an *induced Markov chain $\mathcal{M}^{\sigma}$*.

Definition 6.3 (Induced MC): For an MDP $\mathcal{M} = (S, s_I, Act, \mathcal{P})$ and strategy $\sigma \in Str^{\mathcal{M}}$, the MC induced by $\mathcal{M}$ and σ is $\mathcal{M}^{\sigma} = (S, s_I, Act, \mathcal{P}^{\sigma})$, where

$$\mathcal{P}^{\sigma}(s, s') = \sum_{\alpha \in Act(s)} \sigma(s, \alpha) \cdot \mathcal{P}(s, \alpha, s') \text{ for all } s, s' \in S.$$

In our solution, we use the occupancy measure of a strategy to compute an autonomy protocol, which is introduced below.

Definition 6.4 (*Occupancy Measure*): (Puterman 2014) The occupancy measure x_σ of a strategy σ for an MDP $\mathcal{M}$ is defined as

$$x_\sigma(s, \alpha) = \mathbb{E}\left[\sum_{t=0}^{\infty} P^\sigma(s_t = s, \alpha_t = \alpha | s_0 = s_I)\right], \tag{6.1}$$

where s_t and α_t denote the state and action in $\mathcal{M}$ at time t.

The occupancy measure $x_\sigma(s, \alpha)$ is the expected number of times to take action α at state s under the strategy σ, and the strategy gives the probability of taking action α at state s. Occupancy measures for states can also be similarly defined.

6.2.2 Partially Observable Markov Decision Processes

We now introduce the background for POMDPs. POMDPs generalize MDPs by allowing the agent to not directly observe the underlying state of the system.

Definition 6.5 (*POMDP*): A *POMDP* is a tuple $\mathcal{M}_Z = (\mathcal{M}, Z, O)$ with a finite set S of states, an initial state $s_I \in S$, a finite set *Act* of actions, an *transition function* $P : S \times Act \times S \to \mathbb{R}$, a finite set Z of *observations*, an *observation function* $O : S \times Z \to \mathbb{R}$. The observation function $O(s, z)$ defines the probability perceiving the observation z at state s.

Similar to MDPs, POMDPs also have *nondeterministic choices* of actions at states and the successors are determined similarly.

Definition 6.6 (*Observation-Based Policy*): An *observation-based policy* $\sigma^o : Z \to Distr(Act)$ for an uncertain POMDP maps observations to distributions over actions. Note that such a policy is referred to as memoryless and randomized. More general (and powerful) types of policies take an (in)finite sequence of observations and actions into account. $\Sigma^{\mathcal{M}_{Z,P}}$ is the set of observation-based strategies for $\mathcal{M}_{Z,P}$. Applying $\sigma^o \in \Sigma^{\mathcal{M}_{Z,P}}$ to $\mathcal{M}_{Z,P}$ resolves all choices and partial observability and an induced (uncertain) MC $\mathcal{M}_{Z,P}^{\sigma^o}$ results.

Definition 6.7 (*Finite-Memory Policies*): Recall that an *observation-based policy* $\sigma^o : (Z \times Act)^* \times Z \to Distr(Act)$ for an POMDP maps a *trace*, i.e. a sequence of observations and actions, to a distribution over actions. A *finite-state controller* (FSC) consists of a finite set of memory states and two functions. The *action mapping* $\gamma(n, z)$ takes an FSC memory state n and an observation z and returns a distribution over uncertain POMDP actions. To change a memory state, the *memory update* $\eta(n, z, \alpha)$ returns a distribution over memory states and depends on the action α selected by γ. An FSC induces an observation-based policy by following a joint execution of these functions upon a trace of the POMDP. An FSC is *memoryless* if there is a single memory state. Such FSCs encode policies $\sigma^o : Z \to Distr(Act)$.

6.2.3 Specifications

For an MDP $\mathcal{M}$ or a POMDP $\mathcal{M}_{Z,P}$, the *reachability specification* $\varphi_r = \mathbb{P}_{\geq \beta}(\Diamond T)$ states that a set $T \subseteq S$ of *target states* is reached with probability at least $\beta \in [0,1]$. The *synthesis problem* is to find one particular strategy σ for an MDP $\mathcal{M}$ or an observation-based policy σ^o for an POMDP $\mathcal{M}_{Z,P}$ such that given a reachability specification φ_r and a threshold $\beta \in [0,1]$, the induced MC $\mathcal{M}^\sigma$ satisfies $\mathbb{P}_{\mathcal{M}^\sigma}(s_I \vDash \varphi_r) \geq \beta$, which implies that the strategy σ satisfies the specification φ_r. We note that linear temporal logic specifications can be reduced to reachability specifications (Baier and Katoen 2008); therefore, we omit a detailed introduction. We also consider the expected cost properties $\varphi_c = \mathbb{E}_{\leq \kappa}(\Diamond G)$, that restrict the expected cost to reach the set $G \subseteq S$ of goal states by an upper bound κ. The expected cost of a strategy σ for MDPs is given by

$$\sum_{s \in S} \sum_{\alpha \in Act(s)} x_\sigma(s, \alpha) C(s, \alpha). \tag{6.2}$$

Intuitively, for strategy σ, the cost and the expected number of taking action α at state s are multiplied. This multiplication is summed up for all states and actions. For POMDPs, this cost can be defined similarly for observation-based policies:

$$\sum_{s \in S} \sum_{\alpha \in Act(O(s))} x_\sigma(o, \alpha) C(s, \alpha). \tag{6.3}$$

Example 6.1 Figure 6.2 depicts an MDP $\mathcal{M}$ with initial state s_0. In state s_0, the available actions are a and b. Similarly, for state s_1, the two available actions are c and d. If action a is selected in state s_0, the agent transitions to s_1 and s_3 with probabilities 0.6 and 0.4.

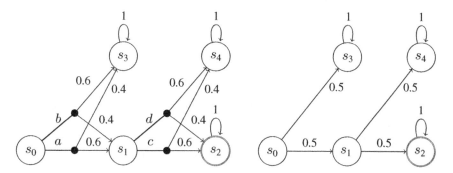

Figure 6.2 MDP $\mathcal{M}$ with target state s_2 and induced MC for strategy σ_{unif}.

For a reachability specification $\varphi_r = \mathbb{P}_{\geq 0.30}(\Diamond s_2)$, the deterministic strategy $\sigma_1 \in Str^{\mathcal{M}}$ with $\sigma_1(s_0, b) = 1$ and $\sigma_1(s_1, s) = 1$ induces a probability of 0.16 to reach s_2. Therefore, the specification is not satisfied, see the induced MC in the right picture of Figure 6.2. Likewise, the randomized strategy $\sigma_{unif} \in Str^{\mathcal{M}}$ with $\sigma_{unif}(s_0, a) = \sigma_{unif}(s_0, b) = 0.5$ and $\sigma_{unif}(s_1, c) = \sigma_{unif}(s_1, d) = 0.5$ violates the specification, as the probability of reaching s_2 is 0.25. However, the deterministic strategy $\sigma_{safe} \in Str^{\mathcal{M}}$ with $\sigma_{safe}(s_0, a) = 1$ and $\sigma_{safe}(s_1, c) = 1$ induces a probability of 0.36, thus $\sigma_{safe} \vDash \varphi_r$.

6.3 Conceptual Description of Shared Control

We now detail the general shared control concept adopted in this chapter and state the formal problem. As inputs, we have a set of task specifications, a model $\mathcal{M}_{Z,P}$ for the robot behavior, and a blending function b. The given robot task is described by certain reachability and expected cost specifications $\varphi = \varphi_1 \wedge \varphi_2 \ldots \wedge \varphi_n$. For example, it may not be safe to take the shortest route because there may be too many obstacles in that route. In order to satisfy performance considerations, the robot should prefer to take the shortest route possible while not violating the safety specifications. We model the behavior of the robot inside a stochastic environment as an POMDP $\mathcal{M}_{Z,P}$.

It may be unrealistic that a human grasps an POMDP that models a realistic shared control scenario. Indeed, a human will likely have difficulties interpreting a large number of possibilities, and the associated probability of paths and payoffs (Fryer and Jackson 2008). We obtain a human strategy as an abstraction of a sequence of human's commands using inverse reinforcement learning (Abbeel and Ng 2004; Ziebart et al. 2008). Specifically, we compute a formal human strategy σ_h based on specific inputs of a human.

Due to the complexity of task specifications, an imperfect input interface or the lack of detailed information, the human strategy may not satisfy the specifications. Put differently, at design time, we compute an abstract strategy σ_h from the human's inputs using inverse reinforcement learning. Using learning to compute human's strategy allows us to account for a variety of imperfections, including interface imperfections and possibly lack of knowledge of human about the underlying POMDP and the specifications.

The *shared control synthesis problem* is then computing a repaired strategy σ_{ha} such that it holds $\sigma_{ha} \models \varphi$ while *deviating minimally* from σ_h. The deviation between the human strategy σ_h and the repaired strategy σ_{ha} is measured by the maximal difference between the two strategies in each state of the POMDP. We state the problem that we study as follows:

Problem 1 Let $\mathcal{M}_{Z,P}$ be an POMDP, φ be an LTL specification, σ_h be a human strategy, and β be a constant. Synthesize a repaired strategy $\sigma_{ha} \in Str * \mathcal{M}_{Z,P}$ that solves the following problem:

$$\underset{\sigma_{ha} \in Str^{\mathcal{M}_{Z,P}}}{\text{minimize}} \quad \underset{o \in Z, \alpha \in Act}{\max} |\sigma_h(o, \alpha) - \sigma_{ha}(o, \alpha)| \tag{6.4}$$

$$\text{subject to} \quad \mathbb{P}_{\mathcal{M}_{Z,P}^{\sigma_{ha}}}(s_I \models \varphi) \geq \beta. \tag{6.5}$$

Using the repaired strategy, and the blending function b, we compute the autonomy strategy σ_a. The blending function reflects the preference over the human strategy or the autonomy strategy in all states of the POMDP. At runtime, we can then blend commands of the human and the autonomy strategy. The resulting "blended" commands will induce the same behavior as the blended strategy σ_{ha}, and the specifications are satisfied. Note that blending commands at runtime according to predefined blending function and autonomy protocol requires a linear combination of real values and is thus very efficient.

6.4 Synthesis of the Autonomy Protocol

In this section, we describe our approach to synthesize an autonomy protocol for the shared control synthesis problem. We start with the concepts of strategy blending and strategy repair. We then show how to synthesize a repaired strategy that deviates minimally from the human strategy based on sequential convex programming. Finally, we discuss how to include additional specifications and discuss other measures for the human and the repaired strategy that induce a similar behavior.

6.4.1 Strategy Blending

Given the human strategy σ_h and the autonomy strategy σ_a, a blending function computes a weighted composition of the two strategies by favoring one or the other strategy in each state of the POMDP at runtime (Javdani et al. 2015; Dragan and Srinivasa 2012, 2013). Put differently, the blending function should assign a *low confidence* to the actions of the human if they may lead to a violation of the specifications. Recall Figure 6.1 and the example in the introduction. In the cells of the gridworld, where some actions may result in a collusion with the vacuum cleaner with a high probability, it makes sense to assign a higher weight to the autonomy strategy. As a design choice, the blending function weighs the confidence in the human strategy and autonomy strategy at each state of the POMDP.

Definition 6.8   (*Linear Blending*):   Given the POMDP $\mathcal{M}_Z = (\mathcal{M}, Z, O)$, two observation-based strategies σ_h, σ_a, and a *blending function* $b : O \to [0,1]$, *the blended strategy* $\sigma_{ha} \in Str^{\mathcal{M}}$ for all observations $o \in Z$, and actions $\alpha \in Act$ is

$$\sigma_{ha}(o, \alpha) = b(o) \cdot \sigma_h(o, \alpha) + (1 - b(o)) \cdot \sigma_a(o, \alpha). \tag{6.6}$$

For each $o \in Z$, the value of $b(o)$ represents the "weight" of σ_h at o, meaning how much emphasis $b(o)$ puts on the human strategy at state o. For instance, if $b(o) = 0.1$, we infer that the blended strategy given an observation o puts more emphasis on σ_a. Note that in our experiments, we design the blending function by making use of the inherent verification result. For each observation o, we will compute the probability of satisfying the objective under the human strategy σ_h. If that probability is too low, we will put higher emphasis on the autonomy.

6.4.2   Solution to the Shared Control Synthesis Problem

We propose an algorithm to solve the shared control synthesis problem. Our solution is based on sequential convex programming, which is done by solving a number of convex optimization problems. We show that the result is the repaired strategy as in Problem 1. The strategy satisfies the specifications and deviates minimally from the human strategy. We use that result to compute the autonomy strategy σ_a.

6.4.2.1   Nonlinear Programming Formulation for POMDPs

In this section, we recall a nonlinear programming (NLP) formulation to compute a strategy that maximizes the probability of satisfying a reachability specification φ_r in an POMDP. The formulation is an straightforward extension of the dual LP formulations for MDPs (Puterman 2014; Forejt et al. 2011).

The variables of the NLP formulation are the following:

- $x_{\sigma_{ha}}(s, \alpha) \in [0, \infty)$ for each state $s \in S \backslash T$ and action $\alpha \in Act$ defines the occupancy measure of a state-action pair for the strategy σ_{ha}, i.e. the expected number of times of taking action α in state s.
- $x_{\sigma_{ha}}(s) \in [0, \infty]$ for each state $s \backslash T$ defines expected number of states of reaching the state s.
- $x_{\sigma_{ha}}(s) \in [0,1]$ for each state in the target set, T denotes the probability of reaching a state s in the target set T.
- $\sigma_{ha}(o, \alpha) \in [0,1]$ for each observation $s \in S$ and action $\alpha \in Act$ defines the probability of taking an action α given an observation s.

$$\text{maximize} \quad \sum_{s \in T} x_{\sigma_{ha}}(s) \tag{6.7}$$

$$\text{subject to}$$

$$\forall s \in S \backslash T. \quad \sum_{\alpha \in Act} x_{\sigma_{ha}}(s, \alpha) = \sum_{s' \in S \backslash T} \sum_{\alpha \in Act} \mathcal{P}(s', \alpha, s) x_{\sigma_{ha}}(s', \alpha) + \alpha_s \tag{6.8}$$

$$\forall s \backslash T. \quad x_{\sigma_{ha}}(s) = \sum_{s' \in S \backslash T} \sum_{\alpha \in Act} P(s', \alpha, s) x_{\sigma_{ha}}(s', \alpha) + \alpha_s \tag{6.9}$$

$$\forall s \backslash T. \quad \forall \alpha \in Act \quad x_{\sigma_{ha}}(s, \alpha) = x_{\sigma_{ha}}(s) \sum_{o \in Z} O(s, o) \sigma_{ha}(o, \alpha) \tag{6.10}$$

$$\sum_{s \in T} x_{\sigma_{ha}}(s) \geq \beta \tag{6.11}$$

where $\alpha_s = 1$ if $s = s_I$ and $\alpha_s = 0$ if $s \neq s_I$. The constraints in (6.8) and (6.9) ensure that the expected number of times transitioning to a state $s \in S$ is equal to the expected number of times to take action α that transitions to a different state $s' \in S$. The nonlinear and nonconvex constraint in (6.10) ensures that the strategy is observation-based, as opposed to state based. The constraint in (6.11) ensures that the specification φ_r is satisfied with a probability of at least β. We determine the states with probability 0 to reach T by a preprocessing step on the underlying graph of the POMDP. To ensure that the variables $x_{\sigma_{ha}}(s)$ encode the actual probability of reaching a state $s \in T$, we then set the variables of the states with probability 0 to reach T to zero.

For any optimal solution $x_{\sigma_{ha}}$ to the NLP in (6.7)–(6.11),

$$\sigma_{ha}(s, \alpha) = \frac{x_{\sigma_{ha}}(s, \alpha)}{\displaystyle\sum_{\alpha' \in Act} x_{\sigma_{ha}}(s, \alpha')} \tag{6.12}$$

is an optimal strategy, and $x_{\sigma_{ha}}$ is the occupancy measure of σ_{ha}, see Puterman (2014) and Forejt et al. (2011) for details.

6.4.2.2 Strategy Repair Using Sequential Convex Programming

Given σ_h, the aim of the autonomy protocol is to compute the blended strategy or the repaired strategy σ_{ha} that induces a similar behavior to the human strategy and yet satisfies the specifications. We compute the repaired strategy by *repairing* the human strategy resulting in the following formulation:

Lemma 6.1 The shared control synthesis problem can be formulated as the following nonlinear programming program with the following variables:

- $x_{\sigma_{ha}}$ and σ_{ha} as defined for the optimization problem in (6.7)–(6.11).
- $\hat{\delta} \in [0,1]$ gives the maximal deviation between the human strategy σ_h and the repaired strategy σ_{ha}.

$$\text{minimize} \quad \hat{\delta} \tag{6.13}$$

$$\text{subject to} \quad (6.8)\text{–}(6.11), \text{ and} \tag{6.14}$$

$$\forall s \in S \backslash T. \forall \alpha \in Act$$
$$\left| x_{\sigma_{ha}}(s, \alpha) - \sum_{\alpha' \in Act} x_{\sigma_{ha}}(s, \alpha') O(s, o) \sigma_h(o, \alpha') \right| \leq \hat{\delta} \sum_{\alpha'' \in Act} x_{\sigma_{ha}}(s, \alpha''). \tag{6.15}$$

Proof: For any solution to the optimization problem abovementioned, the constraints in (6.14) ensure that the strategy computed by (6.12) satisfies the specification. We now show that by minimizing $\hat{\delta}$, we minimize the maximal deviation between the human strategy and the repaired strategy. We perturb σ_h to σ_{ha} by

$$\forall o \in Z \backslash T. \alpha \in Act. \quad \sigma_{ha}(o, \alpha) = \sigma_h(o, \alpha) + \delta(o, \alpha).$$

where $\delta(o, \alpha)$ for $o \in Z$, $\alpha \in Act$ is an perturbation function subject to σ_{ha} being a well-defined strategy. Using (6.12), we reformulate the above constraint into

$\forall s \in S \backslash T.\alpha \in Act.$

$$x_{\sigma_{ha}}(s, \alpha) = \sum_{o \in Z} O(s, o) \sum_{\alpha' \in Act} \left(x_{\sigma_{ha}}(o, \alpha') \left(\sigma_h(o, \alpha') + \delta(o, \alpha') \right) \right). \tag{6.16}$$

Since we are interested in minimizing the maximal deviation, we assign a common variable $\hat{\delta} \in [0,1]$ for all state-action pairs in $\mathcal{M}_{Z,P}$ to put an upper bound on the deviation by

$\forall s \in S \backslash T.\alpha \in Act.$

$$|x_{\sigma_{ha}}(s, \alpha) - \sum_{o \in Z} O(s, o) \sum_{\alpha' \in Act} x_{\sigma_{ha}}(o, \alpha')\sigma_h(o, \alpha')| \leq \hat{\delta} \sum_{o \in Z} O(s, o) \sum_{\alpha'' \in Act} x_{\sigma_{ha}}(o, \alpha''). \tag{6.17}$$

Therefore, by minimizing $\hat{\delta}$ subject to the constraints in (6.14)–(6.15) ensures that σ_{ha} deviates minimally from σ_h. The constraint in (6.17) is a quadratic and nonconvex constraint due to multiplication of $\hat{\delta}$ and x_{ha}. Recall that the constraint in (6.10) is also nonconvex. In Section 6.4.3, we describe our sequential convex programming-based method to repair the strategy.

6.4.3 Sequential Convex Programming Formulation

In this section, we describe our algorithm, adapting a sequential convex programming (SCP) scheme to efficiently solve the policy repair problem. The algorithm involves a *verification step* to compute sound policies and visitation counts, which is not present in the existing off-the-shelf SCP schemes.

6.4.4 Linearizing Nonconvex Problem

The algorithm iteratively linearizes the nonconvex constraints in (6.10) and (6.15) around a previous solution. However, the linearization may result in an infeasible or unbounded linear subproblem (Mao et al. 2018). We first add *slack variables* to the linearized constraints to ensure feasibility. The linearized problem may not accurately approximate the nonconvex problem if the solutions to this problem deviate significantly from the previous solution. Thus, we utilize trust region constraints (Mao et al. 2018; Chen et al. 2013; Yuan 2015) to ensure that the linearization is accurate to the nonconvex problem. At each iteration, we introduce a *verification step* to ensure that the computed policy and visitation counts satisfy the nonconvex policy constraint (6.10), improve the realized deviation $\hat{\delta}$ over past iterations, and satisfy the temporal logic specifications, if available.

6.4.4.1 Linearizing Nonconvex Constraints and Adding Slack Variables

We linearize the nonconvex constraint (6.10), which is quadratic in $x_{\sigma_{ha}}$ and σ_{ha}, around the previously computed solution $\hat{\sigma}_{ha}$, $x_{\hat{\sigma}_{ha}}$, and $x_{\hat{\sigma}_{ha}}$. However, the linearized constraints may be infeasible. We alleviate this drawback by adding *slack variables* $k(s, \alpha) \in \mathbb{R}$ for state $s \in S$ and $\alpha \in Act$ which results in the constraint:

$$x_{\sigma_{ha}}(s, \alpha) + k(s, \alpha) = x_{\hat{\sigma}_{ha}}(s) \sum_{o \in O} O(s, o)\sigma_{ha}(o, \alpha) \tag{6.18}$$

$$+ \left(x_{\sigma_{ha}}(s) - x_{\hat{\sigma}_{ha}}(s) \right) \sum_{o \in Z} O(s, o)\hat{\sigma}_{ha}(o, \alpha).$$

We linearize the constraint (6.15) similarly.

6.4.4.2 Trust Region Constraints

The linearization may be inaccurate if the solution deviates significantly from the previous solution. We add following *trust region* constraints to alleviate this drawback:

$$\forall o \in O. \forall \alpha \in Act \quad \hat{\sigma}_{ha}(o, \alpha)/\rho \leq \sigma_{ha}(o, \alpha) \leq \hat{\sigma}_{ha}(o, \alpha)\rho, \tag{6.19}$$

where ρ is the size of the trust region to restrict the set of allowed policies in the linearized problem. We augment the objective $\hat{\delta}$ with the term $-\tau \sum_{s \in S} \sum_{\alpha \in Act} k(s, \alpha)$ to ensure that we minimize the violation of the linearized constraints, where τ is a large positive constant.

6.4.4.3 Complete Algorithm

We detail our SCP method in Algorithm 6.1. Its basic idea is to find a solution to the NLP in (6.13)–(6.15) such that this solution is feasible to the nonconvex optimization problem. We do this by an iterative procedure. We start with an initial guess of a policy and a trust region radius with $\rho > 0$, and (Step 1) we verify the resulting MC that combines the POMDP and the fixed policy σ to obtain the resulting values of the occupation variables $x_{\sigma_{ha}}$ and an objective value $\hat{\delta}$. This step is not present in existing SCP methods, and this is our main improvement to ensure that the returned policy satisfies the specification and minimally deviates from the human policy. If the uncertain MC indeed satisfies the specification (Step 2 – can be skipped in the first iteration), we return the policy σ. Otherwise, we check whether the deviation $\hat{\delta}$ is improved compared to δ_{old}. In this case, we accept the solution and enlarge the trust region by multiplying ρ with $\gamma > 1$. If not, we reject the solution and contract the trust region by γ and resolve the linear problem at the previous solution. Then (Step 3) we solve the resulting LP with the current parameters. We linearize around previous policy variables $\hat{\sigma}$ and occupancy measures $x_{\sigma_{ha}}$ and solve with updated trust region ρ to get an optimal solution. We iterate this procedure until the radius of the trust region is below a threshold $\omega > 0$. If the trust region size is below ω, we return the policy σ.

6.4.4.4 Additional Specifications

The NLP in (6.13)–(6.15) computes an optimal strategy for a single LTL specification φ. Suppose that we are given another reachability specification $\varphi_r = \mathbb{P}_{\geq \lambda}(\Diamond B)$ with $B \in S$. We can handle this

---

Algorithm 6.1 Sequential convex programming with trust region for strategy repair in POMDPs.

Input: uPOMDP, specification with threshold β, human strategy σ_h, $\gamma > 1$, $\omega > 0$

Initialize: trust region δ, weight τ, repaired strategy σ_{ha}, $\beta_{old} = 0$.

1: **while** $\delta > \omega$ **do**
2:     *Verify* the uncertain MC with strategy σ_{ha}:   ▷ **Step 1**
3:     *Extract* occupancy values $x_{\sigma_{ha}}$, objective value $\hat{\delta}$.
4:     **if** $\hat{\delta} < \hat{\delta}_{old}$ **then**   ▷ **Step 2**
5:         $\hat{x}_{\sigma_{ha}} \leftarrow x_{\sigma_{ha}}, \hat{\sigma}_{ha} \leftarrow \sigma_{ha}, \beta_{old} \leftarrow \beta$ ▷ Accept iteration
6:         $\rho \leftarrow \rho \cdot \gamma$ ▷ Extend trust region
7:     **else**
8:         $\rho \leftarrow \rho/\gamma$ ▷ Reject iteration, reduce trust region
9:     **end if**
10:    *Linearize* (1.10) and (1.15) around $\langle \hat{x}_{\sigma_{ha}}, \hat{\sigma} \rangle$ ▷ **Step 3**
11:    *Solve* the resulting LP
12:    *Extract* optimal solution for strategy σ_{ha}
13: **end while**
14: **return** the repaired strategy σ_{ha}

specification by appending the constraint

$$\sum_{s \in B} x_{\sigma_{ha}}(s) \geq \lambda \tag{6.20}$$

to the NLP in (6.13)–(6.15). The constraint in (6.20) ensures that the probability of reaching T is greater than λ.

We handle an *expected cost specification* $\mathbb{E}_{\leq \kappa}(\Diamond G)$ for $G \subseteq S$, by adding the constraint

$$\sum_{s \in S \setminus (T \cup G)} \sum_{\alpha \in Act} C(s, \alpha) x_{\sigma_{ha}}(s, \alpha) \leq \kappa \tag{6.21}$$

to the NLP in (6.13)–(6.15). The constraint in (6.21) ensures that the expected cost of reaching G is less than κ.

6.4.4.5 Additional Measures

We discuss additional measures that can be used to render the behavior between the human and the autonomy protocol similar based on the occupancy measure of a strategy. Instead of minimizing the maximal deviation between the human strategy and the repaired strategy, we can also minimize the maximal difference of occupancy measures of the strategies. In this case, the difference between the human strategy and the repaired strategy will be smaller in states, where the expected number of being in a state is higher and will be higher if the state is not visited frequently. We can minimize the maximal difference of occupancy measures by minimizing $||x_{\sigma_{ha}} - x_{\sigma_h}||_\infty$.

The occupancy measure of the human strategy can be computed by finding a feasible solution to the constraints in (6.14) for the induced MC $\mathcal{M}^{\sigma_{ha}}$. We can also minimize other convex norms of the occupancy measures of the human strategy and the repaired strategy, such as 1-norm or 2-norm.

6.5 Numerical Examples

We present a numerical example that illustrates the efficacy of the proposed approach. We require a representation of the human's commands as a strategy to use our synthesis approach in a shared control scenario. We discuss how such strategies are obtained using inverse reinforcement learning for POMDPs and report on case study results.

We demonstrate the proposed shared control method in a 3D robot simulation that follows. This simulation has two main components, a virtual ground robot with a full autonomy stack built using ROS (Quigley et al. 2009) and the simulation itself providing an environment and physics engine built using Unity. A robot in this simulation has a set of virtual sensors, including lidar, a monocular camera, and an inertial measurement unit. These sensors are placed on a virtual ground robot modeling a Clearpath Warthog with an accompanying software stack for simultaneous localization and mapping (SLAM) and visual perception. The SLAM system outputs binary occupancy grid feature maps based on the OmniMapper (Trevor et al. 2014). During operation, keyframes are collected to build binary occupancy feature maps, build an obstacle map, and establish the pose of the monocular camera. At each keyframe, a hierarchical inference machine visual classifier (Munoz 2013) takes an image and outputs a bounding box of grid cells which are projected onto a general grid. Each cell in the grid is placed into a unique binary occupancy grid using majority voting. Binary occupancy grid feature maps are disjoint, so their union produces a single map such that each cell in the map is classified as a single terrain.

A screenshot of the robot operating in this unity environment and its corresponding trajectory can be seen in Figure 6.3. This environment contains a variety of obstacles, including buildings, trees, and vehicles, as well as three terrain types describing our features, ϕ, grass, gravel, and road.

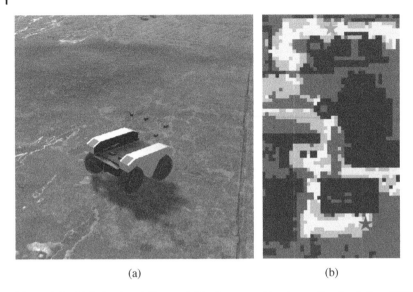

(a)   (b)

Figure 6.3 (a) A simulated Clearpath Warthog operating in a unity simulation. (b) The 2D occupancy grid map of the environment with buildings and other obstacles denoted by dark blue, tarmac and woody roads denoted by light blue and yellow, and grass denoted by green. We denote the initial and final positions by purple and magenta stars on the map. (Readers are requested to refer to the online version for color representation.)

The simulated environment operates in a state space consisting of 6168 states, 56,128 transitions, and 1542 total observations. This simulation is used to gather data for training from simulated human trajectories and test the learned and repaired policies' effectiveness.

6.5.1 Modeling Robot Dynamics as POMDPs

From a ground truth map of the environment in the simulation, we obtain a high-level MDP abstraction of the learner's behavior on the entire state space. Then, we impose partial observability of the robot as follows: the robot does not see the entire map of the world but only sees a fixed radius $r = 4$ (in terms of the number of grid cells) around its current position. Furthermore, we also incorporate uncertainty on the sensor classification of terrain features such that with probability $p = 0.9$, the prediction is correct.

6.5.2 Generating Human Demonstrations

In this section, we describe how we generate human demonstrations and encode the cost function, which describes the desired behavior of the robot. In the left map of Figure 6.4, the ideal trajectory is denoted from the initial location to the final location. Note that the trajectory follows mostly the tarmac roads, denoted by yellow command. In this case study, we assume that a human operator is aiming to follow the trajectory. We incorporate a POMDP framework from Akash et al. (2019) to capture dynamic changes in human trust and workload. The POMDP human trust and workload model estimates how dynamics of the human trust and workload probabilistically evolve during human–robot collaboration.

We incorporate this POMDP model by incorporating a probability p of giving an incorrect command to the robot in the environment. The probability is higher with a lower level of trust and a higher level of workload. In this example, we consider $p = 0.05$, when the human's current state

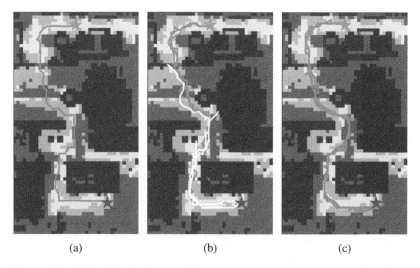

(a)   (b)   (c)

Figure 6.4 (a) The ideal trajectory of the robot to follow during the mission. (b) Sample trajectories generated by human strategy while trying to follow the ideal trajectory. The trajectories that failed to reach the target is denoted by white. (c) Sample trajectories generated by the repaired strategy reaching the target. Note that both the learned human and robot trajectories follow mostly tarmac roads, denoted by yellow cells. (Readers are requested to refer to the online version for color representation.)

indicates a high level of trust and a low level of workload. The probability increases to $p = 0.1$, if the human's current state indicates either a low level of trust or a high level of workload. We set $p = 0.2$, if the human's trust is low and workload is high. Based on this description, we generate 20 simulated human trajectories, shown in the middle map of Figure 6.4. Some of the simulated trajectories fail to safely reach the target and collide with the building in the environment. We denote these trajectories with white color. The (estimated) probability of safely reaching the target is 0.533.

6.5.3 Learning a Human Strategy

Given the human trajectories, we learned a human strategy σ_h using maximum–causal–entropy inverse reinforcement learning method for POMDPS (Djeumou et al. 2022). This method maximizes the likelihood of demonstrated trajectories by learning a cost function and strategy directly from demonstrations. The key notion, intuitively, is that an agent acts to optimize an unknown cost function, which is assumed to be linear in the features, and the objective of maximum entropy inverse reinforcement learning (MEIRL) is to find cost weights that maximize the likelihood of demonstrated paths. We encode the unknown cost function as a linear combination of known features: $C = \theta_1 \phi^{\text{road}} + \theta_2 \phi^{\text{gravel}} + \theta_3 \phi^{\text{grass}} + \theta_4 \phi^{\text{time}} + \theta_5 \phi^{\text{goal}} + \theta_6 \phi^{\text{building}}$, where ϕ^i are feature functions returning a value of 0 when the feature of the corresponding state is not feature i, or 1 otherwise, and θ^i are the weights of the feature functions. In order to incentivize the shortest path, the feature time penalizes the number of actions taken in the environment before reaching the waypoint. Furthermore, goal provides a positive reward upon reaching the waypoint, and building provides a negative reward if the robot collides with a building.

6.5.4 Task Specification

The task specification that the repaired strategy σ_{ha} needs to satisfy is Pr ($\neg$ building **U** goal) ≥ 0.90, where building is an atomic proposition that is true for states having building as its feature, and goal is an atomic proposition that is true at each target state.

6.5.5 Results

The right map of Figure 6.4 demonstrates the behavior of the repaired strategy σ_{ha}. The repaired strategy satisfies the task specification with a probability of 0.92 with a maximum deviation of $\hat{delta} = 0.12$. The strategy also prefers to traverse the tarmac roads. The computation time using Algorithm 6.1 is 2142.33 seconds and 42 iterations on a machine with Intel Core i9-9900u 2.50 GHz CPU and 64 GB of RAM. The repaired strategy also takes more time on average to reach the target than the human strategy, 360 and 312 steps, respectively, showing that the repaired strategy takes a safer path to satisfy the specification with a high probability.

6.6 Conclusion

We introduced a formal approach to synthesize an autonomy protocol in a shared control setting subject to probabilistic temporal logic specifications. The proposed approach utilizes inverse reinforcement learning to compute an abstraction of a human's behavior as a randomized strategy in a Markov decision process. We designed an autonomy protocol such that the resulting robot strategy satisfies safety and performance specifications. We also ensured that the resulting robot behavior is as similar to the behavior induced by the human's commands as possible. We synthesized the robot behavior using sequential convex programming. We showed the practical usability of a 3D simulation involving ground robots.

There is a number of limitations and also possible extensions of the proposed approach. First of all, we assumed that the transition and observation functions of the POMDP are known to the algorithm. Future work will investigate removing this assumption and developing model-free-based approaches. We will also integrate the framework with more expressive neural-network-based reward functions while learning the policy of the human.

We assumed that the human's commands are consistent through the whole execution. This assumption implies the human does not adapt the strategy to the assistance. In the future, we will handle nonconsistent commands by utilizing additional side information, such as task specifications.

Finally, in order to generalize the proposed approach to other task domains, it is worth to explore transfer learning (Pan and Yang 2010) techniques that allow to handle different scenarios without requiring to relearn the human strategy. from the human's commands.

Acknowledgments

This chapter incorporates the results from the previous publications (Cubuktepe et al. 2021a; Djeumou et al. 2022). The work was done while M. Cubuktepe was with the Department of Aerospace Engineering and Engineering Mechanics at the University of Texas at Austin. This work was partially supported by the grants NSF CNS-1836900, and NSF 1652113.

References

Abbeel, P. and Ng, A.Y. (2004). Apprenticeship learning via inverse reinforcement learning. *ICML*, 1. ACM.

Abbeel, P., Coates, A., and Ng, A.Y. (2010). Autonomous helicopter aerobatics through apprenticeship learning. *The International Journal of Robotics Research* 29 (13): 1608–1639.

Akash, K., Polson, K., Reid, T., and Jain, N. (2019). Improving human-machine collaboration through transparency-based feedback–Part I: Human trust and workload model. *IFAC-PapersOnLine* 51 (34): 315–321.

Bai, H., Hsu, D., and Lee, W.S. (2014). Integrated perception and planning in the continuous space: a POMDP approach. *The International Journal of Robotics Research* 33 (9): 1288–1302.

Baier, C. and Katoen, J.-P. (2008). *Principles of Model Checking*. The MIT Press.

Bellare, M. and Rogaway, P. (1993). The complexity of approximating a nonlinear program. In *Complexity in numerical optimization*, 16–32. World Scientific. https://link.springer.com/article/10.1007/BF01585569.

Bogert, K. and Doshi, P. (2014). Multi-robot inverse reinforcement learning under occlusion with interactions. *Proceedings of the 2014 International Conference on Autonomous Agents and Multi-Agent Systems*, 173–180. Citeseer.

Bogert, K. and Doshi, P. (2015). Toward estimating others' transition models under occlusion for multi-robot IRL. *24th International Joint Conference on Artificial Intelligence*.

Bogert, K., Lin, J.F.-S., Doshi, P., and Kulic, D. (2016). Expectation-maximization for inverse reinforcement learning with hidden data. *Proceedings of the 2016 International Conference on Autonomous Agents & Multiagent Systems*, 1034–1042.

Boularias, A., Krömer, O., and Peters, J. (2012). Structured apprenticeship learning. *Joint European Conference on Machine Learning and Knowledge Discovery in Databases*, 227–242. Springer.

Boyd, S. and Vandenberghe, L. (2004). *Convex Optimization*. New York: Cambridge University Press. ISBN 0521833787.

Chen, X., Niu, L., and Yuan, Y. (2013). Optimality conditions and a smoothing trust region newton method for nonlipschitz optimization. *SIAM Journal on Optimization* 23 (3): 1528–1552.

Choi, J.D. and Kim, K.-E. (2011). Inverse reinforcement learning in partially observable environments. *Journal of Machine Learning Research* 12: 691–730.

Cubuktepe, M., Jansen, N., Alsiekh, M., and Topcu, U. (2021a). Synthesis of provably correct autonomy protocols for shared control. *IEEE Transactions on Automatic Control*, Volume 66, 3251–3258. IEEE.

Cubuktepe, M., Jansen, N., Junges, S. et al. (2021b). Robust finite-state controllers for uncertain POMDPs. *35th AAAI Conference on Artificial Intelligence*, Volume 35, 11792–11800. AAAI.

Dehnert, C., Junges, S., Katoen, J.-P., and Volk, M. (2017). A STORM is coming: a modern probabilistic model checker. In *Computer Aided Verification. CAV 2017, Lecture Notes in Computer Science*, vol. 10427 (ed. R. Majumdar and V. Kunčak), 592–600. Cham: Springer.

Djeumou, F., Cubuktepe, M., Lennon, C., and Topcu, U. (2022). Task-guided inverse reinforcement learning under partial information. *Proceedings of the International Conference on Automated Planning and Scheduling*, Volume 32, 53–61.

Dragan, A.D. and Srinivasa, S.S. (2012). Formalizing assistive teleoperation. In: *Robotics: Science and Systems*.

Dragan, A.D. and Srinivasa, S.S. (2013). A policy-blending formalism for shared control. *International Journal of Robotics Research* 32 (7): 790–805.

Fagg, A., Rosenstein, M., Platt, R., and Grupen, R. (2004). Extracting user intent in mixed initiative teleoperator control. *Intelligent Systems Technical Conference*, 6309.

Feng, L., Wiltsche, C., Humphrey, L., and Topcu, U. (2016). Synthesis of human-in-the-loop control protocols for autonomous systems. *IEEE Transactions on Automation Science and Engineering* 13 (2): 450–462.

Finn, C., Levine, S., and Abbeel, P. (2016). Guided cost learning: deep inverse optimal control via policy optimization. *International Conference on Machine Learning*, 49–58. PMLR.

Forejt, V., Kwiatkowska, M., Norman, G. et al. (2011). Quantitative multi-objective verification for probabilistic systems. In: *Tools and Algorithms for the Construction and Analysis of Systems*.

TACAS 2011, Lecture Notes in Computer Science, vol 6605, 112–127. Berlin, Heidelberg: Springer-Verlag.

Fryer, R. and Jackson, M.O. (2008). A categorical model of cognition and biased decision making. *The BE Journal of Theoretical Economics* 8 (1): Article 6.

Fu, J. and Topcu, U. (2016). Synthesis of shared autonomy policies with temporal logic specifications. *IEEE Transactions on Automation Science and Engineering* 13 (1): 7–17.

Galán, F., Nuttin, M., Lew, E. et al. (2016). A brain-actuated wheelchair: asynchronous and non-invasive brain-computer interfaces for continuous control of robots. *Clinical Neurophysiology* 119 (9): 2159–2169.

Hadfield-Menell, D., Russell, S.J., Abbeel, P., and Dragan, A.D. (2016). Cooperative inverse reinforcement learning. In *NIPS*.

Hahn, E.M., Perez, M., Schewe, S. et al. (2019). Omega-regular objectives in model-free reinforcement learning. In: *Tools and Algorithms for the Construction and Analysis of Systems. TACAS 2019, Lecture Notes in Computer Science*, vol. 11427 (ed. T. Vojnar and L. Zhang), 395–412. Cham: Springer.

Hauser, K. (2013). Recognition, prediction, and planning for assisted teleoperation of freeform tasks. *Autonomous Robots* 35 (4): 241–254.

Iturrate, I., Omedes, J., and Montesano, L. (2013). Shared control of a robot using EEG-based feedback signals. *Machine Learning for Interactive Systems*, MLIS '13, 45–50. New York, NY, USA: ACM. ISBN 978-1-4503-2019-1. https://doi.org/10.1145/2493525.2493533. http://doi.acm.org/10.1145/2493525 .2493533 (accessed 9 February 2023).

Jansen, N., Cubuktepe, M., and Topcu, U. (2017). Synthesis of shared control protocols with provable safety and performance guarantees. In *ACC*, 1866–1873. IEEE.

Javdani, S., Bagnell, J.A., and Srinivasa, S. (2015). Shared autonomy via hindsight optimization. In: *Robotics: Science and Systems*.

Kim, D.-J., Hazlett-Knudsen, R., Culver-Godfrey, H. et al. (2012). How autonomy impacts performance and satisfaction: results from a study with spinal cord injured subjects using an assistive robot. *IEEE Transactions on Systems, Man, and Cybernetics-Part A: Systems and Humans* 42 (1): 2–14.

Kitani, K.M., Ziebart, B.D., Bagnell, J.A., and Hebert, M. (2012). Activity forecasting. *European Conference on Computer Vision*, 201–214. Springer.

Kofman, J., Wu, X., Luu, T.J., and Verma, S. (2005). Teleoperation of a robot manipulator using a vision-based human-robot interface. *IEEE Transactions on industrial electronics* 52 (5): 1206–1219.

Kragic, D., Marayong, P., Li, M. et al. (2005). Human-machine collaborative systems for microsurgical applications. *The International Journal of Robotics Research* 24 (9): 731–741.

Kwiatkowska, M., Norman, G., and Parker, D. (2011). PRISM 4.0: verification of probabilistic real-time systems. In: *Computer Aided Verification. CAV 2011, Lecture Notes in Computer Science*, vol. 6806 (ed. G. Gopalakrishnan and S. Qadeer), 585–591. Springer.

Leeper, A., Hsiao, K., Ciocarlie, M. et al. (2012). Strategies for human-in-the-loop robotic grasping. *HRI*, 1–8. IEEE.

Liu, X. and Datta, A. (2012). Modeling context aware dynamic trust using hidden Markov model. *Proceedings of the AAAI Conference on Artificial Intelligence*, Volume 26.

Mao, Y., Szmuk, M., Xu, X., and Açıkmese, B. (2018). Successive convexification: a superlinearly convergent algorithm for non-convex optimal control problems. *arXiv*.

Memarian, F., Xu, Z., Wu, B. et al. (2020). Active task-inference-guided deep inverse reinforcement learning. *2020 59th IEEE Conference on Decision and Control (CDC)*, 1932–1938. IEEE.

Muelling, K., Venkatraman, A., Valois, J.-S. et al. (2015). Autonomy infused teleoperation with application to BCI manipulation. *arXiv preprint arXiv:1503.05451*.

Munoz, D. (2013). Inference machines: parsing scenes via iterated predictions. Dissertation. PhD degree. The Robotics Institute, Carnegie Mellon University.

Ong, S.C.W., Png, S.W., Hsu, D., and Lee, W.S. (2009). POMDPs for robotic tasks with mixed observability. *Robotics: Science and Systems* 5: 4.

Osa, T., Pajarinen, J., Neumann, G. et al. (2018). An algorithmic perspective on imitation learning. *Foundations and Trends in Robotics* 7 (1-2): 1–179.

Pan, S.J. and Yang, Q. (2010). A survey on transfer learning. *IEEE Transactions on Knowledge and Data Engineering* 22 (10): 1345–1359.

Papusha, I., Wen, M., and Topcu, U. (2018). Inverse optimal control with regular language specifications. *2018 Annual American Control Conference (ACC)*, 770–777. IEEE.

Pathak, S., Ábrahám, E., Jansen, N. et al. (2015). A greedy approach for the efficient repair of stochastic models. In: *NASA Formal Methods. NFM 2015, Lecture Notes in Computer Science*, vol. 9058 (ed. K. Havelund, G. Holzmann, and R. Joshi), 295–309. Springer.

Puterman, M.L. (2014). *Markov Decision Processes: Discrete Stochastic Dynamic Programming*. Wiley.

Quigley, M., Conley, K., Gerkey, B. et al. (2009). ROS: an open-source robot operating system. *ICRA Workshop on Open Source Software*, number 3.2, 5. Kobe, Japan,.

Shen, J., Ibanez-Guzman, J., Ng, T.C., and Chew, B.S. (2004). A collaborative-shared control system with safe obstacle avoidance capability. *Robotics, Automation and Mechatronics*, volume 1, 119–123. IEEE.

Suilen, M., Jansen, N., Cubuktepe, M., and Topcu, U. (2020). Robust policy synthesis for uncertain POMDPs via convex optimization. *Proceedings of the 29th International Joint Conference on Artificial Intelligence, IJCAI-20*, 4113–4120. International Joint Conferences on Artificial Intelligence Organization.

Trautman, P. (2015). Assistive planning in complex, dynamic environments: a probabilistic approach. *CoRR* abs/1506.06784. http://arxiv.org/abs/1506.06784.

Trevor, A.J.B., Rogers, J.G., and Christensen, H.I. (2014). OmniMapper: a modular multimodal mapping framework. *2014 IEEE International Conference on Robotics and Automation (ICRA)*, 1983–1990. https://doi.org/10.1109/ICRA.2014.6907122.

Vlassis, N., Littman, M.L., and Barber, D. (2012). On the computational complexity of stochastic controller optimization in POMDPs.

Wen, M., Papusha, I., and Topcu, U. (2017). Learning from demonstrations with high-level side information. *Proceedings of the 26th International Joint Conference on Artificial Intelligence*.

Yu, W., Alqasemi, R., Dubey, R., and Pernalete, N. (2005). Telemanipulation assistance based on motion intention recognition. *Proceedings of the 2005 IEEE International Conference on Robotics and Automation, 2005*. ICRA 2005, 1121–1126. IEEE.

Yuan, Y.-x. (2015). Recent advances in trust region algorithms. *Mathematical Programming* 151 (1): 249–281.

Zhang, S., Sinapov, J., Wei, S., and Stone, P. (2017). Robot behavioral exploration and multimodal perception using POMDPs. *AAAI Spring Symposium on Interactive Multisensory Object Perception for Embodied Agents*.

Zhou, Z., Bloem, M., and Bambos, N. (2017). Infinite time horizon maximum causal entropy inverse reinforcement learning. *IEEE Transactions on Automatic Control* 63 (9): 2787–2802.

Ziebart, B.D. (2010). Modeling purposeful adaptive behavior with the principle of maximum causal entropy. PhD degree. Carnegie Mellon University.

Ziebart, B.D., Maas, A.L., Bagnell, J.A., and Dey, A.K. (2008). Maximum entropy inverse reinforcement learning. *AAAI*, Volume 8, 1433–1438.

7

Parallel Intelligence for CPHS: An ACP Approach

Xiao Wang[1], Jing Yang[1,2], Xiaoshuang Li[1,2], and Fei-Yue Wang[1]

[1] *The State Key Laboratory for Management and Control of Complex Systems, Institute of Automation, Chinese Academy of Sciences, Beijing, China*
[2] *The School of Artificial Intelligence, University of Chinese Academy of Sciences, Beijing, China*

7.1   Background and Motivation

With the continuous development and cost reduction of various technologies in control, computer, and communication over the last two decades, many traditional engineering systems become social systems, and vice versa many traditional social systems become engineering systems. As a consequence, human, social, economic, and ecological factors become more and more important, or simply, simple systems became complex systems, and the complexity of those systems has been rapidly and significantly increased to an unpredictable level. For the effective and efficient operations of these complex systems, we need to consider and deal with both engineering complexity and social complexity simultaneously. This involves issues far beyond traditional control methods, which focus mainly on physical systems (PS) or cyber–physical systems (CPS). For complex systems, we need to take a world of differences and make a difference in the world; therefore, new concepts, new theories, and new technologies must be developed (Wang 2004b,c, 2006).

Originally, the concepts of Cyber–Social–Physical (CSP) systems were introduced in the late 1990s as a general framework to deal with control and management of complex systems that involve human and social factors (Wang 1999). One of the first application examples of CSP was for parallel management of transportation systems and autonomous driving that initiated at the The Complex Adaptive Systems for Transportation Laboratory (CAST Lab) of the Intelligent Control and Systems Engineering Center (ICSEC) in Chinese Academy of Sciences (CAS) (Wang and Tang 2004; Wang 2010b). The second example is for social computing (Wang et al. 2007a) and Human Flesh Search (HFS) (Wang et al. 2010). In 2010, CSP was renamed to CPSS (Wang 2010a). Around 2011, Cyber–Physical–Human Systems (CPHS) appeared as a future research priority in the framework of the European Network of Excellence HYCON2 (Highly Complex and Networked Control Systems) (Alberto et al. 2011) where human–machine symbiosis, humans as operators of complex systems, humans as agents in multiagent systems, and the human as a part of a controlled system were for instance considered. The CPHS concepts were then more precisely introduced in Lamnabhi-Lagarrigue et al. (2017) and Netto and Spurgeon (2017). Regarding Social Systems (Smirnov et al. 2013), in terms of their essences and concepts, CPHS contains exactly the

Cyber–Physical–Human Systems: Fundamentals and Applications, First Edition.
Edited by Anuradha M. Annaswamy, Pramod P. Khargonekar, Françoise Lamnabhi-Lagarrigue, and Sarah K. Spurgeon.
© 2023 The Institute of Electrical and Electronics Engineers, Inc. Published 2023 by John Wiley & Sons, Inc.

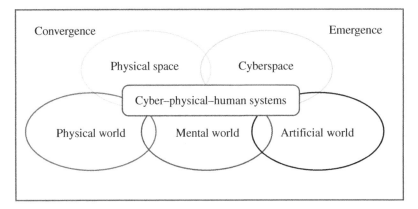

Figure 7.1 The basic philosophy and structure of CPHS (Wang 2010a).

same contents as CPSS. In addition, as pointed out clearly by Wang (2009, 2010a, 2022), and Wang et al. (2022), CPSS or CPHS is the abstract and scientific name for "Metaverse" that has become a hot topic all around the world recently.

Figure 7.1 illustrates the basic philosophy and structure of CPHS. This change from CPS to CPHS is of great philosophical significance because CPHS is based on Karl Popper's theory of reality (Wang 2020a–2021d,c). This theory states that our universe is made up of three interacting worlds: the physical world, the mental world, and the artificial world. Both physical space and cyberspace can be a materialization or reflection of three worlds (Wang et al. 2016a). The structure of CPHS provides a new theory for unifying the contradiction between emergence and convergence, the key concepts in complexity sciences.

The complexity of human and social behaviors in CPHS intensifies a modeling gap between actual systems and artificial systems, which bring a great challenge to the modeling of complex systems. Newton's laws, which are applicable to traditional CPS, are no longer adequate for describing, predicting, and prescripting entities in CPHS (Wang 2013b, 2016). We turn our attention to Merton's laws, which consist of various self-fulfilling prophecy laws (Figure 7.2). A self-fulfilling prophecy is a prediction that directly or indirectly causes itself to become true due to the feedback between

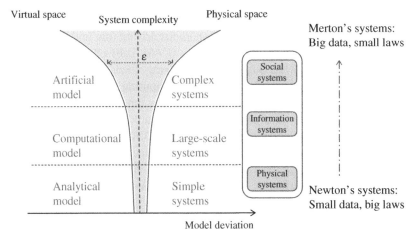

Figure 7.2 The modeling gap between actual systems and artificial systems. Source: Wang et al. (2016a) © [2016] IEEE.

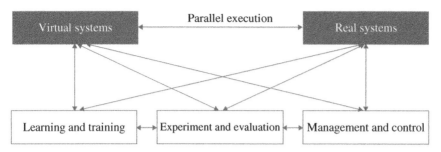

Figure 7.3 The framework and process for systems of parallel intelligence. Source: Wang et al. (2018d)/ Springer Nature.

belief and behaviors (Wang 2012a). Merton's laws can guide and influence Merton's systems, in which human factors must be included in the loop, and machine intelligence as well as human intelligence will work in tandem, think together, and run parallel to each other (Wang 2012a).

The artificial systems, computational experiments, and parallel execution (ACP) approach (Wang 2004b) makes it possible to bridge the modeling gap, which transforms models from system analyzers into data generators to make Merton's systems computable, testable, and verifiable (Figure 7.3). In ACP, "A" denotes "artificial systems, " namely building and generating models for systems (Li et al. 2022); "C" stands for "computational experiments," aiming at data production and analytics, and "P" presents "parallel execution," which targets innovative decision-making. The underlying principle of ACP is that one or more artificial spaces are used to solve issues in complex systems by integrating "artificial systems," "computational experiments," and "parallel execution." In addition, the relationship between actual and artificial systems can be one-to-one, one-to-many, many-to-one, or many-to-many, depending on the nature and complexity of the problem and the goal of the solution (Wang et al. 2018d). Based on the ACP approach, parallel intelligence (PI) (Wang et al. 2016a) is defined as one form of intelligence that is generated from the entanglement and interactions between actual systems and artificial systems.

7.2 Early Development in China

Actually, the biggest drive force for CPHS development is not transportation network but information network. Around 2000, with the rapid development of the Internet as an emerging technology, information security became a significant issue. To address this problem, Fei-Yue Wang (2007) called for research on CSP systems, especially on the topic of CSP for knowledge automation in collecting and processing online information. Around the same period, the phenomenon of HFS gradually emerged in China, which refers to a way to search for information and resources by concentrating the power of many netizens or crowdsourcing as called late in 2006 (Wang 2009; Wang et al. 2010, 2016e). The Microsoft Ziyao Chen Incident in 2001 (Figure 7.4) was the first time that HFS engines attracted widespread attention due to its scale and direct violation of individual privacy. Under the promotion of this event, the research on CPHS was intensified in ICSEC, the Open Source Intelligence Group (OSIG) was established in 2000 and founded for its pioneering work on HFS Cyber Movement Organizations (CMOs) based on CPHS (Wang 2003, 2004a). After that incident, this phenomenon spread to the United States and around the world and crowdsourcing become a norm of internet activity (Wang et al. 2014b,a; Zhang et al. 2016). Ten years later, Fei-Yue Wang and his students completed two books about HFS, but they were not published due to nonacademic reasons. One of these summarizes all the relevant work during these 10 years, and

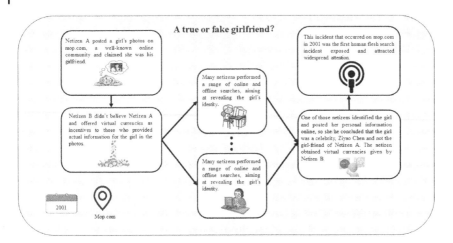

Figure 7.4 The Microsoft Ziyao Chen Incident in 2001.

the other discusses the entire research on HFS in detail (Wang 2011). Fortunately, some of related works have been published by Wang (2010a), Wang et al. (2010), and Zhang et al. (2012).

Fei-Yue Wang considered Wei Zhao of the National Science Foundation (NSF) of US as the originator and biggest promoter of CPS. They were old friends in United States and Zhao visited Wang's new center at Beijing in the earlier 2000s. While Wang's aim was "CSP for knowledge automation (KA)," Zhao's vision was "CPS for knowledge discovery (KD)." Around 2006, Zhao organized several workshops in United States to discuss the related issues and then launched NSF's program on "CPS for KD" and CPS become worldwide research direction and a prominent field today. In view of this new development, Wang changed CSP to CPSS in 2010 (Wang 2010a). In 2007, when the CAS organized a team to develop a strategic roadmap for science and technology to 2050, Fei-Yue Wang, Guojie Li, Tianran Wang, Jiaofeng Pan, Xiaoye Cao, Feng Zhang and others collaborated in promoting CPHS (in Chinese, Ren–Ji–Wu Systems, or Human–Computer–Things Systems) in the roadmap for information technology. They completed 18 books in total, three of which embodied the ideas of CPHS, namely *"National Security," "Information Technology"* as well as *"Smart Manufacturing"* (Wang et al. 2007b).

In 2011, National Natural Science Foundation of China (NNSFC) held a workshop at Beijing University of Technology to discuss the new research directions for information and automation technologies. Fei-Yue Wang brought up the issue of CPSS, and NNSFC's Director of Information Technology Division, Tianyou Chai suggested to use "human" to replace "social," that is, CPHS, but the proposal was not adapted since some of the attendees consider "Social" is more broad than "human." However, in 2019, the term "Human CPS (HCPS)" was promoted again in China. Overall, we consider CPSS, CPHS, HCPS are just different names for the same direction and field of research and development.

In China, CPHS-related research has covered urban transportation, energy, business management, agriculture, and other complex systems (Li et al. 2019; Wang et al. 2017c; Sun and Zheng 2020; Lv et al. 2019a; Wang et al. 2021d). CPHS activities have been promoted and supported by local chapters and technical committees of international, professional associations, such as International Federation of Automatic Control (IFAC), Association for Computing Machinery (ACM), IEEE Systems, Man, and Cybernetics Society (SMC), International Council on Systems Engineering (INCOSE), Institute for Operations Research and the Management Sciences (INFORMS), IEEE Intelligent Transportation Systems Society (IEEE ITSS). CPHS-focused research and development institutes include Qingdao Academy of Intelligent Industries (QAII), Center for Social Computing and Parallel Management, Chinese Academy of Sciences, The State

Key Laboratory for Management and Control of Complex Systems, Chinese Academy of Sciences (SKL-MCCS, CAS), National University of Defense Technology (NUDT), China Center for Economic and Social Security (CCESS), University of Chinese Academy of Sciences and Institute of Philosophy, Chinese Academy of Sciences (CASIP), which have carried out research on CPHS from different perspectives.

7.3 Key Elements and Framework

Karl Popper's theory of reality is the philosophical foundation of CPHS, where physical worlds are consisting of all matters and various phenomena in the objective world, mental worlds include all subjective spiritual activities, and artificial worlds are the worlds of knowledge created by human beings such as language, literary, and artistic works. Thus, the key elements of CPHS are human, cyber-machinery, things, and objects, which are interdependent and developed harmoniously. The intelligent machines represented by computers are no longer just tools used by human beings, and they have been integrated into culture and life, extending human organs and perceptions. As illustrated in Figure 7.5, human beings have been pulled into an information ecosystem that combines cyberspaces and physical spaces, which have a significant impact on human thinking and behaviors. "Intelligent Technology" becomes the third development stage of the concept of IT, addressing the intellectual disparity. The first two stages of IT are "Old IT (Industrial Technology)" and "Past IT (Information Technology)," overcoming the resource disparity and the information disparity, respectively (Wang et al. 2016b).

In comparison with CPS, CPHS, which expansively integrates social systems (Wang 2010a), further strengthens the interactions of cyber systems and physical systems. The core component of CPHS is an artificial system or digital twin or software-defined process, and its full implementation and operation will lead to the parallel era of real-virtual interaction. In CPHS, "C," "P," and "H" cannot be completely separated. However, by and large, "P" denotes the material, natural,

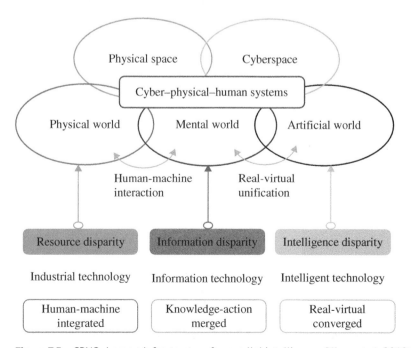

Figure 7.5 CPHS: A smart infrastructure for parallel intelligence (Wang et al. 2018d).

or physical domain of reality, equivalent to physical worlds; "H" represents the social, cognitive, or psychological domain of reality, corresponding to the mental world; "C" stands for the information, knowledge, or artificial domain of reality, equivalent to the artificial world (Wang 2015a). CPHS is the infrastructure for coordinating and integrating Popper's three worlds, which elevate artificial intelligence to hybrid intelligence (Wang 2022).

With the development of CPHS, many researchers have been exploring a wealth of applications, and their solutions only exist in the interactive space, where physical spaces and cyberspaces intersect with each other (Wang 2015b). Moreover, the key is to add intelligent entities that can provide social signals and build professional social networks of knowledge, which aims to perceive society or organizations such as enterprises for intelligent operation and management. The knowledge of CPHS is embedded within a vast number of physical and social intelligent entities. In CPHS, the extraction, analysis, and application of knowledge need to process big data and information that greatly exceeds the bandwidth of the human brains. Therefore, machine intelligence and knowledge engineering techniques are critical, that is, knowledge automation will play an essential role (Cheng et al. 2018).

In the era of big data, knowledge is the set of instructions and rules that guide people and machines to learn and work. Knowledge automation regards knowledge as the controlled object and realizes the cyclic process of automatic generation, acquisition, application, and re-creation of knowledge (Wang 2015b). Knowledge automation has become a new research framework for complex systems, the most urgent task of which is to convert complex systems with the characteristics of uncertainty, diversity, and complexity (UDC) into intelligent systems with the characteristics of agility, focus, and convergence (AFC) (Wang 2015a). To this end, knowledge automation is embedded into an ACP-based PI framework and process for CPHS (Figure 7.6) including three typical connection and operational modes: learning and training, experiment and evaluation, and control and management (Wang 2004c).

As is illustrated in Figure 7.6, the actual system and its artificial counterparts can be connected in a variety of modes for different purposes. By comparing and analyzing real and simulated behaviors, we can learn and predict systems' future actions and correspondingly modify the strategies of control and management. In the learning-and-training mode, artificial systems are primarily utilized as data centers for learning operating processes and for training administrators and operators. It is

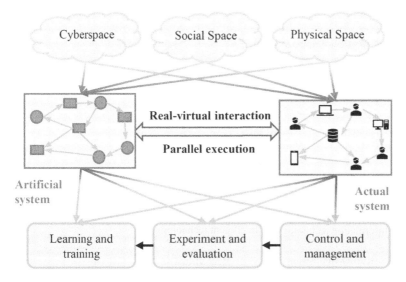

Figure 7.6 ACP-based parallel intelligence framework for CPHS. Source: Wang et al. (2016a) ©[2016] IEEE.

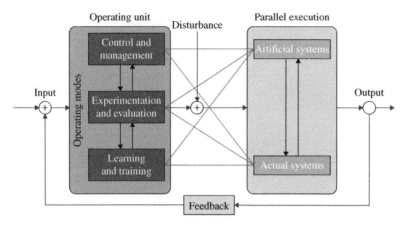

Figure 7.7 The ACP framework for the control and management of complex systems. Source: Wang (2008) ©[2008] IEEE.

not necessary for artificial systems to highly approximate the actual system and an artificial system can be regarded as an alternative to the actual system. In the experiment-and-evaluation mode, artificial systems serve mainly as platforms for performing computational experiments to analyze and predict behaviors of actual systems in various situations. In the control-and-management mode, the actual and artificial systems are required to be connected in real-time and online, and the artificial systems must reproduce actual behaviors with high fidelity. The behavioral differences between actual and artificial systems can be leveraged to determine operating parameters and generate feedback control. This method breaks through the constraints of traditional methods and effectively deals with the contradiction between emergence and convergence in complex systems (Wang 2004b).

The ACP (Wang 2004b) approach builds up the "virtual" and "soft" parts of complex and intelligent systems and makes full use of quantitative and real-time computing to solve practical problems. Figure 7.7 illustrates the ACP framework for the control and management of complex systems. In ACP, "artificial systems" are a generalized knowledge model and can be considered as an extension of traditional mathematical or analytical modeling; "computational experiments" are the way to analyze, predict, and select complex decisions, which can be seen as the sublimation of simulation; "parallel execution" is a new feedback control mechanism composed of real–virtual interactions, which can guide actions and lock targets. The effective management and control of complex systems are achieved by closed-loop feedback, real-virtual interaction, and parallel execution between actual systems and artificial systems. PI is generated during the interaction and execution between these systems.

7.4 Operation and Process

PI is a new theoretical framework of artificial intelligence based on the ACP method, aiming at establishing a mechanism for acquiring, creating, and supporting the intelligence in parallel systems (Li et al. 2017; Wang et al. 2016b–2017b–2018d). PI mainly includes the following three steps: (i) Construct artificial systems corresponding to the actual complex systems; (ii) Use computational experiments to train, predict, and evaluate the complex systems; and (iii) Implement the parallel control and management of the complex systems by setting the interaction and evolution rules between actual systems and artificial systems. Real–virtual interactions in PI can make

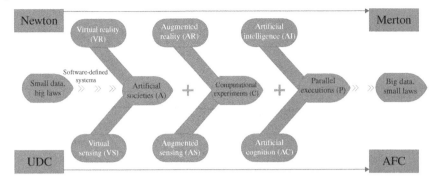

Figure 7.8 ACP-based parallel intelligence: from UDC to AFC. Source: Wang et al. (2016d)/With permission of Elsevier.

physical systems more consistent with artificial systems, which tackles the UDC challenges faced by complex systems to achieve the AFC management and control, as shown in Figure 7.8. Next, we will expound the process of applying PI to CPHS in detail.

7.4.1 Construction of Artificial Systems

Artificial systems are virtual systems defined by software and the digital models of actual systems in cyberspace. The construction of artificial systems can not only provide a reliable experimental environment for subsequent training, evaluation, and interactions but also better guide and support decision-making for complex systems (Wang and Tang 2004). For complex systems including human behaviors and social organizations, artificial systems do not merely digitize actual systems mechanically (Yang et al. 2019), but they integrate knowledge and relationships at social levels such as social units and processes. Therefore, artificial systems adequately model the sum of the relationships formed by humans and their environments.

The construction of artificial systems can overcome the shortcomings that some actual systems are uncontrollable and nonrepeatable (Tian et al. 2020). We can model different virtual entities in cyberspace for different optimization objectives by using some methods such as knowledge representation and knowledge engineering. Every virtual entity corresponds to a possible solution in physical systems. In other words, subject to the constraints of the actual entities and relationships in the physical space, we can also derive multiple artificial systems, providing rich experimental conditions for subsequent predictive intelligence and prescriptive intelligence.

7.4.2 Computational Experiments in Parallel Intelligent Systems

Before real–virtual interactions, computational experiments can be performed in artificial systems based on real data from actual systems (Wang et al. 2016f). Owing to the difficulties in actively testing and evaluating results and limitations of some process factors in experiments that are subjective, uncontrollable, and unobservable, many experimental operations in the physical systems are restricted and the experiments are not reproducible, making it difficult to further apply the experimental results in actual systems.

Computational experiments in parallel systems in Figure 7.9 are a potential perfect solution to the above problems. It is easy and reproducible to design and carry out controllable experiments in diversified artificial systems because artificial systems have many possibilities and broad experimental boundaries. The design of computational experiments follows the principles of replication, randomization, and blocking. Different computational experiments can be performed in different

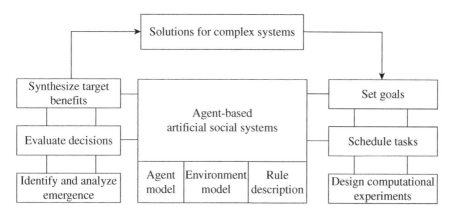

Figure 7.9 Computational experiments based on artificial systems (Wang 2004b).

artificial systems according to its preset optimization goals and control rules, and every result represents a possible evolution direction in the physical system. It means that, with the support of sufficient computing resources, computational experiments can make full use of real data in physical space and maximize the computational potential of artificial systems. Furthermore, it can also provide sufficient choices for actual systems to select an optimal result under the current conditions and then realize prescriptive intelligence through parallel execution.

7.4.3 Closed-Loop Optimization Based on Parallel Execution

After obtaining the prediction results by computational experiments, what needs to be considered is how to better utilize them and promote the iterative optimization of artificial and actual systems. Parallel execution provides a mechanism for managing and controlling actual systems by comparing, evaluating, and interacting with artificial systems, and applies the optimal solution to the actual system to finish decision-making and control actions. Meanwhile, the running state of these systems can be adjusted by analyzing the behavioral differences between actual and artificial systems. In addition, for physical systems, the changes of their states and behaviors can also provide feedback information for the adjustment and evolution of the artificial systems. We can also use the current state of actual systems as the initial condition to build new artificial systems and finally form a close-loop between actual and artificial systems. In short, parallel execution can not only guide the decision-making for actual systems very well but also make full preparations for changes of future scenarios.

7.5 Applications

PI, as a general CPHS theoretical method framework, is able to give full play to the advantages of ACP methods, solve the problems of unpredictability, difficulty in splitting and reducing, and inability to repeat experiments, and translate these advantages into tangible performance gains. Depending on the application domains, PI methods have been widely applied and made great progress in parallel control and intelligent control, parallel robotics and parallel manufacturing, parallel management and intelligent organizations, parallel medicine and smart healthcare, parallel ecology and parallel societies, parallel economic systems and social computing, parallel military systems, and parallel cognition and philosophy. Figure 7.10 shows a wealth of development results of the PI framework and theory.

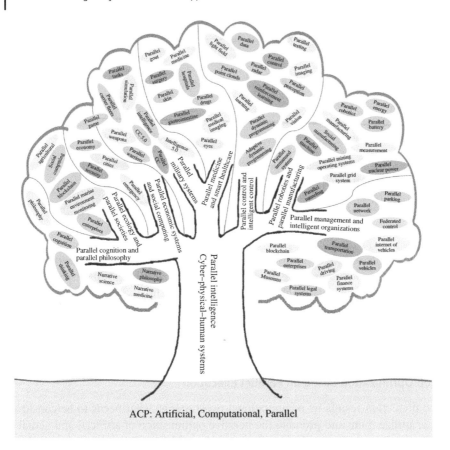

Figure 7.10 Applications tree of parallel intelligence and CPHSs.

7.5.1 Parallel Control and Intelligent Control

Parallel perception (Meng et al. 2017), parallel control (Wei et al. 2022), parallel learning (Li et al. 2018), and parallel testing (Li et al. 2019) provide technical support for perception, modeling, decision-making, control, and testing tasks in complex systems. Parallel perception provides a new theory for intelligent perception and understanding of complex systems (Meng et al. 2017). With the help of the ACP method, artificial sensors and scenes are established in parallel perception, so as to obtain the data of artificial scenes and the actual scenes at the same time. Parallel perception trains and evaluates the sensor (Tian et al. 2020) and scene (Wang et al. 2021b) model parameters through computational experiments, and finally optimizes the perception system through parallel execution to obtain better perception results. Parallel image (Wang et al. 2017c) and parallel vision (Wang et al. 2016c) use artificial image generation methods to obtain a large number of virtual image samples, effectively enhancing the ability of traditional visual perception models.

Parallel learning is a new type of machine learning framework built to address the challenges of inefficient data and difficult strategy selection of existing machine learning methods. The key architecture of parallel learning is a cycle of data, knowledge, and behavior policy. Figure 7.11 shows the basic framework of parallel systems and parallel learning. In the parallel learning framework, the method first generates a large amount of new data using "small data" composed of original data and software-defined artificial systems. The real "small data" and the newly generated data are then mixed to form "big data," which provide data support for the training

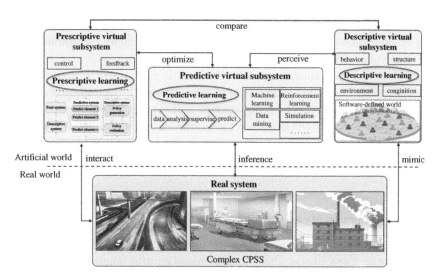

Figure 7.11 Parallel learning based on parallel systems. Source: Yang et al. (2019)/Editorial Department of Journal of Automation.

and parameter optimization of machine learning models. Through the three different learning processes, i.e. descriptive learning, predictive learning, and prescriptive learning, the agent learns precise knowledge that can be used to complete control tasks for complex systems. Those learning forms guide behaviors with the learned knowledge, which in turn generates new data for the update of knowledge, thus enabling the whole learning process to continue operating iteratively. By integrating data, knowledge, and action policies into complete closed-loop optimization systems, parallel learning solves the problems of inefficient data and difficult strategy selection in actual scenarios, and iteratives and optimizes the management policies to reach specific goals and finally achieve the parallel control of complex systems (Figure 7.12) (Li et al. 2018).

Complex system operation processes and industrial products often require a lot of verification and validation to ensure their reliability, but traditional testing methods have encountered huge difficulties to the testing process in terms of robustness and efficiency. Parallel testing technology

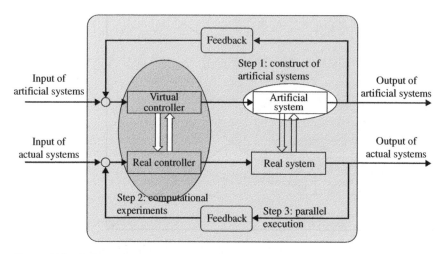

Figure 7.12 Self-learning parallel control framework (Wei et al. 2022).

(Li et al. 2019) proposes a human-in-the-loop testing system, which can combine the advantages of human experts and machine learning systems, so that the system has the ability to continuously upgrade under the guidance of experts, so as to maintain better test ability. Meanwhile, the system introduces an adversarial learning mechanism to get rid of the heavy dependence of the traditional testing process on the prior knowledge and experience of human experts. Through the process of adversarial games, it automatically explores and generates new testing tasks, configures new task scenarios and data, and comprehensively and efficiently performs adequate testing and validation of systems and products. The system provides effective support for the Intelligent Vehicles Future Challenge (IVFC) in China, a national competition from 2009 to 2020 for autonomous vehicles (Wang et al. 2021c).

7.5.2 Parallel Robotics and Parallel Manufacturing

Parallel manufacturing is proposed to meet people's increasing demand for personalized and intelligent products (Wang et al. 2018a; Li et al. 2021). Parallel manufacturing (Figure 7.13) obtains and analyzes manufacturing information existing in social systems via knowledge automation and creates products by gathering collective intelligence so that it can respond to market changes quickly. In addition, manufacturing enterprises rely on the co-evolution of virtual and real systems to optimize process planning and resource scheduling for industrial production. Social manufacturing (Wang 2012b) is a type of parallel manufacturing paradigm which is driven by mass customized production and consumption. In social manufacturing, consumers can directly convert social demands into products, and everyone can participate in the process of product design, production, and promotion in some ways such as crowdsourcing (Wang 2012b; Zhang et al. 2012).

Parallel machine (Bai et al. 2019) is an ACP-based collaborative framework for managing and controlling physical machines. Figure 7.14 illustrates the system architecture of parallel machines.

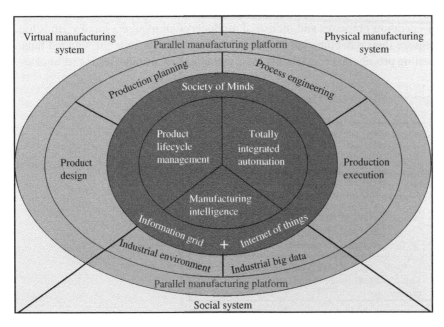

Figure 7.13   The framework of parallel manufacturing. Source: Wang et al. (2018a)/Editorial office of Science & Technology Review.

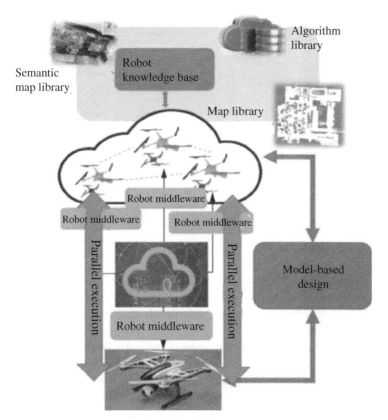

Figure 7.14 The system architecture of parallel machines. Source: Bai et al. (2019)/China InfoCom Media Group.

Compared with traditional machines, parallel machines add an extended loop that is composed of virtual control, virtual execution, and virtual loops. In parallel machines, artificial systems, computational experiments, and parallel execution act as skeleton structures, and knowledge and data circulate as blood, enabling machines to respond quickly to system feedback and administrator instructions in CPHS. In the future, robot workers, digital workers, and natural workers will be coexisting in factories. Robot workers and digital workers are physical machines and software-defined machines in parallel machines, respectively. As driving forces for industrial development, the inherent instincts of energy, i.e. physicality, informatization, and sociality, implies the inevitable and intensive interactions between social systems and energy systems. From this perspective, "social energy" can be defined as a complex sociotechnical system that consists of energy systems, social systems, and derived artificial systems. Parallel energy (Sun and Zheng 2020; Zhang et al. 2017) is a revolutionary theory and method for the investigation of social energy.

7.5.3 Parallel Management and Intelligent Organizations

Currently, organizations are more than ever complex due to the high turnover rate, diversity, and free will of participants. PI framework makes use of the virtual–real interaction to improve the management of various complex systems, such as urban transportation systems, enterprises, etc.

The ACP method is applied to traffic systems, which provide a new idea for their management and control. Based on the artificial transportation systems, parallel traffic (PT) systems (Wang 2008; Xiong et al. 2015; Wang 2010b; Lv et al. 2019a) learn, train, manage, and control the actual traffic

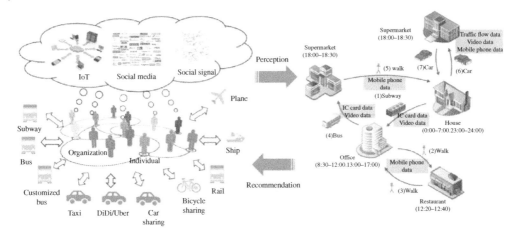

Figure 7.15 Parallel transportation system based on CPHS. Source: Zhang et al. (2018) ©[2018] IEEE.

systems. Figure 7.15 illustrates a parallel transportation system based on CPHS, which implements a closed-loop of intelligent processing including the collection, fusion, and analysis of traffic information, the modeling and prediction of traffic scenes, and the feedback of control measures by building interactive-and-evolutionary parallel systems.

Parallel driving (Wang et al. 2017b) completes complex autonomous driving tasks through simulation and interaction in the artificial world and guides the real world via the knowledge and experience accumulated in the artificial world. The parallel driving system includes a descriptive car, a predictive car, a prescriptive car, and a real car (Yang et al. 2019). Among them, the vehicle, driver, and the external environment are modeled and the dynamic characteristics of the vehicle, the driver's driving strategy as well as the traffic state of the external environment are transformed into software-defined objects. Through the continuous interaction of the above three virtual cars and real car, the safety and comfort of autonomous vehicles can be effectively optimized.

Parallel enterprise (Rui et al. 2017) based on ACP models the processes of purchasing, inventory, production, sales, after-sales, and finance with the help of multiagent and other technologies, while using technologies such as data collection and experimental decision-making to realize the simulation of the whole process of enterprise operation. Through computational experiments, the impact of changes in different positions and functions on enterprise operation costs and benefits can be analyzed quantitatively. With the help of parallel execution, the management strategy is quickly updated and iterated, which eventually realizes the unity of virtual and real interaction of parallel enterprises and promotes the continuous improvement of enterprise operation quality. Figure 7.16 shows the basic operation process of the parallel enterprise.

7.5.4 Parallel Medicine and Smart Healthcare

Smart healthcare refers to the integrated use of Internet of Things(IoTs), artificial intelligence, cloud computing, and big data to build an interactive platform for sharing medical information and facilitate interactions among patients, medical staffs/institutions, and medical equipment (Wang 2020b). Based on artificial intelligence, many researchers have tried to construct smart healthcare systems such as Enlitic and DeepMind to assist doctors in the diagnosis and treatment of diseases. However, the applications of smart healthcare systems are still far from ideal because more and more diseases require cross-border medical knowledge of doctors from different backgrounds, thus it becomes necessary to gather multiple doctors to work together by technological

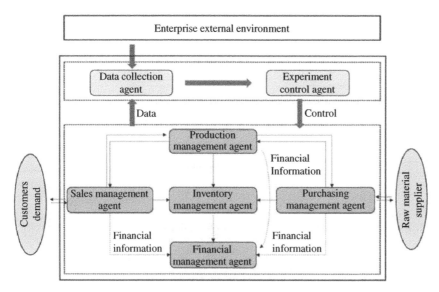

Figure 7.16 The basic workflow of parallel enterprise (Rui et al. 2017).

means. To address these issues, parallel healthcare systems (PHSs) use artificial healthcare systems to model and represent patients' conditions, diagnosis, and treatment process, then apply computational experiments to analyze and evaluate various therapeutic regimens, and implement parallel execution for decision-making support and real-time optimization in both actual and artificial healthcare processes (Wang 2021a; Wang and Wong 2013). In addition, blockchains can be combined with PHS, by building a consortium blockchain that connects patients, health bureaus, hospitals, and healthcare communities for comprehensive healthcare data sharing, medical records review, and care auditability (Wang et al. 2018f). Medical images are important for medical research, but they contain a large amount of information and are difficult to label. In addition, traditional methods lack interpretability in medical imaging analysis. Parallel medical imaging integrates parallel learning, parallel data, and professional knowledge and provides a workable framework to realize the interpretability of medical imaging analysis (Wang et al. 2021a).

The development of smart health will lead to a change in the division of labor among medical staffs. Parallel hospitals can move human doctors into better positions and create new and more medical and healthcare jobs with less tedious and laborious works and more enjoyable creative roles (Wang 2021b). Human monitoring and machine recommendation will be a new norm of medical decision-making and operations. As illustrated in Figure 7.17, parallel hospitals digitalize and parallelize infrastructure and participants in hospitals and introduce digital doctors and robotic doctors, aiming at providing patients with reliable, credible, and efficient services. Digital doctors are decision-makers and commanders that plan and evaluate the patients' treatment schemes. Robotic doctors follow digital doctors' instructions to perform related work. Human doctors are in charge of the entire medical process and have the highest priority that can modify the digital doctors' decisions to interrupt and change the robot doctors' behaviors. A large number of surgeries are performed in hospitals every year. Traditional surgeries are particularly dependent on doctors' experience resulting from a gap between clinical practice and experiments. Parallel surgery constructs parallel systems for surgeries based on CPHS and ACP, which makes full use of artificial data and multiple plans for diagnosis and treatment and considers individual differences to select the optimal surgical plan and guide doctors to complete operations. For the diagnosis and treatment of skin diseases, gout, ophthalmic diseases, and gastrointestinal diseases, parallel skin

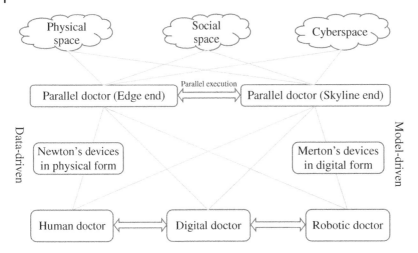

Figure 7.17 The framework of parallel hospital. Source: Wang (2021b)/Editorial Office of Medical Journal of Peking Union Medical College Hospital.

(Wang et al. 2019), parallel gout (Wang et al. 2017a), parallel eyes (Wang et al. 2018e) and parallel gastrointestine (Zhang et al. 2019) are proposed to break through uneven professional levels of different medical staff.

7.5.5 Parallel Ecology and Parallel Societies

Ecological systems consist of natural ecological systems, social-ecological systems as well as artificial or cyber or digital or knowledge ecological systems (Wang and Wang 2020). The agriculture systems are prominent components of natural ecological systems. Agricultural production is featured by strong uncertainty, diversity, and complexity, and its benefit is dependent on natural conditions, market environment, and national policies. For smart control on full production chain including management and service, parallel agriculture (Figure 7.18) (Kang et al. 2019) models the crop growth, farmland, environments, etc., to construct one or more artificial agricultural systems, and then predicts the growth trend of crops under different environment and management. As a consequence, we can adjust the management strategies of agricultural to guide farmers' planting behaviors.

A city is a social-ecological system, including social relationships between human and environment. In parallel cities (Figure 7.19) (Lv et al. 2019b) and parallel societies (Wang et al. 2020),

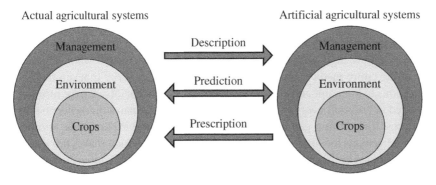

Figure 7.18 Parallel agriculture intelligence technology. Source: Kang et al. (2019).

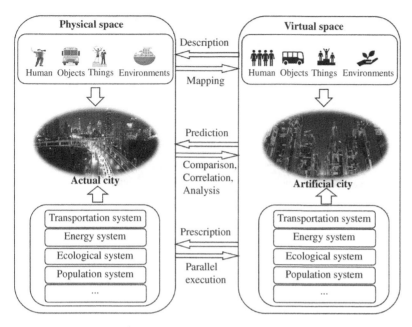

Figure 7.19 Parallel cities with real and virtual interaction. Source: Lv et al. (2019b)/China InfoCom Media Group.

every physical entity has its digital counterparts living or running in parallel such as persons, devices, and macroscopic systems (e.g. transportation, manufacturing, and agriculture). We can combine these digital assets to simulate the operation process of cities or societies based on different plans and choose an optimal solution to achieve the goal of effective operations of cities or societies, including the construction of urban infrastructure, the control and management of citizens' activities, etc.

Blockchain can be considered as a novel architecture for artificial or cyber or digital or knowledge ecological systems, which can store and analyze cyber big data and social signals (Yuan and Wang 2017). The blockchain systems play an irreplaceable role in data privacy and security with the characteristics of decentralization, nontampering as well as the open and transparent acquisition of data. However, due to the inherent features, namely, uncertainty, diversity, and complexity, in blockchain systems, there is still a significant need of an effective approach in the assessment, improvement, and innovation of the existing blockchain framework. We believe parallel blockchains based on ACP is one of the future trends in this line of thinking (Wang et al. 2018b). Parallel blockchains can construct one or more artificial blockchain systems, and then design and conduct diversified computational experiments to evaluate and verify specific behaviors, strategies, and mechanisms.

7.5.6 Parallel Economic Systems and Social Computing

The mechanism and framework of parallel economics based on ACP method and computational economic systems is presented in Figure 7.20. The basic idea is to use artificial economic systems (or software-defined or digital twin economic systems, depending on the complexity of the particular economic problem) to form parallel systems with actual economic systems, the key mechanism is the procedure of computational economic experiments, which produce big economic data from small ones represented by the parallel economic systems and then extracts

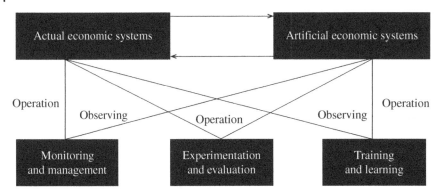

Figure 7.20 Mechanism and framework of parallel economics based on ACP method and computational economic systems (Wang 2004b).

"small" or deep economic intelligence from big data for a specific economic problem using artificial intelligence methods, and, finally, the qualitative novelty is the parallel execution or parallel driven through the real–virtual interaction which creates feedback and builds a closed loop between actual and artificial economic systems (Wang 2020c). Such new feedback and closed loops facilitate the unification of contradiction between *Jean–Baptiste Say's Law* (i.e. supply creates its own demand) and *John Maynard Keynes' law* (i.e. demand creates its own supply) of markets, and develop a new technological philosophy for supply and demand, that is, supply creates largely its own demand actually in real markets, whereas demand creates largely its own supply artificially in virtual markets (Wang 2020c).

Social computing has become the technical foundation for future computational smart societies, with the potential to improve the effectiveness of open-source big data usage, systematically integrating a variety of elements including time, human, resources, scenarios, and organizations in the current cyber–physical–social world, and establish a novel social structure with fair information, equal rights, and a flat configuration (Wang et al. 2016d).

Social computing emphasized technology development for society on the one hand and incorporating social theories and practices into Information and Communications Technology (ICT) development on the other (Wang et al. 2007a) (Figure 7.21). Nowadays, with the prosperity of all kinds of mobile Apps and social media platforms, human and social needs can be easily detected captured by open-source intelligence, which provides inputs for the parallel economic systems. Besides, the new social infrastructures, which are equipped with blockchain, smart contacts, and Decentralized Autonomous Organizations (DAOs), will enable trust and attention to be produced and circulated massively as new commodities, becoming the starting points and basis of a new intelligent economy with a new level of efficiency and effectiveness. While the real and virtual social and economic systems must be coordinated and controlled in parallel, and human operators must be in control through legal enforcement and technological prescription, social computing enables human-in-the-loop for all important real–virtual or actual–artificial interactions (Wang 2020c). Accordingly, we think the future economic systems will march from Adam Smith's division of labor to the current division of human and machinery, further to the coming division of virtual and real, that is, the age of intelligent economy, where "invisible hand" becomes "smart hand," and where parallel economics will be vital for its service, safety, security, sensitivity, and sustainability (Wang 2020c).

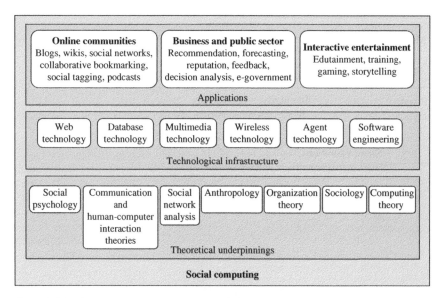

Figure 7.21 The theoretical underpinnings, infrastructure, and applications of social computing. Source: Wang et al. (2007a) ©[2007a] IEEE.

7.5.7 Parallel Military Systems

In recent years, the typical form of warfare has been witnessed to evolve from the nuclear deterrence in the physical world to the so-called "intelligence deterrence" in cyber–physical–social spaces, typical examples include the Iraq war, the Syrian war, and the recent happened Russia–Ukraine conflict, all resulting in a critical need for new intelligent technologies in military application scenarios. During Russia–Ukraine war (RUW), people expressing their opinions actively via Twitter, Chinese Weibo, Facebook, etc. Ukraine used social media platforms to recruit soldiers, mobilize international resources, attract attention all over the world, which definitely extended the war space into both cyber and social spaces. Especially, the scale and intensity of the opinion battles in cyberspace on the RUW, in particular, have even launched a new chapter in the history of international warfare. This is a fantastic demonstration of social cognitive warfare with CPHS that has the potential to significantly influence our humankind today and in the future.

The parallel military systems (Figure 7.22) are proposed based on the "three wars in one" concept, which believed "instead of isolating components in cyber–physical & Social spaces, artificial military systems underline the importance of interrelationship, interactivities, and integration among them (Wen et al. 2012)." By leveraging artificial military systems in actual, simulated, or hybrid conflict scenarios, this approach can generate and develop complex and varied interactions and behavior patterns. As a result, the commanders can observe, analyze, and comprehend complex systems through emergence.

Parallel weapon systems (Bai et al. 2017) such as parallel tanks (Xing et al. 2018) and parallel aircraft carriers (Yang et al. 2018) as well as parallel control and command (Wang 2015a) are developed based on the parallel military systems framework in the past few years. Among them, the core idea of PI is to achieve specific intelligence processes and objectives by building software-defined digital intelligence agencies that interact with real intelligence agencies in a virtual and real way through three forms: learning and training, testing and evaluation, as well as management

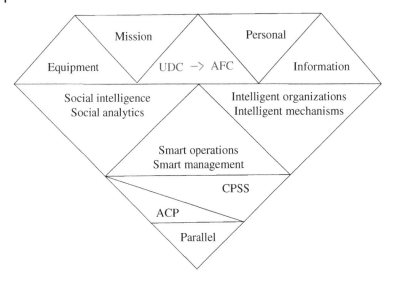

Figure 7.22 The framework of parallel military systems (Wang 2015c).

and control. PI forms a new intelligent system paradigm, led by intelligent intelligence through software-defined intelligence systems and extensive participation of social signals. The key idea is parallel through virtual–real duality, supported by ACP method and CPHS, as infrastructure, accomplished by tasks of intelligent army, smart management, and social intelligence, and finally, integration of personnel information, equipment, and mission.

7.5.8 Parallel Cognition and Parallel Philosophy

The huge success of AlphaGo has led to intelligent technology rapidly attracting a great deal of attention (Wang et al. 2016b). The rapid progress of intelligent science call for the simultaneous development of philosophy of science. On the basis of Karl Popper's three worlds, Fei-Yue Wang proposed the concept of parallel philosophy (Wang 2020a) for intelligent technology, promoting the existing philosophical concepts from process philosophy to parallel philosophy, forming a shift from "Being" and "Becoming" to "Believing" (Figure 7.23). Parallel philosophy establishes

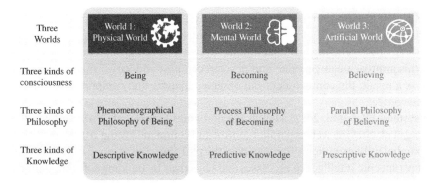

Figure 7.23 Parallel philosophy: Process of parallel interaction and entanglement between virtual and real and correspondent prescriptive knowledge system. Source: Wang (2021d)/Bulletin of Chinese Academy of Sciences.

philosophical systems about descriptive knowledge for "being," predictive knowledge for "becoming," and prescriptive knowledge about "believing" (Wang 2021d).

Parallel thinking and parallel cognition are the basis of parallel philosophy. Parallel thinking requires participants to present different viewpoints on parallel "tracks" from different perspectives as much as possible, thus avoiding the negative effects of adversarial thinking. Based on parallel thinking, we can combine computational thinking, parallel learning, parallel decision making, parallel management, and other intelligent methods into CPHS space to build the culture and behavior of knowledge automation, forming a parallel cognitive science (Wang 2007, 2013a).

7.6 Conclusion and Prospect

According to Karl Popper's reality model of three interacting worlds, we need a Generalized Gaia Hypothesis that extends the natural or physical earth to include both mental earth and artificial earth, and a new thinking and a new theory of ecology for three earths or three worlds correspondingly (Wang and Wang 2020). In intelligent science technology or artificial intelligence, the concept of CPHS has been introduced to integrate three worlds through interactions between physical space and cyberspace as the basic infrastructures for the construction of intelligent systems, accordingly, a new theory of ecology could be developed by investigating CPHS-based interacting ecosystems and corresponding ecological issues through real–virtual interactions.

The remarkable advancement of the next-generation information technologies has profoundly altered the relationship between humans and their physical environments. The resulting emergence of big data, diversified social networks, and online community, and increasingly coupled CPHS spaces, is resulting in a more computable society and thus calling for the parallel-system-based research methodology for social computing and computational social systems (Wang et al. 2018c).

In this new era, we can expect that everything will have its digital avatars that live or run in parallel with the physical entity in the future. For example, every microscopic person, physical equipment, or macroscopic systems will have its online artificial counterparts. As a result, in the future parallel societies, artificial digital assets will become the most vital strategic resources. Possession of how many artificial digital assets rather than how much data, might become the primary criterion for assessing the strength of organizations, enterprises, or countries. We believe these parallel societies (Wang et al. 2020) will be realized in this coming new decade.

References

Alberto, B., Antonio, B., Eduardo, C. et al. (2011). Systems and Control Recommendations for a European Research Agenda Towards Horizon 2020. https://mycore.core-cloud.net/index.php/s/NDggrBynNVmoZjq (accessed February 2023).

Bai, T., Wang, S., Zhao, X., and Qin, J. (2017). Parallel weapons: weapons towards intelligent warfare. *Journal Of Command and Control* 3 (2): 89–98.

Bai, T., Shen, Z., Liu, Y., and Dong, X. (2019). Parallel machine: a framework for the control and management for intelligent machines. *Chinese Journal of Intelligent Science and Technology* 1 (2): 181–191.

Cheng, L., Yu, T., Zhang, X., and Yang, B. (2018). Parallel cyber-physical-social systems based smart energy robotic dispatcher and knowledge automation: concepts, architectures, and challenges. *IEEE Intelligent Systems* 34 (2): 54–64.

Kang, M., Wang, X., Hua, J. et al. (2019). Parallel agriculture: intelligent technology toward smart agriculture. *Chinese Journal of Intelligent Science and Technology* 1 (02): 107–117.

Lamnabhi-Lagarrigue, F., Annaswamy, A., Engell, S. et al. (2017). Systems & control for the future of humanity, research agenda: current and future roles, impact and grand challenges. *Annual Reviews in Control* 43: 1–64.

Li, L., Lin, Y., Zheng, N., and Wang, F.-Y. (2017). Parallel learning: a perspective and a framework. *IEEE/CAA Journal of Automatica Sinica* 4 (3): 389–395.

Li, L., Zheng, N.-N., and Wang, F.-Y. (2018). On the crossroad of artificial intelligence: a revisit to Alan Turing and Norbert Wiener. *IEEE Transactions on Cybernetics* 49 (10): 3618–3626.

Li, L., Wang, X., Wang, K. et al. (2019). Parallel testing of vehicle intelligence via virtual-real interaction. *Science Robotics* 4 (28): https://doi.org/10.1126/scirobotics.aaw4106.

Li, S., Wang, Y., Wang, X., and Wang, F. (2021). Mechanical design paradigm based on ACP method in parallel manufacturing. *2021 IEEE 1st International Conference on Digital Twins and Parallel Intelligence (DTPI)*, 1–4. IEEE.

Li, X., Ye, P., Li, J. et al. (2022). From features engineering to scenarios engineering for trustworthy AI: I& I, C& C, and V& V. *IEEE Intelligent Systems* 37 (4): 18–26.

Lv, Y., Chen, Y., Jin, J. et al. (2019a). Parallel transportation: virtual-real interaction for intelligent traffic management and control. *Chinese Journal of Intelligent Science and Technology* 1 (1): 21.

Lv, Y., Wang, F.-Y., Zhang, Y., and Zhang, X. (2019b). Parallel cities: framework, methodology, and application. *Chinese Journal of Intelligent Science and Technology* 1 (3): 311–317.

Meng, X.-B., Wang, R., Zhang, M., and Wang, F.-Y. (2017). Parallel perception: an ACP-based approach to visual SLAM. *Journal of Command and Control* 3 (004): 350–358.

Netto, M. and Spurgeon, S.K. (2017). Special section on cyber-physical & human systems (CPHS). *Annual Reviews in Control* 44: 249–251.

Rui, Q., Shuai, Z., Juanjuan, L., and Yong, Y. (2017). Parallel enterprises resource planning based on deep reinforcement learning. *Acta Automation Sinica* 43 (9): 1588–1596.

Smirnov, A., Kashevnik, A., Shilov, N. et al. (2013). Context-aware service composition in cyber-physical-human-system for transportation safety. *2013 13th International Conference on ITS Telecommunications (ITST)*, 139–144. IEEE.

Sun, W. and Zheng, Y. (2020). Energy 5.0: stepping into a parallel era of virtuality and reality interaction. *Process Automation Instrumentation* 41 (01): 1–9.

Tian, Y., Shen, Y., Li, Q., and Wang, F.-Y. (2020). Parallel point clouds: point clouds generation and 3D model evolution via virtual-real interaction. *Acta Automatica Sinica* 46 (12): 2572–2582.

Wang, F.-Y. (1999). CAST Lab: A Cyber-Social-Physical Approach for Traffic Control and Transportation Management. *ICSEC Technical Report*. Beijing, China: Intelligent Control and Systems Engineering Center in Chinese Academy of Sciences. http://www.sklmccs.ia.ac.cn/1999reports.html.

Wang, F.-Y. (2003). From nothing to everything: an investigation on research of artificial societies and complex systems. *China Science Times*, 03–17.

Wang, F.-Y. (2004a). Social computing and prototype systems development for key projects in CAS. Chinese Academy of Sciences Reports of Major Projects.

Wang, F.-Y. (2004b). Artificial societies, computational experiments, and parallel systems a discussion on computational theory of complex social-economic systems. *Complex Systems and Complexity Science* 1 (4): 25–35.

Wang, F.-Y. (2004c). Parallel system methods for management and control of complex systems. *Control and Decision* 19 (5): 484–489+514.

Wang, F.-Y. (2006). On the modeling, analysis, control and management of complex systems. *Complex Systems and Complexity Science* 3 (02): 26–34.

Wang, F.-Y. (2007). From computional thinking to computational culture. *Communications of the CFF* 3 (11): 78–82.

Wang, F.-Y. (2008). Toward a revolution in transportation operations: AI for complex systems. *IEEE Intelligent Systems* 23 (6): 8–13.

Wang, F.-Y. (2009). Beyond X 2.0: where should we go? *IEEE Intelligent Systems* 24 (3): 2–4.

Wang, F.-Y. (2010a). The emergence of intelligent enterprises: from CPS to CPSS. *IEEE Intelligent Systems* 25 (4): 85–88.

Wang, F.-Y. (2010b). Parallel control and management for intelligent transportation systems: concepts, architectures, and applications. *IEEE Transactions on Intelligent Transportation Systems* 11 (3): 630–638.

Wang, F.-Y. (2011). A brief history of human flesh search: from crowdsourcing to cyber movement organizations. Beijing, China: State Key Laboratory for Management and Control of Complex Systems, Institute of Automation, Chinese Academy of Sciences. *SKL-MCCS Tech Report*.

Wang, F.-Y. (2012a). A big-data perspective on AI: Newton, Merton, and analytics intelligence. *IEEE Intelligent Systems* 27 (5): 2–4.

Wang, F.-Y. (2012b). From social computing to social manufacturing: the coming industrial revolution and new frontier in cyber-physical-social space. *Bulletin of Chinese Academy of Sciences* 27 (6): 658–669.

Wang, F.-Y. (2013a). General computational education for computational society: computational thinking and computational culture. *Industry and Information Technology Education* 6: 4–8.

Wang, F.-Y. (2013b). A framework for social signal processing and analysis: from social sensing networks to computational dialectical analytics. *Science China Information Sciences* 43 (12): 1598–1611.

Wang, F.-Y. (2015a). CC 5.0: intelligent command and control systems in the parallel age. *Journal of Command and Control* 1 (1): 107–120.

Wang, F.-Y. (2015b). Software-de?ned systems and knowledge automation: a parallel paradigm shift from Newton to Merton. *Acta Automatica Sinica* 41 (1): 1–8.

Wang, F.-Y. (2015c). Intelligence 5.0: parallel intelligence in parallel age. *Journal of the China Society for Scientific and Technical Information* 34 (6): 563–574.

Wang, F.-Y. (2016). Control 5.0: from Newton to Merton in popper's cyber-social-physical spaces. *IEEE/CAA Journal of Automatica Sinica* 3 (3): 233–234.

Wang, F.-Y. (2020a). Parallel philosophy and intelligent science: from Leibniz's Monad to Blockchain's DAO. *Pattern Recognition and Artificial Intelligence* 33 (12): 1055–1065.

Wang, F.-Y. (2020b). Parallel healthcare: robotic medical and health process automation for secured and smart social healthcares. *IEEE Transactions on Computational Social Systems* 7 (3): 581–586.

Wang, F.-Y. (2020c). Parallel economics: a new supply–demand philosophy via parallel organizations and parallel management. *IEEE Transactions on Computational Social Systems* 7 (4): 840–848.

Wang, F.-Y. (2021a). Parallel medicine: from warmness of medicare to medicine of smartness. *Chinese Journal of Intelligent Science and Technology* 3 (1): 1–9.

Wang, F.-Y. (2021b). Digital doctors and parallel healthcare: from medical knowledge automation to intelligent metasystems medicine. *Medical Journal of Peking Union Medical College Hospital* 12 (6): 829–833.

Wang, F.-Y. (2021c). Parallel philosophy and intelligent technology: dual equations and testing systems for parallel industries and smart societies. *Chinese Journal of Intelligent Science and Technology* 3 (3): 245–255.

Wang, F.-Y. (2021d). Parallel philosophy: origin and goal of intelligent industries and smart economics. *Bulletin of Chinese Academy of Sciences* 36 (3): 308–311.

Wang, F.-Y. (2022). Parallel intelligence in metaverses: Welcome to Hanoi! *IEEE Intelligent Systems* 37 (1): 16–20.

Wang, F.-Y. and Tang, S. (2004). Concepts and frameworks of artificial transportation systems. *Complex Systems and Complexity Science* 1 (2): 52–59.

Wang, F.-Y. and Wang, Y. (2020). Parallel ecology for intelligent and smart cyber–physical–social systems. *IEEE Transactions on Computational Social Systems* 7 (6): 1318–1323.

Wang, F.-Y. and Wong, P.K. (2013). Intelligent systems and technology for integrative and predictive medicine: an ACP approach. *ACM Transactions on Intelligent Systems and Technology (TIST)* 4 (2): 1–6.

Wang, F.-Y., Carley, K.M., Zeng, D., and Mao, W. (2007a). Social computing: from social informatics to social intelligence. *IEEE Intelligent Systems* 22 (2): 79–83.

Wang, T., Zhang, Y., Yu, H., and Wang, F.-Y. (2007b). *Advanced Manufacturing Technology in China: A Roadmap to 2050*. Springer.

Wang, F.-Y., Zeng, D., Hendler, J.A. et al. (2010). A study of the human flesh search engine: crowd-powered expansion of online knowledge. *IEEE Computer Architecture Letters* 43 (08): 45–53.

Wang, F.-Y., Zeng, D., Zhang, Q. et al. (2014a). The Chinese "Human Flesh" Web: the first decade and beyond. *Chinese Science Bulletin* 59 (26): 3352–3361.

Wang, T., Liu, Z., Cui, K. et al. (2014b). On mobilizing processes of cyber movement organizations. *The World Congress of the International Federation of Automatic Control (IFAC), IFAC PAPERSONLINE*, Volume 47, 9853–9857.

Wang, F.-Y., Wang, X., Li, L., and Li, L. (2016a). Steps toward parallel intelligence. *IEEE/CAA Journal of Automatica Sinica* 003 (004): 345–348.

Wang, F.-Y., Zhang, J.J., Zheng, X. et al. (2016b). Where does AlphaGo go: from church-turing thesis to AlphaGo thesis and beyond. *Acta Automatica Sinica* 3 (2): 113–120.

Wang, K., Gou, C., and Wang, F.-Y. (2016c). Parallel vision: an ACP-based approach to intelligent vision computing. *ACTA Automatica Sinica* 42 (10): 1490–1500.

Wang, X., Li, L., Yuan, Y. et al. (2016d). ACP-based social computing and parallel intelligence: societies 5.0 and beyond. *CAAI Transactions on Intelligence Technology* 1 (4): 377–393.

Wang, X., Zheng, X., Zhang, Q. et al. (2016e). Crowdsourcing in ITS: the state of the work and the networking. *IEEE Transactions on Intelligent Transportation Systems* 17 (6): 1596–1605.

Wang, X., Zheng, X., Zhang, X. et al. (2016f). Analysis of cyber interactive behaviors using artificial community and computational experiments. *IEEE Transactions on Systems, Man, and Cybernetics: Systems* 47 (6): 995–1006.

Wang, F.-Y., Li, C., Wang, J. et al. (2017a). Parallel gout: an ACP-based system framework for gout diagnosis and treatment. *Pattern Recognition & Artificial Intelligence* 30 (12): 1057–1068.

Wang, F.-Y., Zheng, N.-N., Cao, D. et al. (2017b). Parallel driving in CPSS: a unified approach for transport automation and vehicle intelligence. *IEEE/CAA Journal of Automatica Sinica* 4 (4): 577–587.

Wang, K., Lu, Y., Wang, Y. et al. (2017c). Parallel imaging: a new theoretical framework for image generation. *Pattern Recognition and Artificial Intelligence* 30 (07): 577–587.

Wang, F.-Y., Gao, Y., Shang, X., and Zhang, J. (2018a). Parallel manufacturing and industries 5.0: from virtual manufacturing to intelligent manufacturing. *Science & Technology Review* 36 (21): 10–22.

Wang, F.-Y., Yuan, Y., Rong, C., and Zhang, J.J. (2018b). Parallel blockchain: an architecture for CPSS-based smart societies. *IEEE Transactions on Computational Social Systems* 5 (2): 303–310.

Wang, F.-Y., Yuan, Y., Wang, X., and Qin, R. (2018c). Societies 5.0: a new paradigm for computational social systems research. *IEEE Transactions on Computational Social Systems* 5 (1): 2–8.

Wang, F.-Y., Zhang, J.J., and Wang, X. (2018d). Parallel intelligence: toward lifelong and eternal developmental AI and learning in cyber-physical-social spaces. *Frontiers of Computer Science* 12 (003): 401–405.

Wang, F.-Y., Zhang, M., Meng, X. et al. (2018e). Parallel eyes: an ACP-based smart ophthalmic diagnosis and treatment. *Pattern Recognition and Artificial Intelligence* 31 (6): 495–504.

Wang, S., Wang, J., Wang, X. et al. (2018f). Blockchain-powered parallel healthcare systems based on the ACP approach. *IEEE Transactions on Computational Social Systems* 5 (4): 942–950.

Wang, F.-Y., Gou, C., Wang, J. et al. (2019). Parallel skin: a vision-based dermatological analysis framework. *Journal of Pattern Recognition and Artificial Intelligence* 7: 577–588.

Wang, F.-Y., Qin, R., Li, J. et al. (2020). Parallel societies: a computing perspective of social digital twins and virtual–real interactions. *IEEE Transactions on Computational Social Systems* 7 (1): 2–7.

Wang, F.-Y., Jin, Z.-Y., Guo, C., and Shen, T. (2021a). ACP-based parallel medical imaging for intelligent analytics and applications. *Chinese Journal of Radiology* 55 (3): 309–315.

Wang, F.-Y., Meng, X., Du, S., and Geng, Z. (2021b). Parallel light field: the framework and processes. *Chinese Journal of Intelligent Science and Technology* 3 (01): 110–122.

Wang, F.-Y., Zheng, N., Li, L. et al. (2021c). China's 12-year quest of autonomous vehicular intelligence: the intelligent vehicles future challenge program. *IEEE Intelligent Transportation Systems Magazine* 13 (2): 6–19.

Wang, S., Xiao, P., Chai, H. et al. (2021d). Research on construction of supply chain financial platform based on blockchain technology. *2021 IEEE 1st International Conference on Digital Twins and Parallel Intelligence (DTPI)*, 42–45. IEEE.

Wang, X., Yang, J., Han, J. et al. (2022). Metaverses and DeMetaverses: from digital twins in CPS to parallel intelligence in CPSS. *IEEE Intelligent Systems* 37 (4): 97–102.

Wei, Q., Wang, L., Lu, J. et al. (2022). Discrete-time self-learning parallel control. *IEEE Transactions on Systems, Man, and Cybernetics: Systems* 52 (1): 192–204. https://doi.org/10.1109/TSMC.2020.2995646.

Wen, D., Yuan, Y., and Li, X.-R. (2012). Artificial societies, computational experiments, and parallel systems: an investigation on a computational theory for complex socioeconomic systems. *IEEE Transactions on Services Computing* 6 (2): 177–185.

Xing, Y., Liu, Z., Liu, T. et al. (2018). Parallel tanks: defining a digital quadruplet for smart tank systems. *Journal of Command and Control* 4 (2): 111–120.

Xiong, G., Zhu, F., Liu, X. et al. (2015). Cyber-physical-social system in intelligent transportation. *IEEE/CAA Journal of Automatica Sinica* 2 (3): 320–333.

Yang, D., Wang, K., Chen, D. et al. (2018). Parallel carrier fleets: from digital architectures to smart formations. *Journal of Command and Control* 4 (2): 101–110.

Yang, L., Chen, S., Wang, X. et al. (2019). Digital twins and parallel systems: state of the art, comparisons and prospect. *IEEE/CAA Journal of Automatica Sinica* 45 (11): 2001–2031.

Yuan, Y. and Wang, F.-Y. (2017). Parallel blockchain: concept, methods and issues. *Acta Automatica Sinica* 43 (10): 10.

Zhang, J.J., Gao, D.W., Zhang, Y. et al. (2017). Social energy: mining energy from the society. *IEEE/CAA Journal of Automatica Sinica* 4 (3): 466–482.

Zhang, J.J., Wang, F.-Y., Wang, X. et al. (2018). Cyber-physical-social systems: the state of the art and perspectives. *IEEE Transactions on Computational Social Systems* 5 (3): 829–840.

Zhang, M., Chen, L., Wang, F.-Y. et al. (2019). Parallel gastrointestine: an ACP-based approach for intelligent operations. *Pattern Recognition and Artificial Intelligence* 32 (12): 1061–1071.

Zhang, Q., Wang, F.-Y., Zeng, D., and Wang, T. (2012). Understanding crowd-powered search groups: a social network perspective. *PLoS ONE* 7 (6): e39749.

Zhang, Q., Zeng, D.D., Wang, F.-Y. et al. (2016). Brokers or bridges? Exploring structural holes in a crowdsourcing system. *Computer* 49 (6): 56–64.

Part II

Transportation

8

Regularities of Human Operator Behavior and Its Modeling

Aleksandr V. Efremov

Department of Aeronautical Engineering, Moscow Aviation Institute, National Research University, Moscow, Russian Federation

8.1 Introduction

The design of man-controlled machines requires an understanding of how a person behaves in well-defined tasks required for operation and control. The correct way of solving this problem is by developing a model of the human operator's behavior. The development of such a model is an extremely difficult task for a number of reasons. One of them is the fact that there are different types of pilot behavior demonstrated during their interaction with the machine. There is decision-making, in which the human operator functions as a supervisor; monitoring, where the human operator performs information processing; and manual control, in which the behavior of the human operator is close to the dynamics of a controller. Manual control is typical of many piloting tasks (approach and landing, aim-to-aim tracking, refueling, etc.) requiring high tracking accuracy and appropriate flight safety. Unsatisfactory harmonization between pilot actions and other elements of the pilot aircraft system can lead to the deterioration of flight safety and even to flight accidents. According to statistics (Britain and Authority 1997; Anonymous, International air Transport Association 2019; Anonymous, International air Transport Association 2021), pilot error accounts for 55–65% of all flight accidents. Depending on the type of occurrence (hard landing, loss of control, tail strike, etc.), manual control as a primary factor causing the accident due to pilot error has ranged from 30% to almost 90% over the years. Manual control is also characteristic of spaceship docking, car driving, and berthing of watercraft. This chapter is dedicated to human operator behavior in manual control tasks.

Arnold Tustin was the pioneer in human operator dynamic measurements during World War II. He extended the required feedback control theory framework by introducing the concept of "describing function" and "remnant" measures and quasi-linear systems in general, then applying these concepts to actual human operators (Tustin 1944).

After the war, interest in the study of human operator–machine systems grew, especially in aviation, due to accidents involving supersonic aircraft. A number of notable studies into human operator dynamics for different input spectrum characteristics and plant dynamics were performed by various American organizations.

The publication of "Dynamic Response of Human Operators" by McRuer and Krendel (1957) marked the end of the pioneering era in the study of human dynamic characteristics. This report codified and correlated the available human response data, developed models compatible with it, and prescribed preferred forms for the human operator dynamics. The follow-up studies allowed

Cyber–Physical–Human Systems: Fundamentals and Applications, First Edition.
Edited by Anuradha M. Annaswamy, Pramod P. Khargonekar, Françoise Lamnabhi-Lagarrigue, and Sarah K. Spurgeon.
© 2023 The Institute of Electrical and Electronics Engineers, Inc. Published 2023 by John Wiley & Sons, Inc.

to specify the ideal controlled element dynamics and to develop the so-called "McRuer's pilot" describing function models with different levels of accuracy, widely used in applied research. Intensive work on the creation of fixed- and motion-base simulators allowed to obtain fundamental results on the effects of motion cues on the pilot dynamics, mathematical models of the vestibular sensors, and to develop the technique for the measurement of human operator dynamic characteristics to study human behavior in multiloop and multimodal situations. During the same period (the late 1960s and early 1970s), a new approach to pilot modeling – the optimal pilot model based on modern optimal control theory – was introduced by Kleiman et al. (1970). At the end of the century, the structural approach to pilot behavior modeling based on taking into account an additional inner loop used by the pilot was suggested by Hess (1979). Several modifications to this model have been made in the last few decades (Hess 1990, 1999; Efremov et al. 2010). Besides the compensatory tracking task, the pursuit and preview tasks, typical of manual control of an aircraft or automobile have been studied around the globe. More recently, techniques based on fuzzy logic and neural networks (Efremov et al. 2010) have been gaining increasing attention in problems of human operator modeling. The main motivation behind this chapter was to give a concise and critical review of the research results obtained in different countries in the following areas:

- Regularities of human operator behavior in different manual control tasks;
- Mathematical modeling of human response characteristics and its applications in solving engineering problems.

The majority of publications dedicated to these areas is related specifically to the pilot-aircraft system. Therefore, the main focus hereinafter is on the results of research into human behavior in this system. Additionally, some results of studies of human operator-lunar rover and astronaut–spacecraft systems in manual control tasks are considered.

Knowledge of the regularities of pilot behavior and its mathematical models are widely used in the development of flight control system algorithms, whereby the functions and laws of adaptation generated by the human central nervous system (CNS) are transferred to the controller. It significantly simplifies pilot behavior, improves all pilot-aircraft system characteristics, including accuracy, and decreases pilot workload. In this case, the "human-in-the-loop control system" can be conditionally related to the "human-in-the-controller" category according to the terminology given in Chapter 1. Besides the adaptation carried out by the CNS in the closed-loop system, human behavior is determined by the human sensory system perceiving information and the neuromuscular system transmitting the control signal. This knowledge has allowed to create displays (indicators, flight directors, and primary flight display) and inceptors (side stick, yoke, and central stick) and to determine their best characteristics.

A clear understanding of the sensory and neuromuscular system regularities currently allows to work on a novel type of display (the so-called "predictive display") and to implement active inceptors, as well as controllers, based on the use of new approaches to flight control system design.

In this chapter, in addition to the results of studies on human operator behavior and its modeling, a brief overview of their effective use for the solution of practical tasks is given as well.

8.2 The Key Variables in Man–Machine Systems

Control theory is widely used in the design of different technical devices. Many of its aspects are dedicated to the development of controllers acting in the closed-loop system. The human operator (pilot, astronaut, and driver) is also an integral part of the man–machine system (Figure 8.1),

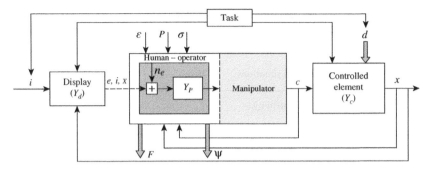

Figure 8.1 Man–machine system and its variables.

engaged as a controller trying to satisfy the overall system purposes. However, this system has a number of peculiarities determined by four kinds of variables:

- task variables;
- environmental variables, ε;
- operator-centered variables, σ;
- procedural variables, P.

Environmental variables include such factors as vibration, temperature, acceleration;

Operator-centered variables include training, motivation, physical condition (e.g. fatigue);

Procedural variables include such aspects as experimental procedure, training schedule, the order of presentation of trials, and so forth.

Because of the limited size of the chapter, only the first group – task variables, which influence the man–machine system characteristics significantly – is considered below. Several of the variables related to this group do not pertain to automatic technical systems. They are the display showing the information perceived by the human operator and the manipulator (inceptor) through which the human operator transmits command signals. A peculiarity of the man–machine system is the dependence of these variables on the piloting task (landing, including the glide tracking and flare, refueling, terrain following, etc.). Each of these tasks is characterized by a specific set of task variables, namely, the display, the controlled element dynamics, and the input signal, for which the human operator and the man–machine system characteristics will be different. This demonstrates the ability of the human operator to adapt their behavior. All manual control tasks can be subdivided into time-limited subtasks. In the case where the dynamic characteristics of the aircraft dynamics remain constant in the process of performing such tasks or change insignificantly, which is true for most precise manual control tasks, the mathematical model of the aircraft dynamics is presented as a linear differential equation system with constant coefficients or transfer function. It determines the vehicle response to the human operator command signals. A series of transfer functions of the controlled element dynamics $Y_c(s)$ in several manual control tasks is presented in Table 8.1.

As will be shown later, these different controlled element dynamics impose different requirements on the human operator action characteristics. In the case of a drastic change in plant characteristics that could happen because of a control subsystem failure, the human operator actions are being restructured. The entire man–machine system thereby becomes significantly nonstationary. Hereinafter, this special case will not be considered in the mathematical description of the vehicle dynamics. Instead, it will be assumed that the characteristics of the vehicle dynamic are constant.

Table 8.1 Influence of the piloting task on the controlled element dynamics.

Transfer function of the controlled element $Y_c(s) = \left\{ \frac{x}{c} \right\}$	Related vehicle control situations	Notes
$\dfrac{K}{s(T_R \cdot s + a)}$	Aircraft roll angle control by ailerons Automobile heading angle response to wheel steer angle deflection	T_R – time constant.
$\dfrac{k(s - Z_w)e^{-s\tau}}{s\left(s^2 + 2\xi_{sp}\omega_{sp}s + \omega_{sp}^2\right)}$	Aircraft pitch angle, θ, control by elevator	$e^{-s\tau}$ represents the time delay approximating FCS dynamics Z = normal force; w = vertical speed perturbation; m = mass; ξ_{sp} = short-period damping ratio; ω_{sp} = short-period frequency. $Z_w = \dfrac{dZ}{dw}\dfrac{1}{m}$
$\dfrac{k\left(s^2 - Z_W - Z_W \cdot \frac{V}{L}\right)}{s^2\left(s^2 + 2\xi_{sp}\omega_{sp}s + \omega_{sp}^2\right)}$	The sight angle, $\varepsilon \cong \theta + \frac{H}{L}$, (case then $L \gg H$), control by elevator	L = distance to the target; V = airspeed; H = aircraft height relative to the target
$\dfrac{k}{s^2}$	Space vehicle attitude control by control jets	

k is the aircraft dynamics gain coefficient.

Manipulators, designed for transferring the pilot's control signals, are divided into

- **displacement sensing control** (DSC) type (Klyde and McRuer 2009), whose output signal is proportional to the inceptor displacement (for the aircraft $c = \delta_e$ in pitch control and $c = \delta_a$ in roll control);
- **force sensing control** (FSC) type (Klyde and McRuer 2009), whose output signal is proportional to the force $c = F$ applied by the pilot.

Usually, the manipulator used in nonmaneuverable aircraft is a yoke, while maneuverable aircraft are outfitted with a center stick. In recent years, small-sized side sticks have been used widely for both types of aircraft.

All of inceptors listed above differ in size and characteristics (force and displacement gradients, preload, etc.).

Usually, displays are specially made technical devices. The most common of them are visual displays: primary flight display, instruments, flight directors, and other possible variants of information display systems. In recent past, considerable attention has been paid to developing the so-called "tactile" and "auditory displays," transferring encoded information through tactile and auditory human sensors. The moving-base system of modern flight simulators can also be considered as a display, affecting the vestibular sensors and transmitting the filtered information about linear and angular acceleration. The command information necessary for task performance can be obtained by the pilot not only through specially made technical devices but also in a natural way as well. For example, in visual flight, the stimulus $e(t)$ is generated by the pilot by matching the

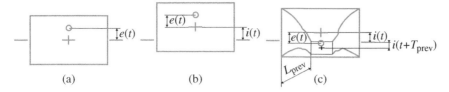

Figure 8.2 Types of displays. (a) Compensatory, (b) pursuit, (c) preview.

marks on the aircraft's windshield or fuselage with external visual cues (elements of the landscape, cloud, and other aircraft outlines). The resulting marks and visual cues system can be considered as a visual flight display.

Depending on the composition of command signal, all displays are divided into compensatory (Figure 8.2a), pursuit (Figure 8.2b), and preview displays (Figure 8.2c). The last one is a typical display for the driver-automobile system.

If the effects of nonlinear properties of the compensatory display are not manifested in the sensor perception frequency range, then the mathematical model of the display in the simplest case is a gain $Y_d = K_d$. This coefficient correlates the state variable to the scale of a device (in instrument flight) or to its projection on the windshield (in visual flight). In the general case, $Y_d(s)$ is a transfer function. The output signal, usually, is the sum of the aircraft state variables passed through corresponding filters. In this case, $Y_d(s)$ could be ascribed to the vehicle dynamics. Then, the effective dynamics of the plant controlled by the pilot is represented by the product of the transfer functions, $Y_d Y_c$. If $Y_d(s) = 1$, the compensatory display transmits the error signal $e(t) = i(t) - x(t)$.

Compensatory tracking corresponds to a certain number of manual control tasks but not all of them. For example, in the tracking task of a target flying against the background of clouds or the terrain, and in the refueling task, aside from the error signal, the pilot also perceives the input signal $i(t)$. In these cases, the pilot-aircraft system corresponds to the so-called "pursuit system." In the case of flying through a mountain gorge or while driving a car, the pilot (driver) sees the target trajectory $i(t + T_{prev})$, where T_{prev} is the preview time.

The properties of an input signal $i(t)$ introduced into the man–machine system or a disturbance $d(t)$ also significantly affect the human operator control actions characteristics. In manual control, the input signal $i(t)$ (or $d(t)$) is close to a random stationary process and can be described by the spectral densities $S_{ii}(\omega)$ or $S_{dd}(\omega)$.

The equations of these spectral densities and their parameters are different for different manual control tasks. For example, the spectral density $S_{ii}(\omega)$ corresponding to the input signal $i(t)$, used in Neal and Smith (1971) for aircraft flying qualities evaluation in a pitch tracking task, has the following form (Efremov et al. 1996): $S_{ii}(\omega) = \frac{k^2}{(\omega^2 + \omega_i^2)}$, where $\omega_i = 0.5$ (1/s), $\sigma_i^2 = 4$ cm^2.

Analysis of a path tracking task (in particular, terrain following) shows that the spectral density of $i(t)$ corresponding to this task has the same form, where the bandwidth frequency ω_i is close to 0, 2 (1/s).

8.3 Human Responses

There are three groups of human responses: control, psychophysiological, and physiological responses.

Control responses (actions) are the most important aspect of human behavior in a control task. Just as in automatic regulators, they characterize the transformation of the input signal into an output signal and are described by the describing function $Y_p(j\omega)$ and remnant $n_e(t)$ that accounts

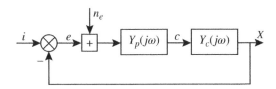

Figure 8.3 Single-loop compensatory man–machine system.

for all nonlinear and nonstationary effects of human behavior. Several studies have demonstrated that $n_e(t)$ is a random stationary process characterized by the spectral density $S_{n_e n_e}(\omega)$. Usually, the remnant is referred to as the error signal (see Figure 8.3). When $Y_p(j\omega)$ is a matrix, and $n_e(j\omega)$ is a vector, it is called a multiaxis system. They are several types of multiaxis systems:

- multiloop system, in which the human operator closes several loops (for example altitude and pitch angle in aircraft glide slope tracking task or path and heading angle in automobile lateral steering control);
- multichannel system, in which the human operator uses several manipulators for control (for example throttle for speed control and inceptor for pitch control);
- multimodal system, in which the pilot perceives signals of different modalities (visual, vestibular, tactile, etc.).

In the simplest case of a single-loop system (Figure 8.3), the matrix Y_p and vector $S_{n_e n_e}$ both consist of one element each.

The measurements of these characteristics can be realized using different techniques. The more commonly used technique is the Fourier coefficients method. For this method, the input signal consists of the sum of harmonics with uncorrelated frequencies $\omega_k = k\omega_0$, where $\omega_0 = \frac{2\pi}{T}$, T is the trial duration. The technique for the selection of frequencies and amplitudes of the input signal harmonics providing agreement with the random spectrum is considered in Efremov et al. (1996). In the case of a multiaxis system several such input signals with sets of uncorrelated frequencies are needed (Efremov et al. 1992).

Psychophysiological responses are determined by the human operator workload required in performing the manual control task. Various rating scales are used as indicators of psychophysiological responses. For the evaluation of aircraft and spacecraft flying qualities or automobile steering characteristics, the Cooper–Harper Scale (Cooper and Harper 1969) is widely used (Figure 8.4).

This scale contains questions, the answers to which help the human operator evaluate the handling (flying) qualities. Moreover, the demand on the human operator in a given task is presented in two metrics: task performance (desired and adequate) and different human operator compensation levels.

In addition to the Cooper–Harper scale, the *pilot induced oscillation* (PIO) scale (Figure 8.5) is used in ground and in-flight studies of PIO tendencies resulting from the pilot–aircraft interaction.

Analysis of in-flight investigations (Neal and Smith 1971; Smith 1978; Bjorkman et al. 1986) in which 117 dynamic configurations were assigned, allowed to determine the relationship between the Cooper–Harper and PIO ratings:

$$PIOR = 0.5PR + 0.25 \qquad (8.1)$$

According Eq. (8.1), a rating of PR $= 6.5$ determining the boundary between levels 2 and 3 of handling (flying) qualities corresponding to a PIO rating of 3.5.

The latter might be considered as the metric, dividing the configurations that are PIO-prone (PIOR > 3.5) and non-PIO-prone (PIOR ≤ 3.5).

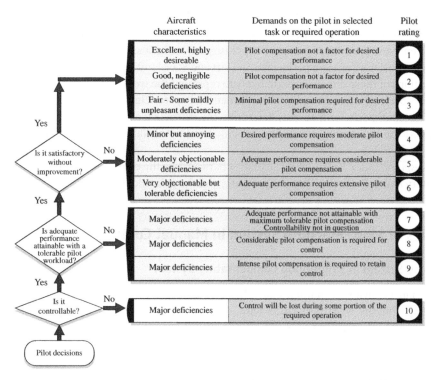

Figure 8.4 Cooper–Harper rating scale. Source: Cooper and Harper (1969)/NASA/Public Domain.

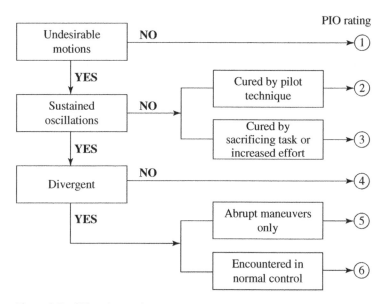

Figure 8.5 PIO rating scale.

In recent years, the so-called "NASA Task Load Index (NASA TLX)" has become the widely used assessment tool for evaluation of human operator workload. This multidimensional scale consists of six subscales that represent somewhat independent clusters of variables: mental, physical, and temporal demands, frustration, effort, and performance (Hart and Staveland 1988). The assumption is that some combinations of these dimensions are likely to represent the workload experienced by most human operators performing most tasks. These dimensions were selected after an extensive analysis of primary factors that do (and do not) determine the subjective experience of workload for different people performing a variety of activities ranging from simple laboratory tasks to flying an aircraft.

The *physiological response* index F (Figure 8.1) usually characterizes the cardiovascular parameters, e.g. the respiratory and pulse rates, body temperature, and the galvanic skin response.

8.4 Regularities of Man–Machine System in Manual Control

8.4.1 Man–Machine System in Single-loop Compensatory System

Most studies on human behavior in manual control were conducted for the single-loop compensatory system. Experiments were performed with simplified controlled element dynamics $\left(Y_c = K; \frac{K}{s}; \frac{K}{(Ts+1)}; \frac{K}{s^2}\right)$ and with rectangular spectrum of input signals consisting of a set of orthogonal harmonics characterized by the bandwidth $\omega_i \geq 1$ (1/s) (McRuer et al. 1965). The obtained experience provided a trim foundation for subsequent extension of the research field.

Copious experimental research into the characteristics of the man–machine system revealed a remarkable property of human behavior – their *adaptation* to task variables. This peculiarity manifests itself both in the characteristics of the human operator describing function and the remnant spectral density.

The adaptation is exposed brightly in the crossover frequency region and is the most important because the frequency response of the open-loop system in it determines the stability and quality of the closed-loop system. It is the region where the amplitude response slope of the open-loop system $|Y_{OL}|$ (Figure 8.6) is relatively constant and equals approximately -20 dB dec^{-1}. This is achieved in the varying dynamics of the controlled element by changing the pilot action properties. Therefore, an increase in the pole order at the origin of the s-plane of the controlled element dynamics causes

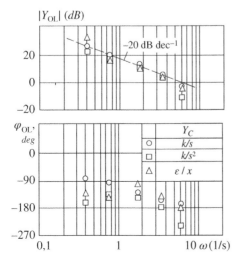

Figure 8.6 Influence of $Y_C(j\omega)$ on $Y_{OL}(j\omega)$.

Figure 8.7 Influence of the controlled element dynamics on $Y_{CL}(j\omega)$.

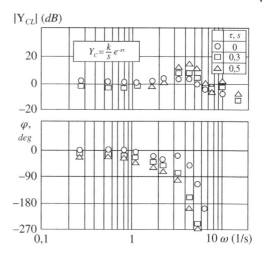

an increase in pilot lead compensation (Figure 8.7). Maintaining a $-20\,\mathrm{dB\,dec^{-1}}$ slope of the $|Y_{OL}|$ is a typical property of the pilot-aircraft system not only for simplified controlled element dynamics but also for real control tasks with more complicated dynamics too.

Analysis of experimental results allows to conclude that an increase in the controlled element's transfer function pole order at the origin of the s-plane causes a decrease in the system crossover frequency ω_c. A similar effect is created by a decrease in the input signal spectral bandwidth if it has a rectangular shape or a sharp decrease in the power spectrum (when $m > 2$; m is the exponent in the expression $\left(S_{ii} = \frac{K_i}{(\omega^2 + \omega_i^2)^m}\right)$. When $m \leq 2$, the decrease in ω_i causes the reverse effect (Efremov et al. 1992). At significant ω_i, values close to ω_c, increasing ω_i in case of rectangular shape of input spectrum causes a decrease in the ω_c ("crossover frequency regress") (McRuer et al. 1965).

While controlling the controlled element whose transfer function includes the $e^{-s\tau}$ element, (τ is the time delay approximating the control loop delays), there is a significant decrease in crossover frequency. For example, for $Y_c = \left(\frac{k}{s}\right)e^{-s\tau}$ transfer function, an increase in τ from 0 to 0.3 seconds causes a decrease in ω_c from 4 to 3 (1/s).

The presence of the time delay in plant dynamics causes a decrease in the amplitude $|Y_{OL}|$ slope up to -14–$16\,\mathrm{dB\,dec^{-1}}$ in the crossover frequency region due to the human operator's actions features trying to compensate the $e^{-s\tau}$ element by including a significant lead. It decreases the amplitude margin δL of $|Y_{OL}|$ and as a consequence, causes an increase in the resonant peak of the closed loop system $Y_{CL}(j\omega)$ (Figure 8.7). Wherein in the middle frequency region (2–5 (1/s)), the amplitude response of the human operator describing function $Y_p(j\omega)$ is characterized by an increase in its slope up to $40\,\mathrm{dB\,dec^{-1}}$ (Efremov et al. 1996, 1992).

The bandwidth of the closed-loop system ω_{CL}, which can be determined as the frequency at which the frequency phase characteristic of the closed-loop system is equal to $-90°$, depends on the dynamics of the control element and the input spectrum. In particular, for the input spectrum $S_{ii} = \frac{K^2}{(\omega^2 + 2^2)^2}$, ω_{CL} obtained at $Y_c = \frac{K}{s}$ is 1.4–1.5 times higher in comparison with ω_{CL} obtained at $Y_c = \frac{K}{s^2}$ or $\frac{K}{s}e^{-0.3s}$.

In the low-frequency region, the characteristics of the human operator's actions considerably depend on the bandwidth ω_i of the input signal spectrum. Studies have shown that the concentration of the power of the input signal in the low-frequency region ($\omega_i < 1.5$ (1/s)) causes a significant increase in amplitude frequency response of the open-loop man–machine system $|Y_{OL}|$ (Efremov et al. 1992), which is provided by an additional low-frequency correction generated by the human

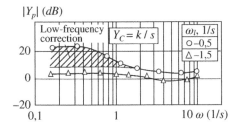

Figure 8.8 Influence of the input signal spectrum bandwidth.

operator (Figure 8.8). The results of numerous experiments show that this regularity manifests itself fully in highly trained operators performing the tracking task with maximum attention. The possibilities of introducing low-frequency corrections are determined, in particular, by the type of inceptor and is more noticeable for the side stick.

In the high-frequency region, the resonant peak of the frequency response of the human operator's actions has specific peculiarities. Its magnitude and the frequency at which it takes place depend on the dynamics of the controlled element, the type of control stick, and the bandwidth of the input spectrum ω_i. Almost all known experiments have in common that the control of Y_c whose transfer function has a high pole order at the origin of the s-plane or a significant phase delay is accompanied by a decrease in the frequency of this resonant peak (Figure 8.9) (Efremov 2017). An increase in the bandwidth of the input signal spectrum and the use of an inceptor with low stiffness causes an increase in the resonant peak, as well as its frequency (Efremov et al. 1992).

Experiments performed with the FSC type of inceptor demonstrated a considerably lower human operator phase delay at high frequencies in comparison with DSC type of inceptor. This leads to an increase in the bandwidth of the closed-loop system ω_{CL} and an improvement of the pilot rating by 1–1.5 units.

The characteristics of the remnant are mainly influenced by the dynamics of the control element. If, for its correction, the human operator is required to introduce lead compensation, such as in the case when the transfer function of the control element is $Y_C = \frac{k}{s^2}$, then the level of normalized spectral density $\overline{S}_{n_e n_e} = \frac{S_{n_e n_e}}{\sigma_e^2}$ becomes noticeably higher than in the case when $Y_C = \frac{k}{s}$ and the shift of the point of intersection of the asymptotes of approximating dependence $\overline{S}_{n_e n_e}$ in the lower-frequency region (here, σ_e^2 is the variance of error).

All the properties considered above are for linear-controlled element dynamics. The nonlinearities in the control system's (for example, actuator's rate limiting) influence on all the characteristics of the man–machine system cause an increase in the remnant power density, a nonsmooth character to all measured characteristics including frequency response of the controlled element Y_c (Figure 8.10), an increase in the spectral density $S_{e_n e_n}(\omega)$ (e_n is the error signal caused by the remnant) (Figure 8.11), and a considerable increase in the variance of error σ_e^2 (Efremov 2017).

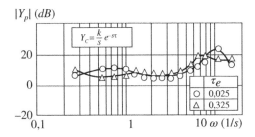

Figure 8.9 Influence of the controlled element dynamics delay on $|Y_p(j\omega)|$.

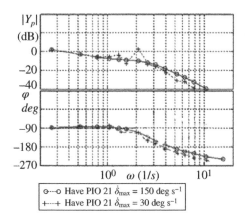

Figure 8.10 Frequency response characteristics of $Y_c(j\omega)$.

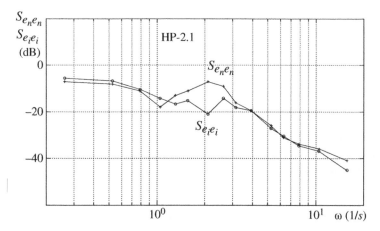

Figure 8.11 Components of spectral density $S_{ee}(\omega)$.

The results discussed above were obtained during experiments in which the pilot tries to keep an error signal at zero for the entire duration of a trial. If the goal of the pilot changes to keeping the error signal within a specified interval "d", then the characteristics of the man–machine system will change as well. The expansion of this interval causes a decrease in the human operator lead compensation crossover frequency, bandwidth of the closed-loop system and its resonant peak. The experiments demonstrated that in the case where the human operator can keep the error signal within the interval d during 95% of the experiment time, the interval d and the root mean square of error σ_e^2 are related by the following formula (Efremov 2017):

$$d = 4\sigma_e \tag{8.2}$$

The deterioration of flying or steering qualities causes not only a higher human operator compensation but also the deterioration of the other man–machine system characteristics: variance of error and human operator control actions. All these leads to an increase in the subjective Cooper–Harper rating, PR. A specific feature of the PR rating is its variability in the evaluation of the same dynamic configuration by the same or different human operators. The random nature of pilot ratings was

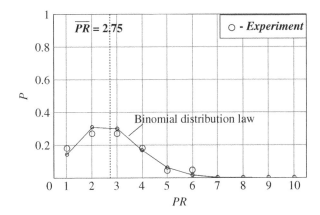

Figure 8.12 Distribution of pilot rating (Configuration HP-4.1).

studied in different papers (Efremov et al. 1992; Efremov 2017; Hodgkinson et al. 1992). It was shown in Efremov et al. (1992) that PR is a random integer from 1 to 10, distributed following the binomial law. In this case, the probability of a given PR can be calculated as given in following equation:

$$P(PR) = C_9^{PR-1} p^{PR-1}(1-p)^{10-PR} \tag{8.3}$$

where $p = \frac{\overline{PR}-1}{9}$; $\tilde{N}_9^{PR-1} = \frac{9!}{(PR-1)!(10-PR)!}$; $\overline{PR}$ is the mean of PR.

The agreement between Eq. (8.3) and the results of experimental research for the same $\overline{PR}$ were checked in a special set of experiments for different configurations. For one of them, the result is shown in Figure 8.12. According to the Cooper–Harper scale, the rating PR = 10 corresponds to the situation of loss of control. The probability of such rating for different mean ratings $\overline{PR}$ can be evaluated with the help of the dependence (8.2) shown in Figure 8.13. It is seen that in case of a $\overline{PR}$ that is close to 3.5 (the boundary between levels 1 and 2 of flying qualities), the probability of loss of control is less than 10^{-5} per flight. This value corresponds to the unreliability allowance (p) for the entire flight control system of airplanes of class I, II, and IV regardless of being manual or automatic.

The 14 CFR part 25 (Federal Aviation Administration [FAA] 2009) stipulates a stricter requirement of an extremely low failure probability of $<10^{-9}$ per flight hour.

If this requirement applies to a pilot, then corresponding flying qualities $\overline{PR}$ must be <2.

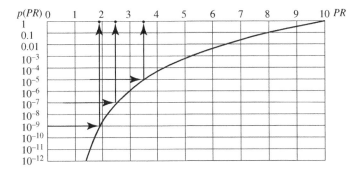

Figure 8.13 Probability of accidents due to incorrect pilot actions.

One of the Cooper–Harper scale metrics is the task performance. There are two levels of the performance: desired and adequate. It is obvious that in a precision tracking task the performance is determined by piloting accuracy. Supposing that a pilot rating is the pilot's response on the stimulus – the flying qualities parameter I, and applying the fundamental psychophysiological Weber–Fechner law (Efremov et al. 1996; Efremov 2017) to the relationship between PR and I, it is possible to get the following equation:

$$PR = A \ln I + B$$

Supposing that the pilot rating PR $= 4$ corresponds to the desired value of parameter I_{des} and PR $= 1$ (minimal rating) corresponds to I_{opt} and introducing the assumption about the feasibility of using the range of error d, which will not be exceeded by the current error $e(t)$, as the flying qualities parameter, it is possible to get the following equation (Efremov 2017):

$$PR = 1 + \frac{3}{\ln \dfrac{d_{des}}{d_{opt}}} \ln \frac{d}{d_{opt}}$$

Here, d_{des} and d_{opt} are the values of the range d determined in experiments in which the human operator performed the tracking task with a dynamic configuration corresponding to the PR $= 4$ and configuration $Y_{c_{opt}}(j\omega)$. The latter is the optimal dynamic configuration providing the minimal variance of error $\sigma^2_{e\,opt}$ and the simplest type of pilot behavior $Y_p(j\omega) = K_p e^{-j\omega\tau}$. It was determined according to the technique given in Efremov et al. (1996, 1992). For the pilot model parameters $\tau = 0.25$ s, $K_{n_e} = 0.01$, spectral density $S_{ii} = \frac{K}{(\omega^2 + 0.5^2)}$, and $\sigma^2_i = 4$ cm, it was obtained in Efremov (2017) that $d_{opt} = 1$ cm, $d_{des} = 1.75$ cm and $PR = 1 + 5.36 \ln \frac{d}{d_{opt}}$.

Taking into account (8.2), this equation can be changed to the following:

$$PR = 1 + 5.36 \ln \frac{\sigma_e}{\sigma_{e\,opt}}$$

This equation allows to calculate the pilot ratings by mathematical modeling of pilot–aircraft system.

8.4.2 Man–Machine System in Multiloop, Multichannel, and Multimodal Tasks

Many of the real manual control tasks are much more complicated than compensatory tracking task. They require the consideration of the man–machine system as multiloop, multichannel, or multimodal system. Several versions of such a system are shown in Figure 8.14.

All this makes research significantly more difficult and complicates the mathematical description of human operator behavior. To carry out such studies, there is a need for more complicated data processing algorithms. Because of this, there is only a limited amount of experimental research in mulitaxis tasks. Despite these difficulties, the resulting generalization still allows to determine the general properties of human operator behavior to some extent.

8.4.2.1 Man–Machine System in the Multiloop Tracking Task

Virtually all known results on multiloop tracking tasks are obtained for the case of dual-loop system (Figure 8.14a or Figure 8.14b), in which the human operator perceives not only the error between the command signal $i(t)$ and the vehicle response $y_1(t)$ but also the additional state variable $y_2(t)$ which is linked to $y_1(t)$ by a transfer function $Y_{c_1}(j\omega)$. In particular, the task of aircraft altitude control in the approach phase or lateral steering control of an automobile are examples of dual-loop systems. In this case, the matrix of describing functions $Y_p(j\omega)$ contains only two elements.

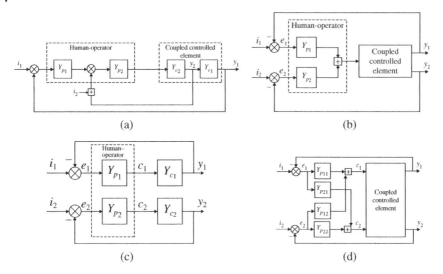

Figure 8.14 Different versions of the multiloop system (* Classification of the multiloop system given in McRuer and Krendel (1974)). (a) Dual-loop system (Single Point Controller, series configuration)*. (b) Dual-loop system (Single Point Controller, parallel configuration)*. (c) Dual-channel system with independent channels (multiple single-loop controller)*. (d) Dual-channel control with cross-coupled channels (cross-fed multipoint controllers)*. Source: Adapted from McRuer and Krendel (1974).

The differences in the properties of dual-loop and single-loop tracking systems can be conveniently analyzed by considering the character of change in the amplitude-phase frequency characteristics of the equivalent describing function of the open-loop system $Y_{OL}(j\omega) = \frac{y_1(j\omega)}{e(j\omega)}$ calculated in the conditions where the human operator perceives the state variable y_2. Experimental research on dual-loop man–machine systems demonstrated that the human operator reacts to an additional state variable if it leads to the extension of the system stability region and improvement of tracking accuracy (Efremov et al. 1992). This feature is sufficiently exposed when the controlled element dynamics is characterized by a high negative slope of the amplitude-frequency response in the region of the crossover frequency $\left(\frac{d|Y_c|}{d \lg \omega} \gg -20 \text{ dB dec}^{-1} \right)$ here $Y_c = Y_{c1} Y_{c2}$. It is typical for $Y_c = \frac{K}{s^2}, Y_{c1} = \frac{K_1}{s}, Y_{c2} = \frac{K_2}{s}$. The fact that the human operator closes the inner loop causes a significant change in the equivalent describing function of the open-loop system $Y_{OL}(j\omega)$ in the region of low and crossover frequencies, to an increase in the phase margin of the open-loop system $\varphi(\omega)$ (Figure 8.15) and in tracking accuracy.

The effects of additional information are more pronounced for plants, whose transfer function has a third order pole at the origin of the s-plane $\left(Y_c = \frac{K}{s^3} \right)$. A lack of this additional information in the case of a third-order pole doesn't allow the operator to realize a stable tracking (Figure 8.16). However, the human operator's perception of the state variable $x_2(t)$, associated with the output $x_1(t)$ through the transfer function $Y_{c1} = \frac{1}{s^2}$, makes the task execution possible.

The identification of two human operator describing functions Y_{p11} and Y_{p12}, with two input signals for the scheme shown in Figure 8.14 a demonstrated that for the case, where $Y_c = \frac{K}{s^2(Ts+1)}$, $Y_{c1} = \frac{K}{s}, Y_{c2} = \frac{K}{s(Ts+1)}$, the amplitude of the describing function $Y_{p12}(j\omega)$ increases as the parameter T increases. It also indicates an increased degree of the human operator's use of the state variable $y_2(t)$ as they form the inner loop.

The pilot's choice of a state variable for inner loop formation is determined by the condition of reaching the maximum accuracy with the minimum workload and also the requirement of maximum possible variability of the pilot's parameters without noticeable change in piloting accuracy

Figure 8.15 Influence of additional information on the open-loop frequency response characteristics.

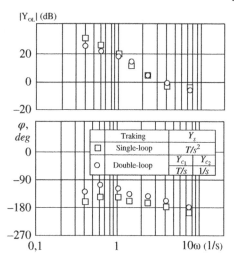

Figure 8.16 Effect of additional information on the tracking process.

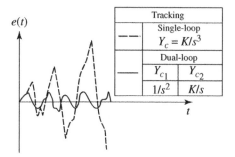

and the stability of the system. Thus, for example, ground-based simulations of a hovering task show that in the case where the pilot perceives the sight angle $\varepsilon = \theta + \arcsin \frac{H}{L}$ (H is the hover height, L is the distance between any object located on the earth surface and the pilot's eye) in the inner loop, the tracking error is reduced by a factor of 5 in comparison with the case where the inner loop is formed with the pitch angle (Efremov et al. 1992). This clearly indicates that the first version of inner loop is more preferred. This conclusion is consistent with the piloting technique used by pilots in the considered task.

8.4.2.2 Man–Machine System in the Multichannel Tracking Task

In multichannel control, the human operator's attention is distributed between several tasks. When performing each of them, the human operator can use several separate manipulators, or the same one tilted in different directions. Depending on whether the movement of the manipulator in one of the directions causes a change in one or more state variables, there are systems with decoupled channels (Figure 8.14c) and systems with cross-coupled channels (Figure 8.14d).

The results of experiments considered in Efremov et al. (1992) demonstrate the human operator ability to decouple the channels in the case where it does not require generating sufficient lead compensation in the transfer functions $Y_{p_{21}}(s)$ and/or $Y_{p_{12}}(s)$ describing pilot actions in the multichannel system with cross-coupled channels (Figure 8.14d).

An increase in the number of control channels in the case of decoupled state variables implies an increase in the number of single-loop tracking tasks that the human operator has to perform simultaneously. As of the fact that the human operator divides their attention between several stimuli, the capacity of each information perception channel, determined by the attention share allocated to this channel f_i, decreases in comparison to the case of single-loop tracking. It leads to an increase in the human operator remnant spectral density. Naturally, an improvement in the handling qualities of one channel should lead to an increased share of attention being allocated to the other channel, and vice versa. When the information is distributed across various displays located outside of the central field of view, the signals can be perceived by the peripheral vision, or the task performance will be accompanied by visual scanning causing a considerable increase in the remnant spectral density and time delay in each channel ($\Delta\tau = 0.05 \div 0.15$ s). Therefore, in multichannel tracking, cross-channel influence is caused by the division of attention between the perceived information and by the differences in the perception process.

The coupling between channels can also arise when using single two-channel inceptor (for example, central stick). In this case, inaccurate human operator actions in one channel can cause a command signal in the other channels and appearance of cross-channel interference.

An increase in the number of channels also causes a degradation of the subjective pilot ratings. This effect can be evaluated by the following expression (McRuer and Krendel 1974):

$$PR_m = 10 + \frac{1}{(-8.3)^{m-1}} \prod_{i=1}^{m} (PR_i - 10) \tag{8.4}$$

Here, PR_m is the total pilot rating of the dynamic properties of m control channels.

PR_i is a partial pilot's rating of flying qualities obtained in experiments in a condition of single-loop tracking (pitch, roll, etc.).

It is obvious that the rating PR_m has to be higher than the partial ratings PR_i. At the same time, analysis of experimental data showed that with some combinations of ratings PR_i, the use of Eq. (8.4) gives inaccurate results. Because of it, a different equation was recommended for the evaluation of PR_m in dual channel tracking task (roll and pitch control) (Efremov 2017):

$$PR_m = B + \sqrt{B^2 - PR_\theta PR_\gamma + B}$$

where $B = \frac{PR_\varphi + PR_\theta}{2}$, and PR_θ, PR_φ are the partial pilot ratings obtained in experiments with single-loop tracking tasks (pitch or roll control).

When the pilot performs the simultaneous multichannel task, the total pilot rating PR_m can be determined according to the following equation:

$$PR_m^* = \max\{PR_i, PR_{j,\dots}\} \tag{8.5}$$

where PR_i, PR_j, … are the pilot ratings of the flying qualities in each control channel.

The agreement of this equation with results of experiments in which the pilot was asked to assign the total flying qualities rating PR_m^t and the rating PR_m^* calculated following Eq. (8.5), where PR_φ, PR_θ are the partial flying qualities ratings of each channel assigned by the pilot just after performing the dual channel task is shown in Figure 8.17.

8.4.2.3 Man–Machine System in Multimodal Tracking Tasks

The difference between multimodal and multiloop tracking is that in multimodal tracking, different command stimuli are perceived by the human operator through different sensors. A multimodal man–machine system can be formed artificially by using special devices – modal displays – aimed at stimuli perception through a specific channel, in parallel to the visual channel. The research performed in the latter half of the last century demonstrated that their use is typically accompanied by improvements in man–machine system performance. For example, transmitting information about the tracking error through the visual and tactile channels allows a decrease of the variance of error by 20% and the open-loop crossover frequency by 30–50% (Efremov et al. 1992;

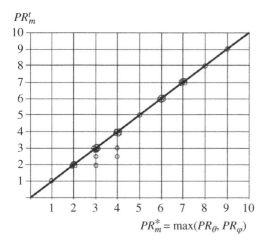

Figure 8.17 Correlation of ground-based simulation results with Eq. (8.5).

Bliss et al. 1967). The simultaneous transmission of the tracking error signal through an audio display and a visual indicator allows to receive both a higher crossover frequency (by approximately 20%) and higher accuracy (Vinje and Pitkin 1972). The use of an audio display in conjunction with visual ones is most effective in multimodal control (Uhlemann and Geiser 1975). Using it for a glide slope-tracking task leads to a decrease in the variance of error of 15% and 25% in the longitudinal and lateral channels, respectively (Efremov et al. 1992). The most significant effect is achieved by using an additional kinesthetic display (Schmidt et al. 1976). In this case, switching to controlling even an unstable vehicle is accompanied by virtually no increase in workload and tracking error. Moreover, their values are significantly lower than those in experiments with only a visual display. In spite of all these positive effects, the modal displays have not yet found practical use.

A multimodal system is formed naturally as the pilot is subjected to angular and linear acceleration. The influence of these factors is complex and nuanced and depends on the task being performed, as well as other variables. It can be presented in terms of equivalent pilot describing function $Y_p^e = \frac{c(j\omega)}{e(j\omega)}$ or the pilot describing function Y_p^{vis}, determining their response to the visual signal $e(t)$, and Y_p^{vest}, determining the pilot response to the rotational motion (Figure 8.18).

The relationship between these describing functions for different piloting tasks is given as follows:

- For the stabilization task, when the input signal is the disturbance, $d(t)$, $Y_p^e = Y_p^{\text{vis}} + Y_p^{\text{vest}}$.
- For the target tracking task, when the input signal is the command signal, $i(t)$, $Y_p^e = \frac{Y_p^{\text{vis}}}{1+Y_c Y_p^{\text{vest}}}$.

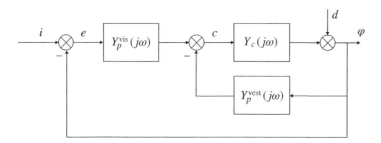

Figure 8.18 Pilot-aircraft multimodal system.

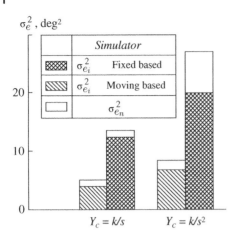

Figure 8.19 Effect of motion information on tracking accuracy (Efremov et al. 1992).

These equations explain the difference in the results of experiments. For the stabilization of the roll angle, the prediction of rotational motion causes a decrease in the pilot phase delay and increase in the crossover frequency and bandwidth of the close-loop system. All these effects lead to a considerable decrease in the variance of error (Figure 8.19) and its components correlated with the input signal $\left(\sigma_{e_i}^2\right)$ and remnant $\left(\sigma_{e_n}^2\right)$.

The effectiveness of using motion cues can be seen in the results of measurements of the describing function Y_p^{vest}, which is virtually the same as Y_p^e in the middle and high-frequency ranges (Figure 8.20).

The positive effects of motion cues perception are noticeable in other piloting tasks as well. In the presence of simulator cockpit motion, the time interval between the moment of a failure in the control system and the pilot response is reduced by a factor of 2. The vertical speed at touchdown when using motion simulators decreases by factors of 1.35 and 3.5 in aircraft with satisfactory and unsatisfactory handling qualities, respectively (Efremov et al. 1992). When tracking the command signal $i(t)$, a different effect of motion on the pilot-aircraft system characteristics is observed. In this case, the perception of rotational motion by the pilot causes an insignificant decrease in the crossover frequency of the open-loop system due a decreased gain of the equivalent describing function $Y_p^e(j\omega)$ (Efremov et al. 1996, 1992).

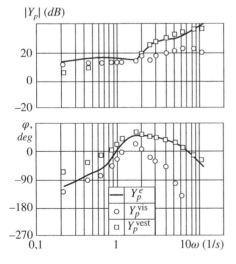

Figure 8.20 Effect of flight simulator rotational motion on the frequency response of pilot actions (Efremov et al. 1992).

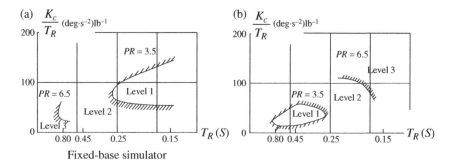

Figure 8.21 Difference between (a) ground-based and (b) in-flight experiments.

More significant effects of the motion cues on the pilot actions take place in the longitudinal target tracking task when the pilot perceives a considerable normal load factor $\left(\frac{a_z}{g} \geq 4 \div 5\right)$. In this case, the pilot gains coefficient and piloting accuracy decreases because of physiological reasons.

The motion cues also influence pilot ratings. The results of flying qualities evaluation in a roll-tracking task obtained in experiments conducted using a fixed-base simulator and an in-flight simulator are shown in Figure 8.21 (Wood 1983). The dynamics of aircraft was presented by the transfer function $Y_c = \frac{K_c}{s(T_R s+1)}$.

The dissimilarity between these results can be explained by the influence of the lateral load factor perceived by the pilot when performing in-flight tests. This load factor is perceived at the level of the pilot's head which is located at a distance from the axis of rotation $\frac{a_y}{g} = \frac{\dot{p}l}{g}$, where $\dot{p}$ is the rolling acceleration. A decrease in the roll time constant T_R causes an increase in this acceleration.

To summarize, under the influence of motion cues, the pilot forms a multimodal system if it helps improve piloting accuracy. Therefore, in this case, the actions of the pilot differ significantly from those in a single-loop compensatory system. Neglecting this fact may lead to inaccurate and sometimes incorrect results of applied research.

8.4.2.4 Human Operator Behavior in Pursuit and Preview Tracking Tasks

Compensatory tracking corresponds to a variety of manual control tasks, but not all of them. In some of the manual control tasks considered in Section 8.2, the man–machine system can be interpreted as a pursuit or preview system in which the human operator perceives the current input signal $i(t)$ (pursuit system) or the planned trajectory $i(t + T_{prev})$ (preview system).

Studies of human operator behavior in preview and pursuit tracking tasks have a long history. First, investigations in this area were performed with simple controlled element dynamics $\left(Y_c = \frac{K}{s}, Y_c = \frac{K}{s^2}, Y_c = \frac{K}{s(s-\lambda)}, Y_c = \frac{1}{s+2}\right)$ and an input signal characterized by the spectral density $S_{ii}(\omega)$ with a bandwidth $\omega_i \geq 1.5$ (1/s) (Reid and Drewell 1972; Wasicko et al. 1966; Tomizuka and Whitney 1973). The results of these studies demonstrated the high potential of preview tracking for improving performance in comparison with compensatory tracking. Such an improvement is achieved if the preview time does not exceed 0.5–1 second. Its subsequent increase does not cause a decrease in tracking error.

Extensive investigations of human operator behavior in preview manual control have been carried out at Delft University since the beginning of the 1990s. Practically, all of them were performed for simple controlled element dynamics (gain coefficient, single/double integrator) and with different task variables: preview time T_{prev}, input bandwidth ω_i, dimensions of the tunnel, and ways of presenting the input signal (Van der El et al. 2018a, 2020, 2016; Mulder and Mulder 2005; Mulder 1999). Several of the Delft studies were aimed at the identification of human operator frequency response characteristics, including those whose outputs are the so-called "near/far view responses" (Van der El et al. 2018a, 2018b). The results of experiments analyzed in Efremov et al. (2022) demonstrated that "increasing" the preview time results in better tracking performance, but beyond a certain "critical" preview time, the effects on both the task performance and human operator control behavior stabilize and the effects of additional preview time are small. The critical preview time is not invariant in tasks with single- and double-integrator controlled elements and is around 0.6 and 1.15–1.4 second, respectively. All this research revealed that the effect of preview on pilot behavior characteristics and its potential for improving task performance for an input signal is characterized by a bandwidth $\omega_i \geq 1.5$ rad s^{-1}. However, since such bandwidths are not typical of vehicle path motion, refinement of these results was required for an input with smaller bandwidths $\omega_i = 0.2 \div 0.5$ rad s^{-1} which are closer to the real planned path trajectory. Experiments with such bandwidths were performed recently at MAI (Efremov et al., 2022) with different bandwidths ω_i and controlled element dynamics for **compensatory** (comp), **pursuit** (pur), and **preview** (prev) tracking tasks, with the results given in Tables 8.2 and 8.3.

The transfer function $Y_c(s) = \frac{\varepsilon_{\text{pr}}(s)}{c(s)}$, where $\varepsilon_{\text{pr}} = \gamma + \dot{\gamma}\frac{T_{\text{pr}}}{2} + \frac{h}{L_{\text{pr}}}$ is the predictive path angle shown in Figure 8.22, corresponds to

$$Y_c = \frac{\varepsilon_{\text{pr}}(s)}{c(s)} = \frac{K_c\left(T_{\text{pr}}s^2 + 2s + \frac{2}{T_{\text{pr}}}\right)}{s^2\left(s^2 + 2\xi_{\text{sp}}\omega_{\text{sp}}s + \omega_{\text{sp}}^2\right)} \tag{8.6}$$

where $T_{\text{pr}} = \frac{L_{\text{pr}}}{V} = 0.9$ seconds, $\omega_{\text{sp}} = 2.4$ rad s^{-1}, $\zeta_{\text{sp}} = 0.64$, h is the deviation in height, and γ is the path angle.

Analysis of all these results demonstrates that the variance of human operator control output σ_c^2 is considerably lower in pursuit and especially preview tracking tasks in comparison with

Table 8.2 Influence of task variables on the variance of pilot control input, σ_c^2.

Y_c	σ_c^2 (comp/pur/prev)		
	$\omega_i = 0.2$ rad s^{-1}	$\omega_i = 0.5$ rad s^{-1}	$\omega_i = 1.0$ rad s^{-1}
$\frac{K}{s}$	2.24/0.94/0.53	3.6/1.9/1.5	12.4/8.9/5.0
$\frac{K}{s^2}$	18.34/6.82/2.42	34.3/22.6/11.2	49.8/46.2/43.5
$\frac{\varepsilon_{\text{pr}}}{X_e}$	5.5/2.11/0.43	17.4/7.9/2.5	32.7/25.4/14.1

Table 8.3 Influence of task variables on the variance of error, σ_c^2.

Y_c	σ_c^2 (comp/pur/prev)		
	$\omega_i = 0.2$ rad s^{-1}	$\omega_i = 0.5$ rad s^{-1}	$\omega_i = 1.0$ rad s^{-1}
$\frac{K}{s}$	0.16/0.15/0.14	0.34/0.30/0.28	0.84/0.69/0.45
$\frac{K}{s^2}$	0.21/0.19/0.16	0.79/0.46/0.34	6.09/4.26/1.08
$\frac{\varepsilon_{\text{pr}}}{X_e}$	0.20/0.19/0.18	0.97/0.62/0.48	2.81/2.29/1.12

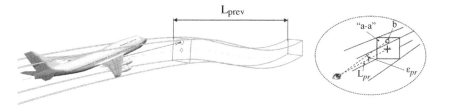

Figure 8.22 Predictive display.

compensatory tracking. The effect of σ_c^2 decreasing depends on the bandwidth ω_i and controlled element dynamics. Such effect is highest for the low bandwidth ω_i (for $\omega_i = 0.2$; 0.5 rad s^{-1}) for all investigated dynamic configurations Y_c (Table 8.2). In addition to the decrease in the human operator control output, the rate of this signal $\dot{c}(t)$ decreases considerably as well. As an example, the root mean square of $\dot{c}(t)$ obtained in experiments with preview and compensatory displays and the dynamics $\frac{\varepsilon_{pr}(s)}{c(s)}$ is shown in Figure 8.23. In a pursuit-tracking task, the effect of σ_c^2 decreasing is several times smaller, but it is still noticeable.

The bandwidth ω_i has an opposite influence on the variance of error σ_e^2. An increase in bandwidth in trials with pursuit and preview displays causes a more significant decrease in σ_e^2 (Table 8.3).

Experiments preformed with the spectrum bandwidth $\omega_i = 0.5$ (1/s) and Y_c corresponding to Eq. (8.5) and with a different preview time T_{prev} demonstrate that the optimum value of T_{prev} close to 2.4 seconds provides the minimum variance of error σ_e^2 (Figure 8.24). This value is higher than the critical preview time obtained in Van der El et al. (2018b) because the bandwidth of the input signal was three times less in comparison with the spectrum used in Van der El et al. (2018a).

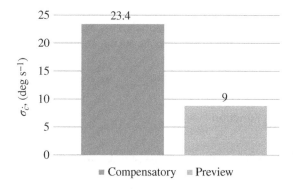

Figure 8.23 Variance $\sigma_{\dot{c}}$ for preview and compensatory tracking tasks.

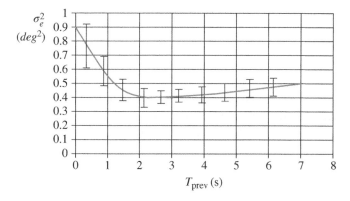

Figure 8.24 $\sigma_e^2 = f(T_{prev})$.

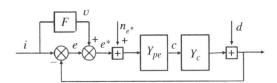

Figure 8.25 Man–machine system in a preview/pursuit tracking task.

The experiments performed with two input signals $i(t)$ and $d(t)$ (Figure 8.25) allowed the measuring of two describing functions $F(s)$ and $Y_{pe}(s)$, determining the pilot's response to the input and error signals. These experiments demonstrated the independence of the describing function $Y_{pe}(j\omega)$ of the preview time T_{prev} (Van der El et al. 2018b).

8.5 Mathematical Modeling of Human Operator Behavior in Manual Control Task

One of the most complicated components of the man–machine system is the human operator. In the general case, their response is determined by the reaction on the visual, vestibular, and proprioceptive cues. One of the possible models, proposed in McRuer et al. (1990) and shown in Figure 8.26, reflects the major processes taking place in the sensory, motor, and CNSs. The most complicated element of this model is the CNS describing the process of adaptation of pilot behavior.

In spite of its straightforwardness, the full model is too complicated to be used as a means of solving engineering tasks. Other, simpler models are used for that. Some of them, which are used widely in solving applied problems, are considered below.

These models consist of

- a controller reflecting a major human operator property – their adaptation to task variables;
- the other so-called "psychological limitations" describing the human operator's sensory and motor systems;
- a technique for the selection of the control law and/or its parameters.

8.5.1 McRuer's Model for the Pilot Describing Function

8.5.1.1 Single-Loop Compensatory Model

This model was the result of the experimental studies on the influence of task variables on the human operator describing function performed at Systems Technology, Inc. in the 1950s and 1960s. The major property of the man–machine system considered in Section 8.4 is the approximately constant slope of $|Y_{OL}(j\omega)|$ close to $-20\,\mathrm{dB\,dec^{-1}}$ in the crossover frequency range of different controlled element dynamics and input signal spectrums. It allows to approximate the open-loop system describing the function by using the so-called "crossover model" (McRuer et al. 1965; McRuer and Krendel 1974).

$$Y_p Y_c(j\omega) = \frac{\omega_c}{j\omega} e^{-j\omega\tau_e} \tag{8.7}$$

where $\omega_{\tilde{n}} = \omega_{\tilde{n}}(\omega_c) + \Delta\omega_{\tilde{n}}(\omega_i)$, $\tau_e = \tau_e(\omega_c) - \Delta\tau_e(\omega_i)$, $\Delta\omega_c = 0.18\omega_i$; $\Delta\tau_e = 0.08\omega_i$.

This model was obtained in experiments with a rectangular input spectrum with a rather high bandwidth ($\omega_i \geq 1.5$ (1/s)). The experiments performed with a lower bandwidth $\omega_i = 0.5 \div 1$ (1/s)

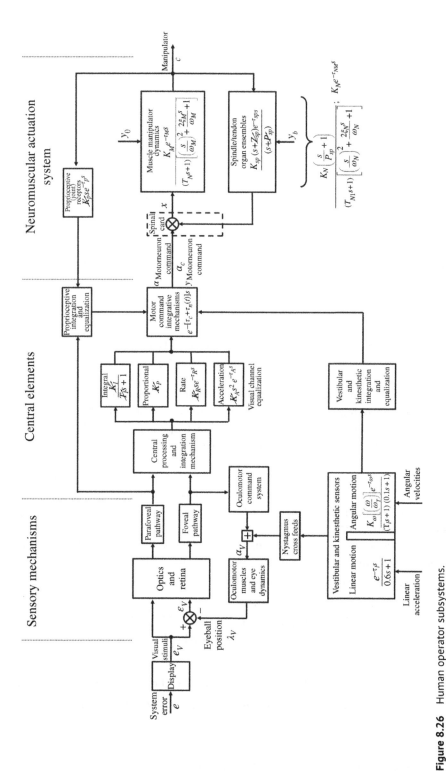

Figure 8.26 Human operator subsystems.

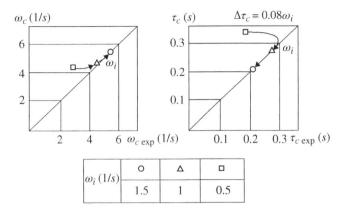

Figure 8.27 Parameters ω_c and τ as the functions of input spectrum bandwidth.

(Efremov et al. 1992) demonstrated the difference in $\omega_{c\,\text{exp}} = f(\omega_i)$ and $\tau_{e\,\text{exp}} = f(\omega_i)$ in comparison with Eq. (8.7) (see Figure 8.27).

The simplest human operator describing function, approximating the experiments in the crossover frequency range with reasonable accuracy, is

$$Y_p = K_p \frac{(T_L j\omega + 1)}{T_I j\omega + 1} e^{-j\omega\tau_{\text{eff}}}$$

where K_p is the pilot static gain; T_L is the lead time constant; T_I is the lag time constant; τ_{eff} is the effective time delay including transport delays and high-frequency neuromuscular lags.

The lead and lag constants T_L and T_I are adjusted by the human operator to achieve the $-20\,\text{dB dec}^{-1}$ slope of the combined $Y_p Y_c$ response, as required by the crossover model, while K_p is adjusted to put ω_c where required. It appears that the human operator attempts to select a lead or lag value such that the sensitivity of the closed-loop, low-frequency characteristics to variations in T_L or T_I is small, leaving the gain and effective time delay as the primary means of adjusting the closed-loop stability and other requirements for the closed-loop system. The resulting equalization is realized according to the so-called "adjustment" rules proposed in McRuer and Krendel (1974).

The simplest human operator describing function model for the single-loop, man–machine system is best suited to conventional stable controlled element which has smooth amplitude and phase characteristics in the crossover region.

The more satisfactory form for approximating the phase and delay in the low-frequency range, $\frac{T_k j\omega+1}{T'_k j\omega+1}$, and the approximation of the neuromuscular system were added in the precision model of the pilot describing function (McRuer and Krendel 1974):

$$Y_p(j\omega) = K_p e^{-j\omega\tau} \left(\frac{T_L j\omega + 1}{T_I j\omega + 1} \right) \left(\frac{T_k j\omega + 1}{T'_k j\omega + 1} \right) \frac{1}{(T_{N_1} j\omega + 1)\left[\left(\dfrac{j\omega}{\omega_N}\right)^2 + \dfrac{2\xi_N}{\omega_N} j\omega + 1 \right]}$$

In the crossover frequency range, this model can be simplified by neglecting the $\left(\frac{T_k j\omega+1}{T'_k j\omega+1} \right)$ element and replacing the dynamics of the neuromuscular system by its effective time delay

$$\tau_{\text{eff}} = \tau + T_{N_1} + \frac{2\xi}{\omega_N}$$

In that case, the precise model is transformed into the simplest one.

8.5.1.2 Multiloop and Multimodal Compensatory Model

The human operator model for multiloop tasks is the extension of the quasi-linear describing function model for single-loop tasks, but with different parameters operating in each loop.

The selection of the loops and human operator equalization in each of the loops corresponds to the adjustment rules used by the engineer during the control system design, taking into account the specificities of human behavior. In particular (McRuer and Krendel 1974),

- – The preferred feedback loops are those which
    - (a) Can be closed with minimum pilot equalization;
    - (b) Require minimum scanning to sense the feedback quantity;
    - (c) Permit wide latitude in the human operator's adapted characteristics.
- – Where distinct inner- and outer-loop closures can be determined by ordering the bandwidths (e.g. the higher the bandwidth, the more inner the loop), a series of multiloop structure applies.
- – Human operator equalization for the outer loop of multiloop systems is adjusted per the crossover model, with the proviso that the effective controlled element transfer functions include the effects of all the inner-loop closures. The crossover model is also directly applicable to many inner-loop closures.
- – The adjustment of the variable gain in each of the loops is, in general, such as to achieve basically simple (i.e. effectively second- or third-order) well-damped dominant modes and nearly uncoupled sets of plant responses. Outer-loop gains, in particular, may be lower than the maximum bandwidth for this reason.

The quasilinear human operator models for single- and multiloop systems developed by McRuer and his colleagues at STI have been used widely in the development of flying qualities criteria, analysis of man–machine system properties, and other applications.

8.5.2 Structural Human Operator Model

The structural human operator model developed by Hess was derived from a theory introduced by Smith (1975). The key idea of this model is to simulate the feedback paths from various sensory modalities. The human equalization occurs through that feedback referred to as proprioceptive feedback (Hess 1979). Such feedback parameters are tuned to match the performance of the crossover model near the crossover frequency (Hess 1979). The structural model offered a compensatory tracking model of the human operator to provide a more realistic representation of the human signal processing, than those human operator models in use.

Hess made several modifications to his model. One of the versions is shown in Figure 8.28.

The parameter K_e is the gain representing the human operator adaptation to the visual information, and τ is the delay of CNS processes. The element Y_m describes the human operator inner-loop feedback, whose key parameter is k. This parameter reflects the adaptive characteristics of the human operator to the controlled element and will mainly depend upon its transfer function around the crossover frequency, ω_c, which is selected as a measure of the specified performance level characterizing the task. The parameter k in the transfer function Y_m can be interpreted as the human operator "internal" model of the vehicle dynamics. That is, in the range of the crossover frequency, the parameter $k = 0$ if the controlled element is a gain coefficient, $k = 1$ if the controlled element is an integrator, and $k = 2$ if the controlled element is a double integrator. The element Y_{pn} is a model of the open-loop neuromuscular + inceptor dynamics system $Y_{pn}(s) = \frac{\omega_{nm}^2}{s^2 + 2\xi_{nm}\omega_{nm}s + \omega_{nm}^2}$, where $\omega_{nm} \cong 10 \ (1/s)$, $\xi = 0.7$, and Y_f is the model of the muscle spindle.

In spite of the human operator's ability to vary the crossover frequency in the analysis of man–machine system properties and other manual control tasks, a constant crossover frequency

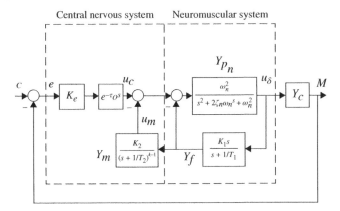

Figure 8.28 Structural model of the human operator for compensatory tracking.

$\omega_c = 2\ \mathrm{rad\,s^{-1}}$ was chosen in Hess (1979). Knowing these parameter values and knowing the controlled element dynamics, the form of Y_m is then obtained by using the crossover model:

$$\frac{Y_c(j\omega)}{Y_m(j\omega)} \cong \frac{K_1}{j\omega}$$

Then the gain coefficient of Y_m is chosen so that the minimum damping ratio of any quadratic closed-loop poles of proprioceptive loop is $\xi_{\mathrm{pr\,min}} = 0.15$. The gain K_e is then selected to achieve a crossover frequency of $2\ \mathrm{rad\,s^{-1}}$. In Hess (1979), it is further stated that the fixed parameters were an analytical simplification. But the so derived human operator model would be of sufficient accuracy to justify its use in the man–machine system analysis, and it would reflect important properties of human behavior.

The Hess model had a number of extensions with the goal to take into account motion cues (Hess 1999) and to add a description of the human operator behavior in pursuit tracking task (Hess 2007). All these modifications and extensities gained popularity as they were able to provide solutions to various engineering problems.

One of the deepest modifications of the structural model was offered in Efremov et al. (2021). It consists of two novelties:

a. The modified structure of the model;
b. The adjustment rules for the selection of the model's parameters based on minimization of the task performance.

The modified model is shown in Figure 8.29.

Here, Y_m is the human operator equalization in the inner-loop, Y_{pf} is the dynamic model of the inceptor, and Y_{pn} is the dynamic model of the neuromuscular system. The observation noise (remnant) n_e is added to the error signal.

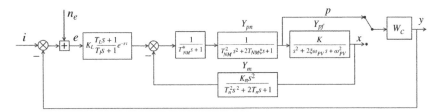

Figure 8.29 Modified version of the structural model.

The model of the remnant spectral density is based on the Levinson model (Levison et al. 1969). Its modified version (Efremov et al. 1992) is

$$S_{n_e n_e} = \frac{K^2}{1 + T_i^2 \omega^2}$$

where $K^2 = \frac{\sigma_{e_i}^2 + T_L^2 \sigma_{\dot{e}_i}^2}{\frac{1}{K n_e} - \int_0^\infty \frac{|Y_{CL}|^2}{1 + T_L^2 \omega^2} d\omega}$, $K_{n_e} = \frac{0.01}{f}$; f is the fraction of attention paid to the piloting task

$$\sigma_{e_i}^2 = \frac{1}{2\pi} \int_{-\infty}^\infty S_{ii} \frac{1}{|1 + Y_{OL}|^2} d\omega; \quad \sigma_{\dot{e}_i}^2 = \frac{1}{2\pi} \int_{-\infty}^\infty S_{ii} \frac{\omega^2}{|1 + Y_{OL}|^2} d\omega$$

One peculiarity worth noting is that the dynamics of the neuromuscular system and inceptor are separated in the proposed model. This allows to investigate the performances of the man–machine system with the DSC and FSC types of inceptors and to study the influence of inceptor characteristics (damping, stiffness) on the man–machine system.

The adjustment rule proposed in this model is $I = \min_{T_L, K_p, T, k} \sigma_e^2$, where

$$\sigma_e^2 = \sigma_{e_i}^2 \left[1 + \frac{1 + T_L^2 \frac{\sigma_{\dot{e}_i}^2}{\sigma_{e_i}^2}}{\frac{1}{K_{n_e}} - \int_0^\infty |Y_{CL}|^2 d\omega} \int_0^\infty \frac{|Y_{CL}|^2}{1 + T_L^2 \omega^2} d\omega \right] \tag{8.8}$$

The mathematical modeling performed with different controlled element dynamics demonstrated good agreement with the results of experiments.

From Eq. (8.8), the following requirements can be obtained: $\sigma = \frac{1}{K_{n_e}} - \int_0^\infty |Y_{CL}|^2 d\omega > 0$. Assuming that the crossover model is valid for the open-loop system, and restricting the time delay element to a first-order approximation in the Padé series, it is possible to obtain the maximum value of $\overline{\omega}_{c_{max}} = \omega_c \tau_e$, for which $\sigma = 0$, $\overline{\omega}_{c_{max}} = -(\overline{\tau}_e + 1) + \sqrt{4\overline{\tau} + (\overline{\tau} - 1)}$.

This value is less than 2, which is the maximum $\overline{\omega}_c$ obtained from the Hurwitz–Gauss criteria of stability.

Several applications of this model are considered in Section 8.6.

8.5.3 Pilot Optimal Control Model

An alternative approach to the estimation and description of human control behavior has been the application of optimal control theory (Kleiman et al. 1970). An optimal system aims to minimize the quadratic performance index in the presence of various system inputs and noises. The current approach takes into account the property of human adaptation and their limitations – effective time and neuromuscular delays, and the observation and motor noises (V_y and V_{u_c}) (Figure 8.30).

The observation noise is added to the display output $y(t)$. For the single-loop system, the output is characterized by two states: error y_e and error rate $\dot{y}_e$. The observation noise is modeled using a white noise source $V_y(t)$ with intensity matrix V_y.

This matrix is made of elements $V_{y_i} = \frac{\rho_{y_i} \pi}{f} \frac{\sigma_{y_i}^2}{E[\sigma_{y_i} a_i]^2}$, where $\rho_{y_i} = 0.01$. The attention traction $0 \le f \le 1$.

The indifference thresholds a_i are 0.05 deg for error perception and 0.1 deg s^{-1} for error rate perception (McRuer et al. 1990) for human eye perception, and

$$E(\sigma_{y_i} a_i) = \text{ertc} \frac{a_i}{\sigma_{y_i} \sqrt{2}}$$

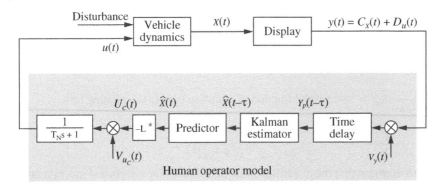

Figure 8.30 Optimal pilot model.

Since V_{y_i} is proportional to the variance of the perceived signal, the solution of the Kalman filter and predictor requires the development of an iterative procedure.

The motor noise, like the observation noise, is assumed to be a zero mean white noise with a spectral density proportional to the variance of operator output, $V_{u_c} = \pi \rho_{u_c} \sigma_{u_c}^2$, $\rho_{u_c} = 0.003$ (McRuer et al. 1990). The neuromuscular system is represented by the lag matrix, T_N.

The remaining elements of the human operator are adaptive to the system characteristics and to changes in the explicit human operator limitations described above. The estimation of the delayed state vector is accomplished with the help of a Kalman filter. This delayed state estimate is fed to a least-mean-squared predictor to yield the estimated state vector, $\hat{x}(t)$. The optimal gain matrix, L^*, is generated by solving the optimal regulator problem for a quadratic cost function of the form:

$$J(u) = E \left\{ \lim_{T \to \infty} \frac{1}{T} \int_0^T (y'Qy + u'Ru + \dot{u}'G\dot{u})dt \right\}$$

Because weights of the cost function preordain the details of the controller gain matrix, L^*, the selection of these weights is critical to the model's success. This is particularly the case when the model's purpose is to simulate human operator responses. For simple single-loop control situations, excellent agreement with experimental measurements has been obtained with a cost function of the extremely simple form:

$$J(u) = E \left\{ \lim_{T \to \infty} \frac{1}{T} \int_0^T (e^2 + G\dot{c}^2)dt \right\}$$

where e is the compensatory system error, and $\dot{c} = \dot{u}$ is the operator's control signal rate. The value of G is selected to yield an appropriate neuromuscular delay, T_N. For more complex situations, the relative weights are determined based either on maximum allowable deviations or limits, or from knowledge of human preferences and capabilities. This is similar to the technique suggested in Bryson and Ho (1969), where the weight of each quadratic term is simply the inverse of the square of the corresponding allowable deviation. The solutions for this modified Kalman filtering prediction and optimal control problem are given by Baron et al. (1970), Curry et al. (1976), Kleinman and Baron (1973) among others.

A number of modifications of the optimal pilot model have been proposed in the last few decades (Doman 1999; Davidson and Shmidt 1992; Hess 1976). In the majority of them, the time delay in the form of a Padé approximation is placed at the human operator's output and is treated as part of the plant dynamics. In all these models, the predictor is eliminated, and so the model has a lower-order representation than when the delay is placed at the human operator input. This kind of implementation allows for rapid calculation of human operator and system transfer function.

8.5.4 Pilot Models in Preview and Pursuit Tracking Tasks

All the mathematical models of pilot behavior considered above were obtained for the compensatory tracking tasks. There is significantly less research dedicated to the development of the human operator's model in pursuit or preview-tracking tasks. Two of these few studies allowed to get not only the structure of the model but also to calculate the parameters of the said structure according to a developed off-line procedure. The mathematical models suggested in Van der El et al. (2018a, 2018b) consist of three describing functions. Two of them describe the pilot's "far viewpoint" response H_{of} and "near viewpoint" response H_{on} to the input signal. The third describing function simulates the pilot response H_{oe*} to the error signal (Figure 8.31); the equations for $H_{of}(j\omega), H_{on}(j\omega)$, and $H_{oe*}(j\omega)$ are the following $H_{of}(j\omega) = Kf\frac{1}{1+T_jj\omega}$; $H_{on}(j\omega) = K_n j\omega$; and $H_{oe*}(j\omega) = K_e^*\left(1 + T_{L_{e*}} j\omega\right)$.

The model takes into account the psychological limitations – time delay and the neuromuscular dynamics $H_{nms}(j\omega) = \frac{\omega_{nm}^2}{(j\omega)^2 + 2\xi_{nm}\omega_{nm} + \omega_{nm}^2}$.

The off-line modeling was performed for the controlled element dynamics $H_{\tilde{n}e}(j\omega) = \frac{K}{j\omega}$ and $\frac{K}{(j\omega)^2}$.

The selection of parameters $\tau_p, \tau_f, \tau_1, K_n, K_f, K_e^*$ of the human operator's model was performed in Van der El et al. (2018a) by minimization of σ_e^2. The human limitation parameters and lead time were constant in the optimization procedure. The possibility to accept the lead time as the constant parameter was proposed in Van der El et al. (2018b), where experiments were performed for different preview times. During the experiments and mathematical modeling, the values of preview times obtained were very close to the optimal ("critical") value of ~0.6 second for the single integrator dynamics and 1.15–1.4 seconds for the double integrator.

However, the comparison of the human operator's frequency response characteristics obtained in experiments and mathematical modeling demonstrated that experimental data matches the predictions reasonably well, but not perfectly.

Another human operator's model (see Figure 8.25) was used in Efremov et al. (2022). Here, the modified Hess's model (see Figure 8.29) for the modeling the Y_{pe} was used in the inner loop. The following equation was used for the remnant spectral density:

$$S_{n_{e^*}n_{e^*}}(\omega) = 0.01\pi\frac{\sigma_{e^*}^2 + \sigma_{e^*}^2 T_L^2}{1 + T_L^2\omega^2}$$

The signal e^* is shown in Figure 8.25.

It was assumed that the output of the model $F(s)$ is

$$v(t) = \alpha_1(t) + \alpha_2(t) \tag{8.9}$$

where $\alpha_1(t)$ is the pilot response in a pursuit-tracking task ($T_{prev} = 0$) and $\alpha_2(t)$ is the human operator response in a preview-tracking task. The following equation for the transfer function $\frac{\alpha_1(s)}{i(s)}$ in a pursuit task was proposed:

$$\frac{\alpha_1(s)}{i(s)} = \frac{K_0(T_1 s + 1)}{T_2 s + 1}$$

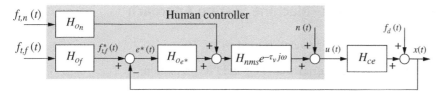

Figure 8.31 Control diagrams of the human operator model for preview tracking tasks.

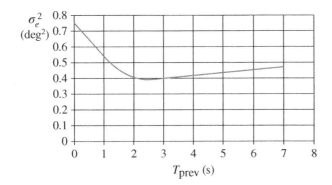

Figure 8.32 Comparison of mathematical modeling and ground-based simulation results (Efremov et al. 2022).

The parameters (K_0, T_1, and T_2) of this transfer function are determined by the minimization of the variance σ_e^2. In the process of the minimization, the structural model Y_{pe} is used.

Due to the frequency response characteristics $Y_{pe}(j\omega)$ being independent of the human operator response to the preview time constant (T_{prev}) (see Section 8.4), the parameters of the structural model $Y_{pe}(j\omega)$, can be determined by considering the compensatory system ($v(t) = 0$) according to the technique outlined in Efremov and Tiaglik (2011) and Efremov (2017). In the case of preview-tracking tasks, the signal $v(t)$ corresponds to Eq. (8.9).

Equation (8.10) was proposed for the signal $\alpha_2(t)$:

$$\alpha_2(t) = \sum_{n=1}^{m} K_n \frac{i(t + n\Delta t) - i(t + (n-1)\Delta t)}{\Delta t \cdot V} \tag{8.10}$$

which is a weighted sum of the planned trajectory slopes. Here, $i(t + n\Delta t)$ is the length of the segment of the planned trajectory, and V is the aircraft velocity.

According to Efremov et al. (2022) the determination of coefficients K_n in Eq. (8.10) is implemented sequentially: first, the coefficient K_1 is determined by minimizing the variance of error, then the coefficient K_2 is determined for the obtained value of K_1. The remaining coefficients K_3, K_4, K_5, etc., are selected in a similar fashion. The details of this procedure are given in Efremov et al. (2022).

Figure 8.32 shows the relation between piloting accuracy and the time T_{prev} for the calculated values of K_i, demonstrating that similarly to the experiments whose results are shown in Figure 8.26.

Besides, the considered pilot behavior modeling techniques, others have been developed lately. In particular, one of them, based on the neural network approach, allowed to obtain the so-called "composite pilot model" (Efremov et al. 2010), characterized by the best predictive properties and agreement with the results of experiments in comparison to the widely used optimal and structural models.

8.6 Applications of the Man–Machine System Approach

Knowledge of human operator behavior models is used widely for

- Development of criteria for **flying qualities** (FQ) and PIO prediction;
- Interface design;
- Flight control system design.

8.6.1 Development of Criteria for Flying Qualities and PIO Prediction

Two types of criteria based on the performance of pilot aircraft systems were developed. One type comprises the criteria for the FQ and PIO tendency prediction (Section 8.6.1.1) as a requirement for the parameters of the pilot–aircraft system. The other type covers the criteria for pilot ratings prediction (Section 8.6.1.2).

8.6.1.1 Criteria of FQ and PIO Prediction as a Requirement for the Parameters of the Pilot-Aircraft System

The first criterion of this type was proposed in Neal and Smith (1971) as a result of in-flight tests of different, so-called "Neal–Smith" dynamic configurations. Knowledge of the transfer functions of each configuration, **pilot opinion ratings** (PR), and the use of McRuer's model allowed the authors to get the parameters of the pilot and pilot–aircraft closed-loop system correlated with PR.

Two parameters were selected: resonant peak for the closed-loop system r, and pilot phase compensation parameters $\Delta\varphi$ calculated as the phase characteristics of the pilot lead/lag equalization $\frac{T_L j\omega+1}{T_I j\omega+1}$ in McRuer's model at the bandwidth of the closed-loop system $\omega_{CL} = 3.5 \text{ rad s}^{-1}$. Correlation between these two parameters and pilot ratings allowed to get the Neal–Smith criterion for the prediction of flying qualities levels (Figure 8.33).

According to Mitchell and Klyde (1998), this criterion has a rather low potentiality in correct prediction of FQ. This led to the modification of the criterion made in Efremov et al. (1996). A different rule for pilot compensation parameter $\Delta\varphi$ was proposed. It was found to be the maximum difference between pilot-phase response characteristics $\varphi|_{Y_p}$ and $\varphi|_{Y_{p\,\text{opt}}}$ ($\Delta\varphi_{\max} = \max_{\Delta\omega}(\varphi|_{Y_p} - \varphi|_{Y_{p\,\text{opt}}})$) that corresponds to the considered dynamic configuration $W_c(j\omega)$ and the optimal controlled element dynamics $Y_{c\,\text{opt}}(j\omega)$ within the any frequency band $\Delta\omega$ that is under consideration. The optimal controlled element dynamics was defined in Efremov et al. (1996) as the dynamics that ensure the simplest type of pilot behavior ($Y_p = K_p e^{-j\omega t}$) within a broad frequency band and a minimum variance σ_e^2 in compensatory tracking task. For the case of $\omega_i = 0.5 \text{ rad s}^{-1}$ and spectrum $S_{ii} = \frac{K^2}{(\omega^2+0.5^2)^2}$, $\tau = 0.18 \div 0.2$ seconds and pilot phase frequency response $\varphi|_{Y_{p\,\text{opt}}} = 57.3\,(0.18 \div 0.2) \cdot \omega$ deg. Several versions of this criterion are considered in Efremov et al. (1996, 2020). The latest modification of this "MAI criterion" (Efremov et al. 2020) is shown in Figure 8.34. For its development, the boundaries of the parameters $\Delta\varphi_{\max}$ and r were determined experimentally using different Neal–Smith, Have PIO, and LAHOS dynamic configurations. The analysis given in Efremov et al. (2020) demonstrated that using this criterion provides

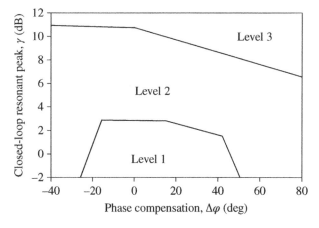

Figure 8.33 Neal–Smith criteria.

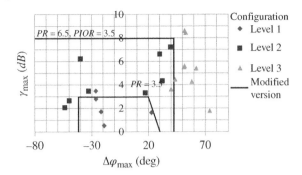

Figure 8.34 Modified MAI criterion (r and $\Delta\varphi$) with boundaries found in experiments.

a percentage of correct FQ predictions equals to 90.9%. Because of the relationship between of PR and PIOR (see Eq. (8.1)), the boundary of the third level of FQ can be considered as the boundary dividing the PIO-prone and non-PIO-prone configurations. This allows to consider the modified MAI criterion as a criterion for PIO prediction as well. The analysis of the potentiality of the criterion for PIO prediction showed that 88.9% of the prone configurations were correctly predicted.

Using the pilot optimal model to calculate the parameters r and $\Delta\varphi_{max}$ required to slightly change the boundaries of the flying qualities levels 1 and 2 to improve the predictive potentiality of the criteria and provide a percentage of correct predictions equal to 89% (Efremov et al. 2020).

The attempt to use the Hess model or any of its modifications when calculating the resonant peak r (a key parameter of the MAI criterion) yielded a poor correlation between the results and boundaries of the initial criterion. According to Efremov et al. (2020) only 52.1% configurations were predicted correctly with none of the first-level configurations and only 31.3% of the third-level configurations being correct. An attempt was made to find another parameter for the pilot-aircraft closed-loop system featuring an improved correlation with pilot's rating. It was found that the correct parameter is the bandwidth ω_{CL} of the pilot-aircraft closed-loop system. This parameter was defined in Section 8.4.1. The calculation of this parameter and the pilot compensation parameters $\Delta\varphi_{max}$ performed by mathematical modeling of the pilot-aircraft system including the modified Hess model allowed to obtain a new criterion (Figure 8.35). This criterion has a high probability of making correct predictions (91.6%) (Efremov et al. 2020).

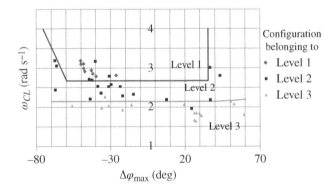

Figure 8.35 New MAI criterion.

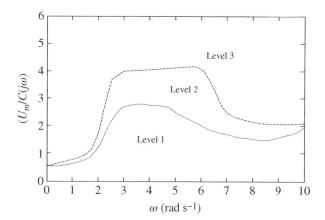

Figure 8.36 Handling qualities boundaries for HQSFs.

Another criterion proposed by Hess is based on the proposition that the central importance for the human pilot to assess a vehicle's handling qualities is the power in the proprioceptive feedback signal $u_m(t)$ when the crossover model (Eq. (8.7)) is satisfied (Hess 1997). It was found that the function $|(U_m/C)(j\omega)|$ could be utilized to predict handling qualities levels because the power of the signal $u_m(t)$ is dependent on this function. The notations of the signals $u_m(t)$ and $c(t)$ are given in Figure 8.28. This way the **handling qualities sensitivity function** (HQSF) was determined:

$$HQSF = \left| \frac{U_m}{C}(j\omega) \right|$$

By analyzing the various aircraft dynamics configurations, obtaining the HQSF and grading them by handling quality-rating levels obtained in flight tests, handling qualities level boundaries were defined as shown in Figure 8.36. This technique showed good results since almost all the HQSFs of the three pilot rating levels from flights tests could be summarized in three areas to form pilot rating level boundaries.

8.6.1.2 Calculated Piloting Rating of FQ as the Criteria

It is apparent from the Cooper–Harper scale (Figure 8.4) that pilot compensation and task performance are the key factors in the rating scale. Consequently, we can expect some connections between these factors and subjective pilot ratings. These connections are intrinsically empirical, and the connections are made somewhat awkward because the rating scale is ordinal. The latter point that can be circumvented as the Cooper–Harper scale can be related to the interval scale. At least in Efremov et al. (1992), the relationship between the Cooper–Harper scale and one of the interval scales (ψ scale (McDonnell 1969)) was determined: $\psi = 3.41 \ln PR + 1.34$. It permits the use of parametric statistics when using ratings. Several functions connecting ratings task performances and pilot's equalization parameters were defined in the past in the framework of the so-called "paper pilot technique" (Anderson 1970). One of them, obtained in the study of the hovering task, is shown in Figure 8.37. The pilot-vehicle model corresponding to this task is shown also in Figure 8.37.

Instead of summarizing the key factors in forming the PR equation, an approach based on Eq. (8.5) was proposed in Efremov (2017) and Efremov et al. (2020) for the development of several

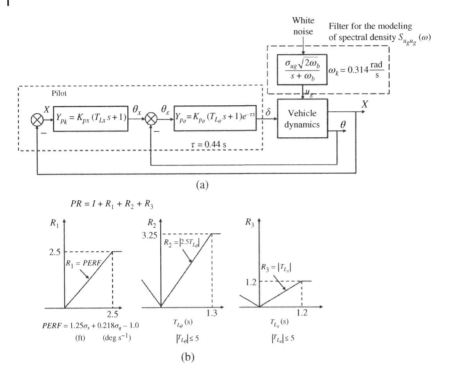

(a)

(b)

Figure 8.37 (a) Typical pilot/vehicle model and form of the (b) pilot rating functional. Source: Adapted from Anderson (1970).

criteria. This technique was used for the flying qualities prediction criteria in a pitch-tracking task, as well as multichannel and multimodal tasks.

8.6.2 Interfaces Design

The display is one of the major elements of a man–machine system. Its role is especially important in the pilot-vehicle system. Here, the pilot continuously perceives the visual information indicated by the instruments (display). This way, the pilot perceives more state variables for each piloting task and closes the corresponding loops.

The display design based on man–machine system modeling has been applied widely for the synthesis of the flight director law:

$$e(s) = Y_1(s)y_1(s) + Y_2(s)y_2(s) + \dots$$

providing the best dynamics of the controlled element, i.e. the dynamics of the display + vehicle system. These goals were realized in Hoh et al. (1977), Hess (1981), and Hoffman et al. (1975) by using the traditional McRuer's or optimal pilot models. In both the cases, the selection of the state variables y_i and filters Y_i is based on the requirement of minimal pilot workload, approaching the controlled element dynamics to the integral in the crossover frequency range, necessary amplitude, and phase margins of the open-loop system, and achieving of the best task performance.

The obtained filters $(Y_i(s))$ are incorporated into the director control law, enabling the pilot to operate as a proportional controller with regard to the director signals with a significantly low workload. Avionics development allowed to incorporate the flight director into the widely used now primary flight display. Its screen displays complex information without requiring the pilot to perform any scanning.

The advantages of preview tracking in comparison with compensatory tracking gave rise to the interest in development of displays generating preview information on future trajectory. A number of publications were dedicated to determining the best way of presenting the preview input signal (Mulder 1999; Wilckens and Schattenmann 1968; Grunwald 1985) and developing the "tunnel-in-the-sky" concept. The suggestion to project the vector of velocity on the cross-section located inside the tunnel at a distance L_{pr} a head of the vehicle and moving at the aircraft velocity transforms the preview display into a predictive display with a preview of the planned trajectory. In this case, the piloting task is the compensation of the error between the predictor and the center of the cross-section. The selection of the parameter $L_{pr} = VT_{pr}$ (Figure 8.22) was performed via experimental studies using a ground-based simulator (Wilckens and Schattenmann 1968) or by using traditional feedback control theory (Sachs 2000). The independence of the describing function Y_{pe} in relation to the preview time T_{prev} (see Section 8.4) allows to select the parameters T_{pr} and T_{prev} separately. The predictive time can be selected by considering the pilot–aircraft–display system shown in Figure 8.38, by minimization of the mean square error $\sigma^2_{\Delta H}$. Here, $\Delta H(t) = h(t) - i(t)$ is the current height error between the aircraft height $h(t)$ and the input signal $i(t)$.

In the general case, the predictive angle $\varepsilon_{pr}(t)$ shown in Figure 8.22 is different in each manual control task.

For example, in an ISS docking task, it has to be equal to $\varepsilon_{pr}(t) = \gamma(t) + \frac{h(t)}{L_{pr}}$ (Efremov et al. 2016).

In the case, where docking with the **International Space Station** (ISS) is executed in **teleoperator mode** (the so-called "TORU"), the control process is accompanied by a considerable time delay ($\tau = 1 \div 1.5$ second) due to the coding and decoding of the signals transmitted from the ISS, where the astronaut is located and back. To compensate for this time delay, it was proposed to use a predictive display implemented with a flight path angle (γ) calculated by using the vehicle mathematical model with the assumption that there is no time delay. In this case, the controlled element dynamics becomes $Y_c(s) = \frac{\varepsilon_{pr}(s)}{\delta_c(s)} = \frac{K_c(e^{-s\tau} + sT_{pr})}{T_{pr}s^2(T_{en}s+1)}$, where T_{en} is the time constant in the jet dynamics. The mathematical modeling of the human–operator–vehicle system with the pilot structural model allowed to obtain the optimal predictive time $T_{pr} = 17$ s. The predictive display corresponding to the considered task is shown in Figure 8.39. The projection of the predictive path angle on the cross-section MN moving inside of the tunnel with the velocity of the spacecraft motion relative to the ISS is the represented by the symbol "a."

The center of such tunnel is indicated by the rhombus "b." The crossing of its axis lies on the line passing the target "d" located at the ISS. Besides these symbols, the center of the screen "c" is indicated on the predictive display as well.

Thus, the task of the operator is to combine the predictive symbol "a" with the center of the predictive window "b" and with the symbol "c" by using jets. During the motion along the planned trajectory, these symbols have to coincide and be directed to the target ("d") and the projection of the velocity vector will demonstrate the direction of the further motion of the vehicle relative to the planned trajectory. This principle of time delay compensation is shown in Figure 8.40.

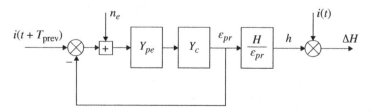

Figure 8.38 Pilot–aircraft–display system.

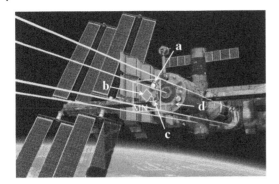

Figure 8.39 Predictive display.

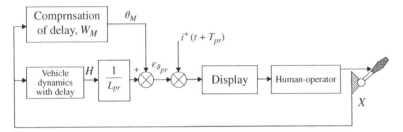

Figure 8.40 Principle of time delay compensation.

Here, $i^*(t + T_{pr})$ is the prediction of the planned trajectory, projected onto the display.

The ground-based simulation confirmed the effectiveness of such display. The root square errors of linear coordinates Z, Y at the moment spacecraft touches the ISS in experiments with such display was 0.2–0.5 cm. This is six to seven times less in comparison with trials performed without such a display.

The same idea was proposed for a lunar rover controlled by an operator located on Earth. In this case, there is a considerably higher time delay ($\tau \cong 4$ s) in the control process. It leads to a "step and go" process of lunar rover control, which deteriorates its research potentialities. A ground-based simulation was performed for simplified rover yaw angle dynamics $\left(Y_c = \frac{K}{s(Ts+1)} \right)$ and the predictive time ($T_{pr} = 7$ s) calculated preliminarily via mathematical modeling of the operator – lunar rover system. These experiments demonstrated that the discrete control process which is typical when using standard displays was transformed into a continuous control process when the predictive display was used (Figures 8.41 and 8.42).

Moreover, the speed of the rover on the lunar surface increases by a factor of 2–2.5 in the last case.

In the aircraft landing task, the equation for predictive path angle ε_{pr} depends on the sign of poles and zeros of the aircraft transfer function $\frac{\gamma}{\delta_e}$. In the case, where all the zeros and poles are negative $\varepsilon_{pr}(t) = \gamma(t) + \dot{\gamma}(t)\frac{T_{pr}}{2} + \frac{h(t)}{L_{pr}}$ (Efremov and Tiaglik 2011). The simplified dynamics of the display-aircraft system in the latter case corresponds to Eq. (8.6).

It was shown in Efremov and Tiaglik (2011) that for the input spectral density $S_{ii}(\omega) = \frac{K^2}{(\omega^2+0.5^2)^2}$ and aircraft with dynamics corresponding to the HP-21 dynamic configuration from the Have PIO database ($\xi = 0.64$, $\omega = 2.4$ rad s^{-1}) (Bjorkman et al. 1986), the predictive time is $T_{pr} = 0.9$ seconds. For the determination of optimal preview time, the technique used for pilot modeling in preview tracking task can be used here as well. For the considered conditions of modeling (S_{ii} and Y_c), the optimal preview time is $T_{prev} \cong 2.4$ s (Efremov et al. 2022). Experiments demonstrate that using

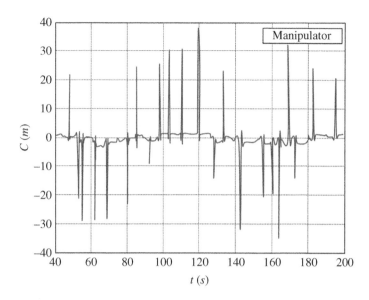

Figure 8.41 Manipulator deflection for the rover control without the predictive display.

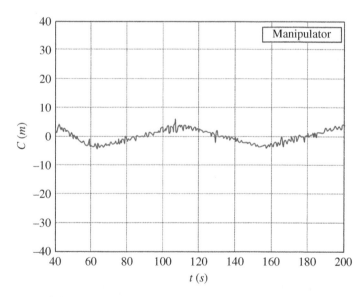

Figure 8.42 Manipulator deflection for rover control with the predictive display.

such a predictive display with a preview of the planned trajectory allows to decrease the variance of error and the mean square stick deflection, in comparison with a primary flight indicator by factors of 2 and 7, respectively (see Section 8.4).

However, in the case where the aircraft is characterized by speed instability or reversible control, the predictive path angle law has to be changed (Efremov et al. 2022), and it has the following form

$$\varepsilon_{pr} = \theta + \frac{h}{L}$$

For that case, another version of predictive display shown in Figure 8.43 was proposed.

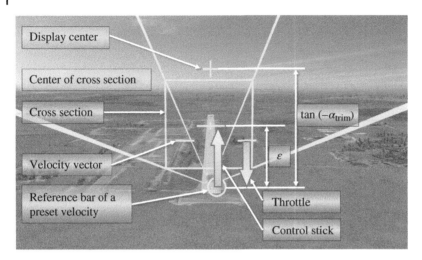

Figure 8.43 Predictive display with a preview of the planned trajectory.

It indicates the predictive path angle ε_{pr} as the difference between the symbols of the center of the tunnel's cross-section located at the distance L_{pr} and the preset velocity vector symbol. With the help of the inceptor, the pilot has to align these symbols. An additional symbol demonstrated on the screen is the velocity vector. With the help of the throttle, the pilot has to minimize the difference between the position of this symbol and the reference bar of the preset velocity vector. When this difference is zero, the position of the display center relative to the reference bar of the preset velocity vector will correspond to the trim angle of attack α_{trim}. Experiments performed using a ground-based simulator demonstrated the high effectiveness of the proposed display in suppressing the effects of wind shear. In this case, the deviations from the glide slope are 5–10 times less in comparison with experiments performed with the standard primary flight display. In case of wake vortex encounter, the deviation in height or lateral variables are no more than 2 m when the proposed display is used.

8.6.3 Optimization of Control System and Vehicle Dynamics Parameters

The catalog of publications dedicated to control system design includes thousands of papers and books. Many of them use the man–machine system approach to solving various problems. One group of these publications is based on the "paper pilot technique" used for the optimization of the flight control system gain coefficients of various vehicles (Anderson 1970; Teper 1972).

The other group of research is founded on parametric optimization by minimization of different cost functions as is discussed in Stapleford et al. (1970), Hess and Peng (2018), and Efremov and Ogloblin (2006). The next approach is the traditional feedback which is used to meet the requirements for a simple type of human operator behavior and the provision of the necessary man–machine system properties or flying qualities criteria formulated in the terms of pilot–aircraft system parameters. This approach is used for the selection of the loops closed by the pilot, development of general methodology of control system design and flight control system synthesis (McRuer and Myers 1988; Ashkenas 1988).

Several examples of the application of pilot-aircraft system approach are given below. The reason for the excessive pilot lead compensation $\left.\frac{d\lg|Y_p(j\omega)|}{d\lg\omega}\right|_{\omega=\omega_c} \geq 40\,\text{dB}\,\text{dec}^{-1}$ in the case of significant time delay in the controlled element dynamics or high negative slope of $|Y_c(j\omega)|$ in the crossover frequency range (see Section 8.4) was studied in Efremov et al. (1992). Here, the experiments with two input signals allowed to obtain the operator frequency characteristics describing the operator's responses to visual and proprioceptive cues. The experiments demonstrated the pilot's ability to react actively to proprioceptive cue in case of such controlled element. Using the modified structural pilot model, this peculiarity was confirmed. The idea to form other information channels impacting on the human operator behavior was offered with the goal to develop the means of preventing the human's reflex stereotype from reacting to proprioceptive cues, increasing the probability of a PIO event. This was implemented by regulating the force (F)/displacement (c) gradient: $F^c = \frac{dF}{dc}$, in the frequency range ($\omega = 2.5 \div 3.5$ rad s^{-1}) in which the resonant peak of the closed-loop system takes place. This implementation is shown in Figure 8.44. Experiments performed using a ground-based simulator confirmed the hypothesis that the pilot actively uses the variable stiffness cues. Such an organization of their behavior leads to the suppression of the resonant peak and a decrease in the slope amplitude of the human operator describing function (Figure 8.45).

The goal to provide the simplest, proportional type of human behavior in the tracking task is implemented by the application of the "inverse dynamics" technique and its modifications (for example NDI, INDI) to the control system design. This technique was used in Efremov et al. (2021) for rotorcraft flight control system design. The results of mathematical modeling of the pilot-vehicle system performed with the modified Hess model demonstrated high effectiveness of this technique

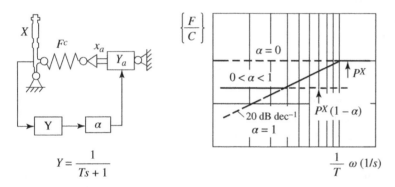

Figure 8.44 System for a dynamical change in inceptor stiffness.

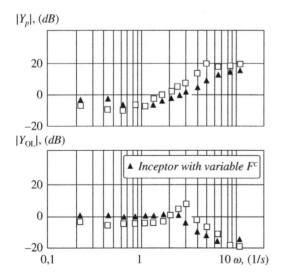

Figure 8.45 Influence of inceptor stiffness.

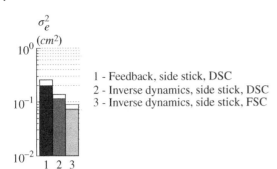

Figure 8.46 Variance of error for different control laws and types of sensing control.

1 - Feedback, side stick, DSC
2 - Inverse dynamics, side stick, DSC
3 - Inverse dynamics, side stick, FSC

in comparison with a control system designed using only feedback gains. It allows to provide the first level of flying qualities for different operating speeds including hovering, effectively increasing piloting accuracy, to decouple the control channels, and to suppress the effects of uncertain aerodynamics in the synthesis of inverse dynamics. The latter is provided by a PI controller in the flight control system. Experiments confirm these results (Figure 8.46).

It was also shown that the effectiveness of the inverse dynamics is higher when using a side stick, in comparison with the central stick. Additionally, a much better synergetic effect was obtained when it was integrated with the FCS type of inceptor output. In the last case, the variance of error decreases by a factor 2.7 in comparison with the case where traditional feedbacks are used and the DSC type of inceptor is used (Figure 8.46).

Man–machine system modeling has a lot of other applications. One of them is understanding the cause of PIO events. In particular, in Ashkenas (1988), a quasilinear pilot model in the dual-loop system was used to determine the possible cause of PIOs taking place during the approach and landing of the Space Shuttle. The vehicle characteristics were represented by the augmented pitch attitude response to the control input $\frac{\theta(s)}{\delta_e(s)}$ and the vehicle response at the pilot station to attitude changes $\frac{H_p(s)}{\theta(s)}$.

The boundaries of the pilot–aircraft system stability demonstrates (Figure 8.47) that while trying to achieve maximum performance, a pilot will increase their gain coefficients $K_{p\theta}$ and K_{pH} in both loops. Increasing these coefficients is accompanied by the narrowing of the stability region. For high pilot gain in the inner-loop $K_{p\theta}$, the range of the stable path gain coefficient K_{ph} is only 1.2 dB. An insignificant variability of the gain coefficient can cause instability in the closed-loop system. Such extreme sensitivity to small increase or decrease in individual pilot control characteristics is the cause of this particular PIO situation (Ashkenas 1988).

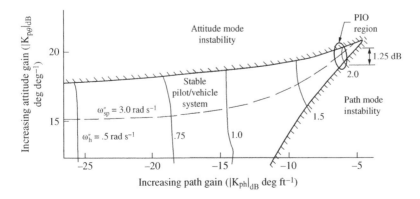

Figure 8.47 Closed-loop path/attitude stability boundaries, pilot/Space-Shuttle ALT system.

8.7 Future Research Challenges and Visions

In spite of certain successes in the studies of regularities of human operator behavior, its modeling and practical use in different applied manual control problems requires conducting additional research. The description of human operator behavior in complex stationary and nonstationary multiloop, multichannel, and multimodal manual control tasks where the human operator simultaneously perceives stimuli of different modalities requires extending the investigations to expose the peculiarities of pilot behavior in these tasks. These include the distribution of human operator attention, coupling arising due to inaccurate human actions in different control channels, disorientation observed when using a moving-base ground simulator, influence of motion, tactile, and proprioceptive cues. Due to the increased role of automatic on-board systems, additional problems have arisen: shared control between the pilot and automatic systems, pilot behavior (their adaptation and ability to generate correct actions) in nonstationary piloting task prompted by failures in automatic systems, sudden switching of flight control system modes, or other anomalies. This time-varying effect causes a considerable change in controlled element dynamics and knowledge of pilot behavior regularities is necessary for the correct reconfiguration of control system laws, minimizing the degradation of man–machine system characteristics.

This knowledge is also necessary for human operator behavior modeling. A modification of the Hess structural model was offered in Hess (2009, 2012) for the case of time-varying aircraft dynamics and a sudden change in vehicle dynamics. This model provides a theoretical framework for the study of nonlinear PIO events. However, a more promising approach to the modeling of pilot behavior in multiaxis and suddenly changing (or time-varying) controlled element dynamics is based on artificial intelligence techniques, several of which have been proposed. One of them is the fuzzy logic technique used for modeling pilot behavior in compensatory and pursuit tracking tasks (Xu et al. 2017) and studying multiaxis manual control in case of flight control system anomalies Xu and Wu (2021). The other techniques are based on the neural network approach. It is shown in Tan et al. (2014) that a pilot model based on this approach provides excellent agreement between the results of experimental studies and the mathematical modeling of pilot control response characteristics in a single-loop tracking task. The widespread development of the neural network approach and its successful application in various engineering tasks allows to assert that this approach will find practical use in the modeling of pilot behavior in complex multiaxis, stationary, and nonstationary manual control tasks.

In addition to the pilot control response characteristics, another group of pilot behavior characteristics – psychophysiological responses – is widely used. One of such widespread characteristics is the rating assigned by the pilot according to the Cooper–Harper flying qualities scale. The use of this scale for flying qualities evaluation demonstrates:

- Disagreement between the PRs assigned by the pilot in-flight and during ground-based simulation (usually, in the range of PR > 6, the ratings assigned in-flight are higher in comparison with the ratings assigned in ground-based simulations, and the opposite tendency is observed in the range of PR < 3÷4).
- High variability of PRs assigned by the pilot in ground and in-flight simulations.

This requires increasing the number of experiments and pilots, which would increase the duration of studies and their cost.

These problems require solving a series of tasks:

- Improvement of the technique used when performing experiments;
- More accurate definition of the key metrics of the Cooper–Harper scale (task performance and especially pilot compensation) for each piloting task;

- Accurate modeling of aircraft dynamics and the flight control system, improvement of the visual and motion cues simulated by their respective simulator subsystems;
- Selection of piloting tasks and maneuvers to be used in the evaluation of flying qualities in ground-based simulations.

The influence of these factors on the agreement between ground-based and in-flight evaluations of flying qualities is considered in Efremov et al. (2014).

With respect to the applications of human operator modeling, several tasks are given below. As a rule, the design of any element of a man–machine system is carried out for preselected characteristics of other elements of this system. Preliminary studies have shown that simultaneous optimization of all elements of the man–machine system (vehicle + control system + display + inceptor) allows to obtain a synergetic effect which improves system performance. This fact allows to recommend the use of such an integrated approach for the optimization of man–machine systems in a wider range of engineering problems.

In spite of the predictive display with a preview of the planned trajectory considered in this chapter being highly effective, it has seen limited use in aviation. The displays developed by the Garmin Corporation show the planned trajectory (or as they call it, "Highway in the Sky"), which allows the pilot to better understand the spatial orientation of the aircraft. However, it does not show the key parameters of the predictive display: the predictive law, cross-section, and lengths of the tunnel in front of and behind the cross section. Showing such symbols and parameters on the display would improve its effectiveness. The predictive display would be very useful for spacecraft control (for docking or landing on the Moon's surface, for example) and UAV control from a ground-based station, especially in the case of a significant time delay in the transition of the command signal. This type of display can prove very effective for car steering (especially in racecars).

Analysis of the different criteria developed for FQ prediction shows that only several of them are based on the requirements for pilot-aircraft system parameters. All of them are criteria for the evaluation of FQs in the longitudinal control channel, and none presents a criterion for FQ evaluation in the lateral channel. Additionally, there are only several studies on developing criteria for predicting PIO Category 2. The disagreement between the results of FQ evaluation obtained using fixed-base and in-flight (or moving-base) simulators considered in Section 8.4.2.3 demonstrate the necessity to take motion cues into account when developing the criteria or at least to modify the existing FQ criteria, taking this factor into consideration more accurately.

Significant progress in the development of handling qualities requirements for rotorcraft (Anonymous, United States Army Aviation and Missile Command, Aviation Engineering Directorate 2000) was achieved at the beginning of the twenty-first century. In that document, all requirements are divided into different piloting tasks, types of control response, and levels of visual cues. As for fixed-wing aircraft, such a document (Anonymous, Department of Defense, Handbook 1997) was published at approximately the same time, but all piloting tasks requiring high accuracy and aircraft maneuverability were grouped into the same category despite each of them being characterized by different controlled element dynamics. Because of the significant influence of this task variable on the pilot-aircraft system characteristics, the principle of division of requirements into categories has to be changed to division according to piloting task.

8.8 Conclusion

The attempts of a human operator to maintain the relative constancy of man–machine system properties are accompanied by his or her adaptation to system variables – the formation of the necessary control loops and corrective actions in each of them. Understanding the adaptation properties considered in this chapter, along with the processes of perception and command action execution, is integral to the mathematical modeling of human operator behavior in control tasks.

The primary models currently employed are the quasilinear model and its extensions – the structural model, as well as the optimal model of the human operator. Their successful use in solving applied tasks instills optimism about the models' relevance in creating new highly augmented human-operated machines.

References

Anderson, R.O. (1970). A new approach to the specification and evaluation of flying qualities, AFFDL-TR-69-120.

Anonymous, Department of Defense, Handbook. Flying qualities of piloted aircraft, MIL-1979A, Department of defense, 1997, 849 pp.

Anonymous, International air Transport Association. A Safety Report 2019, Edition 55, IATA, 2019, 262 pp.

Anonymous, International air Transport Association. A Safety Report 2020, Edition 57, IATA, 2021, 244 pp.

Anonymous, United States Army Aviation and Missile Command, Aviation Engineering Directorate. *Aeronautical design standard performance specification handling qualities requirements for military aircraft*, ADS-33E-PRF, 2000, 106 pp.

Ashkenas, I.L. (1988). *Pilot Modeling Applications*. Systems Technology, Inc., AGARD Lecture Series No. 157, Advances in Flight Qualities.

Baron, S., Kleinman, D.L., Miller, D.C. et al. (1970). Application of optimal control theory to the prediction of the human performance in a complex task, AFFDL-TR-69-81.

Bjorkman, E.A., Eidsaune, D.W., Wilkinson, O.C. et al. (1986). NT-33. Pilot induced oscillation prediction evaluation. In: *Air Force Inst. of Technology USAFTPS-TR-85B-S4*, 1–165. Dayton, OH: Wright–Patterson AFB.

Bliss, I.C., Crane, H.D., Townsend, I.T. et al. (1967). Tactual perception – Experiments and models, NASA-CR-623, 172 p.

Britain, G. and Authority, C.A. (1997). Review of General Aviation Fatal Accidents 1985–1994, CAP 667.

Bryson, A.E. and Ho, Y.C. (1969). *Applied Optimal Control: Optimization, Estimation and Control*. Waltman MA: Taylor & Francis, 481 pp.

Cooper, G.E. and Harper, R.P. (1969). *The Use of Pilot Rating in Evaluation of Aircraft Handling Qualities*. NASA TND-5153, 52 pp.

Curry, R., Hoffman, W., and Young, L. (1976). Pilot modeling for manned simulation, AFFDL-TR-76-124, Vol. 1.

Davidson, J. and Shmidt, D. (1992) Modified optimal control pilot model for computed edit design and analysis, NASA-TN-4384.

Doman, D. (1999). Projection methods for order reduction of optimal human operator models, Ph.D. Thesis.

Efremov, A.V. (2017). *Pilot-Aircraft System. The Regularities and Mathematical Models of Pilot Behavior*, 193. Moscow: Moscow Aviation Institute Press.

Efremov, A.V. and Ogloblin, A.V. (2006). Progress in pilot in the loop investigations for flying qualities prediction and evaluation. In: *25th ICAS Congress*, Hamburg.

Efremov, A.V. and Tiaglik, M.S. (2011). *The Development of Perspective Displays for Highly Precise Tracking Task in the Book*, 163–174. Springer: Advances in Aerospace Guidance, Navigation and Control.

Efremov, A.V., Ogloblin, A.V., Predtechensky, A.N., and Rodchenko, V.V. (1992). *Pilot as a Dynamic System*, 332. Moscow: Mashinostroenie.

Efremov, A.V., Rodchenko, V.V., and Boris, S. (1996). Investigation of Pilot Induced Oscillation Tendency and Prediction Criteria Development, Final report WL-TR, 138 pp.

Efremov, A.V., Alexandrov, V.V., Koshelenko, A.V. et al. (2010). Development of pilot behavior modeling and its application to manual control task. In: *27th Congress of the International Council of the Aeronautical Sciences 2010*, 3319–3326. Nice: ICAS Congress.

Efremov, A.V., Koshelenko, A.V., Tjaglik, M.S., and Tjaglik, A.S. (2014). The ways for improvement of agreement between in-flight and ground-based simulation for evaluation of handling qualities and pilot rating. In: *29th Congress of ICAS*, St. Petersburg.

Efremov, A.V., Tyaglik, M.S., Tyaglik, A.S., and Aleksandrov, V.V. (2016). The flying qualities improvement for the vehicles with time delay in their dynamics, 28 ICAS Congress.

Efremov, A.V., Efremov, E.V., and Tiaglik, M.S. (2020). Advancements in predictions of flying qualities, pilot induced oscillation tendencies and flight safety. *Journal of Guidance, Control and Dynamics* 4–14.

Efremov, A.V., Mbkayi, Z., and Efremov, E.V. (2021). Comparative study of different algorithms for flight control system design and the potentiality of their integration with a side stick. *Aerospace* 8 (10): 290.

Efremov, A.V., Tiaglik, M.S., and Irgaleev, I.K. (2022). *Pilot behavior model in pursuit and preview tracking task*. San Diego: AIAA SciTech Forum.

Federal Aviation Administration (FAA) (2009). 14 CFR part 25 Airworthiness standards: Transport category airplanes.

Grunwald, A.J. (1985). Predictor Laws for Pictorial Flight Displays. *Journal of Guidance and Control* 8 (5): 545–552.

Hart, S. and Staveland, L. (1988). Development of NASA-TLX (Task Load Index): results of empirical and theoretical research. In: *Human Mental Workload* (ed. P. Hancock and N. Meshkati), 139–183. North Holland, Amsterdam: Elsevier.

Hess, R. (1976). A method for generating numerical pilot opinion ratings using the optimal model, NASA-TM-X7310.

Hess, R. (1979). Structural model of the adaptive human pilot. *Journal of Guidance and Control* 3 (5): 416–423.

Hess, R. (1981). Aircraft control display analysis and design using the optimal control model of the human pilot. *IEEE Transactions on MMS* 11 (7): 165–480.

Hess, R. (1990). Analyzing manipulator and Feel System Effects in Aircraft Flight Control. *IEEE Transactions on Systems Man and Cybernetics* 20 (4): 923–931.

Hess, R. (1997). *Unified theory for aircraft handling qualities and adverse pilot-aircraft coupling. Journal of Guidance, Control and Dynamics* 20 (6): 1141–1148.

Hess, R. (1999). A model for the human use of motion cues in vehicular control. *Journal of Guidance, Control and Dynamics* 13 (3): 476–482.

Hess, R. (2007). Obtaining multiloop pursuit control pilot models from computer generation. *Journal of Aerospace Engineering* 22 (G2): 187–200.

Hess, R.A. (2009). Modeling pilot control behavior with sudden changes in vehicle dynamics. *Journal of Aircraft* 45 (5): 1584–1592.

Hess, R. (2012). Modeling pilot definition of time-varying aircraft dynamics. *Journal of Aircraft* 49 (6): 2100–2104.

Hess, R. and Peng, C. (2018). *Design for robust aircraft flight control. Journal of Aircraft* 55 (2).

Hodgkinson, J., Page, M., Preston, J., and Gillette, D. (1992). Continuous flying qualities improvement the measure and the payoff, , AIAA Paper: AIAA-92-413227CP, pp. 172–180.

Hoffman, W., Curry, R., Kleinman, D., and Young, L. (1975). Display control requirements for VTOL aircraft, NASA-CR-14026, 276 p.

Hoh, R., Klein, R., and Johnson, W. (1977). Development of integrated configuration management/flight director system for piloted STOL approach, NASA-CR-2883.

Kleiman, D., Baron, S., and Levison, W. (1970). An optimal control model of human behavior. *Automatica* 6: 357–369.

Kleinman, D. and Baron, S. (1973). Manned vehicle system analysis by modern control theory, NASA-CR-17-53.

Klyde, D. and McRuer, D. (2009). Smart cue and smart gain concepts to alleviate loss of control. *Journal of Guidance, Control and Dynamics* 32: 1409–1417.

Levison, W., Baron, S., and Kleinman, D. (1969). A model for human controller remnant. *IEEE Transactions on MMS*, MMS-10 4 (1): 101–108.

McDonnell, J. (1969). An application of measurement methods to improve the quantitative nature of pilot rating scales. *IEEE Transactions on MMS*, MMS-10 3: 81–92.

McRuer, D.T. and Krendel, E.S. (1957). *Dynamic Response of Human Operators*, WADC-TR-56-524. OH: Wright Air Development Center, Wright-Patterson Air Force Base.

McRuer, D. and Krendel, E. (1974). Mathematical models of human pilot, AGARD, AGD-188, p. 72.

McRuer, D.T. and Myers, T.T. (1988). *Advanced Piloted Aircraft Flight Control System Design Methodology*, vol. I. Hawthorne, CA 90250, Contract No. NAS1-17987: Knowledge Base, System Technology, Inc.

McRuer, D., Graham, D., Krendel, E., and Reisener, W. (1965). Human Pilot Dynamics in Compensatory Systems, AFFDL-TR-65-15, p. 194.

McRuer, D., Clement, W., Thompson, P. and Magdaleno, R., Minimum flying qualities, Vol. II, WRDC-TR-89-3125, Jan., 1990, 132 pp.

Mitchell, D. and Klyde, D. (1998). *A critical examination of PIO prediction criteria*, AIAA-98-4335, A98-37245, pp. 415-427

Mulder, M. (1999). Cybernetics of tunnel-in-the-sky displays, Ph.D. dissertation, Aerospace Engineering, TU Delft, Delft, The Netherlands.

Mulder, M. and Mulder, J.A. (2005). Cybernetic analysis of perspective flight-path display dimensions. *Journal of Guidance, Control, and Dynamics* 28 (3): 398–411.

Neal, T.P. and Smith, R.E. (1971). A flying qualities criteria for the design of fighter-control system. *Journal of Aircraft* 8 (10): 803–809.

Reid, L.D. and Drewell, N.H. (1972). A pilot model for tracking with preview in process. In: *Proceedings of the 8th Annual Conference on Manual Control*, 191–204. Michigan: University of Michigan Ann Arbor.

Sachs, G. (2000). Perspective predictor flight – path display and minimum pilot compensation. *Journal "Guidance, control and dynamics"* 23 (3).

Schmidt, H., Bekey, D., and Reswick, J. (1976). Two dimensional compensatory tracking with tactile displays. In: *Proceedings of the Twelve NASA-University Annual Conference on Manual Control*, 332–354. USA: NASA.

Smith, R. (1975). A theory for handling qualities with application to MIL 8785B WPAFB, AFFDL-TR-75-119.

Smith, R.E. (1978). *Effects of Control System Dynamics on Fighter Approach and Landing Longitudinal Flying Qualities*, vol. 1. Calspan Advanced Technology Center.

Stapleford, R.L., McRuer, D.T., Hofmann L.G., and Teper G.L. (1970). A practical optimization design procedure for stability augmentation systems, AFFDL-TR-70-11.

Tan, W., Wu, Y., Qu, X., and Efremov, A.V. (2014). A methods for predicting aircraft flying qualities using neural network pilot model. In: *2nd International Conference on Systems and Informatics*, 258–263. IEEE.

Teper, G.L. (1972). An assessment of the "Paper Pilot" – An analytical approach to the specification and evaluation of flying qualities, AFFDL-TR-71-147.

Tomizuka, M. and Whitney, D.E. (1973). The preview control problem with application to man–machine system analysis. In: *Proceedings of the 9th Conference on Manual Control*, 429–441. MIT.

Tustin, A. (1944). *An Investigation of the Operator's Response in Manual Control of a Power Driven Gun.* Metropolitan – Vickers Electrical Co. Ltd., C.S. Memorandum №169, Attercliffe Common Works, Sheffield, England.

Uhlemann, H. and Geiser, G. (1975). Multivariable manual control with simultaneous visual and auditory presentation of information. In: *Proceedings of the Eleven NASA-University Annual Conference on Manual Control*, 3–18. NASA.

Van der El, K., Pool, D.M., Damveld, H.J. et al. (2016). *An Empirical Human Controller Model for Preview Tracking Tasks. IEEE Transactions on Cybernetics* 46 (11): 2609–2621.

Van der El, K., Pool, D.M., van Paassen, M.M., and Mulder, M. (2018). Effects of linear perspective on human use of preview in manual control. *IEEE Transactions on Human–Machine Systems* 48 (5): 496–508.

Van der El, K., Pool, D.M., van Paassen, M.M., and Mulder, M. (2018a). *Effects of Preview on Human Control Behavior in Tracking Tasks with Various Controlled Elements. IEEE Transactions on Cybernetics* 48 (4): 1242–1252.

Van der El, K., Padmos, S., Pool, D.M. et al. (2018b). *Effects of Preview Time in Manual Tracking Tasks. IEEE Transactions on Human–Machine Systems* 48 (5): 486–495.

Van der El, K., Pool, D.M., van Paassen, M.M., and Mulder, M. (2020). *Effects of Target Trajectory Bandwidth on Manual Control Behavior in Pursuit and Preview Tracking. IEEE Transactions on Human–Machine Systems* 50 (1): 68–78.

Vinje, E. and Pitkin, E. (1972). Human operator dynamics for aural compensatory tracking. *IEEE Transactions on Systems, Man and Cybernetics* SMC-2: 504–512.

Wasicko, R., McRuer, D., and Magdaleno, R. (1966). Human dynamics in single-loop system with compensatory and pursuit display, AFFDL-TR-66-137, 65 pp.

Wilckens, V. and Schattenmann, W. (1968). *Test Results with New Analog Displays for All Weather Landing.* In: *AGARD Conference Proceedings "Problems of the Cockpit Environment"*, CP-55, 10.1–10.33.

Wood, J.R. (1983). Comparison of fixed-base and in-flight simulation results for lateral high order systems. In: *Proceedings of AIAA Atmospheric Flight Mechanics Conference and Exhibit.* USA.

Xu, S. and Wu, Y. (2021). Modeling multi-loop intelligent pilot control behavior for aircraft-pilot coupling analysis. *Aerospace Science and Technology* 112: 106651.

Xu, S., Tan, W., Efremov, A. et al. (2017). *Preview of control models for human pilot behavior. Annual Reviews in Control* 44: 274–291.

9

Safe Shared Control Between Pilots and Autopilots in the Face of Anomalies

Emre Eraslan[1], Yildiray Yildiz[2], and Anuradha M. Annaswamy[3]

[1]*Department of Mechanical Science & Engineering, University of Illinois at Urbana-Champaign, Urbana-Champaign, IL, USA*
[2]*Department of Mechanical Engineering, Bilkent University, Ankara, Turkey*
[3]*Department of Mechanical Engineering, Massachusetts Institute of Technology, Cambridge, MA, USA*

9.1 Introduction

The twenty-first century is witnessing large transformations in several sectors including energy, transportation, robotics, and healthcare related to autonomy. Decision-making using real-time information over a large range of operations as well as the ability to adapt online in the presence of various uncertainties and anomalies is the hallmark of a resilient system. In order to design such a system, a variety of challenges needs to be addressed. Uncertainties may occur in several forms, both structured and unstructured. Anomalies may often be severe that require rapid detection and swift action to minimize damage and restore normalcy. This chapter addresses the difficult task of making decisions in the presence of severe anomalies. While the specific application we focus on is flight control, the overall solutions we propose are applicable for general complex dynamic systems.

The domain of decision-making is common to both human experts and feedback control systems. Human experts routinely make several decisions when faced with anomalous situations. In the specific context of flight control systems, pilots often take several decisions based on the sensory information from the cockpit, situational awareness, their expert knowledge of the aircraft, and ensure a safe performance. Autopilots in fly-by-wire aircraft are programmed to provide the appropriate corrective input to help the requisite variables follow the specified guidance commands accurately. Advanced autopilots ensure that such a command following occurs even in the presence of uncertainties and anomalies. However, the process of assembling various information that may help detect the anomaly may vary between the pilot and autopilot. Once the anomaly is detected, the process of mitigating the impact of anomaly may also differ between them. Nature of perception, speed of response, intrinsic latencies may all vary significantly between the two decision-makers. It may be argued that perception and detection of the anomaly may be carried out efficiently by the human pilot, whereas fast action following a command specification may be best accomplished by an autopilot. Our thesis in this chapter is that an approach is needed that designs cyber–physical–human systems (CPHS) with a shared control architecture where the decision-making of the human pilot is judiciously combined with that of an advanced autopilot in flight control problems when severe anomalies are present.

Cyber–Physical–Human Systems: Fundamentals and Applications, First Edition.
Edited by Anuradha M. Annaswamy, Pramod P. Khargonekar, Françoise Lamnabhi-Lagarrigue, and Sarah K. Spurgeon.

The larger rubric of CPHS consists of humans, physical plants, and cyber technologies that are interconnected to accomplish a certain goal (Yildiz 2020). Despite the concise name, the level of interaction and the distribution of tasks between the components, and the extent to which these components contribute to the system are still a complex issue to be resolved. In order to understand the optimal manner in which human operators and automation coordinate their decisions in a given context, two important dimensions emerge, that is, task allocation and timeline (Kun et al. 2016; Paternò et al. 2021). The roles of humans and automation can then be delineated using these dimensions. A successful CPHS can be designed using a granularity assignment, which we define to denote the appropriate assignment of humans and automation to various tasks and with an appropriate timeline (see Figure 9.1 for a schematic).

Existing frameworks that combine humans and automation generally rely on human experts supplementing flight control automation, i.e. the system is semiautonomous, with the automation doing all of the work to handle uncertainties and disturbances, and human taking over control once the environment imposes demands that exceeds the automation capabilities (Hess 2009, 2014, 2016). This approach causes bumpy and late transfer of control from the machine to the human, causing the shared-control architecture to fail, as it is unable to keep pace with the cascading demand which may cause actual accidents (Woods and Hollnagel 2006; Woods and Branlat 2011; National Transportation Safety Board NTSB/AAR-96/01,Washinton, D.C., 1996; Woods 2018). This suggests that alternative architectures of coordination between human expert and automation may be needed. In other words, an architecture associated with a suitable granularity assignment as in Figure 9.1 has to be determined that delineates who performs what tasks and when.

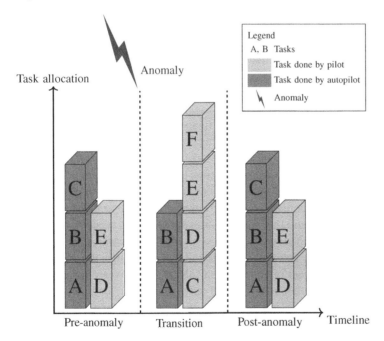

Figure 9.1 A successful CPHS carries out an appropriate granularity assignment of tasks and timelines and is depicted in this figure. The granularity assignment can be divided into three phases according to the timeline of events, that is, pre-anomaly, transition region, and post-anomaly. In the pre-anomaly phase, both the autopilot and pilot carry out their own tasks. In the presence of an undesired event, the autopilot and pilot might be assigned different tasks to mitigate the anomaly. Thanks to his/her critical decision-making, the pilot could have more responsibilities than the autopilot and even undertake a task from the autopilot. When the anomaly is handled by the collaboration of both agents, they might assume their own tasks again depending on the nature of the shared control architecture.

In Section 9.2, we articulate three different architectures of coordination in a CPHS so as to successfully respond to severe anomalies. Recent results that correspond to two of these architectures are then outlined. In each case, the corresponding granularity assignment figure is presented thereby underscoring the usefulness of such an exercise in determining a successful CPHS.

9.2 Shared Control Architectures: A Taxonomy

Various shared control architectures have been proposed in the literature over the years, which can be broadly classified into three forms: a trading action where humans take over control from automation under emergency conditions (Abbink et al. 2018), a supervisory action where the pilot assumes a high-level role and provides the inputs and setpoints for the automation to track (Sheridan 2013), and a combined action where both automation and human expert participate at the same time scale (Mulder et al. 2012). We label all three forms under a collective name of Shared Control Architecture (SCA), with SCA1 denoting the trading action, SCA2 denoting the supervisory action, and SCA3 denoting the combined action (see Figure 9.2 for a schematic of all three forms of SCAs). The adjective *shared* in the phrase of "shared control architectures" (SCAs) is used in a broad sense, where a human-expert shares the decision-making with automation in some form or the other in the CPHS (Abbink et al. 2018). These SCAs can be grouped into three categories: SCA1 as traded (Abbink et al. 2018), SCA2 as supervised (Sheridan 2013), and SCA3 as combined (Mulder et al. 2012).

In SCA1, the authority shifts from the automation to the human-expert as an emergency override, when dictated by the anomaly. For example, in a teleoperation task, if a robot autonomously travels on the remote environment and at certain time-intervals a human operator takes over the control and provides real time speed and direction commands, then this type of interaction can be considered as traded control.

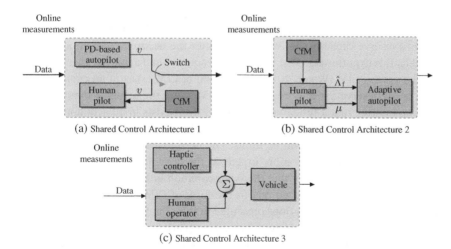

(a) Shared Control Architecture 1

(b) Shared Control Architecture 2

(c) Shared Control Architecture 3

Figure 9.2 An overview of SCAs. (a) represents SCA1, in which the autopilot takes care of the control task until an anomaly, which is then transferred to the pilot based on capacity for maneuver (CfM). (b) represents the SCA2, in which the pilot undertakes a supervisory role. CfM values approaching to dangerously small numbers may trigger the pilot to assess the situation and convey necessary auxiliary inputs μ and $\hat{\Lambda}_f$ to the autopilot. (c) represents SCA3, which is based on the combined control effort continuously transmitted by both haptic controller and human operator who has the choice of abiding by or dismissing the control support of the automation.

In SCA2, the human-expert is on-the-loop, and provides a supervisory, high-level input to the automation. One such example may occur in the teleoperation scenario mentioned above, where the human operator can provide reference points for the mobile robot and the control system on the robot can autonomously navigate the vehicle, help it avoid obstacles and reach the reference points.

In SCA3, different from SCA1 and SCA2, the human and automation are both active simultaneously. For example, in the discussed teleoperation task, if the human controls the robot via a joystick in real time and at the same time the robot controller continuously sends force inputs to the joystick to keep the robot collision-free on the terrain, then this can be considered as combined control.

The main distinction between different classifications originates from the timeline associated with the human and automation control inputs. The commonality between the classes, however, is that both the human and the automation share the load of control, albeit with different time-scales. Therefore, for ease of exposition, we use "shared control" as the collective rubric for both of the interaction types investigated in this chapter.

It is noted that the use of the adjective *shared* in this article is different than what has been proposed by Abbink et al. (2018) and Sheridan (2013), where the term *shared* is reserved exclusively for actions that both humans and automation simultaneously exert, at the same time-scale. We denote this as a "combined action" (SCA3) and the collective human–automation control architectures as *shared*. As we better understand the ramifications of these different types of CPHS, these definitions may evolve as well.

9.3 Recent Research Results

In this section, we provide recent results involving SCA1 and SCA2. In these studies, in order to have an efficient coordination between humans and automation, two principles from cognitive engineering are utilized as to how humans add resilience to complex systems (Woods and Hollnagel 2006; Woods and Branlat 2011; Woods 2018). The first principle is capacity for maneuver (CfM), which denotes a system's reserve capacity that will remain after the occurrence of an anomaly (Woods and Branlat 2011). It is hypothesized that resiliency is rooted in and achieved via monitoring and regulation of the system CfM (Woods 2018). The notion of CfM exists in an engineering context as well, which is the reserve capacity of an actuator. Viewing the actuator input power as the system capacity, and noting that a fundamental capacity limit exists in all actuators in the form of magnitude saturation, one can define CfM for an autopilot-controlled system as the distance between the control input and the saturation limits. The need to avoid actuator saturation, and therefore increasing CfM, becomes even more urgent in the face of anomalies which may push the actuators to their limits. This implies that there is a common link between a pilot-based decision-making with that of an autopilot-based one in the form of CfM. This commonality is utilized in both of the SCA1 and SCA2 applications presented in this chapter.

The foundations for SCA1 and SCA2 come from the results of (Farjadian et al. 2016; Thomsen et al. 2019; Farjadian et al. 2020), with Farjadian et al. (2016) and Thomsen et al. (2019) corresponding to SCA1 and Farjadian et al. (2020) corresponding to SCA2. In Farjadian et al. (2016), the assumption is that the autopilot is designed to accommodate satisfactory operation under nominal conditions, with the requisite tracking performance. An anomaly is assumed to occur in the form of loss of effectiveness of the actuator (which may be due to a damage to the control surfaces caused by a sudden change in environmental conditions or a compromised engine due to bird strikes). The trading action proposed in Farjadian et al. (2016) is for the pilot to step in and take over from the autopilot, based on the pilot perception that the automation is unable to cope with the anomaly. This perception is based on the CfM of the actuator, and when it exceeds a certain threshold, the pilot is proposed to take over control. A well-known model of the pilot in Hess (2009, 2014, 2016) is utilized to propose a specific sequence of control actions that the pilot takes once the control authority has changed. It is shown through simulation studies that when the pilot carries out this sequence of perception and control tasks, the effect of anomaly is contained, and the tracking performance is maintained at a satisfactory level. A slight variation of the above trading role is reported in Thomsen et al. (2019), where instead of the "{autopilot is active, anomaly occurs, pilot takes over}" sequence, the pilot transfers control from one autopilot to a more advanced autopilot following the perception of an anomaly. In the presented studies, the attention is limited to the SCA1 architecture in Farjadian et al. (2016) and its performance validation using human-in-the-loop simulations.

The SCA2 architecture utilized in Farjadian et al. (2020) differs from SCA1, as mentioned above, as the trading action from the autopilot to the pilot is replaced by a supervisory action by the pilot. In addition to using CfM, this architecture utilizes a second principle from cognitive engineering denoted as graceful command degradation (GCD). GCD is proposed as an inherent metric adopted by humans (Woods and Hollnagel 2006) that allows the underlying system to function so as to retain a target CfM. As the name connotes, GCD corresponds to the extent to which the system is allowed to relax its performance goals. When subjected to anomalies, it is reasonable to impose a command degradation; the greater this degradation, the larger the reserves that the system possesses when recovered from the anomaly. A GCD can then be viewed as a control variable that is tuned so as to permit a system to reach its targeted CfM. The role of tuning this variable so that the desired CfM is retained by the system is relegated to the pilot in Farjadian et al. (2020). In particular, a parameter μ is transferred to the autopilot from the autopilot, which is shown to result in an ideal trade-off between the CfM and GCD by the overall closed-loop system with SCA2. In the discussed results, SCA2 is validated with a high-fidelity model of an aircraft and with human-in-the-loop simulations.

The subjects used in the human-in-the-loop simulations include an airline pilot, who is also a flight instructor, with 2600 hours of flight experience, as well as human subjects who were trained in a systematic manner by the experiment designer. The details of the training is explained in Section 9.3.4. It is shown that the SCA1 in Farjadian et al. (2016) and SCA2 in Farjadian et al. (2020) indeed lead to better performance as conjectured therein when the human expert carries out their assigned roles in the respective architectures. The resulting solution therefore is an embodiment of an efficient cyber–physical human system, a topic that is of significant interest of late.

In addition to the aforementioned references, SCAs have been explored in haptic shared control (HSC), which has applications in both aerospace (Smisek et al. 2016) and automotive (van der Wiel et al. 2015; Mulder et al. 2012) domains. In HSC, both automation and human operator exert forces on a common control interface, such as a steering wheel or a joystick, to achieve their individual goals. In this regard, HSC can be considered to have a SCA3 architecture. In Smisek et al. (2016), the goal of the haptic feedback is to help a unmanned aerial vehicle (UAV) operator avoid an obstacle. In van der Wiel et al. (2015), one of the goals is lane-keeping and in Mulder et al. (2012), the goal is assisting the driver while driving around a curve. All of these approaches provide situational awareness to the human operator, and the human has the ability to override the haptic feedback by exerting more force to the control interface. An interesting HSC study is presented in Smisek et al. (2014), where the haptic guidance authority is modified adaptively according to the intensity of the user grip, in which, challenging scenarios are created by introducing force disturbances or incorrect guidance. Another SCA approach is to enable the automation to affect the control input directly instead of using a common interface (Rossetter and Gerdes 2006). In this approach, the human operator can override the automation by disengaging it from the control system. Since SCA1 and SCA2 architectures help with anomaly mitigation and SCA3 is not shown to be relevant in this respect, in this chapter, we mainly focus on SCA1 and SCA2 and their validations with human-in-the-loop simulation studies.

In the specific area of flight control, SCAs have also been proposed in order to lead to greater situational awareness. A bumpless transfer between the autopilot and pilot is proposed to occur in Ackerman et al. (2015) using a more informative pilot display. It is argued in Ackerman et al. (2015) that a common problem in pilot–automation interaction is the unawareness of the pilot of the automation state and the aircraft during the flight, which makes the automation system opaque to the pilot, prompting the use of such a pilot display. The employment of the display is discussed in Ackerman et al. (2015), in the presence of automation developed in Tekles et al. (2014) and Chongvisal et al. (2014) that provides flight envelope protection together with a loss of control logic. A review of recent advances on human–machine interaction can be found in Kaber (2013), where the focus is on adaptive automation, in which the automation monitors the human pilot to adjust itself accordingly, so as to lead to improved situational awareness. No severe anomalies are however considered in Ackerman et al. (2015), Tekles et al. (2014), Chongvisal et al. (2014), and Kaber (2013) which is the focus of this chapter. The goal of the shared control architectures discussed herein consisting of human and automation is to lead to a bumpless efficient performance in the presence of severe anomalies.

Sections 9.3.1 and 9.3.2 introduce the two main components of SCA, respectively: the autopilot and the human pilot. Section 9.3.1 provides the details of the employed controllers. Section 9.3.2 first reviews the main mathematical models of pilots proposed in the literature in the absence of a severe anomaly. The human model used in the development of the proposed SCAs is then proposed. The working principles of these SCAs are later presented in detail in Section 9.3.3. Finally, Section 9.3.4 shows how the underlying ideas are tested with human subjects.

9.3.1 Autopilot

In this section, we describe the technical details related to the autopilot discussed in Farjadian et al. (2016, 2020), Eraslan et al. (2019), and Eraslan (2021).

9.3.1.1 Dynamic Model of the Aircraft

Since the autopilot in Farjadian et al. (2020) is assumed to be determined using feedback control, the starting point is the description of the aircraft model, which has the form

$$
\begin{aligned}
\dot{x}(t) &= Ax(t) + B\Lambda_f u(t) + d + \Phi^T f(x) \\
y(t) &= Cx(t)
\end{aligned}
\tag{9.1}
$$

where $x \in \mathbb{R}^n$ and $u \in \mathbb{R}^m$ are deviations around a trim condition in aircraft states and control input, respectively, d represents uncertainties associated with the trim condition, and the last term $\Phi^T f(x)$ represents higher-order effects due to nonlinearities. A is a $(n \times n)$ system matrix and B is a $(n \times m)$ input matrix, both of which are assumed to be known, with (A, B) controllable, and Λ_f is a diagonal matrix that reflects a possible actuator anomaly with unknown positive entries λ_{f_i}. C is a known matrix of size $(k \times n)$ chosen so that $y \in \mathbb{R}^k$ corresponds to an output vector of interest. It is assumed that the anomalies occur at time t_a, so that $\lambda_{f_i} = 1$ for $0 \le t < t_a$, and λ_{f_i} switches to a value that lies between 0 and 1 for $t > t_a$. It is finally postulated that the higher-order effects are such that $f(x)$ is a known vector that can be determined at each instant of time, while Φ is an unknown vector parameter. Such a dynamic model is often used in flight control problems (Lavretsky and Wise 2013).

As the focus of this chapter is the design of a control architecture in the context of anomalies, we explicitly accommodate actuator constraints. In particular, we assume that the u is assumed to be position/amplitude limited and modeled as

$$u_i(t) = u_{\text{max}_i} \text{ sat}\left(\frac{u_{c_i}(t)}{u_{\text{max}_i}}\right) = \begin{cases} u_{c_i}(t), & |u_{c_i}(t)| \le u_{\text{max}_i} \\ u_{\text{max}_i}(t)\text{sgn}\left(u_{c_i}(t)\right), & |u_{c_i}(t)| > u_{\text{max}_i} \end{cases} \tag{9.2}$$

where u_{max_i} for $i = 1, \ldots, m$ are the physical amplitude limits of actuator i, and $u_{c_i}(t)$ are the control inputs to be determined by the shared control architecture (SCA). The functions sat($\cdot$) and sgn($\cdot$) denote saturation and sign functions, respectively.

9.3.1.2 Advanced Autopilot Based on Adaptive Control

The control input $u_{c_i}(t)$ will be constructed using an adaptive controller. To specify the adaptive controller, a reference model that specifies the commanded behavior from the plant is constructed and is of the form (Narendra and Annaswamy 2005):

$$\dot{x}_m(t) = A_m x_m(t) + B_m r_0(t),$$
$$y_m(t) = C x_m(t) \tag{9.3}$$

where $r_0 \in \mathbb{R}^k$ is a reference input, A_m ($n \times n$) is a Hurwitz matrix, $x_m \in \mathbb{R}^n$ is the state of the reference model, and (A_m, B_m) is controllable, and C is defined as given in (9.1), that is, $y_m \in \mathbb{R}^k$ corresponds to a reference model output. The goal of the adaptive autopilot is then to choose $u_{c_i}(t)$ in (9.5) so that if an error e is defined as

$$e(t) = x(t) - x_m(t) \tag{9.4}$$

where all signals in the adaptive system remain bounded with error $e(t)$ tending to zero asymptotically.

The design of adaptive controllers in the presence of control magnitude constraints is first addressed in Karason and Annaswamy (1993), with guarantees of closed-loop stability through modification of the error used for the adaptive law. The same problem is addressed in Lavretsky and Hovakimyan (2007), using an approach termed *μ-mod adaptive control*, where the effect of input saturation is accommodated through the addition of another term in the reference model. Yet another approach based on a closed-loop reference model (CRM) is derived in Lee and Huh (1997) and Gibson et al. (2013) in order to improve the transient performance of the adaptive controller. The autopilot we propose in this chapter is based on both the μ-mod and CRM approaches. Using the control input in (9.2), this controller is compactly summarized as

$$u_{c_i}(t) = \begin{cases} u_{\text{ad}_i}(t), & |u_{\text{ad}_i}(t)| \le u_{\text{max}_i}^{\delta} \\ \frac{1}{1+\mu}\left(u_{\text{ad}_i}(t) + \mu\text{sgn}\left(u_{\text{ad}_i}(t)\right) u_{\text{max}_i}^{\delta}\right), & |u_{c_i}(t)| > u_{\text{max}_i}^{\delta} \end{cases} \tag{9.5}$$

where

$$u_{\text{ad}_i}(t) = K_x^T(t)x(t) + K_r^T(t)r_0(t) + \hat{d}(t) + \hat{\Phi}^T(t)f(x) \tag{9.6}$$

$$u_{\text{max}_i}^{\delta} = (1 - \delta)u_{\text{max}_i}, \quad 0 \le \delta < 1 \tag{9.7}$$

A buffer region in the control input domain $[(1 - \delta)u_{\text{max}_i}, u_{\text{max}_i}]$ is implied by (9.5) and (9.7) and the choice of μ allows the input to be scaled somewhere in between. The reference model is also modified as

$$\dot{x}_m(t) = A_m x_m(t) + B_m\left(r_0(t) + K_u^T(t)\Delta u_{\text{ad}}(t)\right) - Le(t) \tag{9.8}$$

$$\Delta u_{\text{ad}_i}(t) = u_{\text{max}_i} \text{ sat}\left(\frac{u_{c_i}(t)}{u_{\text{max}_i}}\right) - u_{\text{ad}_i}(t) \tag{9.9}$$

and $L < 0$ is a constant or a matrix selected such that $(A_m + L)$ is Hurwitz. Finally, the adaptive parameters are adjusted as

$$
\begin{aligned}
\dot{K}_x(t) &= -\Gamma_x x(t) e^T(t) PB, \\
\dot{K}_r(t) &= -\Gamma_r r_0(t) e^T(t) PB, \\
\dot{\hat{d}}(t) &= -\Gamma_d e^T(t) PB, \\
\dot{\hat{\Phi}}(t) &= -\Gamma_f f(x(t)) e^T(t) PB, \\
\dot{K}_u(t) &= \Gamma_u \Delta u_{ad} e^T(t) PB_m
\end{aligned}
\tag{9.10}
$$

where $P = P^T$ is a solution of the Lyapunov equation (for $Q > 0$)

$$
A_m^T P + PA_m = -Q
\tag{9.11}
$$

with $\Gamma_x = \Gamma_x^T > 0, \Gamma_r = \Gamma_r^T > 0, \Gamma_u = \Gamma_u^T > 0$.

The stability of the overall adaptive system specified by (9.1), (9.2), (9.3)–(9.11) is established in Lavretsky and Hovakimyan (2007) when $L = 0$. The stability of the adaptive system, when no saturation inputs are present, is also established in Gibson et al. (2013). A very straightforward combination of the two proofs can be easily carried out to prove that when $L < 0$, the adaptive system considered in this chapter has globally bounded solutions if the plant in (9.1) is open-loop stable and bounded solutions for an arbitrary plant if all initial conditions and the control parameters in (9.10) lie in a compact set. The proof is skipped due to page limitations.

The adaptive autopilot in (9.5)–(9.11) provides the required control input, u, in (9.1) as a solution to the underlying problem. The autopilot includes several free parameters including μ in (9.5), δ in (9.7), the reference model parameters A_m, B_m, L in (9.8) and the control parameters $K_x(0)$, $K_r(0)$, $K_u(0)$, $\hat{d}(0)$, $\hat{\Phi}(0)$ in (9.10). In what follows, a brief description is presented as to how the parameters δ and μ are related to CfM and GCD.

The control input u_{c_i} in (9.5) is shaped by two parameters δ and μ, both of which help tune the control input with respect to its specified magnitude limit $u_{\max_i}$. We use these two parameters in quantifying CfM, GCD, and the trade-offs between them as follows:

CfM: As mentioned earlier, qualitatively, CfM corresponds to a system's reserved capacity, which we quantitatively formulate in the current context as the distance between a control input and its saturation limits. In particular, we define CfM as

$$
\text{CfM} = \frac{\text{CfM}^R}{\text{CfM}_d}
\tag{9.12}
$$

where

$$
\begin{aligned}
\text{CfM}^R &= \text{rms} \left(\min_i (c_i(t)) \right) \Big|_{t_a}^{T}, \\
c_i(t) &= u_{\max_i} - |u_i(t)|
\end{aligned}
\tag{9.13}
$$

$c_i(t)$ is the instantaneous available control input of actuator i, CfM^R is the root mean squared CfM variation and CfM_d, which denotes the desired CfM is chosen as

$$
\text{CfM}_d = \max_i \left(\delta u_{\max_i} \right)
\tag{9.14}
$$

In the above equations, min and max are the minimum and maximum operators over the ith index, rms is the root mean square operator, and t_a and T refer to the time of anomaly and final simulation time, respectively. From (9.13), we note that (*i*) CfM^R has a maximum value $u_{\max}$ for the trivial case when all $u_i(t) = 0$, (*ii*) a value close to $\delta u_{\max}$ if the control inputs approach the buffer region, and (*iii*) zero if $u_i(t)$ hits the saturation limit $u_{\max}$. Since CfM$_d = \delta u_{\max}$, it follows that CfM, the corresponding normalized value, is greater than unity when the control inputs are small and far away from saturation, unity as they approach the buffer region, and zero when fully saturated.

GCD: As mentioned earlier, the reference model represents the commanded behavior from the plant being controlled. In order to reflect the fact that the actual output may be compromised if the input is constrained, we add a term that depends on $\Delta u_{\mathrm{ad}_i}(t)$ in (9.9) to become nonzero whenever the control input saturates, that is, when the control input approaches the saturation limit, Δu_{ad_i} becomes nonzero, thereby suitably allowing a graceful degradation of x_m from its nominal choice as in (9.8). We denote this degradation as GCD$_i$ and quantify it as follows:

$$\mathrm{GCD}_i = \frac{\mathrm{rms}(y_{m,i}(t) - r_{0,i}(t))}{\mathrm{rms}(r_{0,i}(t))}, \quad t \in T_0 \tag{9.15}$$

where T_0 denotes the interval of interest, and $y_{m,i}$ and $r_{0,i}$ indicate the *i*th elements of the reference model output and reference input vectors, respectively. It should be noted that once μ is specified, the adaptive controller automatically scales the input into the reference model through Δu_{ad} and K_u, in a wjay so that $e(t)$ remains small and the closed-loop system has bounded solutions.

μ: The intent behind the introduction of the parameter μ in (9.5) is to regulate the control input and move it away from saturation when needed. For example, if $|u_{\mathrm{ad}_i}(t)| > u_{\max_i}^\delta$, the extreme case of $\mu = 0$ will simply set $u_{c_i} = u_{\mathrm{ad}}$, thereby removing the effect of the virtual limit imposed in (9.7). As μ increases, the control input would decrease in magnitude and move toward the virtual saturation limit $u_{\max_i}^\delta$, that is, once the buffer δ is determined, μ controls $u_i(t)$ within the buffer region $[(1 - \delta)u_{\max_i}, u_{\max_i}]$, bringing it closer to the lower limit with increasing μ. In other words, as μ increases, CfM increases as well in the buffer region.

It is easy to see from (9.8) and (9.9) that similar to CfM, as μ increases, GCD increases as well. This is due to the fact that an increase in μ increases $\Delta u_{\mathrm{ad}_i}(t)$ which, in turn, increases the GCD. While a larger CfM improves the responsiveness of the system to future anomalies, a lower bound on the reference command is necessary to finish the mission within practical constraints. In other words, μ needs to be chosen so that GCD remains above a lower limit while maintaining a large CfM. We relegate the task of selecting the appropriate μ to the human pilot.

In addition to μ and δ, the adaptive controller in (9.2), (9.3)-(9.11) requires the reference model parameters A_m, B_m, L, and the control parameters $K_x(0), K_r(0)$, and $K_u(0)$ at time $t = 0$. If no anomalies are present, then $\Lambda_{\mathrm{nom}} = \Lambda_f = I$ which implies that A_m and B_m as well as the control parameters can be chosen as

$$\begin{aligned} A_m &= A + BK_x^T(0), \\ K_r^T(0) &= -(A_m^{-1}B)^{-1} \\ B_m &= BK_r^T(0) \\ K_u^T(0) &= -A_m^{-1}B \end{aligned} \tag{9.16}$$

where $K_x(0)$ is computed using a linear quadratic regulator (LQR) method and the nominal plant parameters (A, B) (Bryson 1996) and $K_r(0)$ in (9.16) is selected to provide unity low-frequency DC gain for the closed-loop system. When anomalies occur, $\Lambda_f \neq I$, at time $t = t_a$ and supposing that an estimate $\hat{\Lambda}_f$ is available, a similar choice as in (9.16) can be carried out using the plant parameters $(A, B\hat{\Lambda}_f)$ and the relations

$$
\begin{aligned}
A_m &= A + B\hat{\Lambda}_f K_x^T(t_a), \\
K_r^T(t_a) &= -(A_m^{-1} B\hat{\Lambda}_f(t_a))^{-1} \\
B_m &= B\hat{\Lambda}_f K_r^T(t_a) \\
K_u^T(t_a) &= -A_m^{-1} B\hat{\Lambda}_f(t_a)
\end{aligned}
\tag{9.17}
$$

with the adaptive controller specified using (9.2)–(9.11) for all $t \geq t_a$. Finally, L is chosen as in Gibson et al. (2013) and lower parameters $\hat{d}(0)$, $\hat{\Phi}(0)$ are chosen arbitrarily. Similar to μ, we relegate the task of assessing the estimate $\hat{\Lambda}_{f_p}$ to the human pilot as well.

9.3.1.3 Autopilot Based on Proportional Derivative Control

To investigate shared control architectures, another autopilot that is employed in the closed loop system is the proportional derivative (PD) controller. Assuming a single-control input (Farjadian et al. 2016), the goal is to control the dynamics (Hess 2009)

$$
Y_p(s) = \frac{1}{s(s+a)}
\tag{9.18}
$$

which represents the aircraft transfer function between the input u and an output $M(t)$ and can be assumed to be a simplified version of the dynamics in (1). The input u is subjected to the same magnitude and rate constraints as in (9.2). Considering the transfer function (9.18) between the input u and the output $M(t)$, the PD controller can be chosen as

$$
u(t) = K_p(M(t) - M_{\text{cmd}}(t)) + K_r(\dot{M}(t))
\tag{9.19}
$$

where M_{cmd} is the desired command signal that M is required to follow. Given the second-order structure of the dynamics, it can be shown that suitable gains K_p and K_r can be determined so that the closed-loop system is stable, and for command inputs at low frequencies, a satisfactory tracking performance can be obtained.

It is noted that during the experimental validation studies certain anomalies are introduced to (9.18) in the form of unmodeled dynamics and time delays. Therefore, according to the crossover model (McRuer and Krendel 1974), the human pilot has to adapt themselves to demonstrate different compensation characteristics, such as pure gain, lead or lag, based on the type of the anomaly. This creates a challenging scenario for the shared control architecture.

9.3.2 Human Pilot

In this section, we discuss mathematical models of human pilot decision-making on the basis of absence and presence of flight anomalies. A great deal of research has been conducted on mathematical human pilot modeling assuming that no failure in the aircraft or no severe disturbances in the environment are present. Since decision-making differs significantly whether the aircraft is under nominal operation or subjected to severe anomalies, the corresponding models are entirely different as well and discussed separately in what follows.

9.3.2.1 Pilot Models in the Absence of Anomaly

In the absence of anomalous event(s), mathematical models of human pilot control behavior can be classified according to control-theoretic, physiological, and more recently, machine learning methods (Lone and Cooke 2014; Xu et al. 2017). One of the most well-known control-theoretic method in the modeling of human pilot, namely, *the crossover model*, is presented in McRuer and Krendel (1974) as an assembly of the pilot and the controlled vehicle, for single-loop control systems. The open loop transfer function for the crossover model is

$$Y_h(j\omega)Y_p(j\omega) = \frac{\omega_c e^{-\tau_e j\omega}}{j\omega} \tag{9.20}$$

where $Y_h(j\omega)$ is a transfer function of the human pilot, $Y_p(j\omega)$ is a transfer function of the aircraft, ω_c is the crossover frequency, and τ_e is the effective time delay pertinent to the system delays and human pilot lags. The crossover model is applicable for a range of frequencies around the crossover frequency ω_c. When a "remnant" signal is introduced to $Y_h(j\omega)$ to account for the nonlinear effects of the pilot-vehicle system, the model is called a quasilinear model (McRuer and Jex 1967).

Other sophisticated quasilinear models can be found in the literature as *the extended crossover model* (McRuer and Jex 1967), which works especially for conditionally stable systems, that is, when a pilot attempts to stabilize an unstable transfer function of the controlled element, $Y_p(j\omega)$, and *the precision model* (McRuer and Jex 1967), which treats a wider frequency region than the crossover model. The single-loop control tasks are covered by quasilinear models. They can be extended to multiloop control tasks by the introduction of *the optimal control model* (Wierenga 1969; Kleinman et al. 1970).

Another approach in modeling the human pilot is employing the information of sensory dynamics relevant to humans to extract the effect of motion, proprioceptive, vestibular, and visual cues on the control effort. An example of this is *the descriptive model* (Hosman and Stassen 1998) in which a series of experiments are conducted to distinguish the influence of vestibular and visual stimuli from the control behavior. Another example can be given as *the revised structural model* (Hess 1997) where the human pilot is modeled as the unification of proprioceptive, vestibular and visual feedback paths. It is hypothesized that such cues help alleviate the compensatory control action taken by the human pilot.

It is noted that due to their physical limitations, models such as in (9.20) are valid for operation in a predefined boundary or envelope where the environmental factors are steady and stable. In the case of an anomaly, they may not perform as expected (Woods and Sarter 2000).

9.3.2.2 Pilot Models in the Presence of Anomaly

When extreme events and failures occur, human pilots are known to adapt themselves to changing environmental conditions, which overstep the boundaries of automated systems. Since the hallmark of any autonomous system is its ability to self-govern even under emergency conditions, modeling of the pilot decision-making upon occurrence of an anomaly is indispensable, and several examples are present in the literature (Hess 2009, 2014, 2016; Farjadian et al. 2016, 2017, 2020; Thomsen et al. 2019; Tohidi and Yildiz 2019; Eraslan et al. 2019, 2020; Habboush and Yildiz 2021). In this chapter, we assume that the anomalies can be modeled either as an abrupt change in the vehicle dynamics (Hess 2009, 2014, 2016; Thomsen et al. 2019; Farjadian et al. 2016; Tohidi and Yildiz 2019), or a loss of control effectiveness in the control input (Farjadian et al. 2017, 2020). In either case, the objective is the modeling of the decision-making of the pilot so as to elicit a resilient performance from the aircraft and recover rapidly from the impact of anomaly. Since the focus of this chapter is shared control, among the pilot models that are developed for anomaly response, we exploit the ones that explains the pilot behavior in relation to the autopilot. These models are developed using a concept known as the capacity for maneuver (CfM), which is a recent method in modeling of human pilot under anomalous events.

A recent method in modeling of human pilot under anomalous events utilizes the capacity for maneuver (CfM) concept (Farjadian et al. 2016, 2017, 2020). As discussed in Section 9.3.1,

CfM refers to the remaining range of the actuators before saturation, which quantifies the available maneuvering capacity of the vehicle. It is hypothesized that the surveillance and regulation of a system's available capacity to respond to all events help maintain the resiliency of a system, which is a necessary merit to recover from unexpected and abrupt failures or disturbances (Woods 2018). We propose two different types of pilot models, both of which use CfM, but in different ways.

Perception Trigger. The pilot model is assumed to assess the CfM and implicitly compute a perception gain based on the CfM. The quantification of this gain K_t is predicated on CfM^R with the definition as in (9.13). The perception trigger is associated with the gain K_t, which is implicitly computed as in Farjadian et al. (2016). The perception algorithm for the pilot is

$$K_t = \begin{cases} 0, & |F_0| < 1 \\ 1, & |F_0| \geq 1 \end{cases} \tag{9.21}$$

where

$$F_0 = G_1(s)[F(t)],$$

$$F(t) = \frac{\frac{d}{dt}(CfM^{R_m}) - \mu_p}{3\sigma_p} \tag{9.22}$$

$$CfM^{R_m} = u_{max} - rms(u(t))$$

$G_1(s)$ is a second-order filter introduced as a smoothing and lagging operator into human perception algorithm, $F(t)$ is the perception variable, CfM^{R_m} is a slightly modified version of CfM^R in (9.13), μ_p is the average of $\frac{d}{dt}(CfM^{R_m})$ and σ_p is the standard deviation of $\frac{d}{dt}(CfM^{R_m})$, both of which are measured over a nominal flight simulation. The computation of these statistical parameters is further elaborated in Section 9.3.5. The hypothesis here is that the human pilot has such a perception trigger K_t, and when this trigger reaches unity, the pilot takes over control from the autopilot. In Section 9.3.5, we validate this perception model.

CfM-GCD Trade-off. The pilot is further assumed to implicitly assess the available (normalized) CfM when an anomaly occurs and decide on the amount of GCD that is allowable so as to let the CfM become comparable to the CfM_d. In other words, we assume that the pilot is capable of assessing the parameter μ and input this value to the autopilot following the occurrence of an anomaly. That is, the pilot model takes the available CfM as the input and delivers μ as the output. In Section 9.3.6, we validate this assessment.

9.3.3 Shared Control

We now propose a CPHS with the appropriate granularity in the microstructure of control and temporality that combines the decision-making of a pilot and autopilot in flight control. The architecture is invoked under alert conditions, with triggers in place that specify when the decision-making is transferred from one authority to another. The specific alert conditions that we focus on in this chapter correspond to physical anomalies that compromise actuator effectiveness. To that end, in Sections 9.3.1.2 and 9.3.1.3, autopilots based adaptive control and PD control are described, with former designed to accommodate parametric uncertainties including loss of control effectiveness in the actuators, and the latter to ensure satisfactory command following under nominal conditions. In Section 9.3.2.2, two different models of decision-making in pilots are proposed, both based on the monitoring of CfM of the actuators. We take this opportunity to suggest two different shared control architectures using the aforementioned models of the autopilots and pilots. We then assemble these architectures with their granularity assignments where, in each case, we outline the corresponding granularity assignment figure and discuss the details.

9.3.3.1 SCA1: A Pilot with a CfM-Based Perception and a Fixed-Gain Autopilot

The first shared control architecture can be summarized as a sequence {autopilot runs, anomaly occurs, pilot takes over}. That is, it is assumed that an autopilot based on PD control as in (9.19) is in place, ensuring a satisfactory command tracking under nominal conditions. The human pilot is assumed to consist of a perception component and an adaptation component. The perception component monitors CfM^{R_m}, through which a perception trigger F_0 is calculated using (9.22). The adaptation component keeps track of the control gain in (9.21) and takes over control of the aircraft when $K_t = 1$. The details of this shared controller and its evaluation using a numerical simulation study can be found in Farjadian et al. (2016). Figure 9.3 illustrates the schematic of SCA1.

Granularity Assignment: The overall traded control architecture presented above can be described using the granularity assignment as in Figure 9.4. It can be seen that the sequence of decisions taken in SCA1 are delineated when faced with an anomaly, elaborated below.

In the pre-anomaly period, the automation is assigned the tasks of determining the control input and overall monitoring of the system and any anomalies that can occur. During this time, the pilot monitors the situation but otherwise remains idle. Immediately following the anomaly, the pilot is given the task of perceiving the presence of an anomaly and takes over the position tracking. In order to render a swift perception of the anomaly and provide a smoother transition of the control task, the pilot utilizes Capacity of Maneuver (CfM). The pilot then takes over the control action from the automation. Such a coordination based on the monitoring of CfM helps in

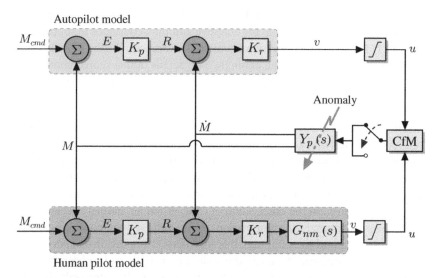

Figure 9.3 Block diagram of the shared control architecture 1 (adapted from Farjadian et al. (2016)). The autopilot model consists of a fixed gain controller, whereas the human pilot model comprehends a perception part based on capacity for maneuver (CfM) concept and an adaptation part governed by empirical adaptive laws (Hess 2016). The neuromuscular transfer function $G_{nm}(s)$ (Hess 2006), which corresponds to a control input formed by an arm or leg is given as $G_{nm}(s) = \frac{100}{s^2+14.14s+100}$. When an anomaly occurs, the plant dynamics $Y_p(s)$, undergoes an abrupt change by rendering the autopilot insufficient for the rest of the control. At this stage, the occurrence of an anomaly is captured by the CfM^{R_m} such that the authority is handed over to the human pilot model for a resilient flight control.

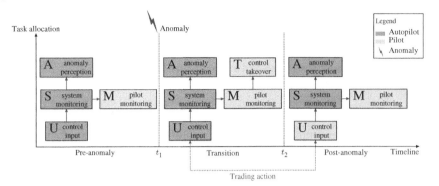

Figure 9.4 Granularity assignment of SCA1. The allocation of control tasks in SCA1 can be divided into three phases according to the timeline of events, that is, pre-anomaly, transition region, and post-anomaly. In the pre-anomaly phase, low-level tasks are conducted under the pilot monitoring. When an anomaly strikes the system, the anomaly perception is triggered and it is signaled to the pilot, who assumes control in the system. The post-anomaly phase is carried out similarly to the pre-anomaly phase.

overriding the automation's role in a bumpless manner. This overall task allocation and timeline capture the granularity assignment of this particular CPHS architecture in SCA1.

9.3.3.2   SCA2: A Pilot with a CfM-Based Decision-Making and an Advanced Adaptive Autopilot

In this shared control architecture, the role of the pilot is a supervisory one while the autopilot takes on an increased and more complex role. The pilot is assumed to monitor the CfM of the resident actuators in the aircraft following an anomaly. In an effort to allow the CfM to stay close to the CfM_d in (9.14), the command is allowed to be degraded; the pilot then determines a parameter μ which directly scales the control effort through (9.5) and indirectly scales the command signal through (9.8) and (9.9). Once μ is specified by the pilot, then the adaptive autopilot continues to supply the control input using (9.5)–(9.11). If the pilot has a high situational awareness, she/he provides $\hat{\Lambda}_{f_p}$ as well, which is an estimate of the severity of the anomaly. The details of this shared controller and its evaluation using a numerical simulation study can be found in Farjadian et al. (2017, 2020). Figure 9.5 shows the schematic of SCA2.

Granularity Assignment: Figure 9.6 illustrates the granularity assignment of tasks and timelines corresponding to SCA2. As the underlying architecture includes a supervisory action, the specific tasks for the pilot and the type of coordination are quite different from SCA1. The pre-anomaly roles of automation remain the same as before, with the control input and system monitoring carried out by the automation. During this time, the pilot monitors the overall system and also assesses the system preparedness through the observation of CfM. When an anomaly occurs, the pilot monitoring task perceives the anomaly, the automation, through the system monitoring task, provides the CfM input to the pilot, and the pilot uses this information to provide a supervisory input to the automation. The pilot continues to assess the system preparedness, by observing the CfM and the GCD, so as to provide the appropriate supervisory inputs as time proceeds. In the post-anomaly, the automation and pilot return to their original tasks as before.

9.3.4   Validation with Human-in-the-Loop Simulations

The goal in this section is to validate two hypotheses of the human-pilot actions, namely SCA1 and SCA2. Despite the presence of obvious common elements to the two SCAs, of a human pilot, an autopilot, and a shared controller that combines their decision-making, the details

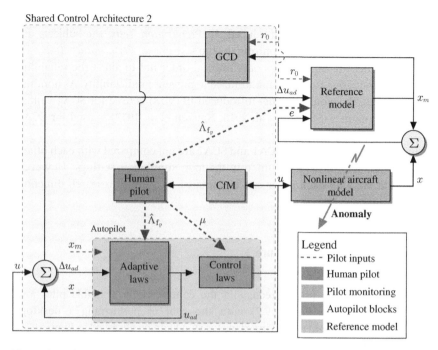

Figure 9.5 Block diagram of the proposed shared control architecture (SCA) 2. The human pilot undertakes a supervisory role by providing the key parameters μ and $\hat{\Lambda}_{f_p}$ to the adaptive autopilot. The blocks are expressed in different colors based on their functions in the proposed SCA2. The numbers in parentheses in each block correspond to the related equations.

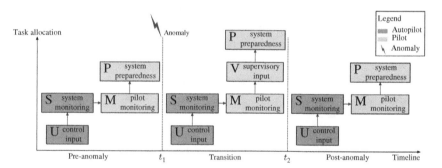

Figure 9.6 Granularity assignment in SCA2. Here the granularity assignment can be divided into three phases according to the timeline of events, that is, pre-anomaly, transition region, and post-anomaly. In the pre-anomaly phase, low-level tasks are conducted under the pilot monitoring. The pilot is in a state of readiness and stays alert to changes in the system. When an anomaly occurs in the system, the anomaly perception is triggered and is signaled to the pilot, who supplements some supervisory inputs to the autopilot to mitigate the effects of the anomaly. The task allocation returns to its initial state in the post-anomaly region.

differ significantly. In each case, we present the validation separately in the following order: the experimental setup, the type of anomaly, the experimental procedure, details of the human subjects, the pilot-model parameters, results, and observations.

Although the experimental procedures used for the validations of SCA1 and SCA2 differ, and therefore are explained in separate sections; they use similar principles. To minimize repetition, main tasks can be summarized as follows: the procedure comprises three main phases, namely,

Pilot Briefing, *Preparation Tests*, and *Performance Tests*. In *Pilot Briefing*, the overall aim of the experiment and the experimental setup are introduced. In *Preparation Tests*, the subjects are encouraged to gain practice with the joystick controls. In the last phase, *Performance Tests*, the subjects are expected to conduct the experiments only once in order not to affect the reliability due to learning. Anomaly introduction times are randomized to prevent predictability. As nominal (anomaly free) plant dynamics, the simpler model introduced in (9.18) is used for SCA1 validation, and a nonlinear F-16 dynamics (Stevens et al. 2015; Nguyen et al. 1979) is used for SCA2 validation.

It is noted that during the validation studies, SCA1 and SCA2 are not compared with each other. They are separately validated with different human-in-the-loop simulation settings. However, comparison of SCA1 and SCA2 with each other can be pursued as a future research direction (Eraslan 2021).

9.3.5 Validation of Shared Control Architecture 1

9.3.5.1 Experimental Setup

The experimental setup consists of a pilot screen, and the commercially available pilot joystick Logitech Extreme 3D Pro (see Figure 9.7) with the goal of performing a desktop, human-in-the-loop, simulation. The flight screen interface for SCA1 can be observed in Figure 9.7a and separately illustrated in Figure 9.8. The line with a circle in the middle shows the reference to be followed by the pilot. This line is moved up and down according to the desired reference command, M_{cmd} (see Figure 9.3). The sphere in Figure 9.8 represents the nose tip of the aircraft and is driven by the joystick inputs. When the subject moves the joystick, she/he provides the control input v, which goes through the aircraft model and produces the movement of the sphere. The control objective is to keep the sphere inside the circle, which translates into tracking the reference command. Similarly, the flight screen for SCA2 is shown in Figure 9.7b and separately illustrated in Figure 9.11. The details on this screen are provided in Section 9.3.6.

 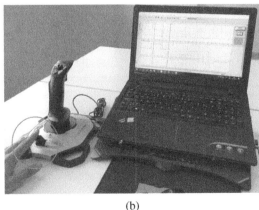

(a)   (b)

Figure 9.7 Experimental setups for shared control architecture (SCA) 1 (a) and 2 (b). SCA1 experiment consists of a pilot screen (see Figure 9.8) and a commercially available pilot joystick, whose pitch input is used. SCA2 experiment consists of a different pilot screen (see Figure 9.11) and the joystick. In SCA2, subjects use the joystick lever to provide input.

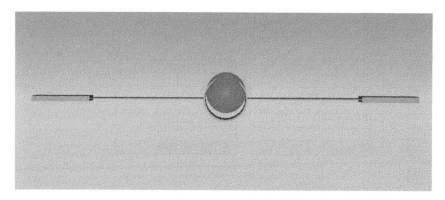

Figure 9.8 Flight screen interface for Shared Control Architecture 1. The sphere represents the aircraft nose and is controlled in the vertical direction via joystick movements. The orange line with a circle in the middle, which also moves in the vertical direction, represents the reference command to be followed. The control objective is to keep the sphere inside the circle. At the beginning of the experiments, the aircraft is controlled by the autopilot. When an anomaly occurs, the subjects are warned to take over the control, via a sound signal. The time of initiating this signal is determined using different alert times.

9.3.5.2 Anomaly

The anomaly is modeled by a sudden change in vehicle dynamics, from $Y_p^{\text{before}}(s)$ to $Y_p^{\text{after}}(s)$ as in Hess (2016), and illustrated in Figure 9.3. Two different flight scenarios, S_{harsh}, and S_{mild}, are investigated, which correspond to a harsh and a mild anomaly, respectively. In the harsh anomaly, it is assumed that

$$Y_{p,h}^{\text{before}}(s) = \frac{1}{s(s+10)}, \quad Y_{p,h}^{\text{after}}(s) = \frac{e^{-0.2s}}{s(s+5)(s+10)} \tag{9.23}$$

whereas in the mild anomaly, it is assumed that

$$Y_{p,m}^{\text{before}}(s) = \frac{1}{s(s+7)}, \quad Y_{p,m}^{\text{after}}(s) = \frac{e^{-0.18s}}{s(s+7)(s+9)} \tag{9.24}$$

The specific numerical values of the parameters in (9.23) and (9.24) are chosen so that the pilot action has a distinct effect in the two cases based on their response time. More details of these choices are provided in Section 9.3.5.5.

The anomaly is introduced at a certain instant of time, t_a, in the experiment. The anomaly alert is conveyed as sound signal at t_s, following which, the pilots take over control at a time t_{TRT}, after a certain reaction time t_{RT}. Denoting $\Delta T = t_s - t_a$ as the alert time, the total elapsed time from the onset of anomaly to the instant of hitting the joystick button is defined as

$$t_{\text{TRT}} = t_{\text{RT}} + \Delta T \tag{9.25}$$

9.3.5.3 Experimental Procedure

The experimental procedure consists of three main parts, which are *Pilot Briefing, Preparation Tests,* and *Performance Tests* (see Figure 9.9). The first part of the procedure is the *Pilot Briefing,* where the subjects are required to read a pilot briefing to have a clear understanding of the experiment. The briefing consists of six main sections, namely, *Overall Purpose, Autopilot, Anomaly, Experimental Setup, Flight Screen,* and *Instructions.* In these sections, the main concepts and experimental hardware such as the pilot screen, and the joystick lever are introduced to the subject.

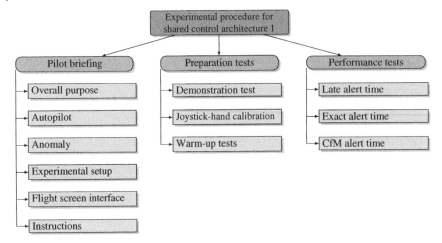

Figure 9.9 Experimental procedure breakdown for the Shared Control Architecture 1. Three main tasks constituting the procedure are seen. In *Pilot Briefing*, the subjects read the pilot briefing, review it with the experiment designer and have a question and answer session. In *Preparation Tests*, the subjects get familiarized with the test via demonstration runs and warm-up tests. Finally, in *Performance Tests*, the real tests are conducted using three different alert times.

The second part is the *Preparation Tests* in which the subjects are introduced to a demonstration test conducted by the experiment designer to familiarize the subjects to the setup. In this test, the subjects observe the experiment designer follow a reference command using the joystick (see Figure 9.7a). They also watch the designer to respond to control switching alert sounds by taking over the control via the joystick. Following these demonstration tests, the subjects are requested to perform the experiment themselves. To complete this part, three preparation tests, each with a duration of 90 seconds are conducted. At the end of each test, the root mean squared error e_{rms} of the subjects is calculated as

$$e_{\text{rms}} = \sqrt{\frac{1}{T_p} \int_{t_a}^{T_p} e(\tau)^2 d\tau} \tag{9.26}$$

where

$$e(t) = M_{\text{cmd}}(t) - M(t) \tag{9.27}$$

and $T_p = 90$s. It is expected that the e_{rms} in each trial decreases as a sign of learning.

The third is the *Performance Tests*, each with a duration of 180 seconds, which aim at testing the performance of the proposed SCA, in terms of tracking error e_{rms}, CfM^{R_m} and a bumpless transfer metric ρ, which is calculated by taking the difference of e_{rms} values that are obtained using the 10 second intervals before and after the anomaly. This calculation is performed as

$$\rho = \sqrt{\frac{1}{t_a + 10} \int_{t_a}^{t_a+10} e(\tau)^2 d\tau} - \sqrt{\frac{1}{t_a} \int_{t_a-10}^{t_a} e(\tau)^2 d\tau} \tag{9.28}$$

9.3.5.4 Details of the Human Subjects

The experiment with the harsh anomaly was conducted by 15 subjects (including 1 flight pilot), whereas the one with the a mild anomaly was conducted by three subjects (including 1 flight pilot). All subjects were over 18 years old, and four of the subjects were left-handed, yet this did not bring about any problems, since an ambidextrous joystick was utilized. Some statistical data pertaining to the subjects are given in Table 9.1.

Table 9.1 Statistical data of the subjects in the SCA1 experiment. $\mu()$ and $\sigma()$ represent the average and the standard deviation operators, respectively.

Scenario	P	F	μ(Age)	σ(Age)	LH
S_{harsh}	15	1	22.9	3.6	3
S_{mild}	3	0	26.0	5.2	0

P: # of participants, F: female, LH: left-handed.

9.3.5.5 Pilot-Model Parameters

The pilot-model in (9.21)–(9.22) includes statistical parameters μ_p and σ_p, the mean and the standard deviation of the time derivative of CfM^{R_m}, respectively, and the parameters of the filter $G_1(s)$. The filter is chosen as $G_1(s) = \frac{2.25}{s^2 + 1.5s + 2.25}$ so as to reflect the bandwidth of the pilot stick motion. To obtain the other statistical parameters, several flight simulations were run with the PD-control based autopilot in closed-loop, and the resulting CfM^{R_m} values were calculated for 180s, both for the harsh and mild anomalies. The time-averaged statistics of the resulting profiles were used to calculate the statistical parameters as $\mu_p = 0.028$, $\sigma_p = 0.038$ for the harsh anomaly, and $\mu_p = 0.091$, $\sigma_p = 0.077$ for the mild anomaly.

The pilot model in (9.21)–(9.22) implies that the pilot perceives the presence of the anomaly at the time instant when K_t becomes unity, following which the control action switches from the autopilot to the human pilot. The action of the pilot, based on this trigger, is introduced in the experiment by choosing t_s, the instant of the sound signal, to coincide with the perception trigger. The corresponding $\Delta T = t_s - t_a$, where t_a is the instant when anomaly is introduced, is denoted as a "CfM-based" one. In order to benchmark, this CfM-based switching action, two other switching mechanisms are introduced, one which we define to be "exact," where $t_s = t_a$, so that $\Delta T = 0$, and another to be "late," where ΔT is chosen to be significantly larger than the CfM-based one. These choices are summarized in Table 9.2.

9.3.5.6 Results and Observations

Scenario 1: Harsh Anomaly We present the results related to the harsh anomaly defined in (9.23) for various alert times transmitted to the subjects. In order to compare the results obtained from the subjects, we also carried out numerical simulation results, where the participants are replaced with the pilot model (9.21)–(9.22), using the same alert times. The results obtained are summarized Table 9.3 using both the tracking error e_{rms} and the corresponding CfM^{R_m}.

Table 9.2 Timeline of anomalies.

Switch	S_{harsh}			S_{mild}		
	$t_a[s]$	$t_s[s]$	$\Delta T[s]$	$t_a[s]$	$t_s[s]$	$\Delta T[s]$
Late	50	55.5	5.5	64	74	10
Exact	50	50	0	64	64	0
CfM-based	50	51.1	1.1	64	70.2	6.2

Scenarios with the harsh and mild anomaly, respectively.
Based on of the switching mechanism, the anomaly is reported to the subject with a sound signal at t_s. ΔT, the alert time, is defined as the time elapsed between the sound signal and the occurrence of anomaly.

Table 9.3 Averaged e_{rms} and CfM^{R_m} values for S_{harsh}. "Sim" refers to simulation and "Exp" refers to experiment.

S_{harsh}	Auto	Sim_{late}	Exp_{late}	Sim_{exact}	Exp_{exact}	$Sim_{CfM\text{-based}}$	$Exp_{CfM\text{-based}}$
e_{rms}	478	363	383	319	354	318	348
CfM^{R_m}	8.92	7.93	6.84	7.83	6.80	7.83	7.06

The autopilot shows the worst tracking error performance. The reason for a high CfM^{R_m} amount for the autopilot is the inability to effectively use the actuators to accommodate the anomaly.

Table 9.4 Mean, μ and standard error, $\sigma_M = \sigma / \sqrt{n}$, where σ is standard deviation, and n is the subject size, of e_{rms} (on the left) and CfM^{R_m} (on the right), for a harsh anomaly.

	e_{rms}		CfM^{R_m}	
Experiment	μ	σ_M	μ	σ_M
Exp_{late}	383	16	6.84	0.10
Exp_{exact}	354	15	6.80	0.12
$Exp_{CfM\text{-based}}$	348	14	7.06	0.11

Table 9.5 Averaged bumpless transfer metric, ρ, standard error, $\sigma_M(\rho) = \sigma(\rho)/\sqrt{n}$, where σ is standard deviation of ρ, and n is the subject size, and averaged reaction times t_{RT} and total reaction times t_{TRT} for S_{harsh}.

Switch	ρ	$\sigma_M(\rho)$	t_{RT} [s]	t_{TRT} [s]
Late	216	4.27	1.07	6.64
Exact	82	1.37	0.98	0.98
CfM-based	26	0.72	0.99	2.12

The least amount of bumpless transfer of control happens to be in the case of CfM-based shared control architecture.

All numbers reported in Table 9.3 are averaged over all 15 subjects. e_{rms} was calculated using (9.26) with $T_p = 180s$, while CfM^{R_m} was calculated using (9.22) with $u_{max} = 10$. The statistical variations of both e_{rms} and CfM^{R_m} over the 15 subjects are quantified for all three SCA experiments, for the late, exact, and CfM-based alert times are summarized in Table 9.4. We also calculate the average bumpless transfer metric ρ and its standard error $\sigma_M(\rho)$ for these three cases in Table 9.5. Table 9.5 also provides the average reaction times t_{RT} and average total reaction times t_{TRT} of the subjects.

Observations The first observation from Table 9.3 is that the e_{rms} for the CfM-based case is at least 25% smaller than the case with the autopilot alone. The second observation is that among the

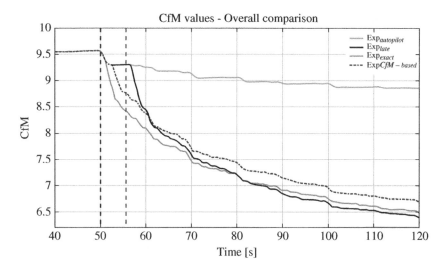

Figure 9.10 CfM^{R_m} variation for the autopilot and all the alert timing mechanisms. CfM^{R_m}s of Exp$_{late}$ and Exp$_{exact}$ show changing trends during the simulation, but CfM^{R_m} of Exp$_{CfM\text{-}based}$ prominently stays at the top especially after the transient effects of the anomaly ($t > t_a + 10$).

shared control architecture experiments, the one with the CfM-based alert time has the smallest tracking error, while the one with the late alert time has the largest tracking error. However, it is noted that the difference between the exact and CfM-based cases is not significant.

The third observation from Table 9.3 is that CfM-based control switching from the autopilot to the pilot not only provides the smallest tracking error but also the largest CfM^{R_m} value compared to other switching strategies. This can also be observed in Figure 9.10, where CfM^{R_m} values, averaged over all subjects, are provided for different alert times. As noted earlier, the large CfM^{R_m} value of the autopilot results from inefficient use of the actuators which manifest itself with a large tracking error. The final observation, which comes from Table 9.5, is that among different alert times, the one with the CfM-based alert time provides the smoothest transfer of control, with a bumpless transfer metric of $\rho = 26$, while the late-alert provides $\rho = 216$, and the exact alert gives $\rho = 82$.

These observations imply the following: The pilot re-engagement after an anomaly should not be delayed until it becomes too late, which corresponds to a late-alert switching strategy. At the same time, immediate pilot action right after the anomaly detection might not be necessary either, which is indicated by the fact that the bumpless transfer metric and the CfM^{R_m} values for the CfM-based switching case is better than the exact-switching case. Instead, monitoring the CfM^{R_m} information carefully may be the appropriate trigger for the pilot to take over.

Scenario 2: Mild Anomaly The averaged e_{rms} (scaled by 10^4) and CfM^{R_m} values for this case are given in Table 9.6. The averaged bumpless transfer metric ρ, its standard error $\sigma_M(\rho)$, the average reaction times t_{RT}, and the average total reaction times t_{TRT} are presented in Table 9.7. The statistical variations of these metrics over the three subjects are shown in Table 9.8.

Observations Similar to the harsh anomaly case, pure autopilot control results in the largest tracking error in the case of a mild anomaly. Also, similar to the harsh anomaly case, pilot engagement based on CfM^{R_m} information produces the smallest tracking error, although the difference between the exact switching and CfM-based switching is not significant. However, in terms of preserving

Table 9.6 Averaged e_{rms} and CfM^{R_m} values for S_{mild}.

S_{mild}	Auto	Sim_{late}	Exp_{late}	Sim_{exact}	Exp_{exact}	$Sim_{CfM-based}$	$Exp_{CfM-based}$
e_{rms}	408	281	259	297	236	243	217
CfM^{R_m}	8.78	7.90	7.75	7.64	7.55	8.13	7.79

The autopilot still shows the worst performance yet, this time, the error introduced is less than the one of the harsh anomaly.

Table 9.7 Averaged bumpless transfer metric ρ, standard error $\sigma_M(\rho) = \sigma(\rho)/\sqrt{n}$, where σ is standard deviation of ρ, and n is the subject size, averaged reaction time t_{RT}, and total reaction time t_{TRT} for the experiment with a mild anomaly, S_{mild}.

Switch	ρ	$\sigma_M(\rho)$	t_{RT} [s]	t_{TRT} [s]
Late	196	3.23	1.06	11.06
Exact	209	8.88	1.02	1.02
CfM-based	203	2.87	0.95	7.20

The nature of the anomaly has a considerable effect on the bumpless transfer metric, that is, the difference between the exact and CfM-based alert timings is not readily noticeable as the one of the harsh anomaly.

Table 9.8 Mean, μ and standard error, $\sigma_M = \sigma/\sqrt{n}$, where σ is the standard deviation, and n is the subject size, of e_{rms} (on the left) and CfM^{R_m} (on the right), for Scenario 2 with a mild anomaly.

Experiment	e_{rms}		CfM^{R_m}	
	μ	σ_M	μ	σ_M
Exp_{late}	259	19	7.75	0.15
Exp_{exact}	236	21	7.55	0.17
$Exp_{CfM-based}$	218	14	7.61	0.19

CfM^{R_m}, Table 9.8 shows that no significant differences between different switching times can be detected due to wide spread (high standard error) of the results. The same conclusion can be drawn for the bumpless transfer metric. One reason for this can be the low sample size, 3, in this experiment. One conclusion that can be drawn from these results is that although CfM-based switching shows smaller tracking errors compared to the alternatives, the advantage of the proposed SCA in the presence of mild anomaly is not as prominent as in the case of harsh anomaly.

9.3.6 Validation of Shared Control Architecture 2

Unlike the SCA1, where the pilot took over control from the autopilot when an anomaly occurred, in SCA2, the pilot plays more of an advisory role, directing the autopilot that remains operational throughout. In particular, the pilot provides appropriate values μ and sometimes $\hat{\Lambda}_{f_p}$ as well.

9.3.6.1 Experimental Setup

Contrary to the SCA1 case, here the subjects use the joystick only to enter the μ and $\hat{\Lambda}_{f_p}$ values using the joystick lever. The flight screen that the subjects see is shown in Figure 9.11. There are three subplots, which are normalized $c_i(t)$ (top), reference command tracking r_0 (middle) and the evolution of graceful command degradation (GCD) (bottom). The horizontal black line in the CfM variation subplot corresponds to the upper bound of the virtual buffer, $[0, \delta] = [0, 0.25]$. The small rectangle at the upper right serves the purpose of showing the amount of the μ input, entered via the joystick lever. In this rectangle, the title "Anti-Locking" is used to emphasize the purpose of the μ input, which is preventing the saturation/locking of the actuators. There is also another region in this rectangle called the "Range," which shows the limits of this input. In the snapshot of the pilot screen shown in Figure 9.11, a scenario with two anomalies introduced at $t_{a_1} = 32$s and $t_{a_2} = 68$s is shown. The instant of anomaly occurrences are marked with vertical lines, the colors of which indicate the severity of the anomaly. The subjects are trained to understand and respond to the severity and the effect of the anomalies by monitoring the colors, CfM information, and the tracking performance. The details of subject training are provided in Section 9.3.6.3.

9.3.6.2 Anomaly

The anomaly considered in this experiment is loss of actuator effectiveness indicated by Λ_f in (9.1). Λ_f is a (2×2) diagonal matrix with equal entries that are between 0 and 1, where 0 corresponds to complete actuator failure and 1 corresponds to no failure. In the experiments, two anomalies are introduced at times $t = t_{a_1}$ and $t = t_{a_2}$. Consequently, the diagonal entries of Λ_f vary as

$$\lambda_{f_i} = \begin{cases} 1, & t < t_{a_1} \\ 0 < \lambda_{f_1} < 1, & t_{a_1} \leq t < t_{a_2} \\ 0 < \lambda_{f_2} < 1, & t_{a_2} \leq t \end{cases} \tag{9.29}$$

where λ_{f_i} refers to the diagonal entries for the ith anomaly introduction. The anomaly injections are communicated to the subjects with colored vertical lines appearing on the pilot screen as shown in

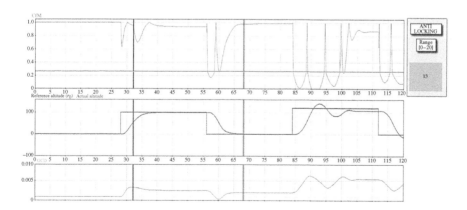

Figure 9.11 Flight screen interface for Shared Control Architecture 2. The top subfigure shows the normalized $c_i(t)$ in (9.13), which can considered as the instantaneous CfM. The vertical lines at $t_{a_1} = 32$s and $t_{a_2} = 68$s, respectively, show the instants of anomaly introduction. The horizontal black line is the anti-locking border below which the μ input becomes effective. It is explained to the subjects, as well as demonstrated during training, that setting μ to high values, where CfM variation is over this border has no influence on CfM, which is apparent from (9.5). The middle subfigure shows reference tracking, and the bottom one is the time variation of the graceful command degradation during flight.

Figure 9.11, together with sound alerts. Three anomalies with different severities are used during the experiments. The anomaly severity quantities are $\Lambda_f = 0.3$ (low), 0.2 (medium), and 0.15 (high), and these are communicated to the subjects using different colored vertical lines, respectively. (See Figure 9.11, where a sample experiment is shown with medium and high severity anomalies.)

9.3.6.3 Experimental Procedure

As in SCA1, the experimental procedure consists of three parts, *Pilot Briefing, Input Training* and *Performance Test* (see Figure 9.12 for a schematic). The first part of the procedure is the *Pilot Briefing*, in which the subjects are demanded to read a pilot briefing to acquire knowledge about the experiment. The briefing consists of four main sections, namely, *Overall Purpose, Flight Scenario, Flight Screen Interface*, and *Instructions*. In these sections, the main concepts and the experimental setup are covered by the experiment designer. The second part is the *Input Training*, in which the subjects are first introduced to the joystick lever, which they are required to use to enter the μ input (see Figure 9.7). By properly moving the lever, an integer value of μ, ranging from 1 to 20, can be given to the controller. Following this, a demonstration test is conducted by the experiment designer. In this test, a sample scenario with two anomalies is run, where the input μ is fixed to its nominal value of $\mu = 1$, throughout the flight. It is explicitly shown that (i) the actuators reaches their saturation limit, and thus the CfM becomes zero, many times during the flight, which jeopardizes the aircraft stability, (ii) it takes for the altitude h, a long time to recover to follow the reference command, and (iii) a certain graceful command degradation occurs to relax the performance goals.

Following the demonstration test, another sample scenario is run by the designer, in which suitable μ values are provided upon occurrence of the anomalies. Different from the previous demonstration, it is pointed out that (i) with a proper μ, CfM can be kept away from zero and (ii) GCD is kept minimal. By this demonstration, the subjects are expected to appreciate that suitable μ inputs help the autopilot recover from severe anomalies in an efficient and swift manner.

Finally, a μ-input-training is performed on the subjects. Six scenarios are introduced to the participants in an interactive manner, by which they learn how to be involved in the overall control architecture.

As a first step, three scenarios with single anomalies are considered. In each severity of the anomaly, an optimal μ, which trades off CfM with GCD, is conveyed to the subjects. Upon doing this, attention is drawn to the fact that each anomaly causes a certain sharp drop in the CfM

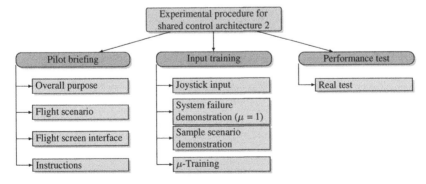

Figure 9.12 Experimental procedure breakdown for the Shared Control Architecture 2. Three main tasks constituting the procedure are observed. In *Pilot Briefing*, the subjects read the pilot briefing, review it with the experiment designer, and have a question and answer session. In *Input Training*, the subjects learn how to provide the auxiliary inputs to the autopilot using the joystick. They also undergo a μ-input training, the details of which are covered in Section 9.3.6.3. Finally, in *Performance Test*, a real test with two successive anomalies Λ_{f_i} is conducted.

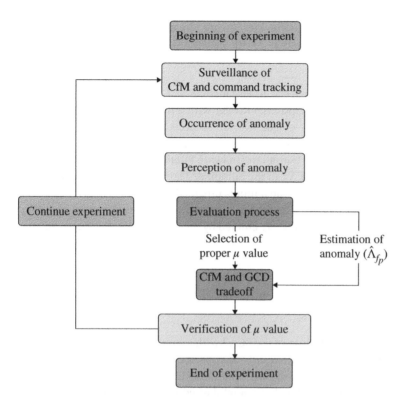

Figure 9.13 Algorithm of pilot tasks. This presents the step-by-step procedure which should be followed by the subject.

variation. In other words, more drastic drops occur in CfM with the increase in the severity of anomaly. Following this, three other scenarios with two successive anomalies are studied. It is noted by the combination of different anomalies that nonlinear effects are present in the flight simulation; that is, the μ values corresponding to single anomalies are no longer effective when the anomalies are combined. In both the training and the performance tests, the anomaly severity estimation $\hat{\Lambda}_{f_p}$ input is automatically fed to the controller with an error of 0.2 in absolute value. Then, to compare the effect of this estimation input, we compare two pilot types where the estimation is not provided in one case and provided in the other.

The third part is the *Performance Test*, in which the subjects are expected to handle a combination of a highly severe anomalies ($\Lambda_{f_1} = 0.20$ and $\Lambda_{f_2} = 0.15$) in a flight simulation. They are expected to give suitable μ values based on their training. Also, in both the training and performance tests, the introduction of anomaly times are chosen to be completely random to prevent the subjects from making a guess whether or not the anomaly is about to happen. The set of tasks to be tackled by the subjects is summarized as a flowchart in Figure 9.13. These tasks are explained interactively during the sample scenario demonstration (see Figure 9.12), where the subjects have the opportunity to practice the steps in controlling the flight simulation.

9.3.6.4 Details of the Human Subjects

The experiment was performed by 10 subjects all of whom attended the previous experiment. This choice was made deliberately since the subjects of the previous experiment had an acquaintance with the autopilot and shared control concepts. For this reason, the pilot briefing regarding this experiment was written in a tone that the subjects were already familiar with these basic notions. The average age of the participants was 24.2 with a standard deviation of 1.8.

9.3.6.5 Results and Observations

The autopilots consist of an adaptive controller as in Farjadian et al. (2020), which has no apparent interface with the autopilot, a μ-mod adaptive controller with a fixed value of $\mu = 50$, and an optimal controller. The numerical results averaged over 10 subjects are presented in Table 9.9. The indices in curly brackets $\{h, v\}$ denote the altitude and velocity, respectively. SAP and SUP refer to the "Situation Aware Pilot" and the "Situation Unaware Pilot", respectively. SUP provides only a μ input and SAP provides both μ and $\hat{\Lambda}_{f_p}$ to the autopilot. The SUP results are obtained by simulating the performance tests by using only the μ inputs provided by the subjects, without their severity estimation inputs $\hat{\Lambda}_{f_p}$. Therefore, to obtain the SUP results, (9.16) is used instead of (9.17), while still incorporating the same μ values that were entered by the participants. Furthermore, the tracking performance γ_i is calculated as

$$\gamma_i = \text{RMSE}_i^+ - \text{RMSE}_i^- \tag{9.30}$$

where

$$\text{RMSE}_i^- = \text{rms}(e_i)\big|_0^{t_{a_1}}, \quad \text{RMSE}_i^+ = \text{rms}(e_i)\big|_{t_{a_1}}^T \tag{9.31}$$

In Figure 9.14, a comprehensive comparison of the SCA2 with autopilot-only cases is given as a matrix of 4×4 plots. Each column presents the results of a specific controller, whereas each row shows the comparison of these controllers based on altitude tracking h, velocity tracking V, CfM and elevator control input δ_{el}, respectively. The horizontal dashed line in the 3rd row shows the buffer limit. The horizontal dashed lines in the 4th row show the limitations posed by u_{max} and u_{max}^δ given in (9.7).

The results given in Table 9.9 and Figure 9.14 can be summarized as follows: first, it is observed that SCA2, whether with SAP and SUP, outperforms all the other autopilot only cases. SCA2 provides a higher tracking performance γ, a higher CfM and a lower GCD. Second, SAP shows a better performance than other controllers, including SUP, throughout the simulation run by not only showing a higher tracking performance but also preventing CfM from reaching zero (saturation point). Third, a quick inspection of the first row of Figure 9.14 shows that both optimal and μ-mod autopilots fail to respond to the second anomaly. This can be explained by the fact that CfM reaches to zero (saturation point) many times, especially in the case of optimal controller.

Table 9.9 Shared Control Architecture (SCA) 2 versus autopilot-only cases. SCA, both with a "Situation Aware Pilot" (SAP) and a "Situation Unaware Pilot" (SUP), results in higher tracking performance γ, higher CfM values, and smaller GCD.

Method	$\text{RMSE}_{\{h,v\}}^-$	$\gamma_{\{h,v\}}$	CfM	GCD
SAP	$\{60,23\} \times 10^{-4}$	$\{36,135\} \times 10^{-4}$	1.21	24.8×10^{-4}
SUP	$\{60,23\} \times 10^{-4}$	$\{51,152\} \times 10^{-4}$	1.16	24.8×10^{-4}
Adaptive	$\{24.1, 0.4\}$	$\{0.51, 1.06\}$	0.92	NA
μ-mod	$\{60,23\} \times 10^{-4}$	$\{0.10, 0.27\}$	0.81	342×10^{-4}
Optimal	$\{24.1, 0.4\}$	$\{160.8, 13.8\}$	0.84	NA

CfM: capacity for maneuver, GCD: graceful command degradation, based on altitude h, NA: Not Applicable.
Note that SAP performs better than SUP in all these metrics.

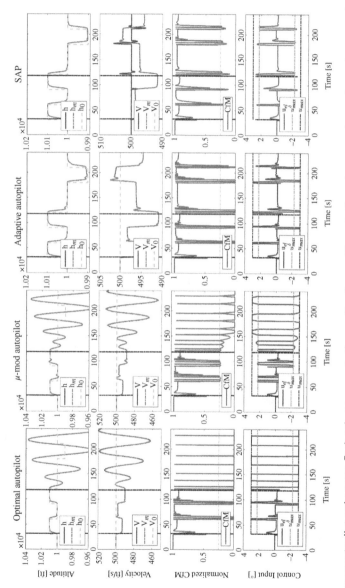

Figure 9.14 SCA2 versus autopilot-only cases. Each column presents the results of a specific controller, whereas each row shows the comparison of these controllers based on altitude tracking h, velocity tracking V, CfM, and elevator control input δ_{el}, respectively. The horizontal dashed line in the 3rd row shows the buffer limit ($\delta = 0.25$).

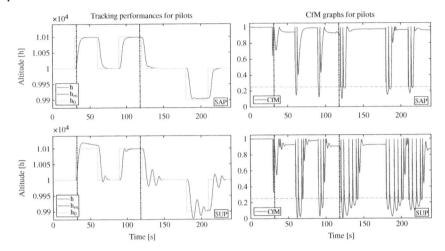

Figure 9.15 Comparison of the SAP and the SUP for the same scenario. The pure difference between these two pilots is based on the additional anomaly estimation input $\hat{\Lambda}_{f_p}$. It greatly attenuates the oscillatory tracking behavior around the reference model and contributes to a fast recovery from the buffer region.

Fourth, the adaptive autopilot shows an acceptable performance up to the second anomaly, but demonstrates degraded behavior with repeated saturation. It is noted that the elevator in the case of adaptive autopilot also hits the saturation point many times and shows a more oscillatory response compared to SAP.

To emphasize the effect of anomaly estimation input $\hat{\Lambda}_{f_p}$, the performances of SCA2 with SAP and SUP are shown separately in Figure 9.15. It is seen that SAP performs better than the SUP in terms of both tracking and CfM metrics. The selection of a suitable μ and the introduction of anomaly estimation, even with an error of 0.2, contribute dramatically to resilient performance, which can also be numerically verified by performance metrics given in Table 9.9.

9.4 Summary and Future Work

Although increased automation has made it easier to control aircraft, ensuring a safe interaction between the pilots and the autopilots is still a challenging problem, especially in the presence of severe anomalies. The domain of decision-making is common to both human experts and feedback control systems (Eraslan et al. 2020). However, the process of detecting and mitigating the anomaly, the speed of response, intrinsic latencies, and the overall decision-making process may vary between the pilot and the autopilot. Considering autonomous systems as fully functional isolated units separate from their connections with the human user may not be appropriate. Current approach often consists of autopilot solutions that disengage themselves in the face of an anomaly and the pilot takes over. This may cause reengagement of the pilot at the worst possible time, which can result in undesired consequences, and a bumpy transfer. We continue to see the consequences of bumpy transfer, for example in accidents where the vehicles with the autonomous driving capabilities are involved. Some unsubstantiated and exaggerated expectations are further contributing to this undesirable behavior by promising a future where all critical tasks are handled by fully autonomous agents without the need of human supervision, control, or collaboration. This promise is fundamentally flawed since automation capabilities are built using finite resources, and hence, they have a design envelope, out of which they break down. The edges of this envelope are exactly where the human intervention is needed, since humans are capable of adaptation to unforeseen uncertainties. This is further elaborated in Woods (2021), where the ideas that favor human-less automation is called zombie beliefs due to their potential to persist in the face of disconfirmation. We believe that a responsible design process should consider both the human and the automation as indispensable and inseparable elements of decision-making under uncertainty. Our thesis in this chapter is that a shared control architecture where the decision-making of the human pilot is judiciously coordinated with that of an advanced autopilot is highly attractive in flight control problems when severe anomalies are present. A granularity assignment of both task distribution among humans and automation and timelines at which these tasks are carried out was articulated and argued to be necessary for a successful CPHS. We presented two distinct shared control architectures (SCAs) through which such coordination can take place and validate these architectures through human-in-the-loop simulations. Though the type of coordination varies between these two architectures, both employ a common principle from cognitive engineering, namely capacity for maneuver (CfM). While these architectures were proposed in recent publications (Farjadian et al. 2016; Thomsen et al. 2019; Farjadian et al. 2020), a common framework within which to view them as well as their validation using human subject data are the main contributions of this chapter.

As automation increases in engineered systems, creation of new CPHS is inevitable. There will be a variety of scenarios where humans and machines will need to interact and engage in combined decision-making in order to lead to resilient autonomous behavior. These interactions will be complex, distinct, and will require new tools and methodologies. Deeper engagement with the social science community so as to get better insight into human decision-making and advanced modeling approaches is necessary. The results reported here including the articulation of a granularity taxonomy should be viewed as a first of several steps in this research direction.

References

Abbink, D.A., Carlson, T., Mulder, M. et al. (2018). A topology of shared control systems - finding common ground in diversity. *IEEE Transactions on Human-Machine Systems* 48 (5): 509–525.

Ackerman, K., Xargay, E., Talleur, D.A. et al. (2015). Flight envelope information-augmented display for enhanced pilot situational awareness. In *AIAA Infotech@ Aerospace*, 1112.

Bryson, A.E. (1996). Optimal control-1950 to 1985. *IEEE Control Systems Magazine* 16 (3): 26–33.

Chongvisal, J., Tekles, N., Xargay, E. et al. (2014). Loss-of-control prediction and prevention for NASA's transport class model. *AIAA Guidance, Navigation, and Control Conference*, 0784.

Eraslan, E. (2021). Shared control in aerial cyber-physical human systems. Master thesis. Bilkent Universitesi (Turkey).

Eraslan, E., Yildiz, Y., and Annaswamy, A.M. (2019). Shared control between pilots and autopilots: illustration of a cyber-physical human system. *arXiv preprint arXiv:1909.07834*.

Eraslan, E., Yildiz, Y., and Annaswamy, A.M. (2020). Shared control between pilots and autopilots: an illustration of a cyberphysical human system. *IEEE Control Systems Magazine* 40 (6): 77–97.

Farjadian, A.B., Annaswamy, A.M., and Woods, D.D. (2016). A resilient shared control architecture for flight control. *Proceedings of the International Symposium on Sustainable Systems and Technologies*. http://aaclab.mit.edu/publications.php (accessed 10 February 2023).

Farjadian, A.B., Annaswamy, A.M., and Woods, D. (2017). Bumpless reengagement using shared control between human pilot and adaptive autopilot. *IFAC-PapersOnLine* 50 (1): 5343–5348.

Farjadian, A.B., Thomsen, B., Annaswamy, A.M., and Woods, D.D. (2020). Resilient flight control: an architecture for human supervision of automation. *IEEE Transactions on Control Systems Technology* 29 (1): 29–42.

Gibson, T.E., Annaswamy, A.M., and Lavretsky, E. (2013). On adaptive control with closed-loop reference models: transients, oscillations, and peaking. *IEEE Access* 1: 703–717.

Habboush, A. and Yildiz, Y. (2021). An adaptive human pilot model for adaptively controlled systems. *IEEE Control Systems Letters* 6: 1964–1969.

Hess, R.A. (1997). Unified theory for aircraft handling qualities and adverse aircraft-pilot coupling. *Journal of Guidance, Control, and Dynamics* 20 (6): 1141–1148.

Hess, R.A. (2006). Simplified approach for modelling pilot pursuit control behaviour in multi-loop flight control tasks. *Proceedings of the Institution of Mechanical Engineers, Part G: Journal of Aerospace Engineering* 220 (2): 85–102.

Hess, R.A. (2009). Modeling pilot control behavior with sudden changes in vehicle dynamics. *Journal of Aircraft* 46 (5): 1584–1592.

Hess, R.A. (2014). A model for pilot control behavior in analyzing potential loss-of-control events. *Proceedings of the Institution of Mechanical Engineers, Part G: Journal of Aerospace Engineering* 228 (10): 1845–1856.

Hess, R.A. (2016). Modeling human pilot adaptation to flight control anomalies and changing task demands. *Journal of Guidance, Control, and Dynamics* 39 (3): 655–666.

Hosman, R. and Stassen, H. (1998). Pilot's perception and control of aircraft motions. *IFAC Proceedings Volumes* 31 (26): 311–316.

Kaber, D.B. (2013). Adaptive automation. In: *The Oxford Handbook of Cognitive Engineering* (ed. J.D. Lee and A. Kirlik), 594–609. Oxford: Oxford University Press.

Karason, S.P. and Annaswamy, A.M. (1993). Adaptive control in the presence of input constraints. *1993 American Control Conference*, 1370–1374. IEEE.

Kleinman, D.L., Baron, S., and Levison, W.H. (1970). An optimal control model of human response Part I: Theory and validation. *Automatica* 6 (3): 357–369.

Kun, A.L., Boll, S., and Schmidt, A. (2016). Shifting gears: user interfaces in the age of autonomous driving. *IEEE Pervasive Computing* 15 (1): 32–38.

Lavretsky, E. and Hovakimyan, N. (2007). Stable adaptation in the presence of input constraints. *Systems & Control Letters* 56 (11–12): 722–729.

Lavretsky, E. and Wise, K.A. (2013). *Robust and Adaptive Control*. London: Springer.

Lee, T.-G. and Huh, U.-Y. (1997). An error feedback model based adaptive controller for nonlinear systems. *ISIE'97 Proceeding of the IEEE International Symposium on Industrial Electronics*, 1095–1100. IEEE.

Lone, M. and Cooke, A. (2014). Review of pilot models used in aircraft flight dynamics. *Aerospace Science and Technology* 34: 55–74.

McRuer, D.T. and Jex, H.R. (1967). A review of quasi-linear pilot models. *IEEE Transactions on Human Factors in Electronics* (3): 231–249.

McRuer, D.T. and Krendel, E.S. (1974). Mathematical Models of Human Pilot Behavior. *Technical Report*. AGARD Report 188. Advisory Group for Aerospace Research and Development.

Mulder, M., Abbink, D.A., and Boer, E.R. (2012). Sharing control with haptics: seamless driver support from manual to automatic control. *Human Factors* 54 (5): 786–798.

Narendra, K.S. and Annaswamy, A.M. (2005). *Stable Adaptive Systems*. Courier Corporation.

National Transportation Safety Board (1996). In-Flight Icing Encounter and Loss of Control, Simmons Airlines, D.B.A. American Eagle Flight 4184 Avions De Transport Regional (ATR) model 72-212, N401am Roselawn, Indiana October 31, 1994. NTSB Aircraft Accident Report. *NTSB/AAR-96/01*. Washinton, DC.

Nguyen, L.T., Ogburn, M.E., Gilbert, W.P. et al. (1979). Simulator study of stall/post-stall characteristics of a fighter airplane with relaxed longitudinal static stability. [F-16].

Paternò, F., Burnett, M., Fischer, G. et al. (2021). Artificial intelligence versus end-user development: a panel on what are the tradeoffs in daily automations? *IFIP Conference on Human-Computer Interaction*, 340–343. Springer.

Rossetter, E.J. and Gerdes, J.C. (2006). Lyapunov based performance guarantees for the potential field lane-keeping assistance system. *Journal of Dynamic Systems, Measurement, and Control* 128 (3): 510–522.

Sheridan, T.B. (2013). *Monitoring Behavior and Supervisory Control*, vol. 1. Springer Science & Business Media.

Smisek, J., Mugge, W., Smeets, J.B.J. et al. (2014). Adapting haptic guidance authority based on user grip. *2014 IEEE International Conference on Systems, Man, and Cybernetics (SMC)*, 1516–1521. IEEE.

Smisek, J., Sunil, E., van Paassen, M.M. et al. (2016). Neuromuscular-system-based tuning of a haptic shared control interface for UAV teleoperation. *IEEE Transactions on Human-Machine Systems* 47 (4): 449–461.

Stevens, B.L., Lewis, F.L., and Johnson, E.N. (2015). *Aircraft Control and Simulation: Dynamics, Controls Design, and Autonomous Systems*. Wiley.

Tekles, N., Holzapfel, F., Xargay, E. et al. (2014). Flight envelope protection for NASA's transport class model. *AIAA Guidance, Navigation, and Control Conference*, 0269.

Thomsen, B.T., Annaswamy, A.M., and Lavretsky, E. (2019). Shared control between adaptive autopilots and human operators for anomaly mitigation. *IFAC-PapersOnLine* 51 (34): 353–358.

Tohidi, S.S. and Yildiz, Y. (2019). Adaptive human pilot model for uncertain systems. *2019 18th European Control Conference (ECC)*, 2938–2943. IEEE.

van der Wiel, D.W.J., van Paassen, M.M., Mulder, M. et al. (2015). Driver adaptation to driving speed and road width: exploring parameters for designing adaptive haptic shared control. *2015 IEEE International Conference on Systems, Man, and Cybernetics*, 3060–3065. IEEE.

Wierenga, R.D. (1969). An evaluation of a pilot model based on Kalman filtering and optimal control. *IEEE Transactions on Man-Machine Systems* 10 (4): 108–117.

Woods, D.D. (2018). The theory of graceful extensibility: basic rules that govern adaptive systems. *Environment Systems and Decisions* 38 (4): 433–457.

Woods, D. (2021). How to Kill Zombie Ideas: why do people tenaciously believe myths about the relationship between people & technology?

Woods, D.D. and Branlat, M. (2011). Basic patterns in how adaptive systems fail. In: *Resilience Engineering in Practice*, (eds. E. Hollnagel, J. Pariès, and J. Wreathall), 127–144. Farnham: Ashgate.

Woods, D.D. and Hollnagel, E. (2006). *Joint Cognitive Systems: Patterns in Cognitive Systems Engineering*. CRC Press.

Woods, D.D. and Sarter, N.B. (2000). Learning from automation surprises and going sour accidents. In: *Cognitive Engineering in the Aviation Domain*, (eds. N.B. Sarter and R. Amalberti), 327–353. Boca Raton: Taylor and Francis.

Xu, S., Tan, W., Efremov, A.V. et al. (2017). Review of control models for human pilot behavior. *Annual Reviews in Control* 44: 274–291.

Yildiz, Y. (2020). Cyberphysical human systems: an introduction to the special issue. *IEEE Control Systems Magazine* 40 (6): 26–28.

10

Safe Teleoperation of Connected and Automated Vehicles

Frank J. Jiang, Jonas Mårtensson, and Karl H. Johansson

Division of Decision and Control Systems, Department of Intelligent Systems, EECS, KTH Royal Institute of Technology, Stockholm, Sweden

10.1 Introduction

Since the DARPA Grand and Urban challenges, automated driving systems have developed rapidly. Nowadays, one can even see several deployments of automated vehicles on public roads, e.g. Einride (2020) and Nobina (2019). Automated vehicles are lauded to be a technological unlock that will make our roads more safe, our transport system more sustainable, and enable new services. The potential benefits of introducing automated vehicles are vast. Since automated vehicles have access to precise sensor suites and can be designed to reliably avoid accidents, if we can deploy automated vehicles at a large scale, then we can significantly reduce the number of accidents on the road. Furthermore, since automated vehicles will likely be electric and can optimize fuel efficiency, they could potentially improve the sustainability of our transport system significantly. Additionally, if we remove the driver onboard the vehicle, then the operational costs of the vehicle significantly decreases and allows for new services that might have not been considered otherwise; for example, Harper et al. (2016) proposes services for taxiing elderly or under-served individuals. However, as is outlined by Koopman and Wagner (2017), before reaping the benefits of automated vehicles, there are several safety challenges lying ahead of us that will require an interdisciplinary approach to handle; an effort that aligns well with the call from Lamnabhi-Lagarrigue et al. (2017) to develop systems under a cyber–physical–human systems (CPHS) perspective.

Currently, in many experimental automated vehicle deployments, these safety challenges are largely alleviated by the presence of onboard human safety drivers. While the presence of onboard drivers is permissible for experimentation, having an onboard human safety driver partially defeats the original purpose of automating our transport system. To address this issue, several automated vehicle companies have been setting up teleoperation systems, e.g. Davies (2017) and Einride (2020) (Figure 10.1). In most cases, teleoperation is a semi-automated mode used in scenarios where the automated driving system is unable to complete the task on its own. The idea is that by introducing teleoperation systems, we will be able to deploy automated vehicles onto routes that cannot be fully handled by automation, allowing for a wider spread deployment of automated vehicles. For tasks where it is still difficult to automate vehicles, we can start by assigning the task to a remote, human operator to handle the task. Then, after some time, we can determine how we could automate

Cyber–Physical–Human Systems: Fundamentals and Applications, First Edition.
Edited by Anuradha M. Annaswamy, Pramod P. Khargonekar, Françoise Lamnabhi-Lagarrigue, and Sarah K. Spurgeon.
© 2023 The Institute of Electrical and Electronics Engineers, Inc. Published 2023 by John Wiley & Sons, Inc.

Figure 10.1 A snapshot of Einride's Remote Driving Station. Source: Used with permission from Einride.

the vehicle to do the same task or if we should keep the task for humans to handle. In this way, teleoperation serves as a technology that allows us to transition our transport system from low to high levels of automation.

To help the community reach a consensus on the relationship between a remote human operator and automated driving systems, Society of Automotive Engineers (SAE) International (formerly known as the Society of Automotive Engineers) recently updated their standard (SAE International 2021) to include definitions for remote assistance and remote driving (teleoperation). In these definitions, they relate teleoperation to the well-known Levels of Driving Automation. Specifically, teleoperation by a remote human operator is described as a possible fallback choice for a Level 4 or 5 automated driving system either when the vehicle exhibits a dynamic driving task-related system failure or if the vehicle exits the operational design domain of the automated driving system. The authors provide an example where a Level 4 automated driving system may request assistance from a remote operator after encountering an unannounced road construction. Additionally, Bogdoll et al. (2021) break down the new SAE description of teleoperation and survey the different implementations in industry and some of the practical experiments conducted in research. However, the survey does not study the different safety aspects of teleoperation or the work done to enhance the safety of teleoperation systems. In any case, it is clear from SAE International's new standard and related experimental deployments of teleoperation systems that the safe inclusion of human operators as fallback support for automated driving systems will require a deep understanding and integration of human capabilities and behaviors.

While the inclusion of teleoperation into the automation of road vehicles is still being heavily discussed, the basic idea of teleoperating a cyber–physical system (CPS) is not a new one. Already in the 2000s, Lichiardopol (2007) surveyed the history of teleoperation systems and introduced some early technical issues in teleoperation: sensors and actuator setup, communication media selection, and latency issues. Upon close inspection of the different systems highlighted by Lichiardopol (2007), it also becomes clear that the evolution of teleoperation systems is intertwined with the evolution of telecommunication technology. For example, after the widespread establishment of the Internet, Fong et al. (2001) presented an early prototype for teleoperation that even included designs for collaborative control, sensor fusion, and a mobile device interface. Similarly, Fong et al. (2003) also proposed early designs for teleoperation that take advantage of innovations of the Internet at the time. After a new generation of telecommunication technology is released, we often start out with designing new teleoperation systems to take advantage of the new technology. We attribute this phenomenon to the fact that implementing a human-in-the-controller system, as is described in Chapter 1, is a natural first step before we have had time to integrate a fully automated controller. Many of the early teleoperation systems were developed to allow humans to operate machinery in inhospitable environments, such as deep sea, within nuclear reactors, and in space. This has long served as an important application of teleoperation systems. However, as latency over wired and wireless connections has decreased, there have been an increase in proposals to use teleoperation for applications such as connected and automated vehicles. Especially with the current deployments of 5G systems, many researchers are beginning to investigate how the design of society's systems will be affected by the upcoming implementation of a new Internet that Fettweis (2014) calls the "tactile internet."

Since 5G networks are capable of decreasing the latency of wireless communication to a level where humans could perform tactile tasks without noticing significant differences between remote interactions and local interactions, this presents new opportunities to re-think the implementation of teleoperation systems. Furthermore, as is reported by Alriksson et al. (2020), 5G networks will also be able to support Critical Internet of Things (IoT) connectivity. Meaning that, for the first time, data can be easily transferred over wireless channels with the necessary quality of service that is required for safety-critical systems, i.e. 1 ms latency and 99.999% reliability. When designing a teleoperation system for connected and automated vehicles, the support for technologies such as the tactile internet and critical IoT connectivity is important for improving the safety of the teleoperated vehicles. Previously, one of the fundamental safety concerns for the teleoperation of vehicles is the fact that controlling a vehicle over a remote communication channel is significantly less reliable than controlling a vehicle from the driver's seat. Thus, the properties of the tactile internet and critical IoT connectivity are promising for closing this gap, since we can start to safely move toward teleoperation systems where the tactile experience is indiscernible, in terms of latency and reliability, from driving the vehicle from the driver's seat.

In many of the setups where a human operator can teleoperate and remotely assist a road vehicle, the human operator will most likely be seated in a "control tower." This control tower will be a central node in the overall communication network for supporting connected and automated vehicles. Specifically, the control tower will have access to data streams from connected vehicles, the infrastructure edge cloud, and other network resources. When we design teleoperation systems within the context of this control tower, we can utilize these data streams to further enhance the situational awareness of human operators. Moreover, by designing under a CPHS perspective, we can ensure the data are effectively used to both maximize the safety of the teleoperation system and to enhance the human operator's teleoperation performance. For example, by combining the video and 3D data from a collection of cooperative vehicles and infrastructure sensors nearby the teleoperated vehicle, it is possible to provide a human operator "x-ray" vision while teleoperating. By designing and implementing such applications, we can potentially set up teleoperation systems where remote driving becomes more safe than local driving.

The remaining chapter is organized as follows: In Section 10.2, we introduce the concept of safe teleoperation of connected and automated vehicles in more detail and overview the innovations brought in by 5G systems to enable the integration of teleoperation into safety-critical systems. Then, in Section 10.3, we will highlight the CPHS design challenges of operator perception and shared autonomy for safe teleoperation. In Section 10.4, we overview the different research contributions toward solving the operator perception and shared autonomy problems. In Section 10.5, we discuss the future research directions that have the potential to enhance the safety of vehicle teleoperation even further. Finally, in Section 10.6, we conclude the chapter with a summary of the state of safe teleoperation research for connected and automated vehicles.

10.2 Safe Teleoperation

The problem of ensuring the overall safety of teleoperation systems is closely related to the problem of ensuring the overall safety for automated vehicles. Koopman and Wagner (2017) outlines this problem as the interdisciplinary need to improve the safety of computing hardware, software, robotics, security, testing, human–computer interaction, social acceptance, legal, and safety engineering for automated vehicles. These topics also have relevance in the pursuit of safe teleoperation. A possible difference lies in the level of activity in each area. In particular, the safety requirement on the human–computer interaction is higher in teleoperation systems compared to automated driving systems, as we are putting a human directly into the control loop, making the human a critical decision-maker (Samad 2022); Chapter 1 in this book. For similar reasons, the safety engineering for teleoperation systems is also different than automated vehicles. We have to include a human operator as a component in our engineered system and apply safety engineering principles to the human operator. In particular, we need to architect a system where the human operator can act as a fallback for automated driving systems, while mitigating any risk the human operator poses on the overall system.

In this section, we start by introducing the architecture of a typical teleoperation system and discuss the opportunities to improve the safety of the overall system. Generally, there are three approaches to maximize the safety of a teleoperated system: (i) make remote operation feel like local operation, (ii) present information that maximizes the human operator's situational awareness, and (iii) design on-board automation to collaborate with the remote human in a way that synergizes the safety capabilities of both human and artificial intelligence.

The human–computer interaction and safety engineering challenges for connected and automated vehicles are similar to the ones the aerospace community has been tackling for several decades. In this work, we take inspiration from the following excerpt from Wickens et al. (1997), which captures the overall philosophy the aerospace community has upheld for their safety engineering:

> In short, we believe that it is impossible to bring the reliability of any system up to infinity, or even to accurately estimate that level (without variability) when it is quite high, and this has profound implications for the introduction of automation. That is, one must introduce automation under the assumption that somewhere, sometime, the system may fail; system design must therefore accommodate the human response to system failure

This perspective motivates the inclusion of pilots onboard and air traffic controllers offboard airplanes to perform fallback tasks when automation fails. Illustrated in the left panel of Figure 10.2, the pilots are ready to provide fallback actions when there is an onboard system failure and the air traffic controllers are ready to provide fallback routing when the air traffic routing system fails. The inclusion of humans into the air traffic systems, along with advances in safety automation itself, are important contributors to the high levels of safety in aviation. For connected and automated vehicles, we must architect a relationship between automation and human operators that enhance the safety of road transportation in a similar fashion. However, as is illustrated on the right-side of Figure 10.2, one big differentiator between aerial vehicle and road vehicle fallback is the missing

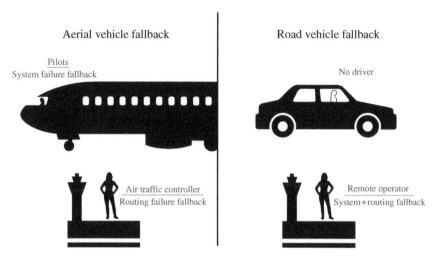

Figure 10.2 An illustration of the aerial and road vehicle fallback options.

onboard driver. Since automated road vehicles will not have an onboard driver, the responsibility of performing fallback actions when vehicles exhibit system failures lies with the remote human operator. This responsibility is in addition to the responsibility of performing other tasks, such as fallback routing. That said, these different responsibilities and fallback tasks could be distributed to different human operators who have different jobs. Nevertheless, to achieve similarly high levels of safety to the aviation industry, we need to develop novel teleoperation systems to mitigate possible automation failures that could happen onboard a connected and automated vehicle.

We start by inspecting the different teleoperation systems that have been developed for connected and automated vehicles. Many of the earlier designs for remote driving, such as the one presented by Gnatzig et al. (2013), are designed as direct teleoperation systems. In these direct teleoperation systems, the human operator interacts with a user interface consisting of screens for viewing the video stream broadcasted from the vehicle and steering wheel and pedals for giving control signals to the vehicle. The human operator's control signals are communicated to the vehicle for direct implementation onto the vehicle's actuators. We refer to this setup as a direct teleoperation system since the human's decisions are directly implemented onto the vehicle. From this work, we know a teleoperation system should include a human operator, a user interface, a vehicle, and a sensor suite. However, direct teleoperation setups do not take into account the automation onboard the vehicle. From the standard set by SAE International (2021), it is clear that the higher a vehicle's automation level, the more dependent a remote operator's teleoperation will be on the decisions of the automated driving system. This perspective is in line with the work done by the robotics community on teleoperation, where experiments, such as in Takayama et al. (2011), show that onboard automation can be used to improve the safety of teleoperation systems. Thus, it is important to include the automated driving system into the teleoperation loop. Additionally, we must consider the algorithms that are used to generate the data for the user interface display. For example, in teleoperation systems, haptic signals and video streams should systematically collected or generated from the vehicle's onboard sensor suite and communicated to the remote operator's user interface. Thus, we should consider a module that we will refer to as the teleoperation "observer," where the main function of this module is to perform local observation on the status of the vehicle for the sake of the human operator's perception. We integrate and illustrate all of these modules in Figure 10.3.

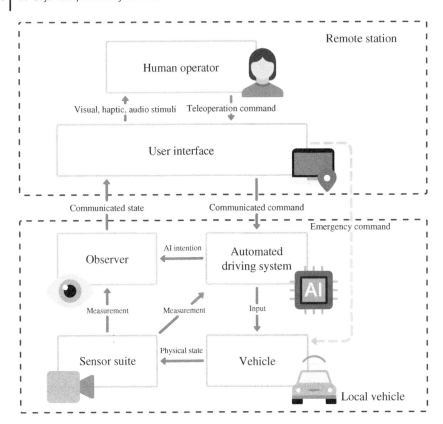

Figure 10.3 Our synthesized architecture for the teleoperation of connected and automated vehicles. We have chosen the modules and their connections based on the overlap of different architectures that exist in the literature.

Among the different modules, we will focus most of the discussion for the rest of this chapter on the challenges and research around the design of the automated driving system, the observer, and the user interface.

For teleoperation of connected and automated vehicles, a critically important factor in the safety of the vehicle is the way the automated driving system receives and handles the communicated command from the remote operator. This is especially true for vehicles that are equipped with automated driving systems with a high level of driving automation. SAE International (2021) specifically states that "Level 4 and 5 [automated driving system]-equipped vehicles that are designed to also accommodate operation by a driver (whether in-vehicle or remote) may allow a user to perform the [dynamic driving task] fallback, when circumstances allow this to be done safely, if s/he chooses to do so." In other words, if a remote human operator decides to perform a fallback action, such as teleoperation, then this is permissible by the level 4 or 5 automated driving system if the teleoperation can be done safely. This means that automated driving systems need to be designed to understand whether or not a remote human operator can perform a teleoperation task safely, presenting the challenge of giving the automated driving system the predictive capabilities to make this decision, a point we will discuss further in Sections 10.4 and 10.5. SAE International (2021) goes on to specify that "a Level 4 or 5 [automated driving system] need not be designed to allow a user to perform [dynamic driving task] fallback and, indeed, may be designed to disallow it in order to reduce crash risk." They highlight the fact that a level 4 or 5 automated driving system does not necessarily need to be designed to handle remote operator fallback. However, if this is the case, then to comply with the principles outlined by the aviation industry, this means that the remote operator must be able to bypass the automated driving system in emergency scenarios where the automated driving system is exhibiting failure. We have included this emergency bypass command into Figure 10.3. In cases where the emergency bypass command is used to control the vehicle, then the architecture becomes a direct teleoperation one. When the communicated command from the human operator's user interface is instead handled by the automated driving system, then the automated driving system needs to safely integrate both the human operator's control input and its own control input when computing the final control input to implement on the vehicle; in the coming Section 10.3, we will address this as a shared autonomy problem.

For shared autonomy setups, one module that is very important is the observer module. The primary concern of the observer module is to select and aggregate measurement data from the vehicle's sensor suite into a communicated state for the human operator's user interface. As we will discuss in Section 10.3, one challenge of designing good teleoperation observers is to develop algorithms that can extract signals from the sensor suite that can give remote operators the "feeling" of driving a real vehicle. Strictly speaking, the objective of these algorithms is to give the operator at least the same information they would have if they were making safety-related decisions during real driving. This can include extra video feeds that allow the operator to "look around" the vehicle as if they were sitting in the drivers seat or haptic signals that inform the driver that the wheels of the vehicle have bumped into an obstacle. Another function of the observer module is to communicate the intentions of the automated driving system. An early example of this idea is implemented by Fong et al. (2003). In their teleoperation set up, they use a dialog system where the automated system could query and communicate with the operator. By allowing the robot to communicate its intention and status, the operator is able to incorporate this information into their decision-making. In this early example, the authors already show that communication can facilitate more effective collaboration between humans and automated systems. Since today's automated driving systems have to make a multitude of decisions while driving, the communication of the driving system's intention is as important as ever. The design of the observer has significant impact on the remote operator's perception of the local status of the vehicle, which is paramount for making safe decisions remotely.

Finally, once the observer has been designed to help collect safety-related information for the human operator, the final module that must be designed with safety in mind is the user interface, which expresses the communicated state from the observer. The user interface typically consists of control hardware, a visual interface, and sometimes an audio interface. For example, the control hardware can be a steering wheel and pedals or the control knobs shown in Einride's setup in Figure 10.1. The control hardware should be haptically enabled to allow for user interface designs that express parts of the communicated state of the vehicle through haptic feedback. The visual interface can consist of monitors, projector screens, or even mixed reality headsets. Sometimes an audio interface is included into a user interface; however, in most teleoperation setups for connected and automated vehicles, audio is not one of the communicated states from the observer module, thus we will not cover this design option. That said, audio can be an important source of information that allow human operators assess the safety of the vehicle and should likely be included in future user interface design. The user interface, designed in conjunction with the observer, has direct impact on the remote operator's understanding of the communicated state from the observer.

In Section 10.3, we will present the design challenges of the user interface, observer, and share autonomy; however, we first must discuss the requirements on the wireless channel that connects them together. The quality of service of the wireless channel that connects the remote user interface to the automated driving system and connects the observer to the user interface changes what we can and cannot implement for these modules. More specifically, the latency, bandwidth, and reliability of the wireless channel affect the design decisions for the three modules, especially from a safety standpoint. The two most common wireless communication mediums, cellular LTE and Wi-fi, might be suitable for certain teleoperation use cases; however, they are not reliable enough for the use in safety-critical use cases such as the teleoperation of vehicles. Studies performed by Neumeier et al. (2019) show that high latency in the wireless channel between the remote station and the vehicle negatively affects remote driving performance, and, in turn, safety. Thus, we claim that the implementation of safe teleoperation systems require the integration of recently deployed 5G technologies. One of the ambitious objectives of safe teleoperation is to allow a remote operator teleoperate a vehicle with a greater level of safety than if they were driving the vehicle onboard. To achieve this, we need to make sure that the latency between the user interface and the vehicle is minimized to the point that the latency is indiscernible by the human operator. Also, the interaction must be reliable to the point that there is minimal difference between the reliability of the teleoperation and real driving. These objectives are almost unreachable; however, as we will present in Section 10.2.1, the technologies introduced by the 5G standard will help us come closer to reaching them.

10.2.1 The Advent of 5G

In this section, we give a brief overview of the technologies deployed with 5G networks. In particular, we will focus in on the technologies that are most relevant for making teleoperation systems more safe. For a general overview of 5G, we refer readers to Dahlman et al. (2020).

The requirements that we seek for safe teleoperation of connected and automated vehicles match the requirements set out for the "tactile internet" and critical IoT connectivity. As described by Fettweis (2014), the tactile internet is the new evolution of the internet that will satisfy the requirements to support tactile, or haptic, services. Specifically, the tactile internet will be able to support services that require around 1 ms round-trip latency. According to Sachs et al. (2019), this new internet's requirements align well with the 5G ultra-reliable and low-latency communication (URLLC) standard, which requires that a data packet can be sent with a one-way latency of 1 ms and can reliably achieve that latency with a success rate of 99.999%. The URLLC standard also fits in with the critical IoT connectivity specified by Alriksson et al. (2020). Even though the URLLC standard is a target of 5G network deployments, achieving this standard and supporting the tactile internet on real use cases is not trivial. In light of this, Sachs et al. (2019) overview and present the 5G technologies required to support a tactile internet. Most importantly, they point out that the tactile internet will rely on an automated and customized configuration of the 5G network that can only be done by leveraging software-defined network designs and network slicing paradigms. Software-defined network architectures are a shift toward being able to program a network and easily configure it, through software updates instead of hardware updates, for new applications and services (Akyildiz et al. 2015). Network slicing paradigms are a specific function of software-defined networks that allow for one network to support multiple service types at the same time (Foukas et al. 2017). In summary, even though 5G networks are capable of supporting use cases such as the teleoperation of connected and automated vehicles, they must be configured and programmed to handle teleoperation services to ensure that the correct network resources are available nearby the vehicle and can handle the mobility of the vehicle.

Since the work done by Sachs et al. (2019) indicate that 5G networks are capable of supporting teleoperation services, and, at the same time, the assessment presented by 5GAA Automative

Association (2017) are pushing for the use of 5G networks for cellular vehicle-to-everything communication, it is becoming clear that 5G is going to be the primary wireless communication medium for connected and automated vehicles. However, we also know from evaluations done by Saeed et al. (2019), who performed thorough analysis of the remote driving potential within the networks of three different carriers, that 5G-based remote driving can currently only be supported for select roads, meaning that we cannot initially expect to perform safe teleoperation throughout all of our road networks.

10.3 CPHS Design Challenges in Safe Teleoperation

As automated systems continue to rapidly advance, especially in the automotive industry, the need to carefully consider how to integrate humans has become increasingly important. Along these lines, a group of experts report that this is one of the key automation challenges that lies ahead of us (Lamnabhi-Lagarrigue et al. 2017). In particular, we have seen in recent developments that the integration of humans in cyber–physical systems needs to be done in a way that is human-centric and not in the traditional approach of treating humans as an external actor of a system. For example, we have seen in the work done by Dextreit and Kolmanovsky (2014) that by including the human driver in a game theoretic model along with the power train of a vehicle, one can create a more fuel-efficient controller. Then, in Shia et al. (2014), authors show that by including driver models into the construction of safe, semi-autonomous controllers, then the controller becomes less conservative and does not falsely intervene as often. These successes exemplify the need to develop teleoperation systems under the CPHS lens, as opposed to a traditional CPS one. Specifically, when we design the automated driving system, teleoperation observer, and user interface module from Figure 10.3, we need to design them in a human-centric fashion, instead of considering the human as an external actor.

While one could claim that teleoperation systems have long been inspected under the lens of the CPHS perspective, we argue that other than the user interface, the CPHS principles have not been well integrated into the design of automated driving systems and teleoperation observers, especially for connected and automated vehicles; this is apparent by the lack of explicit standards for driver behavior model integration into automated driving systems in SAE International (2021). Moreover, as we indicate in Section 10.2, we have a new opportunity to re-design all three modules based on the changes introduced by 5G networks. Leveraging 5G connectivity, we will be able to codesign observers and user interfaces in a way that allows an operator to perceive remote driving as the same as normal driving. In simulation work done by Hosseini and Lienkamp (2016b), authors showed that by improving a remote operator's perceived presence during teleoperation, the safety of their driving also improves. Furthermore, since 5G connections can be configured to be 99.999% reliable, we can reliably integrate teleoperation into the design of automated driving systems without compromising the safety of the overall system. Namely, two of the major CPHS design challenges for making teleoperation systems safe is:

1. the co-design of observers and user interfaces to enhance human operator perception,
2. the design of safe, shared automated driving systems.

For the first challenge, we clarify that an operator's perception, in the context of teleoperation, is the remote operator's ability to visually, haptically, or auditorily perceive and make sense of the situation of the local vehicle. In the literature, some operator perception problems that are referred to are "haptic awareness" or "situational awareness." In the architecture shown in Figure 10.3, the two modules whose design directly affect the operator's perception are the observer and user interface. Since the observer's main function is to collect and synthesize the communicated state from the sensor measurements and the communicated intention from the automated driving system, the design of the aggregated state enables or limits what the human operator's user interface is able to communicate to the human operator. Moreover, the hardware and software capabilities of the user interface will also enable or limit what the human operator can perceive. One of the central issues in the codesign of the observer and user interface is the unclear relationship between the operator's perception and the safety of their teleoperation. This issue lies with the lack of models for how operator's perceive risk. As is presented by Brown et al. (2020), there are models for performing risk assessment of an operator's driving behavior; however, this is not the same as modeling an operator's own perception of risk. There is some initial work by Stefansson et al. (2020) that propose models for how driver's perceive risk in overtaking scenarios. However, since this is still a nascent modeling approach, we refer to this body of work as a future research direction in Section 10.5. Instead, we will focus our discussion in Section 10.4 around the empirical work that has been conducted to find correlations between different observer and user interface codesigns and the safety of teleoperation.

For the second challenge, we define shared autonomy as the challenge of maintaining the autonomy of all participating agents who have control over the same vehicle. The challenge is to make sure the control implemented on the shared system completes the objective and synergizes the abilities of all participating agents. For safe, shared automated systems, the objective and synergy are oriented around highlighting the best safety-aspects of each agent. In the architecture we described previously, the participating agents are a remote human operator and the onboard automated driving system. In this scenario, our objective is to set up a safe, shared automated system that synergizes the human operator's capability to react to safety decisions that our automated driving systems fail to handle and the automated driving system's capability to consistently react to previously well-known safety issues. To highlight a human operator's decision-making capability that automated driving systems still struggle with, SAE International (2021) describes a scenario where a human operator correctly identifies that a bag in the middle of the road is empty, where an automated driving system might falsely identify the bag as a safety threat. In contrast, in scenarios that require fast and precise collision avoidance, automated driving systems have high success rates due to their precise sensor suite and fast-reacting computations. That said, outside of specific examples, there is little consensus on what types of tasks humans are generally better at than automation and vice versa. However, in preparation for when this consensus is reached and standards are issued, several approaches have been proposed for shared autonomy in teleoperation. For example, Jiang et al. (2020) proposed a shared autonomy approach that supports different policy blending approaches, while still providing safety guarantees for the overall system. However, to make matters more complex, another issue for shared autonomy systems is knowing how to evaluate them. From a pure engineering perspective, the success of a shared autonomy system can be measured by comparing the performance, or efficiency, of the integrated system compared to the performance of the individual agents. Some typical engineering, or objective, metrics for measuring teleoperation performance are

- **Total execution time**: The time it takes to complete a given task,
- **Total operator control input**: The amount of effort the operator needs to complete a given task,
- **Success rate**: The rate a given task is completed successfully.

From a social or a human-centric perspective, the success of a shared autonomy system can also be measured based on the human operator's preference and their perspective on usability. Some typical social, or subjective, metrics for measuring teleoperation performance are

- **Predictability**: The operator's ability to predict the behavior of the teleoperated system,
- **Perceived assistance**: The operator's feeling of how much assistance they received,
- **Satisfaction**: The operator feeling about whether their performance was satisfactory.

As can be seen in the experiments conducted by Javdani et al. (2018), human operators sometimes prefer a less-efficient shared autonomy system because they preferred the usability of the system. We will refer to the evaluation and the further experimentation of teleoperation with shared autonomy as an important future research direction in Section 10.5.

In Section 10.4, we will focus on presenting an overview of the research advances on the challenges of enhancing operator perception and designing shared autonomy systems. We have chosen these two topics as they are the two central open questions when it comes to updating teleoperation systems from a safety perspective.

10.4 Recent Research Advances

In this section, we will overview the research advances with regards to both enhancing remote human operator perception and improving shared autonomy. We start by introducing the work that has been done to help a human operator feel like they are operating a vehicle locally instead of remotely. The aim of this body of work is to help a human stimulate the safety-related intelligence they would normally have while operating a vehicle locally. We continue on to introduce the recent developments toward shared autonomy and the integration of both a human operator and an automated driving system into one, shared control system.

10.4.1 Enhancing Operator Perception

In teleoperation, the two forms of perception that have received the most attention is haptic and visual perception. When enhancing a remote operator's perception, there are two types of approaches: (i) provide the remote operator with as many of the raw signals they would have access to normally and (ii) assist the remote operator's perception by filtering signals or adding additional signals using artificial intelligence. The former approach lends itself to the idea that we still do not fully understand human intelligence and have seen, empirically, that we can improve the safety of their teleoperation by providing them with signals more closely resembling what they are used to (Hosseini and Lienkamp 2016b). The latter approach lends itself to situations where we might have an idea of how to help focus a human operator's attention by assisting their perception (Reddy et al. 2020). For both approaches, the key design challenge is the codesign of the observer and the user interface. Most of the work around codesigning the observer and the user interface centers around the preparation and expression of information through visual displays and haptic feedback.

We will start by surveying the different types of observers that have been developed for visual displays. Since we know that a significant amount of our situational awareness is derived from sight, visual display is often the focus when designing teleoperation systems. The simplest form of visual display is a video stream from the vehicle's onboard cameras. In this setup, the design of the observer is trivial, as it simply forwards video streams grabbed from cameras in the sensor suite. However, by adding more intelligence to the observer, one can set up more elaborate visual displays, such as the one shown in Figure 10.1, where additional information, such as the vehicle's predicted future positions, current speed, current speed limit, and fuel status is gathered by the observer, communicated, and rendered on top of the video stream by the user interface. As we indicate in Figure 10.3, observers can also collect information from the automated driving system to be visually displayed to the remote operator. To this end, authors in Fong et al. (2003) develop an early system for performing collaborative control between a remote human and a local vehicle that relies on a dialog system for communication between human and vehicle. In this approach, an observer is collecting intent in the form of information and queries from the automated system. Then, the observer communicates this intent in a text display to the remote operator. Several more recent research approaches have been developed to design more advanced observers that enhance the safety properties of teleoperation visual displays. Hosseini and Lienkamp (2016a) propose a safety-motivated visual reporting system. The system helps analyze the visual situation and reports possible hazards along with some data about hazards such as time-to-collision. Then, to help an operator best handle latency changes throughout a driving task, authors in Neumeier and Facchi (2019) proposed new services for supporting remote teleoperation with routing and vehicle speed suggestions based on network measurements. Using a service that suggests routes and vehicle speeds based on the latency of the wireless communication can allow operators to adapt to variations in network quality of service. Finally, authors in Reddy et al. (2020) propose and evaluate assistive state estimation using recurrent neural networks and partial observations to assist users in their observations. The assistance is proposed for assisting visually impaired humans; interestingly enough, since a vehicle may need to be teleoperated in areas with poor network quality, such assistance can help assist the operator teleoperate with a poor video feed. In these examples, the focus is on the observer's aggregation of information and what the operator needs to see. However, the design of the observer not only depends on what we think the human operator needs to see but also on the capabilities of the user interface's visual display.

The research contributions to the design of the visual display in the user interface has mostly been focused around the use of extended reality (XR) headsets. As is alluded to in Figure 10.1, many implementations of teleoperation for connected and automated vehicles simply use one or more screens for displaying the visual stimuli prepared by the observer. In this case, standard video streaming methods are suitable and the observer design does not receive additional complexity from the user interface. However, there have been several proposals to replace screen displays with XR headsets due to claims that the headsets can improve the perception of the remote operators, and, in turn, enhance the safety of remote driving. Since the advent of 5G, there has been more interest in XR approaches to teleoperation, since XR streams require both high bandwidth and low latency. XR streams can include raw visual signals, such as video streams, and also include virtualized visual elements, such as virtual heads-up displays for different statistics about the vehicle. However, XR feeds that are naively setup might not be cohesive and may contain redundant information that the operator has to comprehend on their own, such as multiple camera feeds that overlap each other. In fact, one of the central issues of using XR headsets as the visual display for the user interface is making the feed cohesive. An early example of an XR-based implementation is proposed and presented in Nielsen et al. (2007), where authors develop "ecological interfaces," which is an interface that synthesizes video, map, and robot pose information into a 3D XR display for improving the situational awareness of a remote operator. However, since software has significantly improved for generating virtualized driving environments, authors in Hosseini and

Lienkamp (2016b) alternatively proposed a design for a teleoperation user interface that cohesively mixes a complex, 3D virtual environment representing the obstacles detected by the vehicle's LiDAR and carefully placed, raw video frames from the vehicle's camera system. They conduct an experiment where they ask a user to perform both precision tasks and urban driving in a simulator. In their experiments, they show that by using XR, they were able to improve the safety of the teleoperation system, the overall precision of the driving, and even decrease the task load of the operator, according to the NASA Task Load Index. These experiments indicate that XR setups not only improve a human operator's immersion but also empirically show that safety of the teleoperation is improved.

In addition to designs for visual displays, another body of work proposes codesigns of observers and user interfaces for haptic feedback. Haptic stimuli is also considered a critically important signal for humans while performing driving tasks. Also inspired by the advent of 5G and the conceptualization of the tactile internet, haptic feedback has received renewed interest for teleoperation tasks. When the round trip latency between the operator's haptic hardware and a connected vehicle's observer is reduced to several milliseconds, haptic stimuli becomes a very effective mode of perception for the operator. For example, with such low latency, when an observer detects that the local vehicle's wheels came into contact with a curb, we can let the remote operator feel this and react to it immediately. As is pointed out by Fettweis (2014), using haptic signals for these types of physical interactions is important since a human's reaction to tactile events is an order of magnitude faster than a human's reaction to visual events. However, as latency increases to tens or hundreds of milliseconds, the utility of these haptic signals decrease significantly, since the vehicle might have already done something unsafe by the time the operator begins to react to the haptic stimuli. Although the advent of 5G only recently enhanced the utility of haptic signals, researchers have been exploring haptic feedback concepts for teleoperation for several decades already. The earliest prototypes for remote driving that use haptics to help a remote driver better understand the situation of the operated vehicle is presented in Brady et al. (2002). In this case, the observer is simply a multiplexer that aggregates accelerometer data, which is then transmitted using a digital RF transmission system. These early prototypes do not attempt to model the human's interaction with the user interface's haptic feedback hardware, and only try to communicate the data to generate forces in the user interface. Then, a decade later, authors in Hirche and Buss (2012) present the first formal treatment of haptic teleoperation as a control problem and introduced dissipativity-based modeling of the human's haptic closed-loop dynamics. They go further to also outline a new evaluation criterion that they call "perceived transparency." This evaluation criterion defines a metric for haptic setups that measure whether a human operator is able to distinguish between direct interactions and remote interactions. This transparency metric is particularly useful for evaluating the codesign of the observer and the user interface's haptic hardware. To maximize the transparency metric, all of the interaction forces a human would normally experience while driving should be re-created. However, in the case of driving a road vehicle, this includes both the forces generated from the steering system, the pedals, *and* the cabin of the vehicle. In other words, to maximize their proposed transparency metric, teleoperation systems will likely need to be set up in full vehicle motion simulators, which are normally used to add haptic transparency to driving simulators. However, the relation between transparency and safety is still unclear, and the empirical studies for highlighting this relationship is an important future research direction. In a more recent work, Ghasemi et al. (2019) jointly model the biomechanics of a driver's arms with the mechanics of a vehicle's control system with the objective of analyzing different haptic feedback models' promise to reduce cognitive load and providing situation awareness. However, there is little empirical evidence showing that these designs improve safety.

To summarize, for both visual and haptic interfaces, there is a dire need to run experiments to identify a more precise relationship between these forms of stimuli and the safety of the overall systems. The experimentation for most of the mentioned work does not focus on measuring the change in safety when introducing new visual or haptic interfaces. In the future, for the sake of developing safe teleoperation systems, experimentation should center around measuring the reduction of safety violations in safety-critical scenarios.

10.4.2 Safe Shared Autonomy

In teleoperation, control over the connected vehicle is shared between the remote human operator and the automated driving system. Previously, before automated driving systems reached maturity, most work focused on the development of direct teleoperation systems. However, now that automated driving systems are becoming increasingly safe, it is become more difficult to determine when a human operator or the automated driving system should be in control. Thus, the development of safe shared autonomous systems has begun to emerge. The development of safe shared autonomous systems sees two main developments: (i) allowing the automated agent to successfully understand and predict the human operator's intent, and (ii) embedding safe policy blending into the automated driving system that synergizes the capabilities of the remote human and the automated driving system.

The first development has seen significant attention. A key contribution that operator intent predictions can have is to make safe shared autonomous systems that are less conservative. If the automated driving system is designed to be defensive toward the worst possible human behavior all the time, then it could be the case that even simple tasks cannot be completed due to conservativeness. Authors in Brown et al. (2020) overview and classify 200 models for driver behavior modeling, which all have different combinations of modeling techniques. However, it is important to note that there is currently no standard approach to applying these 200 models in CPHS's such as teleoperation systems. The utility of predicting and categorizing human operator intent is exemplified in the work of Vasudevan et al. (2012), where authors develop a semi-autonomous framework, where a k-means clustering is performed to extract out less-conservative predictions about the vehicle's driving behavior, which are incorporated into the automated system's intervention mechanism.

To better incorporate the automated driving system's control input with the human's control input, an approach called policy blending emerged. Muslim and Itoh (2019) gives a general overview of different policy blending architectures that exist for human-centric, shared automation systems. In this work, they provide a more system-level perspective on shared autonomy and relate the architectures to the Levels of Driving Automation. However, they do not provide a formal treatment for policy blending. In an earlier work, Dragan and Srinivasa (2013) formally treat the concept policy-blending, where authors present one of the first policy-blending formalisms and performed user studies revealing the importance of confidence measures and information-gathering actions in teleoperation systems. In Dragan and Srinivasa (2013), they propose a predict-then-act approach, where the policy blending function is adjusted based on the confidence the automated system has in its prediction of the human operator's final goal. Several different kinds of blending equations presented in Dragan and Srinivasa (2013) could be considered for combining the inputs of the automated driving system and the human driver. In a follow-up study, authors in Javdani et al. (2018) developed a novel shared autonomy approach that incorporates distributions of the possible goals into the assistance, leading to improved assistance compared to previous predict-then-act methods. In the approach developed by Javdani et al. (2018), the assistance will try to help the operator with teleoperation, even when the automated system is not yet sure about what the operator's final objective is. Both of these approaches focus on the automated system's confidence in predicting the human operator's intent and adjusting the blending of the automated driving system and the human operator's control inputs. They attempt to achieve the best of both worlds: they allow the operator to do the task in the way they prefer, and then seamlessly add in assistance from the automated system to help the operator more efficiently complete the task. Notably, these approaches have the benefit of being more resistant to latency issues. Since the assistance is computed onboard the automated system, the assistance can act to smooth out or filter out control artifacts created by latency. Moreover, in cases where the automated system correctly predicts what the human operator is trying to do, the operator may not notice that the assistance is active and will feel like they completed the task easily, even while using an unstable wireless communication channel. However, these approaches do not provide policy blending approaches that are safety-oriented.

When blending the policies of the human operator and the automated system, it is possible to blend the policies in a way that is oriented around safety. For example, in Takayama et al. (2011), authors implement and study an assisted teleoperation system, where they find that the assistance helps the system stay safe. In this work, the teleoperated system constantly predicts its future trajectories using a dynamic window approach. Whenever the predicted trajectories lead to collisions, then the system filters the operator's teleoperation command and finds the closest alternative that does not result in a collision. This policy-blending approach is simple, but functional. However, the authors do not provide formal safety guarantees that the system will always be safe, even under uncertainty. Interestingly, from their experiments, they found that the background of the operator makes a significant difference on the performance and how taxing the task is, based on the NASA Task Load Index. Authors in Shia et al. (2014) extend the work done in Vasudevan et al. (2012), by adding a safety element to the framework and implementing a more efficient testbed that allowed them to experiment with the approach more extensively than in Vasudevan et al. (2012). In this work, they show that they can guarantee the safety of the vehicle, while reducing the conservatism of the vehicle's safety-controller by including data-driven driver-intent predictions. This approach serves as promising approach to explore more extensively as driver modeling approaches improve.

The safe shared-autonomy approach presented by Shia et al. (2014) is set up for obstacle avoidance while following a road. Obstacle avoidance while following a road is one of the fundamental safety-critical tasks to solve for both automated driving and teleoperation. However, we already know that teleoperation will be needed for tasks that are more complex in both their maneuvering and requirements. For example, currently, a commonly proposed use for teleoperation is for driving vehicles during their first and last mile of transport. The reason for these proposals is that the first and last mile of transport typically involve complex tasks that may be hard to automate, such as parking in a parking lot where humans or human drivers might be moving around at the same time. To address this, there have been recent developments where researchers have begun to use temporal logic specifications to help the automated system better understand the teleoperation task.

Temporal logic is a system of logic formulae that are able to compactly express logical statements over timelines. Temporal logic has been used to describe rich specifications for robots for several decades (Fainekos et al. 2005). In Gao et al. (2020), authors propose one of the earliest frameworks for using linear temporal logic specifications to describe complex teleoperation tasks. Additionally, they also develop a shared autonomy framework that blends the policies of the operator and the automated driving system in a way that guarantees that the vehicle completes the specification safely. The policy blending is also based on a predict-then-act type approach, but uses a geometric confidence measure to infer the intent of the driver. Following this work, authors in Jiang et al. (2020) also proposed the use of linear temporal logic-specifications, but designed a more complete

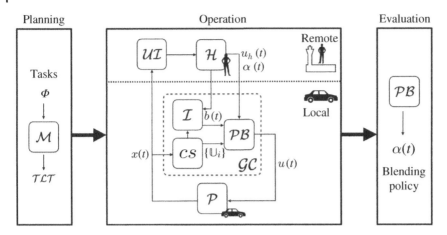

Figure 10.4 A safe teleoperation framework from Jiang et al. (2020). The drawn modules are the model-checking ($\mathcal{M}$), user interface ($\mathcal{UI}$), human operator ($\mathcal{H}$), inference ($\mathcal{I}$), control synthesis ($\mathcal{CS}$), policy-blending ($\mathcal{PB}$), guiding controller ($\mathcal{GC}$), and vehicle ($\mathcal{P}$) modules. For more details about the framework, we refer readers to Jiang et al. (2020). Source: Jiang et al. (2020)/With permission of Elsevier.

framework, shown in Figure 10.4, that gives the human operator a largest amount of freedom throughout the teleoperation, while still providing strong safety guarantees.

In the first stage (on the left of Figure 10.4), a plan is established for the teleoperation task. When this plan is established, it first goes through a model checking module that analyzes the reachability of the task and computes the least-restrictive reachable sets for completing the entire task. The reachable sets for the entire task are then communicated to the local vehicle from which it synthesizes its guiding controller for the teleoperation task. By interfacing with the guiding controller, the human operator can teleoperate the vehicle by sending the steering and acceleration inputs and selecting a policy blending ratio. The human operator's control input and preferred blending ratio are communicated to the guiding controller, which then infers the human operator's preferred approach to completing the specified task, automatically computes its own control input, and then blends its control input with the human's control input based on the preferred blending ratio. Then, it filters out the blended control input to ensure the planned specification will be satisfied safely. If the blended control input is safe, it will be selected; otherwise, the closest alternative control input will be selected instead. Then the selected control input will be implemented on the vehicle. Finally, once the teleoperation task is completed, the blending policy is extracted out in the evaluation phase. In this framework, the human operator is given a great deal of control freedom, while still being robust against human error.

To summarize, although there is a body of literature that develops shared autonomy approaches for teleoperation, few of them include mechanisms that guarantee the safety of the teleoperated system. In light of this, formal verification techniques, such as reachability analysis and temporal logic, are being introduced into shared autonomy teleoperation to provide formal safety guarantees. However, these new safe shared autonomy approaches only utilize operator intent prediction algorithms that are based on completing tasks optimally. In safety-critical scenarios, human operators do not only consider the optimality of their actions but also the future risk associated with their actions. In the future, improved operator prediction algorithms that are able to predict the operator's perception of risk associated with different actions or goals need to be integrated into safe shared autonomy frameworks in order to better assist the operator in keeping the teleoperated system safe.

10.5 Future Research Challenges

Much of the work that we surveyed in Section 10.4, contributes to safe teleoperation by engineering the interaction between the computational resources onboard the vehicle and the remote human operator. This work has been important and has indicated that the design of safe teleoperation systems under the CPHS lens is promising. However, there is still a significant amount of work that will be necessary for bringing safe teleoperation systems to maturity. There are three main future research challenges that need to be addressed to reach fully safe teleoperation for connected and automated vehicles: (i) full utilization of vehicle-to-everything (V2X) networks, (ii) integration of mixed autonomy modeling, and (iii) conducting experiments on newly deployed 5G networks.

10.5.1 Full Utilization of V2X Networks

In all of the methods presented in this chapter so far, the teleoperation systems assume the existence of only two intelligent agents, the human operator and the intelligence onboard the vehicle. However, there is an increasing number of proposals for implementing V2X networks and fog architectures for supporting automated vehicles. An example of a V2X network is illustrated in Figure 10.5. In such networks, many devices, including traffic lights, other connected vehicles, and resources in the cloud will be able to communicate and share their intelligence with each other. Furthermore, as is exemplified in Figure 10.5, the devices will also have different levels of awareness and authority. For example, a vehicle might only have awareness about a couple of meters of road ahead of it and authority over its own actions, whereas a intelligent traffic light might have awareness of 10 s of meters around itself and authority over all connected

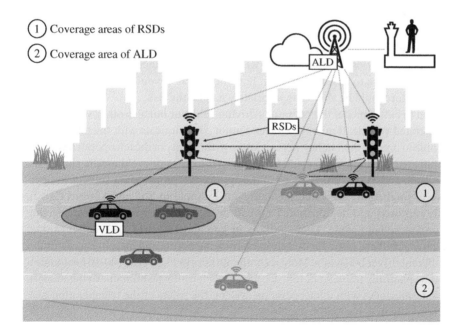

Figure 10.5 Illustration of a V2X network and the different coverage areas of three types of decision-makers: vehicle-level decision (VLD), road-side decision (RSD), and area-level decision (ALD) makers.

vehicles within its range. In these networks, it is most likely that a control tower where human operators will work and teleoperate from will be set up such that they have a radius of awareness of tens, or even hundreds, of kilometers. Furthermore, depending on the access given to the control tower, the human operator may have authority over the intelligent traffic lights and a fleet of vehicles. By sharing the intelligence across the V2X network, a human operator could teleoperate a vehicle, while having "super-situational awareness." Moreover, the shared autonomy problem could be extended to include autonomy from intelligent traffic lights and other vehicles. In this case, while the operator is teleoperating the vehicle, the control authority might shift to the intelligent traffic light when the vehicle is passing through the intersection it has authority over, since the traffic light might know of safety hazards that neither the vehicle nor the operator know about. This means that the policy-blending will no longer be between just the vehicle and the operator but also with other computing resources on the network. Narri et al. (2021) proposes a similar idea where shared situational awareness across V2X networks is used to safely handle occlusions in vehicle sensor suites. Dong and Li (2021) also proposes similar ideas of using V2X networks to detect unexpected safety hazards by collaborating with other intelligent agents.

10.5.2 Mixed Autonomy Traffic Modeling

In addition to better utilizing the V2X network, there is still significant work to be done when it comes to integrating both models of the remote human driver, and the behaviors of the vehicles around the teleoperated vehicle. As is pointed out by authors in Carvalho et al. (2015), by improving forecasting capabilities of human driving behavior, we can improve the assistance capabilities of automated driving systems by providing a more customized assistance. Authors in Brown et al. (2020) overview and classify 200 models for driver behavior modeling, which all have different combinations of modeling techniques. These behaviors models can be used to further experiment with predicting the intent of the operator and for assessing the risk of the operator's driving. However, as we pointed out earlier, one behavioral aspect that needs to be explored is the operator's perception about risk. For example, Stefansson et al. (2020) propose a model for the operator's perception of risk during overtaking scenarios. By doing so, we will be able to enhance the decision-making of the automated driving system to take into account how the operator would perceive a risk, and whether or not the operator would be able to appropriately handle the risk.

Not only do we need to better model the operator's driving behavior but also other vehicles that are nearby the teleoperated vehicle. If nearby vehicles cannot communicate with the teleoperated vehicle, then being able to predict their behaviors can improve the safety of the teleoperated vehicle. Specifically, the automated driving system could better assist the operator avoid unsafe situations. As is described in Albaba and Yildiz (2019), many of the approaches proposed for solving and performing these types of predictions are game-theoretic. For example in Yoo and Langari (2012), authors successfully captured and demonstrated human decision processes in highway driving using a three-person Stackelberg game. Then, authors in Yoo and Langari (2013) demonstrated the use of two-person Stackelberg game for determining merging points and merging acceleration profiles based on different driver behaviors. More recently, authors in Fisac et al. (2019) formulate a general game-theoretic planning framework that accepts nondeterministic human driving models. By integrating these frameworks with a teleoperation system, the safety of the teleoperation task can be increased, especially in dense traffic scenarios.

10.5.3 5G Experimentation

Currently, 5G networks are being deployed across the world for various services. Some of these 5G networks that are being deployed are suitable for the testing and development of teleoperation services, since they provide the required the ultra-low latency and ultra-high reliability necessary for safe teleoperation. In light of these new deployments, we can finally begin field testing and data collection for teleoperation. As shown in Georg et al. (2020), authors setup a teleoperation platform and report the delays of each component of the teleoperation platform identifying that the network delay is the biggest bottleneck to solve. With 5G networks, this bottleneck will be resolved in specialized areas.

Furthermore, many of the experiments conducted on remote driving for LTE systems need to be re-done for 5G networks, as some of the results may change if the tactile internet is realized. Interestingly, if the tactile internet is realized and safe teleoperation systems are matured, it will be important to start measuring the difference in the safety of local driving and remote driving. By fully utilizing the V2X network for assisting safe teleoperation, it could be the case that remote driving will become a safer mode of driving than local driving, due to the super-situational awareness provided by the networked system.

10.6 Conclusions

In general, both technology and research is beginning to come together to support the implementation of safe teleoperation systems for connected and automated vehicles. This bodes well for current and future plans to deploy automated vehicles at scale, since the deployed vehicles will initially require support from human operators in the form of teleoperation. There have been many different implementations of teleoperation systems over the last couple of decades, all of which contribute to improving the usability of teleoperation systems; however, the approaches mostly have not been evaluated for their contribution to improving. We attribute this to the fact that teleoperation is not considered as a viable option for safety-critical applications since the wireless channels were previously too unreliable. However, due to the advancements that have been brought by the introduction of 5G networks and bundled technologies such as network slicing, teleoperation can now be considered a service that is reliable enough to be integrated into safety-critical systems. Moreover, under the paradigm of the tactile internet, the tactile experience of teleoperating a vehicle can potentially become close enough to the tactile experience of driving a vehicle, presenting a unique CPHS design opportunity to develop teleoperation systems that enhance safety by improving operator perception and through implementing safe, shared autonomy. We survey different research approaches to these two design opportunities and highlight the safety improvements they can bring to the teleoperation of connected and automated vehicles. Then, we introduce two important future research directions for safe teleoperation: (i) full utilization of the V2X network and (ii) mixed autonomy traffic modeling. The former is an extension of the shared autonomy formulations that have been presented in recent literature that expands the shared autonomy problem to include other agents on the V2X network. In other words, implementing teleoperation systems as a human-in-multiagent-loops system (Samad 2022); Chapter 1 in this book. By solving this problem, we can potentially allow an operator "see through" buildings by connecting to nearby visual sensors onboard other vehicles or mounted on the side of the road. Then, by improving mixed autonomy traffic models and integrating them into teleoperation systems, we can also improve the predictive capabilities of the automated driving system, leading to better and safer assistance for the operator. Finally, we make a call for more experimentation on newly deployed 5G networks. Through more data-collection on the performance of 5G networks and on the performance of human operators using safe teleoperation systems, we can better understand how to deploy and set up control towers for supporting large-scale deployments of connected and automated vehicles.

References

5GAA Automative Association (2017). An Assessment of Direct Communications Technologies for Improved Road Safety in the EU. Munich, Germany: 5GAA Automotive Association *Technical report number 2017-12-05*.

Akyildiz, I.F., Wang, P., and Lin, S.-C. (2015). SoftAir: A software defined networking architecture for 5G wireless systems. *Computer Networks* 85: 1–18.

Albaba, B.M. and Yildiz, Y. (2019). Modeling cyber-physical human systems via an interplay between reinforcement learning and game theory. *Annual Reviews in Control* 48: 1–21.

Alriksson, F., Boström, L., Sachs, J. et al. (2020). Critical IoT Connectivity. Munich, Germany: Ericsson Technical report, Ericsson Technology Review,.

Bogdoll, D., Orf, S., Töttel, L., and Zöllner, J.M. (2021). Taxonomy and Survey on Remote Human Input Systems for Driving Automation Systems. *arXiv preprint arXiv:2109.08599*.

Brady, A., MacDonald, B., Oakley, I. et al. (2002). Relay: a futuristic interface for remote driving. *Proceedings of EuroHaptics*, 8–10. Citeseer.

Brown, K., Driggs-Campbell, K., and Kochenderfer, M.J. (2020). A taxonomy and review of algorithms for modeling and predicting human driver behavior. *arXiv preprint arXiv:2006.08832*.

Carvalho, A., Lefévre, S., Schildbach, G. et al. (2015). Automated driving: the role of forecasts and uncertainty–a control perspective. *European Journal of Control* 24: 14–32.

Dahlman, E., Parkvall, S., and Skold, J. (2020). *5G NR: The Next Generation Wireless Access Technology*. Academic Press.

Davies, A. (2017). Nissan's Path to Self-Driving Cars? Humans in Call Centers. *Wired Transportation*. https://www.wired.com/2017/01/nissans-self-driving-teleoperation/ (accessed 10 February 2023).

Dextreit, C. and Kolmanovsky, I.V. (2014). Game theory controller for hybrid electric vehicles. *IEEE Transactions on Control Systems Technology* 22 (2): 652–663.

Dong, L. and Li, R. (2021). Collaborative computation in the network for remote driving. *2021 IEEE Symposium Series on Computational Intelligence (SSCI)*, 1–8.

Dragan, A.D. and Srinivasa, S.S. (2013). A policy-blending formalism for shared control. *International Journal of Robotics Research* 32 (7). https://doi.org/10.1177/0278364913490324.

Einride (2020). Einride Will Hire Its First Remote Autonomous Truck Operator in 2020. https://www.einride.tech/news/einride-will-hire-its-first-remote-autonomous-truck-operator-in-2020 (accessed 10 February 2023).

Fainekos, G.E., Kress-Gazit, H., and Pappas, G.J. (2005). Temporal logic motion planning for mobile robots. *Proceedings of IEEE International Conference on Robotics and Automation*, 2020–2025.

Fettweis, G.P. (2014). The tactile internet: applications and challenges. *IEEE Vehicular Technology Magazine* 9 (1): 64–70.

Fisac, J.F., Bronstein, E., Stefansson, E. et al. (2019). Hierarchical game-theoretic planning for autonomous vehicles. *2019 International Conference on Robotics and Automation (ICRA)*, 9590–9596.

Fong, T., Thorpe, C., and Baur, C. (2001). Advanced interfaces for vehicle teleoperation: collaborative control, sensor fusion displays, and remote driving tools. *Autonomous Robots* 11 (1): 77–85.

Fong, T., Thorpe, C., and Baur, C. (2003). Multi-robot remote driving with collaborative control. *IEEE Transactions on Industrial Electronics* 50 (4): 699–704.

Foukas, X., Patounas, G., Elmokashfi, A., and Marina, M.K. (2017). Network slicing in 5G: survey and challenges. *IEEE Communications Magazine* 55 (5): 94–100.

Gao, Y., Jiang, F.J., Ren, X. et al. (2020). Reachability-based human-in-the-loop control with uncertain specifications. *Proceedings of 21st IFAC World Congress*.

Georg, J.-M., Feiler, J., Hoffmann, S., and Diermeyer, F. (2020). Sensor and actuator latency during teleoperation of automated vehicles. *2020 IEEE Intelligent Vehicles Symposium (IV)*, 760–766.

Ghasemi, A.H., Jayakumar, P., and Gillespie, R.B. (2019). Shared control architectures for vehicle steering. *Cognition, Technology and Work* 21 (4): 699–709.

Gnatzig, S., Chucholowski, F., Tang, T., and Lienkamp, M. (2013). A system design for teleoperated road vehicles. *ICINCO (2)*, 231–238.

Harper, C.D., Hendrickson, C.T., Mangones, S., and Samaras, C. (2016). Estimating potential increases in travel with autonomous vehicles for the non-driving, elderly and people with travel-restrictive medical conditions. *Transportation Research Part C: Emerging Technologies* 72: 1–9.

Hirche, S. and Buss, M. (2012). Human-oriented control for haptic teleoperation. *Proceedings of the IEEE* 100 (3): 623–647.

Hosseini, A. and Lienkamp, M. (2016a). Predictive safety based on track-before-detect for teleoperated driving through communication time delay. *2016 IEEE Intelligent Vehicles Symposium (IV)*, 165–172.

Hosseini, A. and Lienkamp, M. (2016b). Enhancing telepresence during the teleoperation of road vehicles using HMD-based mixed reality. *2016 IEEE Intelligent Vehicles Symposium (IV)*, 1366–1373.

Javdani, S., Admoni, H., Pellegrinelli, S. et al. (2018). Shared autonomy via hindsight optimization for teleoperation and teaming. *The International Journal of Robotics Research* 37 (7): 717–742.

Jiang, F.J., Gao, Y., Xie, L., and Johansson, K.H. (2020). Human-centered design for safe teleoperation of connected vehicles. *IFAC Conference on Cyber-Physical & Human-Systems*, Shanghai, China.

Koopman, P. and Wagner, M. (2017). Autonomous vehicle safety: an interdisciplinary challenge. *IEEE Intelligent Transportation Systems Magazine* 9 (1): 90–96.

Lamnabhi-Lagarrigue, F., Annaswamy, A., Engell, S. et al. (2017). Systems & control for the future of humanity, research agenda: current and future roles, impact and grand challenges. *Annual Reviews in Control* 43: 1–64.

Lichiardopol, S. (2007). A Survey on Teleoperation. *Technische Universitat Eindhoven, DCT report*, 20:40–60.

Muslim, H. and Itoh, M. (2019). A theoretical framework for designing human-centered automotive automation systems. *Cognition, Technology and Work* 21 (4): 685–697.

Narri, V., Alanwar, A., Mårtensson, J. et al. (2021). Set-membership estimation in shared situational awareness for automated vehicles in occluded scenarios. *2021 IEEE Intelligent Vehicles Symposium (IV)*, 385–392.

Neumeier, S. and Facchi, C. (2019). Towards a driver support system for teleoperated driving. *2019 IEEE Intelligent Transportation Systems Conference (ITSC)*, 4190–4196. IEEE.

Neumeier, S., Wintersberger, P., Frison, A.-K. et al. (2019). Teleoperation: the holy grail to solve problems of automated driving? Sure, but latency matters. *Proceedings of the 11th International Conference on Automotive User Interfaces and Interactive Vehicular Applications*, 186–197.

Nielsen, C.W., Goodrich, M.A., and Ricks, R.W. (2007). Ecological interfaces for improving mobile robot teleoperation. *IEEE Transactions on Robotics* 23 (5): 927–941.

Nobina (2019). Autonomous buses. https://www.nobina.com/nobina-technology/autonomous-buses/ (accessed 10 October 2023).

Reddy, S., Levine, S., and Dragan, A. (2020). Assisted perception: optimizing observations to communicate state. *Proceedings of the 2020 Conference on Robot Learning, PMLR*, Volume 155, 748–764.

Sachs, J., Andersson, L.A.A., Araújo, J. et al. (2019). Adaptive 5G low-latency communication for tactile internet services. *Proceedings of the IEEE* 107 (2): 325–349.

SAE International (2021). Taxonomy and Definitions for Terms Related to Driving Automation Systems for On-Road Motor Vehicles. *Technical Report number J3016_202104*. SAE International.

Saeed, U., Hämäläinen, J., Garcia-Lozano, M., and González, G.D. (2019). On the feasibility of remote driving application over dense 5G roadside networks. *2019 16th International Symposium on Wireless Communication Systems (ISWCS)*, 271–276.

Samad, T. (2022). Human-in-the-loop control and cyber–physical–human systems: applications and categorization. In: *Cyber–Physical–Human Systems: Fundamentals and Applications* (ed. A. Annaswamy, P. Khargonekar, F. Lamnabhi-Lagarrigue, and S.K. Spurgeon), UK: Wiley (in Press).

Shia, V.A., Gao, Y., Vasudevan, R. et al. (2014). Semiautonomous vehicular control using driver modeling. *IEEE Transactions on Intelligent Transportation Systems* 15 (6): 2696–2709.

Stefansson, E., Jiang, F.J., Nekouei, E. et al. (2020). Modeling the decision-making in human driver overtaking. *IFAC-PapersOnLine* 53 (2): 15338–15345.

Takayama, L., Marder-Eppstein, E., Harris, H., and Beer, J.M. (2011). Assisted driving of a mobile remote presence system: system design and controlled user evaluation. *2011 IEEE International Conference on Robotics and Automation*, 1883–1889.

Vasudevan, R., Shia, V., Gao, Y. et al. (2012). Safe semi-autonomous control with enhanced driver modeling. *2012 American Control Conference (ACC)*, 2896–2903.

Wickens, C.D., Mavor, A.S., and McGee, J.P. (1997). *Flight to the Future Human Factors in Air Traffic Control*. National Academy Press.

Yoo, J.H. and Langari, R. (2012). Stackelberg game based model of highway driving. *Dynamic Systems and Control Conference, volume* 45295, 499–508. American Society of Mechanical Engineers.

Yoo, J.H. and Langari, R. (2013). A stackelberg game theoretic driver model for merging. *Dynamic Systems and Control Conference*, Volume 56130, V002T30A003. American Society of Mechanical Engineers.

11

Charging Behavior of Electric Vehicles

Qing-Shan Jia and Teng Long

Department of Automation, Center for Intelligent and Networked Systems (CFINS), Beijing National Research Center for Information Science and Technology (BNRist), Tsinghua University, Beijing, China

Since the outbreak of the first industrial revolution, the traditional fossil energy represented by oil, coal, and natural gas has injected a steady stream of power into the prosperity and development of the world. Energy has become a worldwide proposition not only related to national prosperity but also related to people's livelihood. According to the data of "BP world energy statistical yearbook (2021 Edition)," the total global primary energy consumption reached 556.63 EJ in 2020, of which more than 83% is fossil energy (BP 2021). Figure 11.1 shows the changes and growth rate of world primary energy consumption from 2011 to 2020. However, the world energy structure dominated by traditional energy is facing unprecedented challenges. On the one hand, due to the nonrenewable nature of fossil energy itself, the energy crisis caused by resource shortage has become increasingly prominent. On the other hand, the combustion of fossil energy is accompanied by a large number of greenhouse gas emissions. The environmental pollution problems such as global warming and ecological deterioration caused by fossil energy have become the focus of attention of all countries. The Intergovernmental Panel on Climate Change (IPCC) special report on global warming of $1.5\,°C$ points out that due to human activities and greenhouse gases, the global temperature is increasing by about $0.2\,°C$ every decade. If this trend is not curbed, it will seriously endanger the health of the earth's ecosystem. Therefore, energy crisis and environmental issues have held center stage recently, energy conservation and emission reduction has become a general consensus of the international community.

In order to achieve high-quality economic and social development and deal with possible energy crisis and environmental problems, China, the United States, and many European countries have set carbon emission reduction targets. From the demand side, the main battlefield of the energy consumption revolution will be the comprehensive electrification of the automotive industry. It is reported that 62.3% of the world's oil consumption comes from the transportation sector, which is considered to be one of the main reasons for the increase in carbon dioxide emissions (International Energy Agency 2021b). Electric vehicles (EVs) provide an alternative energy for fuel-based vehicles, shifting energy demand from fossil fuels to electricity. As electricity can be obtained from a variety of renewable resources, such as wind energy, solar energy, and thermal energy, it can alleviate people's concern about the increasing depletion of oil resources. At the same time, in the smart grid, EVs can provide valuable auxiliary services, rather than pure power consumers.

EVs are an application of CHPS since humans, EVs, and external management control terminals will carry out information interaction and energy flow in the process of driving and charging.

Cyber–Physical–Human Systems: Fundamentals and Applications, First Edition.
Edited by Anuradha M. Annaswamy, Pramod P. Khargonekar, Françoise Lamnabhi-Lagarrigue, and Sarah K. Spurgeon.
© 2023 The Institute of Electrical and Electronics Engineers, Inc. Published 2023 by John Wiley & Sons, Inc.

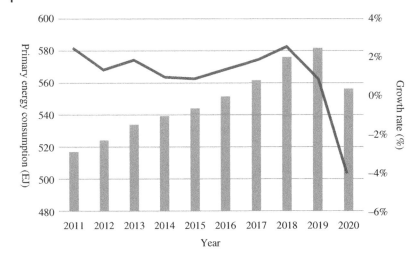

Figure 11.1 Changes and growth rate of world primary energy consumption from 2011 to 2020.

More specifically, charging control of large-scale EVs often involves multiple EVs and humans in the area. They are often coupled and cannot be analyzed separately. Thus, this scenario is a typical case of the humans-in-the-multiagent-loops category. In this chapter, we will introduce the opportunities and challenges brought by the charging behavior of EVs, as well as the related technologies and research development of their interaction with power system and renewable energy.

11.1 History, Challenges, and Opportunities

11.1.1 The History and Status Quo of EVs

Transportation industry is a key area to alleviate energy problems, conserve energy, reduce pollution emissions, and promote sustainable development. According to the data of the International Energy Agency, the greenhouse gas emissions of the transportation industry accounted for 24.5% of the total global emissions (International Energy Agency 2021b). A fuel car that travels about 15 000 mi a year will emit an average of 13 000 lb of exhaust pollutants into the atmosphere. To this end, many countries in the world have taken action. Britain, France, and Norway have successively issued a timetable or proposal for banning the sale of fuel vehicles. Canada, the United States, and other countries have also issued a series of policies to expand the industrial scale of EVs, such as tax relief, purchase subsidies, pollutant emission restrictions, license issuance, and so on (United Nations Environment Programme 2021). In China, vigorously developing renewable energy vehicles represented by EVs and realizing the "replacing oil with electricity" in the transportation industry have also become a national strategy and an important realization path to reduce carbon emissions.

The first generation of EVs came out in 1997 (Al-Alawi and Bradley 2013). After more than 20 years of development, many types of EV products have emerged, as shown in Figure 11.2 (Jia and Long 2020a). It can be divided into three categories according to the mixing degree and the structure of the propulsion system, including plug-in hybrid electric vehicle (PHEV), battery electric vehicle (BEV), and fuel cell electric vehicle (FCEV). PHEV is an upgraded hybrid electric vehicle (HEV), which is equipped with an on-board battery charger and a drive system supplied by electricity and fuel, in which the stored electricity can drive PHEV for a very limited distance. BEV is only driven

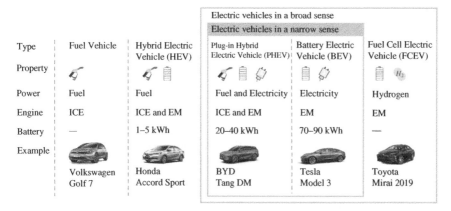

Type	Fuel Vehicle	Hybrid Electric Vehicle (HEV)	Plug-in Hybrid Electric Vehicle (PHEV)	Battery Electric Vehicle (BEV)	Fuel Cell Electric Vehicle (FCEV)
Power	Fuel	Fuel	Fuel and Electricity	Electricity	Hydrogen
Engine	ICE	ICE and EM	ICE and EM	EM	EM
Battery	—	1–5 kWh	20–40 kWh	70–90 kWh	—
Example	Volkswagen Golf 7	Honda Accord Sport	BYD Tang DM	Tesla Model 3	Toyota Mirai 2019

Figure 11.2 Classification of fuel vehicles and EVs. Source: Jia and Long (2020a)/Springer Nature.

by motor, so it needs to be charged by power grid or renewable energy. The energy of FCEV comes from hydrogen fuel cell. However, due to the high cost, production, and storage of hydrogen, the permeability is still low so far.

Compared with traditional fuel vehicles, EVs have many advantages:

- First of all, EVs are an environment-friendly means of transportation with the characteristics of high-energy efficiency and low emissions. Compared with fuel vehicles, EVs can reduce the equivalent greenhouse gas emissions to one-fifth of the original value (Li et al. 2015). Besides, the transportation cost of EVs can be reduced to one-tenth of that of fuel vehicles through energy driving methods such as hybrid power, pure electric power, and hydrogen fuel cell power (Xiao et al. 2019).
- Second, the power sources of EVs can be diverse, especially they can be generated by various renewable energy sources, such as wind power, solar power, and biomass power, which can help the transportation sector get rid of its dependence on fossil fuels, further reduce carbon emissions from the source, and help the development of the renewable energy industry.
- Finally, with the continuous advancement of the networked and intelligent process of EVs, EVs can not only be used as consumers of electricity but also as terminals of distributed energy consumption and storage, deeply participate in the integration of local power grid, transportation, information communication, and other fields, and provide valuable auxiliary services.

In recent years, with concerns about climate change, technical improvement and cost reduction of EVs, the production, sales, and ownership of EVs have shown a rapid upward trend (Calearo et al. 2021). As shown in Figure 11.3, the global ownership of EVs has exceeded 10 million by the end of 2020, of which the global sales volume in 2020 is about 3 million. It is estimated that by 2030, the global stock of EVs will reach 145 million, accounting for 7% of the total number of transportation vehicles, and the sales volume of EVs will exceed that of fuel vehicles (International Energy Agency 2021a). China has the largest EV market in the world. Since 2015, the production, sales, and ownership of EVs in China have always ranked first in the world, and the proportion has increased year by year, as shown in Figure 11.3 (International Energy Agency 2021a).

With the vigorous development of EV industry, the charging infrastructure represented by charging pile is gradually popularized. By 2020, the number of public charging piles in the world has reached 1.3 million, of which 30% are fast-charging piles. In China, the number of slow-charging piles in public places is about 500 000, and the number of fast charging piles (charging power more than 22 kW) is about 310 000, both of which rank first in the world. The extensive deployment of

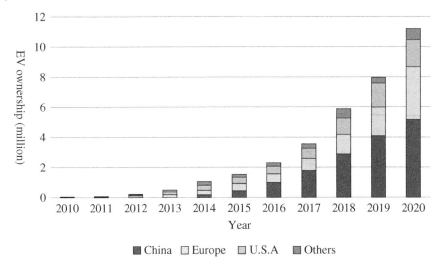

Figure 11.3   Global EV ownership from 2010 to 2020. Source: International Energy Agency (2021a)/IEA/CC BY-4.0.

"new infrastructure construction" such as charging piles makes it possible to integrate EVs with local power grids.

11.1.2   The Current Challenge

Despite the advantages, the disordered charging of large-scale EVs also has an impact on the stable operation of power system. Even at a low permeability, uncontrolled EV charging poses a significant challenge to the power quality of the distribution infrastructure. Firstly, the time period for EV users to access the power grid is often consistent with the peak load period, which will further increase the peak valley difference, thus increasing the operation burden of the power grid. A grid connected charging station that can fully charge the battery of 36 kWh (providing charging power of more than 100 kW) in 20 minutes will cause certain energy loss to the power grid if it charges 10 vehicles (load of 1000 kW) at the same time during peak hours (Ghosh 2020). Meanwhile, the randomness of charging behavior of EVs also brings new issues to power grid planning and dispatching control decision-making. In addition, the disordered charging behavior of EVs will have an impact on voltage distribution, power quality, power grid balance, transformer life, and harmonic distortion (Muratori 2018). The results in Godina et al. (2016) show that the power demand and line current may exceed the transformer ratings considering the uncontrolled EV charging. Mozafar et al. (2018) investigated the impact of EV charging loads enabling vehicle-to-grid (V2G) and grid-to-vehicle (G2V) systems on the stability and reliability of the power grid. Therefore, reliable EV charging control will be the key to its successful integration with the power system.

At the same time, the charging control of EVs often involves the interaction and coupling of multiple intelligent objects, including the vehicle itself, charging pile, charging station, and regional and urban operation managers. The coordination of these intelligent terminals with different spatial scales often has differences in interests and decision-making, and the balance of efficiency and performance. Therefore, it also brings new problems to the charging scheduling optimization of large-scale EVs.

On the other hand, EVs may not be as clean as expected. The carbon emission index of EVs using electricity as energy source depends heavily on the power generation combination during charging.

For most EVs, they often receive power supply from the power grid, which still uses fossil fuels as its main energy source. Moreover, many greenhouse gas measurement indicators of EVs usually only calculate the emission reduction effect compared with the direct combustion of fossil fuels, ignoring the indirect emissions generated in the process of power grid transmission and power generation (Kim et al. 2020). Thus, the energy revolution on the supply side becomes imperative, which means to vigorously develop renewable energy represented by wind power, photovoltaic power, and hydrogen power, and strengthen the integration of renewable energy and power grid.

11.1.3 The Opportunities

The large-scale EV charging not only has a negative impact to the power grid but also brings important opportunities for the development of the next generation of the smart grid in the future, which mainly includes the following three aspects:

- As a typical flexible load, EVs has outstanding charging elasticity. Its charging elasticity is mainly reflected in two dimensions. One is the elasticity in the time dimension. For general private EVs, about 95% of the time is in the shutdown state, with sufficient time to integrate with the power grid access (Lu and Hossain 2015). Simultaneously, less than 40% of the time when EVs are connected to the power grid is in the charging state, and most of the time they are still in the idle state of full power, and this vacancy proportion will continue to rise due to the further development of fast charging and other technologies (Lucas et al. 2019). The second is the elasticity in the spatial dimension. For an EV user, when selecting appropriate charging facilities, he often has multiple adjacent choices, which will be affected by many factors such as charging price, distance, and so on. Therefore, EVs can be regarded as a mobile "big battery," which can realize transregional load regulation through appropriate scheduling methods and incentive means. Thus, the charging scheduling of large-scale EVs can reduce the charging cost, stabilize the operation of the power grid and offer various auxiliary services such as peak cut, frequency regulation, and reserve (Teng et al. 2020).
- The synergy between large-scale EVs and renewable energy can contribute to the development of renewable energy technology. Renewable energy have the characteristics of high randomness, volatility, and intermittence. If large-scale real-time integration into the power system through the distribution network, it may endanger the stable operation of the power grid and cause problems such as voltage rise and power flow reversal. Therefore, the grid connection of uncertain renewable energy often requires a lot of standby and frequency regulation resources. Meanwhile, renewable energy such as wind power often has the characteristics of reverse peak shaving and reverse consumption, that is, the peak time of wind power output is consistent with the load trough. Thus, a large number of renewable energy are difficult to be consumed in real time, resulting in the phenomenon of abandoning wind and light. On the contrary, there may also has the problem of insufficient supply of renewable energy at the peak of power consumption. EVs have considerable energy storage potential and offer a new way for this issue. Taking the battery capacity of the mainstream EVs on the market at present as 50–70 kwh as an example, if the number of EVs continues to increase to more than 300 million, the total power they can store will exceed the total power consumed in China every day. Through charging scheduling, vehicle to grid (V2G) and other technologies, it can effectively alleviate the volatility of renewable energy, improve the local and global consuming capacity, reduce the carbon emission in the charging portfolio of EVs, and further promote the low-carbon development of electric systems (Longo et al. 2018).
- As an intelligent terminal in road traffic, EVs can promote the rapid development of energy Internet. The deep integration of energy and information is the core feature of energy Internet. EVs

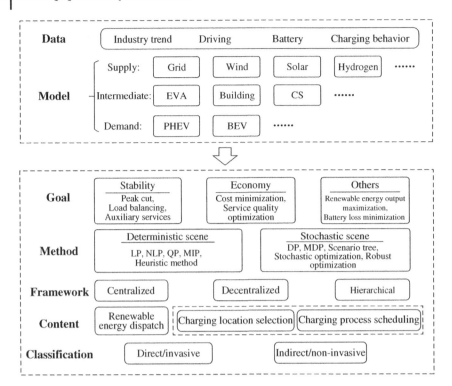

Figure 11.4 State of art of large-scale EV charging scheduling optimization.

and their charging and discharging facilities will establish a perfect communication network, so as to realize the deep integration of vehicle networking, energy network, and Internet, and become an important part of energy Internet.

Therefore, a large number of researchers have paid extensive attention to the charging scheduling optimization of large-scale EVs, including electric propulsion system (Hannan et al. 2014), charging infrastructure (Rubino et al. 2017), problem modeling and optimization (Richardson 2013; García-Villalobos et al. 2014), supply and demand coordination (Jia 2018), fast charging technology (Ashique et al. 2017), and integration with renewable energy (Aghaei and Alizadeh 2013) have been deeply studied. Some key problems and challenges in this field are systematically reviewed in Rahman et al. (2016), Jia and Long (2020a), and Al-Ogaili et al. (2019). Figure 11.4 summarizes and arranges the current research status in the field of charging scheduling optimization of large-scale EVs. Based on this, the rest of this chapter will introduce and analyze the existing research from two aspects of this field: EV charging behavior data sets and problem modeling, charging scheduling control and optimization.

11.2 Data Sets and Problem Modeling

11.2.1 Data Sets of EV Charging Behavior

The research on EV charging scheduling optimization mostly depends on the abstract mathematical model and the actual data collected from the real scene. It is a typical simulation-based learning optimization method. Therefore, the data are particularly important for a comprehensive and accurate understanding of the charging behavior and driving mode of EVs. At present, with the increase of the scale of EVs, the completion of infrastructure deployment and the data connection of intelligent terminals, countries all over the world have begun to start data collection projects around the behavior of EVs. The data collected are mainly concentrated in four aspects, namely, industry trend, driving, battery, and charging behavior. Some typical data set items and acquisition methods are sorted out in Table 11.1. These collated data sets are very limited, and it is likely that important public data sets are not included in the table.

11.2.1.1 Trend Data Sets

The development of EV industry can be reflected by the sales data of EVs. At the same time, it can indirectly represent the technical maturity and penetration in the field of EV, which is also the data support for the necessity of large-scale EV charging scheduling research. Projects such as statista and EV volumes in the United States provide statistical data of EV industry in various countries, while the Alternative Fuels Data Center (AFDC) of the U.S. Department of Energy's Office collects and analyzes the sales trend of EVs in the United States from 2011 to 2019. Websites such as NEVI and Evpartner provide annual and monthly sales data of EVs in China, which can be obtained online.

11.2.1.2 Driving Data Sets

Driving data, including the departure, destination, start, and end time and travel purpose of each trip, can truly reflect the overall statistics of drivers' behavior habits. Such data are often mature and easy to obtain. Therefore, assuming that the driving habits of EV users are consistent with those of ordinary fuel vehicle users, such data are often used to simulate the driving characteristics and charging requirements of EVs. In Germany, investigation, statistics, and data collection on vehicle driving behavior have been carried out for many years to help researchers better understand the behavior habits of German drivers, including Motor Vehicle Traffic in Germany 2010 (KiD 2010), Mobility Panel Germany (MOP), Mobility in Germany (MiD), and Mobility in Cities 2008 (SrV 2008). In the United States, local governments such as the Federal Highway Administration (FHWA), New York, and Chicago city governments have also conducted similar investigations. Italy, China, and other countries and some private enterprises also contributed some relevant data sets.

11.2.1.3 Battery Data Sets

As the core component of EVs, battery pack plays an important role in the simulation and modeling of charging behavior of EVs. Battery capacity, life cycle, working temperature, and self-discharge rate are important factors affecting the performance of EV battery. Lithium battery has become the first choice for common EVs because of its lightweight, small volume, and high power density (Sbordone et al. 2015). The Prognostics CoE at NASA Ames and Advanced Vehicle project provide key data on the charge and discharge of lithium batteries at different temperatures. On the other hand, NASA also provides a detailed random battery usage data set, which can provide a reference benchmark for battery health.

11.2.1.4 Charging Data Sets

The data of charging power, charging frequency, charging location, and charging time are directly related to the charging behavior of EVs. The Office of Energy Efficiency and Renewable Energy provide a complete data set of 12 500 (public and private) charging piles and 8650 EVs in 18 cities. The UK launched an 18-month intelligent charging experiment involving more than 700 EVs. The Electric Nation Project collected more than 2 million hours of charging data for researchers to use. However, due to privacy reasons, it is generally difficult to directly collect and obtain

Table 11.1 Some data sets on EV charging behavior.

Category	Database	Country	Website	References
Trend	Statista	USA	https://www.statista.com/markets/419/topic/487/vehicles-road-traffic/	
Trend	U.S. PEV Sales	USA	http://www.ev-volumes.com/datacenter/	
Trend	EV Volumes	USA	https://afdc.energy.gov/data/	
Trend	Evpartner	China	http://www.evpartner.com/daas/	
Trend	NEVI	China	https://nevi.bbdservice.com/global/analysis	
Drive	KiD 2010	Germany	http://www.kid2010.de	
Drive	MOP	Germany	http://mobilitaetspanel.ifv.kit.edu/	
Drive	MiD	Germany	http://www.mobilitaet-in-deutschland.de/	
Drive	SrV 2008	Germany	https://idp2.tu-dresden.de/	
Drive	NHTS	USA	https://nhts.ornl.gov/	Kelly et al. (2012)
Drive	UP 2013	USA	https://www.kaggle.com/fivethirtyeight/uber-pickups-in-new-york-city	
Drive	Puget Sound Trends	USA	https://www.psrc.org/puget-sound-trends	Wu et al. (2014, 2015)
Drive	Volt Stats	USA and Canada	https://www.voltstats.net/	Plötz et al. (2015, 2018)
Drive	NYC's Taxi	USA	https://chriswhong.com/open-data/foil\LY1\textbackslash_nyc\LY1\textbackslash_taxi/	
Drive	Chicago's Taxi Trip 2013	USA	https://data.cityofchicago.org/Transportation/Taxi-Trips/	
Drive	AUTO21	Canada	http://auto21.uwinnipeg.ca/	Smith et al. (2011)
Drive	Roma's Taxi Trip 2014	Italy	http://crawdad.org/roma/taxi/20140717/	Amici et al. (2014)
Drive	Shanghai's Taxi Trip 2007	China	https://www.cse.ust.hk/scrg/	Liu et al. (2010)
Battery	Battery Charging	USA	https://ti.arc.nasa.gov/tech/dash/groups/pcoe/prognostic-data-repository/#battery	
Battery	Randomized Battery Usage	USA	https://ti.arc.nasa.gov/tech/dash/groups/pcoe/prognostic-data-repository/#batteryrnddischarge	
Battery	VCS Test	USA	https://avt.inl.gov/project-type/data	
Charge	The EV Project	USA	https://www.energy.gov/eere/vehicles/avta-ev-project	Smart and Schey 2012 and Smart et al. (2013, 2014) and Schey et al. (2012)
Charge	ACN-Data	USA	https://ev.caltech.edu/dataset	
Charge	CCVRP	USA	http://www.arb.ca.gov/msprog/aqip/cvrp.html	Tal et al. (2014)

Table 11.1 (Continued)

Category	Database	Country	Website	References
Charge	Unknown	Germany	https://ars.els-cdn.com/content/image/1-s2.0-S2590116820300369-mmc1.xlsx	Hecht et al. (2020)
Charge	Chargepoint Analysis		https://www.gov.uk/government/statistics/electric-chargepoint-analysis-2017-local-authority-rapids	
Charge	Electric Nation Project	England	https://www.westernpower.co.uk/electric-nation-data	
Charge	StarCharge	China	http://www.starcharge.com/	

data about the frequency, time length, the state-of-charge (SoC), and location of charging events. In fact, most of these charging data are held by some large private charging operators and EV manufacturers and can only be obtained through payment.

11.2.2 Problem Modeling

In the charging scheduling scenario of large-scale EVs, it often involves information interaction and energy transmission among multiple intelligent objects, including intelligent buildings, charging stations, power grids, renewable energy, EV aggregator (EVA), battery management system (BMS), charging points (CPs), and so on. Figure 11.5 shows the energy and information exchange relationship between different participating objects. Within the EV, BMS is responsible for monitoring and reporting battery health and SoC. As the information interface with the outside world, the central controller becomes the actual operator in the charging process of EVs. The inverter and connector are used as the transfer station of energy flow to ensure the stability of charging behavior. In particular, the power converters of AC power supply are generally on-board. However, due to the limitations of on-board weight, charging power and space, fast charging is usually realized through external DC power converter and power supply (Haghbin et al. 2012). Externally, local operator and grid also serve as terminals of information flow and energy flow, respectively. Therefore, the

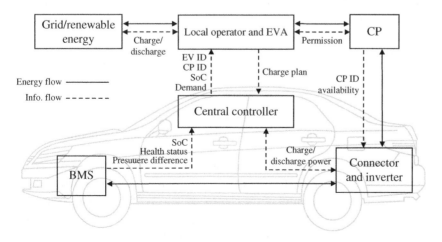

Figure 11.5 EV charging framework. Source: Adapted from García-Villalobos et al. (2014).

problem modeling of EV charging scheduling often needs to consider the joint modeling of multiple stakeholders. In terms of type, the modeling objects in this scene can often be divided into three categories, and they have different optimization objectives.

Firstly, the supplier mainly includes power system and renewable energy. It often hopes to minimize the power operation cost (Su et al. 2013; Khodayar et al. 2012), stabilize the power grid operation (Liu et al. 2017a; Tabatabaee et al. 2017), provide auxiliary services (Ziras et al. 2018; Sortomme and El-Sharkawi 2011), and improve the consumption ratio of renewable energy (Schill and Gerbaulet 2015; Dixon et al. 2020) through coordination with EVs. Morstyn et al. 2020 solved the problems of voltage rise and power limitation when scheduling EV charging, which is usually ignored by other works. When considering the participation of renewable energy, Soares et al. (2017) uses real distribution networks to prove that demand response can indeed play an important role in alleviating the volatility of renewable energy and EV charging. The research results in Mesarić and Krajcar (2015) show that the utilization rate of renewable energy can be further improved by using the power flexibility of flexible loads such as EVs. At the same time, Guo et al. (2015) proposes a two-stage framework to realize the supply–demand matching of microgrid with renewable energy. In the first stage, the optimal charging price is determined according to the power generation of renewable energy and the daily predicted value of EV charging demand, while in the second stage, the real-time scheduling problem of EV charging is further solved based on the determined price.

Secondly, due to the multispatial scale characteristics of large-scale EV charging scheduling, there is often an intermediate coordinator between supply and demand. Such objects often appear as transportation company, regional coordinators, charging stations, or EVA. They form influential load clusters by integrating the charging demand of a certain number of EVs, so as to gain the position to directly talk with the power supplier. In order to avoid affecting the travel needs of EV users, EVA needs to manage the charging process in advance, while maintaining load balance to ensure that the charging needs are met on time. Such objects are studied and analyzed in Zhao et al. (2015), Bessa and Matos (2010), and Bessa and Matos (2013). Vagropoulos and Bakirtzis (2013) established an EVA optimal bidding strategy model based on two-stage stochastic linear programming to achieve overall charging cost reduction. In Xu et al. (2015), the intermediate coordinator plays a role in regulating the balance between supply and demand. Through the mediation of the coordinator, the calculation and communication pressure of the problem can be greatly reduced (Al-Ogaili et al. 2019), and the overall charging cost can be reduced (Liu et al. 2012).

Third, from the demand side, EVs and intelligent buildings with basic load are common demand objects. For such objects, common optimization objectives include minimizing charging cost (González-Garrido et al. 2019) or peak purchase (Moghaddass et al. 2019), battery safety protection (Fang et al. 2016), and improving charging service quality (Kim et al. 2010). Thomas et al. (2018) considered a scheduling scenario including multiple buildings and EVAs to achieve economic objectives by optimizing the basic load of smart appliances and the charging process of EVs. Similar scenarios are also analyzed in Yang et al. (2018) and Wu et al. (2017). Zhang et al. (2013) focused on a charging station scenario with renewable energy power generation and energy storage devices and achieved the goal of minimizing the average waiting time by optimizing the charging queuing process of EVs.

For the charging scheduling control of EVs, it can often be divided into invasive (direct) or noninvasive (indirect) methods. Intrusive methods often require EV users to transfer their power appropriately, so as to realize the direct control of the control center over the charging behavior of EVs. The nonintrusive method is to guide the behavior of EVs through indirect factors (price, waiting time, etc.). For example, EVs will decide whether to charge in real time according to the dynamically changing price in Zeng et al. (2015). No matter which way, in order to promote

EV users to participate in load dispatching, it depends on good market mechanism design to realize the incentive.

Common incentive methods include coordination mechanisms based on additional return (Sundstrom and Binding 2011; Deilami et al. 2011) and dynamic price or bidding (Ghijsen, and D'hulst, R. 2011). Liu and Etemadi (2017b) and Ghosh and Aggarwal (2017) based on the assumption that EV users are price-sensitive, encourage EVs to charge off peak through price mechanism and access control. Li et al. (2011) and Samadi et al. (2010) affect EV users to reasonably allocate power consumption time and realize low-cost charging by determining and publishing charging prices in the real-time market. In Huang et al. (2017), an option based pricing strategy is proposed to reduce the risk of charging cost fluctuation in random process.

On the other hand, in order to overcome the problems of mileage anxiety and high initial cost, sharing economy has been put forward, which also enables users to participate in the EV charging market with new incentives. Jia and Wu (2020b) studied the structural characteristics and optimal properties of EV charging problem under sharing mechanism. Figure 11.6 shows the workflow of the shared EV reservation system. Users make reservations for the shared EVs through a platform, e.g. a website or more popularly an APP on the smartphone. There are a number of parking lots that collaborate with the shared EV company to provide the business. During the reservation, the user browses the EVs in each associated parking lot. Then, the user can see the available time period for each EV (only the EVs that can be fully charged will be available for users). When a user specifies an EV and a pickup time, ideally, the EV should be fully charged (by the company) and available for pickup by the beginning of that step. This is referred to as a departure event (from the parking lot). When a user specifies a return time, ideally, the user should return the EV to one parking lot by the beginning of that step. This is referred to as an arrival event (to the parking lot). Based on that, Jia and Wu (2020b) extend the LLLP principle and further study the structural property of the charging scheduling problem. Authors show that the LLLP applies to the formulated shared EVs' charging scheduling problem and may be used to narrow down the action space while preserving the global optimality. A modified LLLP algorithm is proposed to construct a policy in $O(NT)$, where N is the number of the EVs, and T is the number of time steps in the scheduling problem.

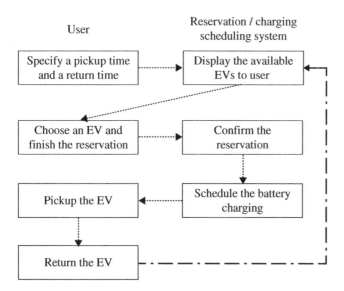

Figure 11.6 Workflow of the reservation system. Source: Jia and Wu (2020b).

11.3 Control and Optimization Methods

For an EV with immediate charging demand, its charging elasticity is mainly reflected in two dimensions: space and time. In terms of space, when the EV puts forward the charging demand on the road, it may face multiple suitable charging stations for selection. Simultaneously, the charging choices of multiple EVs will affect and restrict each other. The optimal scheduling of charging location is the exploration of this spatial elasticity. On the other hand, after driving into the selected charging station, the EV still has the charging elasticity in time, that is, its charging process is partially controllable. Partially controllable means that the premise of charging control is to meet the charging demand of EVs in time. According to the relationship between charging time and parking time, the charging process can be divided into uncontrollable process and controllable process. Uncontrollable process means that when the required charging time is equal to the parking time, the vehicle can only be charged at the maximum power immediately after connecting to the power grid, so it can meet the charging demand before leaving. Therefore, there is no room for optimization. On the contrary, when the required charging time is less than the parking time, the charging process of the vehicle is controllable, that is, the optimization objectives of the system can be achieved by making full use of the low-cost renewable energy and time of use (ToU) electricity price.

11.3.1 The Difficulty of the Control and Optimization

As an important part of the research on stochastic supply and demand matching, the EV charging scheduling optimization with uncertain renewable energy will face the following typical difficulties:

(1) High randomness and partial controllability constraints of supply and demand. On the supply side, the output of renewable energy such as wind and solar power will be affected by weather, light, time, and other factors, so there is high uncertainty and volatility. Meanwhile, if there are no energy storage facilities, the consumption of renewable energy is generally required to be real time, that is to say, it is typically uncontrollable. On the demand side, the charging demand of EVs varies from person to person according to drivers' driving habits and travel plans, which is difficult to predict accurately. On the other hand, although the charging demand of EVs is elastic, its scheduling optimization is also constrained by the conditions that the charging process needs to be completed within the time required by users, so it is partially controllable.

(2) Time-coupled sequential decision process. The charging scheduling of EVs often involves multiple decision time steps, and the decision-making behavior at different time steps will affect each other. In other words, the decision will not only change the current system state but also affect the state transition of each step in the future. This time-coupled characteristic also requires that the policy optimization needs to consider the overall situation of the whole time, rather than just making short-sighted decisions.

(3) The curse of dimensionality. With the increase of the considered spatial scale, the number of EVs in a certain area of a city may reach thousands, which will also bring the size of exponential growth of state space and policy space. Taking the binary decision-making behavior of controlling whether the EV is charged as an example, if we want to optimize the scheduling problem of 1000 EVs, the size of the policy space will reach 2^{1000}.

Moreover, since the charging scheduling of large-scale EVs often involves multiple spatial scales, it may also face the following prominent issues:

(1) Computation efficiency requirements. Facing the frequently changing supply–demand relationship and system state, the solution algorithm needs to have the ability to quickly give high-quality solutions with performance guarantee. Therefore, even if the optimization algorithm can obtain high-precision and high-performance solutions after a long time, it may not be applicable in the real scene.
(2) Constraints of spatial topology. When considering the collaborative optimization of multiregions, the topological constraints such as road network structure and transportation range will further increase the difficulty of solving the problem.
(3) Multiscale optimization objectives and system constraints. The optimization algorithm needs to consider the objective functions and system constraints on different scales and obtain the global optimal solution of the whole system.

11.3.2 Charging Location Selection and Routing Optimization

The research on charging location selection and routing problem is mainly carried out from two aspects.

One aspect is the scheduling optimization of global traffic flow distribution to realize the coordinated operation of power grid and traffic network. In He et al. (2016), the public charging station is optimized to adjust the spatial distribution of vehicle flow and EV charging load, which not only reduces the power loss of power distribution system but also reduces the travel time of passengers. Wei et al. (2016) adopts a two-stage robust optimization method to incorporate the uncertainty of traffic demand into the user equilibrium (UE) model, so as to improve the anti-risk ability of traffic power network cooperative operation. In fact, the implementation of such problems is often carried out by setting dynamic pricing, so that when selecting charging stations, EV users should consider not only traffic conditions but also charging costs, so as to achieve the global impact on macrotraffic flow and strengthen the interdependence between traffic and power grid. He et al. (2013) proposed the retail electricity price to realize the impact on the selection of charging location of EVs. In Wei et al. (2017), the authors combined the traffic allocation problem with the economic dispatching problem of power system, proposed a concept of location marginal price, and derived the equilibrium operation state of the system by using the fixed point method. Alizadeh et al. (2016) guides the charging behavior of EV users through the optimization of road toll, so as to ensure that the traffic distribution is socially optimal.

The other aspect is the location optimization of individual or fleet of EVs based on road topology. Considering the physical constraints in space, it is closer to the category of electric vehicle routing problem (EVRP). Due to the privacy of EVs, such studies mostly exist in the scheduling of public transport vehicles such as passenger/freight EVs and electric buses. Long and Jia (2021a, 2020) consider the joint optimization problem of commercial EV charging location selection and renewable energy dispatch scheduling to achieve the goal of global operation cost minimization. Penna et al. (2016) considered the solution of the multisupply point routing problem with route constraints between multiple charging stations. Yang and Sun (2015) optimized the EVRP in the power exchange scenario under the limitation of battery capacity. On this basis, Erdoğan and Miller-Hooks (2012) and Barco et al. (2013) focus on the problem of minimizing detour in charging location scheduling when the number of charging stations is limited. Not only the number of charging stations may be actually limited but also the charging capacity constraint of each charging station is often one of the challenges in EVRP. Considering the capacity limitation, Ding et al. (2015) proposed a heuristic method for simple charging adjustment. Kchaou-Boujelben and Gicquel (2020) further considers the impact of the randomness of EV charging demand and battery energy state on charging location. Schneider et al. (2014) discusses the EVRP in the context of charging stations with time windows and service requirements constraints. Considering that passenger/freight companies often have many types of fuel vehicles and EVs, their travel range, loading capacity, and cost are different. Therefore, the hybrid fleet of different models can optimize the company's total cost. Goeke and Schneider (2015) studied the EVRP of hybrid fleets of the same type of EVs and fossil fuel vehicles and calculated the corresponding energy consumption cost of each vehicle. Cheng and Szeto (2017) further extended the problem model and studied the solution of EVRP of multiple EV hybrid fleets with different battery capacity and operating cost.

11.3.3 Charging Process Control

Compared with charging location selection and routing optimization, charging process control is a timing control problem for EV charging. Figure 11.7 shows the changes of power grid load under different charging scheduling policies. It can be seen that during disordered charging, the peak power consumption of the power grid often highly coincides with the charging demand of EVs, which will aggravate the peak valley difference of the power system and affect the stable operation. Through the application of smart charging technology, the charging demand can be spread to the low-load period for charging, so as to achieve the purpose of peak shaving and valley filling. Moreover, combined with vehicle-to-grid (V2G) technology and the cooperation of renewable energy, further load balancing can be realized.

According to the optimization scenario, the problem can generally be divided into deterministic scenario and stochastic scenario optimization. Among them, deterministic scenarios often refer to the ability to know the system state information at the current time, as well as the future information or accurate prediction value in advance. Such situations often occur in the planning and operation of the day-ahead market. At this time, the optimal charging policy can be obtained through offline algorithms such as linear programming (Sundstrom and Binding 2011), integer programming (Huang and Zhou 2015), quadratic programming (Clement-Nyns et al. 2009), K-means (Kristoffersen et al. 2011), and branch and bound method (Sun et al. 2016). DeForest et al. (2018) used the mixed integer linear programming method to optimize the EV day-ahead charging plan and bidding strategy, so as to minimize the operating cost and maximize the income of auxiliary services. In Sarker et al. (2015), the optimization of bidding strategy for regional coordinators is considered, and an optimal bidding strategy considering the energy and reserve market and the impact on EV battery degradation is proposed in the day-ahead market.

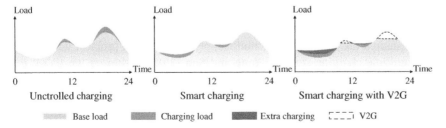

Figure 11.7 Power grid load curve under different charging policies. Source: Adapted from García-Villalobos et al. (2014).

Compared with the day-ahead market, the real-time scheduling of EVs is more realistic and challenging. A simple solution is to convert the online problem into several offline optimization problems and use the above deterministic algorithm for approximate solution (Koufakis et al. 2019). When considering randomness, stochastic optimization (Al-Awami and Sortomme 2011), model predictive control (Guo et al. 2017), robust control (Huang et al. 2016b), game theory (Huang et al. 2013), queuing theory (Gusrialdi et al. 2017), and Markov decision process (MDP) (Widrick et al. 2018) have been widely used and studied. Moghaddass et al. (2019) proposed a mixed integer multiobjective optimization algorithm, and considered two scheduling frameworks including static and dynamic. Wang et al. (2018) combined the real-time access control, charging pricing, and online scheduling of EVs, and introduced a joint optimization method to improve the overall income of charging stations. Scenario tree is another technical means for solving uncertain problems. It uses nodes to represent possible solutions and branches to represent the transformation between different scenarios. A scenario-based stochastic model is proposed in the literature (Aliasghari et al. 2018) to solve the optimal scheduling problem of EV charging under renewable energy supply. In order to overcome the problem of state space explosion, simulation-based method (Zhang and Jia 2017) and reinforcement learning (Long et al. 2019) have attracted great attention in recent years. In real-time scheduling problems, charging stations often have high requirements for solving speed, so heuristic and rule-based methods are often developed and applied. The main idea is to formulate some static and descriptive pricing and charging rules based on subjective knowledge and understanding of the problem. These methods are usually easy to implement and deploy, but they often lack system performance guarantee and have a large space for performance improvement. Ordinal optimization (OO) provides a new idea for solving large-scale simulation-based optimization problems at high speed. It reduces the difficulty of solving the optimal policy by finding a good enough policy set instead. In Long et al. (2021c), an aggregated EV charging model based on the energy curve and incomplete Beta function is proposed, OO-based flexible optimization method is applied, which can solve the problem within seconds.

11.3.4 Control and Optimization Framework

Among the existing large-scale EV charging scheduling optimization frameworks, centralized, distributed, and hierarchical optimization are the most classic and widely used control forms. This subsection will give a brief summary.

11.3.4.1 Centralized Optimization

Centralized optimization refers to the process in which the operators directly establishes communication with each EV, collects global system information such as ID, SoC, charging demand, charging preference, and ToU price, centrally performs back-end calculation and decision-making, and finally directly controls the charging plan of each EV (Sortomme and El-Sharkawi 2010). Zheng et al. (2013) dispatch the charging demand of EVs to reduce the power fluctuation of power grid. Similarly, the centralized policy has also been applied in the battery exchange scenario (Rao et al. 2015) and the integration of V2G technology (Wang and Wang 2013).

The advantage of centralized optimization is that it can get the global optimal solution of the problem, and the structure and implementation of the algorithms are relatively simple. However, for large-scale problems, centralized methods often have high computing cost and poor scalability (Hu et al. 2016). It also relies on distributed global communication network and centralized data integration processor, which also poses new challenges to information privacy and security.

11.3.4.2 Decentralized Optimization

In view of the disadvantages of the above-centralized method, decentralized scheduling has become an attractive alternative. In the distributed control system, each EV can build the charging plan according to its own load and local information. Therefore, personal data will only flow locally, strengthening the protection of privacy (Gan et al. 2012). Garcia Alvarez et al. (2018) proposed an enhanced artificial bee colony algorithm to realize distributed charging scheduling. In Xydas et al. (2016), EVs can learn to adaptively adjust the charging plan and give priority to the use of renewable energy through the distributed multiagent learning policy. Hafiz et al. (2017) proposed a decentralized EV charging method based on energy price optimization, which requires real-time information interaction between EVs and the electric market.

The decentralized optimization framework has the advantages of high computational efficiency and information security and only needs the deployment of locally interconnected communication infrastructure. However, due to these features, the decentralized algorithms often lack understanding of the global information, so the decentralized policy often falls into local optimum (Al-Ogaili et al. 2019). Moreover, this optimization framework will also face problems such as slow convergence and oscillation divergence of optimization results.

11.3.4.3 Hierarchical Optimization

The large-scale EV charging scheduling problem with the participation of renewable energy often involves the interaction between multiple objects of different scales, such as the central control center, distributed renewable energy generation systems, power grid, road network topology, charging stations and regional managers, charging piles, and EV users. Therefore, it has the typical characteristics of multispatial scale. As shown in Figure 11.8, different spatial scales have to achieve trans-regional cooperate in the horizontal direction and exchange energy and information in the vertical direction.

Thus, hierarchical control divides the large-scale EV charging scheduling problem in time or space, so as to reduce the scale of the problem and overcome the uncertainty. Chen et al. (2016) proposed to use EVA as the intermediate coordinator, first optimize the total power supply value by EVA, and then schedule the individual charging behavior of EVs. Huang et al. (2016a) proposed a hierarchical coordinated scheduling algorithm considering both coarse and fine time scales. According to the situation of China, a three-level EV dispatching framework including province, city, and charging station is proposed in Xu et al. (2015), which realizes energy transaction and information exchange between different levels. In addition, by applying

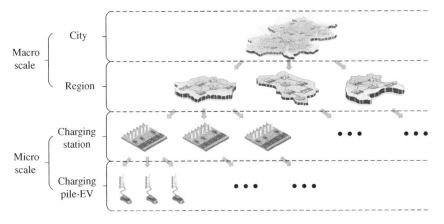

Figure 11.8 Multispatial scale of EV charging scheduling optimization.

the charging demand aggregation technology at the city and station levels, the computing and communication requirements can be effectively reduced. Meanwhile, Rivera et al. (2016) further considers the difference of objective functions between different layers, so as to model the problem as a multiobjective optimization model. Event-based optimization (EBO) realizes further problem compression by defining state transition as event. Long et al. (2017) and Long and Jia (2021b) proposed a bi-level EBO algorithm to handle the transregion EV charging scheduling problem with uncertain wind power, which shows good scalability.

Hierarchical optimization is also called multistage control since it transforms the problem originally solved in one-step into multistep sequential solution. Lam et al. (2014) proposed a two-stage optimization algorithm to solve the problem. In the first stage, the availability of the charging station is transmitted to the EV users through Boolean variables, while in the second stage, the selection of charging points is optimized according to the EV users' own tendency. Pan et al. (2010) consider the optimal battery exchange-scheduling problem and proposes a two-stage stochastic programming solution model.

Although hierarchical control can effectively reduce the difficulty of problem-solving and improve the scalability and operation efficiency of the algorithm, it is often a top-down one-way adjustment process rather than an iterative optimization process. In other words, the optimization problems in the lower layer can only take the existing optimization results in the upper layer as the boundary conditions for subsequent calculation. Although some works consider the multilayer cooperation, they often fall into the local optimal solution because each iteration can only optimize partial variables.

11.3.5 The Impact of Human Behaviors

In the scheduling and charging optimization of EVs, human is an unavoidable role which can strongly affect the decision-making process and results. In general, the impact of human behavior is mainly reflected in the following two aspects.

One the one hand, as the owner and driver of EVs, human driving patterns are always different, including travel time, destination, speed, frequency, parking time, and charging preference. This leads to different distribution of EV charging demand faced by buildings and parking lots in different regions and for different purposes. Furthermore, the uncertainty of human driving behavior will also affect the optimality of charging decisions for the current operation of the charging station, especially in the face of suddenly increased charging demand or sudden early departure time. On the other hand, as a member of the market operators, human beings will participate in the construction of the market mechanism and the formulation of the market price, thus indirectly affecting the charging decision and driving path selection of electric vehicles.

However, in the actual modeling and optimization of large-scale problems, human microindividual behavior is often not directly reflected, but is reflected in the distribution of arrival probability and parking time of EVs in different charging stations and buildings in the form of statistical information.

11.4   Conclusion and Discussion

EVs have become an important part of the infrastructure of the modern society. Despite the promise of EV charging in practice, many researchers have found it difficult to build reliable simulation models to study EV charging and its impact on the power grid and other components of the energy system. This chapter discusses the challenges and opportunities of EV charging and reviews the charging behavior of EVs from data sets, models, control, and optimization methods. Although many excellent research works have been published, the probable future outlook of this research field is worth to be emphasized. We believe that at least the following three aspects of work can be carried out in the future:

(1) Power system integration planning and dispatching combined with V2G technology. In most works, the EV is mainly modeled and analyzed as a flexible load with charging flexibility in time and space. Although through smart charging, EVs can help reduce the power grid operation pressure and fill valley, the charging controllability of EVs has not been fully developed and utilized. In recent years, the vigorous development of V2G has provided a new way for the integration of EVs and the power system. In addition to being a consumer of electricity, EVs can also become energy storage facilities and power supply terminals, and further reduce the peak pressure of power grid and improve the utilization rate of renewable energy. Therefore, large-scale charging scheduling considering the bi-direction charge mechanism of EVs is one of the next feasible research directions in this research field. However, the bi-direction charge mechanism will further expand the decision-making space and solving difficulty of the problem, so it is worth to improve the solving efficiency through methods like charging curve (Long et al. 2021c), event-based optimization (Long and Jia 2021b).

(2) Multiscale smart charging based on swarm intelligence. Although in the multispatial scale scenario, the computational pressure of large-scale EV charging scheduling problem has been alleviated to a certain extent, and part of the computational process has been solved locally. However, the implementation of the algorithm still depends on the centralized control calculation of the central controller and regional coordinators, which still poses a risk to the stable operation and safety of the system. In view of this, using swarm intelligence algorithm like multiagent reinforcement learning methods should become one of the key points of the next research. This method can get rid of centralized data upload and invasive control, which is more conducive to the security of the system. However, the performance instability caused by the confrontation, game and randomness between EVs may become the key problems of this framework. Therefore, how to use hierarchical structure to realize distributed control and to what degree of distributed control will be the focus of researchers in the future.

(3) The design of the noninvasive multiscale market mechanism. Most of the researches on EV charging scheduling are to directly control the charging behavior of EVs. The hypothesis of those researches is to encourage EV users to join the collaborative scheduling in some way and is willing to provide some nonprivate information. Because of this, this intrusive control method is often doubted and criticized by users' privacy and information security. The noninvasive control represented by dynamic price and guidance suggestions can solve this problem as well, which is a useful extension of this research problem. By designing a safe and reasonable market dispatching mechanism, this method can form a strong incentive for the power grid, regional coordinators, and individual EVs in multiscales, and realize the overall optimization goal of the system at the same time. However, the challenge lies in the randomness and uncontrollability of system participants (including charging stations and EV users), which will make it difficult for the algorithm to achieve the theoretical strategic performance.

References

Aghaei, J. and Alizadeh, M.I. (2013). Demand response in smart electricity grids equipped with renewable energy sources: a review. *Renewable and Sustainable Energy Reviews* 18: 64–72.

Al-Alawi, B.M. and Bradley, T.H. (2013). Review of hybrid, plug-in hybrid, and electric vehicle market modeling studies. *Renewable and Sustainable Energy Reviews* 21: 190–203.

Al-Awami, A.T. and Sortomme, E. (2011). Coordinating vehicle-to-grid services with energy trading. *IEEE Transactions on smart grid* 3 (1): 453–462.

Aliasghari, P., Mohammadi-Ivatloo, B., Alipour, M. et al. (2018). Optimal scheduling of plug-in electric vehicles and renewable micro-grid in energy and reserve markets considering demand response program. *Journal of Cleaner Production* 186: 293–303.

Alizadeh, M., Wai, H.T., Chowdhury, M. et al. (2016). Optimal pricing to manage electric vehicles in coupled power and transportation networks. *IEEE Transactions on control of network systems* 4 (4): 863–875.

Al-Ogaili, A.S., Hashim, T.J.T., Rahmat, N.A. et al. (2019). Review on scheduling, clustering, and forecasting strategies for controlling electric vehicle charging: challenges and recommendations. *IEEE Access* 7: 128353–128371.

Amici, R., Bonola, M., Bracciale, L. et al. (2014). Performance assessment of an epidemic protocol in VANET using real traces. *Procedia Computer Science* 40: 92–99.

Ashique, R.H., Salam, Z., Aziz, M.J.B.A., and Bhatti, A.R. (2017). Integrated photovoltaic-grid DC fast charging system for electric vehicle: a review of the architecture and control. *Renewable and Sustainable Energy Reviews* 69: 1243–1257.

Barco, J., Guerra, A., Munoz, L., and Quijano, N. (2013). Optimal routing and scheduling of charge for electric vehicles: a case study. *Mathematical Problems in Engineering* 2017.

Bessa, R.J. and Matos, M.A. (2010). The role of an aggregator agent for EV in the electricity market.

Bessa, R.J. and Matos, M.A. (2013). Optimization models for EV aggregator participation in a manual reserve market. *IEEE Transactions on Power Systems* 28 (3): 3085–3095.

BP (2021). BP world energy statistical yearbook. https://www.bp.com/zh_cn/china/home/news/reports/statistical-review-2021.html (accessed 17 January 2022).

Calearo, L., Marinelli, M., and Ziras, C. (2021). A review of data sources for electric vehicle integration studies. *Renewable and Sustainable Energy Reviews* 151: 111518.

Chen, J., Piao, L., Ai, Q., and Xiao, F. (2016). Hierarchical optimal scheduling for electric vehicles based on distributed control. *Automation of Electric Power Systems* 40 (18): 24–31.

Cheng, Y. and Szeto, W.Y. (2017). Artificial bee colony approach to solving the electric vehicle routing problem. *Journal of the Eastern Asia Society for Transportation Studies* 12: 975–990.

Clement-Nyns, K., Haesen, E., and Driesen, J. (2009). The impact of charging plug-in hybrid electric vehicles on a residential distribution grid. *IEEE Transactions on power systems* 25 (1): 371–380.

DeForest, N., MacDonald, J.S., and Black, D.R. (2018). Day ahead optimization of an electric vehicle fleet providing ancillary services in the Los Angeles Air Force Base vehicle-to-grid demonstration. *Applied energy* 210: 987–1001.

Deilami, S., Masoum, A.S., Moses, P.S., and Masoum, M.A. (2011). Real-time coordination of plug-in electric vehicle charging in smart grids to minimize power losses and improve voltage profile. *IEEE Transactions on Smart Grid* 2 (3): 456–467.

Ding, N., Batta, R., and Kwon, C. (2015). Conflict-free electric vehicle routing problem with capacitated charging stations and partial recharge. SUNY, Buffalo.

Dixon, J., Bukhsh, W., Edmunds, C., and Bell, K. (2020). Scheduling electric vehicle charging to minimise carbon emissions and wind curtailment. *Renewable Energy* 161: 1072–1091.

Erdoğan, S. and Miller-Hooks, E. (2012). A green vehicle routing problem. *Transportation research part E: logistics and transportation review* 48 (1): 100–114.

Fang, H., Wang, Y., and Chen, J. (2016). Health-aware and user-involved battery charging management for electric vehicles: linear quadratic strategies. *IEEE Transactions on Control Systems Technology* 25 (3): 911–923.

Gan, L., Topcu, U., and Low, S.H. (2012). Optimal decentralized protocol for electric vehicle charging. *IEEE Transactions on Power Systems* 28 (2): 940–951.

Garcia Alvarez, J., González, M.Á., Rodriguez Vela, C., and Varela, R. (2018). Electric vehicle charging scheduling by an enhanced artificial bee colony algorithm. *Energies* 11 (10): 2752.

García-Villalobos, J., Zamora, I., San Martín, J.I. et al. (2014). Plug-in electric vehicles in electric distribution networks: a review of smart charging approaches. *Renewable and Sustainable Energy Reviews* 38: 717–731.

Ghijsen, M. and D'hulst, R. (2011). Market-based coordinated charging of electric vehicles on the low-voltage distribution grid. In: *2011 IEEE First International Workshop on Smart Grid Modeling and Simulation (SGMS)* (ed. C. Develder, A. Narayan, and C. Rodine), 1–6. IEEE.

Ghosh, A. (2020). Possibilities and challenges for the inclusion of the Electric Vehicle (EV) to reduce the carbon footprint in the transport sector: a review. *Energies* 13 (10): 2602.

Ghosh, A. and Aggarwal, V. (2017). Control of charging of electric vehicles through menu-based pricing. *IEEE Transactions on Smart Grid* 9 (6): 591jia8–5929.

Godina, R., Rodrigues, E.M., Matias, J.C., and Catalão, J.P. (2016). Smart electric vehicle charging scheduler for overloading prevention of an industry client power distribution transformer. *Applied Energy* 178: 29–42.

Goeke, D. and Schneider, M. (2015). Routing a mixed fleet of electric and conventional vehicles. *European Journal of Operational Research* 245 (1): 81–99.

González-Garrido, A., Thingvad, A., Gaztañaga, H., and Marinelli, M. (2019). Full-scale electric vehicles penetration in the Danish Island of Bornholm—Optimal scheduling and battery degradation under driving constraints. *Journal of Energy Storage* 23: 381–391.

Guo, Y., Xiong, J., Xu, S., and Su, W. (2015). Two-stage economic operation of microgrid-like electric vehicle parking deck. *IEEE Transactions on Smart Grid* 7 (3): 1703–1712.

Guo, L., Gao, B., Li, Y., and Chen, H. (2017). A fast algorithm for nonlinear model predictive control applied to HEV energy management systems. *Science China Information Sciences* 60 (9): 092201.

Gusrialdi, A., Qu, Z., and Simaan, M.A. (2017). Distributed scheduling and cooperative control for charging of electric vehicles at highway service stations. *IEEE Transactions on Intelligent Transportation Systems* 18 (10): 2713–2727.

Hafiz, F., de Quieroz, A.R., Husain, I., and Fajri, P. (2017). Charge scheduling of a plug-in electric vehicle considering load demand uncertainty based on multi-stage stochastic optimization. In: *2017 North American Power Symposium (NAPS)* (ed. J. Solanki), 1–6. IEEE.

Haghbin, S., Lundmark, S., Alakula, M., and Carlson, O. (2012). Grid-connected integrated battery chargers in vehicle applications: review and new solution. *IEEE Transactions on Industrial Electronics* 60 (2): 459–473.

Hannan, M.A., Azidin, F.A., and Mohamed, A. (2014). Hybrid electric vehicles and their challenges: a review. *Renewable and Sustainable Energy Reviews* 29: 135–150.

He, F., Wu, D., Yin, Y., and Guan, Y. (2013). Optimal deployment of public charging stations for plug-in hybrid electric vehicles. *Transportation Research Part B: Methodological* 47: 87–101.

He, F., Yin, Y., Wang, J., and Yang, Y. (2016). Sustainability SI: optimal prices of electricity at public charging stations for plug-in electric vehicles. *Networks and Spatial Economics* 16 (1): 131–154.

Hecht, C., Das, S., Bussar, C., and Sauer, D.U. (2020). Representative, empirical, real-world charging station usage characteristics and data in Germany. *eTransportation* 6: 100079.

Hu, Z., Zhan, K., Zhang, H., and Song, Y. (2016). Pricing mechanisms design for guiding electric vehicle charging to fill load valley. *Applied Energy* 178: 155–163.

Huang, Y. and Zhou, Y. (2015). An optimization framework for workplace charging strategies. *Transportation Research Part C: Emerging Technologies* 52: 144–155.

Huang, J., Leng, M., Liang, L., and Liu, J. (2013). Promoting electric automobiles: supply chain analysis under a government's subsidy incentive scheme. *IIE transactions* 45 (8): 826–844.

Huang, Q., Jia, Q.S., and Guan, X. (2016a). A multi-timescale and bilevel coordination approach for matching uncertain wind supply with EV charging demand. *IEEE Transactions on Automation Science and Engineering* 14 (2): 694–704.

Huang, Q., Jia, Q.S., and Guan, X. (2016b). Robust scheduling of EV charging load with uncertain wind power integration. *IEEE Transactions on Smart Grid* 9 (2): 1043–1054.

Huang, S., Yang, J., and Li, S. (2017). Black-Scholes option pricing strategy and risk-averse coordination for designing vehicle-to-grid reserve contracts. *Energy* 137: 325–335.

International Energy Agency (2021a). Global EV Outlook 2021 – Accelerating ambitions despite the pandemic. https://www.iea.org/reports/global-ev-outlook-2021 (accessed 17 January 2022).

International Energy Agency (2021b). Key world energy statistics 2021. https://www.iea.org/reports/key-world-energy-statistics-2021 (accessed 17 January 2022).

Jia, Q.S. (2018). On supply demand coordination in vehicle-to-grid—a brief literature review. In: *2018 33rd Youth Academic Annual Conference of Chinese Association of Automation (YAC)* (ed. C. Sun and W. He), 1083–1088. IEEE.

Jia, Q.S. and Long, T. (2020a). A review on charging behavior of electric vehicles: data, model, and control. *Control Theory and Technology* 18 (3): 217–230.

Jia, Q.S. and Wu, J. (2020b). A structural property of charging scheduling policy for shared electric vehicles with wind power generation. *IEEE Transactions on Control Systems Technology*.

Kchaou-Boujelben, M. and Gicquel, C. (2020). Locating electric vehicle charging stations under uncertain battery energy status and power consumption. *Computers & Industrial Engineering* 149: 106752.

Kelly, J.C., MacDonald, J.S., and Keoleian, G.A. (2012). Time-dependent plug-in hybrid electric vehicle charging based on national driving patterns and demographics. *Applied Energy* 94: 395–405.

Khodayar, M.E., Wu, L., and Shahidehpour, M. (2012). Hourly coordination of electric vehicle operation and volatile wind power generation in SCUC. *IEEE Transactions on Smart Grid* 3 (3): 1271–1279.

Kim, H.J., Lee, J., Park, G.L. et al. (2010). An efficient scheduling scheme on charging stations for smart transportation. In: *Security-Enriched Urban Computing and Smart Grid. SUComS 2010. Communications in Computer and Information Science*, vol. 78 (ed. T. Kim, A. Stoica, and R.S. Chang), 274–278. Berlin, Heidelberg: Springer.

Kim, I., Kim, J., and Lee, J. (2020). Dynamic analysis of well-to-wheel electric and hydrogen vehicles greenhouse gas emissions: focusing on consumer preferences and power mix changes in South Korea. *Applied Energy* 260: 114281.

Koufakis, A.M., Rigas, E.S., Bassiliades, N., and Ramchurn, S.D. (2019). Offline and online electric vehicle charging scheduling with V2V energy transfer. *IEEE Transactions on Intelligent Transportation Systems* 21 (5): 2128–2138.

Kristoffersen, T.K., Capion, K., and Meibom, P. (2011). Optimal charging of electric drive vehicles in a market environment. *Applied Energy* 88 (5): 1940–1948.

Lam, A.Y., Leung, Y.W., and Chu, X. (2014). Electric vehicle charging station placement: formulation, complexity, and solutions. *IEEE Transactions on Smart Grid* 5 (6): 2846–2856.

Li, N., Chen, L., and Low, S.H. (2011). Optimal demand response based on utility maximization in power networks. In: *2011 IEEE Power and Energy Society General Meeting*, 1–8. IEEE https://urldefense.com/v3/__https://ieeexplore.ieee.org/document/6039082/citations*citations__;Iw!!N11eV2iwtfs!tzAJ67rHs_EnfEBz0MkNY0uMpTZ6nXf2F0D33DuiGRrqa8FPepJUxjs9OjfFc89WioGVo3ml841zTLD4DAG0Sao$.

Li, C., Cao, Y., Zhang, M. et al. (2015). Hidden benefits of electric vehicles for addressing climate change. *Scientific reports* 5 (1): 1–4.

Liu, S. and Etemadi, A.H. (2017b). A dynamic stochastic optimization for recharging plug-in electric vehicles. *IEEE Transactions on Smart Grid* 9 (5): 4154–4161.

Liu, S., Liu, Y., Ni, L.M. et al. (2010). Towards mobility-based clustering. In: *Proceedings of the 16th ACM SIGKDD International Conference on Knowledge Discovery and Data Mining* (ed. B. Rao, B. Krishnapuram, A. Tomkins, and Q. Yang), 919–928. Association for Computing Machinery.

Liu, Z., Wen, F., and Ledwich, G. (2012). Optimal planning of electric-vehicle charging stations in distribution systems. *IEEE transactions on power delivery* 28 (1): 102–110.

Liu, M., Phanivong, P.K., Shi, Y., and Callaway, D.S. (2017a). Decentralized charging control of electric vehicles in residential distribution networks. *IEEE Transactions on Control Systems Technology* 27 (1): 266–281.

Long, T. and Jia, Q.S. (2020). Optimization of large-scale commercial electric vehicles fleet charging location schedule under the distributed wind power supply. In: *2020 IEEE 44th Annual Computers, Software, and Applications Conference (COMPSAC)* (ed. W.K. Chan, B. Claycomb, H. Takakura, et al.), 1398–1404. IEEE.

Long, T. and Jia, Q.S. (2021a). Joint optimization for coordinated charging control of commercial electric vehicles under distributed hydrogen energy supply. *IEEE Transactions on Control Systems Technology.* (Early Access, https://doi.org/10.1109/TCST.2021.3070482).

Long, T. and Jia, Q.S. (2021b). Matching uncertain renewable supply with electric vehicle charging demand—a Bi-level event-based optimization method. *Complex System Modeling and Simulation* 1 (1): 33–44.

Long, T., Tang, J.X., and Jia, Q.S. (2017). Multi-scale event-based optimization for matching uncertain wind supply with EV charging demand. In: *2017 13th IEEE Conference on Automation Science and Engineering (CASE)* (ed. X. Guan and Q. Zhao), 847–852. IEEE.

Long, T., Ma, X.T., and Jia, Q.S. (2019). Bi-level proximal policy optimization for stochastic coordination of EV charging load with uncertain wind power. In: *2019 IEEE Conference on Control Technology and Applications (CCTA)* (ed. J. Chen, R. Smith, and J. Sun), 302–307. IEEE.

Long, T., Jia, Q.S., Wang, G., and Yang, Y. (2021c). Efficient real-time EV charging scheduling via ordinal optimization. *IEEE Transactions on Smart Grid.*

Longo, M., Yaïci, W., and Foiadelli, F. (2018). Electric vehicles integrated with renewable energy sources for sustainable mobility. *New trends in electrical vehicle powertrains* 203–223.

Lu, J. and Hossain, J. (2015). *Vehicle-to-grid: Linking Electric Vehicles to the Smart Grid*. Institution of Engineering and Technology.

Lucas, A., Barranco, R., and Refa, N. (2019). EV idle time estimation on charging infrastructure, comparing supervised machine learning regressions. *Energies* 12 (2): 269.

Mesarić, P. and Krajcar, S. (2015). Home demand side management integrated with electric vehicles and renewable energy sources. *Energy and Buildings* 108: 1–9.

Moghaddass, R., Mohammed, O.A., Skordilis, E., and Asfour, S. (2019). Smart control of fleets of electric vehicles in smart and connected communities. *IEEE Transactions on Smart Grid* 10 (6): 6883–6897.

Morstyn, T., Crozier, C., Deakin, M., and McCulloch, M.D. (2020). Conic optimization for electric vehicle station smart charging with battery voltage constraints. *IEEE Transactions on Transportation Electrification* 6 (2): 478–487.

Mozafar, M.R., Amini, M.H., and Moradi, M.H. (2018). Innovative appraisal of smart grid operation considering large-scale integration of electric vehicles enabling V2G and G2V systems. *Electric Power Systems Research* 154: 245–256.

Muratori, M. (2018). Impact of uncoordinated plug-in electric vehicle charging on residential power demand. *Nature Energy* 3 (3): 193–201.

Pan, F., Bent, R., Berscheid, A., and Izraelevitz, D. (2010). Locating PHEV exchange stations in V2G. In: *2010 First IEEE International Conference on Smart Grid Communications* (ed. G. Arnold, S. Galli, A. Gelman, et al.), 173–178. IEEE.

Penna, P.H.V., Afsar, H.M., Prins, C., and Prodhon, C. (2016). A hybrid iterative local search algorithm for the electric fleet size and mix vehicle routing problem with time windows and recharging stations. *IFAC-PapersOnLine* 49 (12): 955–960.

Plötz, P., Funke, S., and Jochem, P. (2015). Real-world fuel economy and CO_2 emissions of plug-in hybrid electric vehicles (No. S1/2015). Working Paper Sustainability and Innovation.

Plötz, P., Funke, S.Á., and Jochem, P. (2018). Empirical fuel consumption and CO_2 emissions of plug-in hybrid electric vehicles. *Journal of Industrial Ecology* 22 (4): 773–784.

Rahman, I., Vasant, P.M., Singh, B.S.M. et al. (2016). Review of recent trends in optimization techniques for plug-in hybrid, and electric vehicle charging infrastructures. *Renewable and Sustainable Energy Reviews* 58: 1039–1047.

Rao, R., Zhang, X., Xie, J., and Ju, L. (2015). Optimizing electric vehicle users' charging behavior in battery swapping mode. *Applied Energy* 155: 547–559.

Richardson, D.B. (2013). Electric vehicles and the electric grid: a review of modeling approaches, impacts, and renewable energy integration. *Renewable and Sustainable Energy Reviews* 19: 247–254.

Rivera, J., Goebel, C., and Jacobsen, H.A. (2016). Distributed convex optimization for electric vehicle aggregators. *IEEE Transactions on Smart Grid* 8 (4): 1852–1863.

Rubino, L., Capasso, C., and Veneri, O. (2017). Review on plug-in electric vehicle charging architectures integrated with distributed energy sources for sustainable mobility. *Applied Energy* 207: 438–464.

Samadi, P., Mohsenian-Rad, A.H., Schober, R. et al. (2010). Optimal real-time pricing algorithm based on utility maximization for smart grid. In: *2010 First IEEE International Conference on Smart Grid Communications*, October (ed. G. Arnold, S. Galli, A. Gelman, et al.), 415–420. IEEE.

Sarker, M.R., Dvorkin, Y., and Ortega-Vazquez, M.A. (2015). Optimal participation of an electric vehicle aggregator in day-ahead energy and reserve markets. *IEEE Transactions on Power Systems* 31 (5): 3506–3515.

Sbordone, D., Bertini, I., Di Pietra, B. et al. (2015). EV fast charging stations and energy storage technologies: a real implementation in the smart micro grid paradigm. *Electric Power Systems Research* 120: 96–108.

Schey, S., Scoffield, D., and Smart, J. (2012). A first look at the impact of electric vehicle charging on the electric grid in the EV project. *World Electric Vehicle Journal* 5 (3): 667–678.

Schill, W.P. and Gerbaulet, C. (2015). Power system impacts of electric vehicles in Germany: charging with coal or renewables? *Applied Energy* 156: 185–196.

Schneider, M., Stenger, A., and Goeke, D. (2014). The electric vehicle-routing problem with time windows and recharging stations. *Transportation science* 48 (4): 500–520.

Smart, J. and Schey, S. (2012). Battery electric vehicle driving and charging behavior observed early in the EV project. *SAE International Journal of Alternative Powertrains* 1 (1): 27–33.

Smart, J., Powell, W., and Schey, S. (2013). Extended range electric vehicle driving and charging behavior observed early in the EV project (No. 2013-01-1441). SAE Technical Paper.

Smart, J., Bradley, T., and Salisbury, S. (2014). Actual versus estimated utility factor of a large set of privately owned Chevrolet Volts. *SAE International Journal of Alternative Powertrains* 3 (1): 30–35.

Smith, R., Shahidinejad, S., Blair, D., and Bibeau, E.L. (2011). Characterization of urban commuter driving profiles to optimize battery size in light-duty plug-in electric vehicles. *Transportation Research Part D: Transport and Environment* 16 (3): 218–224.

Soares, J., Ghazvini, M.A.F., Borges, N., and Vale, Z. (2017). A stochastic model for energy resources management considering demand response in smart grids. *Electric Power Systems Research* 143: 599–610.

Sortomme, E. and El-Sharkawi, M.A. (2010). Optimal charging strategies for unidirectional vehicle-to-grid. *IEEE Transactions on Smart Grid* 2 (1): 131–138.

Sortomme, E. and El-Sharkawi, M.A. (2011). Optimal scheduling of vehicle-to-grid energy and ancillary services. *IEEE Transactions on Smart Grid* 3 (1): 351–359.

Su, W., Wang, J., and Roh, J. (2013). Stochastic energy scheduling in microgrids with intermittent renewable energy resources. *IEEE Transactions on Smart grid* 5 (4): 1876–1883.

Sun, B., Huang, Z., Tan, X., and Tsang, D.H. (2016). Optimal scheduling for electric vehicle charging with discrete charging levels in distribution grid. *IEEE Transactions on Smart Grid* 9 (2): 624–634.

Sundstrom, O. and Binding, C. (2011). Flexible charging optimization for electric vehicles considering distribution grid constraints. *IEEE Transactions on Smart grid* 3 (1): 26–37.

Tabatabaee, S., Mortazavi, S.S., and Niknam, T. (2017). Stochastic scheduling of local distribution systems considering high penetration of plug-in electric vehicles and renewable energy sources. *Energy* 121: 480–490.

Tal, G., Nicholas, M.A., Davies, J., and Woodjack, J. (2014). Charging behavior impacts on electric vehicle miles traveled: who is not plugging in? *Transportation Research Record* 2454 (1): 53–60.

Teng, F., Ding, Z., Hu, Z., and Sarikprueck, P. (2020). Technical review on advanced approaches for electric vehicle charging demand management, part I: applications in electric power market and renewable energy integration. *IEEE Transactions on Industry Applications* 56 (5): 5684–5694.

Thomas, D., Deblecker, O., and Ioakimidis, C.S. (2018). Optimal operation of an energy management system for a grid-connected smart building considering photovoltaics' uncertainty and stochastic electric vehicles' driving schedule. *Applied Energy* 210: 1188–1206.

United Nations Environment Programme (2021). The emissions gap report. https://www.unep.org/resources/emissions-gap-report-2021 (accessed 25 February 2023).

Vagropoulos, S.I. and Bakirtzis, A.G. (2013). Optimal bidding strategy for electric vehicle aggregators in electricity markets. *IEEE Transactions on power systems* 28 (4): 4031–4041.

Wang, Z. and Wang, S. (2013). Grid power peak shaving and valley filling using vehicle-to-grid systems. *IEEE Transactions on power delivery* 28 (3): 1822–1829.

Wang, S., Bi, S., Zhang, Y.J.A., and Huang, J. (2018). Electrical vehicle charging station profit maximization: admission, pricing, and online scheduling. *IEEE Transactions on Sustainable Energy* 9 (4): 1722–1731.

Wei, W., Mei, S., Wu, L. et al. (2016). Robust operation of distribution networks coupled with urban transportation infrastructures. *IEEE Transactions on Power Systems* 32 (3): 2118–2130.

Wei, W., Wu, L., Wang, J., and Mei, S. (2017). Network equilibrium of coupled transportation and power distribution systems. *IEEE Transactions on Smart Grid* 9 (6): 6764–6779.

Widrick, R.S., Nurre, S.G., and Robbins, M.J. (2018). Optimal policies for the management of an electric vehicle battery swap station. *Transportation Science* 52 (1): 59–79.

Wu, X., Dong, J., and Lin, Z. (2014). Cost analysis of plug-in hybrid electric vehicles using GPS-based longitudinal travel data. *Energy Policy* 68: 206–217.

Wu, X., Aviquzzaman, M., and Lin, Z. (2015). Analysis of plug-in hybrid electric vehicles' utility factors using GPS-based longitudinal travel data. *Transportation Research Part C: Emerging Technologies* 57: 1–12.

Wu, D., Zeng, H., Lu, C., and Boulet, B. (2017). Two-stage energy management for office buildings with workplace EV charging and renewable energy. *IEEE Transactions on Transportation Electrification* 3 (1): 225–237.

Xiao, Y., Zuo, X., Kaku, I. et al. (2019). Development of energy consumption optimization model for the electric vehicle routing problem with time windows. *Journal of Cleaner Production* 225: 647–663.

Xu, Z., Su, W., Hu, Z. et al. (2015). A hierarchical framework for coordinated charging of plug-in electric vehicles in China. *IEEE Transactions on Smart Grid* 7 (1): 428–438.

Xydas, E., Marmaras, C., and Cipcigan, L.M. (2016). A multi-agent based scheduling algorithm for adaptive electric vehicles charging. *Applied energy* 177: 354–365.

Yang, J. and Sun, H. (2015). Battery swap station location-routing problem with capacitated electric vehicles. *Computers & operations research* 55: 217–232.

Yang, Y., Jia, Q.S., Guan, X. et al. (2018). Decentralized EV-based charging optimization with building integrated wind energy. *IEEE Transactions on Automation Science and Engineering* 16 (3): 1002–1017.

Zeng, M., Leng, S., Maharjan, S. et al. (2015). An incentivized auction-based group-selling approach for demand response management in V2G systems. *IEEE Transactions on Industrial Informatics* 11 (6): 1554–1563.

Zhang, Y. and Jia, Q.S. (2017). A simulation-based policy improvement method for joint-operation of building microgrids with distributed solar power and battery. *IEEE Transactions on Smart Grid* 9 (6): 6242–6252.

Zhang, T., Chen, W., Han, Z., and Cao, Z. (2013). Charging scheduling of electric vehicles with local renewable energy under uncertain electric vehicle arrival and grid power price. *IEEE Transactions on Vehicular Technology* 63 (6): 2600–2612.

Zhao, J., Wan, C., Xu, Z., and Wang, J. (2015). Risk-based day-ahead scheduling of electric vehicle aggregator using information gap decision theory. *IEEE Transactions on Smart Grid* 8 (4): 1609–1618.

Zheng, J., Wang, X., Men, K. et al. (2013). Aggregation model-based optimization for electric vehicle charging strategy. *IEEE Transactions on Smart Grid* 4 (2): 1058–1066.

Ziras, C., Zecchino, A., and Marinelli, M. (2018). Response accuracy and tracking errors with decentralized control of commercial V2G chargers. In: *2018 Power Systems Computation Conference (PSCC)*, June (ed. A. Bose and T.C. de Barros), 1–7. IEEE.

Part III

Robotics

12

Trust-Triggered Robot–Human Handovers Using Kinematic Redundancy for Collaborative Assembly in Flexible Manufacturing

S. M. Mizanoor Rahman[1], Behzad Sadrfaridpour[2], Ian D. Walker[3], and Yue Wang[2]

[1]Department of Mechanical Engineering, Pennsylvania State University, Dunmore, PA, USA
[2]Department of Mechanical Engineering, Clemson University, Clemson, SC, USA
[3]Department of Electrical and Computer Engineering, Clemson University, Clemson, SC, USA

12.1 Introduction

Traditional industrial robots are usually separated from humans by placing them in specially designed work cells or restricting their coexistence with barriers or cages, partly for human safety and partly because it is assumed that robot capabilities and human abilities are largely mutually exclusive. This perspective has shifted substantially in recent years, with the shared theme that the skills of people and robots are mainly complementary if they properly used (Ding et al. 2013). Recent advances in robotics (e.g. Kinova, Kuka iiwa) have made it possible for humans to securely share the workspace of robots. Human–robot cooperation (HRC), in which robots and human coworkers share the workspace and collaborate same activities, has sparked a lot of attention (Nicora et al. 2021; Fryman and Matthias 2012). In the current HRC, the strengths of robotic systems (e.g. accuracy, power) and human collaborators (e.g. sensing, decision-making) may be combined to overcome the robots' and humans' inherent limitations (Ding et al. 2013; Fryman and Matthias 2012).

Manufacturing accounts for a sizable component of the country's GDP (Baily and Bosworth 2014). Assembly is a manufacturing stage when the most value may be added (Baily and Bosworth 2014; Rahman et al. 2016). Manual assembly in manufacturing is inefficient, time-consuming, and hazardous to workers' health and safety. For assembly, industrial robotic and automation devices may be less flexible and highly expensive (Fryman and Matthias 2012). The characteristics of robots and human worker capabilities appear to be complementary, and collaboration between robots and humans may make assembly jobs more flexible, adaptable, and productive (Fryman and Matthias 2012). Small-scale, flexible manufacturing processes require more collaboration than large-scale industrial automation because small-scale processes are more unstructured, and resource and equipment requirements are less predictable since operations and product specifications change often (Walker et al. 2015). As a result, there is a rising focus on HRC in industrial manufacturing assembly (Tan et al. 2009). In light of this possibility, HRC in assembly has been a hot topic of research, with researchers delving into many elements of HRC in assembly (Sadrfaridpour and Wang 2017; Kim et al. 2019; Doltsinis et al. 2019; Zanchettin et al. 2018). Despite a few beginning endeavors (Walker et al. 2015; Rahman et al. 2015; Sadrfaridpour et al. 2018), HRC research in the manufacturing industry assembly is still in its infancy.

Cyber–Physical–Human Systems: Fundamentals and Applications, First Edition.
Edited by Anuradha M. Annaswamy, Pramod P. Khargonekar, Françoise Lamnabhi-Lagarrigue, and Sarah K. Spurgeon.
© 2023 The Institute of Electrical and Electronics Engineers, Inc. Published 2023 by John Wiley & Sons, Inc.

For comprehensive, collaborative HRC, payload handovers between robots and humans are required. When a collaborating human requires a tool, an assistance robot may carry it and hand it over (robot to human handover). Such handovers could be used for anything from human hospital care to astronaut extra-vehicular activities (EVAs), according to Diftler et al. (2004). Moreover, human–robot handovers can be from a human to a robot (Sanchez-Matilla et al. 2020), or bidirectional (Strabala et al. 2013). HRC has become an active topic of research as a result of handovers, and substantial contributions have been made to handover developments (Diftler et al. 2004; Strabala et al. 2013). Handovers from robots to humans are becoming increasingly important (Aleotti et al. 2012; Yamada et al. 2020). Strabala et al. (2013) created human–robot handovers based on social signs seen in human–human handoffs. The goal of Aleotti et al. (2012) was to determine the most comfortable stances for robot–human handovers. In Mainprice et al. (2012), motion planning for robot–human handovers was proposed. In Diftler et al. (2004), for example handovers between humanoid robots and astronauts were shown. HRC in the assembly that may need handovers of items (e.g. assembly pieces, tools) requires successful and effective payload handovers between people and robots. As the tasks include human coworkers, the safety considerations in handovers in a collaborative assembly are quite important. When a robot hands over a payload to a person, the problem of safety becomes crucial. In contrast to people, a robot's handover motion and configuration are difficult to modify in changing or uncertain scenarios (Basili et al. 2009).

Considering the human-in-the-loop control systems, as defined in Chapter 1 (Samad 2023), one can list human–robot handovers in the human-in-the-plant category. Recall that the human is part of the "plant" and subject to the control actions in this category. The goal of the robotic system is to either receive or deliver an object from or to the human with a specific set of performance, safety, human factors, and other objectives and constraints. The plant is the surrounding environment of the robot including but not limited to the manipulated object and human. In a simple case, the robot needs to consider the human presence, as part of the plant in the cyber–physical–human system (CPHS), for ensuring collision-free motion planning and possible emergency stop during execution. In a more advanced controller, where additional objectives such as performance, efficiency, and human factors are included, the system still may be categorized as human-in-the-plant. In these cases, the plant still includes the human and the manipulating object. However, if human control was to be added to the control loop, for example through gesture or voice commands, the human-in-the-loop controller category of such a system may switch from human-in-the-plant to human-in-the-controller and even human–machine-control symbiosis.

The majority of human–robot interaction (HRI) safety research has focused on reducing collision impact forces. Designing compliant robotic manipulators with variable stiffness mechanisms (Jujjavarapu et al. 2019) or developing contact-based adaptive switching controllers (Cao et al. 2019) are only a few examples. Although several works focused on collision detection without the need for any additional sensors, such as Zhang et al. (2021), these systems typically require additional sensors to interface the physical interaction and potential consequences. Furthermore, these approaches only evaluate the problem during impact and ignore the handover issue. Handovers between humans and robots, particularly robot-to-human handovers in HRC assembly, are still a work in progress (Walker et al. 2015; Rahman et al. 2016). The issue of trust is overlooked in the state-of-the-art HRC in handovers (Strabala et al. 2013; Diftler et al. 2004). The tendency of a human to rely on or believe in the collaboration of a robot for collaborative tasks (e.g. handovers) is referred as trust (Billings et al. 2012). As the human may not be willing to collaborate with the robot if the robot is not trusted, a satisfactory level of trust in the robot is required. Not only in human–robot systems but also in many agent-based systems, trust is a critical issue. For example, a fuzzy logic technique was used to model customer confidence in vendors in E-commerce transactions (Wang et al. 2015), and a quantitative analysis of security policies based on dynamic trust evaluation on users in a cloud environment was carried out (Yao 2015). Human trust in robots has been reported in many HRI studies (Billings et al. 2012; Hancock et al. 2011) but few HRC studies, e.g. Kaniarasu and Steinfeld (2014). For example Khavas et al. (2021) considered moral trust and performance trust in HRI and Mahani et al. (2020) developed a Bayesian inference model for human trust in multirobot system. However, the current HRC works involving trust are related neither to manufacturing nor to handovers except the preliminary trust-based HRC in assembly in manufacturing proposed in Rahman et al. (2015) and Walker et al. (2015). The well-developed modeling techniques of human trust in robot and methods of real-time measurement of trust are not available in the literature. Apart from a very few preliminary studies, e.g. Rahman et al. (2015), Walker et al. (2015), and Sadrfaridpour and Wang (2017), there is almost no study formally modeling a robot's trust in the collaborating human. Robot trust in a human co-worker is a new concept, and we believe that robot trust in human may enhance the transparency and predictability of the robot's states, behaviors, and actions to the human, which may improve human–robot team fluency for handovers (Walker et al. 2015; Rahman et al. 2016; Hoffman 2019). Knowledge of collaborating robot trust in human may reduce human's cognitive workload, and the human may devote more cognitive resources to handover tasks instead of worrying about or trying to explain the robot behaviors and actions. However, robot trust in collaborating human, especially in payload handovers in assembly, has not received priority (Rahman et al. 2016).

A study on trust in human–human handovers reveals that humans adjust their handover postures based on the trust in the handover collaborator. For an uncertain or unpredictable partner behavior with a possible crashing (collision) at handover, people would adjust their postures to mitigate the effects of impact forces (Basili et al. 2009). Similar incidents were reported in Rahman et al. (2009) where human hand trajectories and applied forces for robot–human manipulation tasks changed when the human faced any uncertainty, doubt, or unusual situation about the objects. A similar approach can be applied for safe and compliant robot–human handovers. The compliant behavior of the robot can be either considered in the design of the robot, for example by variable stiffness mechanisms (Song et al. 2020), or in run-time as we propose in this work. A trade-off between impact force reduction and maximum safe speed by dynamic parameter optimization(mass/inertia and flexibility) of variable stiffness robots was proposed in Song et al. (2020). The proposed compliant handovers can be executed via trust-triggered handover motion planning that makes use of kinematic redundancy (Rahman et al. 2016).

We proposed our preliminary ideas for such a compliant robot handover strategy in Rahman et al. (2015, 2016), Walker et al. (2015), and Sadrfaridpour (2018) and extend them here by providing a more detailed derivation of the proposed motion planning strategy and a comprehensive evaluation through humans-in-the-loop experiments. This chapter develops motion-planning algorithms for robot-to-human handover of payload in the human–robot collaborative assembly tasks based on the trust of robot in human and exploits kinematic redundancy benefits.

12.2 The Task Context and the Handover

We work on human–robot collaborative assembly tasks in flexible small-scale manufacturing processes (Rahman et al. 2016; Sadrfaridpour and Wang 2017; Kim et al. 2019; Doltsinis et al. 2019; Zanchettin et al. 2018). Assumedly, the assembly task is divided into a number of smaller tasks. The subtasks are assigned to the robot, the human, or both based on a subtask allocation optimization scheme (Rahman et al. 2015). Each subtask may be performed in a pre-specified

Figure 12.1 Payload (a screwdriver) hand over to the human coworker in a collaborative (assembly) task.

sequence (Rahman et al. 2016). Once the subtask allocation is determined, the agents are required to complete the subtasks given to them in order, keeping up with one another (Rahman et al. 2016). We assume that human–robot payload (e.g. material, equipment, tool, and accessory) handovers occur during the human–robot collaborative assembly. Though human–robot handover is bidirectional, we here consider the unidirectional handover from a robot assistant to a human, as illustrated in Figure 12.1. It falls in the category where both agents are assigned to perform a single subtask simultaneously as the robot gives the payload (Aleotti et al. 2012; Rahman et al. 2016).

We assume, initially the robot needs to pick the payload from a fixed location, stably grasp the payload before hands over the payload to the human, and the shared workspace between the human and the robot for the collaborative assembly can be suitably used for the handover (Tan et al. 2009; Aleotti et al. 2012; Rahman et al. 2016). The core idea is that the robot becomes able to configure the handover (e.g. moving more cautiously, producing *braced* handover configurations) utilizing the kinematic redundancy so that the robot can avoid dangerous or damaging incidents due to unprepared human actions when robot trust in human diminishes to certain levels.

12.3 The Underlying Trust Model

The trust-based motion planning technique is enabled by a computational representation of the robot's trust in its human coworker. Most trust models study human trust in robots (or systems, machines, humans), e.g. Lee and Moray (1994), Cho et al. (2015), Ting et al. (2021), Pliatsios et al. (2020), Ahmad et al. (2020), Zahi and Hasson (2020), and Lee and Lee (2021). Robot trust in humans is usually not considered except a few initiatives, e.g. Rahman et al. (2015) and Walker et al. (2015). Recently, a unified bi-directional model for natural and artificial trust in human–robot collaboration was proposed in Lee and Lee (2021). The authors in Wang et al. (2022) proposed a model of robot trust in human as a function of the human coworker's performance characterizing various factors including safety, robot singularity, smoothness, physical performance, and cognitive performance. However, the effectiveness of these models needs to be evaluated in actual human–robot collaborative tasks in assembly in flexible manufacturing.

Human trust in robot may be influenced by the factors of robot, task, working environment, and the human (Hoff and Bashir 2015). Trust is a perceptual issue, and the humans have an actual perception of their trust in the robot. However, it is not possible to give the robot a similar perception of its trust in the human. Recent analysis reveals that robot performance and faults are correlated with human trust in the robot (Hancock et al. 2011). Lee and Moray (1994) used a regression model to identify the factors of human trust in robot (automation) and proposed a time-series model as a function of robot performance and fault factors to model the trust. Trust may depend on many factors. We here use human performance and fault factors only to develop a computational model of trust. Such models may not reflect the actual trust but may help incorporate trust in robot–human handover. A general computational model of trust may be expressed as

$$
\begin{aligned}
\text{Trust}(k) = {} & \gamma_0 \, \text{Trust}(k-1) + \gamma_1 \, \text{Performance}(k) \\
& + \gamma_2 \, \text{Performance}(k-1) + \omega_1 \, \text{Fault}(k) \\
& + \omega_2 \, \text{Fault}(k-1) + \tilde{q}
\end{aligned}
\tag{12.1}
$$

with k as the time step, $\gamma_0, \gamma_1, \gamma_2, \omega_1, \omega_2$ as constants dependent to the human–robot system, and $\tilde{q}$ as noise (Lee and Moray 1994). We assign robot trust in the human similar to (12.1). We consider performance and fault as human performance (P_H) and fault status (F). The constant coefficients ($\gamma_0, \gamma_1, \gamma_2, \omega_1$, and ω_2) depend on the collaborative task, the robot, and the human. Equation (12.1) may be updated at every discrete time step k based on measures of P_H and F_H. We normalize the value of T between 0 (no trust) and 1 (full trust).

12.4 Trust-Based Handover Motion Planning Algorithm

This section introduces the trust-based novel handover motion planning algorithm's strategy, explains the novel approaches, and illustrates the strategy for real cases.

12.4.1 The Overall Motion Planning Strategy

If the trust levels drop to below prespecified thresholds, the robot may produce handover motion trajectories different from the "normal or default handover motion" based on the trust levels. The main concern is the generation of impulse forces between human (hand) and robot (end-effector) through the payload during the handover. Unplanned relative motions between the giver (robot) and the receiver (human) during the handover may result in a sudden contact that may generate impulse forces and cause damage to the payload or the robot, and potential injury to the human. To address this issue, we here adopt a hypothesis as follows (see Hypothesis 12.1):

Hypothesis 12.1 *Suppose human makes errors and hence robot trust in human reduces in a human–robot collaborative task (including robot to human handover of payloads). In that case, the human may show premature and unplanned hand motion for receiving the payload from the robot. The amount of reduction in the robot trust in the human may be proportional to the errors, and vice versa.*

In order to make the aforementioned strategy for the robot a reality, we take a two-pronged approach: (i) we loosen the constraints on the robot end-effector trajectory to allow the end-effector to move toward the human for the handover, but the robot adopts a more cautious approach in terms of both space and time; and (ii) we change the robot posture (configuration) to form a *brace*. This approach may ultimately improve the total work productivity and not just the efficiency (Kim et al. 2019).

12.4.2 Manipulator Kinematics and Kinetics Models

The manipulator Jacobian relationship (Chiaverini et al. 2008) is given by (12.2), where x is the end-effector position/orientation in the task space, and q is the corresponding joint space configuration as expressed in (12.3). Here, x, y, z are end-effector position, ϕ_x, ϕ_y, ϕ_z are end-effector orientation, and $J(q) \in \mathbb{R}^{m \times n}$ is the Jacobian relating n-dimensional configuration (shape) to

m-dimensional end-effector coordinates. The Jacobian is known or can be easily calculated for industrial manipulators

$$\dot{x} = J(q)\dot{q} \tag{12.2}$$

$$x = \begin{bmatrix} x & y & z & \phi_x & \phi_x & \phi_z \end{bmatrix}^\mathsf{T} \text{ and}$$

$$q = \begin{bmatrix} q_1 & q_2 & \cdots & q_n \end{bmatrix}^\mathsf{T} \tag{12.3}$$

We presumptively have n independently controllable axes on the robot manipulator. The technique below is easily adaptable to industrial robots with four or five axes or fewer. Manipulator pose defined by x as above can be fully expressed by a 6-dimensional vector. In the motion planning algorithm that follows, we loosen the restrictions on the specification of x by eliminating variables from it, making it generally m-dimensional, where generally $m < 6$ and $m < n$ (Chiacchio 2000; Walker 1994). Under the redundancy assumption ($m < n$), the general solution of (12.2) that minimizes $\|\dot{x} - J\dot{q}\|$ is

$$\dot{q} = J^\dagger(q)\dot{x} + \left(I - J^\dagger(q)J(q)\right)\dot{q}_0 \tag{12.4}$$

where $J^\dagger \in \mathbb{R}^{n \times m}$ is pseudoinverse of J, i.e. an n by m matrix satisfying the Moore–Penrose conditions ($JJ^\dagger J = J$, $J^\dagger JJ^\dagger = J^\dagger$, $\left(JJ^\dagger\right)^\mathsf{T} = JJ^\dagger$, $\left(J^\dagger J\right)^\mathsf{T} = J^\dagger J$), I is the identity matrix, and, $\dot{q}_0$ is an arbitrary joint-space velocity.

12.4.3 Dynamic Impact Ellipsoid

We use an existing body of knowledge in the robotics literature to determine the relation between the magnitude and direction of impulsive forces in the task space and robot posture in joint space due to probable collision at their end-effector (Chiacchio 2000; Walker 1994). This concept is based on the synthesis of a m-dimensional ellipsoid in the end-effector space, which is a function of robot configuration via the Jacobian. The ellipsoid axes and relative magnitudes reveal the robot's exposure to end-effector impact in those directions. The impact ellipsoids for rigid-link robot manipulators are explained in-depth in Walker (1994). These dynamic impact ellipsoids are created by taking into account an impulse force acting at the robot's tip for an infinitesimally brief period of time (the time representing the impact of interest). Then, as shown in (12.5), an expression connecting impulse force and change in joint velocity is derived (see Walker (1994) for derivation details). In (12.5), $D \in \mathbb{R}^{n \times n}$ is the inertia matrix of the manipulator, $F \in \mathbb{R}^m$ is the contact impulse force and $\Delta\dot{q} \in \mathbb{R}^n$ is the vector of instantaneous changes in joint velocities (the units of this term are the same as that of acceleration) caused by the impact. Based on (12.5), we can express the contact impulse force acting at the tip of the manipulator as (12.6).

$$\Delta\dot{q} = D^{-1}(q)J^\mathsf{T}(q)F \tag{12.5}$$

$$F = \left(J^\dagger(q)\right)^\mathsf{T}D(q)\Delta\dot{q} \tag{12.6}$$

The dynamic impact ellipsoid, introduced in Walker (1994), is based on the singular value decomposition (SVD) of the matrix $\left(J^\dagger\right)^\mathsf{T}D$, i.e. $U\Sigma V^\mathsf{T} = \left(J^\dagger\right)^\mathsf{T}D$. Relative magnitudes of the principal axes of the dynamic manipulability ellipsoid given by singular values of $\left(J^\dagger\right)^\mathsf{T}D$ in Σ depict the relative amount of impulse forces that the tip of the manipulator may experience in the corresponding directions (column vectors in U)) for changes in joint velocities ($\Delta\dot{q}$). Thus, the ellipsoid is formally defined using (12.7) in the task space (Walker 1994).

$$\{u \in \mathbb{R}^m : u^\mathsf{T}JD^{-2}J^\mathsf{T}u \leq 1\} \tag{12.7}$$

As shown in Walker (1994), there is a strong correlation between the long axis of the impact ellipsoid (corresponded to the highest singular value in Σ) in (12.7) and the orientation of the robot's last link (typically the wrist/hand). For any two colliding bodies, the highest impact force occurs if they move while facing each other, and it reduces as their directions diverge from each other. Now we define u_l as the unit vector of the long axis of the ellipsoid in (12.7), n_r as the unit vector in the direction of the robot's wrist, n_e as the unit vector in the direction of handover, and ϕ_p as the angle between n_r and n_e. For the payload handover, the highest impact force, $F_{\max}$, happens if the robot moves along the long axis of dynamic ellipsoid, i.e. when u_l aligns with the direction of the handover during impact, n_e, so that $u_l \cdot n_e = 1$. In this case, u_l and $F_{\max}$ are in the same direction and one can write

$$F_{\max} = u_l\|F\| = u_l \cdot F \iff u_l \cdot n_e = 1 \tag{12.8}$$

The direction of the robot's wrist is close to the long axis of the dynamic impact ellipsoid (Walker 1994), i.e. $u_l \approx n_r$. Under the same situation described for (12.8), we can rewrite (12.8) as

$$F_{\max} \approx n_r \cdot F \iff n_r \cdot n_e = 1 \iff \phi_p = 0 \tag{12.9}$$

On the other hand, the possible impact force is close to zero if the robot wrist is close to normal to the handover approach.

$$F \approx 0 \iff n_r \cdot n_e = 0 \iff \phi_p = \pm\frac{\pi}{2} \tag{12.10}$$

This is similar to how people prefer to put their hands up in a *cautious* position when trying to find their way in the dark in order to avoid hurting their wrists in the event of an unexpected impact, as described in Basili et al. (2009). We propose that the robot trajectory should be adjusted to allow the wrist to bend in an analogous *cautious* strategy for low robot trust in the human.

12.4.4 The Novel Motion Control Approach

This article attempts to retain the end-effector's original position trajectory but leave the orientation unspecified (Latombe 1990; LaValle 2006). The strategy proposed here may be tailored to result in a new, modified (reduced dimensional) trajectory in the task space as in (12.11), where $\tilde{x}(t) = [x \ y \ z]^\mathsf{T}$. The trajectory modifies and relaxes the constraints on the robot end-effector path allowing its geometric path to deviate from a direct or most task-efficient course. The net effect may produce a *cautious* movement for the wrist/hand (Walker 1994; Thrun et al. 2005).

$$x_{\text{Modified}} \triangleq \tilde{x}(t) \tag{12.11}$$

This approach makes the robot kinematically redundant (da Graça Marcos et al. 2010; Cheng et al. 2015) to be used to develop the proposed trust-triggered flexible handover configurations (Chiaverini et al. 2008). Equation (12.4) provides the general solution to the inverse kinematics problem (at velocity level). For our problem, we solve the inverse kinematics via the iterative pseudoinverse-based algorithm as expressed in (12.12), where $(\nabla F)^\mathsf{T}$ is the gradient of the magnitude of F in (12.6).

$$\dot{q} = J^\dagger(q)\dot{\tilde{x}}(t) + \alpha \left(I - J^\dagger(q)J(q)\right)(\nabla F)^\mathsf{T} \tag{12.12}$$

It is proved in Walker (1994) that this motion planning algorithm follows the modified end-effector trajectory via the first term in (12.12) and uses the second term in (12.12) to exploit kinematic redundancy to minimize the impulse force magnitudes instantaneously. Here, $\alpha = 0$ indicates no configuration compensation. Thus, only the first term in (12.12) is used that corresponds to robot motions solely concerned with following the modified end-effector trajectory (the least square solution of (12.4). However, increasingly higher values of α produce greater weighing on the second term in (12.12) and result in changes in handover configurations that minimize possible impulse forces (Rahman et al. 2016; Walker et al. 2015; Walker 1994). It has been shown in Walker (1994) that (12.12) can be solved without finding (∇F) as

$$\dot{q} = J^{\dagger}(q)\dot{x}(t) + k_1 \left(I - J^{\dagger}(q)J(q)\right) H \left(q - q_d\right) \tag{12.13}$$

where k_1 is a positive number, H is a diagonal matrix with positive numbers on the diagonal, and q_d is the final configuration of the robot with the minimum possible collision force. Therefore, by letting $k_1 = \alpha$ and $H = I$, one can write

$$\dot{q} = J^{\dagger}(q)\dot{x}(t) + \alpha \left(I - J^{\dagger}(q)J(q)\right) \left(q_d - q\right) \tag{12.14}$$

Hence, we propose the robot trajectory generation strategy for low-trust values as in (12.14). The parameter α in (12.14) may be either taken from a fixed set of values (e.g. based on binary or discrete values of T) or continuously varied values (e.g. based on continuously varying values of T). This will produce a braced robot shape profile in the sense of posturing the robot to minimize the magnitude of impulse forces generated through the human hand's inappropriate movements. This strategy is analogous to the way a skilled human postures their arm(s) to catch a ball (Basili et al. 2009; Walker 1994).

12.4.5 Illustration of the Novel Algorithm

We illustrate the motion planning approach using a robot manipulator, as shown in Figure 12.2. This 6-axis manipulator was aimed to collaborate with a human for collaborative assembly and the proposed robot–human handover during the assembly (Robots 2020). In this case, the robot hands over an object to its human collaborator (Aleotti et al. 2012). The robot is assumed to be handing over to the human on the horizontal direction along the y-axis, $n_e = [0 \ 1 \ 0]^{\mathsf{T}}$ (as shown in Figure 12.2). The potential impulse force magnitude $\|F\|$ may be modeled as (12.15) (Walker 1994), where μ is a constant reflecting the material quantities of the impacting bodies. We see that the unit vector $n_r(q)$ is the aspect the robot can influence with the impulsive force taking its maximum value at μ (last link aligned with handover approach, $\phi_p = 0$) and minimum value at 0 (last link normal to handover approach, $\phi_p = \pm\pi/2$).

$$\|F\| \propto \mu(n_r \cdot n_e) = \mu \cos(\phi_p) \tag{12.15}$$

As shown in Figure 12.2, ϕ_p is the angle of misalignment of the robot last link n_r to the handover approach in the task space n_e, which can be related to the corresponding joint space configuration for the last link via the manipulator Jacobian (Latombe 1990; Thrun et al. 2005). According to (12.15), the potential impact forces would be the maximum for the handover configuration in Figure 12.2a as the final link is aligned with the handover direction, i.e. $\phi_p = 0$. It may be acceptable if the human knows that the robot's trust in the human is high. In this case, the human may move his/her arm toward the robot in the handover direction n_e with a preplanned hand trajectory to receive the payload. Thus, there is almost no possibility of any impact on the human hand by the end-effector. Thus, the impulse forces can be avoided as the humans do not make any error in their trajectory planning. However, if the robot's trust in the human is increasingly low, then the human may approach the robot in the handover direction n_e, but the human's hand trajectory may be unplanned and immature. This may create violent contact between the human hand and the robot's final link as it is directed toward the human direction. This may cause high impulse forces. Therefore, we adopt the approach as introduced in Section 12.4.4 to modify the robot trajectory in the events of lower robot trust in the human.

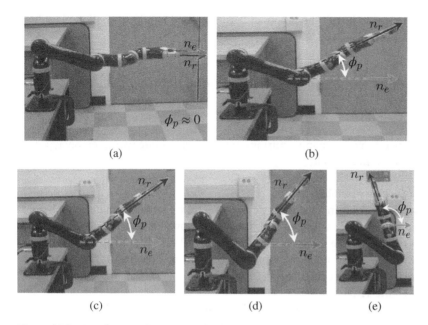

Figure 12.2 The final configurations of the manipulator end-effector for the handover of a payload (a screwdriver) to a human with: (a) high trust, (b) medium trust, (c) low trust, (d) very low trust, and (e) extremely low trust. The arrows along y-axis (n_e) shows the human collaborator receiving the payload (object), and the angular position (in the task space) between the last link holding the object and the handover approach direction is ϕ_p.

For this reason, the trajectory of the end-effector position is unchanged so that the robot can reach the payload to the human properly. However, the end-effector orientation is left free, i.e. the modified end-effector trajectory shown in (12.11) becomes simply the position trajectory $[x\ y\ z]^\mathsf{T}$. In this case, the robot may deviate its final link from n_e so that the contact with the human along n_e either does not take place or reduces. For generating the modified configuration space paths for the robot, (12.11) was used in the first term in (12.14) and the reference configuration, q_d, were used in the second term in (12.14). The parameter α is selected based on robot trust: (i) $\alpha = 0$ for high robot trust in the human (the original robot trajectory, i.e. the first term in (12.12) is chosen), and (ii) $\alpha = \frac{1}{T}$ when robot trust in the human keeps decreasing to below specified thresholds. As shown in Figure 12.2b, in the case of a reduced trust, the final link is misaligned from the y-axis by an angle ϕ_p, which reduces the amount of potential impulse forces based on (12.15). Similarly, based on the varying amount of reduction in the trust, the value of ϕ_p may increase, as shown in Figure 12.2c–e. In Figure 12.2b–d, the handover may occur as the final link is still aligned toward the human with an angle ϕ_p. However, the handover may not take place for Figure 12.2e as the final end-effector direction completely deviates from the direction of the human handover approach, i.e. $\phi_p \geq \pi/2$ due to the least amount of trust of the robot in the human. This configuration can produce minimal impulse forces.

Figure 12.2 shows that the robot handover configuration is transformed to *braced* configurations based on reducing the robot's trust in the human. Such *braced* configurations also indicate the *cautious* movement of the robot to avoid any potential collision. This demonstrates the approach proposed in this article to usefully modify the motion planning of the robot as a function of its trust in the human in the handovers. We know that joint space configuration for the last link is related to task space configuration (e.g. ϕ_p with respect to n_e) via the manipulator Jacobian, and ϕ_p is inversely related to T. An inverse relationship between ϕ_p (measured) and T (computed) in real applications may be sufficient to justify the proposed trust-triggered handover motion planning, given that Hypothesis 12.1 is proven true.

12.5 Development of the Experimental Settings

12.5.1 Experimental Setup

As on our case study in an automotive manufacturing company, we explore two assembly types as feasible representative of HRC in assembly:

12.5.1.1 Type I: Center Console Assembly

To assemble center console products in a laboratory environment, we built a human–robot hybrid cell as shown in Figure 12.3a, where a human and a robot manipulator collaborate to assemble the parts into the final center console products (Tan et al. 2009; Rahman et al. 2016). The robot we used was a Kinova MICO 2-finger manipulator (weight 5.2 kg, max payload 1.3 kg, degrees of Freedom (DOF) 6, reach 0.7 m, power consumption 25 W) (Robots 2020). The human needed to sit on a chair, as shown in Figure 12.3a.

In the hybrid cell, the faceplate parts were kept at "A" and the I-drive and switch row parts were kept at "B" on the table surface. Position "B" was out of reach of the human from their sitting position, but it was within reach of the robot arm. In contrast, position "A" was within reach of the human from their sitting position, but it was out of reach of the robot arm. The robot arm was fastened to the table surface, as shown in the figure. A large computer screen with a key board was placed in front of the human. Taking inspiration from the concepts presented in Rahman et al. (2015), we divided the assembly subtasks in such a way that the human manipulated the faceplate from "A" to "C," and the robot sequentially manipulated (picked and placed) the I-drive and switch row parts from "B" to "C" so that the human could assemble the parts manually at "C" using screws and a screwdriver, and then the human dispatched the assembled center console product manually to "D." This procedure could be repeated to produce many center console parts if the input parts were continuously supplied.

There was an additional screwdriver at "E" (see Figure 12.3a). The robot–human handover occurred as shown in Figure 12.3b when the human needed an alternative screwdriver that the human let the robot know by sending an input to the robot pressing an appropriate key on

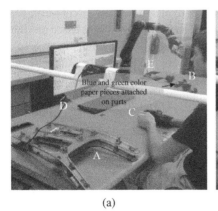

(a)

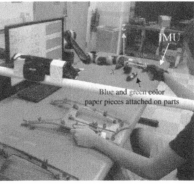

(b)

Figure 12.3 Type I assembly: (a) experimental setup, (b) robot–human handover during the collaborative assembly, and the x-, y-, z-axis directions for the last robot link. (Readers are requested to refer to the online version for color representation.)

the keyboard. Then, the robot stopped the assembly part manipulation, moved to "E" to pick the screwdriver, and handed it over to the human near "C." A Kinect camera was set over the position "C." The human wore an inertial measurement unit (IMU) in his/her dominant hand (wrist) to measure human's assembly speed. See the color paper pieces attached to each part that guided the human to maintain correct orientation of the parts at the time of attaching them with the faceplate (see Figure 12.3).

12.5.1.2 Type II: Hose Assembly

For hose assembly in the automotive industry, the human manually collects the hoses and the fitting parts (end hoses), and assembles them together manually. If the length is determined to be longer than the reference length, a person reduces the length of the hoses by cutting them with a cutter. HRC may address this task, as we consider herein.

We developed another hybrid cell using a Baxter robot (Robots 2022), as shown in Figure 12.4. The hoses and the fitting parts are initially placed in locations that are within the reach of the robot's left arm. The robot picks and manipulates the hoses and the parts to the human who stands in front of the robot, and the human assembles these and then dispatches the assembled products to another section of the table. The human needs to adjust the length of the hoses with a cutter if the lengths are larger than the reference length (the white pipe in Figure 12.4). The cutter is initially placed within the reach of the robot. The robot hands over a cutter to the human. The robot then finishes the current manipulation task and goes to the cutter and picks it up, then moves toward the human for handover. The human wears PhaseSpace motion capture system markers in his/her hand. This system measures the 3D motion of the human hand when the human picks the hoses and fits the parts (end hoses).

12.5.2 Real-Time Measurement and Display of Trust

As seen in (12.1), we need real-time P_H and F_H measurements to obtain a real-time measurement of T. We discuss the human performance and robot fault in the following.

12.5.2.1 Type I: Center Console Assembly

Concerning the setup in Figure 12.3, human performance (P_H) was modeled in (12.16), where V_{Hmn} and V_{Han} are the normalized magnitudes (values) for the human's hand/wrist speed for a part when

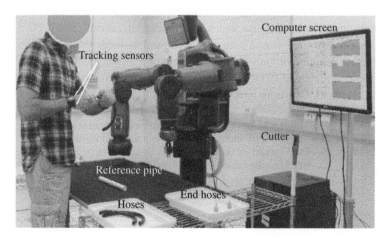

Figure 12.4 Type II assembly: the robot manipulates hoses and parts to the human and hands over a cutter to the human to help the human assemble the hoses and the parts.

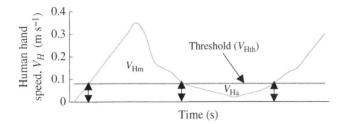

Figure 12.5 Human performance measurement during assembly.

it was manipulated (V_{Hm}), gripped, and released manually, and attached during assembly (V_{Ha}), respectively. w_1 and w_2 are weight factors with $w_2 = 1 - w_1$, $w_1 \in [0,1]$.

$$P_H(k) = w_1 V_{Hmn}(k) + w_2 V_{Han}(k) \tag{12.16}$$

We selected human hand speed during assembly as a human performance measure because achieving high assembly efficiency largely depends on human speed. The IMU (see Figure 12.3) was used to measure human hand speed (V_H), as illustrated in Figure 12.5. V_{Hm} was to be identified when $V_H > V_{Hth}$, where V_{Hth} was a threshold of V_H; otherwise, V_{Ha} was to be identified. V_{Hm} and V_{Ha} were normalized between 0 and 1 to obtain V_{Hmn} and V_{Han}, respectively, which gave the measure of $P_H(k)$ in (12.16) in real-time. w_1 and V_{Hth} were determined based on our experience. P_H varied between 0 (least performance) and 1 (best performance). Effects of human fatigue and the idle time were also reflected in the model indirectly, i.e. the performance could be low if fatigue and idle time were high, and vice versa. Note that the measurements in this article consisted of only the absolute hand speed during the assembly. However, we could consider angular speed, hand positional accuracy, assembly force, and other ergonomic factors for developing the P_H model, making the model more accurate, but we ignored those to avoid complexity.

Incorrectness in part orientation was considered as a measure of human fault status. The incorrectness of the orientation was measured in real-time using the following method. The Kinect ("C" in Figure 12.3a) took the images of the human assembling the parts. Then, the images were used to process in OpenCV. Two pieces of colored paper (see "B" in Figure 12.3a) facilitated the process. OpenCV utilized the colored papers to locate and determine the rough orientation of a part in an image (Suzuki and Abe 1985; Kaplan 1991).

The color blob positions on each assembly part were taken into account to produce a straight line shown in Figure 12.6. The angle (δ) measured between the produced line and a reference was used to indicate incorrect orientation of a part attached with the faceplate. Ideally, $\delta \approx 0$. One of the following three cases could occur. Case 1: one blob was not found (considered as a major fault). Case 2: blobs were believed to be found but not in correct orientations, i.e. $\delta > 0$ (considered as a minor fault). Case 3: blobs were believed to be found in appropriate orientation, i.e. $\delta \approx 0$ (considered as no fault). Case 1 occurred if a wrong part was attached or a right part was attached

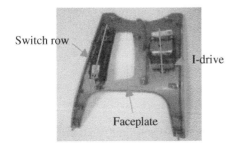

Figure 12.6 The straightline joining two center positions of two color blobs of an assembly part (I-drive or switch row) attached to the faceplate. (Readers are requested to refer to the online version for color representation.)

with wrong sides or views. Case 2 occurred if the right part was attached and, either the screws were put into the wrong holes (the part is attached upside down), or the screws were put into the correct holes but were not tightened enough. Using the above information, $F_H(k)$, the reward score for the human fault status at time step k, was calculated following (12.17).

$$F_H(k) = \begin{cases} 0 & , \quad \text{if case 1 has occurred} \\ 0.5 & , \quad \text{if case 2 has occurred} \\ 1 & , \quad \text{if case 3 has occurred} \end{cases} \tag{12.17}$$

12.5.2.2 Type II: Hose Assembly

For the setup shown in Figure 12.4, human performance (P_H) is modeled as

$$P_H(k) = w_1 V_{\text{Hmn}}(k) + w_2 V_{\text{Han}}(k) + w_3 V_{\text{Hrn}}(k) \tag{12.18}$$

Note that P_H in (12.18) and (12.16) are for two different case studies, but we use the same notation. Generally, P_H and F depend on the task and HRC scenario and need to be defined based on the HRC specifications.

In (12.18), V_{Hmn} and V_{Han} are defined similarly as in (12.16). The normalized magnitude of angular velocity of human hand during fitting an end hose in a hose (V_{Hr}) is V_{Hrn}. Constants w_i are weights between 0 and 1, while $w_1 + w_2 + w_3 = 1$. The linear and rotational speeds are calculated using the position data of the two tracking sensors worn on the hand (see Figure 12.4).

For this task, it is deemed a failure if (i) the fitting components are not installed and fastened appropriately, and (ii) the hose length is not correct. In order to update trust estimation, the experimenter observes the assembly, subjectively, and evaluates $F_H(k)$ using a Likert scale between 0 and 1 with a 0.1 gap between consecutive values, and inputs the score right away into the computer system.

12.5.2.3 Trust Computation

Once F_H and P_H were found in real time, the T could be computed following (12.1). The computed T was displayed on the computer screen (human–computer interface), placed in front of the human subject, as Figure 12.3 shows. The computed trust was updated at every time step k. Figure 12.7 further illustrates the human–computer interface that displayed the computed robot trust in the human T. The graphic user interface provided the trust values from the five most recent time steps

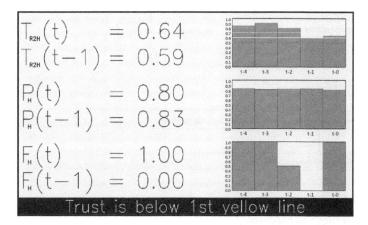

Figure 12.7 Human–computer interface allows for real-time trust display and trust-based alerts to the human coworker. (Readers are requested to refer to the online version for color representation.)

so that the user could see the trend of robot's trust in the human. Depending on the trust values, the green, yellow, and red lines of the trust display (Figure 12.7) displayed varying warning levels to the human.

12.5.3 Plans to Execute the Trust-Triggered Handover Strategy

Two different execution plans were considered for the type I and type II assembly, as described in the following.

12.5.3.1 Type I Assembly
Figure 12.8 shows the plan to execute the proposed strategy. The flowchart considered a few trust-related thresholds to make decisions.

12.5.3.2 Type II Assembly
In this HRC, as described earlier, the human may need to adjust the length of the hoses with a cutter (Figure 12.4), where the cutter is within the reach of the robot. If the human presses the button on the right arm of the robot, the robot will start the hand over process after finishing the current manipulation task. It goes to the cutter and picks it up, then moves toward the human for handover.

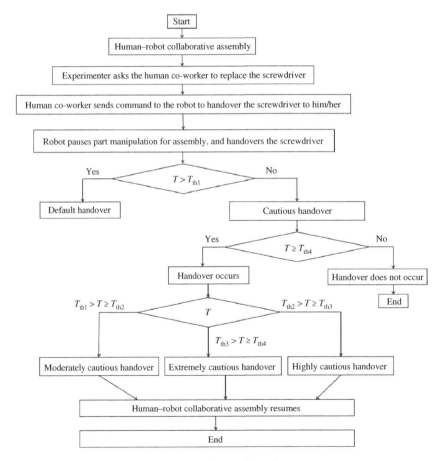

Figure 12.8 Plans to execute the trust-triggered handover.

Contrary to the center console assembly, we do not use discrete values for parameter α in (12.14). For *Type II Assembly*, we define α as

$$\alpha = \begin{cases} \frac{T_{max} - T}{T - T_{min}} \ , & T > \frac{T_{max} + T_{min}}{2} \\ 1 \ , & \text{otherwise} \end{cases} \qquad (12.19)$$

where $T_{max} = 1$ and $T_{min} = 0.5$. If trust is high, then $T = T_{max}$ and $\alpha = 0$. If trust is low, then $T <= \frac{T_{max} + T_{min}}{2}$ and thus $\alpha = 1$.

12.6 Evaluation of the Motion Planning Algorithm

12.6.1 Objective

The experiment's objectives were to evaluate the effectiveness of the novel trust-based collaborative assembly and handover motion planning strategy for improving overall HRI, handover and assembly performance, and safety. We conducted two experimental studies: one for *Type I Assembly* and one for *Type II Assembly*, as described in the following.

12.6.2 Experiment Design

The independent variable was T. The dependent variables were (i) robot handover configuration and motion, (ii) human hand trajectory for receiving payloads during handover, (iii) HRI, (iv) handover success rate, (v) handover and assembly efficiencies, and (vi) safety.

12.6.3 Evaluation Scheme

Human hand trajectories for receiving the payload were captured by the motion capture device worn by the human (Figures 12.3 and 12.4). HRI was classified into the pHRI (meaning physical HRI) and the cHRI (meaning cognitive HRI). Each pHRI criterion was evaluated by human subjects for the assembly including handover using a Likert scale (Carifio and Perla 2007) (from scores 1 to 5 for extremely low to very high). The pHRI criteria given in Table 12.1 guided the subjects to evaluate the pHRI. Two criteria were used to express the cHRI: (i) human trust in robot, (ii) cognitive workload. Human trust was evaluated using the 5-score Likert scale, and the cognitive workload was evaluated applying the NASA TLX (Rahman et al. 2015) by human subjects.

Table 12.1 Physical HRI assessment metrics.

pHRI metric	Explanation
Transparency	Displaying information (warning messages, status) about human's physical performance and fault, and resulting robot trust in the human
Naturalness	The human experienced normalcy and intuitiveness when physically working with the robot for the assembly
Engagement	Amount of human's physical interaction with the robot
Cooperation	The degree of cooperation, sense of teamwork, and team fluency experienced by the human during collaboration in the assembly

For the handover success rate, ϵ_{hsr}, in (12.20), h_f is the total number of failed handover trials and h_t is the total number of handover trials. For the safety in the handover, ϵ_s, in (12.21), h_i is the total number of handover trials when the human co-worker experiences impact forces. For the handover efficiency, λ_h in (12.22), τ_{mtsh} is the targeted time for a handover trial, and τ_{mrh} is the recorded time for a handover trial. The assembly efficiency, λ_a, in (12.23), τ_{mtsa} is the targeted time and τ_{mra} is the recorded time for an assembly (including handover) trial. All criteria were to be evaluated through the experiment.

$$\epsilon_{\text{hsr}} = \left(1 - \frac{h_f}{h_t}\right) \times 100 \tag{12.20}$$

$$\epsilon_s = \left(1 - \frac{h_i}{h_t}\right) \times 100 \tag{12.21}$$

$$\lambda_h = \frac{\tau_{\text{mths}}}{\tau_{\text{mrh}}} \times 100 \tag{12.22}$$

$$\lambda_a = \frac{\tau_{\text{mtsa}}}{\tau_{\text{mra}}} \times 100 \tag{12.23}$$

12.6.4 Subjects

We first conducted the experimental study for *Type I Assembly* recruiting 20 human subjects (university students). We recruit an additional 10 students and conducted the experiments for the *Type II Assembly*. In both studies, the human subjects were divided into two groups/teams (Group I, Group II). The experiments were approved by the Institutional Review Board (IRB).

12.6.5 Experimental Procedures

Each subject independently practiced the subtasks allotted to the human in Phase I, and the robot was trained to carry out the subtasks assigned to it. The purpose of the simulated procedures was to acquaint the subjects with the processes and eliminate any residual learning effects. In order to calculate the standard times for the entire assembly task (including the handover) and the handover only, the time needed by each subject to complete each allocated subtask and handover was recorded separately. To keep the idle time at zero, the robot's manipulation speed was modified to match that of the average human speed (Rahman et al. 2016).

In Phase I, we calculated the reward scores for human performance and fault status so that those score values could be used for P_H and F_H for computing T initially following (12.1) during Phase II. In Phase I, we determined the values of the constants used in the trust model following the ARMA (Sadrfaridpour et al. 2016; Sadrfaridpour and Wang 2017), as given in Table 12.2. Unlike the human trust in the robot (Kaniarasu and Steinfeld 2014), for the robot trust in the human T, we did not consider the prior robot trust (i.e. $\phi_0 = 0$ in (12.1)). As $\Delta t = 30s$, the measurements could be treated

Table 12.2 Constants in (12.1) and trust thresholds.

Constants	Value	Trust thresholds	Value
γ_1	0.472	T_{th1}	0.8
γ_2	0.066	T_{th2}	0.7
ω_1	0.419	T_{th3}	0.6
ω_2	0.043	T_{th4}	0.5

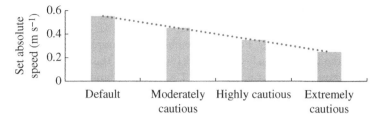

Figure 12.9 Absolute handover speeds set for different strategies.

as "near real-time." We ignored $\tilde{q}$ for z-axis. We determined the trust thresholds ($T_{\text{th}i}$, $i = 1, 2, 3, 4$) used in Figure 12.8, as given in Table 12.2.

We set the absolute handover speeds for different handover configurations for different trust ranges (cautious levels) as shown in Figure 12.9. We extend our effort for more comprehensive implementation of handover motion strategy in (12.12) in *Type II Assembly* by using (12.19).

In Phase II, we performed experiments for the collaborative assembly and handover (Figure 12.3) for two different protocols separately:

  i. Novel motion planning approach based on trust (*assembly and handover with trust*)
 ii. Motion planning approach without considering trust (*assembly and handover without trust*)

The details of these two protocols for the assembly tasks were slightly different. We first describe them for *Type I Assembly* and then for the other type.

12.6.5.1 Type I Assembly

The detailed procedures for this type assembly were presented in Rahman et al. (2016).

12.6.5.2 Type II Assembly

After conducting the experiments for *Type I Assembly*, the benefits of showing the regular trust display were already proven in type I assembly, we here used the regular trust display for both experiment protocols, and focused on handover strategy only.

As discussed earlier, the robot motion trajectory for handover was planned in real-time using (12.14), where α is calculated following (12.19) based on the near real-time trust values. In *assembly and handover without trust*, the trajectory was planned using $\alpha = 0$. The reference pose is the final pose of the end-effector with the minimum collision force, i.e. the cautious pose with minimum trust. Figure 12.10 shows the robot's handover configurations in extremely low- and high-trust conditions for *Type II Assembly* task.

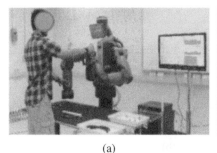
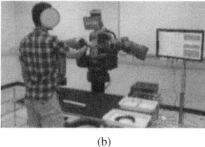

        (a)   (b)

Figure 12.10 Robot's handover configurations in extremely (a) low- and (b) high-trust conditions in representative trials for *Type II Assembly*.

12.7   Results and Analyses, Type I Assembly

We first discuss the results for the *Type I Assembly*. For this assembly type, Figure 12.11a shows that the mean linear position along the y-axis of the last robot link (end-effector) had a direct relationship with trust. For example, when the trust was high (e.g. $T = 0.8$), the handover followed the configuration in Figure 12.2a, where the end-effector tip (screwdriver) was along the y-axis. Thus, the robot used the maximum reach to hand over the payload. Then, with the decreasing trust values, the robot arm attempted to form different *braced* configurations (see Figure 12.2b–e) that caused deviations of the robot arm from the y-axis with the task space angles, ϕ_y. We noted that such deviations from the y-axis caused reductions in the robot's reach along the y-axis, since the handover direction n_e was along the y-axis. Figure 12.11a shows minor fluctuations in linear positions along the z-axis.

It might happen as the end-effector might produce slight upward or downward displacement for the ease of formation of *braced* for different handover configurations for different trust values. The linear position along the x-axis was unaltered as that axis had no contribution to the formation of *braced* configurations. Note that according to our algorithm, the end-effector positions along the x, y, and z-axes should remain unchanged, especially the position along the y-axis should not change. However, it did not happen, as reflected in Figures 12.2 and 12.11. The reason was that the manipulator we used could not form *braced* configurations perfectly due to its structural limitation. Suitably designed manipulators (e.g. hyper-redundant manipulators (da Graça Marcos et al. 2010)) can remove this limitation. The findings partly justify the effectiveness of the proposed motion planning approach.

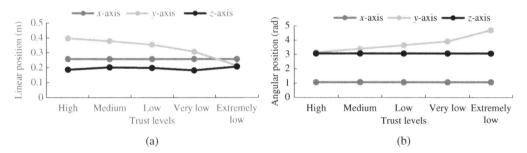

Figure 12.11   Mean (a) linear and (b) angular positions along different axes for the robot's last link for different trust levels.

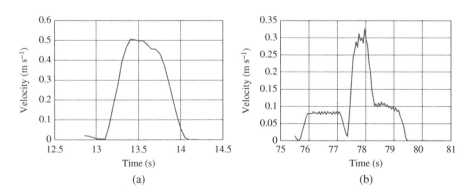

Figure 12.12   Velocity profile of the robot last link for (a) high trust, and (b) low trust situation, as the illustrations (sample from two trials).

Figure 12.11b shows that, along the y-axis, there existed an inverse relationship between ϕ_y and trust, where ϕ_y was related to q_y (joint space configuration for the last link for the y-axis) via direct kinematics equation (Chiaverini et al. 2008). Since the handover direction n_e was along the y-axis, ϕ_y was the angle between robot wrist and handover directions, i.e. $\phi_p = \phi_y$. Thus, the trends in ϕ_y indicated the trends in the impact forces via (9) (Walker 1994). Hence, this also justifies the proposed motion planning algorithm as the potential impulse forces reduced for decreasing trust values through increasing or values (more deviation from the direction of handover, i.e. more *braced* configurations) (Rahman et al. 2016). Figure 12.12 compares the absolute velocities of the last link during the handovers for high-trust and low-trust situations (Table 12.3). The results show that the last link velocity reduced as the robot's trust in the human decreased. We believe that such changes in the velocity profile indicate the robot's *cautious* movement in the low-trust situations, which justifies the proposed motion planning strategy.

Figure 12.13a shows that the human used a very smooth hand trajectory to receive the payload when the T was high. However, as Figure 12.13b–e show, lack of smoothness, jerks, and changes in hand direction (movement of the hand in the backward direction, even a complete U-turn as shown in Figure 12.13e) were observed in the human trajectory as the trust levels decreased. The findings indicate that humans used unplanned, immature, irregular, or hesitating hand trajectories for receiving the payload as the human performance is low with a corresponding low level of robot trust in human, which validates Hypothesis 12.1 (Basili et al. 2009). Such unplanned hand trajectories might cause violent contact between the hand and the robot end-effector that might create impulse forces. In such reduced trust circumstances, as proposed in our motion planning algorithm, the robot took the initiative to save the human from the impulse forces (that might cause injuries) by reducing the impulse forces through generating *braced* configurations of the handover (see Figure 12.2) and *cautious* (slow) movement that is now validated by the information in Figures 12.11 and 12.12.

Figure 12.14a compares the perceived pHRI between *assemblies and handovers with and without trust* (Zheng et al. 2014; Hoffman 2019).

Analysis of variances, ANOVAs (subjects, robot trust), conducted separately on each pHRI criterion (transparency, naturalness, engagement, cooperation, and team fluency) showed that variations in pHRI due to robot trust were statistically significant ($p < 0.05$ for each pHRI criterion, e.g. for transparency $F = 4.66$, $p < 0.05$), which indicated that the differential effects of consideration of robot trust on pHRI. Variations between subjects were not statistically significant ($p > 0.05$ at each pHRI criterion, e.g. for transparency, $F = 3.63$, $p > 0.05$), which indicated the generality of the results.

Table 12.4 indicates a 25.63% reduction in mean cognitive workload and a 37.58% increase in human trust in robot for *assembly and handover with trust* over no trust. Figure 12.14b shows the comparison in the six dimensions of the NASA TLX between assembly and handover with and

Table 12.3 Definitions of trust and cautious levels.

Trust values	Trust levels	Cautious levels	α value
$T \geq 0.8$	High trust	Default/normal	0
$0.8 > T \geq 0.7$	Medium trust	Moderately cautious	0.2
$0.7 > T \geq 0.6$	Low trust	Highly cautious	0.4
$0.6 > T \geq 0.6$	Very low trust	Extremely cautious	0.8
$T < 0.5$	Extremely low trust	N/A	0.8

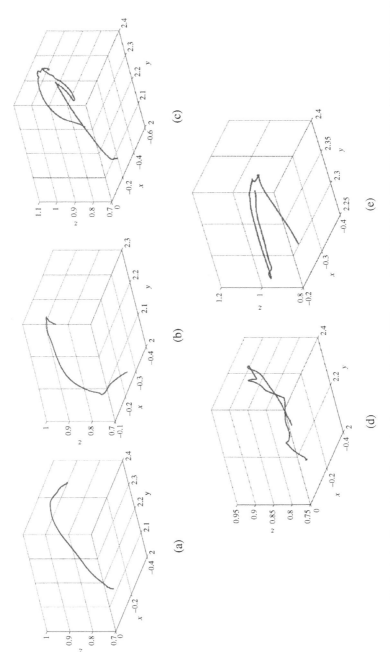

Figure 12.13 Human hand trajectories for receiving the payload, captured by the IMU sensor for (a) high trust, (b) medium trust, (c) low trust, (d) very low trust, and (e) extremely low trust.

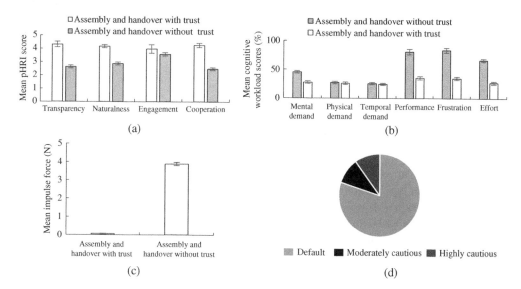

Figure 12.14 Comparison of (a) perceived pHRI, (b) mean ($n = 20$) values of the six dimensions of the NASA TLX, and, (c) mean ($n = 20$) impulse (collision) forces. (d) Statistics ($n = 20$) of different handover trajectories observed.

Table 12.4 Cognitive HRI.

Evaluation criteria	With trust	Without trust
Mean cognitive workload (%)	29.13(1.27)	54.76(2.91)
Mean human's trust (in robot)	4.33(0.45)	3.15(0.27)

without consideration of trust. The results in Table 12.5 indicate 20%, 30%, and 6.73% improvement in handover safety, handover success, and assembly efficiency, respectively, for *assembly and handover with trust* over *assembly and handover without trust*. Figure 12.14c compares the mean impulse forces between assembly and handover with and without trust conditions (Rahman et al. 2016).

The results in Table 12.5 indicate 1.87% less handover efficiency for *assembly and handover with trust*. Figure 12.14d shows that extremely cautious and no handover situations did not occur at all due to the prevailingly high T.

ANOVAs (subjects, robot trust) conducted separately on (i) each cHRI criterion (workload, human trust) and (ii) safety and efficiency showed that variations in cHRI and safety and

Table 12.5 Handover evaluation criteria.

Evaluation criteria	With trust	Without trust
Safety in handover (%)	100	80
Success rate in handover (%)	100	70
Mean efficiency in handover (%)	95.89(1.53)	97.76(2.33)
Mean efficiency in assembly (%)	98.36(2.19)	91.63(1.97)

efficiency due to robot trust were statistically significant ($p < 0.05$ for each criterion, e.g. for cognitive workload, $F = 44.29$, $p < 0.05$), which indicated the effects of consideration of robot trust on cHRI and assembly and handover safety and efficiency. Variations between subjects were not statistically significant ($p > 0.05$ at each criterion, e.g. for workload, $F = 3.58$, $p > 0.05$), which indicated the generality of the results.

12.8 Results and Analyses, Type II Assembly

Figure 12.15 shows the angular and linear positions of the robot end-effector for a typical trial for assembly with trust for type II assembly. The results show that the linear positions along the x, y, and z axis directions were almost unchanged with trust levels, but the angular position (absolute values) along the y-axis (direction of handover, $\boldsymbol{n_e}$, defined in (12.7) significantly changed to produce the braced configuration. The handover speeds also reduced with trust levels as shown in Figure 12.16.

The results thus justify the proposed handover strategy using kinematic redundancy. Figure 12.17 compares the mean impact forces between the assembly with and without trust. We believe that the effectiveness of the handover strategy produced lower impact forces for assembly and handover with trust.

The trends for cognitive HRI and handover evaluation are similar to that of *Type I Assembly* as Table 12.6 shows.

Table 12.7 shows that handover safety, handover success rate, and overall assembly efficiency for the handover with trust are better than that for the handover without trust experiments for

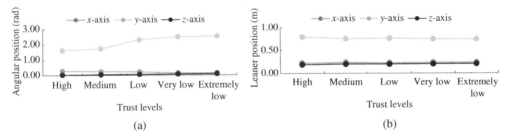

Figure 12.15   Mean (a) linear and (b) angular positions along different axes for the robot's last link for different trust levels during *Type II Assembly*.

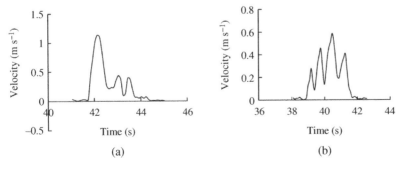

Figure 12.16   Typical absolute velocity profiles of the robot's end-effector during handovers at high (a) and low (b) trust conditions during *Type II Assembly*.

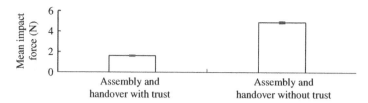

Figure 12.17 The mean impact forces between assembly with trust and without trust in *Type II Assembly*.

Table 12.6 Cognitive HRI for *Type II Assembly*.

Evaluation criteria	With trust	Without trust
Mean cognitive workload (%)	32.1(13.64)	36.46(14.83)
Mean human trust in robot	3.83(0.64)	3.56(0.59)

Table 12.7 Handover evaluation criteria for *Type II Assembly*.

Evaluation criteria	With trust	Without trust
Safety in handover (%)	100	80
Success rate in handover (%)	100	80
Mean efficiency in handover (%)	96.44(0.05)	97.16(0.06)
Mean efficiency in assembly (%)	96.51(0.03)	95.56(0.04)

type II assembly. However, the handover efficiency was reduced slightly for experiments for handover with trust. We also found that pHRI and cHRI for *handover with trust* were satisfactory, and better than that for *handover with no trust*. The results are thus inline with that for *Type I Assembly*.

12.9 Conclusions and Future Work

Handover trajectories for the robot were generated so that the robot could hand over payloads (e.g. a screwdriver) to the human if the human commanded the robot. An algorithm was proposed for the handover motion to adjust the handover configurations through kinematic redundancy based on the robot's trust. Such adjustments in handover configurations were designed to reduce the potential impulse forces between the robot end-effector and the human hand in the handover, thus ensuring reliability and safety. The effectiveness of the proposed trust-triggered assembly and handover motion planning algorithm/strategy was justified for the collaborative assembly through experimental evaluation. The results are novel and may enhance HRI and safety in robot–human handover tasks, especially in the flexible, lightweight assembly in manufacturing.

In the future, we plan to develop data-driven novel control strategies for human–robot collaborative handovers in assembly tasks in manufacturing based on dynamic human–robot bilateral trust.

Acknowledgment

This work is partially supported by the National Science Foundation under Grant No. CMMI-1454139.

References

Ahmad, F., Kurugollu, F., Adnane, A. et al. (2020). Marine: man-in-the-middle attack resistant trust model in connected vehicles. *IEEE Internet of Things Journal* 7 (4): 3310–3322.

Aleotti, J., Micelli, V., and Caselli, S. (2012). Comfortable robot to human object hand-over. *2012 IEEE RO-MAN: The 21st IEEE International Symposium on Robot and Human Interactive Communication*, 771–776. IEEE.

Baily, M.N. and Bosworth, B.P. (2014). Us manufacturing: understanding its past and its potential future. *Journal of Economic Perspectives* 28 (1): 3–26.

Basili, P., Huber, M., Brandt, T. et al. (2009). Investigating human–human approach and hand-over. In: *Human Centered Robot Systems*, Cognitive Systems Monographs, vol. 6 (ed. H. Ritter, G. Sagerer, R. Dillmann, and M. Buss), 151–160. Berlin, Heidelberg, Springer-Verlag.

Billings, D.R., Schaefer, K.E., Chen, J.Y.C., and Hancock, P.A. (2012). Human–robot interaction: developing trust in robots. Proceedings of the 7th Annual ACM/IEEE International Conference on Human–Robot Interaction, 109–110.

Cao, H., He, Y., Chen, X., and Liu, Z. (2019). Control of adaptive switching in the sensing-executing mode used to mitigate collision in robot force control. *Journal of Dynamic Systems, Measurement, and Control* 141 (11): 111003 (12 pages).

Carifio, J. and Perla, R.J. (2007). Ten common misunderstandings, misconceptions, persistent myths and urban legends about Likert scales and Likert response formats and their antidotes. *Journal of Social Sciences* 3 (3): 106–116.

Cheng, C., Xu, W., and Shang, J. (2015). Optimal distribution of the actuating torques for a redundantly actuated masticatory robot with two higher kinematic pairs. *Nonlinear Dynamics* 79 (2): 1235–1255.

Chiacchio, P. (2000). A new dynamic manipulability ellipsoid for redundant manipulators. *Robotica* 18 (4): 381–387.

Chiaverini, S., Oriolo, G., Walker, I.D. (2008). Kinematically redundant manipulators. In: Siciliano, B., Khatib, O. (eds) *Springer Handbook of Robotics*. Springer, Berlin, Heidelberg. https://doi.org/10 .1007/978-3-540-30301-5_12.

Cho, J.-H., Chan, K., and Adali, S. (2015). A survey on trust modeling. *ACM Computing Surveys (CSUR)* 48 (2): 1–40.

da Graça Marcos, M., Machado, J.A.T., and Azevedo-Perdicoúlis, T.-P. (2010). An evolutionary approach for the motion planning of redundant and hyper-redundant manipulators. *Nonlinear Dynamics* 60 (1–2): 115–129.

Diftler, M.A., Ambrose, R.O., Tyree, K.S. et al. (2004). A mobile autonomous humanoid assistant. *4th IEEE/RAS International Conference on Humanoid Robots*, Volume 1, 133–148. IEEE.

Ding, H., Heyn, J., Matthias, B., and Staab, H. (2013). Structured collaborative behavior of industrial robots in mixed human–robot environments. 2013 IEEE International Conference on Automation Science and Engineering (CASE), 1101–1106. IEEE.

Doltsinis, S., Krestenitis, M., and Doulgeri, Z. (2019). A machine learning framework for real-time identification of successful snap-fit assemblies. *IEEE Transactions on Automation Science and Engineering* 17 (1): 513–523.

Fryman, J. and Matthias, B. (2012). Safety of industrial robots: from conventional to collaborative applications. *ROBOTIK 2012; 7th German Conference on Robotics*, 1–5. VDE.

Hancock, P.A., Billings, D.R., Schaefer, K.E. et al. (2011). A meta-analysis of factors affecting trust in human–robot interaction. *Human Factors* 53 (5): 517–527.

Hoff, K.A. and Bashir, M. (2015). Trust in automation: integrating empirical evidence on factors that influence trust. *Human Factors* 57 (3): 407–434.

Hoffman, G. (2019). Evaluating fluency in human–robot collaboration. *IEEE Transactions on Human–Machine Systems* 49 (3): 209–218.

Jujjavarapu, S.S., Memar, A.H., Karami, M.A., and Esfahani, E.T. (2019). Variable stiffness mechanism for suppressing unintended forces in physical human–robot interaction. *Journal of Mechanisms and Robotics* 11 (2): 020915 (7 pages).

Kaniarasu, P. and Steinfeld, A.M. (2014). Effects of blame on trust in human robot interaction. *The 23rd IEEE International Symposium on Robot and Human Interactive Communication*, 850–855. IEEE.

Kaplan, W. (1991). Green's theorem. In: (W. Kaplan) *Advanced Calculus*, Chapter 5, 4e, 286–291. Jones and Bartlett Publishers.

Khavas, Z.R., Perkins, R., Ahmadzadeh, S.R., and Robinette, P. (2021). Moral-trust violation vs performance-trust violation by a robot: which hurts more? *arXiv e-prints*, arXiv–2110.

Kim, W., Lorenzini, M., Balatti, P. et al. (2019). Adaptable workstations for human–robot collaboration: a reconfigurable framework for improving worker ergonomics and productivity. *IEEE Robotics & Automation Magazine* 26 (3): 14–26.

Latombe, J.-C. (1990). *Robot Motion Planning 1991*, vol. 25. Kluwer Academic Publishers.

LaValle, S.M. (2006). *Planning Algorithms*. Cambridge University Press.

Lee, H.-C. and Lee, S.-W. (2021). Towards provenance-based trust-aware model for socio-technically connected self-adaptive system. *2021 IEEE 45th Annual Computers, Software, and Applications Conference (COMPSAC)*, 761–767. IEEE.

Lee, J.D. and Moray, N. (1994). Trust, self-confidence, and operators' adaptation to automation. *International Journal of Human–Computer Studies* 40 (1): 153–184.

Mahani, M.F., Jiang, L., and Wang, Y. (2020). A Bayesian trust inference model for human–multi-robot teams. *International Journal of Social Robotics* 13: 1951–1965.

Mainprice, J., Gharbi, M., Siméon, T., and Alami, R. (2012). Sharing effort in planning human–robot handover tasks. *2012 IEEE RO-MAN: The 21st IEEE International Symposium on Robot and Human Interactive Communication*, 764–770. IEEE.

Nicora, M.L., Ambrosetti, R., Wiens, G.J., and Fassi, I. (2021). Human–robot collaboration in smart manufacturing: robot reactive behavior intelligence. *Journal of Manufacturing Science and Engineering* 143 (3): 031009 (9 pages).

Pliatsios, D., Sarigiannidis, P., Efstathopoulos, G. et al. (2020). Trust management in smart grid: a Markov trust model. *2020 9th International Conference on Modern Circuits and Systems Technologies (MOCAST)*, 1–4. IEEE.

Rahman, S.M.M., Ikeura, R., Nobe, M., and Sawai, H. (2009). Human operator's load force characteristics in lifting objects with a power assist robot in worst-cases conditions. *2009 IEEE Workshop on Advanced Robotics and its Social Impacts*, 126–131. IEEE.

Rahman, S.M.M., Sadrfaridpour, B., and Wang, Y. (2015). Trust-based optimal subtask allocation and model predictive control for human–robot collaborative assembly in manufacturing. *Dynamic Systems and Control Conference*, Volume 57250, V002T32A004. American Society of Mechanical Engineers.

Rahman, S.M.M., Wang, Y., Walker, I.D. et al. (2016). Trust-based compliant robot–human handovers of payloads in collaborative assembly in flexible manufacturing. *2016 IEEE International Conference on Automation Science and Engineering (CASE)*, 355–360. IEEE.

Robots (2020). Jaco. https://robots.ieee.org/robots/jaco/ (accessed 30 January 2022).

Robots (2022). Baxter. https://robots.ieee.org/robots/baxter/ (accessed 30 January 2022).

Sadrfaridpour, B. (2018). Trust-based control of robotic manipulators in collaborative assembly in manufacturing. PhD thesis. Clemson University.

Sadrfaridpour, B. and Wang, Y. (2017). Collaborative assembly in hybrid manufacturing cells: an integrated framework for human–robot interaction. *IEEE Transactions on Automation Science and Engineering* 15 (3): 1178–1192.

Sadrfaridpour, B., Saeidi, H., Burke, J. et al. (2016). Modeling and control of trust in human–robot collaborative manufacturing. In: *Robust Intelligence and Trust in Autonomous Systems* (ed. R. Mittu, D. Sofge, A. Wagner, and W. Lawless), 115–141. Boston, MA: Springer.

Sadrfaridpour, B., Mahani, M.F., Liao, Z., and Wang, Y. (2018). Trust-based impedance control strategy for human–robot cooperative manipulation. *Dynamic Systems and Control Conference*, Volume 51890, V001T04A015. American Society of Mechanical Engineers.

Samad, T. (2023). Human-in-the-loop control and cyber–physical–human systems: applications and categorization. In: *Cyber–Physical–Human Systems: Fundamentals and Applications* (ed. A.M. Annaswamy, P.P. Khargonekar, F. Lamnabhi-Lagarrigue, S.K. Spurgeon). Hoboken, NJ: John Wiley & Sons, Inc.

Sanchez-Matilla, R., Chatzilygeroudis, K., Modas, A. et al. (2020). Benchmark for human-to-robot handovers of unseen containers with unknown filling. *IEEE Robotics and Automation Letters* 5 (2): 1642–1649.

Song, S., She, Y., Wang, J., and Su, H.-J. (2020). Toward tradeoff between impact force reduction and maximum safe speed: dynamic parameter optimization of variable stiffness robots. *Journal of Mechanisms and Robotics* 12 (5): 054503 (8 pages).

Strabala, K., Lee, M.K., Dragan, A. et al. (2013). Toward seamless human–robot handovers. *Journal of Human–Robot Interaction* 2 (1): 112–132.

Suzuki, S. and Abe, K. (1985). Topological structural analysis of digitized binary images by border following. *Computer Vision, Graphics, and Image Processing* 30 (1): 32–46.

Tan, J.T.C., Duan, F., Zhang, Y. et al. (2009). Human–robot collaboration in cellular manufacturing: design and development. *2009 IEEE/RSJ International Conference on Intelligent Robots and Systems*, 29–34. IEEE.

Thrun, S., Burgard, W., Fox, D. et al. (2005). *Probabilistic Robotics*, vol. 1. Cambridge: MIT Press.

Ting, H.L.J., Kang, X., Li, T. et al. (2021). On the trust and trust modeling for the future fully-connected digital world: a comprehensive study. *IEEE Access* 9: 106743–106783.

Walker, I.D. (1994). Impact configurations and measures for kinematically redundant and multiple armed robot systems. *IEEE Transactions on Robotics and Automation* 10 (5): 670–683.

Walker, I.D., Mears, L., Rahman, S.M.M. et al. (2015). Robot–human handovers based on trust. 2015 2nd International Conference on Mathematics and Computers in Sciences and in Industry (MCSI), 119–124. IEEE.

Wang, G., Chen, S., Zhou, Z., and Liu, J. (2015). Modelling and analyzing trust conformity in e-commerce based on fuzzy logic. *WSEAS Transactions on Systems* 14: 1–10.

Wang, Q., Liu, D., Carmichael, M.G. et al. (2022). Computational model of robot trust in human co-worker for physical human–robot collaboration. *IEEE Robotics and Automation Letters* 7 (2): 3146–3153.

Yamada, N., Yani, M., and Kubota, N. (2020). Interactive adaptation of hand-over motion by a robot partner for comfort of receiving. *2020 IEEE Symposium Series on Computational Intelligence (SSCI)*, 1899–1904. IEEE.

Yao, L. (2015). Trusted access control based on FCE of user behavior in cloud environment. *WSEAS Transactions on Computers* 14: 629–637.

Zahi, A.H. and Hasson, S.T. (2020). Trust evaluation model based on statistical tests in social network. *2020 International Conference on Advanced Science and Engineering (ICOASE)*, 1–5. IEEE.

Zanchettin, A.M., Casalino, A., Piroddi, L., and Rocco, P. (2018). Prediction of human activity patterns for human–robot collaborative assembly tasks. *IEEE Transactions on Industrial Informatics* 15 (7): 3934–3942.

Zhang, T., Ge, P., Zou, Y., and He, Y. (2021). Robot collision detection without external sensors based on time-series analysis. *Journal of Dynamic Systems, Measurement, and Control* 143 (4): 041005 (12 pages).

Zheng, M., Moon, A.J. Gleeson, B. et al. (2014). Human behavioural responses to robot head gaze during robot-to-human handovers. *2014 IEEE International Conference on Robotics and Biomimetics (ROBIO 2014)*, 362–367. IEEE.

13

Fusing Electrical Stimulation and Wearable Robots with Humans to Restore and Enhance Mobility

Thomas Schauer[1], Eduard Fosch-Villaronga[2], and Juan C. Moreno[3]

[1]*Department of Electrical Engineering and Computer Science, Control Systems Group, Technische Universität Berlin, Berlin, Germany*
[2]*eLaw Center for Law and Digital Technologies, Leiden University, Leiden, The Netherlands*
[3]*Neural Rehabilitation Group, Translational Neuroscience Department, Cajal Institute, Spanish National Research Council, Madrid, Spain*

13.1 Introduction

Neurological disorders such as stroke, multiple sclerosis (MS), spinal cord injury (SCI), and traumatic brain injury (TBI) can severely limit the mobility and sensation of the affected people. Related lesions of upper motor neurons (UMNs), the source of voluntary movement control, can cause complete or partial paralysis of the skeletal muscles and spasticity. The latter is characterized by increased muscle tone and velocity-dependent resistance to passive joint movements. SCI may also lead to damage to lower motor neurons (LMN) that innervate the muscles. This damage will cause flaccid paralysis linked with ongoing muscle atrophy. In 2017, the prevalence of stroke was ca. 10 million in the 27 European Union countries plus the United Kingdom (EU28), and the incidence was slightly above 1 million (Deuschl et al. 2020). For comparison, the total population in this region was ca. 515 million in 2017. Almost half of the chronic stroke survivors still suffer from motor impairments despite rehabilitation in the acute phase. Prevalence of the progressing disease MS was nearly half a million in the EU28 region in 2017 (Deuschl et al. 2020). One year earlier, the reported prevalence of SCI and TBI in EU28 was 4.9 million and 2.2 million, respectively (James et al. 2019). The reported numbers underline the need for solutions that can increase the quality of life, enable participation in social life, and mitigate economic burdens caused by neurological disorders (Nicolaisen et al. 2020). Given the extraordinary advances in technology, rehabilitation and assistive technologies are currently emerging, including wearable robotics and neuroprostheses, that help patients to recover mobility by generating lost movements or amplifying residual weak functions.

This book section introduces recent technological developments such as neuroprostheses, spinal cord stimulation, and wearable robotics in Section 13.1. We examine various control challenges and sensor technologies to recognize the user's intention, generated movement, muscular fatigue, and environmental disturbances in Section 13.2. Feedback and learning control are essential solutions for user-individual support and assist-as-needed behavior. In addition, we cover hybrid approaches involving multimodal actuation principles to maximize the advantages and counterbalance the disadvantages of the individual support modalities. Examples of hybrid

Cyber–Physical–Human Systems: Fundamentals and Applications, First Edition.
Edited by Anuradha M. Annaswamy, Pramod P. Khargonekar, Françoise Lamnabhi-Lagarrigue, and Sarah K. Spurgeon.
© 2023 The Institute of Electrical and Electronics Engineers, Inc. Published 2023 by John Wiley & Sons, Inc.

systems connecting wearable robots and electrical stimulation (ES) will illustrate in Section 13.3 the current state of the art and research before discussing applicability and transfer to daily practice in Section 13.4.

13.1.1 Functional Electrical Stimulation

The usage of low-frequency (<1 kHz) electrical stimulation has a long history in treating neurological motor disorders and training muscles after orthopedic surgeries or in sports/gyms. The electrical pulses are usually not applied directly to the muscles but to the LMN, which possess a much lower activation threshold than muscle fibers. The action potentials, artificially elicited in the nerves, will travel toward the muscles and cause their contraction. The stimulation is therefore referred to as neuro-muscular electrical stimulation (NMES). Stimuli are predominantly delivered through self-adhesive electrodes attached to the skin, and the setup may involve just pairs of single electrodes for larger muscle groups. To increase selectivity in muscle activation, electrode arrays with many small stimulation contacts have become popular, especially in the forearm, to initiate fine hand and finger movement (Koutsou et al. 2016; Salchow-Hömmen 2020). A demultiplexer connected to a current source can form in real-time virtual electrodes in the array by combining array contacts electrically or by stimulating them sequentially. The training of muscles in gyms and sports by NMES referred to in this application domain as electrical muscle stimulation (EMS) spawns innovative solutions for electrode fixation by garments. First solutions for health care have already emerged (Moineau et al. 2019; Euler et al. 2022).

The generation of functional movements by NMES is called motor neuro-prosthesis (MNP) or functional electrical stimulation (FES). The best known and oldest representative is the peroneal nerve stimulation during the swing phase for foot lifts to prevent a drop foot in stroke patients. Recent applications range from restoration/support of walking (Schauer and Seel 2018), cycling (van der Scheer et al. 2021; Newham and Donaldson 2007), rowing (Ye et al. 2021), grasping and reaching (Kapadia et al. 2020), swallowing (Schauer 2017), bladder and bowel function (Steers et al. 2017), breathing (Chang et al. 2020) to swimming (Wiesener et al. 2020a,b). Most systems involve surface electrodes. Invasive, elaborated implanted neuroprostheses are relatively scarce but exist for some of the listed applications. They allow a much more selective stimulation.

Goal-oriented repetitive movements play an essential role in the motor relearning process based on neuroplasticity. Many studies endorse the application of FES along with the voluntary drive to enhance therapeutic effects (Barsi et al. 2008; Gandolla et al. 2014). The enhanced afferent feedback facilitated by FES modulates motor cortex function and excitability to enable recovery (Ridding et al. 2000). A possible sensory perception of the stimulation, e.g. in stroke patients, can further communicate to the patient the desired timing at which muscles of interest should be activated (Laufer and Elboim-Gabyzon 2011).

A significant limitation of NMES is the associated rapid fatigue of the muscles addressed. This fatigue relates to the following possible issues: (i) The same motor units (MU) (LMN with connected muscle fibers) close to the electrodes are synchronously activated at frequencies of 25–50 Hz to produce tetanic (smooth) contractions (in comparison, physiologically normal recruitment is asynchronous at lower frequencies to allow MUs to recover); (ii) NMES favors fast-fatiguing MUs due to its lower firing threshold rather than fatigue-resisting MUs. Compared with the physiological recruitment order of MU (the Henneman size principle (see Mendell (2005) and references within)), recruitment with NMES is thought to be inverted (Gorman and Mortimer 1983). Sequential stimulation with electrode arrays or multicontact cuff electrodes can emulate physiologically normal asynchronous MU activation to prolong muscular fatigue (Eladly et al. 2020; Gelenitis et al. 2020). Fang and Mortimer (1991) postulated a strategy for implanted tripolar cuff electrodes with particular pulse forms to achieve a proper MU recruitment in line with the Henneman principle. Another restriction when using surface electrodes is that NMES also stimulates nociceptors of the skin, causing discomfort at medium and pain at high stimulation intensities if patients have some remaining sensation in the motor impaired limbs (e.g. after stroke or incomplete SCI). This may limit the tolerable stimulation intensity and producible forces/torques. However, in most subjects, the sensation is weak enough to generate functional movements without discomfort.

Nociceptive electrical stimulation (NES) of the foot's arch at heel-off can support gait training in hemiparetic stroke patients by eliciting the withdrawal reflex. This promotes better initiation and execution of the swing phase and support of the standing phase of the contralateral leg (Spaich et al. 2014). Muscular fatigue, as caused by FES, is less of an issue due to the indirect activation of the muscle's fibers via reflex arches. However, habituation of the reflex responses should be counteracted, e.g. by control of the stimulation intensity and location.

FES is a strongly nonlinear, time-varying, and uncertain dynamic process (Lynch and Popovic 2008; Schauer 2017). The resulting responses heavily depend on the muscular state of fatigue, the patient's additional voluntary contribution, and the level of spasticity. The three latter are hardly predictable and can change continuously during the application of the FES. Even slight changes in surface electrode placements from day to day might cause significant deviations in the motor responses. Precise control of movements and FES-induced torques over more extended periods of time is still not satisfactory.

13.1.2 Spinal Cord Stimulation

Epidural electrical stimulation (EES) of the lumbar spinal cord has recently attracted strong interest among medical practitioners, patients, and society at large because it can restore walking in individuals after severe SCI (Gill et al. 2018; Angeli et al. 2018; Wagner et al. 2018; Rowald et al. 2022). This form of spinal cord stimulation (SCS) alleviates severe lower limb spasticity (Pinter et al. 2000) and produces and enhances rhythmic and locomotor lower limb activity in otherwise paralyzed legs in individuals with SCI (Wagner et al. 2018; Minassian et al. 2004; Angeli et al. 2014; Minassian et al. 2016). This activity is accomplished by recruiting large-to-moderate-diameter proprioceptive and cutaneous afferents within the lumbar and superior sacral posterior roots (Minassian et al. 2004; Ladenbauer et al. 2010). Initially, such EES was administered by re-purposed implants that were originally designed to alleviate pain by targeting the dorsal column of the spinal cord. With this technology, a subgroup of individuals with spinal cord injuries could perform steps with tonic EES after several months of training with the help of multiple physical therapists (Gill et al. 2018; Angeli et al. 2018).

Novel targeted/biometric phasic EES neurotechnologies apply preprogrammed EES sequences in an open-loop fashion or triggered in a closed loop based on feedback from muscle activity and kinematic sensors to a multielectrode array (Wagner et al. 2018). The spatially selective stimulation of posterior lumbosacral roots can be adjusted in real time with precise timing using wireless links and coinciding with the intended movement. It is postulated that such EES activates motor neurons primarily by recruiting proprioceptive circuits within the posterior roots of the spinal cord. However, direct generation of action potentials in efferent nerves cannot be completely excluded as a possible additional source of movement generation. After a high intensive training (several months and almost daily), participants could independently walk with a front-wheel walker or cycle in ecological settings during spatiotemporal stimulation, i.e. with the implant switched on (Wagner et al. 2018). The training consisted of overground walking with multidirectional gravity assistance to compensate for the initially limited weight-bearing capacities and overall performance, and involved multiple therapists. The individuals also regained some voluntary control over previously paralyzed muscles without stimulation, but not their complete natural movements. With a redesign of the paddle-like formed electrode array, Rowald et al. (2022) could target the ensemble of dorsal roots involved in leg and trunk movements more efficiently and restore more diverse motor activities such as walking, cycling, swimming, and trunk control.

The target neural structures of lumbar EES can also be recruited noninvasively by using transcutaneous spinal cord stimulation (tSCS) (Courtine et al. 2007; Minassian et al. 2011; Hofstoetter et al. 2018). Transcutaneous SCS uses surface electrodes placed on the paravertebral and abdominal skin to generate a current flow through the lower trunk, partially crossing the dural sac (Ladenbauer et al. 2010; Hofstoetter et al. 2018). Independent studies have shown the efficacy of tSCS to ameliorate spasticity and augment voluntary motor control, including locomotion in individuals with SCI (Taylor et al. 2021; Hofstoetter et al. 2014, 2015; Estes et al. 2017; Gad et al. 2017; Sayenko et al. 2019; Hofstoetter et al. 2020; Meyer et al. 2020) as well as multiple sclerosis (Hofstoetter et al. 2021; Roberts et al. 2021). As a clinically accessible and noninvasive approach, tSCS was suggested to hold the potential to develop into a widely used neuro-rehabilitation technique and to serve as a screening tool to estimate individually attainable therapeutic outcomes of EES (Hofstoetter et al. 2020).

Typically, tSCS uses monopolar single-channel stimulation (a small electrode on the back and a large electrode on the abdomen). However, recent studies show that using electrodes arrays (or multiple small electrodes) above the spine enables muscle and body-side-specific activation of afferents and efferents within the lumbar and superior sacral posterior roots (Sayenko et al. 2015; Calvert et al. 2019). Hence, tSCS is at least partially equivalent to targeted phasic EES neurotechnologies, allowing a sensor-triggered activation of locomotor circuitry and motor pools. Moshonkina et al. (2021) and Grishin et al. (2021) were the first to demonstrate this for the generation of walking in noninjured subjects on a treadmill with body-weight support.

Spieker et al. (2020) carried out a feasibility study with combined tSCS and FES in a therapeutic aqua pool to promote locomotion in individuals with motor complete and sensory incomplete SCI. After eight weeks of training, twice in the week, the study participant could ambulate independently through the pool without stimulation – exploiting the inherent aquatic weight relief.

The calibration of lumbar EES and tSCS, i.e. determining the suitable electrode positions and intensities for the activation of the target structures, requires neurophysiological expert knowledge. Electromyography (EMG) recordings must be interpreted at the leg musculature after double stimuli. Salchow-Hömmen et al. (2021) proposed algorithms to automate the calibration of tSCS with electrode arrays to facilitate clinical applications.

13.1.3 Wearable Robotics (WR)

Wearable robotics (WR) is still an emerging field of person-oriented devices intimately connected to the human body to achieve a motor function in diverse application scenarios. This type of robot is attractive as a solution to perform action-oriented tasks, delivering augmentation, assistance, or substitution of motor functions. Through mechanical actuation and sensing, WRs are used as assistive technologies due to their ability to mimic the complex motions involved in human movement accurately. As a result, the past few decades have seen an increasing amount of research focused on developing robotic systems intended to interact with the human body.

This interaction (of the human body) with WRs has been established in foundational literature (Pons 2008) as dual, bidirectional physical (physical human–robot interaction (pHRi)) and cognitive (cHRi) interactions. While these systems have proven useful for specific applications, such as in-clinic rehabilitation, current research in pHRi for WRs is more and more focused on developing lightweight and flexible force interactions with hardware solutions more suitable to a broader range of applications. These include adding compliance to rigid exoskeletons (Bortole et al. 2015) or developing "soft exosuits" (Xiloyannis et al. 2019), including injury prevention, telemanipulation, games, and entertainment. On the cHRi side, efforts are focused on developing means to interpret mechanical and neural signals to establish control methods that integrate WRs as parts of human functioning. However, there are many applications in healthcare, civil, and manufacturing domains in which the qualities of self-directness in task performance are expected from a WR as an autonomous, wearable system. Beyond the point of making these machines more active, adaptive, and functional, the point of increasing their proficiency is to make them more capable of interdependent joint activity with the human wearer. In the past years, we are witnessing an unprecedented number of wearable interactive robotics products designed for the clinic environment (e.g. ReWalk, HANK), consumer market (e.g. ARAIG, AxonVR), and human work environments (e.g. EksoVest, FORTIS, shoulderX). In that sense, a reasonable long-term vision is to expect more intelligent architectures for WR to lay human spaces and well-established scientific bases for this new dimension of interaction.

WR devices have been demonstrated to effectively deliver assistance and therapy to the human upper and lower limbs after neurological injuries such as stroke and spinal cord injury. So far, the improvements in clinical outcomes of these interventions have been established with WRs and have been modest compared with traditional therapy. Multiple control modalities have been proposed to precisely control with safety the delivery of robotic guidance that could help mobility or increase motor function. Control design should aim at robustness against both external disturbances and unexpected states from the human user. Control strategies include adaptive control, admittance control, reinforcement learning, EMG-based control, neuromusculoskeletal modeling-based control, and sliding mode control (Baud et al. 2021). A variety of actuation approaches has been tested to establish direct or indirect actuation in WRs, ranging from traditional direct joint motor actuation that can accurately render power to the limb, as mentioned above, to soft approaches based on cable-driven wearable garments that provide less obstructive interaction with voluntary human movements. The former solutions are convenient for applications that require multiple levels of voluntary movement and could, in some applications, achieve reductions in metabolic cost (Chang et al. 2022).

Exoskeletons have been proposed to assist single or multiple joints of the human body in unilateral and bilateral devices that can target joint mobilization of specific functions with inter-joint coordination. In more demanding scenarios with multiple degrees of freedom, the weight and inertia of the entire system become a key challenge. Thus, mechatronic designs have been proposed to relocate the weight of heavy actuators to reduce the mass and inertia of the exoskeleton system, expecting a reduction in metabolic energy cost and wearability. These solutions are often based on cable-driven approaches that allow for this relocation and remote torque delivery. On the other hand, cable-driven systems are not well suited in some applications due to inherent characteristics, such as matching specific human joint motions (wrist rehabilitation movements), low efficiency (parasitic forces exerted by the cables on the human limb), or dynamic performance (including compliant elements has an impact on systems' bandwidth) (Shi et al. 2021).

Conversely, currently available technologies are challenged by several limitations. Solutions based on interaction controllers and sophisticated human–machine interfaces may need a long preparation time, limiting repeatability or requiring difficult and prolonged learning training. A major design problem is the bulkiness and weight of current exoskeletons that may reduce energy efficiency, acceptability, usability, and access (Søraa and Fosch-Villaronga 2020; Pierce and Fosch-Villaronga 2022), as in the case of purely assistive exoskeletons (such as REX exoskeleton or Atalante). In general, robotic exoskeletons need to solve safety issues related to the suboptimal transmission of forces to tissues, risks of failure (that may lead to stumbles and falls), or unexpected behaviors not handled by pre-programmed controllers. Moreover, given that the interaction with the user and the exoskeleton is also cognitive, there will be many other aspects other than physical safety that will have to be addressed, including legal aspects such as cognitive safety, cybersecurity, and privacy (Villaronga 2016; Martinetti et al. 2021); or ethical aspects such as the experience of vulnerability, body and identity impacts, agency, control, and responsibility (Kapeller et al. 2020). As shown under the Cost Action 16116,[1] these aspects relate to a threefold dimension that includes the WR and the self, the interpersonal perspective or how others relate and interact with the exoskeleton wearer, and the broader impacts that these assistive technologies have on society.

13.1.4   Fusing FES/SCS and Wearable Robotics

Combining an exoskeleton with an FES and/or SCS system results in a hybrid exoskeleton or hybrid neuroprosthesis (cf. Figure 13.1). The latter expression also encompasses the combination of FES/SCS with end-effector-based support systems only connected to distal limb parts for movement support or weight compensation. All hybrid systems aim at distributing the movement assistance of the paralyzed or paretic limbs on several types of actuation/support (e.g. drives like motors, springs, artificially activated muscles, and the patient's residual motor activity). Actuators' distribution and degree of co-operation vary depending on the exoskeleton/end effector (passive, semiactive, or active) augmented by electrical stimulation.

In many deployments of hybrid systems, the action of the electrical stimulation and the support provided by the exoskeleton/end-effector are used for independent tasks based on a divide-and-conquer approach. In other words, systems use electrical stimulation to support some limbs' movements and rely on an exoskeleton/end effector to stabilize, support, or actuate other motions (Dunkelberger et al. 2020; Prattichizzo et al. 2021).

Usually, passive mechanical support systems immobilize, stabilize, or compensate for gravity so that the effect of FES/SCS is more repeatable and reliable (Murray et al. 2018; Lee et al. 2020).

Semiactive exoskeletons/end effectors employ controllable brakes, dampers, and springs (single or combined) to hold the limb in place if required (e.g. (Ambrosini et al. 2019)). These relieve the ES from generating holding torques, which usually lead to rapid muscle fatigue. Transitions in limb posture can be induced by electrical stimulation, gravity, or the release of stored energy from springs.

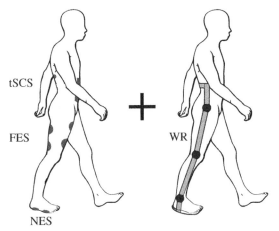

Figure 13.1   Fusion of electrical stimulation techniques and wearable robotics (WR) to form a hybrid exoskeleton/hybrid motor neuro-prosthesis (HMNP) for gait support and restoration (FES, functional electrical stimulation; tSCS, transcutaneous spinal cord stimulation; NES, nociceptive electrical stimulation).

---

1 See https://wearablerobots.eu/.

Active robotic systems can move the joints and limbs through a functional range of motion. This enables more complex integration strategies for electrical stimulation. In exoskeletons, the two actuation strategies can be separately applied on different joints to achieve desired body motion, or co-operating control of the two actuation strategies on the same joint can be realized. As an example of the first approach, many systems use FES for the hand and wrist function to realize grasping and the driven exoskeleton on the arm's proximal joints to generate reaching motions (Choi et al. 2016; Vuckovic et al. 2021). The co-operative approach is currently more established in hybrid exoskeletons for the lower extremities.

Co-operative control at a joint offers several advantages in motion assistance. Firstly, the exoskeleton's drive enables fine control of joint movement and can repeatably deliver power that can compensate for the variant quality of FES/SCS-induced joint movement. On the other hand, muscle action caused by FES/SCS can reduce the energy demand of the wearable exoskeleton device, thereby requiring less powerful joint actuators, which would benefit portability. Secondly, the fusion of WR and FES/SCS may promote functional improvements and neural plasticity because of the intensive, community-based practice involved. For example, such a system would facilitate training in individuals with neurological conditions, something crucial for their motor rehabilitation yet something unachievable with conventional physical-based therapies. Furthermore, the approaches to human–machine interaction put forward by the most recent rehabilitation WRs are not yet taking into account effective strategies for interfacing a MNP and WR to the human body.

In general, FES control systems face the challenge of dealing with time-variant muscle dynamics due to physiological factors such as fatigue. It means that even if the necessary muscle activation patterns for certain motions could be computed (i.e. through whole-body neuromuscular system modeling), how to realize that desired muscle activation patterns in FES is still an open problem, and it is a crucial issue to handle joint torque control in FES. Based on EMG measurements, some researchers have attempted to predict force/torque variations with fatigue. In general, existing techniques rely mainly on physiological findings, which are not yet actively using this information for FES controllers to compensate for muscle fatigue. Fully active hybrid exoskeletons (with powered actuators) allow control of the power delivered at the joint. Moreover, the desired feature enabled by such active hybrid exoskeletons is the delivery of assistance as needed under co-operative controllers that theoretically are thought to adapt to the user's performance, leveraging artificial versus muscle-driven actuation.

SCS provides the opportunity to engage the neuromuscular system of SCI individuals in sustained and active training sessions. It can give even motor complete paralyzed individuals some volitional control over their paralyzed muscles back and triggers locomotion-related pattern generators in the spinal cord – hopefully causing physiologically normal muscle activation with lesser fatigue than caused by FES. Using WRs (and maybe also FES) and weight support under SCS might further improve the therapy outcome and relieve therapists from hard physical work. However, precise control of muscle activation by SCS is still limited compared with FES and currently is under active research.

13.2 Control Challenges

The engineering problems discussed in Sections 13.2.1–13.2.5 challenge the delivery of hybrid actuation in the context of a hybrid FES-robot scenario. The associated human-in-the-loop control problem belongs to the category human–machine control symbiosis, as the patient can be in both "in-the-plant" and "in-the-controller" roles simultaneously (Chapter 1).

13.2.1 Feedback Approaches to Promote Volition

When applied for therapeutic purposes (e.g. after brain damage), assistance with robotic and FES should be adapted in terms of difficulty and challenge of the task to the user's skill. Assistance needs to be shaped to a certain level to promote motor learning via activity-dependent neuroplasticity. For this, robot/FES-mediated training might be properly synchronized and congruent with inputs of the motor system (such as visuomotor and sensory) to efficiently recruit and train nondamaged motor areas for recovery of motor function. Feedback-based methods for neural rehabilitation (including EMG-based biofeedback, virtual reality, sensory-motor stimulation, and others) have been proposed for motor rehabilitation. Feedback modalities could include any combination of haptic, cutaneous, and electrical stimulation at the periphery. Solutions based on virtual reality (VR), augmented reality, and customizable games have been proposed to improve the user interface of WR systems for patient rehabilitation, aiming to increase patients' interest so that they keep performing their exercises (Postolache et al. 2021; Patil et al. 2022). A persistent challenge in these approaches is implementing adaptive visuomotor feedback in a clinically feasible environment to produce a boost of attention to the motor task. Future hybrid WR exoskeletons for neural repair are expected to deliver autonomous training and assistive performances, moving toward optimal visuomotor/sensory feedback delivery.

13.2.2 Principles of Assist-as-Needed

Control layers have been defined as responsible for promoting neural plasticity in the case of neurological injury. At this level, the behavior and response of exoskeletons are programmed to the user's actions, usually based on measured human–robot interaction forces. In this regard, the most widespread control is called the assistive control strategy, in which the exoskeleton provides physical assistance to support the user in achieving a movement (Marchal-Crespo and Reinkensmeyer 2009). It is globally accepted that an assisted-as-needed approach is the most appropriate assistive control strategy to promote neural plasticity and hence motor relearning (Duschau-Wicke et al. 2010). This strategy has been implemented in robotics devices for gait (Duschau-Wicke et al. 2010). This strategy is typically based on impedance-based assistance. It relies on the generation of restoring forces, which are generated using the concept of mechanical impedance when the users deviate from the desired trajectory. As a fixed trajectory can be counterproductive for motor learning (Tucker et al. 2015), and humans present variability in their movement, a dead band often is added to allow more flexibility. Assist-as-needed algorithms encourage effort and concentration and are expected to avoid the slacking hypothesis during the task execution, which favors active user participation (Tucker et al. 2015). On the other hand, another suitable control assistive approach called challenge-based control has been tested for robotic exoskeleton devices. This approach aims to make a task more difficult or challenging by imposing resistance to movement or amplifying the error trajectory during task execution to encourage motor relearning.

13.2.3 Tracking Control Problem Formulation

Some controllers might be optimized to account for tracking objectives that depend on time, while others for tracking phase-dependent variable(s). Hybrid systems incorporating electrical stimulation of muscles often deploy robot controllers with tracking objectives using predefined kinematic patterns. Two extreme scenarios could be sought in applications involving gait pathologies: (i) continuous repetitive training in patients with low volitional control (e.g. body weight-supported treadmill training), and (ii) overground stepping training. Time-dependent trajectory control can be applied for continuous repetitive training. It can rely on fixed reference trajectories or adaptive periodic reference trajectories that can be internally generated into the closed-loop system to achieve steady tracking (Wang 2012). On the other hand, phase-dependent controllers are more suitable for overground stepping training, given the variant gait spatiotemporal characteristics. A suitable approach to synchronize the assistance of the robotic exoskeleton is to use adaptive frequency oscillators (dynamical systems with an oscillatory behavior that can learn the features) with period nature related to several gait features (Qiu et al. 2021). Thus, coupled adaptive oscillators can be entrained to converge with a characteristic periodic signal (e.g. the phase of the adaptive oscillator synchronized with the gait phase).

The objective of the trajectory-tracking controller is to prescribe the robotic joint movement to assist or mobilize the human joint to a given kinematic path. Control approaches are tuned to track joint trajectories; however, these cannot ensure reaching targets in space (targeting a step). In WRs for patients with hemiparesis, a strategy to replay the kinematic trajectory of the sound leg in the affected leg has been proposed. For example, the first and second derivatives of the angular pattern can be computed to derive velocity and acceleration references for the actuators assisting the impaired leg. Moreover, assist-as-needed approaches have been implemented on top of these trajectory-tracking controllers based on a force-tunnel paradigm in which an impedance model is proposed to relate required interaction torque with a prescribed tracking error. Similarly, the force-tunnel paradigm around the desired trajectory has been proposed in hybrid configurations to allow deviation from the kinematic profile based on the response of the FES of muscles around the controlled human joint (del Ama et al. 2014).

Open challenges of such trajectory-tracking controllers are that these controllers are frequently developed and tested mainly for even terrains and can hardly adapt with flexibility to irregular terrains or to respond with safety to unexpected situations, such as obstacles or difficulties in achieving foot clearance. Another critical challenge for more impaired subjects is to ensure reaching targets in space (e.g. targeting a step) instead of targeting a given coordination of internal kinematic trajectories. For these purposes, research can consider environment recognition systems (radar, laser, and camera-based) that could capture the motion of the human–exoskeleton within the environment and set kinematic targets to achieve autonomous motion that could offer both flexibility and safety (Laschowski et al. 2022).

13.2.4 Co-operative Control Strategies

The co-operative control strategy has to cope with a fundamental problem of redundant systems: identical outcomes can be achieved with different involvement of the redundant actuators. In this application, both actuator types (FES and robot) should operate in the same direction and not against each other. First, it seems natural to use the human muscles as much as possible to provoke a training effect. However, this approach will provoke an earlier onset of muscle fatigue. An adaptive distribution of the necessary torque (via a convex combination) between the two actuator types reduces muscle fatigue compared to mere FES. Therefore, the overall co-operative system should be optimized concerning performance, energy consumption, and muscle fatigue while considering both actuators' capabilities and restrictions. The robot yields reliable and fast performance, while increased weight and power consumption need to pay higher values. The artificially activated muscles suffer from a large electromechanical delay (EMD), the time lag between muscle activation and muscle force production, and unreliable performance. In addition, individuals with sensory perception prefer to receive stimulation signals with slow changes in the intensity. Often, optimal control, predictive control, or adaptive feedforward control are used to compensate for the delay in the muscle actuation and to guarantee that the direction of action of the FES aligns with the robotic support. This requires some knowledge of the future reference movement. Such a condition is usually fulfilled for cyclic movements like walking.

Most hybrid MNP can quickly compensate for the effects of muscular fatigue (i.e. drop in torques and joint angles) by simple integral control action that increases motor torques and stimulation intensities up to the technical limits. The stimulation intensities are further limited by the maximally tolerated level of the user. Challenging is to distribute the actuation between robot and muscle to prolong the onset of muscle fatigue and provide recovery periods. It is also possible to change the stimulation strategy (e.g. pulse train composition) upon detection of the first fatigue signs to prevent complete muscle exhaustion (see, e.g. (del Ama et al. 2014; Ha et al. 2016)).

Vallery et al. (2005) reported the first co-operative control strategy for a hybrid MNP which was experimentally validated for the knee joint. Actuated was the joint by an electric motor and the stimulated knee extensor in a sitting position. The concept splits the required net torque profile M_{ref} from a top-level model predictive control (MPC) in the function of its frequency domain characteristics: nearly constant torques are predominantly realized by the muscles and high-frequency torques by the motor. A constraint is the stipulation that both actuators must always produce torque with the same orientation. A simple approach to distribute the torque was chosen, a convex combination using a factor $k \in [0,1]$: the stimulator realizes the part kM_{ref} and the exoskeleton the part $(1 - k)M_{ref}$. The parameter k is adapted based on a frequency analysis of the future M_{ref} profile. Furthermore, k is limited to a certain range to guarantee a certain minimal involvement of each actuator. Figure 13.2 illustrates the control strategy. A trade-off among performance, muscle fatigue, and energy consumption can be achieved by adjustment of the frequency adaptation and the k-limits. However, a detailed analysis was not carried out.

del Ama et al. (2014) realized the first exoskeleton for gait support with a hybrid support of the knee joint involving both knee extensor and flexor. A proportional-derivative (PD) controller of the form

$$\tau(t) = k_s(\theta_{ref}(t) - \theta_{meas}(t)) + k_d(\dot{\theta}_{ref}(t) - \dot{\theta}_{meas}(t)) \tag{13.1}$$

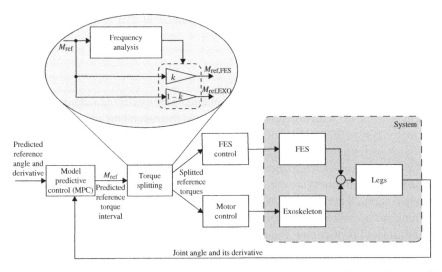

Figure 13.2 Control of a hybrid motor neuroprosthesis for the knee joint in sitting condition. Based on the predicted reference angle trajectory, the model-based torque controller predicts an optimal torque trajectory M_{ref}. This reference torque is split up to the FES-stimulated muscles and the motor-actuated exoskeleton via an adaptive, frequency-dependent convex combination as proposed in Vallery et al. (2005). Source: Adapted from Vallery et al. (2005).

was used to compute the generated torque τ of the electric motor to minimize the error between a reference angle trajectory $\theta_{ref}(t)$ and the measured angle $\theta_{meas}(t)$. In this mechanical setting, the PD controller realizes a desired stiffness and damping of the knee joint (impedance control). The stiffness parameter k_s was modulated dependent on the gait phase – being high at stance and low at swing phase. It was further adjusted to the need of exoskeleton support. Sufficient torque generation by FES will lead to low values of k_s. This enables the mechanical behavior to vary from constrained trajectory control to unhindered motion, allowing to adapt compliance to muscular FES performance. The ES of the knee extensor (quadriceps) was controlled in the stance phase by an empirically tuned proportional–integral–derivative (PID) controller to avoid knee-joint collapse. Iterative learning control (ILC) was applied during the swing phase to adjust the stimulation profile from step to step to minimize the recorded interaction torque profile. After convergence of the ILC, learning stopped and the stimulation intensity profiles were frozen. Then the hybrid motor neuro-prosthesis (NMP) modulated its assistance on a cycle-by-cycle basis. The stiffness was decreased up to the value that allowed a minimum defined knee flexion angle. Furthermore, in the so-called "monitoring state" (ILC off), a muscle fatigue estimation was started. The measured limb-exoskeleton interaction torque was used to calculate the torque–time integral (TTI) during the swing phase to estimate muscle fatigue. A significant muscle fatigue, detected by a certain decay of the TTI, triggers a change in the stimulation parameters to counteract muscle fatigue and to temporarily re-activate the ILC. Figure 13.3 illustrates the entire knee-joint control scheme of this hybrid exoskeleton. The co-operative control approach was successfully validated in healthy subjects (del Ama et al. 2014) and motor incomplete SCI individuals (del Ama et al. 2015).

Ha et al. (2016) proposed a very similar approach for the co-operative control of FES with a powered exoskeleton during level walking for persons with paraplegia. They also used a PD controller to track the desired joint trajectories by the electric motor and an adaptive feedforward controller to adjust the stimulation profiles from step to step to minimize the motor torque contribution required for joint angle trajectory tracking. The FES torque generated at the last step was estimated from the recorded motor torque profiles, and the thus-observed gain in muscle torque generation was used as a fatigue indicator (assuming no volition contribution by the user). When the gain decreased to one-third of the maximum observed gain, stimulation was interrupted for two minutes to allow the muscles to recover. Experimental results from testing the co-operative control system on three motor-complete paraplegics showed that the required motor torque and power could be efficiently reduced due to muscle stimulation.

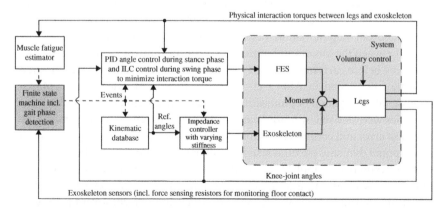

Figure 13.3 Co-operative control approach for the Kinesis hybrid exoskeleton proposed by del Ama et al. (2014). Source: Adapted from del Ama et al. (2014).

In Chang et al. (2022), a cable-driven lower-limb exoskeleton was enhanced by FES of the knee extensors and flexors. Similar to the two previous approaches, the authors used the electric motors of the cable-driven knee-joint to adjust the knee-joint stiffness concerning a knee-joint angle profile throughout the gait cycling using an integral torque feedback controller. However, the FES was now controlled also to achieve kinematic tracking of the knee joint. A Lyapunov-based stability analysis proved exponential tracking of the kinematic and torque closed-loop error systems while guaranteeing that the control input signals remain bounded. The nonlinear kinematic tracking controller is robust for the uncertain and time-varying human–exoskeleton system. The approach was validated in three nondisabled persons instructed to avoid voluntarily contributing to the treadmill walking task. Despite the given guidelines for tuning the control gains, adjusting 10 controller parameters might be challenging for most medical staff and end-users.

Zhang et al. (2017) adopted feedforward control of FES and feedback control of the electric drive as co-operative control strategy in a hybrid active knee exoskeleton. Two muscle groups (quadriceps and hamstrings) were stimulated to generate torque at the knee joint. The method uses a central pattern generator (CPG) with two coupled oscillators to produce reference angles for the knee joint and exoskeleton. The CPG acts as a phase predictor to deal with phase conflicts of motor patterns and realizes synchronization between the two different bodies (shank and exoskeleton). An inverse dynamics model determines the required knee torque to produce the reference knee-joint motion in real time, and an inverse muscle model outputs the nominal stimulation intensity to produce a fraction δ of this torque by the muscles. The exoskeleton compensates for its dynamics (making it transparent to the user) and produces the remaining assistive torque by the feedback of the knee-joint angle and the interaction torque. The innovation is a top-level parameter regulator. It adapts the CPG to synchronize FES and exoskeleton and shifts torque generation between the muscle and the exoskeleton if the muscles fatigue or recover. Under the assumption of cyclic motions, parameter updates occur from cycle to cycle based on the gradients of a defined performance measure. The approach's feasibility was evaluated in experiments with several healthy and paraplegic individuals in a sitting position. One open question is how this concept can be extended to walking tasks.

When controlling multiple joints, more sophisticated strategies are demanded to manage the redundancy of the various actuators. Alibeji et al. (2015) proposed a muscle-synergy-inspired control approach. A subject-specific gait model was used to compute optimal control signals for the multiple actuators and joint angle trajectories to achieve a given step length and duration. The obtained control signals were then dimensionally reduced by principal component analysis (PCA) to extract synergies. For the latter, a low-dimensional controller was designed consisting of an adaptive feedforward controller and a stabilizing robust feedback controller. The approach's feasibility was demonstrated in computer simulations on a 4-degree-of-freedom gait model. The distribution of actuation between the motors and muscles can only be influenced offline by changing the weighting matrices before carrying out the dynamic optimization. The synergy-inspired approach has been further developed in Alibeji et al. (2018). Instead of determining optimal step motion and input signals by dynamic optimization for a complete step for subsequent PCA, dynamic postural synergies were derived by minimizing a convex cost function that minimizes the dynamic posture's position error and control efforts for given reference joint angle trajectories of a half step. Using the computed dynamic postural synergies, dynamic optimizations then compute the optimal synergies' activations to complete a step. Both optimization steps were carried out offline before placing the exoskeleton on a user. During walking, these synergies were used in the feedforward path of the control system. The stimulation levels in the feedforward path were gradually increased based on a model-based fatigue estimate to compensate for the effects of muscle fatigue. A dynamic surface control (DCS) technique with a delay compensation term was used for the feedback controller to address model uncertainty, the activation dynamics, and EMD (Alibeji et al. 2017). It is worth mentioning and surprising that solely adapting the person's height and weight in the generic human–exoskeleton model was sufficient to gain a successful walking trial in one healthy subject and one incomplete SCI individual (Alibeji et al. 2018).

A major disadvantage of the synergy-inspired methods outlined before is the mandatory offline dynamic optimization prior application. To overcome this limitation, Bao et al. (2020) and Molazadeh et al. (2021) proposed a hierarchical control design. A high-level control design was derived to track a time-invariant desired sit-to-stand (STS) movement profile. Bao et al. (2020) used a nonlinear robust feedback controller for the high-level control, while Molazadeh et al. (2021) realized an adaptive feedforward controller by employing neural-network-based iterative learning control (NNILC). The desired standing-up motion of both controllers is governed by a virtual constraint. This approach yields time-invariant joint trajectories that are coupled to a single monotonically increasing state-dependent function. This avoids the design of multiple independent time-based trajectories for the lower-limb joints. Each leg uses an individual controller, but two leg controllers use the same virtual constraint function to maintain coordination between the two legs. The high-level controllers output as control signals the required torques at the hip and knee joint. Then, in Bao et al. (2020) and Molazadeh et al. (2021), low-level MPC controllers allocate the generation of the knee torque to the corresponding electric motors and the FES-activated muscles as initially proposed by Kirsch et al. (2018). A gradient search algorithm was adopted to solve the MPC optimization problem. Weighting parameters in the MPC-related cost function can be adjusted to influence the resulting distribution of the redundant control actions. The MPC employs the user's muscle fatigue and recovery dynamics to determine an optimal ratio between the FES-elicited knee torque and the exoskeleton assist. However, these models need to be identified for an individual user. Unfortunately, the precisely used prediction time horizon and assumptions on the future knee torque profile for the MPC are not apparent in Bao et al. (2020) and Molazadeh et al. (2021). The hybrid concepts by Bao et al. (2020) and Molazadeh et al. (2021) worked successfully in healthy individuals (one SCI individual in Bao et al. (2020) did not respond to FES). The NNILC improved the performance for subsequent STS movements. How realistic such setup is in daily life is questionable, and how NNILC performance will be affected by longer breaks need to be investigated. The power demands for the electric motor decreased compared with pure motor support (Bao et al. 2020), but effects of the weighting parameters in the underlying predictive allocation strategy on avoiding/delaying muscle fatigue for prolonged use of the system have not been explored yet.

One limitation of the above-outlined control strategies for STS is that manual switching of the control between different FES and electric motor allocation levels, as possibly demanded by a user or a therapist, is not foreseen and might lead to the undesired behavior of the MPC allocation controller. Furthermore, detailed model knowledge of the fatigue and recovery behavior is required. To remove these restrictions, Molazadeh et al. (2022) investigated a novel iterative learning neural network-based control law to compensate for structured and unstructured parametric uncertainties in the hybrid exoskeleton model and an MPC-free allocation strategy. Asymptotic stability of the switchable system could be proven and experimentally demonstrated for four healthy and one SCI individual. This co-operative controller will open the way for future fatigue-based update strategies of the allocation levels.

Another hybrid FES-exoskeleton approach to assist STS movements was presented by Alouane et al. (2019). The hip and knee joints were actuated by the exoskeleton using serial elastic actuators (SEAs) that use a torsion spring to enable the measurement of the human–exoskeleton interaction torque. The proposed approach lies within the assist-as-needed control strategy, combining an event-based stimulation trigger of the quadriceps with an admittance control of the exoskeleton.

Based on the detected interaction torque (initiated voluntary by the user or by FES) between the exoskeleton and the human limbs, desired reference angles are calculated by a user-specific admittance model (adapted to the wearer's motor abilities) and tracked by a position controller using the SEAs. Experiments with a healthy subject clearly showed a reduction in the required motor torques when using the hybrid system compared to the robot-only system.

In all experimental studies reported before, surface electrodes were used for artificial muscle activation. In contrast to this, Nandor et al. (2021) presented a hybrid exoskeleton with an implanted stimulation system consisting of 16 chronically indwelling intramuscular electrodes that were connected to two external stimulation boards via percutaneous cables. The exoskeleton's brakes at the hip and knees are locked during stand phases. During swing phases, torque bursts can be administrated synchronously to the reprogrammed stimulation profiles to enhance walking speed, as demonstrated in two SCI individuals (for 10 m walks changing from 0.11 to 0.20 $m\,s^{-1}$ and from 0.11 to 0.41 $m\,s^{-1}$, respectively). Finger switches and weight shift detection sensors control the state machine of the hybrid exoskeleton. A recent simulation study by Makowski et al. (2022) aimed to improve the system's co-operative control by using biologically inspired optimal terminal iterative learning control (TILC) for the swing phase. The objective was to maximize muscular recruitment and activate the motorized exoskeleton bracing only to assist the motion as needed. In TILC, the main learning objective is to control the endpoint of the iteration and not to maximize trajectory performance. There are two learning objectives: (i) The system must be ready to accept weight at the end of the step; (ii) The knee must be sufficiently flexed during the swing to ensure foot clearance. The ILC law was formulated as an optimal control problem to account for redundant actuators in the system. Terms were added to the cost function that minimize the motors' control effort and maximize the muscles' recruitment to ensure the authors' "muscle first" philosophy. The cost function further penalizes rough changes of inputs from step to step. All objectives can be weighted. The simulations show that the desired outcome can be achieved within 15 steps, and the electric drives can compensate for muscle fatigue effects. The avoidance or delay of muscle fatigue was out of the scope of this study.

Romero-Sánchez et al. (2019) explored a theoretical approach based on muscle models to design combined actuation in hybrid exoskeletons. The employed biomechanical model considers 15 muscles in each leg for movement generation in the sagittal plane. Such a quantity of muscles might be realistic only in sophisticated implants. Many muscles act bi-articular and cause movements of two joints. Several muscles (not only one central antagonistic muscle pair) serve each joint of the skeletal system. This increases the complexity of the redundancy problem. The authors developed a sequence of static optimizations to avoid dynamics optimization to solve the torque allocation problem. The pure simulation study by Romero-Sánchez et al. (2019) proposes inverse dynamics analysis (IDA) based on the OpenSim software with normative gait data to determine the net joint torques and joint reaction forces for walking at first. Once the net joint torques and joint reaction forces are known, they are distributed between the orthosis and FES systems. Several cost functions were investigated that incorporate a scalar weight factor δ to tune the torque distribution between muscle activation and motor drives. One cost function also integrates the fatigue state of muscles to enable fatigue-minimizing co-operative support strategies. However, adapting the involved complex neuro-musculoskeletal model to an individual remains a demanding open problem, and an extension of this feedforward control approach by feedback control is further missing.

Co-operative control strategies are still not so intensively explored for the upper extremities. Wolf et al. (2017) controlled the elbow-joint angle for seven healthy subjects in a hybrid fashion. The FES control for the biceps consisted of feedforward control with a small feedback component. The feedforward component was determined empirically for each subject. The exoskeleton utilized a PD controller to track the trajectory (realization of a given joint stiffness and damping).

Simultaneous activation of both controllers yielded much better performance than with FES alone and significantly reduced the motor torque compared with exoskeleton support alone. Participants were instructed to behave passively during all trials.

Nam et al. (2022) presented an EMG-triggered hybrid skeleton with pneumatic muscles and FES to support elbow, wrist, and hand function. FES-induced muscle activation and deflation or inflation of the pneumatic muscles were administrated simultaneously and controlled by a state automaton with EMG-, angle-, and pressure-triggered transitions. No feedback control was considered so far. The clinical benefits of the system have been demonstrated in a pilot trial with 15 chronic stroke patients.

Stewart et al. (2019) describe a portable upper-extremity hybrid exoskeleton for the elbow. The authors aim at an assist-as-need control strategy to distribute the support on the biceps activation by FES and an electric motor that actuates the elbow joint via a pulley mechanism. As stated by the authors, two critical features are desired (Stewart et al. 2019): (i) The assistance provided by the FES and motor should be the minimum that the patient requires to perform the movement at a given time. (ii) The FES should perform the bulk of the movement which the patient is physically unable to. This ensures that most of the movement performed requires effort from the patient's muscles, thus improving muscular strength. The proposed control scheme requires the knowledge of a reference joint-angle trajectory, which restricts the possible usage of the approach. A simple linear FES model needs to be identified every time the system is used. Support and FES parameters will be updated based on the observed error and available input–output values. However, the authors' proposed model updating strategy might fail when mixed FES-generated and voluntary muscle activity occurs. Experimental validation only occurred with one healthy volunteer who was utterly passive during the tests or solved the tracking task entirely without assistance. Therefore, further feasibility studies are missing.

In Passon (2021), a hybrid arm weight compensation was investigated that involved stimulation of the deltoid muscle and a cable-based end-effector robot attached to the forearm. The compensation level (percentage of the shoulder torque required to hold the arm in current position) was set to give the patient the largest possible range of motion by his/her residual volition motion control. Surface EMG measurements were processed to control the desired fraction of lifting torque generated by FES. A low-level control loop automatically adjusted the stimulation intensity to compensate for fatigue effects. A supervisory controller determined the required total support torque based on the measured arm posture (using an inertial sensor) and the set compensation level. By indirectly monitoring the state of muscle fatigue from the gain of FES-induced torque production, the high-level controller continuously adjusted the distribution of torque generation to the robot and FES. This hybrid concept extended an FES-based weight relief controller (Klauer et al. 2016) by a robotic component. Still missing is a concept to adjust the weight compensation level when the residual motor activity of the patient changes. The latter can be assessed from EMG measurements during tasks that challenge the patient above its capacities. EMG processing in the presence of active FES will be discussed in Section 13.2.5 in some detail.

In summary, many co-operative control strategies have already been realized, the majority for the lower limbs. Without any doubt, control performance could be improved by the hybrid systems compared to pure FES solutions. A limitation of the existing studies is the relatively short usage of the hybrid systems for not more than a couple of minutes and walking distances below 20 m. These scenarios are not similar to the use of exoskeletons in the community or in rehabilitation facilities. It is still an open question if hybrid systems can lead to less powerful motors and weight-reduced exoskeletons. If complete muscle fatigue occurs after 20 m or 10 minutes, strong actuators are still required to keep the user from being grounded. Intelligent allocation of muscle and motor power must be investigated over more extended periods. Similar to hybrid cars, an intelligent drive management system must be implemented that takes the predicted usage of the hybrid system over the day and from day to day into account. Another weakness of many existing approaches is insufficient consideration of the patients' residual motor functions. Only a few approaches follow the assist-as-needed principle.

13.2.5 EMG- and MMG-Based Assessment of Muscle Activation

Muscle activity, more specifically the recruitment of motor units, can be recorded in transcutaneous FES systems using surface EMG (Merletti and Farina 2016; Merletti et al. 1992). The EMG records the net electrical activity (action potentials) generated by many contracting motor units. The potential benefits of EMG measurement in FES systems are many (Schauer 2017):

- for evaluating the residual voluntary muscle activity of the muscles to trigger and control the electrical stimulation and/or the WR support,
- for determining how well a muscle responds to stimulation, to control the recruitment of artificially activated motor units, to detect muscle fatigue, and to initiate countermeasures in the event of fatigue,
- for observing how the ES and/or robotics affects the patient's motor coordination and spasticity.

EMG measurement is typically performed via additional electrodes beside the stimulation electrodes on the muscles of interest. Measurement via the stimulation electrodes is also possible, but with some limitations, as explained below.

Figure 13.4 shows a raw surface EMG signal recorded during hybrid muscle activation, i.e. in the presence of active stimulation (at 25 Hz stimulation frequency) and superposing volitional muscle contraction. A high-resolution 24 bit A/D front end was used for EMG recording. The stimulation artifacts caused by the electrical pulses are visible in the EMG, but recovery takes place in less than two milliseconds. These artifacts are usually blanked. When analyzing the remaining EMG signal, one has to distinguish between the FES-evoked EMG portion and patient-induced EMG portion, where the latter includes both intentional (volitional) and unintentional muscle activity (reflexes) (Merletti et al. 1992; Merletti and Farina 2016).

Both quantities can be determined from the raw EMG between stimulation pulses using online signal processing as outlined subsequently (Schauer 2017).

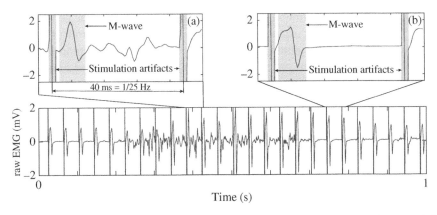

Figure 13.4 EMG recording during active FES with a stimulation frequency of 25 Hz. (a) Stimulation period with clearly visible volitional muscle activity. (b) Stimulation period with almost no volitional muscle activity. Source: Adapted from Schauer (2017).

The FES-evoked EMG is reflected in the so-called "M-wave," a good measure of the number of motor units synchronously activated by the last stimulation pulse. The amplitude of the M-wave can reach several mV. Usually, the 1-norm, root-mean-square, or peak-to-peak value is determined from the current M-wave to quantify the recruitment level. With muscle fatigue, the number of recruited motor units decreases. This is visible in the recorded recruitment level. The power spectrum of the M-wave is another indicator of muscle fatigue. It changes from higher to lower frequencies due to fatigue (Merletti and Farina 2016) due to the early decline of fast-twitch type 2B motor units. As time progresses, mainly the slow and fatigue-resistant motor units are recruited by FES.

For patient-induced muscle activity, the EMG usually has a much smaller amplitude (in the µV range) than the M-wave because the motor units are activated asynchronously. This rather noise-like signal with frequency components in the range of 30–300 Hz (De Luca and Knaflitz 1992) can be extracted from the EMG approximately 20–30 ms after each stimulation pulse either by high-pass filtering or by subtracting the estimated M-wave (see, e.g. (Ambrosini et al. 2014; Klauer et al. 2016) for more details). Both filtering approaches work with a vector of EMG values recorded in the last interpulse interval. Estimating the deterministic M-wave by a linear combination of EMG recordings from previous interpulse intervals was proposed first by Sennels et al. (1997). To reduce unwanted filter transients in the high-pass filter approach, Schauer et al. (2016) applied an energy minimization for the filter output by selecting suitable initial states of the noncausal high-pass filter.

Recording via stimulation electrodes considerably prolongs the recovery time of the stimulation artifacts. Therefore, no evaluation of the M-wave is possible, only the assessment of the patient-induced EMG component (Shalaby et al. 2011).

Drawbacks of EMG recordings are the necessary skin preparation, sensitivity to cable movement artifacts, and power line humming. Furthermore, stimulation artifacts contaminate the biosignals. Mechanomyography (MMG) may be used as an alternative to monitor muscle behavior. MMG represents the mechanical manifestation of the neurophysiological phenomena underlying muscle contraction and is the mechanical counterpart of the electrical activity of the motor units (Gordon and Holbourn 1948). An MMG sensor can be attached via straps directly to the skin or over clothing on the muscle belly. For instance, acceleration is often used as the measurement signal. The MMG is not affected by stimulation artifacts, but cannot distinguish until now the different sources of muscle activity (FES- or patient-induced). Therefore, it may be used to assess the total generated muscular torque or force and fatigue state in hybrid neuroprostheses. Islam et al. (2018) investigated MMG responses to characterize altered muscle function during electrical stimulation-evoked cycling in individuals with SCI.

13.3 Examples

The following four examples of hybrid exoskeletons illustrate the current state of the art and research of assistive technologies for restoring and enhancing mobility.

13.3.1 A Hybrid Robotic System for Arm Training of Stroke Survivors

With the European project RETRAINER,[2] a hybrid robotic system for arm training in stroke patients has been developed and clinically validated (Ambrosini et al. 2019). The chair-mounted system RETRAINER-ARM (see Figure 13.5) consists of a semiactive arm exoskeleton, a

---

2 See https://cordis.europa.eu/project/id/644721.

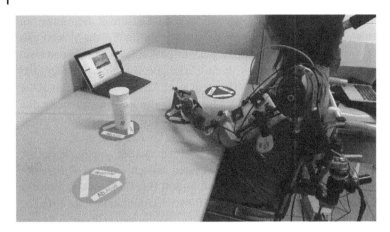

Figure 13.5 The hybrid robotic system RETRAINER-ARM with interactive objects for arm training of stroke survivors. Source: Ambrosini et al. (2019)/IEEE.

current-controlled neuromuscular stimulator, capable of recording EMG signals during stimulation (Valtin et al. 2016), and a set of interactive exercise objects. The system offers a structured exercise program resembling activities of daily life. Six exercises were provided: anterior reaching on a plane, anterior reaching in the space, moving objects on a plane, moving objects in the space, lateral elevation, and hand-to-mouth movement. Each exercise was divided into several subtasks and represented by a state automaton. Task execution is controlled using angular sensor data of the exoskeleton and radio-frequency identification technology inside the interactive objects. For movement support, EMG-triggered FES is combined with the exoskeleton's spring-based gravity compensation. The pre-tension of the upper arm's suspension springs can be electronically controlled. The exoskeleton is characterized by four degrees of freedom (DoFs). An inclination mechanism allows the patient to move the trunk without restriction. The remaining three DoFs (elbow flexion/extension, shoulder rotation, and shoulder elevation) are equipped with a goniometric angle sensor to track the patient's movements and an electromagnetic brake to selectively block DoFs. An adaptive linear filter is used to estimate the volitional EMG contribution during hybrid muscle contractions (Ambrosini et al. 2014). The FES is triggered by the volitional EMG signal using a patient-specific threshold which is set at each session using an automatic procedure. A timer triggers the FES if the threshold is not reached within a predefined period of time. Up to two muscles from a panel of five (biceps, triceps, anterior, medial, posterior deltoid) can be stimulated simultaneously. The system monitors the subject's participation during training. Feedback on voluntary effort is indicated in the graphical user interface (GUI) by a happy/sad emoji to encourage the subject's engagement during training. This system can be enhanced by a wearable multiarray electrode FES system for hand function restoration, which has been developed in parallel within the project RETRAINER (Crema et al. 2018, 2021).

The efficacy of the RETRAINER-ARM system was evaluated in terms of several clinical tests. Furthermore, kinematics- and EMG-based outcome measures were derived directly from data collected during training sessions. A single-blinded randomized controlled trial (RCT) was carried out to evaluate if the hybrid RETRAINER-ARM system is superior to advanced conventional therapy (ACT) of equal intensity in the recovery of arm functions (Ambrosini et al. 2021). For this purpose, 72 patients were randomly assigned to the two groups and exercised three times a week for nine weeks. The experimental group performed task-oriented exercises for 30 minutes using RETRAINER-ARM and ACT (60 minutes), while the control group performed ACT only (90 minutes). Patients considered RETRAINER-ARM moderately usable. The hybrid robotic system allowed the performance of personalized, intensive, and task-oriented training, with enriched sensory feedback, and proved to be superior to ACT in improving arm functions and dexterity after stroke (Ambrosini et al. 2021). The question of the extent to which the combination of FES and robotics had additional clinical benefit compared with the individual procedures remains unanswered.

13.3.2 First Certified Hybrid Robotic Exoskeleton for Gait Rehabilitation Settings

In 2017, the companies Ekso Bionics (Richmond, CA, USA) and HASOMED (Magdeburg, Germany) launched the first CE-marked hybrid robotic gait exoskeleton available for routine clinical use in Europe. The system is based on EksoNR™, a robotic exoskeleton designed specifically for use in rehabilitation facilities to get neuro-rehabilitation patients to the point where they can leave the device and return to their communities (see (Molteni et al. 2021; Edwards et al. 2022) for recent RCTs with stroke and SCI patients, respectively). The EksoNR exoskeleton offers a natural gait to re-teach the brain and muscles to walk again properly. For additional FES, a current-controlled eight-channel stimulator, RehaStim2, is wired to the exoskeleton for information exchange. The stimulator is worn in a special pocket by the therapist who supervises the gait training (cf. Figure 13.6). Self-adhesive stimulation electrodes are attached to the patient's legs and connected via cables to the stimulator. The exoskeleton's motor support adapts for various impairment levels, from full assistance to patient-initiated movement, in the swing and stance phases of walking. Sensors and software continuously monitor and regulate leg movement to minimize compensatory gait patterns. The stimulation timing is determined by the exoskeleton and synchronized to the patient's gait. Stimulation intensities have to be manually set by the therapists – no feedback control is employed. The combination of the exoskeleton technology and FES provides clinicians with the synergistic benefits of early mobility and muscle stimulation to rehabilitate a broader spectrum of patients.

From a clinical point of view, there is only limited evidence for the clinical efficacy of such a hybrid training approach in patients with SCI. For this reason, the Swiss Paraplegic Centre Nottwil currently performs an RCT with chronic, incomplete SCI individuals to investigate the effect of the

Figure 13.6 Gait training with the robotic exoskeleton EksoNR™ in combination with gait-synchronized functional electrical stimulation by means of the eight-channel stimulator RehaStim2. Source: ©Ekso Bionics.

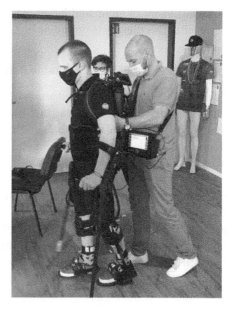

combined application of the EksoNR powered exoskeleton and FES compared to EksoNR therapy alone on functional outcomes and secondary health parameters. Participants of both groups will train for eight weeks, three times per week, for 30 minutes of effective training time per session.

13.3.3 Body Weight-Supported Robotic Gait Training with tSCS

Therapeutic possibilities of tSCS combined with robotic therapy have been explored for motor gait rehabilitation in SCI. tSCS has the potential to improve motor-evoked potentials, voluntary motor activity, trunk stability, standing, gait function, strength, and reduce spasticity in people with SCI. The RECODE Project (coordinated by Neural Rehabilitation Group, Cajal Institute) explores an approach that focuses on determining how to optimize stimulation parameters as an adjunct to intensive robotic therapy with body weight support and determining the clinical effectiveness of this technique. On the one hand, it is needed to determine protocols to deliver effective SCS based on an objective assessment to find the optimal stimulus intensity to promote better motor outcomes with robotic tSCS intervention. From the reflex response, the influence of body weight load (a fundamental variable of robotic therapy) has been characterized to define the gait training rehabilitation program. One of the factors that are still not well understood is the relationship between the changes in the posterior root muscle (PRM) reflex with posture and body weight unloading to optimize the stimulation parameters. It is crucial to establish the possible effect that posture and body weight unloading may have on the PRM reflex as an electrophysiological metric to determine stimulation parameters. A study with SCI patients has determined that it is possible to activate sensorimotor networks with tSCS in different conditions of body weight unloading (Megía-García et al. 2021).

Secondly, a clinical study is set out to analyze the feasibility and safety of tSCS applied to the T11–T12 vertebral level combined with robot-assisted gait training (20 sessions, 30 minutes in the Lokomat robot, 20 minutes tSCS). The first results obtained in patients with incomplete SCI suggest that the application of 30 Hz of tSCS seems safe, and there was no relationship between the level of stimulation intensity and pain perceived. The results, obtained with a small group of patients, have shown positive effects on the muscular strength of the lower limbs. Other functional improvements, while promising, cannot be attributed yet to tSCS as an adjunct to robotic therapy. Further studies involving a control group and larger cohorts will be required to determine the effectiveness and potential of this therapy (Comino-Suárez et al. 2022).

13.3.4 Modular FES and Wearable Robots to Customize Hybrid Solutions

The use of hybrid systems for gait assistance remains somewhat limited due to their complexity and limited effectiveness, which links to the fact that these systems are conceived as generic solutions (one for all) (Valenzuela et al. 2021). With this perspective in mind, the TAILOR project aims to design and develop customizable robotic and neuro-prosthetic technology that could provide an adaptable solution for patient needs.

On the one hand, new functional gait assessment metrics that are not pathology specific are needed. Some of the current metrics have been constructed using a population with a specific condition. An example is the case of the gait deviation index (GDI), which was designed using a gait database from a cerebral palsy population. It has been investigated whether the use of this index is applicable and valid for other populations, such as the incomplete SCI population, and it has been concluded that it is not an appropriate variable (Sinovas-Alonso et al. 2022). Therefore, it is

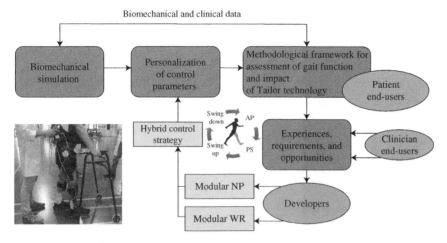

Figure 13.7 Flow diagram of the Tailor project approach.

necessary to improve the metrics used for functional assessment that could also be used to prescribe a customized hybrid solution. One possible solution was introduced by Müller et al. (2020), who adapted reference trajectories of healthy people employing dynamic time warping to assess the gait of individuals with incomplete SCI.

The second main goal of the project is to achieve a highly configurable modular robotic and neuro-prosthetic interface with the human user. In this context, a modular hybrid neuro-prosthesis has been developed that allows for quick and straightforward customization of the solution that could best suit the patient's functional status (gait) for a given prescription. The TAILOR system (cf. Figure 13.7) comprises a wireless network of electrical stimulation nodes and motion sensors and a core module with a software interface responsible for centralizing and managing the information. The sensor network can be configured to have between three and six wireless electrogoniometric sensors.

This network sends the information to the central module of the neuroprosthesis, where the gait is segmented through an event detection algorithm designed to adapt to the patient and variations in gait speed. The core module is responsible for controlling the assistance provided by the peripheral modules at any given moment as programmed in open- or closed-loop control schemes. The number of electrical stimulation nodes can vary between two and four nodes, and each node includes up to 4 FES current-controlled channels. The configuration of the network nodes for stimulation is done via a software interface and integrated with the ABLE exoskeleton to achieve a hybrid solution that assists the patient during gait.

The third main goal of the project is to develop a simulation environment that allows to study and predict the effect of hybrid assistance, considering the patient's biomechanical, anthropometric, and functional characteristics. The aim is to efficiently select the hybrid configuration and its optimal configuration for a specific gait. The computational tool includes an FES simulation model to emulate joint movement in response to FES assistance during gait.

This project is in the execution phase. As part of the work in progress, patients have been recruited and interviewed, deploying the user-centered design methodology. In the group of patients, subjects with and without experience using exoskeletons were included to derive the specifications and user needs. These patients will also be recruited for testing the TAILOR approach (see Figure 13.8) to personalize the configuration and control of the modular system

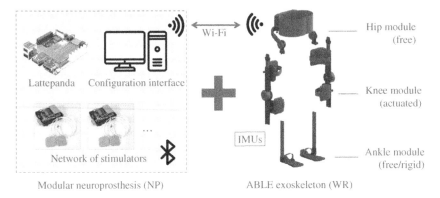

Figure 13.8 Tailor NP and WR modular systems.

based on the proposed methodological framework. Feasibility tests will determine the reliability of the modeling-based approach to tailor control parameters to achieve individually targeted functional goals.

13.4 Transfer into Daily Practice: Integrating Ethical, Legal, and Societal Aspects into the Design

Implanted and wearable systems make the most significant difference in usability and access. Concerning implants, there are substantial risks in any surgery, no predictable guarantee for the effectiveness, time, and effort in rehabilitation, and the high cost of surgical interventions. However, functioning systems are usable at all times, do not require setup time, and enable targeted activation of neural structures. Noninvasive stimulation systems do not carry the risk and cost of surgery and are potentially available to far more users than implanted systems. In a way, systems should strive for an optimal balance between the richness of information needed to perform the intended tasks and the invasiveness of the procedure, which always involves a risk for the user (Tucker et al. 2015; Fosch-Villaronga and Özcan 2020). However, as the need for assistance increases, WRs also require more powerful, heavier, and more drives that often result in bulky devices that are not very user-friendly. The increased functionality and associated complexity require a large amount of time to position and calibrate the electrical stimulation, user interfaces, and wearable robotics. This makes these systems less useful for activities of daily living but more focused on controlled environments such as rehabilitation. Still, given recent advancements toward the development of lightweight devices that would favor the mass-production of these devices, there is an excellent opportunity for technological advances that make donning, doffing, and calibration less time-consuming and simplify the use of experts. Altogether, these aspects make wearable systems more useful in everyday life and can be a more active and healthier alternative to the counter partner wheelchair that is, on the contrary, very limited in providing a healthy lifestyle to disabled populations. One possible way to reduce the need for expert knowledge is to develop intelligent procedures that adapt to the varying electrode or sensor placements and daily changes in muscle response to electrical stimulation. These intelligent procedures would follow the ideals of the principles of assist-as-needed and would adapt more to the real needs of the user and, in that respect, be more user-centered.

Advances in robotic control and design are increasingly focused on improving wearability through adjustable components that fit closely to the body. This may involve using cable-driven systems where actuators are located away from the target joints. However, transmission cables in exosuits pose complex control challenges due to the frictional losses in the cable transmission systems. Another challenge to the robot's portability is the power supply. Power for these actuators must be supplied either by remote batteries (often in a backpack) or by a tethered power supply, limiting mobility, and independence. Therefore, further research into novel devices and gear designs is needed to improve wearability while allowing easy donning and doffing of the robotic system. However, it is important to say that adjustability may be a limiting approach if safety is to be ensured to different users that come with different body shapes, sizes, and anatomies. In this respect, an aspect often forgotten is how intersectional considerations such as age, gender, sex, and health condition play a role in ensuring device safety (Søraa and Fosch-Villaronga 2020). For exoskeleton users, there might be significant differences between users that cannot be accounted for by simply allowing a robotic device to be adjustable. A recent study conducted by Calleja et al. (2022) suggests that the stress levels, calculated based on a WR subject's heart and respiration rate, heart rate variability, and galvanic skin response, showed that female subjects experienced the highest possible stress levels shortly after starting the experiment. The subjects' levels of energy expenditure also stayed at their maximum from the start and throughout the test, these levels being higher than those of the male subjects. According to their preliminary findings, walking with the exoskeleton might be significantly more stressful and tiresome for women than for men. However, more research is needed on the interplay of gender, safety, and exoskeletons to determine the magnitude of the differences. Although it is often said that trade-offs between the function an assistive device provides and the risk, time, and cost a user imposes to use an assistive device need to be made, each user is unique in a way, and a device that ought to serve the needs of that user should, at least, take her specific preferences, including the severity of her disability, into account.

Since these technologies assist users in many new ways, including cognitive, it is unsurprising that many other aspects than mere physical safety are of concern to developers, caregivers, patients, and society. Via different expert workshops, the Cost Action 16116 gathered detailed information on the ethical, legal, and societal issues surrounding the use and development of WR in society (Kapeller et al. 2021). While their work did not focus on neuroprostheses, the issues arising from these technologies correspond to more profound questions that may have an ulterior impact in the design of technology (Kapeller et al. 2021):

- **Wearable robots and the self**: WR may bring about benefits, risks, and harms for the user that may range from body and identity impacts, the experience of vulnerability, and agency, control, and responsibility for the body.
- **Wearable robots and the other**: WR users may suffer from ableism and stigmatization issues fueled by the perception that other people have of the WR-wearer. WR users may also start depending on the device to perform specific tasks and (over)trust the device, leading to potential miscalibrations regarding the device's capabilities (Aroyo et al. 2021).
- **Wearable robots and society**: The existence of WR raises questions about who will benefit from their advantages and what instruments will be available for society to access the technology. In this respect, social justice, nondiscrimination, and the prevention of dehumanizing contexts arise as primordial considerations to ensure in light of the use of technology to extend human caregivers' capabilities.

These considerations may be difficult to anticipate for developers and designers at the moment of creation, especially if the novelty of technology prevented the development of frameworks and guidelines that frame such advances (Villaronga and Virk 2017; Fosch-Villaronga 2019). While current frameworks may lack specific guidance for developers, there are initiatives and resources available to them to anticipate and mitigate related ethical, legal, and societal issues. These include from tools like the Responsible Research and Innovation framework developed by the European Union that moves from technology in society to technology for society (Owen et al. 2012)[3] or initiatives that translate these generic frameworks into specific guidelines for WR developers (Kapeller et al. 2021). Other resources have to do with the development of benchmarking ecosystems such as the H2020 Eurobench project that created a testing zone for WR to ensure that a minimum safeguard baseline is respected and standardized.[4]

These considerations point to the increasing need to establish interdisciplinary teams that can ensure that the device being created serves the purposes of its creation truly to be deployed in research laboratories, clinics, and, potentially, homes. By integrating such considerations, WR and neuroprostheses will be on a better foot to transition to the market.

13.5   Summary and Outlook

The combination of electrical stimulation and wearable robotics is an emerging technology for motor neuro-rehabilitation primary after SCI and stroke. Due to the complexity of the resulting hybrid systems, its use is currently limited to clinical settings with supervised repeated training sessions. Reported home use and support of activities of daily living (ADLs) are rare, although with appropriate research and integration of different aspects concerning physical and cognitive safety and associated ethical and legal considerations, these devices have the potential to be soon deployed in society.

As widely studied in co-operative control concepts, electrical stimulation is applied chiefly peripherally as FES to potentially reduce the energy and power requirements of the WRs and reduce their weight by generating some support using the patients' paralyzed muscles. Drawbacks of FES, such as rapid muscular fatigue and inadequate control performance, shall be mitigated by the almost endless support of the well-controllable electric drives inside the WR. However, the promises of such co-operative hybrid systems have not yet been credibly redeemed. Future supervisory level control should investigate the allocation of FES and robotic support over much more extended periods than in the past to show the genuinely achievable gains of combining FES and WR.

An esteeming alternative to FES in SCI is stimulating primarily posterior (sensory) nerve roots, either epidural with an implant or transcutaneous using noninvasive technology. In people with chronic incomplete SCI, recovery of leg function may be feasible with afferent input to the lumbar spinal cord using electrical stimulation in adjunction with intensive repetitive training. The stimulation may be either FES or SCS, but this should be phasic, synchronized to the flexion and extension of the legs while walking or cycling, and should coincide with descending voluntary drive to optimize neuroplastic recovery (Duffell and Donaldson 2020). Tonic, subthreshold SCS appears to be an effective long-term intervention to enable volitional movements in people with motor complete SCI (American spinal injury association impairment scale (AIS) A and B), at least in part by restoring spinal and descending inhibitory control. However, there is currently little evidence for neuroplastic recovery (Duffell and Donaldson 2020). SCS should be associated with less muscular fatigue than caused by FES, as muscle activation is voluntarily induced and, therefore, more natural. The WR system can support repetitive training. However, co-operative control strategies for targeted/biometric phasic SCS, both epidural and transcutaneous, still need to be developed.

---

3 See for instance https://rri-tools.eu/.
4 See https://eurobench2020.eu/.

Since there is the risk with WR systems that patients behave passively during training exercises, virtual reality, gamification, and biofeedback should be incorporated to motivate the patient. In this respect, the supraspinal drive is significant for the training (Duffell and Donaldson 2020). It has been hypothesized that the combination of ES with descending motor commands is essential for beneficial neuroplastic changes (Duffell and Donaldson 2020).

Besides technological advances, larger RCTs are required to evaluate the clinical effectiveness of hybrid systems and demonstrate the combination's added value. On this note, more research is needed to understand the broader implications of these advances for the user (the self), the persons interacting with the user, and society in general. In this respect, in this book section, we have seen that while differences between male and female anatomy have informed the design of, for instance, bikes with proportional geometry and properly sized components for riders of different sexes, these considerations still need to be implemented in WR and neuroprostheses design. Moreover, while efforts pointing to the integration of ethical, legal, and societal aspects into the design of WR and neuroprostheses are under development (Kapeller et al. 2020, 2021), this reality is far from being a general practice. However, if these devices that intertwine with the human body and have the potential to restore and enhance mobility to stroke, SCI, MS, and TBI patients are meant to hit the market, a holistic approach combining engineering and social sciences will be inevitable.

Acknowledgments

The work presented in the example of the TAILOR project is supported by coordinated grants RTI2018-097290-B-C31/C32/C33 funded by MCIN/AEI/ 10.13039/ 501100011033 and by "ERDF A way of making Europe." This work has also been partially funded by CSIC Interdisciplinary Thematic Platform (PTI+) NEUROAGING+ (PTI-NEURO-AGING+). This work has also been partially supported by the H2020 Cost Action 16116 on Wearable Robotics.

Acronyms

ACT	advanced conventional therapy
ADL	activities of daily living
AIS	American spinal injury association impairment scale
DCS	dynamic surface control
DoF	degree of freedom
EES	epidural electrical stimulation
EMD	electromechanical delay
EMG	electromyography
EMS	electrical muscle stimulation
ES	electrical stimulation
eSCS	epidural spinal cord stimulation
FES	functional electrical stimulation
GDI	gait deviation index
GUI	graphical user interface
HMNP	hybrid motor neuro-prosthesis
ILC	iterative learning control
LMN	lower motor neuron
MNP	motor neuro-prosthesis
MMG	mechanomyography

MPC	model predictive control
MS	multiple sclerosis
MU	motor unit
NES	nociceptive electrical stimulation
NMES	neuro-muscular electrical stimulation
NMP	motor neuro-prosthesis
NNILC	neural-network-based iterative learning control
NP	neuro-prosthesis
PCA	principal component analysis
pHRi	physical human–robot interaction
PID	proportional -integral -derivative
RPM reflex	posterior root muscle reflex
RCT	randomized controlled trial
SCI	spinal cord injury
SCS	spinal cord stimulation
SEA	serial elastic actuator
STS	sit-to-stand
TBI	traumatic brain injury
TILC	terminal iterative learning control
TTI	torque–time integral
tSCS	transcutaneous spinal cord stimulation
UMN	upper motor neuron
VR	virtual reality
WR	wearable robotics/robot

References

Alibeji, N.A., Kirsch, N.A., and Sharma, N. (2015). A muscle synergy-inspired adaptive control scheme for a hybrid walking neuroprosthesis. *Frontiers in Bioengineering and Biotechnology* 3: 203. https://doi.org/10.3389/fbioe.2015.00203.

Alibeji, N., Kirsch, N., and Sharma, N. (2017). An adaptive low-dimensional control to compensate for actuator redundancy and FES-induced muscle fatigue in a hybrid neuroprosthesis. *Control Engineering Practice* 59: 204–219. https://doi.org/10.1016/j.conengprac.2016.07.015.

Alibeji, N.A., Molazadeh, V., Moore-Clingenpeel, F., and Sharma, N. (2018). A muscle synergy-inspired control design to coordinate functional electrical stimulation and a powered exoskeleton: artificial generation of synergies to reduce input dimensionality. *IEEE Control Systems Magazine* 38 (6): 35–60. https://doi.org/10.1109/MCS.2018.2866603.

Alouane, M.A., Huo, W., Rifai, H. et al. (2019). Hybrid FES-exoskeleton controller to assist sit-to-stand movement. *IFAC-PapersOnLine* 51 (34): 296–301. https://doi.org/10.1016/j.ifacol.2019.01.032.

Ambrosini, E., Ferrante, S., Schauer, T. et al. (2014). A myocontrolled neuroprosthesis integrated with a passive exoskeleton to support upper limb activities. *Journal of Electromyography and Kinesiology* 24 (2): 307–317. https://doi.org/10.1016/j.jelekin.2014.01.006.

Ambrosini, E., Zajc, J., Ferrante, S. et al. (2019). A hybrid robotic system for arm training of stroke survivors: concept and first evaluation. *IEEE Transactions on Biomedical Engineering* 66 (12): 3290–3300. https://doi.org/10.1109/TBME.2019.2900525.

Ambrosini, E., Gasperini, G., Zajc, J. et al. (2021). A robotic system with EMG-triggered functional eletrical stimulation for restoring arm functions in stroke survivors. *Neurorehabilitation and Neural Repair* 35 (4): 334–345. https://doi.org/10.1177/1545968321997769.

Angeli, C.A., Edgerton, V.R., Gerasimenko, Y.P., and Harkema, S.J. (2014). Altering spinal cord excitability enables voluntary movements after chronic complete paralysis in humans. *Brain* 137 (5): 1394–1409. https://doi.org/10.1093/brain/awu038.

Angeli, C.A., Boakye, M., Morton, R.A. et al. (2018). Recovery of over-ground walking after chronic motor complete spinal cord injury. *New England Journal of Medicine* 379 (13): 1244–1250. https://doi.org/10.1056/NEJMoa1803588.

Aroyo, A.M., Bruyne, J., Dheu, O. et al. (2021). Overtrusting robots: setting a research agenda to mitigate overtrust in automation. *Paladyn, Journal of Behavioral Robotics* 12 (1): 423–436. https://doi.org/10.1515/pjbr-2021-0029.

Bao, X., Molazadeh, V., Dodson, A. et al. (2020). Using person-specific muscle fatigue characteristics to optimally allocate control in a hybrid exoskeleton–preliminary results. *IEEE Transactions on Medical Robotics and Bionics* 2 (2): 226–235. https://doi.org/10.1109/TMRB.2020.2977416.

Barsi, G.I., Popovic, D.B., Tarkka, I.M. et al. (2008). Cortical excitability changes following grasping exercise augmented with electrical stimulation. *Experimental Brain Research* 191 (1): 57–66. https://doi.org/10.1007/s00221-008-1495-5.

Baud, R., Manzoori, A.R., Ijspeert, A., and Bouri, M. (2021). Review of control strategies for lower-limb exoskeletons to assist gait. *Journal of NeuroEngineering and Rehabilitation* 18 (1): 119. https://doi.org/10.1186/s12984-021-00906-3.

Bortole, M., Venkatakrishnan, A., Zhu, F. et al. (2015). The H2 robotic exoskeleton for gait rehabilitation after stroke: early findings from a clinical study. *Journal of NeuroEngineering and Rehabilitation* 12 (1): 54. https://doi.org/10.1186/s12984-015-0048-y.

Calleja, C., Drukarch, H., and Fosch-Villaronga, E. (2022). Diversity observations in an exoskeleton experiment. *Proceedings of Inclusive HRI Workshop: Equity and Diversity in Design, Application, Methods, and Community.* https://sites.google.com/view/dei-hri-2022/proceedings (accessed 11 February 2023).

Calvert, J.S., Manson, G.A., Grahn, P.J., and Sayenko, D.G. (2019). Preferential activation of spinal sensorimotor networks via lateralized transcutaneous spinal stimulation in neurologically intact humans. *Journal of Neurophysiology* 122 (5): 2111–2118. https://doi.org/10.1152/jn.00454.2019.

Chang, J., Shen, D., Wang, Y. et al. (2020). A review of different stimulation methods for functional reconstruction and comparison of respiratory function after cervical spinal cord injury. *Applied Bionics and Biomechanics* 2020: e8882430. https://doi.org/10.1155/2020/8882430.

Chang, C.-H., Casas, J., Brose, S.W., and Duenas, V.H. (2022). Closed-loop torque and kinematic control of a hybrid lower-limb exoskeleton for treadmill walking. *Frontiers in Robotics and AI* 8: 702860. https://doi.org/10.3389/frobt.2021.702860.

Choi, I., Bond, K., and Nam, C.S. (2016). A hybrid BCI-controlled FES system for hand-wrist motor function. *Proceedings of IEEE International Conference on Systems, Man, and Cybernetics (SMC)*, 2324–2328. https://doi.org/10.1109/SMC.2016.7844585.

Comino-Suárez, N., Gómez-Soriano, J., Serrano-Muñoz, D. et al. (2022). Feasibility of transcutaneous spinal cord stimulation combined with robotic-assisted gait training (Lokomat) for gait rehabilitation of an incomplete spinal cord injury subject. In: *Converging Clinical and Engineering Research on Neurorehabilitation IV*, Biosystems & Biorobotics (ed. D. Torricelli, M. Akay, and J.L. Pons), 735–739. Cham: Springer International Publishing. ISBN 978-3-030-70316-5. https://doi.org/10.1007/978-3-030-70316-5_117.

Courtine, G., Harkema, S.J., Dy, C.J. et al. (2007). Modulation of multisegmental monosynaptic responses in a variety of leg muscles during walking and running in humans. *The Journal of Physiology* 582 (3): 1125–1139. https://doi.org/10.1113/jphysiol.2007.128447.

Crema, A., Malesevic, N., Furfaro, I. et al. (2018). A wearable multi-site system for NMES-based hand function restoration. *IEEE Transactions on Neural Systems and Rehabilitation Engineering* 26 (2): 428–440. https://doi.org/10.1109/TNSRE.2017.2703151.

Crema, A., Furfaro, I., Raschellà, F. et al. (2021). Reactive exercises with interactive objects: interim analysis of a randomized trial on task-driven NMES grasp rehabilitation for subacute and early chronic stroke patients. *Sensors (Basel, Switzerland)* 21 (20): 6739. https://doi.org/10.3390/s21206739.

De Luca, C.F. and Knaflitz, M. (1992). *Surface Electromyography, What's New?*. Turino: CLUT Publishers.

del Ama, A.J., Gil-Agudo, Á., Pons, J.L., and Moreno, J.C. (2014). Hybrid FES-robot cooperative control of ambulatory gait rehabilitation exoskeleton. *Journal of NeuroEngineering and Rehabilitation* 11 (1): 27. https://doi.org/10.1186/1743-0003-11-27.

del Ama, A.J., Gil-Agudo, Á., Bravo-Esteban, E. et al. (2015). Hybrid therapy of walking with Kinesis overground robot for persons with incomplete spinal cord injury. *Robotics and Autonomous Systems* 73 (C): 44–58. https://doi.org/10.1016/j.robot.2014.10.014.

Deuschl, G., Beghi, E., Fazekas, F. et al. (2020). The burden of neurological diseases in Europe: an analysis for the Global Burden of Disease Study 2017. *The Lancet Public Health* 5 (10): e551–e567. https://doi.org/10.1016/S2468-2667(20)30190-0.

Duffell, L.D. and Donaldson, N.N. (2020). A comparison of FES and SCS for neuroplastic recovery after SCI: historical perspectives and future directions. *Frontiers in Neurology* 11: 607. https://doi.org/10.3389/fneur.2020.00607.

Dunkelberger, N., Schearer, E.M., and O'Malley, M.K. (2020). A review of methods for achieving upper limb movement following spinal cord injury through hybrid muscle stimulation and robotic assistance. *Experimental Neurology* 328: 113274. https://doi.org/10.1016/j.expneurol.2020.113274.

Duschau-Wicke, A., von Zitzewitz, J., Caprez, A. et al. (2010). Path control: a method for patient-cooperative robot-aided gait rehabilitation. *IEEE Transactions on Neural Systems and Rehabilitation Engineering* 18 (1): 38–48. https://doi.org/10.1109/TNSRE.2009.2033061.

Edwards, D.J., Forrest, G., Cortes, M. et al. (2022). Walking improvement in chronic incomplete spinal cord injury with exoskeleton robotic training (WISE): a randomized controlled trial. *Spinal Cord*. https://doi.org/10.1038/s41393-022-00751-8.

Eladly, A., Valle, J.D., Minguillon, J. et al. (2020). Interleaved intramuscular stimulation with minimally overlapping electrodes evokes smooth and fatigue resistant forces. *Journal of Neural Engineering* 17 (4): 046037. https://doi.org/10.1088/1741-2552/aba99e.

Estes, S.P., Iddings, J.A., and Field-Fote, E.C. (2017). Priming neural circuits to modulate spinal reflex excitability. *Frontiers in Neurology* 8. https://doi.org/10.3389/fneur.2017.00017.

Euler, L., Guo, L., and Persson, N.-K. (2022). A review of textile-based electrodes developed for electrostimulation. *Textile Research Journal* 92 (7–8): 1300–1320. https://doi.org/10.1177/00405175211051949.

Fang, Z.P. and Mortimer, J.T. (1991). Selective activation of small motor axons by quasi-trapezoidal current pulses. *IEEE Transactions on Biomedical Engineering* 38 (2): 168–174. https://doi.org/10.1109/10.76383.

Fosch-Villaronga, E. (2019). *Robots, Healthcare, and the Law: Regulating Automation in Personal Care*. London: Routledge. ISBN 978-0-429-02193-0. https://doi.org/10.4324/9780429021930.

Fosch-Villaronga, E. and Özcan, B. (2020). The progressive intertwinement between design, human needs and the regulation of care technology: the case of lower-limb exoskeletons. *International Journal of Social Robotics* 12 (4): 959–972. https://doi.org/10.1007/s12369-019-00537-8.

Gad, P., Gerasimenko, Y., Zdunowski, S. et al. (2017). Weight bearing over-ground stepping in an exoskeleton with non-invasive spinal cord neuromodulation after motor complete paraplegia. *Frontiers in Neuroscience* 11: 333. https://doi.org/10.3389/fnins.2017.00333.

Gandolla, M., Ferrante, S., Molteni, F. et al. (2014). Re-thinking the role of motor cortex: context-sensitive motor outputs? *NeuroImage* 91: 366–374. https://doi.org/10.1016/j.neuroimage.2014.01.011.

Gelenitis, K.T., Sanner, B.M., Triolo, R.J., and Tyler, D.J. (2020). Selective nerve cuff stimulation strategies for prolonging muscle output. *IEEE Transactions on Biomedical Engineering* 67 (5): 1397–1408. https://doi.org/10.1109/TBME.2019.2937061.

Gill, M.L., Grahn, P., Calvert, J.S. et al. (2018). Neuromodulation of lumbosacral spinal networks enables independent stepping after complete paraplegia. *Nature Medicine* 24 (11): 1677–1682. https://doi.org/10.1038/s41591-018-0175-7.

Gordon, G. and Holbourn, A.H.S. (1948). The sounds from single motor units in a contracting muscle. *The Journal of Physiology* 107 (4): 456–464. https://doi.org/10.1113/jphysiol.1948.sp004290.

Gorman, P.H. and Mortimer, J.T. (1983). The effect of stimulus parameters on the recruitment characteristics of direct nerve stimulation. *IEEE Transactions on Biomedical Engineering* BME-30 (7): 407–414. https://doi.org/10.1109/TBME.1983.325041.

Grishin, A., Bobrova, E., Reshetnikova, V. et al. (2021). A system for detecting stepping cycle phases and spinal cord stimulation as a tool for controlling human locomotion. *Biomedical Engineering* 54. https://doi.org/10.1007/s10527-021-10029-7.

Ha, K.H., Murray, S.A., and Goldfarb, M. (2016). An approach for the cooperative control of FES with a powered exoskeleton during level walking for persons with paraplegia. *IEEE Transactions on Neural Systems and Rehabilitation Engineering* 24 (4): 455–466. https://doi.org/10.1109/TNSRE.2015.2421052.

Hofstoetter, U.S., McKay, W.B., Tansey, K.E. et al. (2014). Modification of spasticity by transcutaneous spinal cord stimulation in individuals with incomplete spinal cord injury. *The Journal of Spinal Cord Medicine* 37 (2): 202–211. https://doi.org/10.1179/2045772313Y.0000000149.

Hofstoetter, U.S., Krenn, M., Danner, S.M. et al. (2015). Augmentation of voluntary locomotor activity by transcutaneous spinal cord stimulation in motor-incomplete spinal cord-injured individuals. *Artificial Organs* 39 (10): E176–E186. https://doi.org/10.1111/aor.12615.

Hofstoetter, U.S., Freundl, B., Binder, H., and Minassian, K. (2018). Common neural structures activated by epidural and transcutaneous lumbar spinal cord stimulation: elicitation of posterior root-muscle reflexes. *PLoS ONE* 13 (1): e0192013. https://doi.org/10.1371/journal.pone.0192013.

Hofstoetter, U.S., Freundl, B., Danner, S.M. et al. (2020). Transcutaneous spinal cord stimulation induces temporary attenuation of spasticity in individuals with spinal cord injury. *Journal of Neurotrauma* 37 (3): 481–493. https://doi.org/10.1089/neu.2019.6588.

Hofstoetter, U.S., Freundl, B., Lackner, P., and Binder, H. (2021). Transcutaneous spinal cord stimulation enhances walking performance and reduces spasticity in individuals with multiple sclerosis. *Brain Sciences* 11 (4): 472. https://doi.org/10.3390/brainsci11040472.

Islam, Md.A., Hamzaid, N.A., Ibitoye, M.O. et al. (2018). Mechanomyography responses characterize altered muscle function during electrical stimulation-evoked cycling in individuals with spinal cord injury. *Clinical Biomechanics* 58: 21–27. https://doi.org/10.1016/j.clinbiomech.2018.06.020.

James, S.L., Theadom, A., Ellenbogen, R.G. et al. (2019). Global, regional, and national burden of traumatic brain injury and spinal cord injury, 1990–2016: a systematic analysis for the Global Burden of Disease Study 2016. *The Lancet Neurology* 18 (1): 56–87. https://doi.org/10.1016/S1474-4422(18)30415-0.

Kapadia, N., Moineau, B., and Popovic, M.R. (2020). Functional electrical stimulation therapy for retraining reaching and grasping after spinal cord injury and stroke. *Frontiers in Neuroscience* 14: 718. https://doi.org/10.3389/fnins.2020.00718.

Kapeller, A., Felzmann, H., Fosch-Villaronga, E., and Hughes, A.-M. (2020). A taxonomy of ethical, legal and social implications of wearable robots: an expert perspective. *Science and Engineering Ethics* 26 (6): 3229–3247. https://doi.org/10.1007/s11948-020-00268-4.

Kapeller, A., Felzmann, H., Fosch-Villaronga, E. et al. (2021). Implementing ethical, legal, and societal considerations in wearable robot design. *Applied Sciences* 11 (15): 6705. https://doi.org/10.3390/app11156705.

Kirsch, N.A., Bao, X., Alibeji, N.A. et al. (2018). Model-based dynamic control allocation in a hybrid neuroprosthesis. *IEEE Transactions on Neural Systems and Rehabilitation Engineering* 26 (1): 224–232. https://doi.org/10.1109/TNSRE.2017.2756023.

Klauer, C., Ferrante, S., Ambrosini, E. et al. (2016). A patient-controlled functional electrical stimulation system for arm weight relief. *Medical Engineering & Physics* 38 (11): 1232–1243. https://doi.org/10.1016/j.medengphy.2016.06.006.

Koutsou, A.D., Moreno, J.C., del Ama, A.J. et al. (2016). Advances in selective activation of muscles for non-invasive motor neuroprostheses. *Journal of NeuroEngineering and Rehabilitation* 13 (1): 56. https://doi.org/10.1186/s12984-016-0165-2.

Ladenbauer, J., Minassian, K., Hofstoetter, U.S. et al. (2010). Stimulation of the human lumbar spinal cord with implanted and surface electrodes: a computer simulation study. *IEEE Transactions on Neural Systems and Rehabilitation Engineering* 18 (6): 637–645. https://doi.org/10.1109/TNSRE.2010.2054112.

Laschowski, B., McNally, W., Wong, A., and McPhee, J. (2022). Environment classification for robotic leg prostheses and exoskeletons using deep convolutional neural networks. *Frontiers in Neurorobotics* 15. https://doi.org/10.3389/fnbot.2021.730965.

Laufer, Y. and Elboim-Gabyzon, M. (2011). Does sensory transcutaneous electrical stimulation enhance motor recovery following a stroke? A systematic review. *Neurorehabilitation and Neural Repair* 25 (9): 799–809. https://doi.org/10.1177/1545968310397205.

Lee, H., Ferguson, P.W., and Rosen, J. (2020). Lower limb exoskeleton systems–overview. In: *Wearable Robotics – Systems and Applications*, Chapter 11 (ed. J. Rosen and P.W. Ferguson), 207–229. Academic Press. ISBN 978-0-12-814659-0. https://doi.org/10.1016/B978-0-12-814659-0.00011-4.

Lynch, C.L. and Popovic, M.R. (2008). Functional electrical stimulation. *IEEE Control Systems Magazine* 28 (2): 40–50. https://doi.org/10.1109/MCS.2007.914689.

Makowski, N.S., Fitzpatrick, M.N., Triolo, R.J. et al. (2022). Biologically inspired optimal terminal iterative learning control for the swing phase of gait in a hybrid neuroprosthesis: a modeling study. *Bioengineering* 9 (2): 71. https://doi.org/10.3390/bioengineering9020071.

Marchal-Crespo, L. and Reinkensmeyer, D.J. (2009). Review of control strategies for robotic movement training after neurologic injury. *Journal of NeuroEngineering and Rehabilitation* 6 (1): 20. https://doi.org/10.1186/1743-0003-6-20.

Martinetti, A., Chemweno, P.K., Nizamis, K., and Fosch-Villaronga, E. (2021). Redefining safety in light of human–robot interaction: a critical review of current standards and regulations. *Frontiers in Chemical Engineering* 3. https://doi.org/10.3389/fceng.2021.666237.

Megía-García, Á., Serrano-Muñoz, D., Comino-Suárez, N. et al. (2021). Effect of posture and body weight loading on spinal posterior root reflex responses. *The European Journal of Neuroscience* 54 (7): 6575–6586. https://doi.org/10.1111/ejn.15448.

Mendell, L.M. (2005). The size principle: a rule describing the recruitment of motoneurons. *Journal of Neurophysiology* 93 (6): 3024–3026. https://doi.org/10.1152/classicessays.00025.2005.

Merletti, R. and Farina, D. (2016). *Surface Electromyography: Physiology, Engineering, and Applications*, IEEE Press Series on Biomedical Engineering. Wiley. ISBN 978-1-118-98702-5.

Merletti, R., Knaflitz, M., and DeLuca, C.J. (1992). Electrically evoked myoelectric signals. *Critical Reviews in Biomedical Engineering* 19 (4): 293–340.

Meyer, C., Hofstoetter, U.S., Hubli, M. et al. (2020). Immediate effects of transcutaneous spinal cord stimulation on motor function in chronic, sensorimotor incomplete spinal cord injury. *Journal of Clinical Medicine* 9 (11): E3541. https://doi.org/10.3390/jcm9113541.

Minassian, K., Jilge, B., Rattay, F. et al. (2004). Stepping-like movements in humans with complete spinal cord injury induced by epidural stimulation of the lumbar cord: electromyographic study of compound muscle action potentials. *Spinal Cord* 42 (7): 401–416. https://doi.org/10.1038/sj.sc.3101615.

Minassian, K., Hofstoetter, U., and Rattay, F. (2011). Transcutaneous lumbar posterior root stimulation for motor control studies and modification of motor activity after spinal cord injury. In: *Restorative Neurology of Spinal Cord Injury*. Oxford University Press. ISBN 978-0-19-974650-7. https://doi.org/10.1093/acprof:oso/9780199746507.003.0010.

Minassian, K., Hofstoetter, U.S., Danner, S.M. et al. (2016). Spinal rhythm generation by step-induced feedback and transcutaneous posterior root stimulation in complete spinal cord-injured individuals. *Neurorehabilitation and Neural Repair* 30 (3): 233–243. https://doi.org/10.1177/1545968315591706.

Müller, P., del Ama, A.J., Moreno, J.C., and Schauer, T. (2020). Adaptive multichannel FES neuroprosthesis with learning control and automatic gait assessment. *Journal of NeuroEngineering and Rehabilitation* 17 (1): 36. https://doi.org/10.1186/s12984-020-0640-7.

Moineau, B., Chin, C.M., Alizadeh-Meghrazi, M., and Popovic, M. (2019). Garments for functional electrical stimulation: design and proofs of concept. *Journal of Rehabilitation and Assistive Technologies Engineering* 6. https://doi.org/10.1177/2055668319854340.

Molazadeh, V., Zhang, Q., Bao, X. et al. (2021). Shared control of a powered exoskeleton and functional electrical stimulation using iterative learning. *Frontiers in Robotics and AI* 8. https://doi.org/10.3389/frobt.2021.711388.

Molazadeh, V., Zhang, Q., Bao, X., and Sharma, N. (2022). An iterative learning controller for a switched cooperative allocation strategy during sit-to-stand tasks with a hybrid exoskeleton. *IEEE Transactions on Control Systems Technology* 30 (3): 1021–1036. https://doi.org/10.1109/TCST.2021.3089885.

Molteni, F., Guanziroli, E., Goffredo, M. et al., null On Behalf Of Italian Eksogait Study Group (2021). Gait recovery with an overground powered exoskeleton: a randomized controlled trial on subacute stroke subjects. *Brain Sciences* 11 (1): 104. https://doi.org/10.3390/brainsci11010104.

Moshonkina, T., Grishin, A., Bogacheva, I. et al. (2021). Novel non-invasive strategy for spinal neuromodulation to control human locomotion. *Frontiers in Human Neuroscience* 14. https://doi.org/10.3389/fnhum.2020.622533.

Murray, S.A., Farris, R.J., Golfarb, M. et al. (2018). FES coupled with a powered exoskeleton for cooperative muscle contribution in persons with paraplegia. *2018 40th Annual International Conference of the IEEE Engineering in Medicine and Biology Society (EMBC)*, 2788–2792. IEEE. https://doi.org/10.1109/embc.2018.8512810.

Nam, C., Rong, W., Li, W. et al. (2022). An exoneuromusculoskeleton for self-help upper limb rehabilitation after stroke. *Soft Robotics* 9 (1): 14–35. https://doi.org/10.1089/soro.2020.0090.

Nandor, M., Kobetic, R., Audu, M. et al. (2021). A muscle-first, electromechanical hybrid gait restoration system in people with spinal cord injury. *Frontiers in Robotics and AI* 8: 645588. https://doi.org/10.3389/frobt.2021.645588.

Newham, D.J., de, N., and Donaldson, N. (2007). FES cycling. *Acta Neurochirurgica. Supplement* 97 (Pt 1): 395–402. https://doi.org/10.1007/978-3-211-33079-1_52.

Nicolaisen, M., Strand, B.H., and Thorsen, K. (2020). Aging with a physical disability, duration of disability, and life satisfaction: a 5-year longitudinal study among people aged 40 to 79 years. *International Journal of Aging & Human Development* 91 (3): 253–273. https://doi.org/10.1177/0091415019857061.

Owen, R., Macnaghten, P., and Stilgoe, J. (2012). Responsible research and innovation: from science in society to science for society, with society. *Science and Public Policy* 39 (6): 751–760. https://doi.org/10.1093/scipol/scs093.

Passon, A. (2021). Hybrid systems for upper limb rehabilitation after spinal cord injury. Phd thesis. Technische Universität Berlin.

Patil, V., Narayan, J., Sandhu, K., and Dwivedy, S.K. (2022). Integration of virtual reality and augmented reality in physical rehabilitation: a state-of-the-art review. In: *Revolutions in Product Design for Healthcare: Advances in Product Design and Design Methods for Healthcare*, Design Science and Innovation (ed. K. Subburaj, K. Sandhu, and S. Ćuković), 177–205. Singapore: Springer. ISBN 9789811694554. https://doi.org/10.1007/978-981-16-9455-4_10.

Pierce, R.L. and Fosch-Villaronga, E. (2022). Medical robots and the right to health care: a progressive realization. In: *The Cambridge Handbook of Information Technology, Life Sciences and Human Rights*, Cambridge Law Handbooks (ed. M. Ienca, O. Pollicino, L. Liguori), 70–85. Cambridge University Press. https://doi.org/10.1017/9781108775038.008.

Pinter, M.M., Gerstenbrand, F., and Dimitrijevic, M.R. (2000). Epidural electrical stimulation of posterior structures of the human lumbosacral cord: 3. Control of spasticity. *Spinal Cord* 38 (9): 524–531. https://doi.org/10.1038/sj.sc.3101040.

Pons, J.L. (ed.) (2008). *Wearable Robots: Biomechatronic Exoskeletons*. Wiley. ISBN 978-0-470-51294-4.

Postolache, O., Monge, J., Alexandre, R. et al. (2021). Virtual reality and augmented reality technologies for smart physical rehabilitation. In: *Advanced Systems for Biomedical Applications. Smart Sensors, Measurement and Instrumentation*, vol. 39, 155–180. Springer International Publishing. https://doi.org/10.1007/978-3-030-71221-1_8.

Prattichizzo, D., Pozzi, M., Baldi, T.L. et al. (2021). Human augmentation by wearable supernumerary robotic limbs: review and perspectives. *Progress in Biomedical Engineering* 3 (4): 042005. https://doi.org/10.1088/2516-1091/ac2294.

Qiu, S., Guo, W., Zha, F. et al. (2021). Exoskeleton active walking assistance control framework based on frequency adaptive dynamics movement primitives. *Frontiers in Neurorobotics* 15: 672582. https://doi.org/10.3389/fnbot.2021.672582.

Ridding, M.C., Brouwer, B., Miles, T.S. et al. (2000). Changes in muscle responses to stimulation of the motor cortex induced by peripheral nerve stimulation in human subjects. *Experimental Brain Research* 131 (1): 135–143. https://doi.org/10.1007/s002219900269.

Roberts, B.W.R., Atkinson, D.A., Manson, G.A. et al. (2021). Transcutaneous spinal cord stimulation improves postural stability in individuals with multiple sclerosis. *Multiple Sclerosis and Related Disorders* 52. https://doi.org/10.1016/j.msard.2021.103009.

Romero-Sánchez, F., Bermejo-García, J., Barrios-Muriel, J., and Alonso, F.J. (2019). Design of the cooperative actuation in hybrid orthoses: a theoretical approach based on muscle models. *Frontiers in Neurorobotics* 13: 58. https://doi.org/10.3389/fnbot.2019.00058.

Rowald, A., Komi, S., Demesmaeker, R. et al. (2022). Activity-dependent spinal cord neuromodulation rapidly restores trunk and leg motor functions after complete paralysis. *Nature Medicine* 28 (2): 260–271. https://doi.org/10.1038/s41591-021-01663-5.

Salchow-Hömmen, C. (2020). Adaptive hand neuroprosthesis using inertial sensors for real-time motion tracking. PhD thesis. Technische Universität Berlin.

Salchow-Hömmen, C., Schauer, T., Müller, P. et al. (2021). Algorithms for automated calibration of transcutaneous spinal cord stimulation to facilitate clinical applications. *Journal of Clinical Medicine* 10 (22): 5464. https://doi.org/10.3390/jcm10225464.

Sayenko, D.G., Atkinson, D.A., Dy, C.J. et al. (2015). Spinal segment-specific transcutaneous stimulation differentially shapes activation pattern among motor pools in humans. *Journal of Applied Physiology* 118 (11): 1364–1374. https://doi.org/10.1152/japplphysiol.01128.2014.

Sayenko, D.G., Rath, M., Ferguson, A.R. et al. (2019). Self-assisted standing enabled by non-invasive spinal stimulation after spinal cord injury. *Journal of Neurotrauma* 36 (9): 1435–1450. https://doi.org/10.1089/neu.2018.5956.

Schauer, T. (2017). Sensing motion and muscle activity for feedback control of functional electrical stimulation: ten years of experience in Berlin. *Annual Reviews in Control* 44: 355–374. https://doi.org/10.1016/j.arcontrol.2017.09.014.

Schauer, T. and Seel, T. (2018). Gait training by FES. In: *Advanced Technologies for the Rehabilitation of Gait and Balance Disorders*, Biosystems & Biorobotics (ed. G. Sandrini, V. Homberg, L. Saltuari et al.), 307–323. Cham: Springer International Publishing. ISBN 978-3-319-72736-3. https://doi.org/10.1007/978-3-319-72736-3_22.

Schauer, T., Seel, T., Bunt, N.D. et al. (2016). Realtime EMG analysis for transcutaneous electrical stimulation assisted gait training in stroke patients. *IFAC-PapersOnLine* 49 (32): 183–187. https://doi.org/10.1016/j.ifacol.2016.12.211.

Sennels, S., Biering-Sørensen, F., Andersen, O.T., and Hansen, S.D. (1997). Functional neuromuscular stimulation controlled by surface electromyographic signals produced by volitional activation of the same muscle: adaptive removal of the muscle response from the recorded EMG-signal. *IEEE Transactions on Rehabilitation Engineering* 5 (2): 195–206. https://doi.org/10.1109/86.593293.

Shalaby, R., Schauer, T., Liedecke, W., and Raisch, J. (2011). Amplifier design for EMG recording from stimulation electrodes during functional electrical stimulation leg cycling ergometry. *Biomedizinische Technik. Biomedical Engineering* 56 (1): 23–33. https://doi.org/10.1515/bmt.2010.055.

Shi, K., Song, A., Li, Y. et al. (2021). A cable-driven three-DOF wrist rehabilitation exoskeleton with improved performance. *Frontiers in Neurorobotics* 15: 664062. https://doi.org/10.3389/fnbot.2021.664062.

Sinovas-Alonso, I., Herrera-Valenzuela, D., Cano-de-la Cuerda, R. et al. (2022). Application of the gait deviation index to study gait impairment in adult population with spinal cord injury: comparison with the walking index for spinal cord injury levels. *Frontiers in Human Neuroscience* 16: 826333. https://doi.org/10.3389/fnhum.2022.826333.

Søraa, R.A. and Fosch-Villaronga, E. (2020). Exoskeletons for all: the interplay between exoskeletons, inclusion, gender, and intersectionality. *Paladyn, Journal of Behavioral Robotics* 11 (1): 217–227. https://doi.org/10.1515/pjbr-2020-0036.

Spaich, E.G., Svaneborg, N., Jørgensen, H.R.M., and Andersen, O.K. (2014). Rehabilitation of the hemiparetic gait by nociceptive withdrawal reflex-based functional electrical therapy: a randomized, single-blinded study. *Journal of NeuroEngineering and Rehabilitation* 11: 81. https://doi.org/10.1186/1743-0003-11-81.

Spieker, E.L., Wiesener, C., Niedeggen, A. et al. (2020). Motor and sensor recovery in a paraplegic by transcutaneous spinal cord stimulation in water. *Proceedings on Automation in Medical Engineering* 1 (1). https://doi.org/10.18416/AUTOMED.2020.

Steers, W.D., Wind, T.C., Jones, E.V., and Edlich, R. (2017). A review on functional electrical stimulation of bladder and bowel in spinal cord injury. *Journal of Long-Term Effects of Medical Implants* 27 (2-4): 307–317. https://doi.org/10.1615/JLongTermEffMedImplants.v27.i2-4.130.

Stewart, A., Pretty, C., and Chen, X. (2019). A portable assist-as-need upper-extremity hybrid exoskeleton for FES-induced muscle fatigue reduction in stroke rehabilitation. *BMC Biomedical Engineering* 1 (1): 30. https://doi.org/10.1186/s42490-019-0028-6.

Taylor, C., McHugh, C., Mockler, D. et al. (2021). Transcutaneous spinal cord stimulation and motor responses in individuals with spinal cord injury: a methodological review. *PLoS ONE* 16 (11): e0260166. https://doi.org/10.1371/journal.pone.0260166.

Tucker, M.R., Olivier, J., Pagel, A. et al. (2015). Control strategies for active lower extremity prosthetics and orthotics: a review. *Journal of NeuroEngineering and Rehabilitation* 12 (1): 1. https://doi.org/10.1186/1743-0003-12-1.

Valenzuela, D.S.H., Castillo, J.G., Pina, J. et al. (2021). Desarrollo de sistemas modulares robóticos y neuroprotésicos personalizables para la asistencia de la marcha patológica a través del diseño centrado en el usuario: Proyecto TAILOR. *XII Simposio CEA de Bioingeniería. Libro de Actas*, 50–55. ISBN 978-84-09-26469-8.

Vallery, H., Stützle, T., Buss, M., and Abel, D. (2005). Control of a hybrid motor prosthesis for the knee joint. *IFAC Proceedings Volumes* 38 (1): 76–81. https://doi.org/10.3182/20050703-6-CZ-1902.01415.

Valtin, M., Kociemba, K., Behling, C. et al. (2016). RehaMovePro: a versatile mobile stimulation system for transcutaneous FES applications. *European Journal of Translational Myology* 26 (3): 6076. https://doi.org/10.4081/ejtm.2016.6076.

van der Scheer, J.W., Goosey-Tolfrey, V.L., Valentino, S.E. et al. (2021). Functional electrical stimulation cycling exercise after spinal cord injury: a systematic review of health and fitness-related outcomes. *Journal of NeuroEngineering and Rehabilitation* 18 (1): 99. https://doi.org/10.1186/s12984-021-00882-8.

Villaronga, E.F. (2016). ISO 13482:2014 and its confusing categories. Building a bridge between law and robotics. In: *New Trends in Medical and Service Robots*, 31–44. Springer International Publishing. https://doi.org/10.1007/978-3-319-30674-2_3.

Villaronga, E.F. and Virk, G.S. (2017). Legal issues for mobile servant robots. In: *Advances in Robot Design and Intelligent Control* (ed. A. Rodic and T. Borangiu), 605–612. Cham: Springer International Publishing. ISBN 978-3-319-49058-8. https://doi.org/10.1007/978-3-319-49058-8_66.

Vuckovic, A., Osuagwu, B., Altaleb, M.K.H., Czaja, A.Z. et al. (2021). Brain-computer interface controlled functional electrical stimulation for rehabilitation of hand function in people with spinal cord injury. In: *Neuroprosthetics and Brain-Computer Interfaces in Spinal Cord Injury: A Guide for Clinicians and End Users* (ed. G. Müller-Putz and R. Rupp), 281–305. Cham: Springer International Publishing. ISBN 978-3-030-68545-4. https://doi.org/10.1007/978-3-030-68545-4_12.

Wagner, F.B., Mignardot, J.-B., Le Goff-Mignardot, C.G. et al. (2018). Targeted neurotechnology restores walking in humans with spinal cord injury. *Nature* 563 (7729): 65–71. https://doi.org/10.1038/s41586-018-0649-2.

Wang, P. (2012). Gait locomotion generation and leg muscle evaluation for overground walking rehabilitation robots. PhD thesis. Nanyang Technological University.

Wiesener, C., Seel, T., Spieker, L. et al. (2020a). Inertial-sensor-controlled functional electrical stimulation for swimming in paraplegics: enabling a novel hybrid exercise modality. *IEEE Control Systems Magazine* 40 (6): 117–135. https://doi.org/10.1109/MCS.2020.3019152.

Wiesener, C., Spieker, L., Axelgaard, J. et al. (2020b). Supporting front crawl swimming in paraplegics using electrical stimulation: a feasibility study. *Journal of NeuroEngineering and Rehabilitation* 17 (1): 51. https://doi.org/10.1186/s12984-020-00682-6.

Wolf, D., Dunkelberger, N., McDonald, C.G. et al. (2017). Combining functional electrical stimulation and a powered exoskeleton to control elbow flexion. *2017 International Symposium on Wearable Robotics and Rehabilitation (WeRob)*, 1–2. Houston, TX: IEEE. ISBN 978-1-5386-4377-8. https://doi.org/10.1109/WEROB.2017.8383860.

Xiloyannis, M., Annese, E., Canesi, M. et al. (2019). Design and validation of a modular one-to-many actuator for a soft wearable exosuit. *Frontiers in Neurorobotics* 13: 39. https://doi.org/10.3389/fnbot.2019.00039.

Ye, G., Grabke, E.P., Pakosh, M. et al. (2021). Clinical benefits and system design of FES-rowing exercise for rehabilitation of individuals with spinal cord injury: a systematic review. *Archives of Physical Medicine and Rehabilitation* 102 (8): 1595–1605. https://doi.org/10.1016/j.apmr.2021.01.075.

Zhang, D., Ren, Y., Gui, K. et al. (2017). Cooperative control for a hybrid rehabilitation system combining functional electrical stimulation and robotic exoskeleton. *Frontiers in Neuroscience* 11. https://doi.org/10.3389/fnins.2017.00725.

14

Contemporary Issues and Advances in Human–Robot Collaborations

Takeshi Hatanaka[1], Junya Yamauchi[2], Masayuki Fujita[2], and Hiroyuki Handa[3]

[1]Department of Systems and Control Engineering, School of Engineering, Tokyo Institute of Technology, Tokyo, Japan
[2]Graduate School of Information Science and Technology, Department of Information Physics and Computing, The University of Tokyo, Tokyo, Japan
[3]Tsukuba Research Laboratory, YASKAWA Electric Corporation, Tsukuba, Ibaraki, Japan

14.1 Overview of Human–Robot Collaborations

Despite rapid advances in robot technology, most real-world tasks, more or less, require human intervention for various purposes and in various layers. This stems from the human high capability in reasoning and decision-making in unstructured/uncertain environments. A dominant factor in determining the required level of autonomy is the degree of environmental structuring. Taking crop harvesting as an example, harvesting rice or wheat plants in a structured environment may require a human to make only high-level decisions, like sending a sign to start the task. On the other hand, more involvement may be necessary with the harvesting of fruits and vegetables owing to the unstructured nature of the environment in which they grow. Recent groundbreaking advances in artificial intelligence have opened the door for the automation of tasks in unstructured environments. Even though, this approach requires extremely rich data and complicated parameter tuning specific to each scene, which is demanding in many cases. Human–robot collaborations can provide another reasonable solution to this problem.

Driven by the advantage of human–robot collaborations above, a great deal of work has been devoted to this theme in the literature. It was in the period from mid-1990s to the mid-2000s that this theme was established as an independent research field (Goodrich and Schultz 2007), although many studies on human–robot collaboration including pioneering works like (Sheridan and Verplank 1978; Anderson and Spong 1989) had been reported before that. Indeed, many special issues were published in various journals during this period (Murphy and Rogers 2004; Kiesler and Hinds 2004; Laschi et al. 2007; Adams and Skubic 2005). The momentum has not been lost since then, and the field has continued to grow at an accelerated pace, as seen in recent special issues of journals and magazines (Baillieul et al. 2012; Sipahi 2017; Yildiz 2020).

Technological advances in machine learning, networking/communication technology, virtual/augmented reality, and collaborative robots have been a driving force of the field. Machine learning techniques allow one to automate complex tasks that have traditionally been performed manually, specifically by learning the logic behind the unconscious actions of experts and by estimating and predicting human mental states that are difficult to measure, like intentions. Advances in networking technology together with matured distributed control technology

Cyber–Physical–Human Systems: Fundamentals and Applications, First Edition.
Edited by Anuradha M. Annaswamy, Pramod P. Khargonekar, Françoise Lamnabhi-Lagarrigue, and Sarah K. Spurgeon.
© 2023 The Institute of Electrical and Electronics Engineers, Inc. Published 2023 by John Wiley & Sons, Inc.

have enabled collaborations between human(s) and multirobot systems beyond the traditional one-human–one-robot interaction. Virtual and augmented reality enhances 3-D human operability and situational awareness. Collaborative robots can work safely in close proximity to humans, allowing for more flexible collaboration between humans and robots. Another driving force lies in emerging applications. Beyond traditional applications such as factory automation, robot surgery, space exploration, education, and various military uses, any labor-intensive work, including that in agriculture, fishing, the food industry, infrastructure maintenance, and construction, requires technological advances in human–robot collaborations.

In this section, we start by addressing literature reviews highlighting four factors: task architecture, team formation, human modeling in control and decision, and human modeling of other factors. We then present an industrial perspective with real-world use cases of human–robot collaborations. Finally, we clarify how our research results, which will be presented in Sections 14.2 and 14.3, fit within the vast literature on human–robot collaborations.

14.1.1 Task Architecture

Given a robotic mission, a designer is first forced to divide it into subtasks since the overall mission is too complicated to handle in a one-shot manner. For example, let us consider agricultural applications. These involve various tasks, including planting seedlings, watering, crop monitoring, fertilization, pest control, thinning out, weeding, and harvesting. Even if we focus only on harvesting, there are still multiple tasks that need to be completed, such as signaling robots when to start, maneuvering them into place, harvesting the crops, transporting the crops, and providing a signal to end. Moreover, each of these subtasks also contains several subtasks, such as teleoperation, obstacle/collision avoidance, battery management, multirobot coordination, and hardware maintenance. The process of breaking down the overall mission indicates a hierarchical architecture of the subtasks. Regarding the architecture design, Musić and Hirche (2017) presented a six-layer architecture based on the categorization in Parker (1998) that includes task, planning, subtask, action, robot team, and interaction.

Hierarchical task architecture is itself a vital issue, even when considering a fully autonomous system (Parker 1998). Assuming the involvement of human intervention raises a crucial issue when forming the task architecture, namely, determining whether each subtask should be done manually, autonomously, or semiautonomously. The seminal work of (Sheridan and Verplank 1978) specifies which functions should be done autonomously and which should be done manually. They present a graded scale for the required *level of autonomy* with 10 levels ranging from no autonomy to full autonomy. See Table 14.1 for details. Systematically determining the right level of autonomy for a given subtask is far from trivial, but guidelines can be found in the literature. Parasuraman et al. (2000), for example stated that a high level of autonomy is preferable for information acquisition and analysis functions, but not for decision-making functions from the viewpoint of human situational awareness. The main focus of this chapter is on semiautonomous tasks, wherein robot(s) and human(s) are mutually interacting.

Once a set of semiautonomous subtasks is fixed, we then have to determine how humans and robots interact with each other. Among the four architectural templates presented in Samad (2023), this chapter deals primarily with a human-in-the-controller architecture. Musić and Hirche (2017) classified the methods of human–robot interactions for this architecture into three classes: *direct interaction*, *complementary interaction*, and *overlapping (mixed initiative) interaction*. A typical example of direct interaction is (traditional) bilateral teleoperation, where a human operator directly commands a master robot and a remote slave robot follows the motion of the master commanded by the operator.

Table 14.1 Level of autonomy.

Level	
1	The human executes all actions
2	The computer offers complete set of action alternatives
3	The computer offers a selection of action alternatives
4	The computer suggests one alternative
5	The computer executes an action autonomously if the human approves
6	The computer allows the human a restricted time to veto before automatic execution
7	The computer executes an action and informs the human
8	The computer executes an action and informs the human if asked
9	The computer executes an action and informs the human if it decides to
10	The computer executes all actions autonomously

Source: Adapted from Sheridan and Verplank (1978).

In the complementary interaction, a human determines signals for a subset of subtasks, while the remaining ones are managed by the robots. Diaz-Mercado et al. (2017), for example, addressed human–robot collaboration for the coverage control task, where a human specifies a so-called "density function" to determine relative importance over the mission space, and robots run coverage control for the time-varying density to optimally sample environmental data according to the importance specified by the operator (Figure 14.1). Hatanaka et al. (2017b) investigated distributed navigation of a robotic network, wherein robots implement distributed motion synchronization control, while a human operator gives a command to the network so that the group follows motion that is seen as desirable by the human (Figure 14.2). Details of this are highlighted in Section 14.2. Recent work on bilateral teleoperation has also focused on complementary interactions. Rodríguez-Seda et al. (2010) and Franchi et al. (2012b) considered teleoperation of multiple robots where the robots handle subtasks, such as inter-distance control, collision/obstacle avoidance, and network formation, while the operator is in charge of navigation

Figure 14.1 Example of human–robot collaboration for a coverage task where a human specifies the important area with the position of their finger on a touchscreen. The robots then implement coverage control to optimally cover the mission space depending on the specified importance.

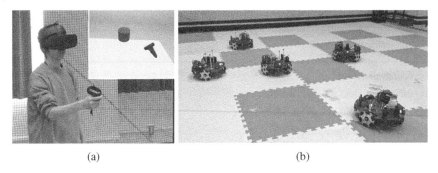

(a)   (b)

Figure 14.2 Multirobot navigation for which a human operator feeds back virtual information that consolidates information from multiple robots (a). The robots then synchronize their motion while following the human command (b).

of the group. Another typical example of the complementary interaction is cooperative payload manipulation (Spletzer et al. 2001; Michael et al. 2011; Erhart and Hirche 2016), where multiple robots grasp and transport a payload. Several studies have investigated scenarios involving a human operator remotely specifying the direction for payload delivery (Lee and Spong 2005; Gioioso et al. 2014; Mohammadi et al. 2016; Staub et al. 2018).

In the overlapping interaction, a human and a robot(s) engage in a common subtask and determine the same control signal, in contrast to the direct and complementary interactions where they determine different control signals. In this paradigm, the issue to be considered is how to blend manual and autonomous control, more specifically, how to determine the gains K_h and K_a in the control signal

$$u(t) = K_h u_h(t) + K_a u_a(t) \tag{14.1}$$

with a manual control u_h and an autonomous control u_a. An example of an overlapping interaction is cooperative manipulation under proximate interactions, where a human holds one end of the payload and robots hold the other(s) (Yokoyama et al. 2003; Medina et al. 2017; Peng et al. 2018).[1] Chipalkatty et al. (2013) investigated a model predictive control-based mixed-initiative control. They showed that a simple zero-order hold human prediction outperforms other complex prediction methods with respect to the human comfort. Saeidi et al. (2017) and Rahman et al. (2023) presented a mixed-initiative bilateral teleoperation scheme, in which manual and autonomous control are blended according to a human–robot mutual trust model. Dragan and Srinivasa (2013) addressed predictions of human intent in a scenario involving shared control teleoperation and presented a prediction method based on inverse reinforcement learning. The learning of human intentions is a key issue in contemporary studies. Geravand et al. (2016) used a decision-making model based on human psychology called the drift diffusion model, to tune the gains in (14.1). They also presented a mixed-initiative control scheme for a reaching task based on the *Gaussian process (GP)* in Hatanaka et al. (2020), the details of which can be found in Section 14.3.

14.1.2 Human–Robot Team Formation

A variety of team formations between humans and robots have been investigated beyond the one-human–one-robot collaborations typical in bilateral teleoperation (Hokayem and Spong 2006; Nu no et al. 2011; Hatanaka et al. 2015).

---

1 The payload manipulation also has an aspect of complementary interactions since the delivery direction is determined by the human, while the robots are devoted solely to payload grasping.

One-human–multiple-robot collaborations have been studied in depth over the last two decades, stimulated by advances in networking and distributed control technology. Early studies investigated bilateral teleoperation of multiple robots (Lee and Spong 2005; Rodríguez-Seda et al. 2010; Liu and Chopra 2012; Franchi et al. 2012b) whose challenge lay in developing a decentralized/distributed control architecture, one in which a human remotely navigates robots while ensuring closed-loop stability and safety certificates in the presence of communication delays. The cooperative payload manipulation mentioned earlier is a typical example of one-human–multiple-robot collaborations both in remote and proximate interactions (Lee and Spong 2005; Gioioso et al. 2014; Mohammadi et al. 2016; Staub et al. 2018; Medina et al. 2017; Peng et al. 2018). The coverage control task in Diaz-Mercado et al. (2017) is also a typical example that inherently needs to be performed by multiple robots. Cummings et al. (2012) studied a search, track, and neutralize mission and showed that human aid enhances task performance.

Teams with multiple humans have also been studied extensively under the name of multilateral teleoperation: see Chong et al. (2000), Sirouspour and Setoodeh (2005), Malysz and Sirouspour (2011), Khademian and Hashtrudi-Zaad (2011), and Franchi et al. (2012a). Formations with multiple humans allow one to separately treat subtasks in a complementary fashion, control robots with a high degree of freedom that is hardly manageable by a single human and blend interventions of humans with different perspectives.

14.1.3 Human Modeling: Control and Decision

Needless to say, human behavior is a key component of human–robot collaboration systems. Having a mathematical model of human behavior opens the door to systematic design and analysis. This is why a great deal of the literature has been devoted to human modeling. In this regard, it is to be noted that the human behavior to be modeled differs depending on the required level of autonomy. For low-level autonomy, the degree of human involvement ranges to the level of robot motion and, accordingly, the human is involved in the control loop. For high-level autonomy, the human makes only high-level decisions while feeding back abstracted task information, which is sometimes described as "human-on-the-loop" (Chen and Barnes 2014). Chen and Barnes (2014) and Mušić and Hirche (2017) call the role formerly assigned to the human the *active role* and the latter one the *supervisory role*.

Early seminal work on human mathematical modeling for the active role can be found in McRuer and Jex (1967), McRuer and Krendel (1974), and McRuer (1980). The authors considered the human-in-the-loop (HIL) system illustrated in Figure 14.3. They built a model, called the *crossover model*, for a scenario involving compensatory control, where the human feeds back only the error between the target r and output y in a manner similar to the standard feedback controller. This model indicates that the open-loop transfer function consisting of the robot $R(s)$ and human $H(s)$ meets

$$H(s)R(s) = \frac{\omega_c}{s}e^{-s\tau} \tag{14.2}$$

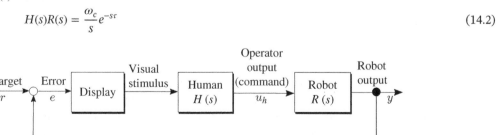

Figure 14.3 A human-in-the-loop system for compensatory control, where the human operator visually feeds back only errors between a reference and the robot output.

where ω_c specifies the gain crossover frequency, and τ describes the human response delay. Equation (14.2) suggests that humans should install the inverse model of $R(s)$, cancel out the dynamical property of $R(s)$, and tune the gain crossover frequency ω_c while taking account of robust stability. A variety of human models beyond the compensatory control have been proposed based on this finding and are summarized well in Mulder et al. (2018). The authors in Hatanaka et al. (2017b) have also validated this model for a scenario of multirobot navigation, at least over a low-frequency domain.

In the field of bilateral teleoperation, human operators are assumed to behave as *passive* systems (Hokayem and Spong 2006; Nu no et al. 2011; Hatanaka et al. 2015), which means that there is a positive scalar β such that the input–output pair (y_h, u_h) of the human always meets

$$\int_0^t u_h^\top(\tau)y_h(\tau)d\tau \geq -\beta \tag{14.3}$$

for any $t \geq 0$. The human passivity assumption together with environmental passivity has allowed researchers to rigorously analyze closed-loop stability for bilateral teleoperation systems in the presence of constant/varying communication delays. On the other hand, relatively few papers have questioned the validity of the human passivity assumption. Dyck et al. (2013) examined the passivity of the human arm for various tasks, and showed that human passivity is task-dependent. Hatanaka et al. (2017b) investigated collaborations between a human operator and multiple networked robots. The authors built a human operator model for various networks and showed not only that the operator tends to approach a passive system through training but also that, depending on the network structure, they may fail to attain passivity even after training. Section 14.2 presents a distributed control architecture under a relaxed assumption for human passivity that is meant to ensure closed-loop stability and synchronization to human reference motions. It has also been pointed out in Lawrence (1993) that passivity (stability) may conflict with the transparency objective.

Human intention is a signal that lies in a higher layer but is also closely related to robot motion. Let us now consider a scenario involving mixed-initiative control for harvesting, one where multiple crops are to be harvested. Without correctly estimating the crop that a human wishes to harvest, the automatic control will never adequately support the reaching and cutting of those crops. In cooperative payload manipulation as well, a correct estimation of intention is key for smooth and safe human–robot collaboration in both the close-proximity interaction and the remote interaction. The intention/emotion has been studied in various fields, and an emotional state representation, termed *valence/arousal*, has emerged in the field of psychophysiology (Bradley 2000). Kulić and Croft (2007) brought this representation to human–robot collaborations as a means of estimating human intention. Other sensory devises for measuring and estimating intention are summarized well in Dani et al. (2020). Various models for estimating human intention have also been presented in the literature. A promising mathematical model for intention estimation is the hidden Markov model (Kulić and Croft 2007; Kelley et al. 2008). Machine learning techniques have also been employed for intention estimation (Li and Ge 2014; Wang et al. 2013; Dani et al. 2020). Li and Ge (2014) estimated the intended trajectories of humans using neural networks. Wang et al. (2013) applied GP over a low-dimensional latent space and demonstrated the effectiveness of their model for a table-tennis robot. Dani et al. (2020) presented an intention estimation scheme combining neural networks, an approximate expectation maximization algorithm, and an extended Kalman filter, together with a safe control strategy based on control barrier functions. Applications of machine learning techniques to estimating high-level human signals like intention and emotion is a promising direction for future human–robot collaborations. One of the authors of this chapter has conducted research related to this topic, and it will be presented in Section 14.3.

In supervisory control, a human is forced to choose an option from a finite number of alternatives (See levels 2 and 3 in Table 14.1). For example, many robotic missions suffer from a dilemma between exploration and exploitation, and the human is expected to moderate these two alternatives. A promising mathematical tool for describing the psychological fluctuations of a human facing the two-alternative forced choice tasks is the *drift-diffusion model* (Bogacz et al. 2006), which was developed in the field of psychology. The model is formulated by the stochastic differential equation:

$$dz = \alpha dt + \sigma dW \tag{14.4}$$

where z is the accumulated evidence in favor of choosing an option, α is the signal intensity of the stimulus acting on z, and σdW is a Wiener process with standard deviation σ. This model chooses an option when z crosses a positive threshold and another when it gets below a negative threshold. Another related model is the *soft-max model* (Stewart et al. 2012), which gives the probability of choosing one of two options at time $t + 1$ based on the anticipated rewards at time t. Stewart et al. (2012) also presented a dynamical model of updating the rewards.

14.1.4 Human Modeling: Other Human Factors

In addition to the above human decision and control models, other human factors can be significant for ensuring successful human–robot collaborations. Chen and Barnes (2014) raised two such factors, *situation awareness* and *trust*, which are mutually related and tightly coupled with the concepts of *transparency* and *human workload*.

Situation awareness is defined in Endsley (2011) as "being aware of what is happening around you and understanding what that information means to you now and in the future." They categorize three levels of situation awareness, perception of elements in the environment (level 1), comprehension of the current situation (level 2), and projection of future status (level 3). In addition to such qualitative models, several mathematical models for measuring situation awareness have also been presented (Nguyen et al. 2019). Chen and Barnes (2014) provided the important insight that automating information acquisition enhances the situation awareness, while automating information analysis for supporting decision-making may degrade awareness (Parasuraman et al. 2000; Rovira et al. 2007; Onnasch et al. 2014). Thus, ensuring a sufficient level of situation awareness for a human operator may put a constraint on the design of a human–robot collaboration system, especially for complementary and overlapping subtasks. The situation awareness is expected to improve if transparency is enhanced (Musić and Hirche 2017).

Human trust is defined in the seminal work of Lee and See (2004) as "the attitude that an agent will help achieve an individual's goals in a situation characterized by uncertainty and vulnerability." It has also been pointed out in Chen and Barnes (2014) that a human operator's trust in automated systems is a critical element that determines the performance of human–robot collaborations. Consequently, many studies have been devoted to modeling human trust against autonomous systems (Lee and Moray 1992, 1994; Moray et al. 2000). Interested readers are recommended to refer to Akash et al. (2020) for a comprehensive survey of this issue, wherein the authors review a variety of traditional human trust models. This includes qualitative and quantitative models, where the quantitative models are further categorized into deterministic and probabilistic models such as hidden Markov models, Markov decision processes, and partially observable Markov decision processes. Saeidi et al. (2017) and Rahman et al. (2023) presented a bilateral teleoperator that scales manual and autonomous controls based on computed human-to–robot mutual trust and demonstrated that the trust-based control strategy improves operator satisfaction. Cognitive feedback control with cognitive modeling has also been studied in detail (Jain 2023).

As demonstrated by that paper, control designs that integrate the trust provide a promising future direction for research in this field. Mercado et al. (2016) reported that transparency again has a positive correlation with trust, as evidenced by Tintarev and Masthoff (2007) and Sinha and Swearingen (2002).

Enhancing transparency has positive impacts for both situation awareness and trust. On the other hand, high transparency increases the amount of information that has to be processed by the human, thus increasing the human workload. The human workload has been a critical factor for human–robot collaborations since the early days of research in this field (Sheridan and Stassen 1979). A well-established metric for human workload is the NASA-TLX (task load index) questionnaire (Hart and Staveland 1988), which has been widely employed to evaluate human–robot collaboration systems (de la Croix and Egerstedt 2012; Egerstedt et al. 2014; Atman et al. 2018a; Hatanaka et al. 2020). The workload issue is even more significant for one-human–multiple-robot interactions since directly controlling multiple robots drastically increases the human workload due to the limited dimensionality of the human commands, as compared with the inherent high dimensionality of multirobot systems (Chen et al. 2011; Musić and Hirche 2017; Goodrich et al. 2005).

The three paragraphs above indicate that transparency is a key quantity that trades off with situation awareness, trust, and workload. In this regard, an interesting problem setup has been addressed in Akash et al. (2020). The authors treated decision support for a reconnaissance mission and presented a novel paradigm in which transparency is viewed as a manipulable quantity and tuned to balance human trust and workload. To this end, they established a partially observable Markov decision process model that describes the relation between transparency, the amount of information, and the human trust-workload behavior. A similar model was also employed in Cubuktepe et al. (2023). To build such models, workload and trust need to be replaced by measurable quantities, specifically the reaction time and compliance, based on the evidence in Newell and Mansfield (2008) and Braithwaite and Makkai (1994). Likewise, human cognitive states are hardly measured by sensors, and it is crucial to link unobservable hidden states with measurable information while making use of insights made in different fields, including the humanities.

Another important research direction was presented in Protte et al. (2020), where the authors investigated a class of bounded rationality revealed in behavioral economics for a particular HIL task. Due to the bilateral interactions between humans and systems inherent in cyber–physical–human systems, the human behavior must be affected by the system. This has great potential to provide novel insights on humans in the loop, whereas behavioral economics basically deals with human behaviors in an open loop.

14.1.5 Industrial Perspective

Human–robot collaborations have not only been studied in academia, but have also been actively developed in industry.

Robot technology has freed workers from dangerous and arduous labor in factories, including assembly, machine tending, palletizing, quality inspection, arc welding, gluing, bonding, polishing, and deburring (Figure 14.4). Robots can be programmed to automatically complete these tasks, but human–robot interactions arise during the teaching phase for the robots. Indeed, robot teaching has been a primary goal from the industrial point of view since the inception of industrial robots (Devol 1961). The main issue here is to teach the robot ideal movements in a way that is intuitive and simple for humans. In typical teaching systems for industrial robots, a human specifies ideal robot motion including way-points, trajectories, and/or link angles using a programming

Figure 14.4 Spot welding of automobiles by multiple industrial robots. Source: Courtesy of Yaskawa Electric Corporation.

Figure 14.5 A scene where a robot is taught with a programming pendant. Source: Courtesy of Yaskawa Electric Corporation.

pendant (Figure 14.5). Many efforts have been made to alleviate the need for this pendant-based teaching, but all still require a substantial workload for human trainers. There is another approach called direct teaching, wherein a human directly handles and operates the robot arm, and the robot learns the correct trajectory from being guided by the trainer. Early work on the development of such a system can be found in Handa et al. (1997), the results of which led to a commercial teaching system. These types of teaching systems have been continuously improved, but there is still room for further development in automating the teaching process. Indeed, in almost all factories robot motion is currently fine-tuned by an operator in the task space due to the inherent difficulty involved in the teaching, which requires expert knowledge of not only the tasks to be automated but also the robot operation. Revolutionary advances of machine learning techniques for mimicking the behavior of experts are expected to produce breakthroughs in teaching systems.

Figure 14.6 A scene showing an application using collaborative robots, MOTOMAN-HC10DTP, where humans and robots cooperate to assemble small-scale robotic arms. Source: Courtesy of Yaskawa Electric Corporation.

Another related technological advances in industry include the development of collaborative robots. Safety is a significant consideration for human–robot collaborations. Traditional industrial robotic arms are fast, heavy, and powerful, and require users to operate them only inside of safety fences in order to avoid danger. Collaborative robots are designed from both a hardware and a software level to ensure safety, eliminating the need for safety fences. Technological advances in this area enable humans to work together with robots while sharing the same space (Figure 14.6). In addition to the safety certificates, collaborative robots bring additional benefits, such as saving space and being easy to install and relocate. Needless to say, they reduce the mental and physical barriers involved with direct teaching. It is because of these benefits that the global sales of collaborative robots have been growing at a steady pace, and it is anticipated, through a survey by the Yaskawa Electric Corporation, that the market will continue to grow at a compound average growth rate of 35% at least by 2025.

Finally, let us introduce some actual use cases of human–robot collaborations in industry. In the assembly task shown in Figure 14.6, a collaborative robot brings assembly parts to their correct positions, where a human receives and assembles them. The robot reduces human workload while leaving the complex assembly task to the human. This can be seen as an example of complementary interactions: repetitive tasks that are simple but require precision are left to the robot while humans focus on complex tasks that are hardly automated. The screw-tightening task shown in Figure 14.7 is another example of this kind of collaboration: a human prepares a workpiece, and the robot performs the tightening of the bolt. This teaming prevents tightening mistakes due to human error. Other examples of this kind of complementary collaboration can be found in scenarios involving inspection and coating. Meanwhile, in the task of loading a workpiece to a press fitting device, a human sets a jig for press fitting to the workpiece and the robot receives and moves it to and from an automatic press, thus preventing any accidents. In the same way, there are many scenarios in

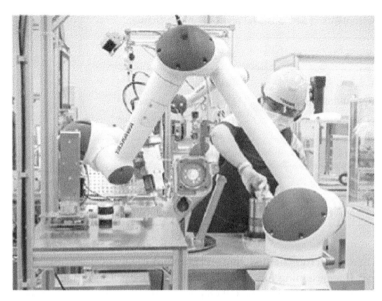

Figure 14.7 Human–robot collaboration for a screw-tightening task. Source: Courtesy of Yaskawa Electric Corporation.

which humans are freed from hard and dangerous work through complementary human–robot collaborations, including machine tending task, air blow task, and palletizing/depalletizing task.

14.1.6 What Is in This Chapter

Summarizing Sections 14.1–14.5, the general procedure for designing a human–robot collaboration system is as follows, though orders may vary depending on the task specifics. We first determine the task architecture, namely subtasks and the relations/interactions among them, which is not exclusive to human–robot collaborations. We then determine whether each subtask should be executed manually, automatically, or semiautonomously. For semiautonomous tasks, we need to further fix

- the team formation including human(s) and the architectural template (Samad 2023) depending on the task goal;
- the required level of autonomy, specifically whether the human plays an active role or a supervisory role and whether we take a HIL or a human-on-the-loop architecture;
- the type of interactions from direct, complementary, and overlapping interactions;
- the required human model of control, decisions, and other factors;
- and the information to be exchanged between robot(s) and human(s).

Finally, we design manipulable signals and designable blocks.

Sections 14.2 and 14.3 present two of the authors' work on human–robot collaborations. Based on the above classification, these studies can be categorized as shown in Table 14.2. Section 14.2 covers multirobot navigation following the line of the passivity-based approaches in Hokayem and Spong (2006), Nu no et al. (2011), and Hatanaka et al. (2015). We address the human passivity shortage, pointed out in Dyck et al. (2013) and Hatanaka et al. (2017b), and present a novel system architecture that ensures stability for the overall system including a possibly passivity-short human

Table 14.2 Categorization of the contents in Sections 14.2 and 14.3 in this chapter.

	Section 14.2	Section 14.3
Subtask	Navigation	Reaching
Team	One-human–multiple-robots	One-human–one-robot
Architecture	Human-in-the-controller	Human-in-the-controller
Human role	Active	Active
Human model	Ppassivity	Gaussian process
Interaction	Complementary	Overlapping
Human → Robot	Velocity command	Velocity command
Robot → Human	Average position/velocity (visually)	Position and vision

operator. Section 14.3 focuses mainly on human modeling and exemplifies the great potential that machine learning methods have for designing human–robot collaborations. There it is demonstrated that the variance information generated by the GP enables ideal operation support for the human operator.

14.2 Passivity-Based Human-Enabled Multirobot Navigation

This section covers a complementary interaction for a one-human–multiple-robot team shown in Figure 14.8. We specifically address the following three challenges:

- designing a distributed control architecture among a human and robots,
- modeling and passivity analysis of a human with an active role,
- and human workload analysis.

The work is motivated by our previous work (Hatanaka et al. 2017b), where we presented a passivity-based distributed control architecture for multirobot navigation. Despite rigorous guarantees of stability/synchronization under human passivity, the user study therein indicated that a

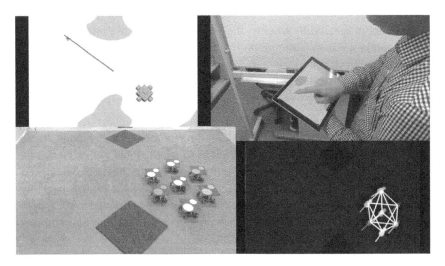

Figure 14.8 A snapshot of an experiment on human-enabled multirobot navigation.

human operator may fail to attain passivity depending on the network structure among a human and robots. In this section, we present a novel architecture that does not require human passivity based on the theory of *passivity-short systems*.

Consider a team formed by a human operator and n mobile robots $\mathcal{V} = \{1, \dots, n\}$. the motion of each robot is assumed to obey the kinematic model

$$\dot{q}_i = u_i, \quad i \in \mathcal{V} \tag{14.5}$$

where $q_i \in \mathbb{R}^2$ and $u_i \in \mathbb{R}^2$ are the position and velocity input of robot i, respectively. The communication network among the robots is modeled by an undirected graph $G = (\mathcal{V}, \mathcal{E})$, $\mathcal{E} \subset \mathcal{V} \times \mathcal{V}$, which is assumed throughout this section to be fixed and connected.[2] Each robot i exchanges information with the neighboring robots defined as $\mathcal{N}_i = \{j \in \mathcal{V} \mid (i, j) \in \mathcal{E}\}$.

The role of the human operator is to decide whether robot positions or velocities are to be controlled, and to navigate all robots to values that he/she deems desired. More precisely, the control objective for position navigation is formulated as

$$\lim_{t \to \infty} \|q_i - r_q\| = 0, \quad \forall i \in \mathcal{V} \tag{14.6}$$

while the objective for velocity navigation is formulated as

$$\lim_{t \to \infty} \|\dot{q}_i - r_v\| = 0, \quad \lim_{t \to \infty} \|q_i - q_j\| = 0, \quad \forall i, j \in \mathcal{V} \tag{14.7}$$

where $r_q \in \mathbb{R}^2$ and $r_v \in \mathbb{R}^2$ are the position and velocity references desired by the operator, respectively. To this end, the human operator visually feeds back a signal y_h associated with the robots' states through the interface, and sends velocity commands u_h to the interface. The interface is connected only to robots belonging to a nonempty subset of robots $\mathcal{V}_h \subseteq \mathcal{V}$. Consequently, the operator may only interact with a limited number of robots.

14.2.1 Architecture Design

In this section, we present a distributed control architecture for stably navigating robots, assuming the complementary interactions between a human and robots. The robots handle a subtask of distributed motion synchronization. To this end, the input u_i to each robot is designed based on a proportional-integral consensus algorithm (Freeman et al. 2006; Bai et al. 2010) as

$$u_i = \delta_i v_s + \sum_{j \in \mathcal{N}_i} a_{ij}(q_j - q_i) + \sum_{j \in \mathcal{N}_i} b_{ij}(\xi_i - \xi_j) \tag{14.8a}$$

$$\dot{\xi}_i = \sum_{j \in \mathcal{N}_i} b_{ij}(q_j - q_i) \tag{14.8b}$$

where $v_s \in \mathbb{R}^2$ denotes an additional input that will be designed later, and $\xi_i \in \mathbb{R}^2$ denotes the internal state of the controller. Furthermore, $a_{ij} = a_{ji} \ \forall i, j \in \mathcal{V}$ with $a_{ij} > 0$ for $(i, j) \in \mathcal{E}$, otherwise $a_{ij} = 0$. The gains b_{ij} obey the same rule. The symbol δ_i takes $\delta_i = 1$ if $i \in \mathcal{V}_h$ and $\delta_i = 0$ otherwise.

The operator intervenes among the coordinated robots through an additional input v_s. By the definition of δ_i, (14.8) meets the communication constraints between a human and robots. The operator takes an active role of navigating the robots toward some desirable motion. The complementary

---

2 The use of communication poses an inherent challenge for robustification against communication delays. Readers interested in this issue are recommended to refer to Yamauchi et al. (2017) and Funada et al. (2020). Experimental studies with long distance communication are also reported in Hatanaka et al. (2017a) and Funada et al. (2020), where the former highlights communication between a human and robots while the latter mainly addresses the inter-robot communication.

division of a task into subtasks drastically alleviates the human workload compared to direct control interactions, which will be explained in the sequel.

Let us now consider the robotic team represented by (14.5) and (14.8) with the graph G. Hatanaka et al. (2017b) showed that by taking the output as the average robot position

$$z_q = \frac{1}{|\mathcal{V}_h|} \sum_{i \in \mathcal{V}_h} q_i \tag{14.9}$$

the robotic team is inherently passive from v_s to z_q. Meanwhile, taking the output as the average robot velocity

$$z_v = \frac{1}{|\mathcal{V}_h|} \sum_{i \in \mathcal{V}_h} \dot{q}_i \tag{14.10}$$

and assuming differentiability of v_s, the robotic team is also shown to be passive from $\dot{v}_s$ to z_v. It is well known that feedback interconnection of passive systems guarantees stability under additional minor assumptions (Hatanaka et al. 2015). These passivity properties suggest that we should connect the operator and robots through the 2-D variables v_s and z_q or z_v, as $v_s = u_h$ and $y_h = z_q$ or $y_h = z_v$, assuming that the operator behaves as a passive system. The block diagram of this system is illustrated in Figure 14.9, where $y_h = z_q$ and $y_h = z_v$ are switched by the selected quantity to be controlled. In this architecture, the human only interacts with a single virtual robot with average position z_q or velocity z_v. Accordingly, the human workload is much smaller than when the human exerts direct control over multiple robots. Moreover, both (14.6) and (14.7) were shown to be ensured under human passivity in Hatanaka et al. (2017b) even without sharing the quantity to be controlled among the robots. However, unfortunately, Hatanaka et al. (2017b) also showed that a human may fail to attain passivity depending on the network structure.

To relax the assumption of human passivity, this section presents another architecture that was originally presented in Atman et al. (2018a). The design of this architecture is based on those for output synchronization of passivity-short systems (Qu and Simaan 2014) and bilateral teleoperation (Hokayem and Spong 2006; Nu no et al. 2011; Hatanaka et al. 2015). Let us first focus on position navigation. In the present architecture, we utilize a virtual master robot that is implemented in the interface. Its position is denoted by q_m, and its dynamics are given by

$$\dot{q}_m = u_h \tag{14.11}$$

This means that the operator only controls the master and none of the robots $\mathcal{V}$, in the same way as bilateral teleoperation. The operator visually feeds back

$$y_h = q_m + k_m(z_q - q_m) = k_m z_q + (1 - k_m) q_m \tag{14.12}$$

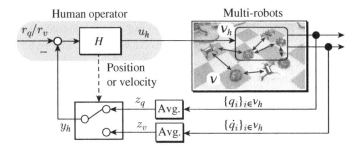

Figure 14.9 Block diagram of passivity-based multirobot navigation system.

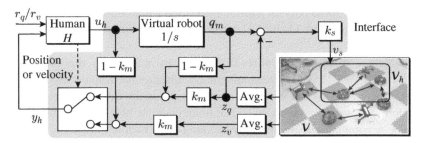

Figure 14.10 Block diagram of a passivity-shortage-based multirobot navigation system.

which blends the master position and the average position of the real robots. We also design v_s in (14.8) as

$$v_s = k_s(q_m - z_q) \tag{14.13}$$

which stems from the architecture in Qu and Simaan (2014).

Let us now consider a system consisting of the robot dynamics (14.5), the robot controller (14.8) with (14.13) and the graph G, the master robot (14.11), and the operator from $r_q - y_h$ to u_h, where y_h is given by (14.12). Assume that r_q is constant and that the cascade system of the operator and the master robot is input passivity-short from $r_q - y_h$ to $q_m - r_q$ with *impact coefficient* $\epsilon \in (0, 1)$, that is, $\exists \beta \geq 0$ such that

$$\int_0^\tau (q_m(t) - r_q)^{\mathrm{T}}(r_q - y_h(t))dt \geq -\beta - \epsilon \int_0^\tau \|r_q - y_h(t)\|^2 dt$$

The system can then be shown to achieve the control goal (14.6) under $k_m \in (0, 1)$, $k_s > 0$, and additional assumptions on the boundedness of the signals (Atman et al. 2018a).

In the case of velocity navigation, we reassign the visual information in (14.12) into

$$y_h = k_m z_v + (1 - k_m)\dot{q}_m \tag{14.14}$$

where y_h is switched at the interface depending on the selected quantity to be controlled. Let us now consider a system consisting of the robot dynamics (14.5), the robot controller (14.8) with (14.13) and the graph G, the master robot (14.11), and the operator from $r_v - y_h$ to u_h, where with y_h is given by (14.14). Assume that r_v is constant and that the operator is input passivity-short from $r_v - y_h$ to $\dot{q}_m - r_v$ with impact coefficient $\epsilon \in (0, 1)$, that is, $\exists \beta \geq 0$ such that

$$\int_0^\tau (\dot{q}_m(t) - r_v)^{\mathrm{T}}(r_v - y_h(t))dt \geq -\beta - \epsilon \int_0^\tau \|r_v - y_h(t)\|^2 dt$$

Similar to the scene for position navigation, the system achieves the control goal (14.7) under $k_m \in (0, 1)$, $k_s > 0$, and additional assumptions on the boundedness of the signals (Atman et al. 2018a) and the time invariance of the human (Atman et al. 2018a).[3] The block diagram for the overall system, including the switch between (14.12) and (14.14), is given in Figure 14.10.

14.2.2 Human Passivity Analysis

As stated at the end of Section 14.2.1, the passivity shortage of the system including the operator and the size of the impact coefficient $\epsilon \in (0, 1)$ are keys in achieving the goals represented by

---

3 The time invariance of the human is discussed in Hatanaka et al. (2017b), where human behavior is shown to be almost time invariant after training for system operation.

(14.6) and (14.7). In this section, we focus only on position navigation and examine the validity of the assumption through closed-loop system identification techniques using human operation data taken from a HIL simulator.

The HIL simulator treats a network of six robots that are interconnected by the network illustrated in Figure 14.11 with $\mathcal{V}_h = \{1, 3, 5\}$. The simulator also displays the window shown in Figure 14.12 on a tablet, where the gray dot is y_h, and the black dot is the externally provided reference r_q, which randomly jumps every 15 seconds. A human operator holds and monitors a tablet and specifies a velocity command u_h (red line) by touching the screen to navigate y_h to r_q. A sample of the time series data is illustrated in Figure 14.13. Throughout this section, we set $k_s = 3$ and $a_{ij} = 1, b_{ij} = 0.4 \; \forall (i,j) \in \mathcal{E}$. We take two options for the gain k_m, $k_m = 0.1$, and $k_m = 0.9$.

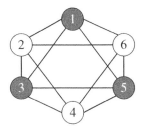

Figure 14.11 Graph *G* in the human-in-the-loop simulator.

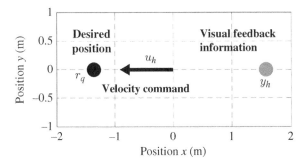

Figure 14.12 Display monitor in the human-in-the-loop simulator.

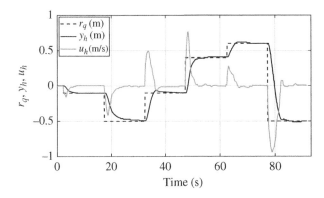

Figure 14.13 A sample of time series data from the human-in-the-loop simulator, where the solid black, solid gray, and dashed black lines show the responses of y_h, u_h, and r_q, respectively.

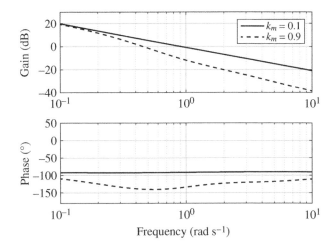

Figure 14.14 Bode diagrams of $T(s)$ for $k_m = 0.1$ (solid black) and $k_m = 0.9$ (dotted black).

By equivalently transforming Figure 14.10 into the standard feedback system with the human from $r_q - y_h$ to u_h and a block from u_h to y_h, the collection of blocks other than the human can be represented by a transfer function:

$$T(s) = \frac{1}{s} - \frac{k_m}{1 + k_s G_{\mathrm{RS}}(s)}$$

where $G_{\mathrm{RS}}(s)$ denotes the transfer function of the robotic team from v_s to z_q. The bode diagrams of $T(s)$ for $k_m = 0.1$ and $k_m = 0.9$ are illustrated in Figure 14.14. While the diagram with $k_m = 0.1$ is close to the single integrator, that with $k_m = 0.9$ has lower gains and additional phase lags.

In the present user study, 10 subjects participated in the trials. The participants received explanations about the system and task in advance, and they had training to familiarize themselves with operation under the $k_m = 0.5$ setting. They then conducted two trials with $k_m = 0.1$ and $k_m = 0.9$, one of which corresponded to 90 seconds of operation and is shown in Figure 14.13.

We applied a closed-loop system identification method, called joint input–output method (Katayama 2005), to the above data, where we took a nonparametric system identification method similarly to Hatanaka et al. (2017b). Approximating the human-operator block with a linear time invariant system, the impact coefficient ϵ is known to be equal to the minimal real part of the frequency response over the frequency domain (Qu and Simaan 2014). An example of a Nyquist plot, for Participant 5, is shown in Figure 14.15. We see from this figure that the impact coefficient ϵ lies within $(0, 1)$ as assumed in theory. Table 14.3 summarizes the impact coefficients of all participants computed in this way. The table does not give any meaningful insight into how k_m affects human passivity, but it does confirm that the assumption $\epsilon < 1$ is satisfied for all cases. We thus conclude from the results that it is reasonable to assume that the human operator will behave as a passivity-short system with $\epsilon < 1$. In combination with the theory, the present control algorithm is expected to achieve the control goal even in the absence of the human passivity assumption.

14.2.3 Human Workload Analysis

Let us get back to (14.12) and (14.14) as perceived by the operator. We immediately see from these equations that a small $k_m \in (0, 1)$ renders y_h close to the master's information, q_m or $\dot{q}_m$. While the

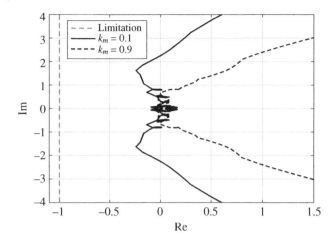

Figure 14.15 Nyquist plot of Participant 5 for $k_m = 0.1$ (solid black) and $k_m = 0.9$ (dotted black).

Table 14.3 Impact coefficient ϵ of 10 participants for $k_m = 0.1$ and $k_m = 0.9$.

	$k_m = 0.1$	$k_m = 0.9$
Participant 1	0.251	0.228
Participant 2	0.172	0.219
Participant 3	0.242	0.088
Participant 4	0.197	0.104
Participant 5	0.146	0.102
Participant 6	0.275	0.278
Participant 7	0.468	0.242
Participant 8	0.240	0.103
Participant 9	0.313	0.694
Participant 10	0.328	0.539

real robots tend to have complicated dynamics, as seen in Figure 14.14, that of the master robot is just a single integrator, which is easier for the operator to control. Consequently, reducing k_m is expected to mitigate the human workload.

To this end, we asked the same subjects in Section 14.2.2 to repeat 20 trials, 10 for $k_m = 0.1$ and 10 for $k_m = 0.9$. After each trial, the participants were required to fill out the prepared NASA TLX questionnaires (Hart and Staveland 1988). The quantified workload scores for all 10 participants are illustrated in Figure 14.16. Each bar consists of six components: the mental demand, physical demand, temporal demand, levels of performance, effort, and frustration. We see from this figure that all participants perceived a greater workload for $k_m = 0.9$ than $k_m = 0.1$, as anticipated.

However, as stated in Atman et al. (2018a,b), taking a small k_m makes the real robots invisible to the operator, which degrades transparency for the operator. This leads to poor high-level decisions and recognition for the operator. For example, if the direction of the human command u_h is blocked by an obstacle, then the operator may fail to recognize and react to the event. Thus, it is expected in

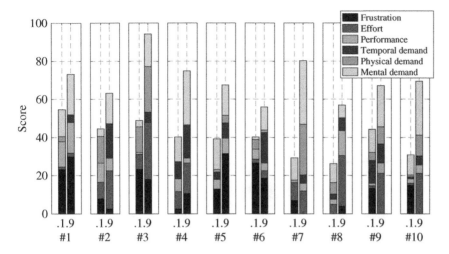

Figure 14.16 Perceived workload of 10 participants for $k_m = 0.1$ (left) and $k_m = 0.9$ (right).

the future that a trade-off between transparency and workload will be managed by appropriately tuning k_m based on the human state like fatigue.

14.3 Operation Support with Variable Autonomy via Gaussian Process

In this section, we consider a scenario with overlapping interactions between a human operator and a robot manipulator, and address the following three challenges:

- designing an operation support system with situation adaptive autonomy,
- modeling and estimating intention of a human with an active role through GP,
- and analyzing human performance and workload.

The work was motivated by remote harvesting in agricultural applications. We specifically address its partial task, namely a reaching task navigating the end-effector to a crop.

We begin by presenting the system components, a robot manipulator, a computer, and a human operator.

We employed the MotoMINI (YASKAWA Electric Corporation) robot manipulator and the YRC1000micro industrial robot controller, both of which are shown in Figure 14.17.[4] The position of the end-effector relative to the robot's coordinate frame Σ is denoted by x. The controller implements a local control algorithm so that the end-effector velocity $\dot{x}$ tracks a velocity reference u generated by the computer. A camera with the capacity for capturing multiple crops is placed in the eye-to-hand configuration, where the camera pose is independent of the robot motion.

We also used a computer, Surface Pro 7 (Microsoft Corporation), running Ubuntu 16.04. The computer exchanges information with YRC1000micro through the Robot Operating System (ROS). The primary role of the computer is to determine the velocity reference u that the local controller follows. The computer also receives image data from the camera and detects the

---

4 The product names mentioned in this paper may be trademarks and/or registered trademarks of their respective companies.

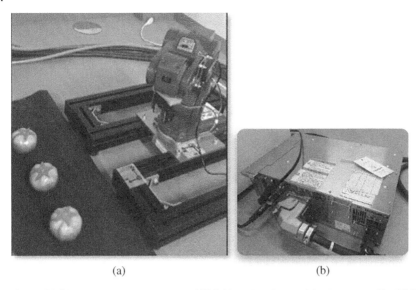

(a)   (b)

Figure 14.17   Robot manipulator MotoMINI (a) and an industrial robot controller YRC1000micro (b).

positions of crops, $r_1, r_2, \dots, r_n$, relative to Σ. A target crop at position r is identified from among these crops by the computer based on a designated rule (e.g. choose the one closest to the current x or the one with the highest similarity to sample images). Note that the target crop may differ from the next crop that the operator wishes to harvest primarily for the following two reasons: In the harvesting task, one wishes to harvest only fully ripe crops, but the ability of the computer to identify ripe crops may be inferior to that of humans. Even if ripe crops are correctly identified by the computer in an image containing multiple candidates, the operator may choose one other than r according to their preference. In other words, uncertain human intention may cause a target mismatch. In what follows, we denote by r_h the position in Σ of the target crop for the operator. Due to the better identifiability of the operator, the goal of the control algorithm on the computer is to ensure that the end-effector x reaches the human reference r_h.

The human operator, who monitors the image data from the camera, first identifies the target crop r_h. Additionally, the operator holds a game controller and determines a velocity command u_h that allows x to be driven to r_h.

In this section, we design an operation support system illustrated in Figure 14.18 for the reaching task. The primary motivation for doing so is that despite the human operator's ability to identify fully ripe crops, it is not assumed that the operator has sufficient robot operating skills. It is by compensating for this assumed deficiency that the computer supports the system operation. In other words, the task inherently requires an overlapping interaction between the operator and the robot, where the velocity reference u is determined by blending the human command u_h and the automatic control u_a. The central issue in the system design is how to blend the inputs depending on the situation.

To address this issue, we also built a training system, shown in Figure 14.19, that simulates idealized situations with only the end-effector of a manipulator and a single clear target. The window includes the target position r determined by the computer, the robot position x, the manual command u_h, and the TimeLimit circle that will be explained later. The operator was asked to follow the externally provided r by manipulating a joystick on the game controller. The time series data for (r, u_h, x) acquired on the training system are denoted by $\mathcal{D}_o = \{r_o(k), u_o(k), x_o(k)\}_{k=0}^{K}$. We then assume that the data $\mathcal{D}_o$ is available for designing the operating support system. We also assume that

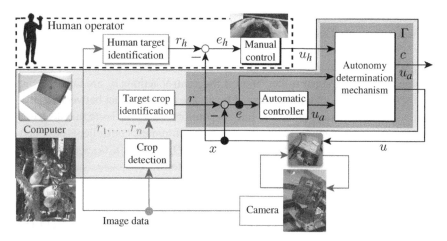

Figure 14.18 Block diagram of the operation support system, where the processes performed by a human are shown in the area enclosed by the dotted line and those by a computer are highlighted in light gray. The main focus of this section lies in the design for the collection of the blocks in dark gray, which determine the velocity reference u from the manual control u_h, the target crop r determined by the computer, and the current robot position x.

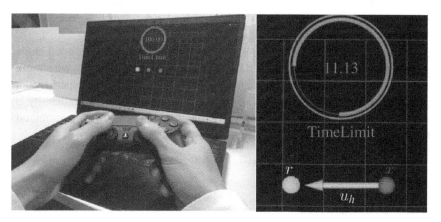

Figure 14.19 The training system in which the operator takes the visual information on the right and specifies the velocity command u_h using a game controller, the goal being to ensure the ensure the dark gray dot, x, tracks the light gray one, r.

data from an expert in the system operation, denoted by $\mathcal{D}_e = \{r_e(k), u_e(k), x_e(k)\}_{k=0}^K$, was available for the system design. In this section, we prepare the data for seven subjects including an expert of the system operation. The expert conducted two experiments, where x was driven to $r = -0.08$ m and $r = 0.08$ m, both from the initial state $x(0) = 0$. The data was sampled with a frequency of 30Hz and was decimated to 40 samples, which were used for $\mathcal{D}_e$. The six remaining nonexpert subjects conducted the experiments for $r = \pm0.06$ m, $r = \pm0.12$ m, and $r = \pm0.18$ m, all from $x(0) = 0$. The data for each nonexpert subject was decimated to 100 samples from which we prepared the data set $\mathcal{D}_o$.

14.3.1 Design of the Operation Support System with Variable Autonomy

In this section, we design an automatic controller for determining the signal u_a and a mechanism for determining autonomy, K_h and K_a in (14.1). The basic policy for designing the latter is to prioritize automatic control u_a if the real scene is close to the idealized training system, and to prioritize manual control u_h if the real scene includes uncertain factors not duplicated in the training system. The latter may involve not only the case, where $r \neq r_h$ but also scenarios where the direct path from x to r_h is blocked by an obstacle and the operator has to take commands to avoid it.

We begin by designing the automatic controller. Now, if the control actions u_a are far from the human actions, then the manual control will be affected by u_a due to the conflict between u_a and u_h, which might negatively impact performance and human comfort. This is an issue that does not exist in the normal control system design. To address the issue, the automatic controller is designed from the data gathered on the expert's use of the training system, $D_e = \{r_e(k), u_e(k), x_e(k)\}_{k=0}^{K}$, the assumption being that the expert's actions are natural to any operator.

Motivated by this design policy, we first build an expert model from the data. In this section, we take the model to have the form

$$u_e(k) = f_e(e_e(k)), \quad e_e(k) = r_e(k) - x_e(k) \tag{14.15}$$

which assumes that the expert determines the control action based on the error $e = r - x$. We employ GP to build the model. The GP provides not only the mean function $\mu_e(e)$ of $f_e(e)$ but also the variance function $\sigma_e^2(e)$ for any error e, which will enable flexible autonomy tuning. Let us now take the mean $\mu_e(e(t))$ as the automatic control $u_a(t) = \mu_e(e(t))$. The GP expert model built from the data D_e is shown in Figure 14.20a, where we separately build two models (μ_e^i, σ_e^i) $(i = 1, 2)$ using data from $x(0)$ to $r = 0.08$ m and from $x(0)$ to $r = -0.08$ m, respectively. Please refer to Hatanaka et al. (2020) behind the construction of these two models.

Let us now imagine the normal scene for which $r = r_h$ and the real environment does not include any factor that is not duplicated in the training system. Then, as mentioned before, we should prioritize the automatic control $u_a(t) = \mu_e(e(t))$ at each time instant t as long as the expert model is reliable around the current error $e(t)$. An advantage of the GP is that the model reliability is quantified by the variance $\sigma_e^2(e)$. To reflect the model reliability, we blend the human command $u_h(t)$ and the automatic control $u_a(t)$ using

$$u_{\text{normal}}(t) = a(\sigma_{et}^1, \sigma_{et}^2)u_a(t) + (1 - a(\sigma_{et}^1, \sigma_{et}^2))u_h(t) \tag{14.16}$$

$$u_a(t) = \sum_{i=1}^{2} \frac{\alpha(\sigma_{et}^i)\mu_{et}^i}{\alpha(\sigma_{et}^1) + \alpha(\sigma_{et}^2)} \tag{14.17}$$

$$\alpha(\sigma_{et}) = \frac{0.98}{1 + \exp(-857(\sigma_{et} - 0.013))} + 0.01 \tag{14.18}$$

$$a(\sigma_{et}^1, \sigma_{et}^2) = 1 - (1 - \alpha(\sigma_{et}^1))(1 - \alpha(\sigma_{et}^2)) \tag{14.19}$$

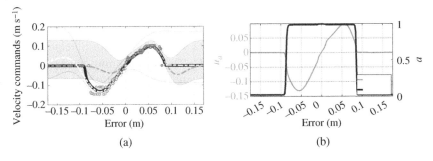

(a)　　　　　　　　　　　　(b)

Figure 14.20 (a) Two GP mean models from the data set with $r_h = -0.08$ (dashed light gray curve) and $r_h = 0.08$ (dashed dark gray curve), where the light and dark gray shaded area describe the 95% con?dence regions. The solid black curve illustrates the support from automatic control in the normal scene. (b) The automatic control for u_a (gray) and the reliability parameter a (black).

where $\mu_{et}^i = \mu_e^i(e(t))$ and $\sigma_{et}^i = \sigma_e^i(e(t))$ $(i = 1, 2)$. The functions $a(\sigma_e^1(e), \sigma_e^2(e))$ and $\mu_e(e)$ are illustrated in Figure 14.20b. Please refer to Hatanaka et al. (2020) for more details on how to design $\alpha(\sigma_{et})$ in (14.18). The amount of support au_a in (14.16) is illustrated by the solid purple curve in Figure 14.20a.

We next consider an unexpected scene for which $r \neq r_h$ and/or there are factors that differ significantly between the real environment and the training system. In this scene, it is desirable to prioritize the manual control u_h. To this end, we need to quantify the degree of anomaly for the operator's actions. The variance function generated by the GP model is also expected to be useful for this purpose. Since it is the operator's anomaly that we need to quantify, we have to refer to the operator's model rather than the expert model.

To identify the operator's anomaly, we build the operator's GP model from the offline data $D_o = \{r_o(k), u_o(k), x_o(k)\}_{k=0}^K$ on the training system, assuming the regression model

$$u_o(k) = f_o(e_o(k), u_o(k-1)), \quad e_o(k) = r_o(k) - x_o(k) \tag{14.20}$$

Note that the command $u_o(k-1)$ is added as an explanatory variable in a different manner than (14.15), thus allowing one to quantify the anomaly of the operator's current actions. Figure 14.21 illustrates the variance function $\sigma_o(e, u_h)$ along with the data for one of the six nonexpert operators. We see from the figure that the data are not uniformly distributed over the domain of (e, u_h), reflecting the operator's normal actions for reducing e. Accordingly, the variance tends to be large over the region with sparse data. In other words, the variance can be regarded as a metric for the operator's anomaly, which can be computed online from the feedback information given by $e(t)$ and $u_h(t-1)$ at each time instant t.

Finally, let us combine the anomaly parameter $\sigma_o(e(t), u_h(t-1))$ with the control (14.16) as

$$u(t) = (1 - b(\sigma_{ot}))u_h(t) + b(\sigma_{ot})u_{normal}(t) \tag{14.21}$$

with $\sigma_{ot} = \sigma_o(e(t), u_h(t-1))$. The weighting coefficient $b(\sigma_{ot})$ is designed in the same way as (14.18) with σ_{et} replaced by σ_{ot}. (14.21) then means that the manual control u_h is prioritized when the anomaly parameter $\sigma_o(e(t), u_h(t))$ is high even if the expert model is reliable around $e(t)$.

Substituting (14.16) into (14.21) yields

$$u(t) = (1 - c(\sigma_{et}, \sigma_{ot}))u_h(t) + c(\sigma_{et}, \sigma_{ot})u_a(t) \tag{14.22}$$

where $c(\sigma_{et}, \sigma_{ot}) = a(\sigma_{et})b(\sigma_{ot})$. The algorithm for determining the autonomy is illustrated in Figure 14.22. The mechanism for producing $c(\sigma_{et})$ and u_a from x and r is represented by the symbol Γ in Figure 14.18 and is used as a module in the sequel.

So far we have investigated a scene in which the computer determines just one candidate r for the human intention r_h. In real crop harvesting, however, multiple crops may be captured by the camera, making it fairly difficult for the computer to estimate which target crop r_h is intended by the operator. When $r \neq r_h$, the mechanism Γ always prioritizes manual control, and the operation

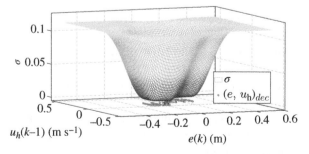

Figure 14.21 The variance function of a nonexpert operator's GP model.

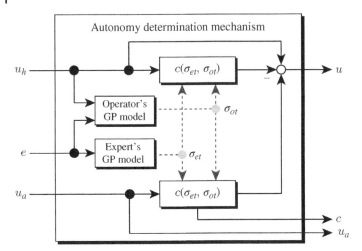

Figure 14.22 Block diagram of the autonomy determination mechanism in Figure 14.18.

support may work only in very limited cases. A way to avoid this problem is to run the module Γ for multiple candidates $r_1, \dots, r_m$ $(m \leq n)$ among n detected targets. Denoting each module by Γ_i $(i = 1, \dots, m)$ and their outputs by $c_i(\sigma_{et})$ and $u_{a,i}$, we need to appropriately blend $u_{a,i}$ $(i = 1, \dots, m)$ depending on $c_i(\sigma_{et})$ $(i = 1, \dots, m)$. To this end, we present the mechanism

$$u(t) = (1 - \bar{c})u_h(t) + \sum_{i=1}^{m} \frac{c(\sigma_{et,i}, \sigma_{ot,i})u_{a,i}(t)}{\bar{c}} \tag{14.23}$$

$$\bar{c} = \sum_{i=1}^{m} c(\sigma_{et,i}, \sigma_{ot,i}) \tag{14.24}$$

$$\sigma_{et,i} = \sigma_e(r_i(t) - x(t)), \quad \sigma_{ot,i} = \sigma_o(r_i(t) - x(t), u_h(t)) \tag{14.25}$$

To clarify, a collection of operator GP models in Γ_i $(i = 1, \dots, m)$ works as a multiple-model intention estimator for the human. A similar concept was also presented in Dani et al. (2020). Note that (14.20) depends only on the error e rather than the reference r, so we only have to run the same m modules simultaneously by changing r_i without building different models specialized to r_i. The overall system is illustrated in Figure 14.23.

14.3.2 User Study

In this section, we demonstrate the effectiveness of the present operation support system.

14.3.2.1 Operational Verification
We begin by presenting the operational verification, where we first conduct two experiments with $r = r_h$ and $r \neq r_h$.

In the first experiment, we demonstrate the case of $r_h = r = -0.14$ m with $x(0) = 0$ m. Note that the initial error $e(0) = r - x(0) = -0.14$ m exceeds the range, ± 0.08 m, of the training data for the expert. The expert model at around $e = -0.14$ is not reliable, and the variance tends to be large. It is thus expected that manual control is initially prioritized but then switches to automatic control once e is within ± 0.08 m. The time series data for e, u_h, u, and u_a are illustrated in Figure 14.24. It is observed here that the velocity reference u to the robot almost coincides with the human input u_h up until one second, during which the error e is outside of ± 0.08 m. Once e is within that range, u starts

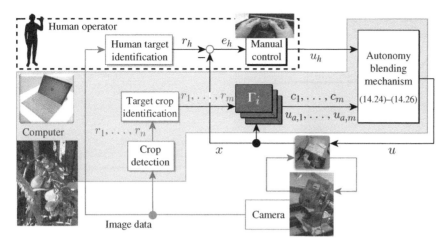

Figure 14.23 Block diagram of the operation support system with a human intention estimator based on multiple GP models.

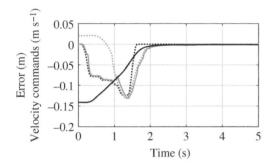

Figure 14.24 Time series data for e (solid black), u_h (dotted black), u (solid gray), and u_a (dotted gray) when $r_h = r = -0.14$ m.

following the automatic control u_a as expected. Consequently, automatic control only supports accurate adjustment of x to the reference $r_h = r$, which is demanding for untrained operators.

Next, we address the case of a target mismatch, where $r_h \neq r$, and we set $r_h = 0$ m, $r = -0.14$ m, and $x = -0.14$ m. The same signals presented in Figure 14.24 are also shown in Figure 14.25. Initially, the automatic control is zero since the controller thinks that the error $e(0) = r - x(0)$ is zero. Meanwhile, the operator has another target crop at $r_h = 0$ m, and he/she provides the positive command u_h for reducing the error $e_h = r_h - x$, which must be significantly different from normal actions to reduce e. The reference u, then, obeys the human command u_h with slight delays while ignoring the automatic control u_a. Consequently, it is concluded that manual control is prioritized in the case of a target mismatch, as expected.

We finally demonstrate the algorithm in Figure 14.23 with multiple Γ_i. In the experiment, we set two target crops $r_1 = 0$ m and $r_2 = -0.14$ m and prepared two mechanisms Γ_1 and Γ_2. The experiment started from $x(0) = 0$ m and the operator drove x to $r_h = -0.14$ m. The resulting signals are illustrated in Figure 14.26. We see from the figure that after the operator starts giving commands u_h, the velocity reference u follows the human command with a slight delay while ignoring the automatic control u_a, similar to Figure 14.25. They eventually coincide at around one second. At around 1.7 seconds, the error between x and $r_h = r_2$ falls into the ± 0.08 m range, and then $c(\sigma_{et,2})$ gets close to 1. Accordingly, the automatic control $u_{a,2}$ is prioritized and u almost synchronizes $u_{a,2}$

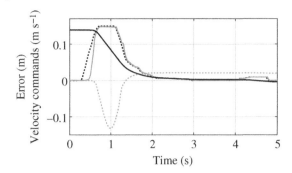

Figure 14.25 Time series data for e, u_h, u, and u_a when $r_h = 0$ m and $r = -0.14$ m.

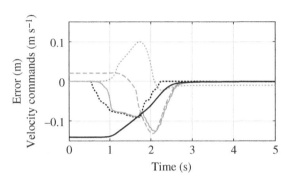

Figure 14.26 Time series data for e (solid black), u_h (dotted black), u (solid gray), $u_{a,1}$ with $r_1 = 0$ m (dotted gray), and $u_{a,2}$ with $r_2 = -0.14$ m (dashed gray).

while ignoring u_h. As a result, the settling of x to $r_h = r_2 = -0.14$ is automated differently from Figure 14.25.

In summary, the results above demonstrate that the present operation support system works as expected.

14.3.2.2 Usability Test

Let us next demonstrate the usability of the present system. To this end, we individually prepared operator models for the six nonexpert trial subjects along with the expert model. Each subject conducted the following four-phase trial twice, once without operation support and once with.

Phase (i): $r = 0$ m and $r_h = 0.14$ m
Phase (ii): $r = 0$ m and $r_h = 0$ m
Phase (iii): $r = -0.14$ m and $r_h = -0.14$ m
Phase (iv): $r = 0$ m and $r_h = 0$ m

Remark that only phase (i) contains a target mismatch. The phases were switched in order from (i) to (iv) at a constant interval. In all phases, the remaining time until the next switch was displayed at the top of the monitor, as can be seen in Figure 14.19, to prevent performance degradation due to insufficient information, an issue that was pointed out in Walker et al. (2018).

We used two metrics to evaluate system performance: the settling time and the NASA TLX (Hart and Staveland 1988). The former is defined as the time after which e_h is within $\pm 2\%$ of 0.14 m. In addition to those, we also took the p-value to assess how the support system affects the human operators.

The average settling time among the six subjects for each phase is shown in Figure 14.27, where the individual variations are illustrated by black bars. Figure 14.27 indicates that the case with operation support outperforms manual control in the absence of a target mismatch, while the support slightly degrades performance in the presence of a mismatch due to the delay observed in

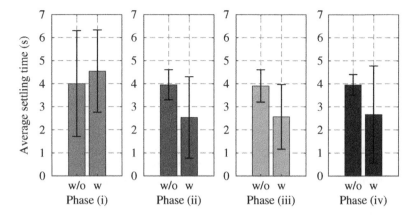

Figure 14.27 Average settling times among the six trial subjects for phases (i)–(iv).

Table 14.4 p-values under the null hypothesis that individual variations are ignorable (top) and that manual control is not affected by the system (bottom).

	Phase (i)	Phase (ii)	Phase (iii)	Phase (iv)
Individuals	0.0140	0.1223	0.0476	0.2806
w. versus w/o	0.3010	0.0527	0.0160	0.1592

Figure 14.25. Table 14.4 shows the p-values used to analyze individual variations and the difference between having and not having operation support. The large p-value in the bottom-left section indicates that the present system does not disturb manual control too much, even in the case of $r \neq r_h$, despite the delay in Figure 14.25. In addition, the larger values for phases (ii)–(iv) in the top row indicate that individual differences are reduced for phases (ii)–(iv) compared to (i), which would be beneficial especially for untrained operators.

Figure 14.28 illustrates the workload perceived by the individual subjects with and without operation support. It was observed that, with the exception of subject 6, the workload was reduced by the present support system, mainly because the precise settling of x to r_h was automated for phases (ii)–(iv). Notice that the score for subject 6 is minimal among all subjects. This indicates the possibility that subjects who are good at manually operating the robot may feel frustrated by the support.

14.4 Summary

In this chapter, we started by reviewing historical and contemporary issues regarding human–robot collaborations focusing on specific key factors, including task architecture, team formation, and human modeling. Although remarkable developments have been made both in academia and industry, real applications for technologies centered around human–robot collaboration are restricted to limited scenarios within structured environments, such as those found in laboratories and factories. However, the potential future applications of the technology are extremely broad, and it is no exaggeration to say that all labor-intensive tasks can be applications. To both improve productivity and free workers from hard and dangerous tasks, increasing the level of autonomy is an important issue. Taking crop harvesting as an example, it is not difficult to imagine the many barriers acting against full automation in real, unstructured, uncertain, and complex environments. On the other hand, a human, even a child, is capable of harvesting a variety of crops, even without expert knowledge. Learning the behavior of humans, as was discussed in Section 14.3, will be significant for pushing up the level of autonomy, but collecting rich data covering all situations is demanding and, at the very least, increases system cost. Innovations in accurate learning from limited amounts of data are inevitable as we progress toward a higher level of autonomy.

Control problems involving a high level of autonomy and taking into account factors such as human intention, situation awareness, trust, workload, and transparency, remain an untapped and lucrative area for research. Despite numerous efforts to resolve such problems in the past, real-world use cases of solutions are rarely reported. Innovations in sensing technologies for high-level cognitive signals are indispensable, but, as seen in Protte et al. (2020), fusing established ideas from other disciplines, including the humanities, with robot control technology is inevitable. As has been the case since the inception of the field (Goodrich and Schultz 2007), interdisciplinary, transdisciplinary, and even convergence research are a major driving force behind it.

We have presented our work on human-enabled multirobot navigation and human operation support with variable autonomy, both of which still possess many issues. In both of these problems, human studies are limited to one dimension. Our preliminary investigations indicated that poor human recognition in three dimensions drastically degrades not only control performance but also human comfort. We need to make more efforts on the interface design, including incorporating VR/AR technology. While the impact of such technology on people has been studied extensively in other fields, it also poses a question to control researchers: how and in what sense are such technologies advantageous from the control theoretic perspective? For example, one of our preliminary investigations suggested that using a VR control interface with a head-mounted display has a positive impact on human passivity in multirobot 3-D navigation (Hatanaka et al. 2022). More thorough investigations like this should be conducted in the future.

We would like to close this chapter by noting that a new motivation for automating or remotely performing tasks has arisen due to the COVID-19 pandemic. For example, the installation and utilization of robotic technology in food processing, as shown in Figure 14.29, has proceeded primarily to enhance labor productivity, but the pandemic has added a new aspect to this application since physical intervention by a person during food processing increases the risk of virus contamination.

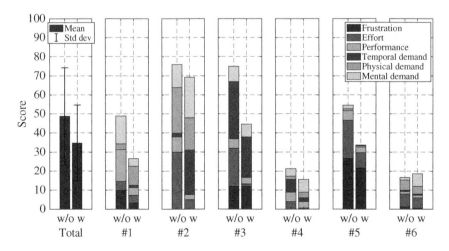

Figure 14.28 Perceived workloads quantified by NASA TLX.

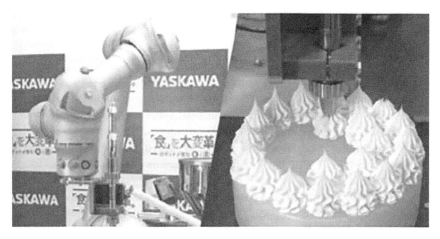

Figure 14.29 Cake decoration by a collaborative robot. Source: Courtesy of Yaskawa Electric Corporation.

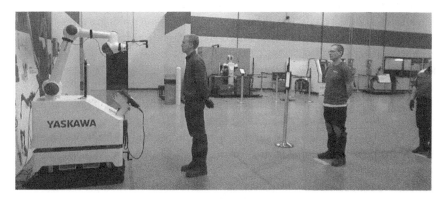

Figure 14.30 A temperature-taking robot. Source: Courtesy of Yaskawa Electric Corporation.

This is also true for medical applications (Figure 14.30), where increasing the level of autonomy is important for preventing the spread of infection. Considering the difficulties associated with fully automating any task, we can confidently say that the demand for human–robot collaborations are rising at an unprecedented rate.

Acknowledgments

We express our sincere thanks to Dr. M. W. S. Atman, Prof. Z. Qu, and Prof. N. Chopra for their fruitful discussions when completing the results in Section 14.2. We would also like to thank Mr. K. Noda, Mr. M. Horikawa, Mr. M. Adachi, Mr. R. Hirata, Mr. K. Sokabe, and Mr. K. Shimamoto for their dedicated support when completing the results in Section 14.3. This work was in part supported by JSPS KAKENHI Grant Number 21K04104.

References

Adams, J.A. and Skubic, M. (eds.) (2005). Special issue on human–robot interaction. *IEEE Transactions on Systems, Man, and Cybernetics: Part A – Systems and Humans* 35.

Akash, K., McMahon, G., Reid, T., and Jain, N. (2020). Human trust-based feedback control: dynamically varying automation transparency to optimize human–machine interactions. *IEEE Control Systems Magazine* 40 (6): 98–116. https://doi.org/10.1109/MCS.2020.3019151.

Anderson, R.J. and Spong, M.W. (1989). Bilateral control of teleoperators with time delay. *IEEE Transactions on Automatic Control* 34 (5): 494–501. https://doi.org/10.1109/9.24201.

Atman, M.W.S., Hatanaka, T., Qu, Z. et al. (2018a). Motion synchronization for semi-autonomous robotic swarm with a passivity-short human operator. *International Journal of Intelligent Robotics and Applications* 2 (2): 235–251. https://doi.org/10.1007/s41315-018-0056-8.

Atman, M.W.S., Hay, J., Yamauchi, J. et al. (2018b). Two variations of passivity-short-based semi-autonomous robotic swarms. *2018 SICE International Symposium on Control Systems*, 12–19. https://doi.org/10.23919/SICEISCS.2018.8330150.

Bai, H., Freeman, R.A., and Lynch, K.M. (2010). Robust dynamic average consensus of time-varying inputs. *49th IEEE Conference on Decision and Control*, 3104–3109. https://doi.org/10.1109/CDC.2010.5717485.

Baillieul, J., Leonard, N.E., and Morgansen, K.A. (eds.) (2012). Interaction dynamics: the interface of humans and smart machines. *Proceedings of the IEEE* 100 (3): 567–570.

Bogacz, R., Brown, E., Moehlis, J. et al. (2006). The physics of optimal decision making: a formal analysis of models of performance in two-alternative forced-choice tasks. *Psychological Review* 113 (4): 700–765. https://doi.org/10.1037/0033-295X.113.4.700.

Bradley, M.M. (2000). Emotion and motivation. In: *Handbook of Psychophysiology* (ed. J.T. Cacioppo, L.G. Tassinary, and G.G. Berntson), 602–642. Cambridge University Press.

Braithwaite, J. and Makkai, T. (1994). Trust and compliance. *An International Journal of Research and Policy* 4 (1): 1–12. https://doi.org/10.1080/10439463.1994.9964679.

Chen, J.Y.C. and Barnes, M.J. (2014). Human–agent teaming for multirobot control: a review of human factors issues. *IEEE Transactions on Human–Machine Systems* 44 (1): 13–29. https://doi.org/10.1109/THMS.2013.2293535.

Chen, J.Y.C., Barnes, M.J., and Harper-Sciarini, M. (2011). Supervisory control of multiple robots: human-performance issues and user-interface design. *IEEE Transactions on Systems, Man, and Cybernetics, Part C - Applications and Reviews* 41 (4): 435–454. https://doi.org/10.1109/TSMCC.2010.2056682.

Chipalkatty, R., Droge, G., and Egerstedt, M. (2013). Less is more: mixed-initiative model-predictive control with human inputs. *IEEE Transactions on Robotics* 29 (3): 695–703. https://doi.org/10.1109/TRO.2013.2248551.

Chong, N.Y., Kotoku, T., Ohba, K. et al. (2000). Remote coordinated controls in multiple telerobot cooperation. *IEEE International Conference on Robotics and Automation*, Volume 4, 3138–3143. https://doi.org/10.1109/ROBOT.2000.845146.

Cubuktepe, M., Jansen, N., and Topcu, U. (2023). Shared control with human trust and workload models. In: *Cyber–Physical–Human Systems: Fundamentals and Applications* (ed. A. Annaswamy, P.P. Khargonekar, F. Lamnabhi-Lagarrigue, and S.K. Spurgeon). UK: Wiley (in Press).

Cummings, M.L., How, J.P., Whitten, A., and Toupet, O. (2012). The impact of human–automation collaboration in decentralized multiple unmanned vehicle control. *Proceedings of the IEEE* 100 (3): 660–671. https://doi.org/10.1109/JPROC.2011.2174104.

Dani, A.P., Salehi, I., Rotithor, G. et al. (2020). Human-in-the-loop robot control for human–robot collaboration: human intention estimation and safe trajectory tracking control for collaborative tasks. *IEEE Control Systems Magazine* 40 (6): 29–56. https://doi.org/10.1109/MCS.2020.3019725.

de la Croix, J.-P. and Egerstedt, M. (2012). Controllability characterizations of leader-based swarm interactions. *AAAI Fall Symposium: Human Control of Bioinspired Swarms*.

Devol, G.C. (1961). Programmed article transfer. *US Patent # 2988237*.

Diaz-Mercado, Y., Lee, S.G., and Egerstedt, M. (2017). Human–swarm interactions via coverage of time-varying densities. In: *Trends in Control and Decision-Making for Human–Robot Collaboration Systems* (ed. Y. Wang and F. Zhang), 357–385. Switzerland: Springer. https://doi.org/10.1007/978-3-319-40533-9_15.

Dragan, A. and Srinivasa, S. (2013). A policy blending formalism for shared control. *International Journal of Robotics Research* 32 (7): 790–805. https://doi.org/10.1177/0278364913490324.

Dyck, M., Jazayeri, A., and Tavakoli, M. (2013). Is the human operator in a teleoperation system passive? *2013 World Haptics Conference*, 683–688. https://doi.org/10.1109/WHC.2013.6548491.

Egerstedt, M., de la Croix, J.-P., Kawashima, H., and Kingston, P. (2014). Interacting with networks of mobile agents. In: *Large-Scale Networks in Engineering and Life Sciences* (ed. P. Benner, R. Findeisen, D. Flockerzi et al.), 199–224. Cham: Springer. https://doi.org/10.1007/978-3-319-08437-4_3.

Endsley, M.R. (2011). *Designing for Situation Awareness An Approach to User-Centered Design*, 2e. CRC Press. ISBN 978-1420063554.

Erhart, S. and Hirche, S. (2016). Model and analysis of the interaction dynamics in cooperative manipulation tasks. *IEEE Transactions on Robotics* 32 (3): 672–683. https://doi.org/10.1109/TRO.2016.2559500.

Franchi, A., Secchi, C., Ryll, M. et al. (2012a). Shared control: balancing autonomy and human assistance with a group of quadrotor UAVs. *IEEE Robotics and Automation Magazine* 19 (3): 57–68. https://doi.org/10.1109/MRA.2012.2205625.

Franchi, A., Secchi, C., Son, H.I. et al. (2012b). Bilateral teleoperation of groups of mobile robots with time-varying topology. *IEEE Transactions on Robotics* 28 (5): 1019–1033. https://doi.org/10.1109/TRO.2012.2196304.

Freeman, R.A., Yang, P., and Lynch, K.M. (2006). Stability and convergence properties of dynamic average consensus estimators. *45th IEEE Conference on Decision and Control*, 338–343. https://doi.org/10.1109/CDC.2006.377078.

Funada, R., Cai, X., Notomista, G. et al. (2020). Coordination of robot teams over long distances from Georgia tech to Tokyo tech and back; an 11,000-km multi-robot experiment. *IEEE Control Systems Magazine* 40 (4): 53–79. https://doi.org/10.1109/MCS.2020.2990515.

Geravand, M., Werner, C., Hauer, K., and Peer, A. (2016). An integrated decision making approach for adaptive shared control of mobility assistance robots. *International Journal of Social Robotics* 8 (5): 631–648. https://doi.org/10.1007/s12369-016-0353-z.

Gioioso, G., Franchi, A., Salvietti, G. et al. (2014). The flying hand: a formation of UAVs for cooperative aerial tele-manipulation. *2014 IEEE International Conference on Robotics and Automation*, 4335–4341. https://doi.org/10.1109/ICRA.2014.6907490.

Goodrich, M.A. and Schultz, A.C. (2007). Human–robot interaction: a survey. *Foundations and Trends in Human–Computer Interaction* 1 (3): 203–275. https://doi.org/10.1561/1100000005.

Goodrich, M.A., Quigley, M., and Cosenzo, K. (2005). Task switching and multi-robot teams. In: *Multi-Robot Systems. From Swarms to Intelligent Automata Volume III: Proceedings from the 2005 International Workshop on Multi-Robot Systems* (ed. L.E. Parker, F.E. Schneider, and A.C. Schultz), 185–195. Dordrecht: Springer. https://doi.org/https://doi.org/10.1007/1-4020-3389-3_15.

Handa, H., Okumura, S., and Nio, S. (1997). The robotic easy teaching system in computer aided welding. *7th International Conference on Computer Technology in Welding*, 562–575.

Hart, S.G. and Staveland, L.E. (1988). Development of NASA-TLX (Task Load Index): results of empirical and theoretical research. In: *Human Mental Workload, Advances in Psychology*, vol. 52 (ed. P.A. Hancock and N. Meshkati), 139–183. North-Holland. https://doi.org/doi.org/10.1016/S0166-4115(08)62386-9.

Hatanaka, T., Chopra, N., Fujita, M., and Spong, M.W. (2015). *Passivity-Based Control and Estimation in Networked Robotics*. Springer. ISBN 978-3-319-15170-0.

Hatanaka, T., Chopra, N., Yamauchi, J. et al. (2017a). A passivity-based system design of semi-autonomous cooperative robotic swarm. *ASME Dynamic Systems and Control Magazine* 5 (2): 14–18. https://doi.org/10.1115/1.2017-Jun-6.

Hatanaka, T., Chopra, N., Yamauchi, J., and Fujita, M. (2017b). A passivity-based approach to human–swarm collaborations and passivity analysis of human operators. In: *Trends in Control and Decision-Making for Human–Robot Collaboration Systems* (ed. Y. Wang and F. Zhang), 325–355. Switzerland: Springer. https://doi.org/10.1007/978-3-319-40533-9_14.

Hatanaka, T., Noda, K., Yamauchi, J. et al. (2020). Human–robot collaboration with variable autonomy via Gaussian process. *3rd IFAC Workshop on Cyber-Physical & Human Systems*, 126–133. https://doi.org/10.1016/j.ifacol.2021.04.091.

Hatanaka, T., Mochizuki, T., Maestre, J.M., and Chopra, N. (2022). Impact of VR technology on a human in semi-autonomous multi-robot navigation: control theoretic perspective. *4th IFAC Workshop on Cyber-Physical Human Systems*, page to appear.

Hokayem, P.F. and Spong, M.W. (2006). Bilateral teleoperation: an historical survey. *Automatica* 42 (12): 2035–2057. https://doi.org/10.1016/j.automatica.2006.06.027.

Jain, N. (2023). Enabling human-aware autonomy through cognitive modeling and feedback control. In: *Cyber–Physical–Human Systems: Fundamentals and Applications* (ed. A. Annaswamy, P.P. Khargonekar, F. Lamnabhi-Lagarrigue, and S.K. Spurgeon). UK: Wiley (in Press).

Katayama, T. (2005). *Subspace Methods for System Identification*. London: Springer. ISBN 978-1-84628-158-7.

Kelley, R., Nicolescu, M., Tavakkoli, A. et al. (2008). Understanding human intentions via hidden Markov models in autonomous mobile robots. *3rd ACM/IEEE International Conference on Human–Robot Interaction*, 367–374. https://doi.org/10.1145/1349822.1349870.

Khademian, B. and Hashtrudi-Zaad, K. (2011). Dual-user teleoperation systems: new multilateral shared control architecture and kinesthetic performance measures. *IEEE Robotics and Automation Magazine* 17 (5): 895–906. https://doi.org/10.1109/TMECH.2011.2141673.

Kiesler, S. and Hinds, P.J. (eds.) (2004). *Special Issue on Human–Robot Interaction*, vol. 19. Human–Computer Interaction.

Kulić, D. and Croft, E.A. (2007). Affective state estimation for human–robot interaction. *IEEE Transactions on Robotics* 23 (5): 991–1000. https://doi.org/10.1109/TRO.2007.904899.

Laschi, C., Breazeal, C., and Nakauchi, Y. (eds.) (2007). Special issue on human–robot interaction. *IEEE Transactions on Robotics* 23.

Lawrence, D.A. (1993). Stability and transparency in bilateral teleoperation. *IEEE Transactions on Robotics and Automation* 9 (5): 624–737. https://doi.org/10.1109/70.258054.

Lee, J.D. and Moray, N. (1992). Trust, control strategies and allocation of function in human–machine systems. *Ergonomics* 35 (10): 1243–1270. https://doi.org/10.1080/00140139208967392.

Lee, J.D. and Moray, N. (1994). Trust, self-confidence, and operators' adaptation to automation. *International Journal of Human–Computer Studies* 40 (1): 153–184. https://doi.org/10.1006/ijhc.1994.1007.

Lee, J.D. and See, K.A. (2004). Trust in automation: designing for appropriate reliance. *The Journal of the Human Factors and Ergonomics Society* 46 (1): 50–80. https://doi.org/10.1518/hfes.46.1.50.30392.

Lee, D. and Spong, M.W. (2005). Bilateral teleoperation of multiple cooperative robots over delayed communication network: theory. *2005 IEEE International Conference on Robotics and Automation*, 360–365. https://doi.org/10.1109/ROBOT.2005.1570145.

Li, Y. and Ge, S.S. (2014). Human–robot collaboration based on motion intention estimation. *IEEE/ASME Transactions on Mechatronics* 19 (3): 1007–1014. https://doi.org/10.1109/TMECH.2013.2264533.

Liu, Y. and Chopra, N. (2012). Controlled synchronization of heterogeneous robotic manipulators in task space. *IEEE Transactions on Robotics* 28 (1): 268–275. https://doi.org/10.1109/TRO.2011 .2168690.

Malysz, P. and Sirouspour, S. (2011). Trilateral teleoperation control of kinematically redundant robotic manipulators. *The International Journal of Robotics Research* 30 (13): 1643–1664. https://doi.org/10 .1177/0278364911401053.

McRuer, D. (1980). Human dynamics in man–machine systems. *Automatica* 16 (3): 237–253. https://doi.org/10.1016/0005-1098(80)90034-5.

McRuer, D.T. and Jex, H.R. (1967). A review of quasi-linear pilot models. *IEEE Transactions on Human Factors in Electronics* HFE-8 (3): 231–249. https://doi.org/10.1109/THFE.1967.234304.

McRuer, D.T. and Krendel, E.S. (1974). Mathematical Models of Human Pilot Behavior. *AGARDograph*, AGARD-AG-188.

Medina, J.R., Lorenz, T., and Hirche, S. (2017). Considering human behavior uncertainty and disagreements in human–robot cooperative manipulation. In: *Trends in Control and Decision-Making for Human–Robot Collaboration Systems* (ed. Y. Wang and F. Zhang), 207–240. Switzerland: Springer. https://doi.org/10.1007/978-3-319-40533-9_10.

Mercado, J.E., Rupp, M.A., Chen, J.Y.C. et al. (2016). Intelligent agent transparency in human–agent teaming for Multi-UxV management. *Human Factors: The Journal of the Human Factors and Ergonomics Society* 58 (3): 3261–3270. https://doi.org/10.1177/0018720815621206.

Michael, N., Fink, J., and Kumar, V. (2011). Cooperative manipulation and transportation with aerial robots. *Autonomous Robots* 30: 73–86. https://doi.org/10.1007/s10514-010-9205-0.

Mohammadi, M., Franchi, A., Barcelli, D., and Prattichizzo, D. (2016). Cooperative aerial tele-manipulation with haptic feedback. *2016 IEEE/RSJ International Conference on Intelligent Robots and Systems*, 5092–5098. https://doi.org/10.1109/IROS.2016.7759747.

Moray, N., Inagaki, T., and Itoh, M. (2000). Adaptive automation, trust, and self-confidence in fault management of time-critical tasks. *Journal of Experimental Psychology: Applied* 6 (1): 44–58. https://doi.org/10.1037/1076-898x.6.1.44.

Mulder, M., Pool, D.M., Abbink, D.A. et al. (2018). Manual control cybernetics: state-of-the-art and current trends. *IEEE Transactions on Human–Machine Systems* 48 (5): 468–485. https://doi.org/10 .1109/THMS.2017.2761342.

Murphy, R.R. and Rogers, E. (eds.) (2004). Special issue on human–robot interaction. *IEEE Transactions on Systems, Man, and Cybernetics: Part C – Applications and Reviews* 34 (1): 101–102.

Musić, S. and Hirche, S. (2017). Control sharing in human–robot team interaction. *Annual Reviews in Control* 44: 342–354. https://doi.org/10.1016/j.arcontrol.2017.09.017.

Newell, G.S. and Mansfield, N.J. (2008). Evaluation of reaction time performance and subjective workload during whole-body vibration exposure while seated in upright and twisted postures with and without armrests. *International Journal of Industrial Ergonomics* 38 (5–6): 499–508. https://doi.org/10.1016/j.ergon.2007.08.018.

Nguyen, T., Lim, C.P., Nguyen, N.D. et al. (2019). A review of situation awareness assessment approaches in aviation environments. *IEEE Systems Journal* 13 (3): 3590–3603. https://doi.org/10 .1109/JSYST.2019.2918283.

Nu no, E., Basa nez, L., and Ortega, R. (2011). Passivity-based control for bilateral teleoperation: a tutorial. *Automatica* 47 (3): 485–495. https://doi.org/10.1016/j.automatica.2011.01.004.

Onnasch, L., Wickens, C.D., Li, H., and Manzey, D. (2014). Human performance consequences of stages and levels of automation: an integrated meta analysis. *Human Factors* 56 (3): 476–488. https://doi.org/10.1177/0018720813501549.

Parasuraman, R., Sheridan, T.B., and Wickens, C.D. (2000). A model for types and levels of human interaction with automation. *IEEE Transactions on Systems, Man, and Cybernetics - Part A: Systems and Humans* 30 (3): 286–297. https://doi.org/10.1109/3468.844354.

Parker, L.E. (1998). Alliance: an architecture for fault tolerant multirobot cooperation. *IEEE Transactions on Robotics and Automation* 14 (2): 220–240. https://doi.org/10.1109/70.681242.

Peng, Y.-C., Carabis, D.S., and Wen, J.T. (2018). Collaborative manipulation with multiple dual-arm robots under human guidance. *International Journal of Intelligent Robotics and Applications* 2 (2): 252–266. https://doi.org/10.1007/s41315-018-0053-y.

Protte, M., Fahr, R., and Quevedo, D.E. (2020). Behavioral economics for human-in-the-loop control systems design: overconfidence and the hot hand fallacy. *IEEE Control Systems Magazine* 40 (6): 57–76. https://doi.org/10.1109/MCS.2020.3019723.

Qu, Z. and Simaan, M.A. (2014). Modularized design for cooperative control and plug-and-play operation of networked heterogeneous systems. *Automatica* 50 (9): 2405–2414. https://doi.org/10.1016/j.automatica.2014.07.003.

Rahman, S.M.M., Sadrfaridpour, I.D., Walker, B., and Wang, Y. (2023). Trust-triggered robot–human handovers using kinematic redundancy for collaborative assembly in flexible manufacturing. In: *Cyber–Physical–Human Systems: Fundamentals and Applications* (ed. A. Annaswamy, P.P. Khargonekar, F. Lamnabhi-Lagarrigue, and S.K. Spurgeon). UK: Wiley (in press).

Rodríguez-Seda, E.J., Troy, J.J., Erignac, C.A. et al. (2010). Bilateral teleoperation of multiple mobile agents: coordinated motion and collision avoidance. *IEEE Transactions on Control Systems Technology* 18 (4): 984–992. https://doi.org/10.1109/TCST.2009.2030176.

Rovira, E., McGarry, K., and Parasuraman, R. (2007). Effects of imperfect automation on decision making in a simulated command and control task. *Human Factors* 49 (1): 76–87. https://doi.org/10.1518/001872007779598082.

Saeidi, H., Wagner, J.R., and Wang, Y. (2017). A mixed-initiative haptic teleoperation strategy for mobile robotic systems based on bidirectional computational trust analysis. *IEEE Transactions on Robotics* 33 (6): 1500–1507. https://doi.org/10.1109/TRO.2017.2718549.

Samad, T. (2023). Human-in-the-loop control and cyber–physical–human systems: applications and categorization. In: *Cyber–Physical–Human Systems: Fundamentals and Applications* (ed. A. Annaswamy, P.P. Khargonekar, F. Lamnabhi-Lagarrigue, and S.K. Spurgeon). UK: Wiley (in press).

Sheridan, T.B. and Stassen, H.G. (1979). Definitions, models and measures of human workload. In: *Mental Workload: Its Theory and Measurement* (ed. N. Moray), 219–233. Switzerland: Springer. https://doi.org/10.1007/978-1-4757-0884-4_12.

Sheridan, T.B. and Verplank, W.L. (1978). Human and Computer Control of Undersea Teleoperators. *Technical Report*. DTIC Document.

Sinha, R. and Swearingen, K. (2002). The role of transparency in recommender systems. *Human Factors in Computing Systems*, 830–831. https://doi.org/10.1145/506443.506619.

Sipahi, R. (ed.) (2017). *Human Machine Interaction: Bridging Humans in The Loop*, vol. 5. ASME Dynamic Systems and Control Magazine.

Sirouspour, S. and Setoodeh, P. (2005). Multi-operator/multi-robot teleoperation: an adaptive nonlinear control approach. *2005 IEEE/RSJ International Conference on Intelligent Robots and Systems*, 1576–1581. https://doi.org/10.1109/IROS.2005.1545353.

Spletzer, J., Das, A.K., Fierro, R. et al. (2001). Cooperative localization and control for multi-robot manipulation. *2001 IEEE/RSJ International Conference on Intelligent Robots and Systems*, 631–636. https://doi.org/10.1109/IROS.2001.976240.

Staub, N., Mohammadi, M., Bicego, D. et al. (2018). The tele-magmas: an aerial-ground comanipulator system. *IEEE Robotics and Automation Magazine* 25 (4): 66–75. https://doi.org/10.1109/MRA.2018.2871344.

Stewart, A., Cao, M., Nedic, A. et al. (2012). Towards human–robot teams: model-based analysis of human decision making in two-alternative choice tasks with social feedback. *Proceedings of the IEEE* 100 (3): 751–775. https://doi.org/10.1109/JPROC.2011.2173815.

Tintarev, N. and Masthoff, J. (2007). A survey of explanations in recommender systems. *2007 IEEE 23rd International Conference on Data Engineering Workshop*, 801–810. https://doi.org/10.1109/ICDEW .2007.4401070.

Walker, M., Hedayati, H., Lee, J., and Szafir, D. (2018). Communicating robot motion intent with augmented reality. *2018 13th ACM/IEEE International Conference on Human–Robot Interaction*, 316–324. https://doi.org/10.1145/3171221.3171253.

Wang, Z., Mülling, K., Deisenroth, M.P. et al. (2013). Probabilistic movement modeling for intention inference in human–robot interaction. *International Journal of Robotics Research* 32 (7): 841–858. https://doi.org/10.1177/0278364913478447.

Yamauchi, J., Atman, M.W.S., Hatanaka, T. et al. (2017). Passivity-based control of human–robotic networks with inter-robot communication delays and experimental verification. *2017 IEEE International Conference on Advanced Intelligent Mechatronics*, 628–633. https://doi.org/10.1109/ AIM.2017.8014087.

Yildiz, Y. (ed.) (2020). *Cyberphysical Human Systems*, vol. 40. IEEE Control Systems Magazine.

Yokoyama, K., Handa, H., Isozumi, T. et al. (2003). Cooperative works by a human and a humanoid robot. *the 2003 IEEE International Conference on Robotics & Automation*, 2985–2991. https://doi.org/ 10.1109/ROBOT.2003.1242049.

Part IV

Healthcare

15

Overview and Perspectives on the Assessment and Mitigation of Cognitive Fatigue in Operational Settings

Mike Salomone[1], Michel Audiffren[2], and Bruno Berberian[3]

[1] *Laboratoire de Psychologie et NeuroCognition, Univ. Grenoble Alpes, Univ. Savoie Mont Blanc, CNRS, Grenoble, France*
[2] *Centre de Recherches sur la Cognition et l'Apprentissage, UMR CNRS 7295, Université de Poitiers, Poitiers, France*
[3] *Information Processing and Systems Department, ONERA, Salon-de-Provence, France*

15.1 Introduction

For about 50 years now, humans have been surrounded by automated systems that assist them in most of their activities. In daily life and at work, we rely on very sophisticated systems able not only to replace us during manual activities but also to make extremely complex decisions (Frey and Osborne 2017). This replacement, whose objective is in many cases an increase in performance or safety, is bound to increase given the considerable expansion of Artificial Intelligence (AI) within the cyber–physical systems (CPS) from increasingly varied domains (production, aeronautics, automobile, economy, to name a few examples).

The transport field in general and, more specifically, aeronautics is a good example of the changes generated by the integration of automated systems. In this field, this integration has been realized for many years with the objective to limit the occurrence of human errors that were one of the main causes of accidents (e.g. Wiener 1989; Diehl 1991; Billings 1996). Therefore, in part due to these new technologies (e.g. autopilots, flight directors, and alerting and warning systems), between 2013 and 2017, an average of about nine fatal accidents occurred per year. One accident for every 559 000 flights (according to the International Air Transport Association). It is effectively to limit human deficiencies or to reduce the constraints that can be placed on humans that the CPS are developed. In healthcare, where medical professionals are often under pressure due to work schedules, technologies such as surgical robots relieve workers of certain tasks to help them provide quality care.

However, this integration did not come without a cost. The development of these CPS has been realized without optimally integrating the human and his functioning, assuming that the automation of systems was only a simple substitution of human activity by a machine called the "myth of substitution" (Woods and Tinapple 1999). This assumption is a distorted reflection of the real impact of automation since it has in fact profoundly transformed the nature of human work. In addition, its limitations are particularly apparent in unexpected situations where its inadequacy to human cognitive functioning can affect the takeover behavior and thus have the opposite effect of that desired. All these changes can lead to degraded cognitive states such as cognitive fatigue. This state is detrimental to safety and performance. It is important to identify the measures not only to detect it but also to define effective countermeasures to develop CPHS, i.e.

Cyber–Physical–Human Systems: Fundamentals and Applications, First Edition.
Edited by Anuradha M. Annaswamy, Pramod P. Khargonekar, Françoise Lamnabhi-Lagarrigue, and Sarah K. Spurgeon.
© 2023 The Institute of Electrical and Electronics Engineers, Inc. Published 2023 by John Wiley & Sons, Inc.

human-centered systems (Lamnabhi-Lagarrigue et al. 2017; chapter 2.1 in this volume by Samad 2022 to have a definition of this concept).

15.2 Cognitive Fatigue

15.2.1 Definition

Cognitive fatigue is considered a subtype of the broader concept of fatigue. Fatigue is defined by the International Civil Aviation Organization (2015) as "A physiological state of reduced mental or physical performance capability resulting from sleep loss or extended wakefulness, circadian phase, or workload." This definition highlights the multidimensional nature of fatigue and differentiates between various subtypes with different causes and effects. Cognitive fatigue is distinct from drowsiness, which is usually the result of a lack of sleep or variation in circadian rhythms, and from physical fatigue generated by physical exercise. Cognitive fatigue occurs after sustained mental effort. It is therefore induced by the execution of a task, even of a few minutes, which solicits cognitive capacities and requires mental effort. It should be noted that these different subtypes of fatigue are not totally independent since interactions have already been observed between them (e.g. Marcora et al. 2009), and they can appear in similar conditions (Salomone et al. 2021). Indeed, it is sometimes difficult to differentiate between drowsiness and cognitive fatigue since both appear, for example, during a monotonous activity or one that is performed for a significant period of time. Cognitive fatigue can therefore be defined as a suboptimal psychophysiological state caused by exertion that results in changes in strategies or use of resources so that the original levels of mental processing or physical activity are maintained or reduced (Phillips 2015). In addition, cognitive fatigue can also be divided according to whether it is chronic or acute in nature. Chronic fatigue usually persists for several days and may be associated with the presence of a pathology (DeLuca 2005) and long-term structural changes in the brain (Chaudhuri and Behan 2004). Acute fatigue can be present for several hours but is usually experienced over a relatively short period of time. Acute cognitive fatigue can be induced within minutes (Borragán et al. 2017). It is generally the result of a temporary activity (Enoka and Duchateau 2016).

15.2.2 Origin of Cognitive Fatigue

The factors at the origin of cognitive fatigue can be directly linked to the activity performed, such as its duration or complexity. In this section, we will therefore review these different factors in order to better understand the causes of cognitive fatigue. Next, some of the current explanations for the origin of cognitive fatigue will be briefly presented.

First, it is necessary to introduce the concept of effort since it is central when it comes to defining what generates cognitive fatigue. Effort can be defined as the set of processes that determine the level of performance that will be achieved compared to what is possible to achieve (Shenhav et al. 2017). Effort can also be viewed as a mechanism for deciding whether or not to act, switching from one activity to another, and managing the level of engagement based on the costs and benefits associated with achieving the task goal (André et al. 2019). The exertion of cognitive effort can be experienced as costly or aversive, but it can also be valued (see Inzlicht et al. 2018). Subjectively, it corresponds to the intensity of mental activity dedicated to the achievement of a goal (Eisenberger 1992).

Time on task is one of the major contributors to the development of cognitive fatigue. Whatever the mechanism(s) at the origin of cognitive fatigue, prolonged mental effort generates fatigue that accumulates and increases over time (Grandjean 1979). The task durations used to induce it vary considerably, ranging from a few minutes to several hours. Borragán et al. (2017), for example, evaluated the evolution of performance during a task lasting just over 20 minutes (Borragán et al. 2017), while for others, the duration used was six hours (Blain et al. 2016). However, not all negative effects of time on task are due to cognitive fatigue. Confounding factors such as boredom or more globally decreased engagement, especially when the task is repetitive, is likely to influence performance or even the experience of fatigue (Charbonnier et al. 2016; Raffaelli et al. 2018). The involvement of circadian (Aeschbach et al. 1997) or ultradian (Lavie 1989; Maffei and Angrilli 2018) rhythms, depending on the length of the task, has also been suggested. In addition, a training effect can sometimes mask the presence of cognitive fatigue (Möckel et al. 2015).

Cognitive fatigue is also modulated by cognitive load. A cognitive load will be present when the task requires cognitive control. This concept corresponds to the ability to coordinate a set of cognitive processes intended to accomplish a goal, allowing not only to go against our automatic behaviors but also to adapt the goal followed according to the context (Miller and Cohen 2001; Koechlin and Summerfield 2007; Braver 2012; Cools et al. 2019). This definition is based on the theoretical distinction between automatic, nonconscious cognitive processes activated by a particular stimulus, and therefore sometimes unwanted, and controlled processes consciously activated to achieve a desired goal. Cognitive control is by definition a process requiring mental effort, unlike automatic processes (Schneider and Shiffrin 1977). There is no consensus on the cognitive processes involved in cognitive control. Nevertheless, inhibition of information or behavior, mental flexibility, planning, and working memory are frequently cited (Luria 1973; Miyake et al. 2000). For example Klaassen et al. (2014) assigned to one group of participants cognitively demanding tasks, while the second (control) group watched a documentary or read a magazine and thus did not exert cognitive control. Both groups were then assessed on the same working memory task. They observed that the number of errors was lower for the control group, and that lower cognitive fatigue was also reported (Klaassen et al. 2014). The level of cognitive load is not necessarily based on the "amount" of cognitive control exerted but can also be dependent of the time available to process information. The less time available, the greater the cognitive load (Barrouillet et al. 2004). Thus, it has been observed that the greater the time pressure to perform a task, the greater the induced cognitive fatigue (Borragán et al. 2017).

It is not necessary to perform a task with a high cognitive load to induce cognitive fatigue. This may seem paradoxical, but it appears that performing a simple monotonous task also requires a cognitive effort to be achieved. Numerous studies report, both in laboratory and ecological contexts, that performing a monotonous task (e.g. driving, surveillance task) generates cognitive fatigue associated with decreased performance (Lim et al. 2010; Kamzanova et al. 2014; Ma et al. 2018).

Finally, what stops us from pursuing a mental effort over time? Cognitive effort is costly and is therefore not exercised in an unlimited way. Although cognitive fatigue is a century-old research topic, there is currently no consensus on the nature of the costs (e.g. metabolic) and the causes of the disengagement of cognitive effort (Hockey 2013; Shenhav et al. 2017, for a review). This could be explained by the need to resolve the question of the existence and nature of cognitive resources. This subject, which is still open in the field of cognitive science, has long been the dominant approach to understanding what might cause difficulty in pursuing a cognitive effort over time (Hockey 2013). Currently, cognitive fatigue is generally presented as resulting either from a decrease in cognitive resources, a decrease in the capability to exert cognitive control because of short-term synaptic mechanisms, or a shift in motivation toward less costly activities. In the first case, the general idea is that sustained effort draws on resources we have in limited amount. When resources are depleted, a decrease in performance is observed. These resources can be metabolic. Glucose, for example, is suggested to be the finite resource (Baumeister et al. 1998; Christie and Schrater 2015, for a different form of glucose, glycogen), but many results argue against this hypothesis (Kurzban et al. 2013). According to the second approach, accumulation of adenosine in specific brain regions associated with brain glucose consumption would be associated with a decrease in the capacity to maintain cognitive control over time (Martin et al. 2018; André et al. 2019). The third approach considers cognitive fatigue as a motivational mechanism that helps to regulate effort according to motivational requirements, meaning to direct it toward relevant and valued activities. The mechanisms invoked by the second and the third approaches could also be viewed as synergistic. These different models of cognitive fatigue do not all deny the existence of resources, but propose other types of costs from structural constraints or our information processing limitations, for example Shenhav et al. (2017). It is the opinion shared by the opportunity cost model developed by Kurzban et al. (2013). This model postulates that cognitive resources are limited. They argue that the prefrontal cortex has limited processing capacity that prevents it from performing several cognitive operations in parallel. The choice of cognitive processing performed will be dependent on its utility and cognitive fatigue would act as an adaptive indicator redirecting the allocation of limited resources to an activity with higher utility (Kurzban et al. 2013).

15.2.3   Effects on Adaptive Capacities

Regardless of the models used to explain the origins of cognitive fatigue, all agree that it impairs adaptive capacities. Adapting to an unexpected situation requires the use of cognitive control and the set of cognitive processes that it includes. An alteration of these processes usually leads to perseverative behaviors or distractibility (Cools et al. 2019). Their solicitation requires cognitive effort, and it is therefore these processes that are impacted by cognitive fatigue, while automatic processes are relatively preserved (Lorist and Faber 2011). Thus, it has been observed that cognitive fatigue alters the ability to detect errors and to adjust behavior accordingly (Lorist et al. 2005), to plan (van der Linden et al. 2003), to inhibit an action (Boksem et al. 2006), or to modify behavior and information processing according to environmental constraints (i.e. cognitive flexibility; Petruo et al. 2018); i.e. an umbrella of high-level cognitive functions called executive functions, which are anchored in a frontoparietal network (Seeley et al. 2007; Menon and Uddin 2010; Diamond 2013).

15.3   Cyber–Physical System and Cognitive Fatigue: More Automation Does Not Imply Less Cognitive Fatigue

Although the scientific literature indicates a deterioration of adaptive capacities through cognitive fatigue, it is poorly considered when it comes to integrating a human agent into the CPS. In the various application domains, fatigue has been studied mainly through drowsiness. Most of the solutions currently proposed are intended to limit the occurrence and effects of cognitive fatigue generated by extended work hours and staggered schedules that are inconsistent with circadian rhythms and can lead to sleep debt. This lack of consideration is problematic because since the introduction of automated CPS, the cognitive abilities of human agents and their motivation have never been so challenged. Although many operations are now handled by the system, the result is that certain aspects of the activity are now more likely to generate cognitive fatigue. It is therefore important to consider it in order to propose adapted solutions limiting its apparition. To do this, it is first essential to specify the determinants of the activity that induce cognitive fatigue.

The automation of aircraft with the integration of artificial intelligence (AI) has significantly changed the activity of human agents. Previously in control of the system, the human agent must now share control with automation technology and to coordinate with it. Critically, adding or expanding the machine's role changes the cooperative architecture, changing the human's role, often in profound ways (Sarter et al. 1997). The introduction of automation into complex systems has led to a redistribution of operational responsibility between human agents and computerized automated systems. This entails new coordination demands for the human agent – they must ensure that their own actions and those of the automated agent are synchronized and consistent (Christoffersen and Woods 2002).

First, this new role entails the need for increased monitoring by the human agent to detect possible failures and requires maintaining a high level of vigilance. Depending on the level of automation implemented, one or more functions initially performed by the human agent will then be performed by the system, relegating the human agent to a supervisory role for the functions in question. The evolution of the human agent's activity in the automotive domain is very illustrative of this role change. These vehicles have become extremely complex CPS. They now have a large number of external (e.g. visual, meteorological environment) and internal (e.g. speed, trajectory) information processing units that allow them to propose driving assistance systems (e.g. braking assistance, trajectory control) up to autonomous driving (this automation is still limited to level 3, requiring the resumption of control by the human agent if necessary). The human agent therefore no longer has to be concerned with many aspects of driving but must pay increased attention to the actions of the system. Research on vigilance has shown that humans are poorly suited for monitoring role (Davies and Parasuraman 1982; Parasuraman 1987; Wiener 1987). There is a long history of cases in which human agents are reportedly unaware of automation failures and do not detect critical system changes when acting as monitors of automated systems (Wickens and Kessel 1979; Ephrath and Young 1981; Kessel and Wickens 1982). Interestingly, simple tasks in which few actions are required from the operators and which require them to maintain their attention for a prolonged duration are perceived as mentally demanding and can generate fatigue (Grier et al. 2003; Warm et al. 2008). The decrease in performance resulting from this effort is attributed by some researchers to a depletion of attentional resources (Grier et al. 2003; Smit et al. 2004).

Second, human agents may meet difficulties to develop sufficient understanding of the behavior of their automated partners. When interacting with automated system, human agent will develop a mental model of the system's behavior and use it to assess the degree of risk of his/her decisions, to predict and verify his/her actions to ensure that the goals of the system and his/her own are in line (Klein et al. 2004). Without this, it is difficult for the human agent to regain control of the system by being out of the control loop. However, with increase in system complexity (for example, the multiplication of the number of possible "modes"), it is sometimes difficult for the human agent to track the activities of their automated partners. The result can be situations where the human agent is surprised by the behavior of the automation asking questions like, what is it doing now, why did it do that, or what is it going to do next (Wiener 1989). These "automation surprises" are particularly well documented (e.g. Van Charante et al. 1992; Palmer 1995; Sarter and Woods 1994, 1995; Degani and Heymann 1999) and have been listed as one of the major cause of incidents (see for example, Federal Aviation Authority 1995). The lack of system predictability is certainly a central point in understanding automation surprises and associated difficulties of takeover (Norman et al. 1990; Christoffersen and Woods 2002; Dekker and Woods 2002; Klein et al. 2004). With the progress of technology, current man-made complex systems tend to develop cascades and runaway chains of automatic reactions that decrease, or even eliminate predictability and cause outsized and unpredicted events (Taleb 2012). This is what we may call the "system opacity": i.e. the difficulty for a human agent to have a clear idea of the system's intentions and to predict the sequence of events that will occur. The increasing use of artificial intelligence (AI) tends to exacerbate these challenges, as AI systems typically provide no explanation for their results and offer a low level of prediction (Biran and Cotton 2017). This is, for example the case in control rooms (e.g. air traffic control, automobile) where the development of information processing and visualization algorithms has considerably reduced the need for human agents to process a large amount of data. The automated CPS, able to integrate a lot of information, ultimately leaves it up to the human agent to decide based on the results provided to him. However, it does not necessarily provide all relevant information, such as the level of certainty of the outcome for example, to human agents who decide without all the elements (Baber et al. 2019; Hullman 2020). Critically, this opacity will, directly or indirectly, increase cognitive load and thus cognitive fatigue (Galster et al. 2001). Besides the increase of the cognitive load, which will be associated with the effort to understand the system, the maladaptive interaction will directly influence the level of trust of the human agent in the system. This level of confidence is indeed a modulator of the cognitive load. If the information presented does not allow for the establishment of an adequate level of confidence between the system and the human agent, its use may be inappropriate and the processing of this information may instead result in an additional cognitive load (Naiseh et al. 2020).

Finally, automation has not completely relieved the human agent of certain tasks. The task performed by the automated system has been replaced by a supervisory task. The human agent must now perform other tasks, in parallel or alternatively to this task. Multitasking refers to a situation where cognitive processes involved in performing two (or more) tasks that overlap in time. This multitasking constraint has a cognitive cost that is added to the cost already attributable to the supervision of the system. Indeed, it is widely accepted that performing two tasks in parallel, or alternating from one to another, requires greater cognitive effort than performing a single task (Kiesel et al. 2010). Many studies have shown through simple experimental paradigms (e.g. memorizing stimuli and deciding in parallel the parity of numbers) that this cognitive cost is accompanied by an alteration of performance that can be measured, for example through a higher number of errors or higher reaction times (see for example Borragán et al. 2017). This cognitive cost has two sources. First, performing task simultaneously needs to maintain several task-sets at the same time, to update the task rules in working memory, to keep in mind the current status of each task, or evaluate the outcome of each task. This dual-task cost has often been explained with a reference to a division of attention, which has been conceptualized as a limited processing capacity (Kahneman 1973). Second, one should be able to switch between tasks. When switching task, it requires a switch in the mental task set. The process of activating and implementing a new task set is often called task-set reconfiguration (e.g. Monsell et al. 2000; Rubinstein et al. 2001) and is assumed to be time- and resources-consuming. Critically, switching costs or dual task costs have been demonstrated (Koch et al. 2018). In addition, "task switching" paradigm has been classically used to induce cognitive fatigue (Rogers and Monsell 1995). Asking a human agent to multitask puts their cognitive system at the limits of their capacity. This cognitive effort, if sustained over time, then results in increased cognitive fatigue exacerbating the decline in performance already present (Petruo et al. 2018).

In summary, while automation was initially thought to relieve the human agent of part of his work and reduce his workload, these changes have created conditions favorable to the development of cognitive fatigue.

15.4 Assessing Cognitive Fatigue

Cognitive fatigue is expressed through different types of responses, i.e. subjective and objective. The latter is divided into behavioral and physiological responses (Bartley and Chute 1947). It is important to evaluate each of these responses given the multidimensional nature of cognitive fatigue and the diversity of the experimental protocols used, which can elicit different responses from the participants. In addition, subjective and objective manifestations are perceived as relatively independent since a relationship between these measures is rarely observed (Kluger et al. 2013). Thus, the subjective experience is not always accompanied by a performance deficit. A nonexhaustive review of the different measures used inside and outside the laboratory is presented below as well as advantages and disadvantages of these different methods.

15.4.1 Subjective Measures

A large number of questionnaires have been developed to assess the presence of cognitive fatigue, but most of them are designed to assess chronic fatigue; so they will not be presented here. Since fatigue is studied mainly when it is caused by sleep deprivation or work schedules, most of the questionnaires used measure the level of alertness/arousal. The Karolinska Sleepiness Scale is, for example one of the most used scales (Åkerstedt and Gillberg 1990). It situates a person between two extremes: "extremely alert" and "very sleepy, great effort to keep alert, fighting sleep." The Samn–Perelli scale is a hybrid scale in the sense that it essentially evaluates, like the Karolinska Sleepiness Scale, the level of alertness of human agents, but some items examine concentration difficulties (Samn and Perelli 1982). There are few scales specifically evaluating cognitive fatigue. The majority of studies (especially those conducted in laboratories) assess cognitive fatigue using a visual analog scale (Lee et al. 1991) or a simple question. The measure obtained with this kind of scale is often dependent on the definition given by the experimenter. For example, Borragán et al. (2017) told participants that cognitive fatigue corresponded to the need to stop the current effort (Borragán et al. 2017). New scales are being developed to specifically extract cognitive fatigue. For example, an Australian team has developed a scale focused on the capacity to provide mental effort. It is similar in length and structure to the Karolinska Sleepiness Scale and has been tested under ecological conditions (Hu and Lodewijks 2021). Finally, the presence of cognitive fatigue is sometimes assessed indirectly through the presence of the effects it may cause or by assessing the factors associated with its occurrence. Thus, modulation of effort or even motivation can sometimes account for its presence. A person who is tired also reports having less motivation as well as a feeling of exhaustion or lack of energy (Boksem and Tops 2008; Gergelyfi et al. 2015). Questionnaires such as the NASA TLX (Hart 2006), widely used to assess cognitive load, or the Rating Scale Mental Effort (Zijlstra 1993) and the Borg's rating of perceived exertion (Borg 1982) can be encountered as indirect measures of cognitive fatigue.

Subjective measures have the advantage of being quickly administered and collected. When assessing cognitive fatigue, the perception of the human agent remains crucial as the behavioral impairment may be compensated and masked by additional cognitive effort while the human agent relates its presence. However, these are disturbing measures as they force human agents to stop their activity. Moreover, the measurement provided is a snapshot of the current time and does not provide a real-time estimation of the state of the human agent. Finally, although the subjective experience is important, it reflects only a portion of the cognitive activity, and it is sometimes difficult to accurately judge our own mental state (Schmidt et al. 2009).

15.4.2 Behavioral Measures

Response time (RT) or number of errors are the main measures used to detect the presence of cognitive fatigue. Studies generally show a poorer ability to detect rare events and a higher detection time (e.g. failure detection; Mackworth 1948; Molloy and Parasuraman 1996). Most of the results are at the individual level, but group-level results have also been observed, including an increase in decision errors within a team (Hollenbeck et al. 1995). The behavior of the human agents on the control system (e.g. lever, brake pedal, steering wheel) shows a sensitivity to this condition (Li et al. 2017). Cognitive fatigue also appears to be associated with greater behavioral variability (Seli et al. 2013). This can directly not only affect RT (Wang et al. 2014) but also accentuate the difficulty in maintaining trajectory (Atchley and Chan 2011).

Cognitive fatigue is also detectable through eye behaviors whose metrics include the evaluation of blinks, saccades, pupil size, or the exploratory pattern of eye movements. Studies show a modulation (decrease or increase) of pupillary diameter during fatigue (Wilson et al. 2007). This contradiction can be explained by the level of engagement of the participants at the time of the evaluation since the pupillary diameter is linked to a modulation of the cognitive load and mental effort (Hopstaken et al. 2015). The percentage of eyelid closure is also higher, given that drowsiness is also experienced with mental fatigue (Guo et al. 2018). The number and duration of eye blinks are also increased when drowsiness is present (Di Stasi et al. 2015) but are instead lower when engagement is high (Wilson and Fisher 1991). Finally, there is ample evidence of fatigue directly accessible through the analysis of head movements and facial expressions. For instance, yawning or inclination of the head can be used to assess fatigue (Ansari et al. 2021). However, these measures are rather focused on the evaluation of drowsiness which can appear after a certain time.

Behavioral measures (i.e. RT and number of errors), although objective, may have a major disadvantage shared with subjective measures, which is the delay between the apparition of cognitive fatigue and its detection. The presence of unwanted behavior and even errors is inconceivable in an environment where security is a major issue. Oculomotor and pupillary behavior may be more relevant options since they could detect cognitive fatigue before it leads to unwanted behavior. The diversity of the metrics and their sensitivity to different cognitive and emotional states make it a relevant tool, given the nature of cognitive fatigue. However, not all of them are as easily exploitable in real-life situations. For example, the variations in luminosity that can be encountered in the cockpit or in a car can modify the pupillary diameter and thus complexify the detection of cognitive fatigue as compared to a measurement made in the laboratory. Some cognitive fatigue assessments, such as the evaluation of the eye-exploratory pattern, certainly suffer less from environmental conditions.

15.4.3 Physiological Measurements

There are many tools available to measure cerebral and peripheral physiological activity (i.e. electroencephalography – EEG, electrocardiography – ECG or functional near-infrared spectroscopy – fNIRS). Usually designed for laboratory research, the improvement of their portability and of data processing techniques has democratized their use in ecological situations. Thus, many measures of cognitive fatigue are now identified (Borghini et al. 2014; Hu and Lodewijks 2021).

The ECG consists of measuring cardiac reactivity. It is more specifically used for investigating autonomic nervous system activity. Inexpensive, easy to use, and with a high signal-to-noise ratio, the ECG is a relevant tool to use in ecological situations. Several metrics seem relevant to detect cognitive fatigue. Decreased heart rate and increased heart rate variability over time are frequently reported (Hidalgo-Muñoz et al. 2018; Matuz et al. 2021). Frequency-domain-based measures like low-frequency (LF) power density of heart rate variability (HRV) also changes with fatigue. For example, Qin et al. (2021) observed an increase in LF spectral power during a flight simulation (Qin et al. 2021); i.e. an increase in parasympathetic activity. The ECG is not the only tool to measure cardiac reactivity. Photoelectric plethysmography, for example provides similar metrics to those collected with the ECG. This tool sends and receives infrared light, which has the particularity of being absorbed by blood. The intensity of the light received therefore informs of the variations in blood flow associated with cardiac activity.

The fNIRS is an interesting tool for measuring cognitive fatigue. Its signal-to-noise ratio, superior to EEG, and its ease of use make this tool suitable for measurement in ecological situations (Dehais et al. 2018). fNIRS measures blood oxygenation level from the surface of the human brain. With better spatial resolution than EEG, it allows for accurate characterization of where variations in oxygen consumption occur. Thus, during a driving task, increased oxygenation in the frontal, parietooccipital, and motor areas has been observed (Chuang et al. 2018) or a decrease of intra-hemispheric connectivity between left anterior frontal and frontal areas (Borragán et al. 2019).

The EEG captures the electrical potentials generated by the pyramidal neurons whose oscillatory activity can be characterized by its frequency. These oscillations can be grouped into different frequency bands with a functional role. Many studies have observed changes in these spectral bands that may be associated with the onset of cognitive fatigue. Some of them show a higher consistency than others regarding the evolution of their direction. This is for example the case of alpha activity whose increase is often reported with time on task (Wascher et al. 2014). This increase may be associated with the presence of mind-wandering, which can be considered as a form of task disengagement (Compton et al. 2019). Theta activity, particularly measured in the frontomedial area, is considered as a correlate of cognitive effort and cognitive control (Smit et al. 2005; Cavanagh and Frank 2014). Its increase and decrease have therefore already been reported, depending on whether participants had already disengaged from the task or not (Borghini et al. 2014). The same is also true for beta activity, which increases with arousal and effort but decreases with cognitive fatigue (Craig et al. 2012). Different ratios can be calculated between these different bands and account for certain states. The engagement index for example, calculated using the ratio of beta band spectral power to the sum of alpha band and theta band spectral powers, is used to estimate cognitive fatigue (Dehais et al. 2018) and decreased over time among air traffic controllers (Berka et al. 2007).

On the theoretical level, several models have highlighted the communication between specific brain structures as markers of cognitive fatigue (Holroyd 2016; André et al. 2019). This form of communication can be assessed through functional connectivity measures representing, for example, the phase synchronization of oscillations (Fries et al. 2001). Several studies have observed a modulation of functional connectivity over different frequency bands (Liu et al. 2020). For example, a decrease on the alpha and theta bands between frontoparietal regions has been reported (Qi et al. 2020; for an opposite result see Lorist et al. 2009) or an increase on the beta band (Liu et al. 2010).

Finally, in addition to describing the strength of functional connectivity, it is also possible to describe the topology of the brain network using measures from graph theory (Bullmore and Sporns 2012). Several studies have shown that topology shifts with mental fatigue, toward more connections between close rather than distant spatial regions for example, demonstrating a shift to a more economical mode of functioning (Sun et al. 2014; Zhao et al. 2017).

Physiological measures, such as eye-tracking, are the most interesting measures, given the variety of metrics available to estimate cognitive fatigue. However, some of them are more invasive and difficult to implement in a real situation because of their sensitivity to environmental noise, such as movements. Nevertheless, these are the measures that demonstrate the greatest sensitivity and some tools, such as the EEG, offer the greatest capacity for development.

15.5 Limitations and Benefits of These Measures

This brief review presents several methods and measures to detect and assess cognitive fatigue, but it should be noted that they are not necessarily specific to cognitive fatigue. The duration or the task performed can be very different, which generates a variety of responses corresponding to different types of fatigue, making the detection of cognitive fatigue more complex. For example, when assessed after a long task, which is done in the majority of studies, the measures will also reflect the presence of sleepiness or a decrease in arousal. The timing of the assessment also matters. For example, cognitive fatigue is not directly associated with disengagement and a drop in performance. An initial phase of intense effort may precede these phenomena without them being present but still with a feeling of cognitive fatigue (Wang et al. 2016). For instance, the maintenance of a stable performance over time instead of the presence of cognitive fatigue can be associated with a compensatory increase of mental effort. It is therefore possible to observe contradictory results from one study to another or between measures, some of them being more appropriate to specify other psychophysiological states than cognitive fatigue (e.g. level of arousal or effort). Besides the need to develop measures that specifically assess cognitive fatigue, considering the use of multiple measures in the same study can improve its detection. By doing so, the variety of measures allows for the disentanglement of what can be attributed to cognitive fatigue or to other responses of the organism. In addition, recording multiple modalities helps to overcome individual tool limitations.

As systems tend to become more automated, a reconsideration of the relevance of tools designed to detect cognitive fatigue is necessary. For example, measurements of the vehicle's behavior should be excluded, since by definition the human agents do not have to deal with the system, except in the case of a takeover. Measurements concerning, for example, trajectory or speed control will therefore become obsolete in this context. One of the major contributions of the use of physiological indicators is the possibility to use them in real time to inform the system of the cognitive state of the human agent to create a direct connection between the human and the machine. This connection can then be used to create a human–machine interface adaptable to the state of the human agent. The development of this type of adaptive system is part of the field of "physiological computing" (Fairclough 2017).

15.6 Current and Future Solutions and Countermeasures

15.6.1 Physiological Computing: Toward Real-Time Detection and Adaptation

Physiological computing consists of inputting not only physiological data but also behavioral data, such as eye data, into the system as a source of information to infer mental state or behavior (Fairclough 2017). For several decades now, this type of system has been developed using, in particular, the electrical activity of the brain measured with EEG in order to replace motor functions that are no longer ensured in the case of handicaps, for example. These brain–computer interfaces (BCI) can allow tetraplegic people to write on a computer screen by concentrating on letters presented on a virtual keyboard (Farwell and Donchin 1988). This type of interface is referred to as active since it requires the involvement of the human agent to perform an action or by asking him to "produce" a specific mental state. Conversely, the passive equivalent of these interfaces detects brain activity without user intervention. They will therefore be employed to detect the affective and cognitive states of users (Zander et al. 2010). The core of the functioning of passive (and active) BCI is therefore the analysis and classification of neurophysiological signals. The steps followed after data acquisition, common to all types of BCI, are generally data preprocessing, which allows to improve the signal-to-noise ratio by applying, among others, temporal or spatial filters in order to facilitate the next step, the extraction of parameters and finally their classification. Machine learning algorithms are frequently used for this classification step. The most used algorithms are supervised learning and unsupervised learning (other types exist, such as reinforcement learning). Deep learning algorithms are used but remain a minority at this time (e.g. Han et al. 2020; Wu et al. 2021). Supervised learning involves providing the algorithm with a database of examples, usually containing valid and invalid examples, to learn to classify a mental state. The classification methods included in this category are, for example neural networks or support vector machines. On the contrary, unsupervised learning methods do not need examples to learn but will rather look for similar patterns in the data by clustering for example. The choice of these algorithms will depend on several criteria, such as the cognitive state and the context in which the measures are recorded or the amount and the characteristics (e.g. distribution) of the data included in the training database if a learning is necessary. Each of these algorithms has its own advantages and disadvantages, which make them relevant to each case.

The detection and classification of cognitive fatigue in real time has already been achieved through different methods. For example, the use of vehicle commands (Li et al. 2017), blink rate, and heart rate variability (Qin et al. 2021), head movements (Ansari et al. 2021), spectral power of EEG activity (Trejo et al. 2015), or functional connectivity within or between neuronal networks have been used (Qi et al. 2019). All these studies show a high classification rate, between 70% and 100% accuracy. A more precise and accurate classification can be obtained by combining the different measures (Han et al. 2020).

Once the data are categorized and the mental state detected, this information will help to improve the interaction between the human agent and the system in two ways. First, the human agent will be able to adapt his mental state and behavior according to the information provided, and second, the system will be able to use this information to provide an adapted and relevant response to the presence of cognitive fatigue.

The real-time transmission of physiological or behavioral data by the system is referred to as biofeedback. The objective in this context is to increase the human agent's awareness of cognitive fatigue so that he can adapt his behavior to reduce it. There are several ways to do this. If we consider the resource models, stopping the task to restore the cognitive resources and the capacity to exert cognitive control should reduce cognitive fatigue. Taking a break when necessary would be a solution. Stopping mental effort to allow recovery is indeed a generally advocated and easily implemented countermeasure (Helton and Russell 2015). Successful results have already been observed. A break performed following detection of cognitive fatigue via eye-tracking data was more effective than if it was performed according to the participants' will (Marandi et al. 2019). However, caution must be exercised, and the conditions under which the pause is performed must be considered. Research is currently limited, and some have shown negative effects. More details on the use of breaks are presented in Section 5.3 entitled "Recovering from Cognitive Fatigue."

Biofeedback can also be used to engage human agents to reallocate cognitive effort. Maksimenko et al. (2019) alerted participants when a decline in attention was observed while they had to sustain their attention during a visual perception task. To do this, they assessed in real-time changes in alpha and beta activity in the parietooccipital region, markers of the presence of cognitive fatigue (Roy et al. 2016) and sent auditory feedback to alert participants of changes in this activity.

They observed with this procedure an increase in the duration of periods of high level of attention (Maksimenko et al. 2019).

These are just a few examples of the use of biofeedback to reduce cognitive fatigue, but there are many other possible solutions (e.g. Bazazan et al. 2019, which used postural adjustment). It could also be used to improve human agent training. Operator training is a way to prevent the occurrence and effects of cognitive fatigue. It could not only guide human agents toward a more adapted use of the systems but also provide measures to objectively assess their progresses (Aricò et al. 2015). A more adapted use could not only improve their reaction ability when they are subjected to this condition but also help them develop compensatory strategies and/or a higher capacity to exert cognitive control under difficult conditions (e.g. long task duration). However, this requires developing our knowledge on the mechanisms impacted by cognitive fatigue (Salomone et al. 2021).

Informing the human agent and relying on his adaptive skills may be inappropriate in some cases. For example, during flight phases with difficult weather conditions (e.g. wind), it would be difficult to rely on the human agent to regulate his own fatigue level. The second solution is therefore to let the system adapt to the human agent's state. With an automated system, it is easy to imagine how the system can adapt to reduce the cognitive effort of the human agent. There are many ways to reduce the effort and here are some examples. The first would be to let the system handle certain tasks to let the human agent take a break or reduce multitasking (e.g. Aricò et al. 2016). It would also be possible to alleviate the information-processing capacities by reducing the amount of information presented to select only the most relevant in the current situation or modulate their salience in order to facilitate their processing. It would also be possible to alleviate information-processing capacities by reducing the amount of information presented to retain only the most relevant in the current situation or modulate its salience to facilitate its processing. Attention can indeed be involuntarily drawn to certain information (Anderson et al. 2011). In this case, inhibition capacities are mobilized, increasing cognitive effort. These different options use the data provided to estimate whether a cognitive state, such as cognitive fatigue, is present. A further step would be to use more sophisticated data processing to anticipate its emergence by "predicting" the evolution of the data (Roy et al. 2020). Moreover, the responses that the automated system can provide in reaction to the occurrence of fatigue can be more or less accurate, depending on the number of categories defined. Thus, a binary categorization (i.e. fatigued or not) corresponds to the simplest of classifications to obtain but may have limitations. A larger number of categories may allow for a greater variety of responses to different degrees of cognitive fatigue.

15.7 System Design and Explainability

A human agent is generally cognitively fatigued when confronted to sustained effortful activities. Cognitive fatigue, like other cognitive states such as stress, emerges from normal human cognitive functioning and therefore cannot be totally suppressed. Reacting to the presence of fatigue is essential, but a complementary solution would be to create the conditions that are not conducive to the emergence of cognitive fatigue by taking into account knowledge about its causes and consequences. These conditions could be implemented by acting directly on the design of systems. The problem with automation is not its presence per se, as evidenced by the many positive effects that have emanated from its presence, but rather its design, which is not always adapted to human functioning (Norman et al. 1990).

We have highlighted the design of opaque automated systems as one of the causes of cognitive fatigue. Addressing this problem would reduce the occurrence of direct and indirect causal factors, such as workload and system trust. Recently, a growing body of research has addressed the issue of opacity, particularly in AI systems. This area of research, called eXplainable Artificial Intelligence (XAI), aims to determine how to improve the interaction between the system and the human agent by making the decisions made by an AI predictable and intelligible.

Several solutions exist to deliver an intelligible explanation from the AI. The first is to provide the human agent with a simplified model of what is being achieved by the AI (Guidotti et al. 2019). This may consist in directly presenting the outputs of what is computed by the AI (e.g. predictions) or the different steps of the decision process. In this case, a preliminary processing is necessary to simplify the presented information. Visual aids also seem to be an adequate solution. However, they must generally be based on the extraction of relevant data to make a decision in progress. A study recently showed that highlighting the information on which the AI based its decision decreased the cognitive load and increased confidence in the AI (Hudon et al. 2021). Yet, we know remarkably little today about how humans perceive and evaluate algorithms and their results, about what makes a human trust or distrust an algorithm, and about how we can empower humans to act – to adopt or challenge an algorithmic decision (Stoyanovich et al. 2020). Thus, a prerequisite for the design of explicability tools is the identification of the information to be provided to enable the human agent to work in cooperation with automation. Research in cognitive sciences offers many insights. Recently, it was shown that when the system communicates metacognitive information (i.e. the confidence it had in its decisions), the system is perceived as more intelligible and acceptable. More importantly, metacognitive information increases the sense of control as well as confidence in the decision and ultimately performance of the human agent (Vantrepotte et al. 2022). Similar results were obtained when the system directly communicates its intentions (Le Goff et al. 2018).

15.8 Future Challenges

15.8.1 Generalizing the Results Observed in the Laboratory to Ecological Situations

The contradictions observed among the measures that can be used highlight the need to generalize the results to ecological situations experienced by human agents. We have seen that slight modifications of the difficulty, time on task, or human agent's effort engagement can generate opposite behavioral and physiological responses. When the choice of the measures is made, it is therefore necessary to first define what exactly is done by the human agent.

Apart from these considerations, the use of physiological instruments and classification algorithms in real-life conditions requires robust measurements. The data processed by the classification algorithms may have differences between those obtained in the laboratory and in the real world. Data obtained in the laboratory may, for example, reflect specific behaviors and thus limit the generalization from the laboratory to the real world. In addition, algorithm development generally has some specificity to the data set (e.g. to the task, to the individual). Progress must be made in the ability to transfer data sets to other data sets. In addition to the difference in the quality of the signal obtained in an ecological situation, there are, as we have just seen, differences in the temporality of the onset of cognitive fatigue and the cognitive states that emerge alongside it.

15.8.2 Determining the Specificity of Cognitive Fatigue

The use of methods, such as physiological computing, implies that the measurement performed allows precise inference of the human agent's cognitive state human agent. However, we have seen that one of the particularities that stands out when it comes to detecting cognitive fatigue is the difficulty of determining a specific signature of its presence. The conditions of its occurrence are also at the origin of other states like drowsiness or stress. Thus, several measures used (e.g. pupil dilation, frontocortical activity) can account for their presence as well as for cognitive fatigue. There is, therefore, a real operational need to characterize measures specific to cognitive fatigue or cognitive effort. To do this, it is necessary to determine their origin (i.e. the mechanisms underpinning their occurrence).

Several models propose alternative theories regarding the nature of cognitive costs and have outlined a fatigue network that opens the door to more directed measures. Trujillo (2019), for example, proposed using free energy, "an information-theoretic system property of the brain that reflects the difference between the brain's current and predicted states, as an index of the information-processing resource costs allocated by mental effort." During a categorization task, he observed that the task-related differences in brain free energy were related to variations in cognitive load and information-processing resource allocations related to task decisions (Trujillo 2019). Holroyd (2016) proposed instead that cognitive fatigue is generated by a problem of elimination of metabolic waste, β-amyloid peptides, whose accumulation within the interstitial fluid could lead to brain damage. According to him, cognitive fatigue would appear in order to reduce the behaviors that generate this accumulation and would therefore also have a protective role. To explain this increase, Holroyd hypothesizes that the anterior cingulate cortex (ACC) and locus coeruleus would inhibit the default mode network. In addition to the structures involved in the task, cognitive fatigue should therefore be reflected by an increase in synchronization between the ACC-locus coeruleus and the default mode network. This synchrony should generate an accumulation of metabolic waste within the structures targeted by the ACC and locus coeruleus, which is thought to cause the sensation of exertion and fatigue (Holroyd and McClure 2015; Holroyd 2016). Müller and Apps (2019) proposed a set of brain structures, which they divide into systems, involved in the onset of fatigue by distinguishing on the one hand the structures solicited by the current activity and on others involved in motivation. It is within the structures solicited in the accomplishment of the task in progress that cognitive fatigue would manifest itself. The structures would therefore be different depending on the task being performed. Conversely, the structures of the motivational system would be solicited independently of the task in progress. They would assess the state of the system involved in the task in order to determine the amount of effort required to perform a task and would perform a cost/benefit assessment to determine whether this effort would be expended. If so, an increase in activity should be observed. Conversely, a decrease or disengagement from the current task to another should be observed. The structures involved in the motivational system are the dorsolateral prefrontal cortex, the anterior insula, and the dorsal ACC. Several studies have observed this pattern of functioning using functional magnetic resonance imaging (Müller and Apps 2019; Aben et al. 2020; Wylie et al. 2020). New measures may therefore be relevant but need to be tested with other tools (e.g. EEG) and in ecological situations.

Finally, one of the remaining challenges, and not the least, is to characterize the physiological and behavioral correlates that would predict the transition from the use of a compensatory mechanism to the drop in performance. As already mentioned in this chapter, a compensatory mechanism is put in place for a certain period of time before eventually giving way to a decline performance explained by a disengagement or a decrease in cognitive resources or in the capacity to exert cognitive control, depending on the explanatory model. Cognitive fatigue is often detected once this stage has been passed and therefore potentially a high level of cognitive fatigue. Questions can be asked about our ability to recover once a significant level has been reached.

15.8.3   Recovering from Cognitive Fatigue

Cognitive fatigue is a vast topic that has been studied for many years and that has focused on many aspects: the theoretical understanding of its origins, its effects on cognition, how to detect it, or how to limit its occurrence. However, a fundamental aspect, both in terms of theory and applicability, has been relatively neglected. Indeed, there is very little knowledge about the recovery of cognitive fatigue. Determining how, by doing what, and in how much time we can recover from cognitive fatigue is of crucial interest. However, some answers have already been provided by observing participants after they have taken a break or by analyzing the evolution of physiological measures during a rest period carried out after a cognitive activity. Theoretically, the different models of cognitive fatigue stipulate that the interruption of cognitive activity can be sufficient to reduce cognitive fatigue (e.g. Hockey 2013). This proposition is broadly accepted across models (Warm et al. 2008; Hockey 2013).

At present, the results are scarce. Even if overall a reduction in fatigue is observed after stopping the activity, the reality seems more complex since negative effects are sometimes observed. This was indeed the case in the study conducted by Lim et al. (2016). In their study, breaks did not only have completely positive effects when participants had to re-engage in a task. Rather, a change in strategy had been reported. In their study, between each block, participants had a 12 or 28-second break. When the duration of the pause was the longest, participants intensified their cognitive effort directly after reengagement, which was accompanied by not only a boost in performance but also a greater decline in performance at the end of the task (Lim et al. 2016). Thus, a minimal delay after stopping the activity seems to be necessary before seeing recovery without a rebound effect. Lim et al. (2013), in a previous study, observed that spectral power in the upper alpha band (10–12 Hz) predicted improved RT after a break. However, significant individual differences were reported, as the positive effect was present in only a portion of the participants (Lim et al. 2013). The break was longer this time, lasting five minutes. Positive results were also observed after a break of 15 minutes or even 30 seconds (Chen et al. 2010). In the latter case, the decrease observed on RT was less than that observed after a break of about 10 minutes (Arnau et al. 2017). Positive effects in any case do not seem to differ according to the age of the participants (Gilsoul et al. 2021). However, persistence of cognitive fatigue has sometimes been reported after 20 minutes (Jacquet et al. 2021).

The delay before recovering from the effects of cognitive fatigue has also been investigated during a resting-state period. Esposito et al. (2014) compared resting-state brain activity after several hours of helicopter training or after a day without training. They observed that activity within the default network or regions involved in visual processing was greater after training. The opposite had been observed in the activity of the frontoparietal network (Esposito et al. 2014). Breckel et al. (2013) compared some measures reporting the topology of the brain network and observed that after 30 minutes of performing an attentional task, the network was composed of shorter connections, with more clusters and overall higher connectivity. This topology was present during the first six minutes of the resting task and for another six minutes for the least resilient participants (Breckel et al. 2013). This organization would be suitable for recovery as it would reduce the metabolic cost associated with, for example, maintaining connections between distant regions (Bullmore and Sporns 2012). In conclusion, it is difficult at present to rule on a minimum length of time before observing the effects of recovery. Numerous factors not only modulate this delay, such as the level of fatigue, but also the evaluation measure. As we will see hereafter, this will also depend on the activity performed during the recovery phase and the cognitive process evaluated.

In addition to the length of the break, other studies have tried to determine the type of activity performed during the break that promotes recovery. For example, several studies have compared breaks during which physical activity was performed (e.g. boxing) or relaxation, with the breaks usually performed. The observed results generally show positive effects of "alternative" breaks through a report of lower fatigue (Blasche et al. 2018) or an evolution of alpha activity (Scholz et al. 2018). But these results seem to show a large interindividual variability. In all cases, it is in line with the postulate that recovery does not necessarily imply stopping the activity to be optimal. This question can also be answered by looking at the literature evaluating the effect of secondary activities (e.g. radio, smartphone) performed while driving an automated system. Some studies show a positive effect on engagement or arousal (Jamson et al. 2013). However, if we refer to the opportunity cost model, performing a second valued activity will increase the costs associated with performing the main activity (Dora et al. 2021). Thus, recovery appears to be more of a regulatory or a motivational process (Inzlicht et al. 2014; Zijlstra et al. 2014).

As it is critical to determine precisely which cognitive processes are disrupted by cognitive fatigue, it is equally important to assess whether some of these processes are more resilient than others. Indeed, differences have been reported. Magnuson et al. (2021) observed a different pattern of recovery from the effects of fatigue on motor and cognitive processes. The motor skills (fine and coarse) they assessed showed a faster recovery of more than about 10 minutes than alpha and theta power, representative of the cognitive functioning of participants (Magnuson et al. 2021).

In summary, it is impossible at this point to make specific recommendations on measures to promote recovery. Although breaks generally demonstrate their effectiveness, it is essential to determine the conditions that can lead to negative effects. Again, generalization to the ecological task is essential given the differences that can be observed depending on the nature of the activities performed during recovery.

15.9 Conclusion

The objective of this chapter was to highlight how system automation has increased the occurrence of cognitive fatigue, what solutions exist to counteract the detrimental effects of system automation, and what challenges exist to reduce the negative effects or occurrence of cognitive fatigue. After defining the concept of cognitive fatigue, by delimiting its specificities compared to other subtypes of fatigue, we first specified the disrupted cognitive processes and explained how these effects were particularly detrimental during unexpected situations. In a second part, we highlighted, through the three following changes, how the increase of automation of systems could cause this cognitive fatigue: (i) the monotonous nature of the activity with an automated system, (ii) the opacity of the automated system for human agents, and (iii) the increase of multitasking. After reviewing different measures used to evaluate it, we briefly presented the concept of physiological computing whose goal is to detect cognitive fatigue in real time. We have seen that countermeasures based on the biofeedback technique or the new adaptation capacities of the system can be deduced from this detection. As this type of countermeasures can be considered as reactive measures, we then proposed proactive solutions focused on the modification of the system design. Finally, the last part presented three topics to be explored so that the results of the research would be beneficial from both a fundamental and an applied point of view, in particular for the design of CPHS. Since cognitive fatigue often occurs in conjunction with other cognitive states, we emphasized that it is crucial to improve the specificity of current measures by clarifying the origin of cognitive fatigue. We then concluded by presenting a brief review of the literature highlighting the contrasting and sparse findings regarding the conditions for recovery from cognitive fatigue.

References

Aben, B., Buc Calderon, C., Van den Bussche, E., and Verguts, T. (2020). Cognitive effort modulates connectivity between dorsal anterior cingulate cortex and task-relevant cortical areas. *The Journal of Neuroscience* 40 (19): 3838–3848. https://doi.org/10.1523/JNEUROSCI.2948-19.2020.

Aeschbach, D., Matthews, J.R., Postolache, T.T. et al. (1997). Dynamics of the human EEG during prolonged wakefulness: evidence for frequency-specific circadian and homeostatic influences. *Neuroscience Letters* 239 (2–3): 121–124. https://doi.org/10.1016/S0304-3940(97)00904-X.

Åkerstedt, T. and Gillberg, M. (1990). Subjective and objective sleepiness in the active individual. *International Journal of Neuroscience* 52 (1–2): 29–37. https://doi.org/10.3109/00207459008994241.

Anderson, B.A., Laurent, P.A., and Yantis, S. (2011). Value-driven attentional capture. *Proceedings of the National Academy of Sciences* 108 (25): 10367–10371. https://doi.org/10.1073/pnas.1104047108.

André, N., Audiffren, M., and Baumeister, R.F. (2019). An integrative model of effortful control. *Frontiers in Systems Neuroscience* 13: 79. https://doi.org/10.3389/fnsys.2019.00079.

Ansari, S., Naghdy, F., Du, H., and Pahnwar, Y.N. (2021). Driver mental fatigue detection based on head posture using new modified reLU-BiLSTM Deep Neural Network. *IEEE Transactions on Intelligent Transportation Systems* 1–13: https://doi.org/10.1109/TITS.2021.3098309.

Aricò, P., Borghini, G., Graziani, I. et al. (2015). Air-traffic-controllers (ATCO): neurophysiological analysis of training and workload. *Italian Journal of Aerospace Medicine* 12: 35.

Aricò, P., Borghini, G., Di Flumeri, G. et al. (2016). Adaptive automation triggered by EEG-based mental workload index: a passive brain-computer interface application in realistic air traffic control environment. *Frontiers in Human Neuroscience* 10: 539. https://doi.org/10.3389/fnhum.2016.00539.

Arnau, S., Möckel, T., Rinkenauer, G., and Wascher, E. (2017). The interconnection of mental fatigue and aging: an EEG study. *International Journal of Psychophysiology* 117: 17–25. https://doi.org/10.1016/j.ijpsycho.2017.04.003.

Atchley, P. and Chan, M. (2011). Potential benefits and costs of concurrent task engagement to maintain vigilance: a driving simulator investigation. *Human Factors* 53 (1): 3–12. https://doi.org/10.1177/0018720810391215.

Baber, C., Morar, N.S., and McCabe, F. (2019). Ecological interface design, the proximity compatibility principle, and automation reliability in road traffic management. *IEEE Transactions on Human-Machine Systems* 49 (3): 241–249. https://doi.org/10.1109/THMS.2019.2896838.

Barrouillet, P., Bernardin, S., and Camos, V. (2004). Time constraints and resource sharing in adults' working memory spans. *Journal of Experimental Psychology: General* 133 (1): 83–100. https://doi.org/10.1037/0096-3445.133.1.83.

Bartley, S.H. and Chute, E. (1947). *Fatigue and Impairment in Man*. McGraw-Hill Book Company https://doi.org/10.1037/11772-000.

Baumeister, R.F., Bratslavsky, E., Muraven, M., and Tice, D.M. (1998). Ego depletion: is the active self a limited resource? *Journal of Personality and Social Psychology* 74 (5): 1252–1265. https://doi.org/10.1037/0022-3514.74.5.1252.

Bazazan, A., Dianat, I., Feizollahi, N. et al. (2019). Effect of a posture correction–based intervention on musculoskeletal symptoms and fatigue among control room operators. *Applied Ergonomics* 76: 12–19. https://doi.org/10.1016/j.apergo.2018.11.008.

Berka, C., Levendowski, D.J., Lumicao, M.N. et al. (2007). EEG correlates of task engagement and mental workload in vigilance, learning, and memory tasks. *Aviation, Space, and Environmental Medicine* 78 (5): B231–B244.

Billings, C.E. (1996). *Human-Centered Aviation Automation: Principles and Guidelines*. Moffett Field, CA: National Aeronautics and Space Administration, Ames Research Center.

Biran, O. and Cotton, C. (2017). Explanation and justification in machine learning: a survey. In: *IJCAI-17 Workshop on Explainable AI (XAI)* (Vol. 8, No. 1, pp. 8–13).

Blain, B., Hollard, G., and Pessiglione, M. (2016). Neural mechanisms underlying the impact of daylong cognitive work on economic decisions. *Proceedings of the National Academy of Sciences* 113 (25): 6967–6972. https://doi.org/10.1073/pnas.1520527113.

Blasche, G., Szabo, B., Wagner-Menghin, M. et al. (2018). Comparison of rest-break interventions during a mentally demanding task. *Stress and Health* 34 (5): 629–638. https://doi.org/10.1002/smi .2830.

Boksem, M.A.S. and Tops, M. (2008). Mental fatigue: costs and benefits. *Brain Research Reviews* 59 (1): 125–139. https://doi.org/10.1016/j.brainresrev.2008.07.001.

Boksem, M.A.S., Meijman, T.F., and Lorist, M.M. (2006). Mental fatigue, motivation and action monitoring. *Biological Psychology* 72 (2): 123–132. https://doi.org/10.1016/j.biopsycho.2005.08.007.

Borg, G.A. (1982). Psychophysical bases of perceived exertion. *Medicine and Science in Sports and Exercise* 14 (5): 377–381. https://doi.org/10.1249/00005768-198205000-00012.

Borghini, G., Astolfi, L., Vecchiato, G. et al. (2014). Measuring neurophysiological signals in aircraft pilots and car drivers for the assessment of mental workload, fatigue and drowsiness. *Neuroscience & Biobehavioral Reviews* 44: 58–75. https://doi.org/10.1016/j.neubiorev.2012.10.003.

Borragán, G., Slama, H., Bartolomei, M., and Peigneux, P. (2017). Cognitive fatigue: a time-based resource-sharing account. *Cortex* 89: 71–84. https://doi.org/10.1016/j.cortex.2017.01.023.

Borragán, G., Guerrero-Mosquera, C., Guillaume, C. et al. (2019). Decreased prefrontal connectivity parallels cognitive fatigue-related performance decline after sleep deprivation. An optical imaging study. *Biological Psychology* 144: 115–124. https://doi.org/10.1016/j.biopsycho.2019.03.004.

Braver, T.S. (2012). The variable nature of cognitive control: a dual mechanisms framework. *Trends in Cognitive Sciences* 16 (2): 106–113. https://doi.org/10.1016/j.tics.2011.12.010.

Breckel, T.P.K., Thiel, C.M., Bullmore, E.T. et al. (2013). Long-term effects of attentional performance on functional brain network topology. *PLoS One* 8 (9): e74125. https://doi.org/10.1371/journal.pone .0074125.

Bullmore, E. and Sporns, O. (2012). The economy of brain network organization. *Nature Reviews Neuroscience* 13 (5): 336–349. https://doi.org/10.1038/nrn3214.

Cavanagh, J.F. and Frank, M.J. (2014). Frontal theta as a mechanism for cognitive control. *Trends in Cognitive Sciences* 18 (8): 414–421. https://doi.org/10.1016/j.tics.2014.04.012.

Charbonnier, S., Roy, R.N., Bonnet, S., and Campagne, A. (2016). EEG index for control operators' mental fatigue monitoring using interactions between brain regions. *Expert Systems with Applications* 52: 91–98. https://doi.org/10.1016/j.eswa.2016.01.013.

Chaudhuri, A. and Behan, P.O. (2004). Fatigue in neurological disorders. *The Lancet* 363 (9413): 978–988. https://doi.org/10.1016/S0140-6736(04)15794-2.

Chen, L., Sugi, T., Shirakawa, S. et al. (2010). Integrated design and evaluation system for the effect of rest breaks in sustained mental work based on neuro-physiological signals. *International Journal of Control, Automation and Systems* 8 (4): 862–867. https://doi.org/10.1007/s12555-010-0419-x.

Christie, S.T. and Schrater, P. (2015). Cognitive cost as dynamic allocation of energetic resources. *Frontiers in Neuroscience* 9: https://doi.org/10.3389/fnins.2015.00289.

Christoffersen, K. and Woods, D.D. (2002). How to make automated systems team players. In: *Advances in Human Performance and Cognitive Engineering Research*, vol. 2, 1–12. Emerald Group Publishing Limited https://doi.org/10.1016/S1479-3601(02)02003-9.

Chuang, C.-H., Cao, Z., King, J.-T. et al. (2018). Brain electrodynamic and hemodynamic signatures against fatigue during driving. *Frontiers in Neuroscience* 12: https://www.frontiersin.org/article/10 .3389/fnins.2018.00181.

Compton, R.J., Gearinger, D., and Wild, H. (2019). The wandering mind oscillates: EEG alpha power is enhanced during moments of mind-wandering. *Cognitive, Affective, & Behavioral Neuroscience* 19 (5): 1184–1191. https://doi.org/10.3758/s13415-019-00745-9.

Cools, R., Froböse, M., Aarts, E., and Hofmans, L. (2019). Dopamine and the motivation of cognitive control. In: *Handbook of Clinical Neurology*, vol. 163, 123–143. Elsevier https://doi.org/10.1016/B978-0-12-804281-6.00007-0.

Craig, A., Tran, Y., Wijesuriya, N., and Nguyen, H. (2012). Regional brain wave activity changes associated with fatigue: regional brain wave activity and fatigue. *Psychophysiology* 49 (4): 574–582. https://doi.org/10.1111/j.1469-8986.2011.01329.x.

Davies, D.R. and Parasuraman, R. (1982). *The Psychology of Vigilance*. Academic Press.

Degani, A. and Heymann, M. (1999). Pilot autopilot interaction: a formal perspective. In: *Proceeding of the 10th Aviation Psychology Symposium* (ed. R. Jensen). Columbus, OH: Ohio State University.

Dehais, F., Dupres, A., Di Flumeri, G. et al. (2018). Monitoring pilot's cognitive fatigue with engagement features in simulated and actual flight conditions using an hybrid fNIRS-EEG passive BCI. In: *2018 IEEE International Conference on Systems, Man, and Cybernetics (SMC)*, 544–549. https://doi.org/10.1109/SMC.2018.00102.

Dekker, S.W. and Woods, D.D. (2002). MABA-MABA or abracadabra? Progress on human–automation co-ordination. *Cognition, Technology & Work* 4 (4): 240–244. https://doi.org/10.1007/s101110200022.

DeLuca, J. (ed.) (2005). *Fatigue as a Window to the Brain*. MIT Press.

Di Stasi, L.L., McCamy, M.B., Pannasch, S. et al. (2015). Effects of driving time on microsaccadic dynamics. *Experimental Brain Research* 233 (2): 599–605. https://doi.org/10.1007/s00221-014-4139-y.

Diamond, A. (2013). Executive functions. *Annual Review of Psychology* 64 (1): 135–168. https://doi.org/10.1146/annurev-psych-113011-143750.

Diehl, A.E. (1991). Human performance and systems safety considerations in aviation mishaps. *The International Journal of Aviation Psychology* 1 (2): 97–106. https://doi.org/10.1207/s15327108ijap0102_1.

Dora, J., van Hooff, M.L.M., Geurts, S.A.E. et al. (2021). The effect of opportunity costs on mental fatigue in labor/leisure trade-offs. *Journal of Experimental Psychology: General* https://doi.org/10.1037/xge0001095.

Eisenberger, R. (1992). Learned industriousness. *Psychological Review* 99 (2): 248–267. https://doi.org/10.1037/0033-295X.99.2.248.

Enoka, R.M. and Duchateau, J. (2016). Translating fatigue to human performance. *Medicine & Science in Sports & Exercise* 48 (11): 2228–2238. https://doi.org/10.1249/MSS.0000000000000929.

Ephrath, A.R. and Young, L.R. (1981). Monitoring vs. Man-in-the-loop detection of aircraft control failures. In: *Human Detection and Diagnosis of System Failures* (ed. J. Rasmussen and W.B. Rouse), 143–154. Springer US https://doi.org/10.1007/978-1-4615-9230-3_10.

Esposito, F., Otto, T., Zijlstra, F.R.H., and Goebel, R. (2014). Spatially distributed effects of mental exhaustion on resting-state fMRI networks. *PLoS One* 9 (4): e94222. https://doi.org/10.1371/journal.pone.0094222.

Fairclough, S.H. (2017). Physiological computing and intelligent adaptation. In: *Emotions and Affect in Human Factors and Human-Computer Interaction*, 539–556. Elsevier https://doi.org/10.1016/B978-0-12-801851-4.00020-3.

Farwell, L.A. and Donchin, E. (1988). Talking off the top of your head: toward a mental prosthesis utilizing event-related brain potentials. *Electroencephalography and Clinical Neurophysiology* 70 (6): 510–523. https://doi.org/10.1016/0013-4694(88)90149-6.

Federal Aviation Authority (1995). *American Airlines Flight 965, B-757, Accident near Cali – Accident Overview*. http://lessonslearned.faa.gov/ll_main.cfm?TabID=3&LLID=43.

Frey, C.B. and Osborne, M.A. (2017). The future of employment: how susceptible are jobs to computerisation? *Technological Forecasting and Social Change* 114: 254–280. https://doi.org/10.1016/j.techfore.2016.08.019.

Fries, P., Reynolds, J.H., Rorie, A.E., and Desimone, R. (2001). Modulation of oscillatory neuronal synchronization by selective visual attention. *Science (New York, N.Y.)* 291 (5508): 1560–1563. https://doi.org/10.1126/science.1055465.

Galster, S.M., Duley, J.A., Masalonis, A.J., and Parasuraman, R. (2001). Air traffic controller performance and workload under mature free flight: conflict detection and resolution of aircraft self-separation. *The International Journal of Aviation Psychology* 11 (1): 71–93. https://doi.org/10.1207/S15327108IJAP1101_5.

Gergelyfi, M., Jacob, B., Olivier, E., and Zénon, A. (2015). Dissociation between mental fatigue and motivational state during prolonged mental activity. *Frontiers in Behavioral Neuroscience* 9: https://doi.org/10.3389/fnbeh.2015.00176.

Gilsoul, J., Libertiaux, V., and Collette, F. (2021). Cognitive fatigue in young, middle-aged, and older: breaks as a way to recover. *Applied Psychology* apps.12358: https://doi.org/10.1111/apps.12358.

Grandjean, E. (1979). Fatigue in industry. *Occupational and Environmental Medicine* 36 (3): 175–186. https://doi.org/10.1136/oem.36.3.175.

Grier, R.A., Warm, J.S., Dember, W.N. et al. (2003). The vigilance decrement reflects limitations in effortful attention, not mindlessness. *Human Factors: The Journal of the Human Factors and Ergonomics Society* 45 (3): 349–359. https://doi.org/10.1518/hfes.45.3.349.27253.

Guidotti, R., Monreale, A., Ruggieri, S. et al. (2019). A survey of methods for explaining black box models. *ACM Computing Surveys* 51 (5): 1–42. https://doi.org/10.1145/3236009.

Guo, Z., Chen, R., Liu, X. et al. (2018). The impairing effects of mental fatigue on response inhibition: An ERP study. *PLoS One* 13 (6): e0198206. https://doi.org/10.1371/journal.pone.0198206.

Han, S.-Y., Kwak, N.-S., Oh, T., and Lee, S.-W. (2020). Classification of pilots' mental states using a multimodal deep learning network. *Biocybernetics and Biomedical Engineering* 40 (1): 324–336. https://doi.org/10.1016/j.bbe.2019.12.002.

Hart, S.G. (2006). NASA-task load index (NASA-TLX); 20 years later. *Proceedings of the Human Factors and Ergonomic Society Annual Meeting* 50 (9): 904–908. https://doi.org/10.1177/154193120605000909.

Helton, W.S. and Russell, P.N. (2015). Rest is best: The role of rest and task interruptions on vigilance. *Cognition* 134: 165–173. https://doi.org/10.1016/j.cognition.2014.10.001.

Hidalgo-Muñoz, A.R., Mouratille, D., Matton, N. et al. (2018). Cardiovascular correlates of emotional state, cognitive workload and time-on-task effect during a realistic flight simulation. *International Journal of Psychophysiology* 128: 62–69. https://doi.org/10.1016/j.ijpsycho.2018.04.002.

Hockey, R. (2013). *The Psychology of Fatigue: Work, Effort, and Control.* Cambridge University Press.

Hollenbeck, J.R., Ilgen, D.R., Tuttle, D.B., and Sego, D.J. (1995). Team performance on monitoring tasks: an examination of decision errors in contexts requiring sustained attention. *Journal of Applied Psychology* 80 (6): 685–696. https://doi.org/10.1037/0021-9010.80.6.685.

Holroyd, C. (2016). The waste disposal problem of effortful control. In: *Motivation and Cognitive Control* (ed. T. Braver), 235–260. New York, NY: Routledge.

Holroyd, C.B. and McClure, S.M. (2015). Hierarchical control over effortful behavior by rodent medial frontal cortex: a computational model. *Psychological Review* 122 (1): 54–83. https://doi.org/10.1037/a0038339.

Hopstaken, J.F., van der Linden, D., Bakker, A.B., and Kompier, M.A.J. (2015). The window of my eyes: task disengagement and mental fatigue covary with pupil dynamics. *Biological Psychology* 110: 100–106. https://doi.org/10.1016/j.biopsycho.2015.06.013.

Hu, X. and Lodewijks, G. (2021). Exploration of the effects of task-related fatigue on eye-motion features and its value in improving driver fatigue-related technology. *Transportation Research Part F: Traffic Psychology and Behaviour* 80: 150–171. https://doi.org/10.1016/j.trf.2021.03.014.

Hudon, A., Demazure, T., Karran, A. et al. (2021). Explainable artificial intelligence (XAI): how the visualization of ai predictions affects user cognitive load and confidence. In: *Information Systems and Neuroscience (Vol. 52, pp. 237–246)* (ed. F.D. Davis, R. Riedl, J. vom Brocke, et al.). Springer International Publishing https://doi.org/10.1007/978-3-030-88900-5_27.

Hullman, J. (2020). Why authors don't visualize uncertainty. *IEEE Transactions on Visualization and Computer Graphics* 26 (1): 130–139. https://doi.org/10.1109/TVCG.2019.2934287.

International Civil Aviation Organization (2015). *Fatigue Management Guide for Airline Operators*. Montreal, Canada: International Civil Aviation Organization.

Inzlicht, M., Schmeichel, B.J., and Macrae, C.N. (2014). Why self-control seems (but may not be) limited. *Trends in Cognitive Sciences* 18 (3): 127–133. https://doi.org/10.1016/j.tics.2013.12.009.

Inzlicht, M., Shenhav, A., and Olivola, C.Y. (2018). The effort paradox: effort is both costly and valued. *Trends in Cognitive Sciences* 22 (4): 337–349. https://doi.org/10.1016/j.tics.2018.01.007.

Jacquet, T., Poulin-Charronnat, B., Bard, P., and Lepers, R. (2021). Persistence of mental fatigue on motor control. *Frontiers in Psychology* 11: 588253. https://doi.org/10.3389/fpsyg.2020.588253.

Jamson, A.H., Merat, N., Carsten, O.M.J., and Lai, F.C.H. (2013). Behavioural changes in drivers experiencing highly-automated vehicle control in varying traffic conditions. *Transportation Research Part C: Emerging Technologies* 30: 116–125. https://doi.org/10.1016/j.trc.2013.02.008.

Kahneman, D. (1973). *Attention and effort*, vol. 1063, 218–226. Englewood Cliffs, NJ: Prentice-Hall.

Kamzanova, A.T., Kustubayeva, A.M., and Matthews, G. (2014). Use of EEG workload indices for diagnostic monitoring of vigilance decrement. *Human Factors: The Journal of the Human Factors and Ergonomics Society* 56 (6): 1136–1149. https://doi.org/10.1177/0018720814526617.

Kessel, C.J. and Wickens, C.D. (1982). The transfer of failure-detection skills between monitoring and controlling dynamic systems. *Human Factors: The Journal of the Human Factors and Ergonomics Society* 24 (1): 49–60. https://doi.org/10.1177/001872088202400106.

Kiesel, A., Steinhauser, M., Wendt, M. et al. (2010). Control and interference in task switching—a review. *Psychological Bulletin* 136 (5): 849–874. https://doi.org/10.1037/a0019842.

Klaassen, E.B., Evers, E.A.T., de Groot, R.H.M. et al. (2014). Working memory in middle-aged males: age-related brain activation changes and cognitive fatigue effects. *Biological Psychology* 96: 134–143. https://doi.org/10.1016/j.biopsycho.2013.11.008.

Klein, G., Woods, D.D., Bradshaw, J.M. et al. (2004). Ten challenges for making automation a "team player" in joint human-agent activity. *IEEE Intelligent Systems* 19 (06): 91–95. https://doi.org/10.1109/MIS.2004.74.

Kluger, B.M., Krupp, L.B., and Enoka, R.M. (2013). Fatigue and fatigability in neurologic illnesses. *Neurology* 80 (4): 409–416. https://doi.org/10.1212/WNL.0b013e31827f07be.

Koch, I., Poljac, E., Müller, H., and Kiesel, A. (2018). Cognitive structure, flexibility, and plasticity in human multitasking—an integrative review of dual-task and task-switching research. *Psychological Bulletin* 144 (6): 557–583. https://doi.org/10.1037/bul0000144.

Koechlin, E. and Summerfield, C. (2007). An information theoretical approach to prefrontal executive function. *Trends in Cognitive Sciences* 11 (6): 229–235. https://doi.org/10.1016/j.tics.2007.04.005.

Kurzban, R., Duckworth, A., Kable, J.W., and Myers, J. (2013). An opportunity cost model of subjective effort and task performance. *Behavioral and Brain Sciences* 36 (6): 661–679. https://doi.org/10.1017/S0140525X12003196.

Lamnabhi-Lagarrigue, F., Annaswamy, A., Engell, S. et al. (2017). Systems & control for the future of humanity, research agenda: current and future roles, impact and grand challenges. *Annual Reviews in Control* 43: 1–64.

Lavie, P. (1989). Ultradian rhythms in arousal-the problem of masking. *Chronobiology International* 6 (1): 21–28. https://doi.org/10.3109/07420528909059139.

Le Goff, K., Rey, A., Haggard, P. et al. (2018). Agency modulates interactions with automation technologies. *Ergonomics* 61 (9): 1282–1297. https://doi.org/10.1080/00140139.2018.1468493.

Lee, K.A., Hicks, G., and Nino-Murcia, G. (1991). Validity and reliability of a scale to assess fatigue. *Psychiatry Research* 36 (3): 291–298. https://doi.org/10.1016/0165-1781(91)90027-M.

Li, Z., Chen, L., Peng, J., and Wu, Y. (2017). Automatic detection of driver fatigue using driving operation information for transportation safety. *Sensors* 17 (6): 1212. https://doi.org/10.3390/s17061212.

Lim, J., Wu, W., Wang, J. et al. (2010). Imaging brain fatigue from sustained mental workload: an ASL perfusion study of the time-on-task effect. *NeuroImage* 49 (4): 3426–3435. https://doi.org/10.1016/j.neuroimage.2009.11.020.

Lim, J., Quevenco, F.-C., and Kwok, K. (2013). EEG alpha activity is associated with individual differences in post-break improvement. *NeuroImage* 76: 81–89. https://doi.org/10.1016/j.neuroimage.2013.03.018.

Lim, J., Teng, J., Wong, K.F., and Chee, M.W.L. (2016). Modulating rest-break length induces differential recruitment of automatic and controlled attentional processes upon task reengagement. *NeuroImage* 134: 64–73. https://doi.org/10.1016/j.neuroimage.2016.03.077.

van der Linden, D., Frese, M., and Meijman, T.F. (2003). Mental fatigue and the control of cognitive processes: effects on perseveration and planning. *Acta Psychologica* 113 (1): 45–65. https://doi.org/10.1016/S0001-6918(02)00150-6.

Liu, J.-P., Zhang, C., and Zheng, C.-X. (2010). Estimation of the cortical functional connectivity by directed transfer function during mental fatigue. *Applied Ergonomics* 42 (1): 114–121. https://doi.org/10.1016/j.apergo.2010.05.008.

Liu, J., Zhu, Y., Sun, H. et al. (2020). Sustaining attention for a prolonged duration affects dynamic organizations of frequency-specific functional connectivity. *Brain Topography* 33 (6): 677–692. https://doi.org/10.1007/s10548-020-00795-0.

Lorist, M.M. and Faber, L.G. (2011). Consideration of the influence of mental fatigue on controlled and automatic cognitive processes and related neuromodulatory effects. In: *Cognitive Fatigue: Multidisciplinary Perspectives on Current Research and Future Applications* (ed. P.L. Ackerman), 105–126. American Psychological Association https://doi.org/10.1037/12343-005.

Lorist, M.M., Boksem, M.A.S., and Ridderinkhof, K.R. (2005). Impaired cognitive control and reduced cingulate activity during mental fatigue. *Cognitive Brain Research* 24 (2): 199–205. https://doi.org/10.1016/j.cogbrainres.2005.01.018.

Lorist, M.M., Bezdan, E., ten Caat, M. et al. (2009). The influence of mental fatigue and motivation on neural network dynamics; an EEG coherence study. *Brain Research* 1270: 95–106. https://doi.org/10.1016/j.brainres.2009.03.015.

Luria, A.R. (1973). The frontal lobes and the regulation of behavior. In: *Psychophysiology of the Frontal Lobes*, 3–26. Elsevier https://doi.org/10.1016/B978-0-12-564340-5.50006-8.

Ma, J., Gu, J., Jia, H. et al. (2018). The relationship between drivers' cognitive fatigue and speed variability during monotonous daytime driving. *Frontiers in Psychology* 9: 459. https://doi.org/10.3389/fpsyg.2018.00459.

Mackworth, N.H. (1948). The breakdown of vigilance during prolonged visual search. *Quarterly Journal of Experimental Psychology* 1 (1): 6–21. https://doi.org/10.1080/17470214808416738.

Maffei, A. and Angrilli, A. (2018). Spontaneous eye blink rate: an index of dopaminergic component of sustained attention and fatigue. *International Journal of Psychophysiology* 123: 58–63. https://doi.org/10.1016/j.ijpsycho.2017.11.009.

Magnuson, J.R., Doesburg, S.M., and McNeil, C.J. (2021). Development and recovery time of mental fatigue and its impact on motor function. *Biological Psychology* 161: 108076. https://doi.org/10.1016/j .biopsycho.2021.108076.

Maksimenko, V.A., Hramov, A.E., Grubov, V.V. et al. (2019). Nonlinear effect of biological feedback on brain attentional state. *Nonlinear Dynamics* 95 (3): 1923–1939. https://doi.org/10.1007/s11071-018-4668-1.

Marandi, R.Z., Madeleine, P., Omland, Ø. et al. (2019). An oculometrics-based biofeedback system to impede fatigue development during computer work: a proof-of-concept study. *PLoS One* 14 (5): e0213704. https://doi.org/10.1371/journal.pone.0213704.

Marcora, S.M., Staiano, W., and Manning, V. (2009). Mental fatigue impairs physical performance in humans. *Journal of Applied Physiology* 106 (3): 857–864. https://doi.org/10.1152/japplphysiol.91324 .2008.

Martin, K., Meeusen, R., Thompson, K.G. et al. (2018). Mental fatigue impairs endurance performance: a physiological explanation. *Sports Medicine* 48 (9): 2041–2051. https://doi.org/10.1007/s40279-018-0946-9.

Matuz, A., van der Linden, D., Kisander, Z. et al. (2021). Enhanced cardiac vagal tone in mental fatigue: analysis of heart rate variability in Time-on-Task, recovery, and reactivity. *PLoS One* 16 (3): e0238670. https://doi.org/10.1371/journal.pone.0238670.

Menon, V. and Uddin, L.Q. (2010). Saliency, switching, attention and control: a network model of insula function. *Brain Structure and Function* 214 (5–6): 655–667. https://doi.org/10.1007/s00429-010-0262-0.

Miller, E.K. and Cohen, J.D. (2001). An integrative theory of prefrontal cortex function. *Annual Review of Neuroscience* 24 (1): 167–202. https://doi.org/10.1146/annurev.neuro.24.1.167.

Miyake, A., Friedman, N.P., Emerson, M.J. et al. (2000). The unity and diversity of executive functions and their contributions to complex "frontal lobe" tasks: a latent variable analysis. *Cognitive Psychology* 41 (1): 49–100. https://doi.org/10.1006/cogp.1999.0734.

Möckel, T., Beste, C., and Wascher, E. (2015). The effects of time on task in response selection—an ERP study of mental fatigue. *Scientific Reports* 5 (1): https://doi.org/10.1038/srep10113.

Molloy, R. and Parasuraman, R. (1996). Monitoring an automated system for a single failure: vigilance and task complexity effects. *Human Factors: The Journal of the Human Factors and Ergonomics Society* 38 (2): 311–322. https://doi.org/10.1177/001872089606380211.

Monsell, S., Yeung, N., and Azuma, R. (2000). Reconfiguration of task-set: is it easier to switch to the weaker task? *Psychological Research* 63 (3): 250–264. https://doi.org/10.1007/s004269900005.

Müller, T. and Apps, M.A.J. (2019). Motivational fatigue: a neurocognitive framework for the impact of effortful exertion on subsequent motivation. *Neuropsychologia* 123: 141–151. https://doi.org/10 .1016/j.neuropsychologia.2018.04.030.

Naiseh, M., Jiang, N., Ma, J., and Ali, R. (2020). Explainable recommendations in intelligent systems: delivery methods, modalities and risks. In: *Research Challenges in Information Science* (ed. F. Dalpiaz, J. Zdravkovic, and P. Loucopoulos), 212–228. Springer International Publishing https://doi .org/10.1007/978-3-030-50316-1_13.

Norman, D.A., Broadbent, D.E., Baddeley, A.D., and Reason, J. (1990). The 'problem' with automation: Inappropriate feedback and interaction, not 'over-automation.' *Philosophical Transactions of the Royal Society of London. B, Biological Sciences* 327 (1241): 585–593. https://doi.org/10.1098/rstb.1990 .0101.

Palmer, E. (1995). Oops, it didn't arm-a case study of two automation surprises. In: *Proceedings of the Eighth International Symposium on Aviation Psychology*, 227–232. Columbus, Ohio: Ohio State University.

Parasuraman, R. (1987). Human-computer monitoring. *Human Factors: The Journal of the Human Factors and Ergonomics Society* 29 (6): 695–706. https://doi.org/10.1177/001872088702900609.

Petruo, V.A., Mückschel, M., and Beste, C. (2018). On the role of the prefrontal cortex in fatigue effects on cognitive flexibility—a system neurophysiological approach. *Scientific Reports* 8 (1): 6395. https://doi.org/10.1038/s41598-018-24834-w.

Phillips, R.O. (2015). A review of definitions of fatigue – and a step towards a whole definition. *Transportation Research Part F: Traffic Psychology and Behaviour* 29: 48–56. https://doi.org/10.1016/j.trf.2015.01.003.

Qi, P., Ru, H., Gao, L. et al. (2019). Neural mechanisms of mental fatigue revisited: new insights from the brain connectome. *Engineering* 5 (2): 276–286. https://doi.org/10.1016/j.eng.2018.11.025.

Qi, P., Hu, H., Zhu, L. et al. (2020). EEG functional connectivity predicts individual behavioural impairment during mental fatigue. *IEEE Transactions on Neural Systems and Rehabilitation Engineering* 28 (9): 2080–2089. https://doi.org/10.1109/TNSRE.2020.3007324.

Qin, H., Zhou, X., Ou, X. et al. (2021). Detection of mental fatigue state using heart rate variability and eye metrics during simulated flight. *Human Factors and Ergonomics in Manufacturing & Service Industries* 31 (6): 637–651. https://doi.org/10.1002/hfm.20927.

Raffaelli, Q., Mills, C., and Christoff, K. (2018). The knowns and unknowns of boredom: a review of the literature. *Experimental Brain Research* 236 (9): 2451–2462. https://doi.org/10.1007/s00221-017-4922-7.

Rogers, R.D. and Monsell, S. (1995). Costs of a predictable switch between simple cognitive tasks. *Journal of Experimental Psychology: General* 124 (2): 207–231. https://doi.org/10.1037/0096-3445.124.2.207.

Roy, R.N., Charbonnier, S., Campagne, A., and Bonnet, S. (2016). Efficient mental workload estimation using task-independent EEG features. *Journal of Neural Engineering* 13 (2): 026019. https://doi.org/10.1088/1741-2560/13/2/026019.

Roy, R.N., Drougard, N., Gateau, T. et al. (2020). How can physiological computing benefit human-robot interaction? *Robotics* 9 (4): 100. https://doi.org/10.3390/robotics9040100.

Rubinstein, J.S., Meyer, D.E., and Evans, J.E. (2001). Executive control of cognitive processes in task switching. *Journal of Experimental Psychology: Human Perception and Performance* 27 (4): 763–797. https://doi.org/10.1037/0096-1523.27.4.763.

Salomone, M., Burle, B., Fabre, L., and Berberian, B. (2021). An electromyographic analysis of the effects of cognitive fatigue on online and anticipatory action control. *Frontiers in Human Neuroscience* 14: 615046. https://doi.org/10.3389/fnhum.2020.615046.

Samad, T. (2022). Human-in-the-loop control and cyber–physical–human systems: applications and categorization. In: *Cyber–Physical–Human Systems: Fundamentals and Applications* (ed. A. Annaswamy et al.). UK: Wiley in Press.

Samn, S.W., and Perelli, L.P. (1982). *Estimating Aircrew Fatigue: A Technique with Application to Airlift Operations (No. SAM-TR-82-21)*. San Antanio, TX: School of Aerospace Medicine Brook Air Force Base.

Sarter, N.B. and Woods, D.D. (1994). Pilot Interaction with cockpit automation II: an experimental study of pilots' model and awareness of the flight management system. *The International Journal of Aviation Psychology* 4 (1): 1–28. https://doi.org/10.1207/s15327108ijap0401_1.

Sarter, N.B. and Woods, D.D. (1995). How in the world did we ever get into that mode? Mode error and awareness in supervisory control. *Human Factors: The Journal of the Human Factors and Ergonomics Society* 37 (1): 5–19. https://doi.org/10.1518/001872095779049516.

Sarter, N.B., Woods, D.D., and Billings, C.E. (1997). Automation surprises. *Handbook of Human Factors and Ergonomics* 2: 1926–1943.

Schmidt, E.A., Schrauf, M., Simon, M. et al. (2009). Drivers' misjudgment of vigilance state during prolonged monotonous daytime driving. *Accident Analysis & Prevention* 41 (5): 1087–1093. https://doi.org/10.1016/j.aap.2009.06.007.

Schneider, W. and Shiffrin, R.M. (1977). Controlled and automatic human information processing: I. Detection, search, and attention. *Psychological Review* 84 (1): 1–66. https://doi.org/10.1037/0033-295X.84.1.1.

Scholz, A., Ghadiri, A., Singh, U. et al. (2018). Functional work breaks in a high-demanding work environment: an experimental field study. *Ergonomics* 61 (2): 255–264. https://doi.org/10.1080/00140139.2017.1349938.

Seeley, W.W., Menon, V., Schatzberg, A.F. et al. (2007). Dissociable intrinsic connectivity networks for salience processing and executive control. *Journal of Neuroscience* 27 (9): 2349–2356. https://doi.org/10.1523/JNEUROSCI.5587-06.2007.

Seli, P., Cheyne, J.A., and Smilek, D. (2013). Wandering minds and wavering rhythms: linking mind wandering and behavioral variability. *Journal of Experimental Psychology: Human Perception and Performance* 39 (1): 1–5. https://doi.org/10.1037/a0030954.

Shenhav, A., Musslick, S., Lieder, F. et al. (2017). Toward a rational and mechanistic account of mental effort. *Annual Review of Neuroscience* 40 (1): 99–124. https://doi.org/10.1146/annurev-neuro-072116-031526.

Smit, A.S., Eling, P.A.T.M., and Coenen, A.M.L. (2004). Mental effort causes vigilance decrease due to resource depletion. *Acta Psychologica* 115 (1): 35–42. https://doi.org/10.1016/j.actpsy.2003.11.001.

Smit, A.S., Eling, P.A.T.M., Hopman, M.T., and Coenen, A.M.L. (2005). Mental and physical effort affect vigilance differently. *International Journal of Psychophysiology* 57 (3): 211–217. https://doi.org/10.1016/j.ijpsycho.2005.02.001.

Stoyanovich, J., Van Bavel, J.J., and West, T.V. (2020). The imperative of interpretable machines. *Nature Machine Intelligence* 2 (4): 197–199. https://doi.org/10.1038/s42256-020-0171-8.

Sun, Y., Lim, J., Kwok, K., and Bezerianos, A. (2014). Functional cortical connectivity analysis of mental fatigue unmasks hemispheric asymmetry and changes in small-world networks. *Brain and Cognition* 85: 220–230. https://doi.org/10.1016/j.bandc.2013.12.011.

Taleb, N.N. (2012). *Antifragile: Things that gain from disorder*, vol. 3. Random House.

Trejo, L.J., Kubitz, K., Rosipal, R. et al. (2015). EEG-based estimation and classification of mental fatigue. *Psychology* 06 (05): 572. https://doi.org/10.4236/psych.2015.65055.

Trujillo, L.T. (2019). Mental effort and information-processing costs are inversely related to global brain free energy during visual categorization. *Frontiers in Neuroscience* 13: https://www.frontiersin.org/article/10.3389/fnins.2019.01292.

Van Charante, E.M., Cook, R.I., Woods, D.D. et al. (1992). Human-computer interaction in context: physician interaction with automated intravenous controllers in the heart room. *IFAC Proceedings Volumes* 25 (9): 263–274.

Vantrepotte, Q., Berberian, B., Pagliari, M., and Chambon, V. (2022). Leveraging human agency to improve confidence and acceptability in human-machine interactions. *Cognition* 222: 105020. https://doi.org/10.1016/j.cognition.2022.105020.

Wang, C., Ding, M., and Kluger, B.M. (2014). Change in intraindividual variability over time as a key metric for defining performance-based cognitive fatigability. *Brain and Cognition* 85: 251–258. https://doi.org/10.1016/j.bandc.2014.01.004.

Wang, C., Trongnetrpunya, A., Samuel, I.B.H. et al. (2016). Compensatory neural activity in response to cognitive fatigue. *Journal of Neuroscience* 36 (14): 3919–3924. https://doi.org/10.1523/JNEUROSCI.3652-15.2016.

Warm, J.S., Parasuraman, R., and Matthews, G. (2008). Vigilance requires hard mental work and is stressful. *Human Factors* 50 (3): 433–441. https://doi.org/10.1518/001872008X312152.

Wascher, E., Rasch, B., Sänger, J. et al. (2014). Frontal theta activity reflects distinct aspects of mental fatigue. *Biological Psychology* 96: 57–65. https://doi.org/10.1016/j.biopsycho.2013.11.010.

Wickens, C.D. and Kessel, C. (1979). The effects of participatory mode and task workload on the detection of dynamic system failures. *IEEE Transactions on Systems, Man, and Cybernetics* 9 (1): 24–34. https://doi.org/10.1109/TSMC.1979.4310070.

Wiener, E.L. (1987). Application of vigilance research: rare, medium, or well done? *Human Factors: The Journal of the Human Factors and Ergonomics Society* 29 (6): 725–736. https://doi.org/10.1177/001872088702900611.

Wiener, E. L. (1989). *Human factors of advanced technology (glass cockpit) transport aircraft* (Tech. Report 177528). Moffett Field, CA: NASA Ames Research Center.

Wilson, G.F. and Fisher, F. (1991). The use of cardiac and eye blink measures to determine flight segment in F4 crews. *Aviation, Space, and Environmental Medicine* 62 (10): 959–962.

Wilson, G.F., Caldwell, J.A., and Russell, C.A. (2007). Performance and psychophysiological measures of fatigue effects on aviation related tasks of varying difficulty. *The International Journal of Aviation Psychology* 17 (2): 219–247. https://doi.org/10.1080/10508410701328839.

Woods, D.D. and Tinapple, D. (1999). W3: Watching human factors watch people at work. In: *43rd Annual Meeting of the Human Factors Ergonomics and Society*. TX: Houston.

Wu, E.Q., Deng, P.-Y., Qiu, X.-Y. et al. (2021). Detecting fatigue status of pilots based on deep learning network using EEG signals. *IEEE Transactions on Cognitive and Developmental Systems* 13 (3): 575–585. https://doi.org/10.1109/TCDS.2019.2963476.

Wylie, G.R., Yao, B., Genova, H.M. et al. (2020). Using functional connectivity changes associated with cognitive fatigue to delineate a fatigue network. *Scientific Reports* 10 (1): 21927. https://doi.org/10.1038/s41598-020-78768-3.

Zander, T.O., Kothe, C., Jatzev, S., and Gaertner, M. (2010). Enhancing human-computer interaction with input from active and passive brain-computer interfaces. In: *Brain-computer Interfaces: Applying Our Minds to Human-computer Interaction* (ed. D.S. Tan and A. Nijholt), 181–199. Springer https://doi.org/10.1007/978-1-84996-272-8_11.

Zhao, C., Zhao, M., Yang, Y. et al. (2017). The reorganization of human brain networks modulated by driving mental fatigue. *IEEE Journal of Biomedical and Health Informatics* 21 (3): 743–755. https://doi.org/10.1109/JBHI.2016.2544061.

Zijlstra, F.R.H. (1993). Efficiency in work behavior. A design approach for modern tools. PhD thesis, Delft University of Technology. Delft, The Netherlands: Delft University Press. Book Citation.

Zijlstra, F.R.H., Cropley, M., and Rydstedt, L.W. (2014). From recovery to regulation: an attempt to reconceptualize 'recovery from work.' *Stress and Health* 30 (3): 244–252. https://doi.org/10.1002/smi.2604.

16

Epidemics Spread Over Networks: Influence of Infrastructure and Opinions

Baike She[1], Sebin Gracy[2], Shreyas Sundaram[3], Henrik Sandberg[4], Karl H. Johansson[4], and Philip E. Paré[3]

[1]Department of Mechanical and Aerospace Engineering, University of Florida, Gainesville, FL, USA
[2]Department of Electrical and Computer Engineering, Rice University, Houston, TX, USA
[3]Elmore Family School of Electrical and Computer Engineering, Purdue University, West Lafayette, IN, USA
[4]Division of Decision and Control Systems, Department of Intelligent Systems, EECS, KTH Royal Institute of Technology, Stockholm, Sweden

16.1 Introduction

As introduced in Samad (2023), Chapter 1 in this book, distinct from cyber–physical systems (CPSs) that are composed of the interactions between cyber and physical factors, cyber–physical–human systems also consider the human factors in modeling, analysis, and decision-making. In frameworks such as epidemic modeling and control, human behaviors and interactions, together with physical environment and cyber-surveillance are critical in determining the epidemic spreading process and decision-making. Further, the human-in-loop framework is one of the most essential pieces in policy making in epidemic mitigation and suppression, since we cannot rely on policy-making knowledge that is automatically generated through the CPS, like the traditional closed-loop feedback control systems. Instead, humans, i.e. policy-makers, will leverage knowledge provided by the system to establish epidemic mitigation strategies. In order to understand the epidemic spreading processes and mitigation strategies in modern society, as well as the critical role that the human-in-loop framework plays in decision-making in epidemic spreading, we first introduce epidemic spreading models. In this section, we present a review of infectious diseases, the history of disease spreading processes, and compartmental models for modeling and analyzing disease spreading processes.

16.1.1 Infectious Diseases

The emergence and spread of infectious diseases, with the first influenza pandemic being pathologically documented in 1510 (Morens et al. 2010), have been concomitant features of human civilization. Due to various transmission and spreading mechanisms, infectious diseases can spread within the same species (e.g. smallpox), but they can also spread across species (e.g. Dengue fever, SARS-CoV-2) (Acheson 2007). Nevertheless, recent headlines have recorded an increasing number of major pandemics and epidemics that have already afflicted humanity extensively. For instance, viruses that lead to well-known pandemics over the past several hundred years include cholera (Finkelstein and Feeley 1973), influenza (Potter 2001), severe acute respiratory syndrome coron-

Cyber–Physical–Human Systems: Fundamentals and Applications, First Edition.
Edited by Anuradha M. Annaswamy, Pramod P. Khargonekar, Françoise Lamnabhi-Lagarrigue, and Sarah K. Spurgeon.
© 2023 The Institute of Electrical and Electronics Engineers, Inc. Published 2023 by John Wiley & Sons, Inc.

avirus (SARS-CoV) (Seto et al. 2003), Middle East respiratory syndrome coronavirus (MERS-CoV) (De Groot et al. 2013), severe acute respiratory syndrome coronavirus 2 (SARS-CoV-2) (Chin et al. 2020), etc. Human beings have paid a huge price including, but not limited to, life, economical and social loss, ecological damage (Benatar 2002), etc., when fighting against infectious diseases (Helms et al. 2020; Davies et al. 2021; Barro et al. 2020). While some infectious diseases that caused enormous damage have faded away over the long course of history due to modern science and human effort (Fenner et al. 1988), numerous viruses that can cause pandemics still coexist with us to this day (e.g. human immunodeficiency virus [HIV] (Friedland and Klein 1987) and influenza (Potter 2001)). Furthermore, new outbreaks and pandemics are inevitable (Fauci et al. 2020). Hence, it is critical for research communities to continue to study epidemic spreading mechanisms and curing strategies to mitigate the potential damage caused by infectious diseases. For example, SARS-CoV-2 (the virus that causes COVID-19) is mainly transmitted by exposure to infectious respiratory fluids (Davies et al. 2020). HIV Type-1 is transmitted by sexual contact across mucosal surfaces, by maternal–infant exposure, and by percutaneous inoculation (Hollingsworth et al. 2008). From a thorough understanding of the disease spreading mechanisms and patterns in communities, regions, and countries, researchers are able to construct mathematical models to track spreading trends and to propose appropriate approaches to mitigate and eradicate the transmission of diseases. These analyses are capable of assisting in early warning of outbreaks, prediction of spreading patterns, further preparations for mitigation strategies, etc. Consequently, mathematical models and analysis of epidemic spreading processes are pivotal for comparing, planning, implementing, evaluating, and optimizing various detection, prevention, therapy, and control programs for disease spreading interventions (Teng 1985; Piot and Seck 2001; Alisic et al. 2022; Chowell 2017).

16.1.2 Modeling Epidemic Spreading Processes

Disease spreading models are of great importance for analyzing and designing epidemiological surveys, suggesting crucial data that should be collected, identifying spreading patterns, making general forecasts, estimating the uncertainty in forecasts, and facilitating policy making for epidemic eradication. The documented history of modeling epidemic spreading dynamics can be traced back to Daniel Bernoulli, who formulated a model to evaluate the effectiveness of variolation[1] of healthy people with the smallpox virus in 1760 (Bernoulli 1760). The trend toward establishing spreading models of epidemics began to take shape in the twentieth century. In 1906, Hamer (1906) formulated and analyzed an epidemic spreading model by assuming that the number of new cases per unit time depends on the product of the densities of the susceptible and infected population. The infection mechanism proposed by Hamer (1906) became one of the most popular ways to model epidemic spreading processes. Later on, Ross developed a model to characterize the spreading of malaria in 1911 (Ross 1911). In order to capture more detailed behaviors of epidemic spreading processes and to explore the spreading mechanism analytically, Kermack and McKendrick developed threshold conditions that can lead to either an outbreak or eradication (Kermack and McKendrick 1927; McKendrick 1925). Other representative works on modeling and analyzing epidemic spreading processes can be found in Dietz (1967, 1988), Kendall (1956), Bailey (1975), and Hethcote (2000).

Considered as one of the most foundational and classic disease spreading models, group compartmental models (usually known as compartmental models) that consider epidemic spreading over

---

1 Variolation was a method of immunizing patients against smallpox by infecting them with substance from the pustules of patients with a mild form of the disease, in the hope that a mild, but protective, infection would result (Stewart and Devlin 2006).

a well-mixed population are popular ways for epidemiologists to analyze and forecast spreading patterns (Hethcote 2000; Anderson and May 1991; Bailey 1957). Group compartmental models are capable of offering insights into detecting epidemic outbreaks and forecasting spreading patterns within a well-mixed population with little outside influence (Nowzari et al. 2016), which are applicable for studying local disease outbreaks and spreading processes within communities.

When building classic compartmental models, one usually splits the whole well-mixed population into different compartments with labels. For instance, one can split a whole population into susceptible (people who are vulnerable to the virus) and infected (people who are infected with the virus). Based on these labels, we introduce one classic compartmental model: susceptible–infected–susceptible (SIS) models. SIS models categorize the population into either susceptible and infected compartments, depending on whether the population members are uninfected or infected. The SIS models lay the foundation for researchers to further develop various group models, when considering more complicated disease properties with detailed labels. For instance, recent models have involved aspects such as passive immunity, gradual loss of immunity, various stages of the infection (presymptomatic, asymptomatic, and symptomatic), vertical transmission, disease vectors, macroparasitic loads, age structure, social and sexual mixing groups, spatial spread, vaccination, quarantine, hospitalization. Hence, it is natural for researchers to formulate particular models for common diseases such as measles, chickenpox, smallpox, malaria, HIV/AIDS, and influenza. As for COVID-19 infections, there is an incubation period during which individuals have been infected but are not yet infectious themselves (Shi et al. 2020). The incubation period can be captured by introducing a compartment E for the exposed population. In addition, recent studies have shown that some individuals who are infected with COVID-19 will never develop any symptoms at all (Pollock and Lancaster 2020). Hence, in order to capture the spread pattern for COVID-19 and to predict the course of the epidemic to help plan effective control strategies, Giordano et al. (2020) proposed a new group compartmental model that considered eight stages of infection: susceptible (S), infected (I), diagnosed (D), ailing (A), recognized (R), threatened (T), healed (H), and extinct (E), collectively termed the SIDARTHE compartmental model. The proposed SIDARTHE model discriminated between infected individuals depending on whether they have been diagnosed and on the severity of their symptoms. For more detailed information about different compartmental models of epidemics, we refer the interested reader to the following papers (Hethcote 2000; Paré et al. 2020; Heesterbeek et al. 2015).

16.1.3 Susceptible–Infected–Susceptible (SIS) Compartmental Models

Among these group compartmental models, SIS group models are widely used for diseases that can infect people repeatedly, e.g. recurrent bacterial and fungal infections, such as *Streptococcus pyogenes*, *Neisseria gonorrhoeae* bacteria, *Trichophyton rubrum* (one cause of athlete's foot), *Microsporum canis* (one cause of ringworm) fungi (Paré et al. 2020). Introduced by Kermack and McKendrick (1932), the continuous-time deterministic form of the SIS model in its simplest form is given by

$$\dot{S}(t) = -\beta S(t)I(t) + \delta I(t)$$
$$\dot{I}(t) = \beta S(t)I(t) - \delta I(t)$$

(16.1)

where $S(t)$ is the proportion of the susceptible population, and $I(t)$ is the proportion of the infected proportion; see also Figure 16.1. Let β denote the rate of infection or contact between susceptible and infected compartments. Similarly, let δ denote the healing or recovery rate of the infected population. From Eq. (16.1), we have $\dot{S}(t) + \dot{I}(t) = 0$, which implies that the size of the population

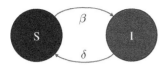

Figure 16.1 SIS model.

remains unchanged. This phenomenon can be explained by assuming a homogeneous population with no vital dynamics, that is, birth and death processes are not included. Hence, we denote the constant population size as N.

16.2 Epidemics on Networks

16.2.1 Motivation

Group compartmental models capture epidemic spreading in a well-mixed population, such as within a city or densely populated country. However, modern transportation systems and means of traveling make it much easier for people to travel between cities and countries. The frequent flow of the population across large distances enables epidemics to spread widely and quickly. Consequently, local outbreaks can evolve into global pandemics. For example, during the COVID-19 pandemic, it took less than a month for the United States to detect its first case within its borders following the COVID-19 outbreak in Wuhan, China. In such cases, a well-mixed population representation might be insufficient both for modeling and studying the spreading pattern. Hence, networked epidemic spreading models that can capture spreading processes between subpopulations become necessary.

We use a more detailed example to illustrate the advantage of leveraging networked epidemic spreading models instead of group models. During the COVID-19 crisis, countries implemented different policies in response to the pandemic. One strategy was to decrease traffic flows between countries. The reduced traffic/population flows between countries during the pandemic influenced virus spreading patterns across the world. For instance, China just recently, has reestablished extremely strict border policies for travelers who would like to enter the country (Hossain et al. 2020). In order to fly to China, all travelers must show negative testing results before boarding. In addition, after landing on China, travelers must be quarantined in hotels for around two weeks and will be released after testing negative. In comparison, it is more challenging for countries with relatively looser border policies to keep the virus out. Therefore, networked epidemic spreading models that reflect epidemic transmission between different countries based on traffic/population flows along with the pandemic policies are important. Further, networked models have advantages over group models, when tackling resource allocation for epidemic mitigation problems. By studying transmission couplings between countries in networked models, policy-makers can adjust epidemic mitigation policies based on the transmission rates between different countries. For example, it is unnecessary for policy-makers to allocate many resources (e.g. canceling flights) to mitigate the disease transmission between countries with zero infection. In contrast, a country with a small infected population might propose extremely strict border policies for travelers from countries with relatively higher infection levels.

In addition, when assuming that individuals are connected in a well-mixed population model, it is impossible to perform contact-tracing analysis. Thus, beyond capturing the spreading dynamics between individual elements, networked epidemic spreading models can also facilitate the trace of origins of epidemic outbreaks and establish epidemic mitigation strategies via construction of contact-tracing networks. In general, contact-tracing networks are wildly used for tracing the origin of viruses and thereby preventing the spread of viruses (Eames and Keeling 2003;

Armbruster and Brandeau 2007). For infections that have an incubation period like COVID-19, it is critical to trace potentially infected individuals and isolate them from interacting with others in order to prevent the virus from spreading further (Munzert et al. 2021). To this end, networked epidemic spreading models can provide researchers more valuable information in modeling, analyzing, and mitigating epidemic spreading processes.

16.2.2 Modeling Epidemics over Networks

Recent breakthroughs in both software and algorithms significantly facilitate the exploration of connected structures and dynamics of complex systems, such as biological, engineering, social systems (Newman 2003). The research area of network science considers distinct elements represented by nodes (or vertices) and the connections between the elements as links (or edges) to capture connected systems. Drawing upon tools such as graph theory, statistics and machine learning, systems and control theory, societal and ecological theory, there has been a resurgence of interest in mathematical modeling and analysis of epidemic spreading processes over networks (Newman 2018). When studying networked epidemic spreading dynamics, we roughly classify research on networked epidemic spreading models into two categories through network structures. One is to assume that epidemics spreading over specific network structures, which are applicable to some particular transmission processes. For instance, modeling epidemics spreading over star graphs can capture outbreaks caused by shared resources, such as water sources (Liu et al. 2019a). Other works that consider epidemics over particular networks study more complex structures like small-world networks (Oleś et al. 2013) and scale-free networks (Pastor-Satorras and Vespignani 2001). More detailed reviews of epidemic spreading over particular networks can be found in Pastor-Satorras et al. (2015).

The analytical tools and results developed for specific disease spreading structures are usually limited. In order to develop models and analytical tools for disease spreading over arbitrary network structures, similar to the introduction of group compartmental models, we focus on research results from one of the classic networked compartmental models, networked SIS models. Researchers have developed and studied the threshold conditions and equilibria of networked SIS models (Dietz 1967; Lajmanovich and Yorke 1976; Fall et al. 2007; Abakuks 1973; Kryscio and Lefévre 1989). Lajmanovich and Yorke (1976) studied the spread of gonorrhea by considering heterogeneous epidemic spreading parameters between subpopulations, and further explored the asymptotic stability properties of the developed model. Then, the work by Fall et al. (2007) established the stability of the classic deterministic SIS models through Lyapunov analysis, while Abakuks (1973) and Kryscio and Lefévre (1989) analyzed the behaviors of the classic stochastic SIS models. Studies on networked SIS models also included considerations of convergence properties (Ahn and Hassibi 2013), effect of the network topology on epidemic spreading process (Masucci and Silva 2013), extensions to time-varying networks (Paré et al. 2017a), discrete-time systems (Liu et al. 2019a; Paré et al. 2018b,a), and multiple viruses spreading over networks while competing with each other (Granell et al. 2014; Paré et al. 2017b; Liu et al. 2019b; Paré et al. 2021). For epidemic mitigation problems, Hota and Sundaram (2019) studied vaccine allocation problems of networked SIS spreading dynamics. Other works analyzing networked SIS epidemic models include the studies by Van Mieghem (2011, 2012), Enyioha et al. (2013), Khanafer et al. (2016), Stegehuis et al. (2016), Li et al. (2017), Zino et al. (2018), Chen et al. (2020), and Cisneros-Velarde and Bullo (2021). For more detailed information about leveraging networked compartmental models to capture and mitigate different epidemic spreading processes over networks, we refer to the review papers of (Nowzari et al. 2016; Paré et al. 2020; Zino and Cao 2021; Mei et al. 2017).

16.2.3   Networked Susceptible–Infected–Susceptible Epidemic Models

We introduce networked SIS models with a detailed derivation in this section. Consider an epidemic spreading over connected communities, where the infection level of each community represents the average infected proportion within that community. Further, it is natural to consider that an epidemic could spread from one arbitrary community to another within the network, either through direct interactions (population flow between two communities) and/or through indirect interactions (contaminated intermediate communities). Mathematically, we define an epidemic spreading process over n strongly connected communities represented by a directed graph $G = (\mathcal{V}, \mathcal{E})$, where the node set $\mathcal{V} = \{v_1, \ldots, v_n\}$ and the edge set $\mathcal{E} \subseteq \mathcal{V} \times \mathcal{V}$ represent the communities and the interconnections between the communities, respectively. The adjacency matrix of G is denoted by $A \in \mathbb{R}^{n \times n}$, where $A_{ij} = a_{ij} \in \mathbb{R}_{\geq 0}$ captures the strength of the interconnection between the ith and jth communities. A directed edge (v_j, v_i) indicates that an individual in community j can infect an individual in community i. If community j can infect community i, $(v_j, v_i) \in \mathcal{E}$ and $a_{ij} > 0$. Otherwise, community j cannot transmit virus to community i directly, and $a_{ij} = 0$.

In order to use networked SIS models to capture epidemic dynamics over n communities, we let $x_i \in [0, 1]$ represent the proportion of the infected population in community i, $i \in [n]$. Note that for any positive integer n, we use $[n]$ to denote the index set $\{1, 2, \ldots, n\}$. Therefore, $1 - x_i \in [0, 1]$ denotes the proportion of the susceptible population in community i, $i \in [n]$. We leverage the networked SIS epidemic dynamics proposed in Fall et al. (2007) to capture the epidemic spreading process over the n communities, given by

$$\dot{x}_i(t) = -\delta_i x_i(t) + (1 - x_i(t)) \sum_{j=1}^{n} \beta_i a_{ij} x_j(t) \tag{16.2}$$

where $\delta_i \in \mathbb{R}_{\geq 0}$ is the average recovering rate of community i, and $\beta_i a_{ij} \in \mathbb{R}_{\geq 0}$ is the average infection rate of community j to community i, $i, j \in [n]$. The deterministic SIS epidemic spreading process in Eq. (16.2) is a mean-field approximation of the stochastic SIS epidemic spreading process. In addition, the model in Eq. (16.2) provides a lower complexity deterministic approximation to the full-dimensional Markov process model of an SIS spread process over a static network. In the rest of this section, we explain how to obtain the networked SIS epidemic spreading model in Eq. (16.2) through two methods. For the first interpretation of the state, the model can be derived from a subpopulation perspective as was done by Fall et al. (2007) in the following manner.

With some abuse of notation, let $S_i(t)$ and $I_i(t)$ denote the proportions of the population that are susceptible and infected within community i, $\forall i \in [n]$, at time step t, $t \geq 0$. Let N_i denote the size of the population in each community i. We introduce epidemic spreading parameters that are associated with each community i: recovery rate ψ_i, birth rate μ_i, death rate $\overline{\mu}_i$, and infection rates α_{ij} $i, j \in [n]$. Similar to the SIS group compartmental model, we introduced in Section 16.1.3, we consider no vital dynamics for all communities, that is, $S_i(t) + I_i(t) = N_i$, $\overline{\mu}_i = \mu_i$, for all $i \in [n]$ and $t \geq 0$. The evolution of the number of infected and susceptible individuals in each community i is as follows:

$$\dot{S}_i(t) = \mu_i N_i - \overline{\mu}_i S_i(t) + \psi_i I_i(t) - \sum_{j=1}^{n} \alpha_{ij} \frac{S_i(t)}{N_i} I_j(t)$$

$$= (\mu_i + \psi_i) I_i(t) - \sum_{j=1}^{n} \alpha_{ij} \frac{S_i(t)}{N_i} I_j(t) \tag{16.3}$$

$$\dot{I}_i(t) = -\psi_i I_i(t) - \overline{\mu}_i I_i(t) + \sum_{j=1}^{n} \alpha_{ij} \frac{S_i(t)}{N_i} I_j(t)$$

$$= (-\psi_i - \mu_i)I_i(t) + \sum_{j=1}^{n} \alpha_{ij} \frac{S_i(t)}{N_i} I_j(t) \tag{16.4}$$

To simplify the model, we define the proportion of infected individuals in community i by

$$x_i(t) = \frac{I_i(t)}{N_i}$$

and let

$$\beta_{ij} = \alpha_{ij} \frac{N_j}{N_i}, \quad \delta_i = \psi_i + \mu_i$$

then,

$$\dot{x}_i(t) = -\delta_i x_i(t) + (1 - x_i(t)) \sum_{j=1}^{n} \beta_{ij} x_j(t) \tag{16.5}$$

which is the same as Eq. (16.2) with $\beta_{ij} = \beta_i a_{ij}$. The derivation can be generalized to derive extensions of SIS models for multi-city epidemics (Lewien and Chapman 2019), competing viruses (Liu et al. 2019b), and other compartmental models and extensions thereof. We leverage this method to derive multivirus models under the impact of shared resources in Section 16.4.3.

Another way to derive the model in Eq. (16.2) is to use a mean field approximation of a 2^n-state Markov chain model that captures the networked SIS dynamics. Van Mieghem et al. (2008) built the connections between networked SIS epidemic models and Markov chain models by introducing a 2^n-state Markov chain. Each state of the chain, $\Sigma_k(t)$, $k \in [2^n]$, corresponds to a binary-valued string w of length n, where the ith agent is either infected or susceptible, indicated by $w_i = 1$ or $w_i = 0$, respectively, and the state transition matrix, $\overline{Q}$, is defined by

$$\overline{q}_{kl} = \begin{cases} \delta, & \text{if } w_i = 1, k = l + 2^{i-1} \\ \beta \sum_{j=1}^{n} a_{ij} w_j, & \text{if } w_i = 0, k = l - 2^{i-1} \\ -\sum_{l \neq j} \overline{q}_{jl}, & \text{if } k = l \\ 0, & \text{otherwise} \end{cases} \tag{16.6}$$

for $i \in [n]$. Here, a virus is propagating over a network structure defined by a_{ij} (nonnegative, with $a_{ii} = 0$, for $j \in [n]$), with n agents, β is the homogeneous (same for each node) infection rate, δ is the homogeneous recovery rate, and, again, $w_i = 1$ or $w_i = 0$ indicates that the ith agent is either infected or susceptible, respectively. The state vector $\sigma(t)$ is defined by

$$\sigma_k(t) = Pr[\Sigma_k(t) = k] \tag{16.7}$$

with $\sum_{k=1}^{2^n} \sigma_k(t) = 1$. The Markov chain evolves according to

$$\frac{d\sigma^{\mathsf{T}}(t)}{dt} = \sigma^{\mathsf{T}}(t)\overline{Q} \tag{16.8}$$

Let $v_i(t) = Pr[\mathcal{W}_i(t) = 1]$, where $\mathcal{W}_i(t)$ is the random variable representing whether the ith agent is infected (not to be confused with w_i, which is the ith entry of the binary string associated with each state of the 2^n Markov chain).

Then

$$v^{\mathsf{T}}(t) = \sigma^{\mathsf{T}}(t)M \tag{16.9}$$

where $M \in \mathbb{R}^{2^n \times n}$ with the rows being lexicographically ordered binary numbers, bit reversed.[2] That is, $v_i(t)$ reflects the summation of all probabilities, where $x_i = 1$; therefore, giving the mean, $E[X_i]$, of the infection, X_i, of agent i. Note that the first chain state of the chain, which corresponds to $x = \mathbf{0}$, the vector of zeros, or the "disease free" equilibrium, for $\delta > 0$, is the absorbing, or sink, state of the chain. This means that the Markov chain will never escape the state once in it, and further, since it is the only absorbing state, the system will converge to the healthy state with probability one (Norris 1998).

Now, consider the probabilities associated with node i being healthy ($\mathcal{W}_i = 0$) or infected ($\mathcal{W}_i = 1$) at time $t + \Delta t$:

$$Pr(\mathcal{W}_i(t + \Delta t) = 0 | \mathcal{W}_i(t) = 1, \mathcal{W}(t)) = \delta \Delta t + o(\Delta t)$$

$$Pr(\mathcal{W}_i(t + \Delta t) = 1 | \mathcal{W}_i(t) = 0, \mathcal{W}(t)) = \beta \sum_{j=1}^{n} a_{ij} \mathcal{W}_j \Delta t + o(\Delta t)$$

$$\vdots$$

Letting Δt go to zero, and taking expectations, leads to

$$\dot{E}(\mathcal{W}_i(t)) = E\left((1 - \mathcal{W}_i(t)) \beta \sum_{j=1}^{n} a_{ij} \mathcal{W}_j(t) \right) - \delta E(\mathcal{W}_i(t)) \tag{16.10}$$

Using the above equation, the identities $Pr(z) = E(1_z)$, $x_i(t) = Pr(\mathcal{W}_i(t) = 1)$, $(1 - x_i(t)) = Pr(\mathcal{W}_i(t) = 0)$, and approximating $Pr(\mathcal{W}_i(t) = 1, \mathcal{W}_j(t) = 1) \approx x_i(t) x_j(t)$ (which again inaccurately assumes independence) gives

$$\dot{x}_i(t) = (1 - x_i(t)) \beta \sum_{j=1}^{n} a_{ij} x_j(t) - \delta x_i(t)$$

which is the same as Eq. (16.2) for the homogeneous virus case (Paré 2018), i.e. homogeneous transmission and recovering rates. It can be verified that for both interpretations of the model to be well defined, each state $x_i(t)$ must remain in the domain $[0, 1]$ for all $t \geq 0$, $i \in [n]$.

Now that we have introduced two different ways of deriving networked SIS models in Eq. (16.2), these techniques can be used for more complicated networked models with multiple compartments, such as networked susceptible-infected-recovered (SIR) and susceptible-exposed-infected-recovered (SEIR) compartmental models.

16.3 Epidemics and Cyber–Physical–Human Systems

The recent breakthroughs from control, communications, automation, and computing in CPSs have led to substantial progress (Baheti and Gill 2011). There is an extensive body of work devoted to designing CPSs and improving their performance, spanning from smart grids to autonomous vehicles, to medical systems, etc. (Wolf 2009). Among these systems, humans are considered as independent entities outside of the system (e.g. operators) in terms of functionality. However, when modeling and analyzing complex systems like epidemic spreading processes, it is necessary to consider humans as parts of these systems, since human factors will have an impact on both the epidemic modeling and analysis as well as policy-making for epidemic control and mitigation (Li et al. 2021). On the one hand, infamous pandemics caused by human carriers such as malaria, SARS, COVID-19 have spread and evolved based on human behaviors and interactions.

---

2 Matlab code: $M = fliplr(dec2bin(0 : (2^n) - 1) - $ "0").

Hence, modeling and analyzing epidemic spread without factoring in human behavior may lead to insufficient conclusions. On the other hand, unlike traditional control systems, where feedback loops can generate signals for actuators of the physical systems, for epidemic mitigation problems, it is critical to generate policies through human decision-making instead of leveraging the analysis of epidemic spreading directly, i.e. human-in-the-plant. Therefore, it is essential to build models and analysis tools that establish the connections between epidemic spread and human factors.

Inspired by cyber–physical human systems (CPHS), where human factors are implemented within the system modeling and analysis, in this section, we focus on the background of how to characterize the influence of human factors in epidemic spreading and mitigation processes. In contrast to modeling, prediction, and control of epidemic spreading processes and outbreaks through epidemic models that are introduced in Section 16.2.3, research communities are motivated to study complex transmission behaviors of epidemic spreading processes through integrating human factors and infrastructure. Human behaviors and infrastructure can potentially exacerbate epidemic outbreaks, postpeak dynamics, resurgence of cases, multiple spreading waves, etc. Since recent research results are pushing the boundaries even further by incorporating human factors when studying epidemic spreading processes (Bauch and Galvani 2013), we address two aspects to capture human factors in epidemic spreading processes, namely, human opinions/behaviors and infrastructure.

16.3.1 Epidemic and Opinion Spreading Processes

During the ongoing COVID-19 pandemic, the effectiveness of nonpharmaceutical interventions (NPIs), such as social distancing policies, wearing face masks, and self-isolation, and pharmaceutical interventions like vaccinations, are heavily affected by beliefs and behaviors of people toward the seriousness of the epidemic. Higher beliefs toward the seriousness of the epidemic and higher trust in vaccines will lead people to follow appropriate intervention strategies and to seek out treatment and vaccines. On the other hand, people with lower beliefs in the seriousness of the epidemic are less likely to follow appropriate instructions in response to outbreaks. Further, people who oppose the use of vaccines or regulations mandating vaccination will not get vaccinated to protect themselves from the virus.

By investigating a population's awareness toward the seriousness of the COVID-19 pandemic, Weitz et al. (2020) showed that incorporating fatigue and long-term behavior into spreading models can potentially lead to a resurgence of infected cases during the COVID-19 pandemic. In order to study the interplay between the epidemic spreading and awareness diffusion on time-varying network structures, Hu et al. (2018) considered a model that captured behavior changes of susceptible individuals influenced by awareness of people toward the epidemic. In the developed model, susceptible populations will become alert and then adopt a preventive behavior under the local risk perception, which could change the model parameters such as transmission rate and recovery rate. Similarly, Granell et al. (2014) explored the critical relation between the awareness spreading and infection. The work by Granell et al. (2014) also analyzed the consequences of a massive broadcast of awareness (mass media) on the final level of infection. In addition, Funk et al. (2009) revealed that behavior changes in response to a disease outbreak in human populations can alter the progression of the infectious individuals. Beyond computational and numerical analysis, Funk et al. (2009) also formulated and analyzed a mathematical model in order to understand the correlation between the spread of awareness in a host population and the spread of a disease. Funk et al. (2009) found that more informed hosts will likely have lower susceptibility to the disease. Likewise, She et al. (2021) also built a mathematical model to capture a coupled epidemic and opinion spreading dynamics over networks, in order to study how the beliefs of a population impact the peak infection time of a networked epidemic spreading process. These findings can be used to interpret epidemic spreading behaviors, as well as in the prediction of future outbreaks. Other works that consider coupled epidemic and human behavior/opinions/awareness can be found in Granell et al. (2014), Paré et al. (2021), Hu et al. (2018), Funk et al. (2009), Wang et al. (2015), and Paarporn et al. (2017).

16.3.2 Epidemic and Infrastructure

We now introduce the research concerning the impact of infrastructure on epidemic spreading processes. Examples of unsuccessful interventions in epidemic mitigation are typically attributed to several factors, among which are expansion of urban centers with poor sanitation and inadequate vector control infrastructure (Wilder-Smith et al. 2017). In May 1983, an estimated 865 cases of epidemic gastrointestinal disease occurred in Greenville, Florida (Sacks et al. 1986). It was subsequently confirmed that the virus fecal coliforms found in water samples from the city water plant was accountable for the epidemic outbreak: the city water plant, a deep well system with an unlicensed operator, a failure of chlorination, and open-top treatment towers, was actually contaminated by birds. From 1 January through 30 June 1997, 8901 cases of typhoid fever were reported in Dushanbe, Tajikistan (Mermin et al. 1999). Researchers found that Salmonella typhi infection was associated with drinking unboiled water that showed fecal coliform contamination as well (Threlfall 2002). Likewise, the lack of chlorination, equipment failure, and back-siphonage in the water distribution system led to contamination of the drinking water. Both examples illustrate the influence of infrastructure on disease spreading and epidemic outbreaks. Further, these cases also indicate the importance of protecting shared resources, i.e. water distribution systems, from contamination. Other examples that are considered as shared resources, such as water distribution systems, bathrooms, markets, are major factors for epidemic outbreaks and spreading (Water and Organization 2002; Troesken 2004; Lee 2003; Hung 2003; Weissman et al. 1976). Furthermore, it may be noted that modern infrastructure networks, such as water distribution networks, often have a coupling between the cyber-space and the physical space. For example, the contamination levels of a water resource could be monitored remotely (cyber) while said resource could be connected with other resources through a network of pipes (physical). Sensors that measure contamination levels could malfunction, either occurring naturally or as the result of a deliberate action by a hostile entity. Assuming the latter case, the spread of diseases in such settings may be thought of in the following sense: an adversary injects a biological virus into the water distribution network; to prevent it from being detected on time, he/she also compromises (via the spread of a computer virus) the sensor readings of said water distribution network, resulting in a large number of the human population becoming sick – a topic that falls within the realm of **CPHS**. This part of the chapter, however, focuses on the case where only the biological virus is present in the population and the network of shared resources. Thus, this chapter may be seen as a first step toward the analysis of such complex systems.

In order to capture the phenomena of disease spread in the presence of shared resources, research communities have integrated the impact of shared resources into epidemic spreading models. Paré et al. (2019) and Liu et al. (2019a) proposed a layered networked susceptible–infected–water–susceptible (SIWS) model, for an SIS-type waterborne disease spreading over a human contact network connected to a water distribution network that has a pathogen spreading in it. Gracy et al. (2021) developed a time-varying SIWS model, with the water compartment representing the contamination level in the shared resource. Gracy et al. (2021) also established an on/off lockdown strategy to eradicate the infection spread. By leveraging these mathematical models and corresponding analysis, researchers can forecast the epidemic spreading trends and develop effective intervention strategies for epidemic eradication.

16.4 Recent Research Advances

After introducing the history of disease spreading processes and how to leverage mathematical models to further capture epidemic spreading behaviors, in particular, epidemics under the influence of human factors (opinions/awareness) and infrastructure (shared resources), we introduce some of our research on epidemic modeling and analysis over networks by considering epidemic spreading under the impact of opinion dynamics and infrastructure, respectively. One can find more detailed information on opinion diffusion processes over social networks through Chapter 3 in this book (Zino and Cao 2023).

16.4.1 Notation

First, we introduce the notation we use in this section. For any positive integer n, we use $[n]$ to denote the index set. We view vectors as column vectors and write $x^\top$ to denote the transpose of a column vector x. For a vector x, we use x_i to denote the ith entry. We use $M = \text{diag}\{m_1, \ldots, m_n\}$ to represent a diagonal matrix $M \in \mathbb{R}^{n \times n}$ with $M_{ii} = m_i$, $\forall i \in [n]$. We use $\mathbf{0}$ and e to denote the vectors whose entries all equal 0 and 1, respectively, and I to denote the identity matrix. The dimensions of the vectors and matrices are to be understood from the context.

For a real square matrix M, we use $\rho(M)$ and $s(M)$ to denote its spectral radius and spectral abscissa (the largest real part among its eigenvalues), respectively. For any two vectors $v, w \in \mathbb{R}^n$, we write $v \geq w$ if $v_i \geq w_i$, and $v \gg w$ if $v_i > w_i$, $\forall i \in [n]$. The comparison notations between vectors are applicable for matrices as well, for instance, for $A, B \in \mathbb{R}^{n \times n}$, $A \gg B$ indicates that $A_{ij} > B_{ij}$, $\forall i, j \in [n]$. Given a matrix A, $A \prec 0$ (resp. $A \preccurlyeq 0$) indicates that A is negative definite (resp. negative semidefinite), whereas $A \succ 0$ (resp. $A \succcurlyeq 0$) indicates that A is positive definite (resp. positive semidefinite). For the sets P and Q, we denote a subset by $P \subseteq Q$, a proper subset by $P \subset Q$, and set difference by $P \backslash Q$.

We also employ a modified sign function:

$$sgnm(x) = \begin{cases} 1, & \text{if} \quad x \geq 0 \\ -1, & \text{if} \quad x < 0 \end{cases}$$

Note that we use $sign(\cdot)$ to represent the original sign function, where $sign(x) = sgnm(x)$, $\forall x \neq 0$, and $sign(0) = 0$. We use the modified sign function $sgnm(\cdot)$ to classify non-negative and negative opinion states, and use the original sign function $sign(\cdot)$ to distinguish between positive and negative edge weights.

Consider a directed graph $G = (\mathcal{V}, \mathcal{E})$, with the node set $\mathcal{V} = \{v_1, \ldots, v_n\}$ and the edge set $\mathcal{E} \subseteq \mathcal{V} \times \mathcal{V}$. Let matrix $A = [a_{ij}] \in \mathbb{R}^{n \times n}$ denote the adjacency matrix of $G = (\mathcal{V}, \mathcal{E})$, where $a_{ij} \in \mathbb{R} \backslash \{0\}$ if $(v_j, v_i) \in \mathcal{E}$ and $a_{ij} = 0$, otherwise. Graph G does not allow self-loops, i.e. $a_{ii} = 0$, $\forall i \in [n]$. Let $k_i = \sum_{j \in \mathcal{N}_i} |a_{ij}|$, where $\mathcal{N}_i = \{v_j | (v_j, v_i) \in \mathcal{E}\}$ denotes the neighbor set of v_i and $a_{ij}|$ denotes the absolute value of a_{ij}. The graph Laplacian of G is defined as $L \triangleq K - A$, where $K \triangleq \text{diag}\{k_1, \ldots, k_n\}$ denotes the degree matrix of the graph G.

16.4.2 Epidemic and Opinion Spreading Processes

Social media significantly increases the rate at which opinions spread through communities. During the ongoing SARS-CoV-2 pandemic, polarized opinions about the seriousness of the epidemic and the attitudes toward vaccines have significantly impacted the epidemic spreading processes (Czeisler et al. 2020). The polarized opinions toward the seriousness of the epidemic enable shifts in people's reactions toward the virus in different ways, leading to drastic differences in daily infected proportions over different communities/areas. A significant drawback of the research reviewed in Section 16.3.1 is that it lacks an explicit dynamical model to capture the impact of the population's perception on the severity of the epidemics over time, and vice versa. In order to provide a foundational understanding regarding epidemic spreading processes under the effect of polarized opinions, we build a coupled epidemic and opinion spreading model.

16.4.2.1 Opinions Over Networks with Both Cooperative and Antagonistic Interactions

In order to capture the coupled epidemic spreading process with opinion dynamics over networks, we leverage the networked SIS model defined in Eq. (16.2). First, we propose an opinion spreading model to capture opinion exchanging processes with both cooperative and antagonistic interactions.

Consider an epidemic spreading over a group of connected communities captured by the networked SIS model in Eq. (16.2). The disease spreading graph $G = (\mathcal{V}, \mathcal{E})$ and the corresponding adjacency matrix A are defined in Section 16.2.3. In order to capture polarized opinion spreading over the same n connected communities, we use a signed graph $\overline{G} = (\mathcal{V}, \overline{E})$ to represent the opinion spreading network. The signed graph $\overline{G}$ allows both positive and negative edge weights, representing cooperative and antagonistic exchanges of opinions between the communities, respectively. Note that we use the same node set $\mathcal{V}$ in the opinion spreading network to represent the identical n communities in the epidemic spreading network. However, we use a different edge set $\overline{E}$ for opinions than for epidemic spreading. In addition, in order to capture the structure of the opinion spreading graph, we construct the adjacency matrix $\overline{A}_u \in \mathbb{R}^{n \times n}$ with $\left[\overline{A}_u\right]_{ij} = |\overline{a}_{ij}|$, where

$\overline{a}_{ij}$ is the ijth entry of the adjacency matrix of $\overline{G}$. Therefore, $\overline{A}_u$ captures the coupling strength of the opinion dynamics without considering the signs. We use $\overline{K}$ and $\overline{L}$ to denote the degree matrix and Laplacian matrix of $\overline{G}$, respectively.

Let $o(t) \in \mathbb{R}^n$ $\forall t \geq 0$ be the opinion vector of the n communities, $o_i(t) \in [-0.5, 0.5]$, $i \in [n]$. The state $o_i(t) \in [-0.5, 0.5]$ denotes the belief of community i about the severity of the epidemic at time t. The opinion $o_i(t) \in [0, 0.5]$ indicates that community i considers the epidemic to be relatively serious, while $o_i(t) \in [-0.5, 0)$ implies that community i has a relatively lower belief regarding the seriousness of the epidemic. It is natural to assume that communities with higher beliefs toward the seriousness of the epidemic exchange their opinions cooperatively. Likewise, communities with lower beliefs toward the seriousness of the epidemic exchange their opinions cooperatively. However, communities with relatively higher beliefs and communities with relatively lower beliefs in the seriousness of the epidemic will exchange their opinions antagonistically. Hence, based on the communities' beliefs toward the epidemic at any given time, we allow the edge signs of the opinion graph $\overline{G}$ to switch. We achieve this behavior by partitioning the node set of the communities $\mathcal{V}$ into two groups, $\mathcal{V}_1(o(t)) = \{v_i \in \mathcal{V} \mid sgnm(o_i(t)) = 1, i \in [n]\}$ and $\mathcal{V}_2(o(t)) = \{v_i \in \mathcal{V} \mid sgnm(o_i(t)) = -1, i \in [n]\}$. Then we construct the adjacency matrix $\overline{A}$ of graph $\overline{G}$ as $\overline{A}(o(t)) = \Phi(o(t))\overline{A}_u\Phi(o(t))$, where $\Phi(o(t)) = \text{diag}\{sgnm(o_1), \ldots, sgnm(o_n)\}$ is a gauge transformation matrix. The entries of the gauge transformation matrix $\Phi\,(o\,(t))$ are chosen as $\phi_i\,(o_i\,(t)) = 1$ if $i \in \mathcal{V}_p\,(t)$ and $\phi_i\,(o_i\,(t)) = -1$ if $i \in \mathcal{V}_q\,(t)$, $p \neq q$, and $p, q \in \{1, 2\}$, $\forall i \in [n]$. Through the construction of $\overline{A}\,(o\,(t))$, $|\overline{a}_{ij}|$ is fixed and $sign|\overline{a}_{ij}|$ is switchable. In particular, for all nonzero entries in $\overline{A}\,(o\,(t))$, $sign|\overline{a}_{ij}| = 1$ if $sgnm\,(o_j\,(t))$ and $sgnm\,(o_i\,(t))$ are the same; otherwise, $sign|\overline{a}_{ij}| = -1$. Therefore, $\overline{A}\,(o\,(t))$ captures the switch of the opinion interactions through the attitude changing toward the epidemic. Note that an opinion graph $\overline{G}$ constructed in this manner is always structurally balanced; the following definition and lemma explains why it is so.

Definition 16.1 (***Structural Balance (Cartwright and Harary 1956)***) A signed graph $\overline{G} = \left(\mathcal{V}, \overline{\mathcal{E}}\right)$ is structurally balanced if the node set $\mathcal{V}$ can be partitioned into $\mathcal{V}_1$ and $\mathcal{V}_2$ with $\mathcal{V}_1 \cup \mathcal{V}_2 = \mathcal{V}$ and $\mathcal{V}_1 \cap \mathcal{V}_2 = \emptyset$, where $\overline{a}_{ij} \geq 0$ if $v_i, v_j \in \mathcal{V}_q$, $q \in \{1, 2\}$, and $\overline{a}_{ij} \leq 0$ if $v_i \in \mathcal{V}_q$ and $v_j \in \mathcal{V}_r$, $q \neq r$, and $q, r \in \{1, 2\}$.

Lemma 16.1 (***Zaslavsky 1982***) A connected signed graph $\overline{G}$ is structurally balanced if and only if there exists a gauge transformation matrix $\Phi \in \mathbf{R}^{n \times n}$, with $\Phi = \text{diag}\,\{\phi_1, \dots, \phi_n\}$ and $\phi_i \in \{\pm 1\}$, such that $\Phi \overline{A} \Phi \in \mathbf{R}^{n \times n}$ is nonnegative.

After defining the opinion interaction matrix $\overline{A}\,(o\,(t))$, we modify the opinion dynamics model with both cooperative and antagonistic interactions in Altafini (2013) to capture opinions evolving over the n communities with sign-switching structures:

$$\dot{o}_i\,(t) = \sum_{j=1}^{n} |\overline{a}_{ij}\,(o\,(t))| \,\left(sign\,(\overline{a}_{ij}\,(o\,(t)))\, o_j\,(t) - o_i\,(i)\right) \tag{16.11}$$

with the compact form

$$\dot{o}\,(t) = -\Phi\,(o\,(t)) \overline{L}_u \Phi\,(o\,(t))\, o\,(t) \tag{16.12}$$

where $\overline{L}_u$ represents the Laplacian matrix of $\overline{A}_u$. The model in Eq. (16.12) states that the belief of each community regarding the seriousness of the epidemic evolves as a function of the community's infected proportion and the opinions of neighboring communities.

16.4.2.2 Coupled Epidemic and Opinion Dynamics

Now that we have introduced the networked SIS epidemic spreading model in Section 16.1.3 and the opinion spreading model with both cooperative and antagonistic interactions in this section, we construct the coupled epidemic and opinion spreading model. People's beliefs about a health problem can affect their behaviors in response to the problem, and vice versa (Glanz et al. 2008). We propose an epidemic spreading model that integrates the beliefs of communities in their perceived severity of an outbreak and/or illness for the purpose of capturing the impact of opinions on the chance of being infected by and effectively healing from a virus. Assume that a community's opinion/belief toward the severity of an epidemic will affect its behaviors which lead to the variation of the community's average recovering rate and infection rate. For instance, a community with a relatively higher belief regarding the seriousness of an epidemic will react to the epidemic spreading process actively, in terms of monitoring and prediction, policy-making, resource allocation, etc., thus result in the community having a lower average infection rate and a higher average healing rate. On the other hand, a community with a relatively lower belief regarding the seriousness of an epidemic will react to the virus passively, which can lead to a higher average infection rate and a lower average healing rate. Following this idea, we incorporate the opinion dynamics in Eq. (16.11) into the epidemic dynamics in Eq. (16.2), in the following manner:

$$\dot{x}_i(t) = -\left[\delta_{\min} + \left(\delta_i - \delta_{\min}\right) o_i'(t)\right] x_i(t) \tag{16.13}$$

$$+ \left(1 - x_i(t)\right) \sum_{j=1}^{n} \left[\beta_{ij} - \left(\beta_{ij} - \beta_{\min}\right) o_i'(t)\right] x_j(t)$$

where $\delta_{\min}$ and $\beta_{\min}$ are the possible minimum average healing rate and infection rate for all communities, respectively. Note that $o_i'(t) = o_i(t) + 0.5$ shifts the opinion into the range $[0, 1]$. In particular, the term $o_i'(t)$ scales the average healing and infection rates between their maximum and minimum values. In the case when $o_i(t) = -0.5$, which implies that community i has the lowest belief in the seriousness of the epidemic at time t, community i will take no action to protect itself and thus is maximally exposed to the infection. Hence, communities i has $\delta_{\min}$ and $\sum_{j=1}^{n} \beta_{ij}, i, j \in [n]$ as its spreading parameters. In the case when $o_i(t) = 0.5$, which implies that community i believes the epidemic is extremely serious, it will implement policies and limitations to decrease the chance of its population being infected and seek out all the possible medical treatment options and hospitalizations to increase the healing of its infected population. Thus, communities i has $\delta_i \geq \delta_{\min}$ and $\sum_{j=1}^{n} \beta_{\min}, i, j \in [n]$, as its spreading parameters. By introducing the model in Eq. (16.13), we allow the communities' opinions to affect how susceptible they are and how effectively they heal from the virus, capturing the health belief model (Glanz et al. 2008), as explained in Rosenstock (1974).

Recall that we introduced Eq. (16.11) by capturing the opinion of each community evolving as a function of the community's own opinion and the opinions of neighboring communities. Similarly, inspired by the health belief model (Glanz et al. 2008), it is reasonable to assume that the opinion of each community is also affected by the infected level within the community, i.e. the proportion of the infected population of itself. Thus, we incorporate the opinion dynamics in Eq. (16.11) with Eq. (16.2) as follows:

$$\dot{o}_i = \left(x_i(t) - o_i'(t)\right) + \sum_{j=1}^{n} \left|\bar{a}_{ij}(o(t))\right| \left(sign\left(\bar{a}_{ij}(o(t))\right) o_j(t) - o_i(t)\right) \tag{16.14}$$

The first term on the right-hand side of Eq. (16.14) captures how the proportion of infections of a community affects its own opinion. If $o_i'(t)$ is small, but the community is heavily infected, i.e. if $x_i(t)$ is larger than $o_i'(t)$, $o_i'(t)$ will increase. If $o_i'(t)$ is large, but the community has few infections, i.e. if $x_i(t)$ is smaller than $o_i'(t)$, $o_i'(t)$ will decrease. This behavior reflects the fact that a community's infection level should affect its belief in the severeness of the virus. The second term on the righthand side of Eq. (16.14) is from Eq. (16.11). The neighbors of community i affect its opinion cooperatively $(sign\left(\bar{a}_{ij}(o(t))\right) = 1)$ or antagonistically $(sign\left(\bar{a}_{ij}(o(t))\right) = -1)$.

Now that we have presented the epidemic-opinion model in Eqs. (16.13) and (16.14), we write the system in Eqs. (16.13) and (16.14) in the following compact form:

$$\dot{x}(t) = -D(o(t))x(t) + (I - X(t))B(o(t))x(t) \tag{16.15}$$

$$\dot{o}(t) = Ix(t) - \left(\Phi(o(t))\overline{L}_u\Phi(o(t)) + I\right)o(t) - 0.5e \tag{16.16}$$

where $D(o(t)) = D_{\min} + \left(D - D_{\min}\right)(O(t) + 0.5I)$, and $B(o(t)) = B - (O(t) + 0.5I)\left(B - B_{\min}\right)$ capture the opinion-dependent healing and infection matrix, respectively. The opinion-dependent healing and infection matrices describe the impact of opinions of the n communities on their epidemic spreading parameters (healing and transmission rates), thus are the key features for analyzing the epidemic spreading behavior. In addition, we have $O(t) = \text{diag}\left\{o_1(t), \ldots, o_n(t)\right\}$, $X = \text{diag}\left\{x_1, \ldots, x_n\right\}$, $D = \text{diag}\left\{\delta_1, \ldots, \delta_n\right\}$, $D_{\min} = \delta_{\min}I$, $B = \left[\beta_{ij}\right] \in \mathbb{R}^{n \times n}$ and $B_{\min} = \beta_{\min}\tilde{A}$, with $\tilde{A} \in \mathbb{R}^{n \times n}$ being the unweighted adjacency matrix of graph G (with $\tilde{A}_{ij} \in \{0, 1\}, \forall i, j \in [n]$).

16.4.2.3 Opinion-Dependent Reproduction Number

After defining the compact form of the model in Eqs. (16.15) and (16.16), we introduce our results building upon the developed model. We are interested in exploring the mutual influence between the epidemic spreading over n communities captured by graph G in Eq. (16.13) and the opinions of the n communities about the epidemic captured by graph $\overline{G}$ in Eq. (16.14). A key epidemiologic metric in studying epidemic spreading models is the basic reproduction number (R_0), which describes the expected number of cases directly generated by one case in an infection-free population; similarly, the effective reproduction number (R_t) characterizes the average number of new infections caused by a single infected individual at time t in the partially susceptible population (Nishiura and Chowell 2009). Contemporary properties of disease spreading and threshold conditions for epidemic outbreak and eradication are often associated with reproduction numbers (Nishiura and Chowell 2009; Liu et al. 2019b; Van Mieghem et al. 2008; Khanafer et al. 2016; Fall et al. 2007). If the reproduction number is greater than 1, the epidemic will keep spreading. If the reproduction number is less than 1, the infection level will decrease. Inspired by leveraging the next-generation matrix to derive reproduction numbers of group compartmental models (Diekmann et al. 2010), we define a reproduction number of the model in Eq. (16.15) through the following definition.

Definition 16.2 (*Opinion-Dependent Reproduction Number*) We leverage $R_t^o = \rho\left(D(o(t))^{-1}B(o(t))\right)$ to denote the Opinion-Dependent Reproduction Number, where $D(o(t))$ and $B(o(t))$ are defined in Eq. (16.15).

We have the following theorem to capture the spreading process of the coupled epidemic and opinion model through the proposed opinion-dependent reproduction number R_t^o.

Theorem 16.1 *If $R_t^o < 1$ $\forall o_i(t) \in [-0.5, 0.5]$, $\forall t \geq 0$, $i \in [n]$, then for any initial condition, the epidemic state $x(t)$ in Eq. (16.15) will asymptotically converge to zero.*

The behavior of the opinion-dependent epidemic spreading process in Eq. (16.15) is captured by R_t^o. Aligning with the epidemic spreading behaviors captured by reproduction numbers, Theorem 16.1 states that the epidemic state in Eq. (16.15) will decrease to zero when the defined opinion-dependent reproduction number is less than 1. Hence, the epidemic will eventually fade away when the defined opinion-dependent reproduction number is less than 1. The proof and more detailed analysis of the opinion-dependent reproduction number and the coupled epidemic-opinion dynamics can be found in She et al. (2022).

After studying the behavior of the opinion-dependent epidemic spreading process when R_t^o is less than 1 $\forall o_i(t) \in [-0.5, 0.5]$, $\forall t \geq 0$, $i \in [n]$, we explore the behavior of the epidemic spreading process when $R_t^o > 1$ $\forall o_i(t) \in [-0.5, 0.5]$, $\forall t \geq 0$, $i \in [n]$.

Theorem 16.2 *Suppose that $R_t^o > 1$ $\forall o_i(t) \in [-0.5, 0.5]$, $\forall t \geq 0$, $i \in [n]$. Then, Eq. (16.15) has at least one endemic equilibrium.*

A disease outbreak becomes an endemic when it is consistently present and never fades away. Since we consider a networked SIS epidemic spreading process, where the recovered population can become infected again, it is possible for the disease to continue spreading over the network and never disappear. Theorem 16.2 indicates that if the defined opinion-dependent reproduction number R_t^o is greater than 1, the epidemic will become endemic, which aligns with the definition of reproduction numbers introduced in Fall et al. (2007), Liu et al. (2019b), Khanafer et al. (2016), Van Mieghem et al. (2008), and Nishiura and Chowell (2009). A detailed proof of the existence of the endemic state when $R_t^o > 1$ $\forall o_i(t) \in [-0.5, 0.5]$, $\forall t \geq 0$, $i \in [n]$, can be found in She et al. (2022).

16.4.2.4 Simulations

We illustrate Theorem 16.1 through the following example. Consider an epidemic spreading over 10 communities, with the epidemic and opinions spreading through the same strongly connected network, captured by Figure 16.2. Note that we use the same graph structure in G to capture the epidemic and opinion spreading network to simplify the simulation, and our results still apply to communities with different epidemic and opinion interactions. In addition, for the epidemic spreading network in Figure 16.2a, the node size scales with respect to the value of the infected state $x_i(t)$ $\forall t = 0$, $i \in [n]$. Similarly, for the opinion spreading network in Figure 16.2b, the node size scales with respect to the absolute value of the opinion state $o_i(t)$ $\forall t = 0$, $i \in [n]$. White nodes represent positive values while shaded nodes represent negative values. Meanwhile, solid edges represent positive edges and dashed edges represent negative edges. Figure 16.2 illustrates the initial conditions for the following simulation.

We consider the epidemic spreading process under the condition that $R_t^o < 1$ $\forall$, $o_i(t) \in [-0.5, 0.5]$, $\forall t \geq 0$, $i \in [n]$. Figure 16.3a illustrates that epidemic states of all communities will converge to zero, i.e. the epidemic will eventually fade away, and there will be no infection in each community. Meanwhile, most communities will think the epidemic is not serious eventually, as shown in Figure 16.3b. The only exception is community 6; this community still holds a relatively higher belief in the seriousness of the epidemic, due to the existence of the antagonistic interactions in the opinion graph.

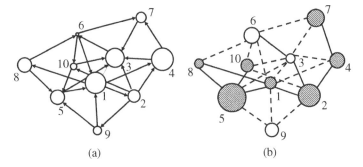

(a)   (b)

Figure 16.2 (a) Epidemic spreading network and (b) opinion spreading network.

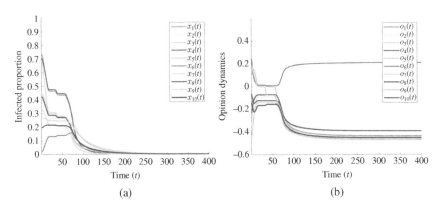

(a)   (b)

Figure 16.3 (a) Epidemic spreading process and (b) opinion spreading process.

To summarize this section, we study the behavior of coupled epidemic and opinion spreading models in Eqs. (16.15) and (16.16) through defining an opinion-dependent reproduction number. We explore the stability of the equilibria of the developed model to infer the behavior of epidemic and opinion spreading over n communities when R_t^o is less than 1. We leverage an example to illustrate the spreading behavior, when the defined reproduction number is less than 1. The results can potentially guide the use of social media to broadcast the severity of the pandemic to appropriate communities to suppress the epidemic. More detailed analysis and results of the work can be found in She et al. (2022).

16.4.3 Epidemic Spreading with Shared Resources

In this section, we introduce our work on modeling epidemic spreading processes with shared infrastructure. We define a model of multi-viral spread across a population network with a shared resource. We analyze the dynamics of the model and leverage the model to establish epidemic spreading behavior under the impact of shared infrastructure. The results can be applied to analyze spreading patterns and to develop control interventions for epidemic spreading when shared infrastructure, like contaminated water systems, is a factor.

16.4.3.1 The Multi-Virus SIWS Model

Consider a population of individuals, subdivided into n population nodes in a network, with a resource W being shared among some or all of the population nodes. Suppose that m viruses are active in the population. An individual can become infected by a virus, either by coming into contact with an infected individual, or due to interaction with the (possibly) contaminated shared resource. We make the assumption that an individual can be infected by no more than one virus at the same time. An infected individual can then recover, returning to the susceptible state (SIS). The model is visualized from an individual's perspective in Figure 16.4. The spread of the m viruses across the population can be represented by a multi-layer network G with m layers, where the vertices correspond to population nodes and the shared resource, and each layer contains a set of directed edges, $\mathcal{E}^k$, specific to each virus k. There exists a directed edge from node j to node i in $\mathcal{E}^k$ if an individual, infected by virus k in node j, can directly infect individuals in node i. Furthermore, the existence of a directed edge in $\mathcal{E}^k$ from node j (resp. from the shared resource W) to the shared resource W (resp. to the node j) signifies that the shared resource W (resp. node j) can be contaminated with virus k by infected individuals in node j (resp. by the shared resource W). We say that the kth layer of G is strongly connected if there is a path via the directed edges in $\mathcal{E}^k$ from each node, and from the shared resource, to every node, and to the shared resource. In real-world scenarios, it is often the case that viral epidemics can spread from each subpopulation to every other subpopulation, in which case we assume that each layer is strongly connected.

Figure 16.4 Visualization of the model for the case when $m = 2$. An individual is either susceptible (S), infected with virus 1 (I^1), or infected with virus 2 (I^2). The shared resource (W) is contaminated by individuals infected with either virus, and in turn augments the corresponding infection rate.

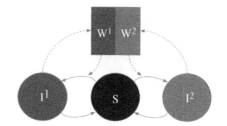

Each population node i contains N_i individuals, with a birth rate μ_i equal to its death rate $\overline{\mu}_i$, i.e. we consider no vital dynamics. We use $S_i(t)$ to represent the number of susceptible individuals in node i, while $I_i^k(t)$ to represent the number of individuals infected by virus k in node i, $\forall t \geq 0$. For each population node, we have $S_i(t) + \sum_{k=1}^m I_i^k(t) = N_i \ \forall t \geq 0$. The rate at which individuals infected by virus k in node j infect susceptible individuals in node i is denoted by α_{ij}^k, where $\alpha_{ij}^k = 0$ corresponds to the absence of a directed edge from node j to node i in $\mathcal{E}^k$. In node i, individuals infected by virus k will recover and become susceptible again at a rate γ_i^k. The shared resource contains a viral mass with respect to each virus k, denoted by $W^k(t)$, representing the level of contamination at time $t \geq 0$. The viral mass of virus k grows at a rate proportional to all $I_i^k(t)$ scaled by their corresponding rates ζ_i^k, and decays at a rate δ_w^k. The resource-to-node infection rate to node i with respect to virus k is denoted by α_{iw}^k. The time evolution of the number of susceptible and infected individuals (with respect to each virus $k \in [m]$) in population node i, $i \in [n]$, is given by

$$\dot{S}_i(t) = \mu_i N_i - \overline{\mu}_i S_i(t) + \sum_{k=1}^m \gamma_i^k I_i^k(t) - \sum_{k=1}^m \left(\alpha_{iw}^k W^k(t) - \sum_{j=1}^n \alpha_{ij}^k \frac{I_j^k(t)}{N_i} \right) S_i(t)$$

$$\dot{I}_i^k(t) = -(\overline{\mu}_i + \gamma_i^k) I_i^k(t) + \left(\alpha_{iw}^k W^k(t) + \sum_{j=1}^n \alpha_{ij}^k \frac{I_j^k(t)}{N_i} \right) S_i(t) \qquad (16.17)$$

$$\dot{W}^k(t) = -\delta_w^k W^k(t) + \sum_{j=1}^n \zeta_j^k I_j^k(t)$$

We define new variables to simplify the system. Let

$$x_i^k(t) = \frac{I_i^k(t)}{N_i}, \quad z^k(t) = \frac{\delta_w^k W^k(t)}{\sum_{j=1}^n \zeta_j^k N_j}, \quad \delta_i^k = \gamma_i^k + \mu_i$$

$$\beta_{ij}^k = \alpha_{ij}^k \frac{N_j}{N_i}, \quad \beta_{iw}^k = \frac{\alpha_{iw}^k}{\delta_w^k} \sum_{j=1}^n \zeta_j^k N_j, \quad c_i^k = \frac{\zeta_i N_i}{\sum_{j=1}^n \zeta_j^k N_j}$$

We interpret these variables as follows: With respect to virus k, $x_i^k(t)$ is the fraction of currently infected individuals in node i, $z^k(t)$ is a scaled contamination level in the shared resource, δ_i^k is the healing rate in node i, β_{ij}^k is the node-to-node infection rate from node j to i, scaled with respect to population ratios, β_{iw}^k is a scaled resource-to-node infection rate from the resource W to node i, and c_i^k is a scaled node-to-resource contamination rate from node i to the resource W. Then, assuming that the birth rates and death rates are equal for each node, we rewrite Eq. (16.17) as

$$\dot{x}_i^k(t) = -\delta_i^k x_i^k(t) + \left(1 - \sum_{l=1}^m x_i^l(t)\right)\left(\beta_{iw}^k z^k(t) + \sum_{j=1}^n \beta_{ij}^k x_j^k(t)\right)$$

$$\dot{z}^k(t) = \delta_w^k\left(-z^k(t) + \sum_{i=1}^n c_i^k x_i^k(t)\right) \qquad (16.18)$$

Using vector notation, we rewrite Eq. (16.18) as

$$\dot{x}^k(t) = \left(\left(I - \sum_{l=1}^m \text{diag}(x^l(t))\right) B^k - D^k\right) x^k(t) + \left(I - \sum_{l=1}^m \text{diag}(x^l(t))\right) b^k z^k(t)$$

$$\dot{z}^k(t) = \delta_w^k\left(-z^k(t) + c^k x^k(t)\right) \qquad (16.19)$$

where B^k is an $n \times n$-matrix with β_{ij}^k as the ijth element, D^k is a diagonal $n \times n$ matrix with $D_{ii}^k = \delta_i^k$ for all $i \in [n]$, b^k is a column vector with β_{iw}^k as the ith element, and c^k is a row vector with c_i^k as the ith element. To simplify notations further, we define

$$y^k(t) = \begin{bmatrix} x^k(t) \\ z^k(t) \end{bmatrix}, \quad y(t) = \begin{bmatrix} y^1(t) \\ \vdots \\ y^m(t) \end{bmatrix}, \quad B_w^k = \begin{bmatrix} B^k & b^k \\ \delta_w^k c^k & 0 \end{bmatrix}$$

$$X(x^k(t)) = \begin{bmatrix} \text{diag}(x^k(t)) & 0 \\ 0 & 0 \end{bmatrix}, \quad D_w^k = \begin{bmatrix} D^k & 0 \\ 0 & \delta_w^k \end{bmatrix}.$$

Based on these defined variables, now that we can rewrite Eq. (16.19) as the following compact form:

$$\dot{y}^k(t) = (-D_w^k + (I - \sum_{l=1}^{m} X(y^l(t)))B_w^k)y^k(t) \tag{16.20}$$

Further, we define $A_w^k(y(t)) = (-D_w^k + (I - \sum_{l=1}^{m} X(y^l(t)))B_w^k)$, the dynamics of the system of all m viruses spreading across the connected population are given by

$$\dot{y}(t) = \begin{bmatrix} A_w^1(y(t)) & 0 & \cdots & 0 \\ 0 & A_w^2(y(t)) & \cdots & 0 \\ \vdots & \vdots & \ddots & \vdots \\ 0 & 0 & \cdots & A_w^m(y(t)) \end{bmatrix} y(t) \tag{16.21}$$

We defined that a virus $k \in [m]$ is eradicated, or in its eradicated state, if $y^k = \mathbf{0}$, which is an equilibrium of Eq. (16.20). When considering the system of m viruses in Eq. (16.21), we state that the system is in the healthy state if all viruses are eradicated, i.e. $y = \mathbf{0}$. If Eq. (16.21) has an endemic (nonzero) equilibrium, it can belong to one of two types: single-virus endemic equilibrium, where $y^k > \mathbf{0}$ for some $k \in [m]$ and all other viruses are eradicated; or coexisting equilibrium, where $y^k > \mathbf{0}$ for multiple $k \in [m]$. After introducing the model, we propose several questions regarding the insight brought by analyzing the spreading behaviors through the model.

16.4.3.2 Problem Statements

Consider the compact form of the coupled population and resources spreading model in Eq. (16.21). In order to study the influence of multi-virus and/or contaminated resources in epidemic spreading processes, we are motivated to tackle the following questions:

(i) Under what conditions does $y^k(t)$ converge exponentially, or asymptotically, to its eradicated state, i.e. $y^k = \mathbf{0}$, for some $k \in [m]$?

(ii) For a single-virus setup, i.e. $m = 1$, under what conditions does the system have a unique single-virus endemic equilibrium, $y^* > \mathbf{0}$, and under such conditions, does the system converge asymptotically to y^* from any nonzero initial condition?

(iii) What is a necessary and sufficient condition for the healthy state, i.e. $y = \mathbf{0}$, to be the unique equilibrium?

Before addressing these questions, we point out two connections between the considered setup and the existing literature. First, note that if $m = 1$, system Eq. (16.20) coincides with the single-virus model proposed in Liu et al. (2019a). Hereafter, when a single-virus system is considered, we drop the superscripts that identify the virus from all variables: $B_w = B_w^1$, $D_w = D_w^1$, etc. Second, note that if $b^k = \mathbf{0}$ for all $k \in [m]$, the influence of the shared resource on the population is nullified. Then, the multi-virus dynamics of the population nodes, $x^k(t)$, in Eq. (16.21) are equivalent to the time-invariant multi-virus setup of (Paré et al. 2017b, 2021). Here, we consider general spreading processes captured by Eq. (16.21).

In order for Eq. (16.21) to be well-defined and realistic, we assume that $\delta_i^k > 0, \delta_w^k > 0, \beta_{ij}^k \geq 0$, $\beta_{iw}^k \geq 0$ and $c_i^k \geq 0$ for all $i,j \in [n]$ and $k \in [m]$, with $c_l^k > 0$ for at least one $l \in [n]$. Then, for all $k \in [m]$, it can be testified that B_w^k is a nonnegative matrix and D_w^k is a positive diagonal matrix. Moreover, recall that a square matrix M is said to be irreducible if, replacing the nonzero elements of M with ones and interpreting it as an adjacency matrix, the corresponding graph is strongly connected. Then, noting that nonzero elements in B_w^k represent directed edges in the set $\mathcal{E}^k$, it is true that B_w^k is irreducible whenever the kth layer of the multi-layer network G is strongly connected. Further, we can restrict our analysis to the sets $\mathcal{D} := \{y(t) : x^k(t) \in [0,1]^n, z^k(t) \in [0,\infty) \; \forall k \in [m]\}$

and $\mathcal{D}^k := \{y^k(t) : x^k(t) \in [0,1]^n, z^k(t) \in [0,\infty)\}$. Since we interpret $x_i^k(t)$ as a fraction of a population, and $z^k(t)$ is a nonnegative quantity, these sets represent the sensible domain of the system. That is, if $y(t) \in D \, \forall t \geq 0$, $\mathcal{D}^k$ is positively invariant with respect to Eq. (16.20), $\forall k \in [m]$. In addition, if $y^k(t), \forall t \geq 0, \forall k \in [m]$, takes values outside of $\mathcal{D}^k$, then those values would lack physical meaning.

16.4.3.3 Analysis of the Eradicated State of a Virus

Through showing that the model and parameters in Eq. (16.20) are well defined, we study sufficient conditions for the exponential (resp. asymptotic) stability of the eradicated state of a virus in Eq. (16.20). Similar to the analysis of the transmission and healing matrix of the model in Eq. (16.15), we explore the property of Eq. (16.20) through analyzing eigenvalue property of $(B_w^k - D_w^k)$, as shown in the following theorem.

Theorem 16.3 *Consider the SIWS model Eq. (16.20) with the initial condition $y(0) \in D$. Suppose that for some virus $k \in [m]$, we have $s(B_w^k - D_w^k) < 0$. Then the eradicated state of virus k is exponentially stable, with domain of attraction containing $\mathcal{D}^k$.*

Theorem 16.3 states that if the linearized state matrix of virus k is Hurwitz, then, for all initial conditions in the sensible domain, virus k is eradicated exponentially fast. Theorem 16.3 answers the first part of question 1 in Section 16.4.3.2. Compared to (Liu et al. 2019b, Proposition 2), Theorem 16.3 is an improvement in the sense that it holds globally (on the sensible domain), and accounts for the multi-virus case, whereas (Liu et al. 2019b, Proposition 2) established local exponential stability for the single-virus case.

Theorem 16.3 guarantees exponential eradication of virus k; however, the condition is quite strict. For certain viruses, Theorem 16.3 suffices to know whether or not the virus will be eradicated, but the speed with which this eradication takes place is of less importance. Indeed, it turns out that a relaxation of the strict inequality of the eigenvalue condition in Theorem 16.3 guarantees asymptotic eradication of a virus, as stated in the following theorem.

Theorem 16.4 *Consider the SIWS model Eq. (16.20) with $y(0) \in D$. Suppose that for some virus $k \in [m]$, we have $s(B_w^k - D_w^k) \leq 0$, and that the matrix B_w^k is irreducible. Then the eradicated state of virus k is asymptotically stable, with domain of attraction containing $\mathcal{D}^k$.*

Recall we assume that the kth layer of the graph is strongly connected. Theorem 16.4 states that if the largest real part of any eigenvalue of the linearized state matrix of virus k is nonpositive, then, for all initial conditions in the sensible domain, virus k is eradicated asymptotically. Theorem 16.4 answers the second part of question 1 in Section 16.4.3.2. Note that for the single-virus case ($m = 1$), Theorem 16.4 improves (Liu et al. 2019a, Theorem 1) by relaxing the requirements $\beta_{iw} > 0$ and $c_i > 0$ for all $i \in [n]$.

Further, the conditions in Theorem 16.3 (Theorem 16.4) are equivalent to $\rho((D_w^k)^{-1}B_w^k) < 1$ $(\rho((D_w^k)^{-1}B_w^k) \leq 1)$ (Liu et al. 2019b, Proposition 1). The fact that Theorem 16.3 (resp. Theorem 16.4) guarantees exponential (asymptotic) eradication of a virus k whenever we have $\rho((D_w^k)^{-1}B_w^k) < 1$ $(\rho((D_w^k)^{-1}B_w^k) \leq 1)$ is consistent with an interpretation of $\rho((D_w^k)^{-1}B_w^k)$ as the basic reproduction number of the virus in the network, as introduced in Section 16.4.2. Thus, Theorem 16.3 (Theorem 16.4) states that whenever the basic reproduction number of a virus is strictly less than (less than or equal to) one, the virus will exponentially (asymptotically) converge to its eradicated state.

16.4.3.4   Persistence of a Virus

After studying the stability of the healthy state, we study the possibility of viruses persisting in the network, corresponding to nonzero equilibria of Eq. (16.21). Naturally, the persistence of a virus must follow from the violation of the conditions of Theorem 16.4. To further capture the behavior of the spreading process, we have the following theorem for a single-virus system, guaranteeing existence of a unique, asymptotically stable, single-virus endemic equilibrium when the eigenvalue condition in Theorem 16.4 is violated, i.e. $s(B_w^k - D_w^k) > 0$.

Theorem 16.5   *Consider the SIWS model Eq. (16.20) with $m = 1$. Suppose that B_w is irreducible and $s(B_w - D_w) > 0$. Then there exists a unique single-virus endemic equilibrium $\tilde{y} \in D$, with $\mathbf{0} \ll \tilde{y} \ll \mathbf{1}$, and it is asymptotically stable with domain of attraction containing $D \backslash \{\mathbf{0}\}$.*

For a single-virus system, Theorem 16.5 states that when the eigenvalue condition in Theorem 16.4 is violated, then as long as some viral infection is present initially, the viral spreading process will converge to a unique infection ratio in each population node and a unique contamination level in the shared resource. Theorem 16.4, thus, answers question 2 in Section 16.4.3.2. Further, the condition in Theorem 16.5 is equivalent to $\rho((D_w)^{-1} B_w) > 1$ (Liu et al. 2019b, Proposition 1). This expression is again consistent with the interpretation of $\rho((D_w)^{-1} B_w)$ as the basic reproduction number of the virus, since a persisting virus should have a basic reproduction number greater than one. In addition, Theorem 16.5 establishes the existence, uniqueness, and asymptotic stability of a single-virus endemic equilibrium, extending (Fall et al. 2007, Theorem 2.4.) to the setting with a shared resource, whereas (Liu et al. 2019a) illustrated this extension in simulations, without providing theoretical guarantees. Note that detailed proofs from Theorems 16.3 to 16.5 can be found in Janson et al. (2020). We can partially extend Theorem 16.5 to the multi-virus case, specifically the existence and uniqueness of a single-virus endemic equilibrium for a virus. Finally, we introduce the following necessary and sufficient condition for the healthy state to be the unique equilibrium of the system in Eq. (16.21).

Theorem 16.6   *Consider the SIWS model Eq. (16.21). Suppose that, for all $k \in [m]$, B_w^k is irreducible. Then the healthy state is the unique equilibrium in D if, and only if, for all $k \in [m]$, $s(B_w^k - D_w^k) \leq 0$.*

Theorem 16.6 states that as long as the largest real part of the eigenvalue of the linearized state matrix of each virus is nonpositive, the healthy state is the only equilibrium of Eq. (16.21). Theorem 16.6 answers question 3 in Section 16.4.3.2. Note that Theorem 16.6 extends (Liu et al. 2019b, Theorem 1) to the setting with more than two viruses and a shared resource, albeit under the assumption that the healing rate of each agent with respect to each virus is strictly positive.

16.4.3.5   Simulations

In this section, we present a number of simulations to illustrate our theoretical findings, using the city of Stockholm as an example setting. In particular, 15 districts in and around Stockholm are taken to be the population nodes of the network, thus, $n = 15$. All of these districts are connected to the Stockholm metro, which we view as the shared resource of the network; see Figure 16.5.

In all simulated scenarios, we consider two competing viruses, namely virus 1 and virus 2. We denote the average infection ratio of virus k, i.e. $\frac{1}{n} \sum_i^n x_i^k(t)$, by $\overline{p}^k(t)$. The contact network of each virus is the same, i.e. $\mathcal{E}^1 = \mathcal{E}^2$, and is represented in Figure 16.5. We set $x_i^k(0) = 0.5$ and $z^k(0) = 0.5$ for all $i \in [15]$ and $k \in [2]$, and use the same in all simulated scenarios throughout this section. Observe that the aforementioned choice of initial states represents the case where half of the population in each node is infected by virus 1, while the other half is infected by virus 2, and

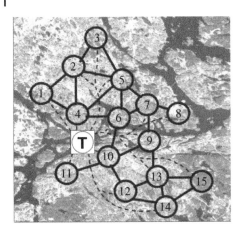

Figure 16.5 A map of 15 city districts in Stockholm. All districts are connected to the metro system. Source: Janson et al. (2020)/arXiv.org.

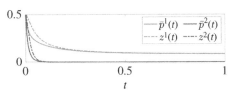

Figure 16.6 Simulation with two viruses, one (virus 2) converges to eradication, while the other (virus 1) converges to its single-virus endemic equilibrium.

the shared resource is contaminated by both viruses. The spread parameters β_{ij}^k are taken to be 1 if district i is adjacent to district j, or if $i = j$, and 0 otherwise, for $k \in [2]$. In all scenarios, we also assume that each district is bi-directionally connected to the shared resource, i.e. the Stockholm metro, with $\beta_{iw}^k = 1$ and $c_i^k = 1/15$, for all $i \in [15]$ and $k \in [2]$. As a consequence, B_w^k is irreducible for $k \in [2]$. The following scenarios differ only in terms of the choice of δ_i^k and δ_w^k.

In the simulation depicted in Figure 16.6, we chose $\delta_i^1 = 4.6$, $\delta_w^1 = 4$ and $\delta_i^2 = 10$, $\delta_w^2 = 10$, for all $i \in [15]$. As a consequence, $s(B_w^2 - D_w^2) = -4.2$, so Theorem 16.3 applies to virus 2. Consistent with the result in Theorem 16.3, virus 2 becomes eradicated exponentially fast. However, since $s(B_w^1 - D_w^1) = 0.3$, virus 1 is eradicated, and the system is essentially a single-virus system. Treated as such, virus 1 satisfies the conditions in Theorem 16.5. In line with the result in Theorem 16.5, virus 1 converges to a single-virus endemic equilibrium. Insofar as it can be determined by varying $y^1(0)$ in $\mathcal{D}^1$, this single-virus endemic equilibrium appears to be unique and asymptotically stable.

16.5 Future Research Challenges and Visions

Epidemic spreading processes are heavily affected by human factors, which significantly adds to the complexity of modeling and forecasting the spreading dynamics. In order to incorporate human factors including opinions, behaviors, infrastructure, etc., it is critical to build models that factor in the aforementioned human factors. Further, in an effort to implement these developed models and mathematical analysis in practice, we need to leverage epidemic spreading data, social network data, human mobility data, infrastructure information, etc. Hence, it is essential to have collaborations between research teams from diverse research communities (epidemiology, ecology, control theory, computer science, data science, civil engineering, sociology, etc.). For instance, communities from different backgrounds have their unique insights and research tools to study epidemic spreading, prediction, and mitigation problems. As far as the automatic control community is concerned, there is a very strong interest in devising models that are more representative of the underlying phenomena, but by using only limited data. To this end, we are interested in taking advantage of recent advances in machine learning toward modeling and prediction of spreading processes. In the meantime, we plan on leveraging rigorous control-theoretical guarantees for proposing mitigation policies that also account for model uncertainties, making them robust to variations in model parameters. Hence, in response to future outbreaks, not only will we be able to model and forecast epidemic outbreaks and spreading pattern under the considerations of spreading mechanism, human factors, and infrastructure information, but we also expect to propose and design rigorous control strategies to mitigate and eradicate epidemic spreading processes with less societal and economic cost.

References

Abakuks, A. (1973). An optimal isolation policy for an epidemic. *Journal of Applied Probability* 10 (2): 247–262.

Acheson, N.H. (2007). *Fundamentals of Molecular Virology*, Wiley.

Ahn, H.J. and Hassibi, B. (2013). Global dynamics of epidemic spread over complex networks. *Proceedings of the 52nd IEEE Conference on Decision and Control (CDC)*, 4579–4585.

Alisic, R., Paré, P.E., and Sandberg, H. (2022). Change time estimation uncertainty in nonlinear dynamical systems with applications to COVID-19. *International Journal of Robust and Nonlinear Control*. https://doi.org/10.1002/rnc.5974.

Altafini, C. (2013). Consensus problems on networks with antagonistic interactions. *IEEE Transactions on Automatic Control* 58 (4): 935–946.

Anderson, R.M. and May, R.M. (1991). *Infectious Diseases of Humans*. Oxford University Press.

Armbruster, B. and Brandeau, M.L. (2007). Contact tracing to control infectious disease: when enough is enough. *Health Care Management Science* 10 (4): 341–355.

Baheti, R. and Gill, H. (2011). Cyber-physical systems. *The Impact of Control Technology* 12 (1): 161–166.

Bailey, N.T.J. (1957). *The Mathematical Theory of Epidemics*. London: Hafner *Technical Report*.

Bailey, N.T.J. (1975). *The Mathematical Theory of Infectious Diseases and its Applications*, 2e. High Wycombe, Bucks HP13 6LE: Charles Griffin & Company Ltd.

Barro, R.J., Ursúa, J.F., and Weng, J. (2020). The Coronavirus and the Great Influenza Pandemic: Lessons from the "Spanish flu" for the Coronavirus's Potential Effects on Mortality and Economic Activity. *Technical Report No. w26866*. National Bureau of Economic Research.

Bauch, C.T. and Galvani, A.P. (2013). Social factors in epidemiology. *Science* 342 (6154): 47–49.

Benatar, S.R. (2002). The HIV/AIDS pandemic: a sign of instability in a complex global system. *The Journal of Medicine and Philosophy* 27 (2): 163–177.

Bernoulli, D. (1760). *Essai d'une nouvelle analyse de la mortalité causée par la petite vérole, et des avantages de l'inoculation pour la prévenir*, 1–45. Histoire de l'Acad., Roy. Sci. (Paris) avec Mem.

Cartwright, D. and Harary, F. (1956). Structural balance: a generalization of Heider's theory. *Psychological Review* 63 (5): 277.

Chen, X., Ogura, M., and Preciado, V.M. (2020). SDP-based moment closure for epidemic processes on networks. *IEEE Transactions on Network Science and Engineering* 7 (4): 2850–2865.

Chin, A.W.H., Chu, J.T.S., Perera, M.R.A. et al. (2020). Stability of SARS-CoV-2 in different environmental conditions. *The Lancet Microbe* 1 (1): e10.

Chowell, G. (2017). Fitting dynamic models to epidemic outbreaks with quantified uncertainty: a primer for parameter uncertainty, identifiability, and forecasts. *Infectious Disease Modelling* 2 (3): 379–398.

Cisneros-Velarde, P. and Bullo, F. (2021). Multi-group SIS epidemics with simplicial and higher-order interactions. *IEEE Transactions on Control of Network Systems* 9 (2): 695–705.

Czeisler, M.E., Tynan, M.A., Howard, M.E. et al. (2020). Public attitudes, behaviors, and beliefs related to COVID-19, stay-at-home orders, nonessential business closures, and public health guidance–United States, New York City, and Los Angeles, May 5–12, 2020. *Morbidity and Mortality Weekly Report* 69 (24): 751.

Davies, N.G., Klepac, P., Liu, Y. et al. (2020). Age-dependent effects in the transmission and control of COVID-19 epidemics. *Nature Medicine* 26 (8): 1205–1211.

Davies, N.G., Jarvis, C.I., Edmunds, W.J. et al. (2021). Increased mortality in community-tested cases of SARS-CoV-2 lineage B.1.1.7. *Nature* 593 (7858): 270–274.

De Groot, R.J., Baker, S.C., Baric, R.S. et al. (2013). Commentary: Middle East respiratory syndrome coronavirus (MERS-CoV): announcement of the coronavirus study group. *Journal of Virology* 87 (14): 7790–7792.

Diekmann, O., Heesterbeek, J.A.P., and Roberts, M.G. (2010). The construction of next-generation matrices for compartmental epidemic models. *Journal of the Royal Society Interface* 7 (47): 873–885.

Dietz, K. (1967). Epidemics and rumours: a survey. *Journal of the Royal Statistical Society: Series A (General)* 130 (4): 505–528.

Dietz, K. (1988). The first epidemic model: a historical note on PD En'ko. *Australian Journal of Statistics* 30 (1): 56–65.

Eames, K.T.D. and Keeling, M.J. (2003). Contact tracing and disease control. *Proceedings of the Royal Society of London. Series B: Biological Sciences* 270 (1533): 2565–2571.

Enyioha, C., Preciado, V., and Pappas, G. (2013). Bio-inspired strategy for control of viral spreading in networks. *Proceedings of the 2nd ACM International Conference on High Confidence Networked Systems*, 33–40.

Fall, A., Iggidr, A., Sallet, G., and Tewa, J.-J. (2007). Epidemiological models and Lyapunov functions. *Mathematical Modelling of Natural Phenomena* 2 (1): 62–83.

Fauci, A.S., Lane, H.C., and Redfield, R.R. (2020). COVID-19–navigating the uncharted. *New England Journal of Medicine* 382 (13): 1268–1269.

Fenner, F., Henderson, D.A., Arita, I. et al. (1988). *Smallpox and Its Eradication*. World Health Organization Geneva.

Finkelstein, R.A. and Feeley, J.C. (1973). Cholera. *CRC Critical Reviews in Microbiology* 2 (4): 553–623.

Friedland, G.H. and Klein, R.S. (1987). Transmission of the human immunodeficiency virus. *New England Journal of Medicine* 317 (18): 1125–1135.

Funk, S., Gilad, E., Watkins, C., and Jansen, V.A.A. (2009). The spread of awareness and its impact on epidemic outbreaks. *Proceedings of the National Academy of Sciences of the United States of America* 106 (16): 6872–6877.

Giordano, G., Blanchini, F., Bruno, R. et al. (2020). Modelling the COVID-19 epidemic and implementation of population-wide interventions in Italy. *Nature Medicine* 26 (6): 855–860.

Glanz, K., Rimer, B.K., and Viswanath, K. (2008). *Health Behavior and Health Education: Theory, Research, and Practice*. Jossey-Bass.

Gracy, S., Morarescu, I.C., Varma, V.S., and Paré, P.E. (2021). Analysis and on/off lockdown control for time-varying SIS epidemics with a shared resource. https://www.diva-portal.org/smash/get/diva2:1622798/fulltext01.pdf (accessed 13 February 2023).

Granell, C., Gómez, S., and Arenas, A. (2014). Competing spreading processes on multiplex networks: awareness and epidemics. *Physical Review E* 90 (1): 012808.

Hamer, W.H. (1906). The Milroy lectures on epidemic disease in England–the evidence of variability and of persistency of type. *The Lancet* 167 (4305): 569–574.

Heesterbeek, H., Anderson, R.M., Andreasen, V. et al., Isaac Newton Institute IDD Collaboration (2015). Modeling infectious disease dynamics in the complex landscape of global health. *Science* 347 (6227): aaa4339.

Helms, J., Kremer, S., Merdji, H. et al. (2020). Neurologic features in severe SARS-CoV-2 infection. *New England Journal of Medicine* 382 (23): 2268–2270.

Hethcote, H.W. (2000). The mathematics of infectious diseases. *SIAM Review* 42 (4): 599–653.

Hollingsworth, T.D., Anderson, R.M., and Fraser, C. (2008). HIV-1 transmission, by stage of infection. *The Journal of Infectious Diseases* 198 (5): 687–693.

Hossain, M.P., Junus, A., Zhu, X. et al. (2020). The effects of border control and quarantine measures on the spread of COVID-19. *Epidemics* 32: 100397.

Hota, A.R. and Sundaram, S. (2019). Game-theoretic vaccination against networked SIS epidemics and impacts of human decision-making. *IEEE Transactions on Control of Network Systems* 6 (4): 1461–1472.

Hu, P., Ding, L., and An, X. (2018). Epidemic spreading with awareness diffusion on activity-driven networks. *Physical Review E* 98 (6): 062322.

Hung, L.S. (2003). The SARS epidemic in Hong Kong: what lessons have we learned? *Journal of the Royal Society of Medicine* 96 (8): 374–378.

Janson, A., Gracy, S., Paré, P.E. et al. (2020). Networked multi-virus spread with a shared resource: analysis and mitigation strategies. *arXiv preprint arXiv:2011.07569*.

Kendall, D.G. (1956). Deterministic and stochastic epidemics in closed populations. *Proceedings of the 3rd Berkeley Symposium on Mathematical Statistics and Probability, Volume 4: Contributions to Biology and Problems of Health*, 149–165. University of California Press.

Kermack, W.O. and McKendrick, A.G. (1927). A contribution to the mathematical theory of epidemics. *Proceedings of the Royal Society of London. Series A, Containing papers of a mathematical and physical character* 115 (772): 700–721.

Kermack, W.O. and McKendrick, A.G. (1932). Contributions to the mathematical theory of epidemics. II. The problem of endemicity. *Proceedings of the Royal Society A* 138 (834): 55–83.

Khanafer, A., Başar, T., and Gharesifard, B. (2016). Stability of epidemic models over directed graphs: a positive systems approach. *Automatica* 74: 126–134.

Kryscio, R.J. and Lefévre, C. (1989). On the extinction of the S–I–S stochastic logistic epidemic. *Journal of Applied Probability* 26 (4): 685–694.

Lajmanovich, A. and Yorke, J.A. (1976). A deterministic model for gonorrhea in a nonhomogeneous population. *Mathematical Biosciences* 28 (3–4): 221–236.

Lee, S.-H. (2003). The SARS epidemic in Hong Kong. *Journal of Epidemiology & Community Health* 57 (9): 652–654.

Lewien, P. and Chapman, A. (2019). Time-scale separation on networks for multi-city epidemics. *Proceedings of the IEEE 58th Conference on Decision and Control (CDC)*, 746–751.

Li, X.-J., Li, C., and Li, X. (2017). Minimizing social cost of vaccinating network SIS epidemics. *IEEE Transactions on Network Science and Engineering* 5 (4): 326–335.

Li, G., Shivam, S., Hochberg, M.E. et al. (2021). Disease-dependent interaction policies to support health and economic outcomes during the COVID-19 epidemic. *iScience* 24 (7): 102710.

Liu, J., Paré, P.E., Du, E., and Sun, Z. (2019a). A networked SIS disease dynamics model with a waterborne pathogen. *Proceedings of the 2019 American Control Conference (ACC)*, 2735–2740. IEEE.

Liu, J., Paré, P.E., Nedić, A. et al. (2019b). Analysis and control of a continuous-time bi-virus model. *IEEE Transactions on Automatic Control* 64 (12): 4891–4906.

Masucci, A.M. and Silva, A. (2013). Information spreading on almost torus networks. *Proceedings of the 52nd IEEE Conference on Decision and Control (CDC)*, 3273–3280.

McKendrick, A.G. (1925). Applications of mathematics to medical problems. *Proceedings of the Edinburgh Mathematical Society* 44: 98–130.

Mei, W., Mohagheghi, S., Zampieri, S., and Bullo, F. (2017). On the dynamics of deterministic epidemic propagation over networks. *Annual Reviews in Control* 44: 116–128.

Mermin, J.H., Villar, R., Carpenter, J. et al. (1999). A massive epidemic of multidrug-resistant typhoid fever in Tajikistan associated with consumption of municipal water. *The Journal of Infectious Diseases* 179 (6): 1416–1422.

Morens, D.M., North, M., and Taubenberger, J.K. (2010). Eyewitness accounts of the 1510 influenza pandemic in Europe. *The Lancet* 376 (9756): 1894–1895.

Munzert, S., Selb, P., Gohdes, A. et al. (2021). Tracking and promoting the usage of a COVID-19 contact tracing app. *Nature Human Behaviour* 5 (2): 247–255.

Newman, M. (2003). The structure and function of complex networks. *SIAM Review* 45 (2): 167–256.

Newman, M. (2018). *Networks*. Oxford University Press.

Nishiura, H. and Chowell, G. (2009). The effective reproduction number as a prelude to statistical estimation of time-dependent epidemic trends. In: *Mathematical and Statistical Estimation Approaches in Epidemiology* (ed. G. Chowell, J.M. Hyman, L.M.A. Bettencourt, and C. Castillo-Chavez), 103–121. Dordrecht: Springer.

Norris, J.R. (1998). *Markov Chains*. Cambridge University Press.

Nowzari, C., Preciado, V.M., and Pappas, G.J. (2016). Analysis and control of epidemics: a survey of spreading processes on complex networks. *IEEE Control Systems Magazine* 36 (1): 26–46.

Oleś, K., Gudowska-Nowak, E., and Kleczkowski, A. (2013). Efficient control of epidemics spreading on networks: balance between treatment and recovery. *PLoS ONE* 8 (6): e63813.

Paarporn, K., Eksin, C., Weitz, J.S., and Shamma, J.S. (2017). Networked SIS epidemics with awareness. *IEEE Transactions on Computational Social Systems* 4 (3): 93–103.

Paré, P.E. (2018). Virus spread over networks: modeling, analysis, and control. PhD thesis. University of Illinois at Urbana-Champaign.

Paré, P.E., Beck, C.L., and Nedić, A. (2017a). Epidemic processes over time-varying networks. *IEEE Transactions on Control of Network Systems* 5 (3): 1322–1334.

Paré, P.E., Liu, J., Beck, C.L. et al. (2017b). Multi-competitive viruses over static and time-varying networks. *Proceedings of 2017 American Control Conference (ACC)*, 1685–1690. IEEE.

Paré, P.E., Kirwan, B.E., Liu, J. et al. (2018a). Discrete-time spread processes: analysis, identification, and validation. *American Control Conference (ACC)*, 404–409. IEEE.

Paré, P.E., Liu, J., Beck, C.L. et al. (2018b). Analysis, estimation, and validation of discrete-time epidemic processes. *IEEE Transactions on Control Systems Technology* 28 (1): 79–93.

Paré, P.E., Liu, J., Sandberg, H., and Johansson, K.H. (2019). Multi-layer disease spread model with a water distribution network. *Proceedings of 2019 IEEE 58th Conference on Decision and Control (CDC)*, 8335–8340. IEEE.

Paré, P.E., Beck, C.L., and Başar, T. (2020). Modeling, estimation, and analysis of epidemics over networks: an overview. *Annual Reviews in Control* 50: 345–360.

Paré, P.E., Liu, J., Beck, C.L. et al. (2021). Multi-competitive viruses over time-varying networks with mutations and human awareness. *Automatica* 123: 109330.

Pastor-Satorras, R. and Vespignani, A. (2001). Epidemic spreading in scale-free networks. *Physical Review Letters* 86 (14): 3200.

Pastor-Satorras, R., Castellano, C., Van Mieghem, P., and Vespignani, A. (2015). Epidemic processes in complex networks. *Reviews of Modern Physics* 87 (3): 925.

Piot, P. and Coll Seck, A.M.C. (2001). International response to the HIV/AIDS epidemic: planning for success. *Bulletin of the World Health Organization* 79: 1106–1112.

Pollock, A.M. and Lancaster, J. (2020). Asymptomatic transmission of COVID-19. *BMJ* 371: m4851.

Potter, C.W. (2001). A history of influenza. *Journal of Applied Microbiology* 91 (4): 572–579.

Rosenstock, I.M. (1974). Historical origins of the health belief model. *Health Education Monographs* 2 (4): 328–335.

Ross, R. (1911). *The Prevention of Malaria*. John Murray.

Sacks, J.J., Lieb, S., Baldy, L.M. et al. (1986). Epidemic campylobacteriosis associated with a community water supply. *American Journal of Public Health* 76 (4): 424–428.

Samad, T. (2023). Human-in-the-loop control and cyber–physical–human systems: applications and categorization. In: *Cyber–Physical–Human Systems: Fundamentals and Applications* (ed. A. Annaswamy, P.P. Khargonekar, F. Lamnabhi-Lagarrigue, S.K. Spurgeon). Hoboken, NJ: John Wiley & Sons, Inc.

Seto, W.H., Tsang, D., Yung, R.W.H. et al., Advisors of Expert SARS group of Hospital Authority (2003). Effectiveness of precautions against droplets and contact in prevention of nosocomial transmission of severe acute respiratory syndrome (SARS). *The Lancet* 361 (9368): 1519–1520.

She, B., Leung, H.C.H., Sundaram, S., and Paré, P.E. (2021). Peak infection time for a networked SIR epidemic with opinion dynamics. *arXiv preprint arXiv:2109.14135*.

She, B., Liu, J., Sundaram, S., and Paré, P.E. (2022). On a networked SIS epidemic model with cooperative and antagonistic opinion dynamics. *IEEE Transactions on Control of Network Systems* 9 (3): 1154–1165.

Shi, Y., Wang, Y., Shao, C. et al. (2020). COVID-19 infection: the perspectives on immune responses. *Cell Death & Differentiation* 27: 1451–1454.

Stegehuis, C., Van Der Hofstad, R., and Van Leeuwaarden, J.S.H. (2016). Epidemic spreading on complex networks with community structures. *Scientific Reports* 6 (1): 1–7.

Stewart, A.J. and Devlin, P.M. (2006). The history of the smallpox vaccine. *Journal of Infection* 52 (5): 329–334.

Teng, P.S. (1985). A comparison of simulation approaches to epidemic modeling. *Annual Review of Phytopathology* 23 (1): 351–379.

Threlfall, E.J. (2002). Antimicrobial drug resistance in Salmonella: problems and perspectives in food-and water-borne infections. *FEMS Microbiology Reviews* 26 (2): 141–148.

Troesken, W. (2004). *Water, Race, and Disease*. MIT Press.

Van Mieghem, P. (2011). The N-intertwined SIS epidemic network model. *Computing* 93 (2): 147–169.

Van Mieghem, P. (2012). Epidemic phase transition of the SIS type in networks. *EPL (Europhysics Letters)* 97 (4): 48004.

Van Mieghem, P., Omic, J., and Kooij, R. (2008). Virus spread in networks. *IEEE/ACM Transactions On Networking* 17 (1): 1–14.

Wang, Z., Andrews, M.A., Wu, Z.-X. et al. (2015). Coupled disease–behavior dynamics on complex networks: a review. *Physics of Life Reviews* 15: 1–29.

Water, S. and World Health Organization (2002). *Emerging Issues in Water and Infectious Disease*. World Health Organization.

Weissman, J.B., Craun, G.F., Lawrence, D.N. et al. (1976). An epidemic of gastroenteritis traced to a contaminated public water supply. *American Journal of Epidemiology* 103 (4): 391–398.

Weitz, J.S., Park, S.W., Eksin, C., and Dushoff, J. (2020). Awareness-driven behavior changes can shift the shape of epidemics away from peaks and toward plateaus, shoulders, and oscillations. *Proceedings of the National Academy of Sciences of the United States of America* 117 (51): 32764–32771.

Wilder-Smith, A., Gubler, D.J., Weaver, S.C. et al. (2017). Epidemic arboviral diseases: priorities for research and public health. *The Lancet Infectious Diseases* 17 (3): e101–e106.

Wolf, W. (2009). Cyber-physical systems. *Computer* 42 (03): 88–89.

Zaslavsky, T. (1982). Signed graphs. *Discrete Applied Mathematics* 4 (1): 47–74.

Zino, L. and Cao, M. (2023). Social diffusion dynamics in cyber-physical-human systems. In: *Cyber–Physical–Human Systems: Fundamentals and Applications* (ed. A. Annaswamy, P.P. Khargonekar, F. Lamnabhi-Lagarrigue, S.K. Spurgeon). Hoboken, NJ: John Wiley & Sons, Inc.

Zino, L. and Cao, M. (2021). Analysis, prediction, and control of epidemics: a survey from scalar to dynamic network models. *IEEE Circuits and Systems Magazine* 21 (4): 4–23.

Zino, L., Rizzo, A., and Porfiri, M. (2018). Modeling memory effects in activity-driven networks. *SIAM Journal on Applied Dynamical Systems* 17 (4): 2830–2854.

17

Digital Twins and Automation of Care in the Intensive Care Unit

J. Geoffrey Chase[1], Cong Zhou[1,2], Jennifer L. Knopp[1], Knut Moeller[3], Balázs Benyo[4], Thomas Desaive[5], Jennifer H. K. Wong[7], Sanna Malinen[6], Katharina Naswall[7] Geoffrey M. Shaw[8], Bernard Lambermont[9], and Yeong S. Chiew[10]

[1] Department of Mechanical Engineering, Centre for Bio-Engineering, University of Canterbury, Christchurch, New Zealand
[2] School of Civil Aviation, Northwestern Polytechnical University, Taicang, China
[3] Department of Biomedical Engineering, Institute of Technical Medicine, Furtwangen University, Villingen-Schwenningen, Germany
[4] Department of Control Engineering and Information Technology, Budapest University of Technology and Economics, Budapest, Hungary
[5] GIGA In Silico Medicine, Liege University, Liege, Belgium
[6] Department of Management, Marketing, and Entrepreneurship, University of Canterbury, Christchurch, New Zealand
[7] School of Psychology, Speech and Hearing, University of Canterbury, Christchurch, New Zealand
[8] Department of Intensive Care, Christchurch Hospital, Christchurch, New Zealand
[9] Department of Intensive Care, CHU de Liege, Liege, Belgium
[10] Department of Mechanical Engineering, School of Engineering, Monash University Malaysia, Selangor, Malaysia

17.1 Introduction

Healthcare and intensive care unit (ICU) medicine, in particular, are facing a devastating tsunami of rising demand due to aging demand multiplied by increasing chronic disease and increasing inequity of access to care. The difficulty arises where increases in demand are not matched by society's ability to pay. In contrast, digital technologies, particularly software, increased data, and automation, have brought significant productivity gains to many industries and manufacturing in particular. However, such productivity gains have not yet come to the field of medicine.

At the cutting edge of modern manufacturing, digital twins, model-based optimization of manufacturing systems, and equipment are a rapidly growing means of further enhancing productivity and quality (Cimino et al. 2019; Kritzinger et al. 2018; Negri et al. 2017), and are a major growth technology trend overall (Panetta 2019, 2020). They represent the next step toward increased automation and improved productivity and quality, as well as the foundation for an emerging range of digital services.

This concept intersects well with the model-based decision support and control just beginning to emerge into clinical use in medicine and diabetes in particular (Lee et al. 2017; Lewis et al. 2016; Thabit et al. 2017) It is especially relevant to the technology-laden ICU (Chase et al. 2018c, 2019; Desaive et al. 2019; Morton et al. 2019a). In medicine, digital twins offer the opportunity to personalize care, as well as providing the foundation digital models and methods upon which automation can be introduced to care.

More specifically, ICU medicine involves a wide range of sensors and devices to measure patient state and further devices to deliver care. These existing technologies, such as infusions pumps and

Cyber–Physical–Human Systems: Fundamentals and Applications, First Edition.
Edited by Anuradha M. Annaswamy, Pramod P. Khargonekar, Françoise Lamnabhi-Lagarrigue, and Sarah K. Spurgeon.
© 2023 The Institute of Electrical and Electronics Engineers, Inc. Published 2023 by John Wiley & Sons, Inc.

mechanical ventilators, are relatively simple devices mechanically and electronically. They thus provide an excellent foundation upon which to develop digital twins as all sensor and actuation technologies exist. In particular, increasing automation will bring these systems into use as cyber–physical–human systems (CPHS), with a human in the control loop, such as the nurse or the clinician, as well as the human physiological system being controlled or managed.

This chapter presents digital twins in a manufacturing concept and translates it into clinical practice. It then reviews the state of the art in key areas of ICU medicine to show how these CPHS represent a future of personalized, productive, and, outside of the ICU, patient-led care. Next, it covers the role of social-behavioral sciences in innovation and technology uptake, where innovation can be stifled during implementation and where the term *disruptive innovation* does not always carry the same context in healthcare settings as in technology fields, and may exclude a range of high-impact frugal or process innovations (Sounderajah et al. 2021). The final section covers future research challenges and visions, delineating the key hurdles to seeing digital twins move *from research bench to clinical bedside*.

17.1.1 Economic Context

Healthcare in developed countries is growing at an annualized rate of 5–9% annually (Economist, T. 2021; Micklethwait 2011) and consumes 10–18% of GDP (OECD 2021), all of which is far faster than GDP growth in these countries. Thus, the ability to provide care is outpaced by the demand for care, creating significant stress on the ability to deliver equity of access to care (Glorioso and Subramanian 2014; Levine and Mulligan 2015; Neilson 2018), and increased rationing of care in response (Bauchner 2019; Keliddar et al. 2017; Williams et al. 2012). Intensive care medicine is one of the most technology-laden areas of healthcare. It consumes ~1–2% of GDP (or ~10% of health budgets) in modern countries, while treating 0.1% of patients or less (Chalfin et al. 1995; Halpern et al. 2004; Kelley et al. 2004).

ICU patients are very complex and highly variable, making management difficult. Aging demographics, chronic disease, and increasing life spans are driving increasing cost and reducing equity of access to care (Baumol and De Ferranti 2012; Dombovy 2002; Halpern 2009, 2011; Orsini et al. 2014; Shorr 2002; Truog et al. 2006; van Exel et al. 2015). This issue has been particularly highlighted during the Covid-19 pandemic, where the ICU bed numbers became a household topic as ICU services were "overrun" with patient demand (ANZICS 2020; Beitler et al. 2020; Cohen et al. 2020; Ricci and Gallina 2020), which in turn required innovative strategies to manage patient load and care (Aziz et al. 2020; Beitler et al. 2020; Buijs et al. 2021; Chase et al. 2020a,b; de Jongh et al. 2020; Holder-Pearson et al. 2021; Lambermont et al. 2021; von Düring et al. 2020; Wunsch 2020), not all of which were proven safe or effective (e.g. SCCM 2020).

Thus, there is a significant need to improve both the productivity and quality of care. In fact, the need for improved productivity of care can be traced to poor and very low uptake of productivity enhancing innovations in healthcare delivery (Baumol and De Ferranti 2012; Economist, T. 2016; Micklethwait 2011; Morris et al. 2011). As a result, there has been increasing discussion of digital forms of healthcare delivery in economics focused discussions (Economist, T. 2016, 2020; Micklethwait 2011), as well as clinical studies and editorials (Bosch et al. 2014; Carville 2013; Glorioso and Subramanian 2014; Halpern 2009; Levine and Mulligan 2015; Warrillow and Raper 2019), with some studies noting the increasing burden on patients themselves.

Personalized care using patient-specific models identified from clinical data offers the means to directly manage the significant intra- and interpatient variability characterizing the ICU patient (Chase et al. 2018c). It adds automation to care, using digital technologies to improve quality and cost, as in many other sectors, but much less in medicine (Baumol and De Ferranti

2012; Micklethwait 2011). Thus, some level of automation will be necessary to improve the productivity, personalization, and quality of care given society's increasing inability to meet rising costs and provide equal equity of access to care (Baumol and De Ferranti 2012; Bosch et al. 2014; OECD 2015).

17.1.2 Healthcare Context

Intensive care unit (ICU) patients are complex and highly variable, making management difficult. Aging demographics, chronic disease, and increasing life spans drive increasing patient complexity and cost, which in turn increases length of stay and reduces equity of access to care (Baumol and De Ferranti 2012; Dombovy 2002; Halpern 2009, 2011; Orsini et al. 2014; Shorr 2002; Truog et al. 2006; van Exel et al. 2015). Reduced equity of access to care increases poor outcomes (Benatar et al. 2018; Gulliford et al. 2002; Neilson 2018), such as increased mortality due to reduced opportunity for timely care (Craxì et al. 2020; Halpern 2011; Orsini et al. 2014; Truog et al. 2006), creating further inequity. This positive feedback loop has negative consequences but highlights the significant need to improve productivity and quality of care to address this issue and applies to both ICU care, as well as healthcare in general.

Personalized care using patient-specific mathematical models, which are personalized to the patient using system identification methods and clinical data, offer the means to directly manage intra- and interpatient variability (Chase et al. 2018c). In particular, ICU patient state is highly variable in key areas such as glycemic control (Du et al. 2020; Egi et al. 2006; Le Compte et al. 2010; Uyttendaele et al. 2017) and mechanical ventilation (Chiew et al. 2015; Kim et al. 2019; Lee et al. 2021) and can include differences due to sex (Dickson et al. 2015; Kim et al. 2020a; Uyttendaele et al. 2021), which can further contribute to inequity of access to both care and outcomes, which disproportionately affect the poorer segments of society (Artiga et al. 2020; Benatar et al. 2018; Essue et al. 2018; Robinson et al. 2014). Thus, such model-based care represents a potential means to improve outcomes by personalizing care and to also improve equity by enabling automation, improved productivity, and (thus) greater access to (better) care.

There is a further advantage to using deterministic models based on directly modeled physics and mechanics to guide care; specifically, the reduction or elimination of personal and systemic bias. Bias in care, particularly racial bias, is increasingly being discovered via analyses of care and outcomes (Hagiwara et al. 2019; Maina et al. 2018; Rumball-Smith 2009; Talamaivao et al. 2020). Racial biases often also mask or duplicate socioeconomic biases. However, a deterministic model-based care approach is strictly numerical and model-driven, a *one method fits all* form of care (Chase et al. 2011a, 2018c).

In particular, racial or sex differences play zero role because the computational model used contains only direct physical mechanics terms identified and personalized using objective measured clinical data. Clinical data, such as blood glucose level or airway pressure and flow, are not biased or able to be biased as they are objective measures. Thus, physics models, objective measured data, and system identification methods offer a potentially powerful means to remove race, sex, socioeconomic status, and other biases from care decisions.

As noted, the key difference is deterministic models. In particular and in contrast, machine learning and artificial intelligence approaches are data-driven and increasingly touted as a healthcare solution. However, the data and resulting data-driven algorithms can have unintentional bias, carrying over into care recommendations (Nelson 2019; Obermeyer et al. 2019; Wiens et al. 2020). Thus, a deterministic, fully objective, model-based approach can significantly reduce, or eliminate, bias in the care delivery, which is a significant gain from implementing a model-based DT approach in its own right, even if no other gain was obtained from their use.

Thus, enabling automation, using digital technologies can improve quality and cost, an outcome which has occurred in many industries, but much less so in medicine (Baumol and De Ferranti 2012; Economist, T. 2011; Micklethwait 2011). Hence, from a healthcare context, automation and personalization enable healthcare futures with improved equity of access to care and outcomes. In particular, improved equity of access is not possible today nor with incremental changes in healthcare productivity.

17.1.3 Technology Context

ICU medicine is one of the most technology-laden areas of medicine, where the doctor or nurse almost always touches some form of technology, such as a ventilator or infusion pump, in treating the patient (Tunlind et al. 2015). In turn, high levels of technology require understanding of the associated human factors and ergonomics for effective use (Carayon and Gurses 2005; Carayon et al. 2014; Chase et al. 2008a). Automation can streamline this process and current ICU technologies could be readily automated with existing wireless and wired technologies (Gatouillat et al. 2018; Joyia et al. 2017). However, there is a significant lack of interoperability between devices, which hinder this process, despite calls for greater interoperability and human-centered automation in ICU care (Dominiczak and Khansa 2018; Gurkan and Merchant 2010; Joseph et al. 2020; Poncette et al. 2019).

Thus, there is no automation technology barrier to automating critical and core areas of ICU care, such as drug and fluid delivery for glycemic control and cardiovascular management or mechanical ventilation. These areas cover 90–100% of ICU patients and are leading causes of ICU admission, length of stay, mortality, and, thus, cost. Equally, the hardware technology itself is often relatively simple (Hatcliff et al. 2019; Lewis et al. 2016; Payne et al. 2021; Poncette et al. 2019). The missing elements are in the area of digital twin modeling for clinical application, or virtual patients, and the difficulty in using them to find the appropriate model-based metrics upon which to titrate care (Chase et al. 2011b, 2018c).

In particular, computational physiological models can combine medical data and model identification methods to generate a "virtual patient" representing a given patient in a particular state and point of time for a given organ or physiological system (e.g. metabolic, cardiovascular, pulmonary). What is missing is a collection of accurate, validated virtual patient or digital twin models for use at the bedside to automate care. In particular, there are no accepted standards for modeling approach, model identifiability, or accepted levels of model validation, although some have been proposed (Chase et al. 2018a,c). Thus, the technological issue is one of linking existing bedside care delivery technologies and known communication and control technologies to their clinical medicine application.

Specifically, the lack of accurate, implementable virtual patient or digital twin models is the primary technical and scientific hurdle to linking measurements and care delivery devices to automate care in an accurate, personalized fashion.

17.1.4 Overall Problem and Need

Hyper-automation and digital twins (DT) capture the essence of this potential model-based approach. They are also a major growth area in manufacturing technology (Cimino et al. 2019; Panetta 2019). Reduced cost and optimization arise from using sensor data to monitor, model, and manage real-world systems. In medicine, the difference is the humans-in-the-loop, both the patients and the clinicians, where the former is the system controlled and the latter is part of the control or patient management. Both introduce cyber–physical–human aspects into DT implementation.

This chapter addresses the key modeling and social science links missing in bringing DTs to use in ICU medicine, including translating these concepts into a medical space. Specifically, how models can be created and effectively integrated into clinically applied DTs in the ICU and how to address barriers to uptake of innovation in medicine (Wong et al. 2021). Both issues hinder the creation of cyber–physical–human systems (CPHS) in medical care, where one is a technical question and the other a social science question.

Given the deep interaction of physiology, medicine, and clinical practice with both modeling and identification methods in creating healthcare delivery digital twins, as well as the significant literature in each area alone, this review provides several added references to ensure suitable supporting citations for any interested reader to follow-up any specific area. The chapter should thus serve as a relatively complete reference to the current state of the art for a reader from any relevant background.

17.2 Digital Twins and CPHS

17.2.1 Digital Twin/Virtual Patient Definition

Digital twins arise at the intersection of Industry 4.0 and the internet of things (IoT). A DT is "a virtual copy of a system able to interact with the physical system in a bi-directional way" (Cimino et al. 2019; Kritzinger et al. 2018). Bi-directional information exchange synchronizes virtual system response to match the physical system to "forecast and optimise the behavior of the physical system in real time."

In the manufacturing context, digital twins sit in the middle, on top of a "control layer" of supporting technologies, and under an "enterprise resource planning" layer integrating organizational functions and goals into how the DT is applied (Figure 17.1). The upper and lower layers both inform the DT and its design and use. The middle layer DT itself is connected through the control layer to its physical counterpart in real time and uses modeling and computation to continually update the virtual digital twin (model) (Negri et al. 2017).

More specifically, DTs are defined by their integration (Kritzinger et al. 2018). A ***digital model*** (DM) does not interact with the physical system. A ***digital shadow*** (DS) has one-direction flow, updated with data from the physical system without returning a control input. A ***digital twin*** or DT arises when the DM is updated from physical system data, and the resulting simulation is used to control the physical system, or specifically, two-directional data flow integrating both the enterprise and control levels to create a full system.

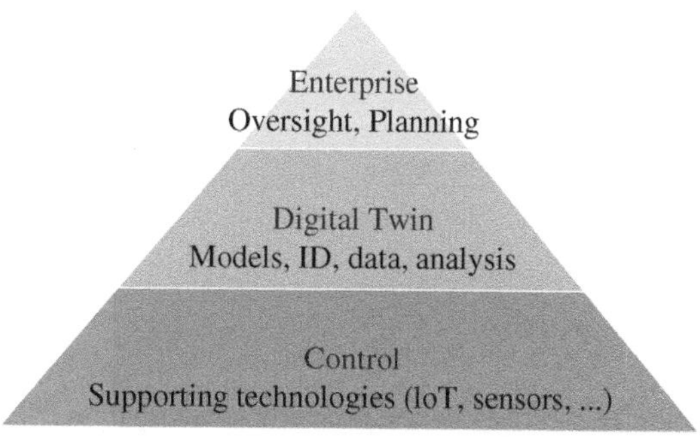

Figure 17.1 The DT lies between supporting technologies and a guiding organization level or protocol. Source: Adapted from Chen (2005).

Enterprise
Oversight, Planning

Digital Twin
Models, ID, data, analysis

Control
Supporting technologies (IoT, sensors, ...)

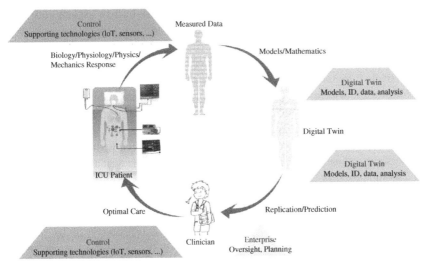

Figure 17.2 DT control loop in medicine with clinical staff in the control loop and main elements or blocks of Figure 17.1 schematically showing where they are applied in this application loop.

In medicine, the physical system is the patient and their particular organ or physiological system to be managed. The control layer is created by communication and command/control technologies, sometimes referred to as the medical Internet of things (Gatouillat et al. 2018; Joyia et al. 2017), and connect the sensor measurements and care delivery technologies to the DT model. The enterprise layer defines how a DT model is applied clinically, typically via an agreed protocol for care using

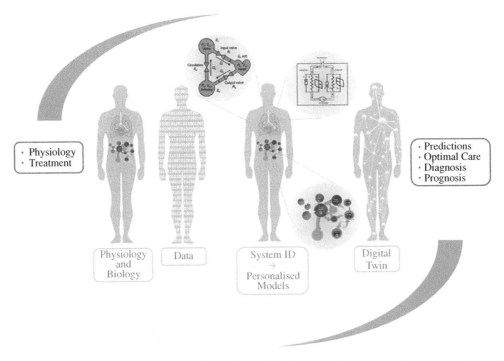

Figure 17.3 DT model creation and updating from clinical data, leading to prediction of response, and subsequent new data.

the DT (Chase et al. 2016, 2018c). The result is the control loop in Figure 17.2, showing also the main elements of Figure 17.1, where bi-directional information to and from the DT is clear.

Specific to the DT model, Figure 17.3 shows the specific steps for model updating and subsequent use to predict patient-specific response in a real-time loop. Figure 17.3 thus defines a "virtual patient" in the context of its application. Virtual patient or DT models are thus the core of a DT application in medicine, where, as noted, control layer technologies connecting the DT model to sensors and delivery devices are within current technology capabilities.

From Figures 17.2 and 17.3, interventions are optimized via prediction and a clinically agreed protocol for using the DT within the clinical workflow (Chase et al. 2008a; Szlavecz et al. 2014; Ward et al. 2012). In this framework, the protocol is derived from clinical standards and local approaches to care, and thus comes from the enterprise level of Figure 17.1. Finally, the clinician in the loop in Figure 17.2 is there primarily for safety, but could be removed, given full automation. Full automation creates a "human-in-the-plant" system when fully automated, and is also a "human-in-the-controller" element if clinical input is required to confirm choices or for safety (Samad 2022).

17.2.2 Requirements in an ICU Context

Given either interoperable devices and communications (e.g. Gatouillat et al. 2018; Ng et al. 2021; Szlavecz et al. 2014) or clinical personnel in the loop, the requirements for a DT model in an ICU context thus include the following:

- Physiological relevance
- Clinical relevance
- Treatment sensitivity, practically identifiable from clinically available data (nonadditionally invasively via no new sensors added is preferred).

Physiological relevance requires a model structure capturing relevant dynamics. To be useful, these models must be able to reproduce the measurable physiological dynamics to the resolution of available sensors (Carson and Cobelli 2001; Ljung and Glad 1994; Wongvanich et al. 2015). *Clinical relevance* emphasizes models simulated in "clinical real-time" used to make predictions about changes in care to guide decisions, where the model inputs and outputs match the clinical situation. These two requirements thus define the feasible range of model structure and complexity.

A *treatment sensitivity identifiable from clinically available data* is the crucial element in titrating care, assessing rate of change of output variable per unit of input administered, for example, insulin sensitivity (Blaha et al. 2016; Chase et al. 2010, 2011b, Dalla Man et al. 2005; Hovorka et al. 2008; Lin et al. 2008, 2011b; Cobelli et al. 1986; Dalla Man et al. 2002; Docherty et al. 2011b; Hann et al. 2005; Hovorka et al. 2002; Langouche et al. 2007; Le Compte et al. 2009; Lin et al. 2006; Mari et al. 2005; Pielmeier et al. 2010b; Pillonetto et al. 2006; Wilinska et al. 2008) in metabolism and recruitment elastance in pulmonary mechanics (Bates 2009; Carvalho et al. 2013; Chiew et al. 2015; Redmond et al. 2014; Rees et al. 2006, 1999, 2002; Sundaresan et al. 2011; van Drunen 2014; Rees 2011; Stenqvist et al. 2012; Sundaresan et al. 2010; van Drunen et al. 2013), where they capture patient's state and are often clinically used in simpler forms to guide care (Benyo et al. 2012; Camilo et al. 2014; Carvalho et al. 2007, 2008, 2013; Garcia-Estevez et al. 2003; Gross et al. 2003; Lambermont et al. 2008; Pintado et al. 2013; Suarez-Sipmann et al. 2007). In cardiovascular management, these sensitivities do not yet exist as the desired outcome metrics of stroke volume or stressed blood volume cannot yet be directly measured, requiring a model-based sensor from which to derive a sensitivity (Jansen et al. 2010; Kamoi et al. 2014; Maas et al. 2009, 2012; Nouira et al. 2005; Pironet et al. 2015; Wujtewicz 2012), where these measures increasingly called for clinically to guide care (Cecconi and Rhodes 2009; Cecconi et al. 2014). The final element is the requirement to be *practically identifiable* using available clinical measurements so further invasive measures are not required (Audoly et al. 1998, 2001; Cobelli et al. 1984; Docherty et al. 2011a; Ljung and Glad 1994; Pillonetto et al. 2003; Schranz et al. 2012). Overall, these requirements limit both minimum and maximum model complexity and structure, within which a DT model and DT system solution can exist for a given problem.

Implicitly, these requirements also include specific CPHS aspects from two perspectives. From the patient's perspective, the modeling captures their human physical state from the available measurements. Equally, from the clinical staff perspective, model inputs and outputs representing only key available metrics place the models and resulting digital twins directly into the same care space that the clinical staff occupy. Thus, these model requirements implicitly include human-centered aspects of care, and with an enterprise-level protocol, they also include the human input to, and control of, care.

17.2.3 Digital Twin Models in Key Areas of ICU Care and Relative to Requirements

Physiological models are very common in the scientific, particularly engineering science, literature. These models cover a range of ICU and other areas of care, physiological systems, and potential clinical applications. All these models can be assessed by their potential use as one of three specific types of system models. In particular, (**1**) digital models (**DM**); (**2**) digital shadows (**DS**); or (**3**) digital twin (**DT**) models. Thus, the clinical requirements in Section 2.2 link these models to their potential clinical uses, and DT models sit within the middle control layer of Figures 17.1–17.3.

In particular, by use, most models are, in fact, DMs, used to analyze information, validated offline, but with no ability to be personalized or updated. Models capable of being personalized from data in a clinically relevant timeframe, or "clinical real-time," are DS. Very few offer the clinical real-time identification and prediction accuracy to optimize patient-specific care as a DT model. More simply, there is a hierarchy comprising many models (DM), relatively very few can be identified and personalized using available clinical data (DS), and only a few of those remaining offer prediction accuracy able to guide care (Chase et al. 2018b).

There is also the ability to call upon a growing range of models, methods, and databases, ranging from simple to detailed and at multiple scales of physiology, space, and time (e.g. Barrett et al. 1998; Ben-Tal 2006; Bergman et al. 1979; Bradley et al. 2011; Callegari et al. 2003; Carson and Cobelli 2001; Chase et al. 2006, 2008b, 2011a; Dalla Man et al. 2006; Hunter et al. 2008; Tawhai et al. 2009; Wilinska et al. 2010; Cohen 2012; Cooling and Hunter 2015; Dalla Man et al. 2002; Garny and Hunter 2015; Hovorka et al. 2003; Hunter et al. 2013; Keener and Sneyd 1998; Man et al. 2009; Nickerson et al. 2014; Pielmeier et al. 2010b; Pillonetto et al. 2006; Pironet et al. 2013; Smith et al. 2011; Tawhai et al. 2006a; Tawhai and Lin 2010; Tawhai and Bates 2011; Toffolo and Cobelli 2003; Vodovotz and Billiar 2013; Wilinska et al. 2008; Yu et al. 2011). However, for deterministic, physiologically relevant digital twin models, the overall approach relies on identifying patient-specific and time-varying parameters capturing all relevant intra- and interpatient variability for use in titrating care to clinically recognized endpoints. These parameters typically relate clinical inputs or care to clinical outputs or metrics defining (successful) response to care.

These "sensitivities" are the key, as they provide an input–output relationship reflecting patient's status and response to care, and can thus be used to titrate delivery. However, this approach thus defines the feasible model structure and complexity. Specifically, it segregates more complex anatomically and biophysically based models (referred to as "physiome models") to an informative role, by requiring simpler models for immediate use at the bedside (referred to as "bedside models").

In more detail, physiome models can provide significant insight into dysfunction at levels bedside models, with their simpler single organ and/or single system dynamics, cannot (Cutrone et al. 2009; Hunter et al. 2005, 2010, 1992; Hunter 2016; Ramachandran et al. 2009; Safaei et al. 2016; Smith et al. 2011; Tawhai et al. 2006b, 2009; Tawhai and Burrowes 2008; Tawhai and Bates

2011; Viceconti and Hunter 2016; Willmann et al. 2007). Physiome models can be patient-specific, but require significant amounts of data, which are often not typically available at the ICU bedside, thus precluding use in real-time care. In contrast, the last 10–15 years has seen growing numbers of model-based sensors and decision support systems in critical care (e.g. Blaha et al. 2005; Blaha et al. 2009; Cochran et al. 2006; Desaive et al. 2013; Dong et al. 2012; Evans et al. 2011; Fisk et al. 2012; Larraza et al. 2015; Le Compte et al. 2012; Lin et al. 2011a; Morris et al. 2008; Pielmeier et al. 2010a; Pielmeier et al. 2012; Pironet et al. 2015; Plank et al. 2006; Rees et al. 2006; Rees et al. 2002; Rees 2011; Revie et al. 2013; Szlavecz et al. 2014; Van Herpe et al. 2009; Van Herpe et al. 2013), including in some very rare cases their implementation as a standard of care. There is thus growing interest in using computational models to guide care of ICU patients.

Thus, in a DT context, medicine and physiology offer many DMs, models informed by data, but not receiving real-time patient data (Hunter et al. 2010; Nickerson et al. 2016; Safaei et al. 2016; Viceconti and Hunter 2016). However, most are too complex to be personalized in real-time with available data (Audoly et al. 2001; Bellu et al. 2007; Chapman et al. 2003; Chase et al. 2018c; Docherty et al. 2011a; Pironet et al. 2016, 2017; Schranz et al. 2012). The vast majority of physiome models fit this category. There are several examples in the metabolic (Afshar et al. 2019; Chase et al. 2019; Dalla Man et al. 2014), pulmonary (Burrowes et al. 2013, 2005; Morton et al. 2019a; Tawhai et al. 2019, 2009), and cardiovascular areas (Desaive et al. 2019; Hunter 2016; Safaei et al. 2016; Smith et al. 2011).

DS models are increasingly common, differentiated by their identifiability from the relatively limited clinical data typically available at the ICU bedside (Chase et al. 2011a, 2019; Morton et al. 2019a), which effectively limits these models complexity. They have been used to assess new medical technology applications (Zhou et al. 2018, 2019) and protocols (Fisk et al. 2012; Uyttendaele et al. 2018). However, very few to none are used in regular care, and thus are not DT models, where it is important to note <1% of model-based decision support systems are implemented beyond testing, let alone as a standard care (Garg et al. 2005; Wears and Berg 2005).

A DT model is further differentiated from a DS by its (critical) ability to accurately predict the outcome to changes in care or dosing, and to do so in clinical real-time for decision support. DT models innately include application and are very rare. In short, while many models exist, few are identifiable, and even fewer can accurately predict patient-specific response to clinically reasonable changes in care well enough to guide care.

In cardiovascular systems, modeling DT models and model-based sensors are just emerging for tracking key clinical and physiological variables like stroke volume and stressed blood volume over major changes in patient state or care (Desaive et al. 2019; Murphy et al. 2020a,b). For pulmonary mechanics, the first accurate predictive models have emerged since 2018 (Morton et al. 2018, 2019b, 2020), culminating in very accurate nonlinear models of mechanics capturing all key metrics in ventilation care (Knopp et al. 2021; Sun et al. 2021; Zhou et al. 2021). Finally, the metabolic area is furthest along with several DS models (Blaha et al. 2016; Evans et al. 2012; Fisk et al. 2012; Knopp et al. 2019; Le Compte et al. 2011; Lonergan et al. 2006; Mesotten et al. 2017; Pielmeier et al. 2010a; Van Herpe et al. 2013; Wilinska et al. 2008). There are also full DT models in standard of care use in ICU and NICU (neonatal ICU) care (Dickson et al. 2016, 2018; Evans et al. 2011; Hovorka et al. 2008; Knopp et al. 2019; Stewart et al. 2016), as well as in outpatient diabetes (Boiroux et al. 2017; Kovatchev et al. 2017; Thabit et al. 2017).

Prediction accuracy requires validation to ensure trust in the methods and approach, based on accuracy relative to clinical metrics and goals, as defined in (Chase et al. 2018c). The gold standard is the ability to capture a cohort of patients with accurate prediction so entire cohorts of predictions match, or a cohort-based cross validation. To date, only one model has achieved this outcome (twice) across multiple cohorts, which was in the metabolic modeling domain (Chase et al. 2010; Dickson et al. 2018). Patient-level prediction accuracy has been demonstrated for any model in standard of care use, as noted above, but has also been recently demonstrated for pulmonary models (Sun et al. 2021; Zhou et al. 2021). Cardiovascular models are not beyond the DS level at this time, and thus not yet validated for DT use in care.

Validation of prediction accuracy defines the level of confidence in a given DT model meeting all other requirements. It is the key element verifying whether a potential DT model can make the last step for use in care, where a model alone is not enough. In short, prediction is the key to turning a model that meets the requirements of a DT, into a DT suitable for guiding care.

17.2.4 Review of Digital Twins in Automation of ICU Care

As stated, there are extremely few DT models in use in care at this time. They are all part of metabolic control systems (Boiroux et al. 2017; Dickson et al. 2016, 2018; Evans et al. 2011; Hovorka et al. 2008; Knopp et al. 2019; Kovatchev et al. 2017; Stewart et al. 2016; Thabit et al. 2017). In particular, the STAR and eMPC glycemic control systems in ICU care, and similar artificial pancreas solution for type 1 diabetes, are the exemplars. There are thus three examples of such models with largely automated control, despite a human control element in the loop (per Figure 17.2). An equally limited selection of truly well-validated models is the limiting factor in greater examples.

Relevant to core ICU care areas, pulmonary models have achieved the level of prediction accuracy necessary for use in care as a DT in all common mechanical ventilation modes (Knopp et al. 2021; Sun et al. 2021; Zhou et al. 2021). However, they are not yet in use as a DT, although clinical trials are planned (Kim et al. 2020b). These DS models do show a pathway to DT automation using accurate prediction of key outcomes to guide care; for example to safely set positive end expiratory pressure to minimize lung elastance (Amato et al. 2015; Goligher et al. 2021), maximize recruited lung volume (Turbil et al. 2020; Wallet et al. 2013), and simultaneously minimize the risk of over distension and ventilator induced lung injury (Bates and Smith 2018). Such multidimensional optimization problems are typical in mechanical ventilation (Major et al. 2018; Morton et al. 2019a) and can be best managed via an objective, automated DT instead of relying on clinical experience and intuition, which can over simplify such problems due to their higher dimensionality.

Equally, cardiovascular management is only nearing the digital shadow, DS, phase. In particular, recent models enable personalized model-based sensors to assess key metrics to guide care, which are unmeasurable without extensive invasive measures otherwise. These measures, such as stroke volume and stressed blood volume (perfusion) (Murphy et al. 2020a,b; Smith et al. 2021a,b), will enable similar predictive accuracy and the ability to personalize and optimize a very difficult, multidimensional clinical optimization problem (Desaive et al. 2019).

The question thus arises as to what are the major aspects hindering adoption and uptake?

The only bi-directional DTs in clinical ICU use are also not fully automated, with a human-in-the-loop (e.g. Knopp et al. 2019; Stewart et al. 2016). Thus, where a proven, validated DT model exists, the main element missing reverts to being technological and is not model or control based. Specifically, there is a need to create greater interoperability and access to data from the range of ventilators and infusion pumps in the ICU.

This issue has been a great source of difficulty for proprietary and other reasons (Hudson and Clark 2018; Jaleel et al. 2020; Joseph et al. 2020; Mavrogiorgou et al. 2019; Poncette et al. 2019; Sandhya et al. 2017; Williams 2017). It will grow in importance and be a greater hurdle as more DT models able to be automated emerge. Whether this is addressed by change within the medical and healthcare industries and infrastructure, or via open-source approaches remains to be seen and is outside the scope of this review.

17.2.5 Summary

The key element to implementing digital twins in ICU, or any area of healthcare, is the need for an accurate DT model, able to be readily personalized and accurate in prediction to guide care. While there are many models in many core areas of care, few are able to be digital shadows (DS), and far fewer have been validated in their prediction accuracy or use as digital twins (DT). There are further issues of technology interoperability hindering full automation, which can be resolved through either added, external sensors, or via greater access. Beyond the technological, there are social factors affecting uptakes, which are addressed in the next section (Wong et al. 2021).

17.3 Role of Social-Behavioral Sciences

17.3.1 Introduction

Having the technologies, models, and protocols to create DTs for use as standard of care is not enough. The "enterprise layer" of Figure 17.1 also includes decision-making on the adoption of new standards of care. Adoption is a decision made at both the ICU and clinician level, as well as higher management and/or a health system level. Worryingly, patients benefit from only 30% to 50% of validated healthcare technologies (Grol 2001; Schuster et al. 2005). If the issues surrounding technology implementation are not remediated, at least half of the DTs that pass rigorous clinical validation will fail to be successfully adopted into ICU healthcare practices.

Consideration of factors other than technical aspects of DTs is the key to enabling successful adoption. From a social-behavioral perspective, technology adoption is more than the acceptance of new technology in an environment. Sustainable adoption requires new technology to be integrated into the everyday processes of the healthcare delivery unit (Anderson et al. 2019), and for the use of technology to be "normalized" as part of protocolized healthcare. This requires consideration of social factors at the individual, the team/unit, and the organizational system levels, and the dynamic interrelationships across these levels.

Socialization of new technology to clinical staff can begin prior to its implementation into workplaces. That is, staff should be engaged at the technology development stage, when aspects of need, ergonomics, and ease-of-use are deliberated. Co-design principles not only ensure compatibility of DTs with existing practices, but also generate a sense of ownership over its final design with staff (Bird et al. 2021; Laurance et al. 2014; Llopis and D'este 2016; Merito and Bonaccorsi 2007), increasing the likelihood of successful adoption.

17.3.2 Barriers to Innovation Adoption

Majority of barriers to technology adoption relate to individual employee perceptions, including their past experience with technology (Gagnon et al. 2012; Koivunen and Saranto 2018; McGinn et al. 2011; Schreiweis et al. 2019), perceptions of the new technology's usability, the expected benefits of the innovation (system usefulness; Kruse et al. 2016), and ease of use (Gagnon et al. 2012). Other factors driving technology adoption decisions relate to individuals' motivation to use the technology (Rumball-Smith 2009), their ability to learn to use the technology (Robinson et al. 2014), and their level of trust toward the technology (Samad 2022).

Emotion-based elements can similarly hinder the adoption process, with the fear of technology being the most commonly cited negative emotion surrounding technology adoption. Clinical staff can be fearful of procedural changes at work and of the implications of technology on their professional identity (Koivunen and Saranto 2018; Kruse et al. 2016), including the possibility of aspects of their role being replaced by technology and their professional credibility undermined (Koivunen and Saranto 2018). The capacity to provide high-standard quality of care is a core value in healthcare professionals (Ko et al. 2018). Therefore, the introduction of new technology can give rise to fears of depersonalization of healthcare, and thereby fears of reduced quality of care (Gagnon et al. 2012; Koivunen and Saranto 2018; Lluch 2011; McGinn et al. 2011).

In addition to individual cognitive and emotional factors related to adoption, consideration should be placed on the hierarchical nature of healthcare and the group-level social dynamics. Healthcare delivery teams rely heavily on conformity through hierarchical decision-making processes to maintain performance and minimize risk (Hughes et al. 2016). Power dynamics across specialists–nurses, clinical–nonclinical staff, and unit-level managers–medical decision-makers likely influence the process of technology adoption. Rigid hierarchical structures within units and teams are known barriers to communication and collaboration (Baker et al. 2011) and likely hinder opportunities for peer support, which could otherwise assist the adoption process. Peer attitudes can also be a barrier to adoption, specifically in the case where an influential peer is resistant to adoption (Gagnon et al. 2012; Greenhalgh and Stones 2010; Greenhalgh et al. 2017; van Deen et al. 2019).

At the organizational system-level, communications from healthcare management about and support for technology use are crucial for sustainable uptake (Gagnon et al. 2012; Ingebrigtsen et al. 2014; Kruse et al. 2016; Schreiweis et al. 2019). Clear communications about the need for new technology and appropriate change management processes (Kotter 2007), in addition to workload management to enable appropriate levels of engagement with technology and training prior to implementation (Eastman and McCarthy 2012), clearly increase the likelihood of adoption.

In addition to these organization-level factors, barriers to technology adoption can be reduced if healthcare practitioners are involved, or in some cases, even drive, the technology development. In particular, practitioners in supervisory roles can play the part of change leaders: offering both instrumental (e.g. consistent and clear communication about the change (Ingebrigtsen et al. 2014)) and relational support (e.g. instilling a positive learning culture supportive of the new technology (Wong et al. 2021) to staff.

17.3.3 Ergonomics and Codesign

The development stage of new medical innovation is an opportunity to engage clinical staff, the intended end-users, to commit to and accept the new technology. This participatory, codesign process involves (i) developers gaining insight into the working lives of the end-users through accessing contextual knowledge, and (ii) a feedback loop between developers and end-users, creating opportunities for mutual learning and understanding of the usability of the innovation under development (Harte et al. 2018).

The codesign process relies on multiple information gathering methods, where insights are gained through observations and interactions with end-users (e.g. via ethnography, interviews). Mutual learning can occur through these informal interactions and more deliberately as part of the design process, such as hosting in-person informational workshops for end-users (Fonseka et al. 2020) and prototyping iterations of the innovation with end-users. Formal and informal interaction and information gathering can be triangulated with data on the usage of various technologies and processes, creating a richer and more nuanced perspective of activities relating to technology implementation. Basing the development of technology on a more comprehensive understanding of how it can work in the healthcare context will increase the likelihood that it is deemed relevant by the end-user.

Through participatory codesign, siloed thinking between technology developers and end-users is reduced (Barber et al. 2019). At its best, codesign involves boundaryless interaction between developer and users, resulting in the final technology being fit for purpose and able to meet the unique needs of the end-users and their healthcare context (O'Kane et al. 2021). When successful, cognitive and emotional barriers preventing successful adoption can be addressed and alleviated at the stage of development. Staff are empowered by their involvement (Desmond et al. 2018), which will allow for quicker buy-in during the implementation period.

Codesign is not without its challenges, however. In any participatory project, barriers can occur at the collaboration (e.g. power hierarchies; lack of psychological safety to raise concerns), organization (e.g. lack of staff training in using the new technology), process (e.g. lacking connection to other developments), dissemination (e.g. reliance on a few insiders for spreading co-created knowledge), and information collection stages (e.g. poor choice of methods (Pirinen 2016)). Further, developers and end-users are typically from different fields and have different expertise, and this diverse understanding of knowledge and values can hinder collaborative codesign (Reay et al. 2017).

Within the context of healthcare, tensions of codesign involve balancing the need for change versus resistance to change, focusing on patient-centered outcomes versus staff-centered needs, and balancing the value of innovation versus financial constraints (Cunningham and Reay 2019). A clear obstacle to successful codesign is resource constraints. Collaborating on technology development is often an extended, time-consuming process, potentially taking staff away from their core tasks, resulting in frustration and work overload. Investing in a sustainable participatory approach that allows for end-user engagement in design is likely to reduce multiple barriers to successful implementation, ultimately leading to the anticipated benefits of technology such as more accurate, objective, and equitable treatment approaches.

17.3.4 Summary (Key Takeaways)

Digital twins offer a unique opportunity to transform ICU care. However, this opportunity carries significant potential change in process and how care is given, along with change in the range and complexity of digital tools employed in care. They are also CPHSs. Both aspects require careful consideration of the social science aspects in both usability and ergonomics, as well as design and implementation directly addressing innovation adoption.

To successfully implement novel technology and achieve end-user uptake, a number of social-behavioral factors need to be considered at the individual, team, and enterprise levels of healthcare system. The existing research on barriers to technology adoption mainly focus on the individual employee perspective and has so far not managed to solve the problem. We propose that the solution lies in understanding and addressing the sociorelational aspects of the system, particularly those centered on the decision-makers at the enterprise level who enable technology adoption and uptake.

Similarly, success in technology adoption requires early engagement of end-users – the medical staff. Codesign principles can enable this and result in the innovation filling a gap in healthcare delivery needs. Yet, the success of codesign is also burdened with challenges, including resource constraints such as end-users' time to engage in the development process.

17.4 Future Research Challenges and Visions

17.4.1 Technology Vision of the Future of CPHS in ICU Care

The science and technology for all necessary elements of DTs in ICU care already exist. There are modeling hurdles to overcome, but those elements are already under development with some models and other elements already in standard of care use. Thus, the primary technological hurdle to a future with significant use of digital twins in care is one of incumbent barriers to entry in making medical devices interoperable, a nascent area in its own right (Hudson and Clark 2018; Jaleel et al. 2020; Mavrogiorgou et al. 2019; Williams 2017), where some have taken to creating their own interoperability (Lee et al. 2017; Lewis et al. 2016).

An interoperable future, enabling digital twins in care would create low cost, commoditized devices, most of which are relatively simple and which is a relatively new field itself (Hatcliff et al. 2019; Niezen et al. 2016; Payne et al. 2021; Pearce 2020; Read et al. 2021). These emerging devices would not only reduce costs but also separate hardware and the data in the device, including command and control. In essence, such a change would create the opportunity for digital twins and full automation where incumbent barriers make this very difficult currently. Figure 17.4 shows how unlocking these elements could create new opportunities for digital twins and CPHSs in healthcare, including new products and services to further optimize care and outcomes.

Importantly, this path leads to interoperability. Equally, separating the hardware and computational aspects of a device is not new or novel in general, although it is a significant disruption to medical devices today. In particular, the well-accepted "Christenson model" of innovative disruption (Christensen et al. 2013) envisaged innovations emerging as novel, or even higher cost, solutions for specific areas, which grow to provide dominant solutions in many fields holding to a given status quo solution.

In this case, digital twins emerging elsewhere may more significantly alter the status quo of ICU care delivery than other areas, which have already made significant productivity gains. The aspect of being wedded to a status quo solution to the detriment of productivity and innovation is thus particularly relevant to healthcare and its need to embrace novel solutions from other fields to enhance productivity and create social, as well as economic, change (Christensen et al. 2006), where personalized care is a core element of this disruption (Christensen et al. 2009; Christensen 2011).

A clear analogy arises in portable memory storage, where larger portable disk drives have given way memory sticks and portable solid-state drives, which in turn are yielding to cloud data storage. A similar virtualization of computational elements, moving hardware to the cloud, has further delinked computation, data, and the hardware or location upon which it occurs. Digital twins in manufacturing already make use of these virtualized solutions, and their arrival into healthcare should not be surprising.

Finally, writing on healthcare and innovation in social change, Christensen et al. wrote: "What's required is expanded support for organizations that are approaching social-sector problems in a fundamentally new way and creating scalable, sustainable, systems-changing solutions" (Christensen et al. 2006). This vision matches ours, where enhanced DT models and interoperable/open devices can create this sustainable change to fundamentally alter ICU care, and health care in general. In particular, where much of this computation and data are decentralized or cloud based (even if locally cloud based for security), there will be an increasing separation of hardware from software, data, and computation. Thus, these types of digital twin solutions will, in turn, fundamentally impact society and its equity of access to care and outcomes.

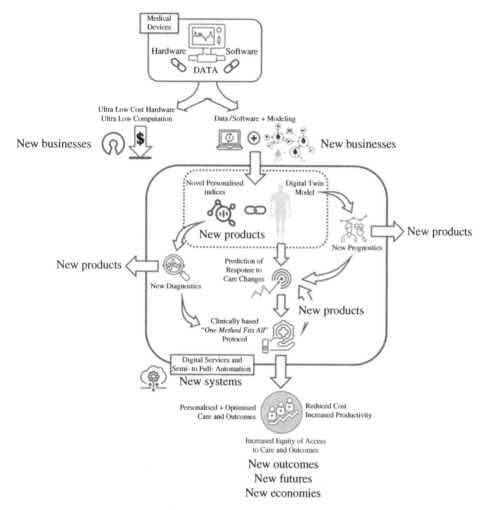

Figure 17.4 Technology vision of a possible future of CPHS in ICU care by separating the hardware of medical devices from the software, control, and data these devices also contain and link together within the device.

17.4.2 Social-Behavioral Sciences Vision of the Future of CPHS in ICU Care

Across development and implementation of healthcare innovation, the common underlying theme is that fostering relationships are required to ensure innovation uptake. Within the implementation stage, this can involve decision-makers engaging with managers, early adopters, and staff of influence. At the development stage, this can involve engaging clinical staff in the development of technology from the outset; i.e. decision-makers prioritizing developers' access to staff who will be the eventual users of the technology. Thereby, the future of CPHS in ICU care from a social-behavioral science perspective includes a commitment from the enterprise-level in the socialization of novel technology, such as DTs, into healthcare workplaces. Ultimately, the buy-in for CPHS needs to occur at the highest enterprise level among the decision-makers themselves.

We present a framework for supporting medical decision-makers in understanding and managing the social dynamics around technology adoption attitudes, which are the cognitive and

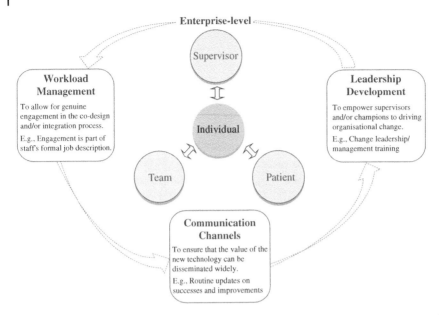

Figure 17.5 A framework around social feedback loops for medical decision-makers to drive innovation adoption in healthcare. Source: Adapted from Wong et al. (2021).

emotional inclinations to "accept, embrace, and adopt a particular plan to purposefully alter the status quo" (Holt et al. 2007, pp. 235). Considering workplace social dynamics as a social feedback loop between the individual and other groups of people in the work system, there are three key relationships within a healthcare context containing social feedback loops: clinical staff and (i) their team members; (ii) their supervisors; and (iii) their patients (Figure 17.5). Taken together, while an individual staff member may have a predisposed inclination toward technology (e.g. history with similar technology), attitudes are further constructed when they engage with or hear of team members, supervisors, and/or patient opinions of new technology.

Equally, without adoption, innovation cannot thrive. Within this framework, it is now possible to consider technology adoption as a dynamic temporal process which requires medical decision-makers' commitment over time. Further, introducing a new technology seldom is a concise event, the acceptance and adoption take place over time, and thus could be considered a manageable process, which can be optimized to maximize impact for patients, caregivers, and healthcare systems.

Therefore, we recommend three specific management areas for medical decision-makers to invest in to socialize technology adoption:

1. We suggest clinical staff workload should be managed at all times to enable genuine engagement in developing and integrating new technology in healthcare workplaces. We further recommend staff's availability to designers (to aid codesign of technology) and for training (to learn how to integrate new technology in the workplace) be part of their official job description that is prioritized.

2. We suggest formal and informal communications are set up across healthcare organization so information about the use and value of the new technology can be disseminated widely. We further recommend medical decision-makers encourage and reinforce feedback from clinical staff regarding the technology (e.g. follow through with suggestions for improvement and celebrate small adoption successes of individual or team champions).

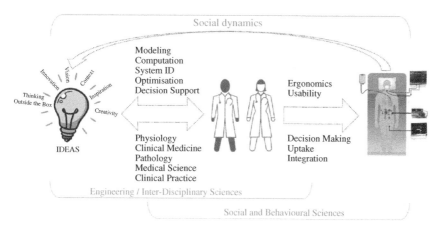

Figure 17.6 The overlapping roles of technical sciences and social and behavioral sciences in taking DT innovation from "science bench to clinical bedside."

3. We suggest clinical staff in positions of power and influence (e.g. supervisory role, champions) be trained and supported in managing organizational change, as technology adoption is a change in the workplace. We further recommend development opportunities focus on both the instrumental (e.g. routine communications about the new technology) and emotional aspects of change leadership (e.g. managing diverse attitudes regarding the new technology).

17.4.3   Joint Vision of the Future and Challenges to Overcome

Given this segregated discussion, there is a significant overlap between technical and social sciences in the creation and implementation, including adoption, of DTs in ICU and in healthcare in general. Figure 17.6 shows how these roles overlap in this timeline, including the impact of social dynamics over all stages of this process. In particular, the technical and scientific development of DTs for healthcare, though nascent, has overlooked the key role social sciences can play in making these innovations a clinical reality, including the role of social dynamics in all these interactions.

As noted, issues of clinical and patient uptake are addressed via social sciences engagement in the technology development. However, ethics have not been explicitly addressed in this framework. However, ethics is implicit in every step of the development and uptake process. In particular, the clinical trials to develop and prove these CPHS and digital twin systems require ethics approval from each center, each of which will reflect their local or national perspectives. Thus, these issues are directly addressed within the process.

Overall, Figure 17.6 summarizes the joint vision for DT innovation and its use to improve care and outcomes. This view envisions that codesign, development, and implementation will be able to optimize innovation outcomes at all stages of development. As a result, it envisions a tighter connection along the engineering–clinical–social science axis, which needs to work in as highly integrated a manner as possible to maximize the potential outcomes and benefits for patients, clinical staff, healthcare systems, and society.

17.5   Conclusions

Digital twins are emerging in manufacturing and for optimizing the use of advanced, complex systems, such as aircraft engines. These overtly technology-focused digital twins can be readily automated, and with the right digital twin model, controlled and optimized with little to no human intervention outside of setting goals for utilization. However, the first key outcome from this state-of-the-art analysis is translating digital twins to healthcare, and the ICU in particular, requires consideration of the human(s)-in-the-loop.

In particular, digital twins in the ICU are CPHSs. There is human input not only at the enterprise level to protocolized care, but also in terms of the patient, where modeling is more nonlinear and less certain than with purely physical, technological systems. Further, there is the clinical staff in the loop, doctors and nurses particularly, who use the digital twin in care, and have to work with it within a highly dynamic, highly variable work environment. These aspects force translation of the digital twin concept and approach to ICU care to significantly consider these human constraints and inputs.

Technologically, the key missing element to fully automate care is the digital twin model, where automating command and control signals from devices presents no specific technical hurdle. Thus, while device interoperability is significantly limited, open devices are emerging to enable digital twins in this space, and new business models can emerge where the data are separated from the hardware used to deliver care, creating new opportunities to improve care and healthcare economics.

Digital twin models are extremely limited. There are significant constraints on their complexity created by the need for real-time identification from limited data, and their need to mesh well within variable clinical workflows so they are able to be effectively integrated into the wider care of the patient. As a result, there is a significant lack of digital twin models in use, but, in contrast, significant opportunity to create and validate them to meet these constraints and emerging validation requirements.

Finally, this analysis recommends far greater codesign and overlap between the engineering sciences leading to digital twin research and modeling, and the clinical end-users. This codesign includes not just frameworks to not only enhancing adoption of what is a significant innovation in the delivery of healthcare but also creating the means to enable clinical staff to see this significant change in a positive manner and to further champion these changes in the clinical space. This codesign would significantly enable faster adoption and uptake of new ideas into ICU care to benefit patients and healthcare systems alike, and clearly highlight the ongoing CPHS aspects of digital twins in healthcare.

References

Afshar, N., Safaei, S., Nickerson, D.P. et al. (2019). Computational modeling of glucose uptake in the enterocyte. *Frontiers in Physiology* 10: 380.

Amato, M.B., Meade, M.O., Slutsky, A.S. et al. (2015). Driving pressure and survival in the acute respiratory distress syndrome. *New England Journal of Medicine* 372 (8): 747–755.

Anderson, J., O'Moore, K., Faraj, M., and Proudfoot, J. (2019). Stepped care mental health service in Australian primary care: codesign and feasibility study. *Australian Health Review* 44 (6): 873–879.

ANZICS (2020). Report on COVID-19 Admissions to Intensive Care in Victoria: 01 January 2020 to 31 August 2020. ANZICS Centre for Outcome and Resource Evaluation (CORE). Melbourne, VIC, Australia: Australian and New Zealand Intensive Care Society (ANZICS). www.anzics.com.au/annual-reports.

Artiga, S., Orgera, K., and Pham, O. (2020). *Disparities in Health and Health Care: Five Key Questions and Answers*. Kaiser Family Foundation.

Audoly, S., D'Angio, L., Saccomani, M.P., and Cobelli, C. (1998). Global identifiability of linear compartmental models – a computer algebra algorithm. *IEEE Transactions on Biomedical Engineering* 45 (1): 36–47.

Audoly, S., Bellu, G., D'Angio, L. et al. (2001). Global identifiability of nonlinear models of biological systems. *IEEE Transactions on Biomedical Engineering* 48 (1): 55–65.

Aziz, S., Arabi, Y.M., Alhazzani, W. et al. (2020). Managing ICU surge during the COVID-19 crisis: rapid guidelines. *Intensive Care Medicine* 46: 1303–1325.

Baker, L., Egan-Lee, E., Martimianakis, M.A., and Reeves, S. (2011). Relationships of power: implications for interprofessional education. *Journal of Interprofessional Care* 25 (2): 98–104.

Barber, T., Sharif, B., Teare, S. et al. (2019). Qualitative study to elicit patients' and primary care physicians' perspectives on the use of a self-management mobile health application for knee osteoarthritis. *BMJ Open* 9 (1): e024016.

Barrett, P.H., Bell, B.M., Cobelli, C. et al. (1998). SAAM II: simulation, analysis, and Modeling software for tracer and pharmacokinetic studies. *Metabolism* 47 (4): 484–492.

Bates, J.H.T. (2009). *The Linear Single-Compartment Model Lung Mechanics*. Cambridge University Press.

Bates, J.H. and Smith, B.J. (2018). Ventilator-induced lung injury and lung mechanics. *Annals of Translational Medicine* 6 (19): 30460252.

Bauchner, H. (2019). Rationing of health care in the United States: an inevitable consequence of increasing health care costs. *JAMA* 321 (8): 751–752.

Baumol, W.J. and De Ferranti, D.M. (2012). *The Cost Disease : Why Computers Get Cheaper and Health Care doesn't*. New Haven: Yale University Press. xxi, 249 p.

Beitler, J.R., Mittel, A.M., Kallet, R. et al. (2020). Ventilator sharing during an acute shortage caused by the COVID-19 pandemic. *American Journal of Respiratory and Critical Care Medicine* 202 (4): 600–604.

Bellu, G., Saccomani, M.P., Audoly, S., and D'Angio, L. (2007). DAISY: a new software tool to test global identifiability of biological and physiological systems. *Computer Methods and Programs in Biomedicine* 88 (1): 52–61.

Benatar, S., Sullivan, T., and Brown, A. (2018). Why equity in health and in access to health care are elusive: insights from Canada and South Africa. *Global Public Health* 13 (11): 1533–1557.

Ben-Tal, A. (2006). Simplified models for gas exchange in the human lungs. *Journal of Theoretical Biology* 238 (2): 474–495.

Benyo, B., Illyes, A., Nemedi, N.S. et al. (2012). Pilot study of the SPRINT glycemic control protocol in a Hungarian medical intensive care unit. *Journal of Diabetes Science and Technology* 6 (6): 1464–1477.

Bergman, R.N., Ider, Y.Z., Bowden, C.R., and Cobelli, C. (1979). Quantitative estimation of insulin sensitivity. *The American Journal of Physiology* 236 (6): E667–E677.

Bird, M., McGillion, M., Chambers, E. et al. (2021). A generative co-design framework for healthcare innovation: development and application of an end-user engagement framework. *Research Involvement and Engagement* 7 (1): 1–12.

Blaha, J., Hovorka, R., Matias, M. et al. (2005). Intensive insulin therapy in critically ill patients: comparison of standard and MPC protocols. *Intensive Care Medicine* 31 (S1): S203.

Blaha, J., Kopecky, P., Matias, M. et al. (2009). Comparison of three protocols for tight glycemic control in cardiac surgery patients. *Diabetes Care* 32 (5): 757–761.

Blaha, J., Barteczko-Grajek, B., Berezowicz, P. et al. (2016). Space GlucoseControl system for blood glucose control in intensive care patients – a European multicentre observational study. *BMC Anesthesiology* 16: 8.

Boiroux, D., Duun-Henriksen, A.K., Schmidt, S. et al. (2017). Adaptive control in an artificial pancreas for people with type 1 diabetes. *Control Engineering Practice* 58: 332–342.

Bosch, X., Moreno, P., and Lopez-Soto, A. (2014). The painful effects of the financial crisis on Spanish health care. *International Journal of Health Services* 44 (1): 25–51.

Bradley, C., Bowery, A., Britten, R. et al. (2011). OpenCMISS: a multi-physics & multi-scale computational infrastructure for the VPH/Physiome project. *Progress in Biophysics and Molecular Biology* 107 (1): 32–47.

Buijs, P., Catena, R., Holweg, M., and van der Vaart, T. (2021). Preventing disproportionate mortality in ICU overload situations: Empirical evidence from the first COVID-19 wave in Europe. medRxiv. https://doi.org/10.1101/2021.05.03.21255735 (accessed 2 March 2023).

Burrowes, K.S., Hunter, P.J., and Tawhai, M.H. (2005). Anatomically based finite element models of the human pulmonary arterial and venous trees including supernumerary vessels. *Journal of Applied Physiology* 99 (2): 731–738.

Burrowes, K., De Backer, J., Smallwood, R. et al. (2013). Multi-scale computational models of the airways to unravel the pathophysiological mechanisms in asthma and chronic obstructive pulmonary disease (AirPROM). *Interface Focus* 3 (2): 20120057.

Callegari, T., Caumo, A., and Cobelli, C. (2003). Bayesian two-compartment and classic single-compartment minimal models: comparison on insulin modified IVGTT and effect of experiment reduction. *IEEE Transactions on Biomedical Engineering* 50 (12): 1301–1309.

Camilo, L.M., Ávila, M.B., Cruz, L.F.S. et al. (2014). Positive end-expiratory pressure and variable ventilation in lung-healthy rats under general Anesthesia. *PLoS One* 9 (11): e110817.

Carayon, P. and Gurses, A. (2005). A human factors engineering conceptual framework of nursing workload and patient safety in intensive care units. *Intensive & Critical Care Nursing* 21 (5): 284–301.

Carayon, P., Wetterneck, T.B., Rivera-Rodriguez, A.J. et al. (2014). Human factors systems approach to healthcare quality and patient safety. *Applied Ergonomics* 45 (1): 14–25.

Carson, E.R. and Cobelli, C. (2001). *Modelling Methodology for Physiology and Medicine*, Biomedical Engineering, xiv–421. Academic Press.

Carvalho, A., Jandre, F., Pino, A. et al. (2007). Positive end-expiratory pressure at minimal respiratory elastance represents the best compromise between mechanical stress and lung aeration in oleic acid induced lung injury. *Critical Care* 11 (4): R86.

Carvalho, A., Spieth, P., Pelosi, P. et al. (2008). Ability of dynamic airway pressure curve profile and elastance for positive end-expiratory pressure titration. *Intensive Care Medicine* 34 (12): 2291–2299.

Carvalho, A.R., Bergamini, B.C., Carvalho, N.S. et al. (2013). Volume-independent elastance: a useful parameter for open-lung positive end-expiratory pressure adjustment. *Anesthesia & Analgesia* 116 (3): 627–633. https://doi.org/10.1213/ANE.0b013e31824a95ca.

Carville, O. (2013). *Busy ICU Spilling Patients*, *The Press*, 2. Christchurch: Fairfax NZ News.

Cecconi, M. and Rhodes, A. (2009). Validation of continuous cardiac output technologies: consensus still awaited. *Critical Care* 13 (3): 159.

Cecconi, M., De Backer, D., Antonelli, M. et al. (2014). Consensus on circulatory shock and hemodynamic monitoring. Task force of the European Society of Intensive Care Medicine. *Intensive Care Medicine* 40 (12): 1795–1815.

Chalfin, D., Cohen, I., and Lambrinos, J. (1995). The economics and cost-effectiveness of critical care medicine. *Intensive Care Medicine* 21 (11): 952–961.

Chapman, M.J., Godfrey, K.R., Chappell, M.J., and Evans, N.D. (2003). Structural identifiability for a class of non-linear compartmental systems using linear/non-linear splitting and symbolic computation. *Mathematical Biosciences* 183 (1): 1–14.

Chase, J., Shaw, G., Wong, X. et al. (2006). Model-based glycaemic control in critical care – a review of the state of the possible. *Biomedical Signal Processing and Control* 1 (1): 3–21.

Chase, J., Andreassen, S., Jensen, K., and Shaw, G. (2008a). The impact of human factors on clinical protocol performance – a proposed assessment framework and case examples. *Journal of Diabetes Science and Technology (JoDST)* 2 (3): 409–416.

Chase, J., LeCompte, A., Shaw, G. et al. (2008b). A benchmark data set for model-based glycemic control in critical care. *Journal of Diabetes Science and Technology (JoDST)* 24 (4): 584–594.

Chase, J.G., Suhaimi, F., Penning, S. et al. (2010). Validation of a model-based virtual trials method for tight glycemic control in intensive care. *Biomedical Engineering Online* 9: 84.

Chase, J.G., Le Compte, A.J., Preiser, J.C. et al. (2011a). Physiological modeling, tight glycemic control, and the ICU clinician: what are models and how can they affect practice? *Annals of Intensive Care* 1 (1): 11.

Chase, J.G., Le Compte, A.J., Suhaimi, F. et al. (2011b). Tight glycemic control in critical care – the leading role of insulin sensitivity and patient variability: a review and model-based analysis. *Computer Methods and Programs in Biomedicine* 102 (2): 156–171.

Chase, J., Desaive, T., and Preiser, J.C. (2016). Virtual patients and virtual cohorts: a new way to think about the design and implementation of personalized ICU treatments. In: *Annual Update in Intensive Care and Emergency Medicine* (ed. J.L. Vincent), 435–448. Springer.

Chase, J.G., Desaive, T., Bohe, J. et al. (2018a). Improving glycemic control in critically ill patients: personalized care to mimic the endocrine pancreas. *Critical Care* 22 (1): 182.

Chase, J.G., Preiser, J.-C., Dickson, J.L. et al. (2018b). Next-generation, personalised, model-based critical care medicine: a state-of-the art review of in silico virtual patient models, methods, and cohorts, and how to validation them. *Biomedical Engineering Online* 17 (1): 1–29.

Chase, J.G., Preiser, J.C., Dickson, J.L. et al. (2018c). Next-generation, personalised, model-based critical care medicine: a state-of-the art review of in silico virtual patient models, methods, and cohorts, and how to validation them. *Biomedical Engineering Online* 17 (1): 24.

Chase, J.G., Benyo, B., and Desaive, T. (2019). Glycemic control in the intensive care unit: a control systems perspective. *Annual Reviews in Control* 48: 359–368.

Chase, J.G., Chiew, Y.-S., Lambermont, B. et al. (2020a). In-parallel ventilator sharing during an acute shortage: too much risk for a wider uptake. *American Journal of Respiratory and Critical Care Medicine* 202 (9): 1316–1317.

Chase, J.G., Chiew, Y.S., Lambermont, B. et al. (2020b). Safe doubling of ventilator capacity: a last resort proposal for last resorts. *Critical Care* 24 (222): 1–4.

Chen, D. (2005). Enterprise-control system integration – an international standard. *International Journal of Production Research* 43 (20): 4335–4357.

Chiew, Y.S., Pretty, C., Docherty, P.D. et al. (2015). Time-varying respiratory system elastance: a physiological model for patients who are spontaneously breathing. *PLoS One* 10 (1): e0114847.

Christensen, C. (2011). A disruptive solution for health care. In: *Harvard Business Review*, 496. Harvard University Press.

Christensen, C.M., Baumann, H., Ruggles, R., and Sadtler, T.M. (2006). Disruptive innovation for social change. *Harvard Business Review* 84 (12): 94.

Christensen, C., Grossman, J., and Hwang, J. (2009). *The Innovator's Prescription: A Disruptive Solution for Healthcare*. USA: McGraw-Hill Education.

Christensen, C., Raynor, M.E., and McDonald, R. (2013). *Disruptive Innovation*. Harvard Business Review.

Cimino, C., Negri, E., and Fumagalli, L. (2019). Review of digital twin applications in manufacturing. *Computers in Industry* 113: 103130.

Cobelli, C., Carson, E.R., Finkelstein, L., and Leaning, M.S. (1984). Validation of simple and complex models in physiology and medicine. *The American Journal of Physiology* 246 (2 Pt 2): R259–R266.

Cobelli, C., Pacini, G., Toffolo, G., and Sacca, L. (1986). Estimation of insulin sensitivity and glucose clearance from minimal model: new insights from labeled IVGTT. *The American Journal of Physiology* 250 (5 Pt 1): E591–E598.

Cochran, S., Miller, E., and Dunn, K. (2006). EndoTool software for tight glucose control for critically ill patients. *Critical Care Medicine* 34 (Suppl 2): A68.

Cohen, M.J. (2012). Use of models in identification and prediction of physiology in critically ill surgical patients. *The British Journal of Surgery* 99 (4): 487–493.

Cohen, I.G., Crespo, A.M., and White, D.B. (2020). Potential legal liability for withdrawing or withholding ventilators during COVID-19: assessing the risks and identifying needed reforms. *JAMA* 323 (19): 1901–1902.

Cooling, M.T. and Hunter, P. (2015). The CellML metadata framework 2.0 specification. *Journal of Integrative Bioinformatics* 12 (2): 260.

Craxì, L., Vergano, M., Savulescu, J., and Wilkinson, D. (2020). Rationing in a pandemic: lessons from Italy. *Asian Bioethics Review* 12 (3): 325–330.

Cunningham, H. and Reay, S. (2019). Co-creating design for health in a city hospital: perceptions of value, opportunity and limitations from 'designing together' symposium. *Design for Health* 3 (1): 119–134.

Cutrone, A., De Maria, C., Vinci, B. et al. (2009). A new library of HEMET model: insulin effects on hepatic metabolism. *Computer Methods and Programs in Biomedicine* 94 (2): 181–189.

Dalla Man, C., Caumo, A., and Cobelli, C. (2002). The oral glucose minimal model: estimation of insulin sensitivity from a meal test. *IEEE Transactions on Biomedical Engineering* 49 (5): 419–429.

Dalla Man, C., Yarasheski, K.E., Caumo, A. et al. (2005). Insulin sensitivity by oral glucose minimal models: validation against clamp. *American Journal of Physiology. Endocrinology and Metabolism* 289 (6): E954–E959.

Dalla Man, C., Camilleri, M., and Cobelli, C. (2006). A system model of oral glucose absorption: validation on gold standard data. *IEEE Transactions on Biomedical Engineering* 53 (12 Pt 1): 2472–2478.

Dalla Man, C., Micheletto, F., Lv, D. et al. (2014). The UVA/PADOVA type 1 diabetes simulator: new features. *Journal of Diabetes Science and Technology* 8 (1): 26–34.

van Deen, W.K., Cho, E.S., Pustolski, K. et al. (2019). Involving end-users in the design of an audit and feedback intervention in the emergency department setting – a mixed methods study. *BMC Health Services Research* 19 (1): 1–13.

Desaive, T., Lambermont, B., Janssen, N. et al. (2013). Assessment of ventricular contractility and ventricular-arterial coupling with a model-based sensor. *Computer Methods and Programs in Biomedicine* 109 (2): 182–189.

Desaive, T., Horikawa, O., Ortiz, J.P., and Chase, J.G. (2019). Model-based management of cardiovascular failure: where medicine and control systems converge. *Annual Reviews in Control* 48: 383–391.

Desmond, D., Layton, N., Bentley, J. et al. (2018). Assistive technology and people: a position paper from the first global research, innovation and education on assistive technology (GREAT) summit. *Disability and Rehabilitation. Assistive Technology* 13 (5): 437–444.

Dickson, J.L., Chase, J.G., Pretty, C.G. et al. (2015). Hyperglycaemic preterm babies have sex differences in insulin secretion. *Neonatology* 108 (2): 93–98.

Dickson, J.L., Pretty, C.G., Alsweiler, J. et al. (2016). Insulin kinetics and the neonatal intensive care insulin-nutrition-glucose (NICING) model. *Mathematical Biosciences* 284: 61–70.

Dickson, J.L., Stewart, K.W., Pretty, C.G. et al. (2018). Generalisability of a virtual trials method for glycaemic control in intensive care. *IEEE Transactions on Biomedical Engineering* 65 (7): 1543–1553.

Docherty, P.D., Chase, J.G., Lotz, T.F., and Desaive, T. (2011a). A graphical method for practical and informative identifiability analyses of physiological models: a case study of insulin kinetics and sensitivity. *Biomedical Engineering Online* 10 (1): 1–20.

Docherty, P.D., Chase, J.G., Morenga, L. et al. (2011b). A spectrum of dynamic insulin sensitivity test protocols. *Journal of Diabetes Science and Technology* 5 (6): 1499.

Dombovy, M.L. (2002). U.S. health care in conflict – Part I. The challenges of balancing cost, quality and access. *Physician Executive* 28 (4): 43–47.

Dominiczak, J. and Khansa, L. (2018). Principles of automation for patient safety in intensive care: learning from aviation. *The Joint Commission Journal on Quality and Patient Safety* 44 (6): 366–371.

Dong, Y., Chbat, N.W., Gupta, A. et al. (2012). Systems modeling and simulation applications for critical care medicine. *Annals of Intensive Care* 2 (1): 18.

van Drunen, E., Chiew, Y.S., Chase, J. et al. (2013). Expiratory model-based method to monitor ARDS disease state. *Biomedical Engineering Online* 12 (1): 57.

van Drunen, E., Chiew, Y.S., Pretty, C. et al. (2014). Visualisation of time-varying respiratory system elastance in experimental ARDS animal models. *BMC Pulmonary Medicine* 14 (1): 33.

Du, Y., Liu, C., Li, J. et al. (2020). Glycemic variability: an independent predictor of mortality and the impact of age in pediatric intensive care unit. *Frontiers in Pediatrics* 8: 403.

von Düring, S., Primmaz, S., and Bendjelid, K. (2020). COVID-19: desperate times call for desperate measures. *Critical Care* 24 (1): 1–2.

Eastman, D. and McCarthy, C. (2012). Embracing change: healthcare technology in the 21st century. *The Journal of Excellence in Nursing Leadership* 43 (6): 52–54.

Economist, T (2011). Patient, heal thyself. In: *The Economist*, 2. London, UK: The Economist.

Economist, T (2016). *Where the smart is*. In: *The Economist*. London, UK: The Economist Newspaper Limited.

Economist, T (2020). *The dawn of digital medicine*. In: *The Economist*. London, UK: The Economist Newspaper Limited.

Economist, T (2021). Governments are not going to stop getting bigger. In: *The Economist*, vol. 7. London, UK: The Economist Newspaper Limited.

Egi, M., Bellomo, R., Stachowski, E. et al. (2006). Variability of blood glucose concentration and short-term mortality in critically ill patients. *Anesthesiology* 105 (2): 244–252.

Essue, B.M., Laba, M., Knaul, F. et al. (2018). Economic burden of chronic Ill health and injuries for households in low- and middle-income countries. In: *Disease Control Priorities: Improving Health and Reducing Poverty*, rd, et al., Editors, 121–143. Washington (DC), https://opus.lib.uts.edu.au/ bitstream/10453/135163/1/TL_DCP3%20Volume%209_Ch%206.pdf.

Evans, A., Shaw, G.M., Le Compte, A. et al. (2011). Pilot proof of concept clinical trials of stochastic targeted (STAR) glycemic control. *Annals of Intensive Care* 1 (1): 38.

Evans, A., Le Compte, A., Tan, C.S. et al. (2012). Stochastic targeted (STAR) glycemic control: design, safety, and performance. *Journal of Diabetes Science and Technology* 6 (1): 102–115.

van Exel, J., Baker, R., Mason, H. et al. (2015). Public views on principles for health care priority setting: findings of a European cross-country study using Q methodology. *Social Science & Medicine* 126: 128–137.

Fisk, L., Lecompte, A., Penning, S. et al. (2012). STAR development and protocol comparison. *IEEE Transactions on Biomedical Engineering* 59 (12): 3357–3364.

Fonseka, T.M., Pong, J.T., Kcomt, A. et al. (2020). Collaborating with individuals with lived experience to adapt CANMAT clinical depression guidelines into a patient treatment guide: the CHOICE-D co-design process. *Journal of Evaluation in Clinical Practice* 26 (4): 1259–1269.

Gagnon, M.-P., Desmartis, M., Labrecque, M. et al. (2012). Systematic review of factors influencing the adoption of information and communication technologies by healthcare professionals. *Journal of Medical Systems* 36 (1): 241–277.

Garcia-Estevez, D.A., Araujo-Vilar, D., Fiestras-Janeiro, G. et al. (2003). Comparison of several insulin sensitivity indices derived from basal plasma insulin and glucose levels with minimal model indices. *Hormone and Metabolic Research* 35 (1): 13–17.

Garg, A.X., Adhikari, N.K., McDonald, H. et al. (2005). Effects of computerized clinical decision support systems on practitioner performance and patient outcomes: a systematic review. *JAMA* 293 (10): 1223–1238.

Garny, A. and Hunter, P.J. (2015). OpenCOR: a modular and interoperable approach to computational biology. *Frontiers in Physiology* 6: 26.

Gatouillat, A., Badr, Y., Massot, B., and Sejdić, E. (2018). Internet of medical things: a review of recent contributions dealing with cyber-physical systems in medicine. *IEEE Internet of Things Journal* 5 (5): 3810–3822.

Glorioso, V. and Subramanian, S.V. (2014). Equity in access to health care services in Italy. *Health Services Research* 49 (3): 950–970.

Goligher, E.C., Costa, E.L., Yarnell, C.J. et al. (2021). Effect of lowering tidal volume on mortality in ARDS varies with respiratory system elastance. *American Journal of Respiratory and Critical Care Medicine* 203 (11): 1378–1385.

Greenhalgh, T. and Stones, R. (2010). Theorising big IT programmes in healthcare: strong structuration theory meets actor-network theory. *Social Science & Medicine* 70 (9): 1285–1294.

Greenhalgh, T., Wherton, J., Papoutsi, C. et al. (2017). Beyond adoption: a new framework for theorizing and evaluating nonadoption, abandonment, and challenges to the scale-up, spread, and sustainability of health and care technologies. *Journal of Medical Internet Research* 19 (11): e367.

Grol, R. (2001). Successes and failures in the implementation of evidence-based guidelines for clinical practice. *Medical Care* 39 (8 Suppl 2): II46–II54.

Gross, T.M., Kayne, D., King, A. et al. (2003). A bolus calculator is an effective means of controlling postprandial glycemia in patients on insulin pump therapy. *Diabetes Technology & Therapeutics* 5 (3): 365–369.

Gulliford, M., Figueroa-Munoz, J., Morgan, M. et al. (2002). What does' access to health care'mean? *Journal of Health Services Research & Policy* 7 (3): 186–188.

Gurkan, D. and Merchant, F. (2010). Interoperable medical instrument networking and access system with security considerations for critical care. *Journal of Healthcare Engineering* 1 (4): 637–654.

Hagiwara, N., Lafata, J.E., Mezuk, B. et al. (2019). Detecting implicit racial bias in provider communication behaviors to reduce disparities in healthcare: challenges, solutions, and future directions for provider communication training. *Patient Education and Counseling* 102 (9): 1738–1743.

Halpern, N.A. (2009). Can the costs of critical care be controlled? *Current Opinion in Critical Care* 15 (6): 591–596.

Halpern, S.D. (2011). ICU capacity strain and the quality and allocation of critical care. *Current Opinion in Critical Care* 17 (6): 648–657.

Halpern, N.A., Pastores, S.M., and Greenstein, R.J. (2004). Critical care medicine in the United States 1985–2000: an analysis of bed numbers, use, and costs. *Critical Care Medicine* 32 (6): 1254–1259.

Hann, C.E., Chase, J.G., Lin, J. et al. (2005). Integral-based parameter identification for long-term dynamic verification of a glucose-insulin system model. *Computer Methods and Programs in Biomedicine* 77 (3): 259–270.

Harte, R., Quinlan, L.R., Andrade, E. et al. (2018). Defining user needs for a new sepsis risk decision support system in neonatal ICU settings through ethnography: user interviews and participatory design. In: *Human Systems Engineering and Design: Proceedings of the First International Conference on Human Systems Engineering and Design (IHSED2018): Future Trends and Applications, 25–27*

October 2018, CHU-Université de Reims Champagne-Ardenne, France 1 2019, 221–227. Springer International Publishing.

Hatcliff, J., Larson, B., Carpenter, T. et al. (2019). The open PCA pump project: an exemplar open source medical device as a community resource. *ACM SIGBED Review* 16 (2): 8–13.

Holder-Pearson, L., Lerios, T., and Chase, J.G. (2021). Physiologic-range three/two-way valve for respiratory circuits. *HardwareX* 10: e00234.

Holt, D.T., Armenakis, A.A., Feild, H.S., and Harris, S.G. (2007). Readiness for organizational change: the systematic development of a scale. *The Journal of Applied Behavioral Science* 43 (2): 232–255.

Hovorka, R., Shojaee-Moradie, F., Carroll, P.V. et al. (2002). Partitioning glucose distribution/transport, disposal, and endogenous production during IVGTT. *American Journal of Physiology. Endocrinology and Metabolism* 282 (5): E992–E1007.

Hovorka, R., Chassin, L.J., and Wilinska, M.E. (2003). Virtual type 1 diabetic treated by CSII: Model description. WC2003. Sydney, Australia.

Hovorka, R., Chassin, L.J., Ellmerer, M. et al. (2008). A simulation model of glucose regulation in the critically ill. *Physiological Measurement* 29 (8): 959–978.

Hudson, F. and Clark, C. (2018). Wearables and medical interoperability: the evolving frontier. *Computer* 51 (9): 86–90.

Hughes, A.M., Gregory, M.E., Joseph, D.L. et al. (2016). Saving lives: a meta-analysis of team training in healthcare. *Journal of Applied Psychology* 101 (9): 1266–1304.

Hunter, P. (2016). The virtual physiological human: the physiome project aims to develop reproducible, multiscale models for clinical practice. *IEEE Pulse* 7 (4): 36–42.

Hunter, P.J., Nielsen, P.M., Smaill, B.H. et al. (1992). An anatomical heart model with applications to myocardial activation and ventricular mechanics. *Critical Reviews in Biomedical Engineering* 20 (5–6): 403–426.

Hunter, P., Smith, N., Fernandez, J., and Tawhai, M. (2005). Integration from proteins to organs: the IUPS physiome project. *Mechanisms of Ageing and Development* 126 (1): 187–192.

Hunter, P.J., Crampin, E.J., and Nielsen, P.M. (2008). Bioinformatics, multiscale modeling and the IUPS Physiome Project. *Briefings in Bioinformatics* 9 (4): 333–343.

Hunter, P., Coveney, P.V., de Bono, B. et al. (2010). A vision and strategy for the virtual physiological human in 2010 and beyond. *Philosophical Transactions. Series A, Mathematical, Physical, and Engineering Sciences* 368 (1920): 2595–2614.

Hunter, P., Chapman, T., Coveney, P.V. et al. (2013). A vision and strategy for the virtual physiological human: 2012 update. *Interface Focus* 3 (2): 20130004.

Ingebrigtsen, T., Georgiou, A., Clay-Williams, R. et al. (2014). The impact of clinical leadership on health information technology adoption: systematic review. *International Journal of Medical Informatics* 83 (6): 393–405.

Jaleel, A., Mahmood, T., Hassan, M.A. et al. (2020). Towards medical data interoperability through collaboration of healthcare devices. *IEEE Access* 8: 132302–132319.

Jansen, J.R., Maas, J.J., and Pinsky, M.R. (2010). Bedside assessment of mean systemic filling pressure. *Current Opinion in Critical Care* 16 (3): 231–236.

de Jongh, F.H., de Vries, H.J., Warnaar, R.S. et al. (2020). Ventilating two patients with one ventilator: technical setup and laboratory testing. *ERJ Open Research* 6 (2).

Joseph, R., Lee, S.W., Anderson, S.V., and Morrisette, M.J. (2020). Impact of interoperability of smart infusion pumps and an electronic medical record in critical care. *American Journal of Health-System Pharmacy* 77 (15): 1231–1236.

Joyia, G.J., Liaqat, R.M., Farooq, A., and Rehman, S. (2017). Internet of medical things (IoMT): applications, benefits and future challenges in healthcare domain. *The Journal of Communication* 12 (4): 240–247.

Kamoi, S., Pretty, C., Docherty, P. et al. (2014). Continuous stroke volume estimation from aortic pressure using zero dimensional cardiovascular model: proof of concept study from porcine experiments. *PLoS One* 9 (7): e102476.

Keener, J.P. and Sneyd, J. (1998). Mathematical physiology. In: *Interdisciplinary Applied Mathematics*, vol. 8. New York: Springer. viii, 766 p.

Keliddar, I., Mosadeghrad, A.M., and Jafari–Sirizi, M. (2017). Rationing in health systems: a critical review. *Medical Journal of the Islamic Republic of Iran* 31: 47.

Kelley, M.A., Angus, D.C., Chalfin, D.B. et al. (2004). The critical care crisis in the United States: a report from the profession. *Critical Care Medicine* 32 (5): 1219–1222.

Kim, K.T., Knopp, J., Dixon, B., and Chase, G. (2019). Quantifying neonatal pulmonary mechanics in mechanical ventilation. *Biomedical Signal Processing and Control* 52: 206–217.

Kim, K.T., Knopp, J., Dixon, B., and Chase, J.G. (2020a). Mechanically ventilated premature babies have sex differences in specific elastance: a pilot study. *Pediatric Pulmonology* 55 (1): 177–184.

Kim, K.T., Morton, S., Howe, S. et al. (2020b). Model-based PEEP titration versus standard practice in mechanical ventilation: a randomised controlled trial. *Trials* 21 (1): 130.

Knopp, J.L., Lynn, A.M., Shaw, G.M., and Chase, J.G. (2019). Safe and effective glycaemic control in premature infants: observational clinical results from the computerised STAR-GRYPHON protocol. *Archives of Disease in Childhood. Fetal and Neonatal Edition* 104 (2): F205–F211.

Knopp, J.L., Chase, J.G., Kim, K.T., and Shaw, G.M. (2021). Model-based estimation of negative inspiratory driving pressure in patients receiving invasive NAVA mechanical ventilation. *Computer Methods and Programs in Biomedicine* 208: 106300.

Ko, M., Wagner, L., and Spetz, J. (2018). Nursing home implementation of health information technology: review of the literature finds inadequate investment in preparation, infrastructure, and training. *INQUIRY: The Journal of Health Care Organization, Provision, and Financing* 55: 0046958018778902.

Koivunen, M. and Saranto, K. (2018). Nursing professionals' experiences of the facilitators and barriers to the use of telehealth applications: a systematic review of qualitative studies. *Scandinavian Journal of Caring Sciences* 32 (1): 24–44.

Kotter, J.P. (2007). Leading change: Why transformation efforts fail. In: *Museum Management and Marketing*, 20–29. Routledge.

Kovatchev, B., Cheng, P., Anderson, S.M. et al. (2017). Feasibility of long-term closed-loop control: a multicenter 6-month trial of 24/7 automated insulin delivery. *Diabetes Technology & Therapeutics* 19 (1): 18–24.

Kritzinger, W., Karner, M., Traar, G. et al. (2018). Digital twin in manufacturing: a categorical literature review and classification. *IFAC-PapersOnLine* 51 (11): 1016–1022.

Kruse, C.S., Kristof, C., Jones, B. et al. (2016). Barriers to electronic health record adoption: a systematic literature review. *Journal of Medical Systems* 40 (12): 1–7.

Lambermont, B., Ghuysen, A., Janssen, N. et al. (2008). Comparison of functional residual capacity and static compliance of the respiratory system during a positive end-expiratory pressure (PEEP) ramp procedure in an experimental model of acute respiratory distress syndrome. *Critical Care* 12 (4): R91.

Lambermont, B., Rousseau, A.-F., Seidel, L. et al. (2021). Outcome improvement between the first two waves of the coronavirus disease 2019 pandemic in a single tertiary-care hospital in Belgium. *Critical Care Explorations* 3 (5).

Langouche, L., Vander Perre, S., Wouters, P.J. et al. (2007). Effect of intensive insulin therapy on insulin sensitivity in the critically ill. *The Journal of Clinical Endocrinology and Metabolism* 92 (10): 3890–3897.

Larraza, S., Dey, N., Karbing, D.S. et al. (2015). A mathematical model approach quantifying patients' response to changes in mechanical ventilation: evaluation in pressure support. *Journal of Critical Care* 30 (5): 1008–1015.

Laurance, J., Henderson, S., Howitt, P.J. et al. (2014). Patient engagement: four case studies that highlight the potential for improved health outcomes and reduced costs. *Health Affairs* 33 (9): 1627–1634.

Le Compte, A., Chase, J., Lynn, A. et al. (2009). Blood glucose controller for neonatal intensive care: virtual trials development and 1st clinical trials. *Journal of Diabetes Science and Technology (JoDST)* 3 (5): 1066–1081.

Le Compte, A.J., Lee, D.S., Chase, J.G. et al. (2010). Blood glucose prediction using stochastic modeling in neonatal intensive care. *IEEE Transactions on Biomedical Engineering* 57 (3): 509–518.

Le Compte, A.J., Chase, J.G., Lynn, A. et al. (2011). Development of blood glucose control for extremely premature infants. *Computer Methods and Programs in Biomedicine* 102 (2): 181–191.

Le Compte, A.J., Lynn, A.M., Lin, J. et al. (2012). Pilot study of a model-based approach to blood glucose control in very-low-birthweight neonates. *BMC Pediatrics* 12: 117.

Lee, J.M., Newman, M.W., Gebremariam, A. et al. (2017). Real-world use and self-reported health outcomes of a patient-designed do-it-yourself mobile technology system for diabetes: lessons for mobile health. *Diabetes Technology & Therapeutics* 19 (4): 209–219.

Lee, J.W.W., Chiew, Y.S., Wang, X. et al. (2021). Stochastic modelling of respiratory system elastance for mechanically ventilated respiratory failure patients. *Annals of Biomedical Engineering* 49: 1–16.

Levine, D. and Mulligan, J. (2015). Overutilization, overutilized. *Journal of Health Politics, Policy and Law* 40 (2): 421–437.

Lewis, D., Leibrand, S., and Open, A.P.S.C. (2016). Real-world use of open source artificial pancreas systems. *Journal of Diabetes Science and Technology* 10 (6): 1411.

Lin, J., Lee, D., Chase, J.G. et al. (2006). Stochastic modelling of insulin sensitivity variability in critical care. *Biomedical Signal Processing and Control* 1 (3): 229–242.

Lin, J., Lee, D., Chase, J.G. et al. (2008). Stochastic modelling of insulin sensitivity and adaptive glycemic control for critical care. *Computer Methods and Programs in Biomedicine* 89 (2): 141–152.

Lin, J., Parente, J.D., Chase, J.G. et al. (2011a). Development of a model-based clinical sepsis biomarker for critically ill patients. *Computer Methods and Programs in Biomedicine* 102 (2): 149–155.

Lin, J., Razak, N.N., Pretty, C.G. et al. (2011b). A physiological intensive control insulin-nutrition-glucose (ICING) model validated in critically ill patients. *Computer Methods and Programs in Biomedicine* 102 (2): 192–205.

Ljung, L. and Glad, T. (1994). On global identifiability for arbitrary model parametrizations. *Automatica* 30 (2): 265–276.

Llopis, O. and D'este, P. (2016). Beneficiary contact and innovation: the relation between contact with patients and medical innovation under different institutional logics. *Research Policy* 45 (8): 1512–1523.

Lluch, M. (2011). Healthcare professionals' organisational barriers to health information technologies – a literature review. *International Journal of Medical Informatics* 80 (12): 849–862.

Lonergan, T., LeCompte, A., Willacy, M. et al. (2006). A simple insulin-nutrition protocol for tight glycemic control in critical illness: development and protocol comparison. *Diabetes Technology & Therapeutics* 8 (2): 191–206.

Maas, J.J., Geerts, B.F., van den Berg, P.C. et al. (2009). Assessment of venous return curve and mean systemic filling pressure in postoperative cardiac surgery patients. *Critical Care Medicine* 37 (3): 912–918.

Maas, J.J., Pinsky, M.R., Aarts, L.P., and Jansen, J.R. (2012). Bedside assessment of total systemic vascular compliance, stressed volume, and cardiac function curves in intensive care unit patients. *Anesthesia and Analgesia* 115 (4): 880–887.

Maina, I.W., Belton, T.D., Ginzberg, S. et al. (2018). A decade of studying implicit racial/ethnic bias in healthcare providers using the implicit association test. *Social Science & Medicine* 199: 219–229.

Major, V.J., Chiew, Y.S., Shaw, G.M., and Chase, J.G. (2018). Biomedical engineer's guide to the clinical aspects of intensive care mechanical ventilation. *Biomedical Engineering Online* 17 (1): 169.

Man, C.D., Breton, M.D., and Cobelli, C. (2009). Physical activity into the meal glucose-insulin model of type 1 diabetes: in silico studies. *Journal of Diabetes Science and Technology* 3 (1): 56–67.

Mari, A., Pacini, G., Brazzale, A.R., and Ahren, B. (2005). Comparative evaluation of simple insulin sensitivity methods based on the oral glucose tolerance test. *Diabetologia* 48 (4): 748–751.

Mavrogiorgou, A., Kiourtis, A., Perakis, K. et al. (2019). IoT in healthcare: achieving interoperability of high-quality data acquired by IoT medical devices. *Sensors* 19 (9): 1978.

McGinn, C.A., Grenier, S., Duplantie, J. et al. (2011). Comparison of user groups' perspectives of barriers and facilitators to implementing electronic health records: a systematic review. *BMC Medicine* 9 (1): 1–10.

Merito, M. and Bonaccorsi, A. (2007). Co-evolution of physical and social technologies in clinical practice: the case of HIV treatments. *Research Policy* 36 (7): 1070–1087.

Mesotten, D., Dubois, J., Van Herpe, T. et al. (2017). Software-guided versus nurse-directed blood glucose control in critically ill patients: the LOGIC-2 multicenter randomized controlled clinical trial. *Critical Care* 21 (1): 212.

Micklethwait, J. (2011). Taming Leviathan. In: *The Economist*, 6. London, UK: The Economist.

Morris, A.H., Orme, J. Jr., Truwit, J.D. et al. (2008). A replicable method for blood glucose control in critically ill patients. *Critical Care Medicine* 36 (6): 1787–1795.

Morris, Z.S., Wooding, S., and Grant, J. (2011). The answer is 17 years, what is the question: understanding time lags in translational research. *Journal of the Royal Society of Medicine* 104 (12): 510–520.

Morton, S.E., Dickson, J., Chase, J.G. et al. (2018). A virtual patient model for mechanical ventilation. *Computer Methods and Programs in Biomedicine* 165: 77–87.

Morton, S.E., Knopp, J.L., Chase, J.G. et al. (2019a). Optimising mechanical ventilation through model-based methods and automation. *Annual Reviews in Control* 48: 369–382.

Morton, S.E., Knopp, J.L., Chase, J.G. et al. (2019b). Predictive virtual patient modelling of mechanical ventilation: impact of recruitment function. *Annals of Biomedical Engineering* 47 (7): 1626–1641.

Morton, S.E., Knopp, J.L., Tawhai, M.H. et al. (2020). Prediction of lung mechanics throughout recruitment maneuvers in pressure-controlled ventilation. *Computer Methods and Programs in Biomedicine* 197: 105696.

Murphy, L., Chase, J., Davidson, S. et al. (2020a). Minimally invasive model based stressed blood volume as an index of fluid responsiveness. *IFAC Papers-Online* 53 (2): 16257–16262.

Murphy, L., Davidson, S., Chase, J.G. et al. (2020b). Patient-specific monitoring and trend analysis of model-based markers of fluid responsiveness in sepsis: a proof-of-concept animal study. *Annals of Biomedical Engineering* 48 (2): 682–694.

Negri, E., Fumagalli, L., and Macchi, M. (2017). A review of the roles of digital twin in CPS-based production systems. *Procedia Manufacturing* 11: 939–948.

Neilson (NZ Herald), M. (2018). Waitangi Tribunal hearings over 'inequity and institutionalised racism' in health system. NZ Herald [cited 2018 November 5, 2018]. www.nzherald.co.nz/nz/news/article.cfm?c_id=1&objectid=12141442 (accessed 5 November 2018).

Nelson, G.S. (2019). Bias in artificial intelligence. *North Carolina Medical Journal* 80 (4): 220–222.

Ng, Q.A., Chiew, Y.S., Wang, X. et al. (2021). Network data acquisition and monitoring system for intensive care mechanical ventilation treatment. IEEE Access. 9: 91859–91873.

Nickerson, D.P., Ladd, D., Hussan, J.R. et al. (2014). Using CellML with OpenCMISS to simulate multi-scale physiology. *Frontiers in Bioengineering and Biotechnology* 2: 79.

Nickerson, D., Atalag, K., de Bono, B. et al. (2016). The human physiome: how standards, software and innovative service infrastructures are providing the building blocks to make it achievable. *Interface Focus* 6 (2): 20150103.

Niezen, G., Eslambolchilar, P., and Thimbleby, H. (2016). Open-source hardware for medical devices. *BMJ Innovations* 2 (2).

Nouira, S., Elatrous, S., Dimassi, S. et al. (2005). Effects of norepinephrine on static and dynamic preload indicators in experimental hemorrhagic shock. *Critical Care Medicine* 33 (10): 2339–2343.

Obermeyer, Z., Powers, B., Vogeli, C., and Mullainathan, S. (2019). Dissecting racial bias in an algorithm used to manage the health of populations. *Science* 366 (6464): 447–453.

OECD (2015). FOCUS on Health Spending @ OECD Health Statistics 2015. OECD Health Statistics 2015, July 2015. pp. 1–8.

OECD (2021). Health at a Glance 2021: OECD Indicators. 2021, Paris: OECD. https://doi.org/10.1787/ae3016b9-en (accessed 2 March 2023).

O'Kane, C., Haar, J., Mangematin, V. et al. (2021). Distilling and renewing science team search through external engagement. *Research Policy* 50 (6): 104261.

Orsini, J., Blaak, C., Yeh, A. et al. (2014). Triage of patients consulted for ICU admission during times of ICU-bed shortage. *Journal of Clinical Medical Research* 6 (6): 463–468.

Panetta, K. (2019). Hyperautomation, blockchain, AI security, distributed cloud and autonomous things drive disruption and create opportunities in this year's strategic technology trends. Smarter with Gartner. (accessed 27 April 2020), https://www.gartner.com/smarterwithgartner/gartner-top-10-strategic-technology-trends-for-2020.

Panetta, K. (2020). Distributed cloud, AI engineering, cybersecurity mesh and composable business drive some of the top trends for 2021. Smarter with Gartner (accessed 4 May 2021).

Payne, M., Pooke, F., Chase, J.G. et al. (2021). The separation of insulin pump hardware and software-a novel and low-cost approach to insulin pump design. *IFAC-PapersOnLine* 54 (15): 502–507.

Pearce, J.M. (2020). A review of open source ventilators for COVID-19 and future pandemics. *F1000Research* 9: 218.

Pielmeier, U., Andreassen, S., Juliussen, B. et al. (2010a). The Glucosafe system for tight glycemic control in critical care: a pilot evaluation study. *Journal of Critical Care* 25 (1): 97–104.

Pielmeier, U., Andreassen, S., Nielsen, B.S. et al. (2010b). A simulation model of insulin saturation and glucose balance for glycemic control in ICU patients. *Computer Methods and Programs in Biomedicine* 97 (3): 211–222.

Pielmeier, U., Rousing, M.L., Andreassen, S. et al. (2012). Decision support for optimized blood glucose control and nutrition in a neurotrauma intensive care unit: preliminary results of clinical advice and prediction accuracy of the Glucosafe system. *Journal of Clinical Monitoring and Computing* 26 (4): 319–328.

Pillonetto, G., Sparacino, G., and Cobelli, C. (2003). Numerical non-identifiability regions of the minimal model of glucose kinetics: superiority of Bayesian estimation. *Mathematical Biosciences* 184 (1): 53–67.

Pillonetto, G., Caumo, A., Sparacino, G., and Cobelli, C. (2006). A new dynamic index of insulin sensitivity. *IEEE Transactions on Biomedical Engineering* 53 (3): 369–379.

Pintado, M.-C., de Pablo, R., Trascasa, M. et al. (2013). Individualized PEEP setting in subjects with ARDS: a randomized controlled pilot study. *Respiratory Care* 58 (9): 1416–1423.

Pirinen, A. (2016). The barriers and enablers of co-design for services. *International Journal of Design* 10 (3): 27–42.

Pironet, A., Desaive, T., Kosta, S. et al. (2013). A multi-scale cardiovascular system model can account for the load-dependence of the end-systolic pressure-volume relationship. *Biomedical Engineering Online* 12: 8.

Pironet, A., Desaive, T., Geoffrey Chase, J. et al. (2015). Model-based computation of total stressed blood volume from a preload reduction manoeuvre. *Mathematical Biosciences* 265: 28–39.

Pironet, A., Dauby, P.C., Chase, J.G. et al. (2016). Structural identifiability analysis of a cardiovascular system model. *Medical Engineering & Physics* 38 (5): 433–441.

Pironet, A., Docherty, P.D., Dauby, P.C. et al. (2017). Practical identifiability analysis of a minimal cardiovascular system model. *Computer Methods and Programs in Biomedicine* 171: 53–65.

Plank, J., Blaha, J., Cordingley, J. et al. (2006). Multicentric, randomized, controlled trial to evaluate blood glucose control by the model predictive control algorithm versus routine glucose management protocols in intensive care unit patients. *Diabetes Care* 29 (2): 271–276.

Poncette, A.-S., Spies, C., Mosch, L. et al. (2019). Clinical requirements of future patient monitoring in the intensive care unit: qualitative study. *JMIR Medical Informatics* 7 (2): e13064.

Ramachandran, D., Luo, C., Ma, T.S., and Clark, J.W. Jr. (2009). Using a human cardiovascular-respiratory model to characterize cardiac tamponade and pulsus paradoxus. *Theoretical Biology & Medical Modelling* 6: 15.

Read, R.L., Clarke, L., and Mulligan, G. (2021). VentMon: an open source inline ventilator tester and monitor. *HardwareX* 9: e00195.

Reay, S., Collier, G., Kennedy-Good, J. et al. (2017). Designing the future of healthcare together: prototyping a hospital co-design space. *CoDesign* 13 (4): 227–244.

Redmond, D., Chiew, Y.S., van Drunen, E. et al. (2014). A minimal algorithm for a minimal recruitment model–model estimation of alveoli opening pressure of an acute respiratory distress syndrome (ARDS) lung. *Biomedical Signal Processing and Control* 14 (0): 1–8.

Rees, S.E. (2011). The Intelligent Ventilator (INVENT) project: the role of mathematical models in translating physiological knowledge into clinical practice. *Computer Methods and Programs in Biomedicine* 104 (Supplement 1): S1–S29.

Rees, S.E., Andreassen, S., Freundlich, M. et al. (1999). Selecting ventilator settings using INVENT, a system including physiological models and penalty functions. *Proceedings of the Joint conference of European societies of Artificial Intelligence in Medicine and Medical Decision Making. Workshop, Computers in Anesthesia and Intensive Care*, Aalborg, Denmark.

Rees, S.E., Kjærgaard, S., Thorgaard, P. et al. (2002). The Automatic Lung Parameter Estimator (ALPE) system: non-invasive estimation of pulmonary gas exchange parameters in 10–15 minutes. *Journal of Clinical Monitoring and Computing* 17 (1): 43–52.

Rees, S., Allerød, C., Murley, D. et al. (2006). Using physiological models and decision theory for selecting appropriate ventilator settings. *Journal of Clinical Monitoring and Computing* 20 (6): 421–429.

Revie, J.A., Stevenson, D.J., Chase, J.G. et al. (2013). Validation of subject-specific cardiovascular system models from porcine measurements. *Computer Methods and Programs in Biomedicine* 109 (2): 197–210.

Ricci, M. and Gallina, P. (2020). COVID-19 – immunity from prosecution for physicians forced to allocate scarce resources: the Italian perspective. *Critical Care* 24 (1): 295.

Robinson, M.R., Daniel, L.C., O'Hara, E.A. et al. (2014). Insurance status as a sociodemographic risk factor for functional outcomes and health-related quality of life among youth with sickle cell disease. *Journal of Pediatric Hematology/Oncology* 36 (1): 51–56.

Rumball-Smith, J.M. (2009). Not in my hospital? Ethnic disparities in quality of hospital care in New Zealand: a narrative review of the evidence. *New Zealand Medical Journal* 122 (1297): 68–83.

Safaei, S., Bradley, C.P., Suresh, V. et al. (2016). Roadmap for cardiovascular circulation model. *The Journal of Physiology* 594 (23): 6909–6928.

Samad, T. (2022). Human-in-the-loop control and cyber–physical–human systems: applications and categorization. In: *Cyber–Physical–Human Systems: Fundamentals and Applications* (ed. A. Annaswamy et al.). UK: Wiley (In-Press).

Sandhya, M., Madhumitha, R., and Sankar, S. (2017). Analysis of threats in interoperability of medical devices. *International Journal of Biomedical and Biological Engineering* 11 (5): 282–285.

SCCM, AARC, ASA, ASPF, AACN, and CHEST (2020). Consensus statement on multiple patients per ventilator. SCCM Website. https://www.sccm.org/Disaster/Joint-Statement-on-Multiple-Patients-Per-Ventilato (accessed 25 March 2020).

Schranz, C., Docherty, P.D., Chiew, Y.S. et al. (2012). Structural identifiability and practical applicability of an alveolar recruitment model for ARDS patients. *IEEE Transactions on Biomedical Engineering* 59 (12): 3396–3404.

Schreiweis, B., Pobiruchin, M., Strotbaum, V. et al. (2019). Barriers and facilitators to the implementation of ehealth services: systematic literature analysis. *Journal of Medical Internet Research* 21 (11): e14197.

Schuster, M.A., McGlynn, E.A., and Brook, R.H. (2005). How good is the quality of health care in the United States? *The Milbank Quarterly* 83 (4): 843–895.

Shorr, A.F. (2002). An update on cost-effectiveness analysis in critical care. *Current Opinion in Critical Care* 8 (4): 337–343.

Smith, N., Waters, S., Hunter, P., and Clayton, R. (2011). The cardiac physiome: foundations and future prospects for mathematical modelling of the heart. *Progress in Biophysics and Molecular Biology* 104 (1–3): 1.

Smith, R., Chase, J.G., Pretty, C.G. et al. (2021a). Preload & Frank-Starling curves, from textbook to bedside: clinically applicable non-additionally invasive model-based estimation in pigs. *Computers in Biology and Medicine* 135: 104627.

Smith, R., Murphy, L., Pretty, C.G. et al. (2021b). Tube-load model: a clinically applicable pulse contour analysis method for estimation of cardiac stroke volume. *Computer Methods and Programs in Biomedicine* 204 (Paper #106062): 10.

Sounderajah, V., Patel, V., Varatharajan, L. et al. (2021). Are disruptive innovations recognised in the healthcare literature? A systematic review. *BMJ Innovations* 7 (1): 208–216.

Stenqvist, O., Grivans, C., Andersson, B., and Lundin, S. (2012). Lung elastance and transpulmonary pressure can be determined without using oesophageal pressure measurements. *Acta Anaesthesiologica Scandinavica* 56 (6): 738–747.

Stewart, K.W., Pretty, C.G., Tomlinson, H. et al. (2016). Safety, efficacy and clinical generalization of the STAR protocol: a retrospective analysis. *Annals of Intensive Care* 6 (1): 24.

Suarez-Sipmann, F., Bohm, S.H., Tusman, G. et al. (2007). Use of dynamic compliance for open lung positive end-expiratory pressure titration in an experimental study. *Critical Care Medicine* 35: 214–221.

Sun, Q., Chase, J.G., Zhou, C. et al. (2021). Over-distension prediction via hysteresis loop analysis and patient-specific basis functions in a virtual patient model. *Computers in Biology and Medicine* 141: 105022.

Sundaresan, A., Chase, J.G., Hann, C.E., and Shaw, G.M. (2010). Model-based PEEP selection in mechanically ventilated patients – first clinical trial results. *UKACC International Conference on Control*, Coventry, UK (07–10 September 2010): IEEE.

Sundaresan, A., Chase, J., Shaw, G. et al. (2011). Model-based optimal PEEP in mechanically ventilated ARDS patients in the intensive care unit. *Biomedical Engineering Online* 10 (1): 64.

Szlavecz, A., Chiew, Y., Redmond, D. et al. (2014). The Clinical Utilisation of Respiratory Elastance Software (CURE Soft): a bedside software for real-time respiratory mechanics monitoring and mechanical ventilation management. *Biomedical Engineering Online* 13 (1): 140.

Talamaivao, N., Harris, R., Cormack, D. et al. (2020). Racism and health in Aotearoa New Zealand: a systematic review of quantitative studies. *The New Zealand Medical Journal* 133 (1521): 55–55.

Tawhai, M.H. and Bates, J.H.T. (2011). Multi-scale lung modeling. *Journal of Applied Physiology* 110 (5): 1466–1472.

Tawhai, M.H. and Burrowes, K.S. (2008). Multi-scale models of the lung airways and vascular system. *Integration in Respiratory Control* 605 (5): 190–194.

Tawhai, M.H. and Lin, C.-L. (2010). Image-based modeling of lung structure and function. *Journal of Magnetic Resonance Imaging* 32 (6): 1421–1431.

Tawhai, M.H., Burrowes, K.S., and Hoffman, E.A. (2006a). Computational models of structure-function relationships in the pulmonary circulation and their validation. *Experimental Physiology* 91 (2): 285–293.

Tawhai, M.H., Nash, M.P., and Hoffman, E.A. (2006b). An imaging-based computational approach to model ventilation distribution and soft-tissue deformation in the ovine lung. *Academic Radiology* 13 (1): 113–120.

Tawhai, M.H., Hoffman, E.A., and Lin, C.-L. (2009). The lung physiome: merging imaging-based measures with predictive computational models. *Wiley Interdisciplinary Reviews: Systems Biology and Medicine* 1 (1): 61–72.

Tawhai, M., Clark, A., and Chase, J. (2019). The lung physiome and virtual patient models: from morphometry to clinical translation. *Morphologie* 103 (343): 131–138.

Thabit, H., Hartnell, S., Allen, J.M. et al. (2017). Closed-loop insulin delivery in inpatients with type 2 diabetes: a randomised, parallel-group trial. *The Lancet Diabetes & Endocrinology* 5 (2): 117–124.

Toffolo, G. and Cobelli, C. (2003). The hot IVGTT two-compartment minimal model: an improved version. *American Journal of Physiology. Endocrinology and Metabolism* 284 (2): E317–E321.

Truog, R.D., Brock, D.W., Cook, D.J. et al. (2006). Task Force on Values, E, and Rationing in Critical, C. Rationing in the intensive care unit. *Critical Care Medicine* 34 (4): 958–963; quiz 971.

Tunlind, A., Granström, J., and Engström, Å. (2015). Nursing care in a high-technological environment: experiences of critical care nurses. *Intensive & Critical Care Nursing* 31 (2): 116–123.

Turbil, E., Terzi, N., Cour, M. et al. (2020). Positive end-expiratory pressure-induced recruited lung volume measured by volume-pressure curves in acute respiratory distress syndrome: a physiologic systematic review and meta-analysis. *Intensive Care Medicine* 46: 1–14.

Uyttendaele, V., Dickson, J.L., Shaw, G.M. et al. (2017). Untangling glycaemia and mortality in critical care. *Critical Care* 21 (1): 152.

Uyttendaele, V., Knopp, J.L., Pirotte, M. et al. (2018). Preliminary results from the STAR-Liège clinical trial: virtual trials, safety, performance, and compliance analysis. *IFAC-PapersOnLine* 51 (27): 355–360.

Uyttendaele, V., Chase, J.G., Knopp, J.L. et al. (2021). Insulin sensitivity in critically ill patients: are women more insulin resistant? *Annals of Intensive Care* 11 (1): 1–14.

Van Herpe, T., De Moor, B., and Van den Berghe, G. (2009). Towards closed-loop glycaemic control. *Best Practice & Research. Clinical Anaesthesiology* 23 (1): 69–80.

Van Herpe, T., Mesotten, D., Wouters, P.J. et al. (2013). LOGIC-insulin algorithm-guided versus nurse-directed blood glucose control during critical illness: the LOGIC-1 single-center randomized, controlled clinical trial. *Diabetes Care* 36 (2): 189–194.

Viceconti, M. and Hunter, P. (2016). The Virtual physiological human: ten years after. *Annual Review of Biomedical Engineering* 18: 103–123.

Vodovotz, Y. and Billiar, T.R. (2013). In silico modeling: methods and applications to trauma and sepsis. *Critical Care Medicine* 41 (8): 2008–2014.

Wallet, F., Delannoy, B., Haquin, A. et al. (2013). Evaluation of recruited lung volume at inspiratory plateau pressure with PEEP using bedside digital chest X-ray in patients with acute lung injury/ARDS. 2013. *Respiratory Care* 58 (3): 416–423.

Ward, L., Steel, J., Le Compte, A. et al. (2012). Interface design and human factors considerations for model-based tight glycemic control in critical care. *Journal of Diabetes Science and Technology* 6 (1): 125–134.

Warrillow, S. and Raper, R. (2019). The evolving role of intensive care in health care and society. *The Medical Journal of Australia* 211 (7): 294–297. e1.

Wears, R.L. and Berg, M. (2005). Computer technology and clinical work: still waiting for Godot. *JAMA* 293 (10): 1261–1263.

Wiens, J., Price, W.N., and Sjoding, M.W. (2020). Diagnosing bias in data-driven algorithms for healthcare. *Nature Medicine* 26 (1): 25–26.

Wilinska, M.E., Chassin, L., and Hovorka, R. (2008). In Silico testing – impact on the Progress of the closed loop insulin infusion for critically ill patients project. *Journal of Diabetes Science and Technology* 2 (3): 417–423.

Wilinska, M.E., Chassin, L.J., Acerini, C.L. et al. (2010). Simulation environment to evaluate closed-loop insulin delivery systems in type 1 diabetes. *Journal of Diabetes Science and Technology* 4 (1): 132–144.

Williams, P. (2017). Standards for safety, security and interoperability of medical devices in an integrated health information environment: standards for safety, security, and interoperability of medical devices in an integrated health information environment. *Journal of American Health Information Management Association* 88 (4): 32–34.

Williams, I., Dickinson, H., and Robinson, S. (2012). *Rationing in Health Care: The Theory and Practice of Priority Setting*. Policy Press.

Willmann, S., Hohn, K., Edginton, A. et al. (2007). Development of a physiology-based whole-body population model for assessing the influence of individual variability on the pharmacokinetics of drugs. *Journal of Pharmacokinetics and Pharmacodynamics* 34 (3): 401–431.

Wong, J.H.K., Naswall, K., Pawsey, F., et al. (2021). Adoption of Technological Innovation in Healthcare Delivery: A Social Dynamic Perspective.

Wongvanich, N., Hann, C.E., and Sirisena, H.R. (2015). Robust global identifiability theory using potentials – application to compartmental models. *Mathematical Biosciences* 262: 182–197.

Wujtewicz, M. (2012). Fluid use in adult intensive care. *Anaesthesiol Intensive Ther* 44 (2): 92–95.

Wunsch, H. (2020). Mechanical ventilation in COVID-19: interpreting the current epidemiology. *American Journal of Respiratory and Critical Care Medicine* 202 (1): 1–21.

Yu, T., Lloyd, C.M., Nickerson, D.P. et al. (2011). The physiome model repository 2. *Bioinformatics* 27 (5): 743–744.

Zhou, T., Dickson, J.L., Shaw, G.M., and Chase, J.G. (2018). Continuous glucose monitoring measures can be used for glycemic control in the ICU: an in-Silico study. *Journal of Diabetes Science and Technology* 12 (1): 7–19.

Zhou, T., Knopp, J.L., and Chase, J.G. (2019). The state of variability: a vision for descriptors of glycaemia. *Annual Reviews in Control* 48: 472–484.

Zhou, C., Chase, J.G., Knopp, J. et al. (2021). Virtual patients for mechanical ventilation in the intensive care unit. *Computer Methods and Programs in Biomedicine* 199: p., Paper #105912: 24.

Part V

Sociotechnical Systems

18

Online Attention Dynamics in Social Media

Maria Castaldo[1], Paolo Frasca[1], and Tommaso Venturini[2,3]

[1] *Université Grenoble Alpes, CNRS, Inria, Grenoble INP, GIPSA-lab, Grenoble, France*
[2] *Medialab, Université de Genéve, Geneva, Switzerland*
[3] *Center for Internet et and Society, CNRS, Paris, France*

18.1 Introduction to Attention Economy and Attention Dynamics

This chapter aims to emphasize a number of questions that, although crucial since the early days of media studies, have not yet been the object of the empirical and computational study that they deserve: how does collective attention concentrate and dissipate in modern communication systems? How do subjects and sources rise and fall in public debates? How are these dynamics shaped by media infrastructures? In the perspective of addressing these questions, this chapter provides a review of the literature on the dynamics of online content dissemination: our goal is to prepare the ground for the necessary study, which should comprise empirical investigation, mathematical modeling, numerical simulation, and rigorous system-theoretic analysis.

The interest in the dynamics of collective attention is as old as sociology. Already in the nineteenth century Tarde (1890, 1893), argued that these fleeting dynamics (rather than the more stable structures and norms) should make up the core of social research (Latour 2002). Attention dynamics rose again in sociological preoccupations in the 1970s and 1980s, when the major problem of the nascent media research was to describe the competition for the limited bandwidth of radio and television broadcasting. Concepts such as "attention cycles" (Downs 1972; Hilgartner and Bosk 1988) and "agenda setting" (McCombs and Shaw 1972; McCombs 2005) became prominent to investigate media schedules and their consequences on public debate. With the advent of digital media, the interest for collective attention shifted from the supply to the demand side. Vindicating Herbert Simon's 1971 prophecy (Simon 1971), media scholars (and commercial actors) realized that in an information-rich environment, attention becomes a scarce and therefore valuable resource. This gave rise to many critical reflections on the consequences of the rise of the "attention economy" and its way of transforming collective attention and debates into a marketable commodity (Crogan and Kinsley 2012; Citton 2014).

The research on attention economy is extremely interesting for its effort to conceptualize a very large phenomenon (the way in which collective attention flows through the media system) through the continuous convergence and divergence of a myriad of individual choices (Terranova 2012). At the same time, and for the same reason, the literature on attention economy has remained largely theoretical. Until recently, the empirical investigation of the dynamics of collective attention has been hindered by the difficulty to procure data sets broad enough to account for an entire media

Cyber–Physical–Human Systems: Fundamentals and Applications, First Edition.
Edited by Anuradha M. Annaswamy, Pramod P. Khargonekar, Françoise Lamnabhi-Lagarrigue, and Sarah K. Spurgeon.
© 2023 The Institute of Electrical and Electronics Engineers, Inc. Published 2023 by John Wiley & Sons, Inc.

population, but rich enough to distinguish each fleeting individual choice (Venturini and Latour 2010; Latour et al. 2012). In the last few years, however, the massive investments by commercial and governmental actors into the surveillance of media interactions (Zuboff 2019) have generated the data necessary for the empirical and computational study of the flows of collective attention, and scholars have begun to seize this possibility.

Based on this growing literature, this chapter has the purpose of introducing the "cyber-social" research on attention dynamics, including its conceptualizations and its empirical findings, in the perspective of proposing useful mathematical models on which the instruments of control systems theory can be effectively deployed. Online media feature a tight integration of human and technological (digital) components and involve the participation of large numbers of users: they thus appear to be a socio-technical system with humans-in-multiagent-loops, as per the taxonomy proposed earlier in this book by Samad (2023). We are hopeful that this perspective will help the readers to appreciate the relevance and the challenges of attention dynamics in online media.

The rest of the chapter is divided into three sections. Section 18.2 summarizes this emerging field of research by considering seven features that characterize collective attention: (i) its limitation, (ii) its skewedness, (iii) its sensitivity to novelty and (iv) popularity, (v) its burstiness, and (vi) its dependency on online platforms' policies, and (vii) their dynamism. Section 18.3 considers the latest preoccupation of computational attention studies: how the increasingly sophisticated recommendation systems, introduced by online platforms, interfere with the expression of individual and thus collective attention. To this purpose, we present a simple model, meant to highlight the growing importance of trendiness in recommendation systems and its potential deleterious consequences on public debate. Section 18.4 contains some concise final remarks.

18.2 Online Attention Dynamics

We can summarize the current status of the knowledge about online attention dynamics in some few key ideas:

- collective attention is limited;
- collective attention is distributed in a highly nonuniform way among contents;
- novelty has a fundamental role in directing collective attention;
- previous popularity influences future popularity;
- people take part in the diffusion of content, and their individual activity is bursty, i.e. highly concentrated in time;
- online platforms' policies and infrastructures have a major influence on the diffusion of content;
- platforms themselves are ever-changing.

In the following sections, we consider each of these ideas separately and discuss the literature that regards them.

18.2.1 Collective Attention Is Limited

When Herbert A. Simon first theorized the concept of attention economy, he built his reasoning on the statement that human attention can be treated as a scarce commodity. Despite the impressive complexity and processing power of the human brain, it is undeniable that its capacities are limited: we can barely attend to over one object at a time, and we can hardly perform two tasks at once (Marois and Ivanoff 2005). Considerable research in the cognitive science has investigated the limitations of our brain, and we refer the interested reader to Marois and Ivanoff (2005). For our purpose, it suffices to point out that these limitations have become standard assumptions in many works not only to analyze the patterns of online engagement (Lorenz-Spreen et al. 2019; Weng et al. 2012; Qiu et al. 2017) but also in modeling opinion dynamics in social groups (Rossi and Frasca 2020; Ceragioli et al. 2021b,a). Crucially, for this chapter, these assumptions have become an essential starting point for the research on collective attention, as the scarcity of individual cognitive resources has turned attention into the object of an increasingly competitive market, in which attention ceases to be an individual feature and becomes a collective commodity that is consumed online.

18.2.2 Skewed Attention Distribution

As a result of the limitedness of individual and collective attention, news items have to compete with each other to gain the consideration of the public. This competition rewards few items which become over-popular, while the vast majority of them remain unnoticed. As largely discussed in the last years, online popularity is highly skewed, with a relatively small number of participants getting most of the public attention. The distribution of popularity among online items has often been found to respect the "80–20 rule," also known as the Pareto rule: the 20% of the online content accounts for the 80% of the popularity. Evidence of this have been brought out on many platforms: it turned out to be true for videos on Metacafe, Yahoo!, Dailymotion, Veoh (Mitra et al. 2009) and YouTube (Cha et al. 2007), and for retweets in Twitter (Bild et al. 2015; Lu et al. 2014).

It is legitimate to wonder what causes this skewness and whether it exists also outside of user-generated content platforms like the ones mentioned above. A first answer to these questions is given by Cha et al. (2009), where they compared the consumption of videos on platforms of User-Generated Content (UGC), like YouTube, and Professionally Generated Content (PGC), like Netflix or Yahoo! Movies. They outlined that in UGC platforms attention is less equally distributed among items. More precisely, at that time, on YouTube 10% of the most popular videos were accounting for nearly 80% of the total views, while on-demand videos presented a less skewed distribution of popularity. In practice, while on the PGC platform, it never happens that a content is left with no public, on YouTube, there is a significant quantity of videos that do not receive any view at all. But that is not the only difference between UGC and PGC: the authors also stress an enormous difference in the quantity of content uploaded on the two kinds of platforms, with UGC platform collecting a significantly higher quantity of material. The massive amount of contents present on YouTube, together with the human cognitive limitations and the peer influence, might be the cause of the stressed skewness in UGC platforms: people, only disposing of a limited attention, have to choose among an excessive variety of contents and may end up relying on imitation for their choices. As a result, collective attention is more concentrated and many items cannot arouse the slightest interest.

Acknowledging that the distribution of attention is skewed is crucial not only to conduct research on attention dynamics but also to contextualize previous works in the field: working with empirical data on social platforms often means dealing with a majority of items that received no or very little attention. When aiming at modeling popularity trends, it then becomes important to remove the nonrelevant observations. For instance, Crane and Sornette (2008) based their model of content diffusion on only the 10% of YouTube videos in their dataset, as the remaining 90% either showed a too low number of views or could be accurately described as (purely random) Poisson processes. Similarly, Kämpf et al. (2012) had to perform a significant filtering of their Wikipedia data set: the vast majority of articles they monitored were rarely accessed and almost never experienced significant bursts of activity. To filter their data, they focused on articles that exhibited a minimum rate of 256 views at least in one hour, over the period of observation. This threshold, that might not seem particularly demanding, was met by only the 0.17% of the Wikipedia articles studied by the authors. Such a low percentage is, again, evidence of the quantity of content available on the web but never or rarely accessed.

18.2.3   The Role of Novelty

Once acknowledged that few items capture most of the public attention, it comes natural to wonder which are the factors that concentrate everyone's interest on specific items. In the "Attention Economy" literature, novelty is often presented as one of the main factors (Simon 1971; Goldhaber 1997). Indeed, as Goldhaber stated, since it is hard to get new attention by repeating exactly what has already been done in the past, a key role in the attention economy is played by novelty.

Online platform managers are well aware of the importance of novelty. Its promotion is expressly sought by platforms and specifically encoded in recommendation systems and in particular in the algorithms that select and suggest content to users, trying to meet their tastes and interests. Covington et al. (2016) developers at YouTube in 2016, include *freshness* among the three major needs of YouTube recommendation system. In particular, they acknowledge that as "many hours' worth of videos are uploaded each second to YouTube," "recommending recently uploaded ('fresh') content is extremely important for YouTube as a product." In fact, they observe that users constantly prefer fresh content and, hence, to keep their engagement high and make them spend time on the platform, YouTube has to satisfy their need for novelty.

The preference for novel content has been observed also in other contexts. In 2012 on Twitter, among the total tweets published in a week, 45% had never been published before (Weng et al. 2012). Similarly, according to Roth et al. (2020) in 2020, two-thirds of the suggestions of YouTube given at a certain moment were not anymore associated with the same video after two days. Novelty can thus be listed as a key factor for capturing people's attention, and it should be considered when modeling popularity trends and shifts of collective attention from one topic to another.

18.2.4   The Role of Popularity

Various models, stemming from the different branches of science, have been adopted to describe the evolution of popularity online. By classifying these models according to the research field that generated them, we could distinguish *epidemic models*, issued from the tradition of mathematical models describing the spread of infectious diseases, *Bass-like diffusion models* stemming from the theory of innovation diffusion, and *self-exciting processes* belonging to the larger family of counting processes in statistics.

Independently from the specific model used, one assumption is recurrent: people influence each other. To describe this occurrence, different terms have been adopted in different fields. When dealing with innovation diffusion models, we usually refer to the tendency of people to be influenced by others as an *imitation processes*. When adopting epidemic models instead, we usually refer to it as a *contagion* effect or a word-of-mouth effect. Despite its different names, the concept is the same: when many people are aware of a content, it becomes more likely for an individual to encounter it. We could also talk of *popularity effects*: the future spread of a piece of information is influenced by its previous success. In the current section, we aim to summarize the various approaches that have been used to study the evolution of popularity in different contexts.

- **Epidemic models**: although originally designed to model the spread of infectious diseases, epidemic models can effectively describe the propagation of content online. Daley and

Kendall (1964), in 1964, first proposed the analogy between epidemics and the spread of rumors, suggesting the same mathematical model might apply to both fields of study. The foundation of these models resides in partitioning a population into different classes of individuals. Among the most common classes we find Susceptible, Infected, and Recovered individuals. Susceptible individuals are those who still have not been in contact with the disease. Infected individuals have caught the disease and can spread it, while recovered individuals healed from the disease and cannot transmit it anymore. Of course, when adapting these models to attention dynamics, the disease is replaced by a piece of information and healing is replaced by forgetting. We can elaborate on different models, depending on the class of individuals considered. For example, the SI model considers only susceptible or infected individuals. In an SI model, the fraction of infected individuals I evolves in time according to

$$\dot{I} = \alpha(1 - I)I$$

Here, the fraction of newly infected individuals is given by the probability of a susceptible individual to meet an infected one, multiplied by a transmission rate α.

Many variations of this basic model have been proposed in literature with the specific aim of explaining information and rumor diffusion. In 2006, Bettencourt et al. (2006) proposed a variation of the SI model (which they called SEIZ model), to fit the spread of the use of Feynman diagrams through the theoretical physics communities in USA, Japan, and USSR in the period immediately after World War II. In their SEIZ model, actors can either be susceptible (S), i.e. still unaware of an idea, exposed (E), i.e. having been in contact with the idea, infected (I), i.e. adopters of the idea, or skeptic (Z), namely aware of the idea but unconvinced by it. They prove that introducing exposed individuals consistently increases the capability of the model to fit the data: in fact, inserting a delay between the moment physicists first met Feynman diagrams and the moment they adopted them brought major improvements to the data explanation.

Jin et al. later applied the same model (Jin et al. 2013) to the spread of rumors online, with similar outcomes: they confirmed the improvement due to the introduction of exposed individuals in a simple SI model. Another confirmation of the importance of introducing an exposure delay between the reception and the adoption of an idea comes from the work of Xiong et al. in 2012 (Xiong et al. 2012). The authors proposed a diffusion model with four different states: susceptible (S), contacted (C), infected (I), and refractory (R). Contacted individuals behaved exactly as the exposed individual in the SEIZ model: they acknowledged the information but have not decided yet whether to spread it or not.

Besides the abovementioned variations of the SI model, there also exists some which do not include the addition of new classes of individuals. In Richier et al. (2014), the authors proposed different biologically inspired models which proved to fit at least the 90% of videos of a conspicuous YouTube dataset. Among the considered models, we find a variation of the SI model called *Gompertz model* and governed by

$$\dot{I} = \alpha I \log(M/I) \tag{18.1}$$

where M represents the potentially interested public and I represents the number of users that viewed a video. They compared it with a simple *exponential model* $\dot{I} = \alpha(M - I)$ and with some variations of the Gompertz and the exponential models, where the authors added a term kt to the temporal evolution of $I(t)$. While the simple Gompertz and exponential models failed at fitting the majority of the videos, the modified models brought a sensible improvement to the fittings: they explained almost 75% of videos popularity evolution. Even though, adding a term kt to $I(t)$ seems rather contrived and difficult to interpret, we can explain the need of adding a further parameter k as the necessity of higher degrees of freedom when explaining complex dynamics.

In this respect, adding a term kt to the evolution of infected individuals, or adding new classes of individuals to the SI model, plays the same role.

- **Bass diffusion models**: the Bass diffusion model owns its name to Frank Bass, an academic pioneer of marketing research in the second half of the twentieth century. It was first introduced in Bass (1969) to model the process of adoption of new products in a market. It is based on a simple classification of individuals into two groups: *innovators* and *imitators*. The Bass model found its main application in forecasting innovations or technology sales. The model formulation is the following:

$$\dot{F}(t) = p(1 - F(t)) + qF(t)(1 - F(t))$$

where $F(t)$ is the fraction of adopters in a population at time t, p is the coefficient of innovation, and q is a coefficient of imitation. As we can see, also in this formulation of the problem, we have a term of contagion $F(t)(1 - F(t))$ that models the influence of users on each other. The analogy between infected individuals in epidemic models and innovators in models of adoption is glaring, and it has been made explicit in many works (Bass 1969; Coleman et al. 1957; Toole et al. 2012). That is one of the reason why, especially when these models are complemented with additional assumptions, it is hard to make a sharp distinction between adoption models and epidemic models. The main difference relies on the terms used and the community the researchers addressed. Most explanations of real data through the Bass model come from its agent-based extension, that was at first studied in Rand and Rust (2011). In this agent-based formulation, the model presents some initially unaware agents connected by relationship links. Over time, agents get the opportunity to become aware of the information through two mechanisms: either by spontaneous adoption, or by influence exerted by their neighbors in the network. In Rand et al. (2015), Rand et al. apply an agent-based Bass model to the diffusion of information in Twitter during four major events happened in the United States in 2011–2012. They obtained pretty satisfactory fitting of the increase in time of the number of people talking about each topic. Many other examples can be discussed, e.g. Chica and Rand (2017), but for a complete review on innovation diffusion process, we refer the reader to Kiesling et al. (2012). Here, our aim is to highlight the similarities between innovation diffusion processes and epidemic models.

- **Self-exciting processes**: the best example of how self-exciting processes can model content diffusion is given by Crane and Sornette (2008). The authors provide a model to fit the evolution of videos on YouTube, and they base it on three essential assumptions: (i) the relevance of human interactions in spreading a piece of information, (ii) the existence of influences external to social media, and (iii) the fact the humans activity follows very specific patterns, which we are going to discuss more in detail in Section 18.2.5. Crane and Sornette's model consists of a self-excited Hawkes' conditional Poisson process (Hawkes and Oakes 1974) with an instantaneous rate of views given by

$$\lambda(t) = V(t) + \sum_{i, t_i < t} \mu_i \varphi(t - t_i)$$

where the term $\sum_{i, t_i < t} \mu_i \varphi(t - t_i)$ models the contagion/imitation process, and $V(t)$ represents an exogenous source of views, which capture all spontaneous engagement not triggered by epidemic effects. The parameter μ_i represents the number of potential viewers who will be influenced by person i that views a video at time t_i. The kernel $\varphi(t)$ is chosen to be equal to

$$\varphi(t) = \frac{1}{t^{1+\theta}} \tag{18.2}$$

and it represents the rate at which individuals consume information. Such a formulation stems from a wider literature investigating the rhythms of individual human activity, which, in many contexts, is characterized by power law distribution of waiting times between consequent actions, for instance, between receiving an e-mail and replying to it (Barabási 2005; Vázquez et al. 2006; Oliveira and Vazquez 2009). Here, the memory kernel $\varphi(t)$ describes the distribution of waiting times between acknowledging the existence of a video and actually watching it. We could consider it as another way to model the *latency* time between acknowledgment and adoption of an idea proposed in the SEIZ model by Bettencourt et al. (2006). The rich literature on human interevents (Deschâtres and Sornette 2005; Johansen and Sornette 2000; Johansen 2001) that justifies Crane and Sornette's choice of $\varphi(t)$ will be discussed in Section 18.2.5.

18.2.5 Individual Activity Is Bursty

Individual actions online (reading, posting, re-posting, or replying) have been the object of a vast literature analyzing the temporal patterns of human activity. A consistent number of empirical evidences has outlined that human activities are often inhomogeneous in time: long periods of inactivity are followed by sudden bursts of subsequent events (Karsai et al. 2018). This behavior affects every task we perform in daily life: the process of editing a scientific paper (Jo et al. 2012; Mryglod et al. 2012; Hartonen and Alava 2013), the loans demanded in a university library (Vázquez et al. 2006), the requests sent to printers (Harder and Paczuski 2006), messages in a chat system (Dewes et al. 2003), page downloads on a news site (Dezsö et al. 2006), e-mails (Johansen 2004). Besides being inhomogeneous in time, interevents time share another constant behavior across most of the examples mentioned above: a heavy tailed distribution.

Vazquez et al. extensively analyze the bursts characterizing human activity in Vázquez et al. (2006). Among many datasets (the number of clicks by the same user on a Hungarian search engine, the number of e-mails by the same researcher in a university, the library book checks by the same student, the trade transactions by the same broker) they outline that the interevent time distribution has a power-law tail $P(\tau) = \tau^{-\alpha}$. They recognize two universal classes of exponent $\alpha = 1$ and $\alpha = 3/2$ in their datasets and proposed two decision-based queuing process to explain them. Decision-based queuing processes are processes in which tasks get ranked and executed on the basis of some perceived priority. Vazquez et al. proved that when there is no limitation to the number of tasks an individual can handle at any time, the distribution of the waiting time of the individual tasks follows a heavy tail distribution characterized by $\alpha = 3/2$. On the other hand, when imposing a limitation on the queue length, the resulting distribution of the waiting time is characterized by $\alpha = 1$. Vazquez's work has been vastly appreciated, even if further studies attenuated the net distinction between the two classes he identified. Nowadays, we know that many other bursty systems can be explained through power laws of various exponents, ranging between 1 and 2. Many other causes of this peculiar distribution of interevents have been proposed in literature, and we refer the interested reader to Karsai et al. (2018). Our general aim here was to provide the reader with the intuition that heavy-tailed interevents in human activity are likely to be the outcome of the way individuals process tasks.

Focusing on the implications of bursty activities rather than on their causes, we are confronted with a vast literature, at a first sight partially contradictory. In fact, the effects of heterogeneous interevents times in dynamical processes have been considerably debated in the last 10 years. On the one hand, burstiness of interevents seemed to slow down diffusion processes and epidemic spreading (Karsai et al. 2011; Vazquez et al. 2007). On the other hand, evidence has been brought to light that, in a certain environment, the speed of contagion might be increased by heavy-tailed interevents (Rocha et al. 2011). A good synthesis of these apparently opposing effects of bursty human activity can be found in Chapter 5 of Karsai et al. (2018). Here, Karsai explains that "heterogeneous inter-event times may have different effects when considering the early and late time behavior of a dynamical process." Summarizing his analysis, we could say that bursty activity on networks has an opposite effect when considering the early stages of the diffusion or the later ones: in the first case, in fact, it speeds up the spread, while, in the second case, it slows it down. Given the evidence that bursty activity can influence the outcome of a diffusion process, it is important to acknowledge their existence and their universality. Mean field models like the basic SI model defined in (18.1) fail in representing users' individual activity. The need for adding a delay to slow down the spread of information in the SEIZ model can be justified by the slowdown bursty individual activity causes in latter stages of a diffusion process.

18.2.6 Recommendation Systems Are the Main Gateways for Information

Knowing how human behave in their daily activities can help us understand their way of posting, forwarding, or liking content on social media and hence, better describe the process of diffusion and consumption of information. However, when dealing with online attention dynamics, we have to keep in mind that influence from neighbors and word-of-mouths spreading are not the only way people get into contact with information. On the contrary, in many online platforms, they appear to be rather marginal when compared to recommendation systems.

Many investigations have been carried out to understand the ratio of content diffused by algorithmic and human recommendation. Already in 2010, Zhou et al. (2010) confirmed that the related video recommendation on YouTube was the main source of views for most of the videos on YouTube. Furthermore, the authors reveal that there is a strong correlation between the view count of a video and the average view count of its suggested "videos to watch next." This implies that a video has a higher chance of becoming popular when it is placed on the related video recommendation lists of popular videos. In 2011, Figueiredo et al. (2011) confirmed that the most likely way to get to watch a video on YouTube is either "the related video" list, which displays a list of 20 videos that the platform suggests based on the previously watched video. These two internal sources of views account for the 32%, 43% of the total views. On the other hand, external sources like links to the video in other social media, or suggestions made by friends, account for only the 8–16% of the total views of a video. This very day, YouTube itself, on its official blog, declares that "recommendations drive a significant amount of the overall viewership on YouTube, even more than channel subscriptions or search." In 2018 YouTube Chief Product Officer Neal Mohan admitted that for over the 70% of the time users spend watching videos on the platform, they are lured in by one of the service's AI-driven recommendations (Solsman 2018).

18.2.7 Change Is the Only Constant

Given the relevance of platform's suggestions in shaping what people watch, it is important to get a better understanding of their evolution, in order to contextualize researches conducted in different periods. Despite being created and designed relatively recently, online platforms have an incredibly rich history in terms of updates and enhancements: the evolution of their policies, their design, and their functionalities is a fast, never-ending process. By keeping in mind that Facebook was created in 2004, YouTube in 2005, Twitter in 2006, and Instagram in 2010, the extent these social media reached in only less than two decades is impressive. Nowadays, over 300 hours of videos get uploaded every minute on YouTube, and the platform gathers over 30 million visitors per day. In October 2021, Facebook had 2.910 billion monthly active users (met), Twitter had 206 million daily active users (twi), Instagram had 500 million daily active users (met).

Along their way to success, these platforms and their engineers had to face constant challenges. For instance, as already discussed, they had to confront the need for sorting unprecedented quantities of user-generated content and ranking it in a personalized way for each user. The tools used to meet this requirement, recommendation systems, are constantly updated, and undergo continuous improvements to keep up with the latest discoveries in artificial intelligence. Some of the changes that recommendation systems have undergone in the last years have been documented by developers (Covington et al. 2016; Zhao et al. 2019; Naumov et al. 2019; Cheng et al. 2016; Gupta et al. 2013). They always aimed at enhancing the efficiency of recommendations, increasing the engagement of the public and the time they spent online.

Among the other arduous challenges platforms had to face, those maybe most known by the wide public concern ethics. Platforms have been accused, over the years, of promoting the diffusion of fake news, of disseminating hate speech, and of polarizing people by suggesting extremist contents. To cite one of the major and earlier events that rose these concerns, we could name the 2016 US Presidential elections: Buzzfeed reported that false news stories at that time outperformed real news on Facebook (Silverman 2016). Throughout the years, platforms have changed their policies to overcome these ethical issues. Videos about discrimination, segregation, or exclusion have been banned by YouTube (Wojcicki 2021), and creators can no longer monetize videos using inappropriate language or dealing with controversial content (Google YouTube Terms of Service 2022). Facebook does not allow objectionable or violent content and hate speech (Meta 2022a) and the same holds for Instagram (Meta n.d.b). Ethical concerns have hence contributed to shape platform policies and did affect the kind of content diffused online.

In a nutshell, online platforms are ever-changing, under the combined forces of technological progress and ethical and political criticism. No man ever steps in the same platform twice. For it is not the same platform, and he is not the same man. That is how we could rephrase Heraclitus to stress that this continuous evolution of platform should be kept in mind when considering the previous literature, to better contextualize the findings and avoid improper generalizations.

18.3 The New Challenge: Understanding Recommendation Systems Effect in Attention Dynamics

So far, we discussed the main features that should be taken into consideration when dealing with online content diffusion. When considering these features together with the models used in literature (discussed in Section 18.2.4), it becomes self-evident that one key variable has been left out when modeling online information spread: the role of recommendation systems.

As we have seen, many models (Crane and Sornette 2008) include in their formulation an exogenous source of popularity. We do believe this analysis to be potentially superficial as, by calling "exogenous" everything not explained by a mechanism of contagion/imitation, we risk including in the term both what is effectively triggered by events external to the platform, and recommendations made by social networks, hiding the potential role of platforms in influencing information spreads. When studying online content diffusion, we cannot avoid considering the means *on* which information circulates: online platforms.

Even though much has been written in the last years about online media, we claim that not much has been done to investigate how they shape the speed of content diffusion and whether they have a responsibility in accelerating our collective attention dynamics (Lorenz-Spreen et al. 2019). Most of the literature about online media has rather focused on other potentially dangerous implications of their use. "Selective exposure" (Sears and Freedman 1967), "eco chambers" (Garrett 2009), and "filter bubbles" (Pariser 2011) fall within the best-known threats of social media. To briefly summarize these concerns: social media may create environments in which "thousands or perhaps millions or even tens of millions of people are mainly listening to louder echoes of their own voices" (Sunstein 2001). This idea has sparked much interest in computational sociology (Colleoni et al. 2014; Geschke et al. 2019) and computer science (Nguyen et al. 2014) despite evidence of a possible overestimation of its impact (Bakshy et al. 2015; Dubois and Blank 2018). Recent work includes both mathematical modeling (Jiang et al. 2019; Rossi et al. 2021) and empirical data analysis, with a focus on e-commerce services (Ge et al. 2020; Anderson et al. 2020).

Instead of focusing on the kind of content suggested to users and on the possible distortion of reality induced by social media, **in this chapter, we would like to provide a formal example of how content diffusion might depend on platforms' policies**. Our models is built on few features and assumption and our aim is, through a simplification of recommendation systems, to investigate the effect they might have on the speed of content diffusion.

18.3.1 Model Description

The first ingredient of our model is a **population of matters of attention** defined as the entities that compete to capture public attention. They can represent tweets on Twitter, posts on Facebook, or videos on YouTube.

The second ingredient of our model is a platform in which attention matters live. On this platform, we assume attention to be limited, given the evidence already discussed in Section 18.2.1. For the sake of simplicity, we do not account for circadian or weekly rhythms that affect the attention availability on platforms. Hence, as discussed in Section 18.2.1, considering a sufficiently small time window on the life of a platform, we can avoid considering the arrival of new users and consider the attention fixed. The third ingredient of our model is a representation of the recommendation system that prizes trendiness. In other words, an item that grew fast in popularity in the previous interval of time is more likely to be suggested and viewed by other users in the next future. This boosting of trendiness is consistent with the way in which online platforms "emphasiz[e] novelty and timeliness… [by] identifying unprecedented surges of activity" and "reward[ing] popularity with visibility" (Gillespie 2016). The model therefore should reward rising items and penalize declining ones.

To shape these features into a formal model, we define a set of x_i, $\{i \in 1, \dots, n\}$ attention matters, and we name π_i^t the share of attention captured by x_i at time t. At every time step, the recommendation system rewards rising attention matters according to

$$p_i^{t+1} = \pi_i^t + \alpha(\pi_i^t - \pi_i^{t-1}) + \varepsilon_i^t \tag{18.3}$$

where ε is a random variable drawn from a normal distribution with mean 0 and standard deviation $1/(\sqrt{c} * n)$, where c controls the noise. The crucial variable of our model consists of α. It represents how much weight is given to trendiness when promoting a content on a platform. It is the boost induced by trendiness.

Equation (18.3) provides us with a *potential* visibility of item x_i at time t. Its *actual* visibility is given by a normalization of the overall attention on the platform and eventual rounding of negative potential visibilities to zero. The resulting model is fully defined by the following set of equations:

$$p_i^{t+1} = \pi_i^t + \alpha(\pi_i^t - \pi_i^{t-1}) + \varepsilon_i^t$$

$$\hat{p}_i^{t+1} = \max(0, p_i^{t+1}) \tag{18.4}$$

$$\pi_i^{t+1} = \frac{\hat{p}_i^{t+1}}{\sum_j \hat{p}_j^{t+1}}$$

18.3.2 Results and Discussion

Our interest lies in understanding the effects of the α parameter modeling a potential recommendation system. When simulating system (18.4) some typical outcomes, shown by Figure 18.1, can be observed.

The comparison between the graphs in Figure 18.1 suggests that as the boost of the trendiness grows, the rise and fall of attention matters become steeper. This relation can be tested by computing the mean increment by unit of time and observing that it increases with α, as analyzed in Figure 18.2. This behavior is independent on the level of noise and on the number of items populating the platform.

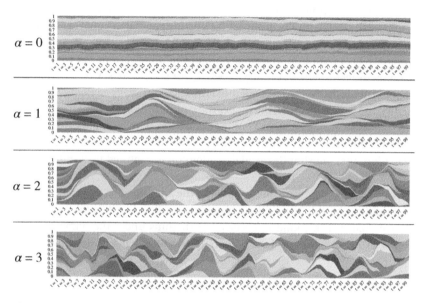

Figure 18.1 Evolution of our model for trendiness boost = 0, 1, 2, and 3 (with $N = 20$ and $c = 12$). Each color area corresponds to the attention received by an item. The first 100 iterations are shown.

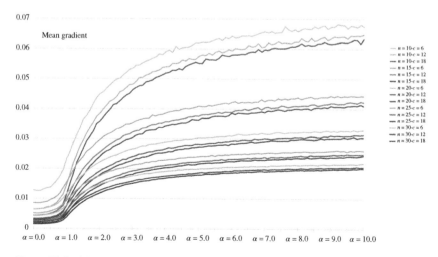

Figure 18.2 Mean increment of attention curves as a function of the trendiness boost (for different values of n and c).

Overall, from Figure 18.1 emerges that prizing trendiness accelerates the dynamics of collective attention. Without the need of a complex representation of recommendation systems, we obtain different regimes of content diffusion depending on the platform policy. Moreover, not only we obtain different regimes but also we could claim that some of them are healthier than others. To make this point clearer Figure 18.3a shows the relation between the parameter α and some key metrics of an attention dynamics. The first one is the average life cycle of an item, namely the time from its first view and its latest one. As a consequence of faster rise and decay of interest in news items, the attention waves shortens. As a consequence, a higher number of attention matters enter and exit the arena, as shown by Figure 18.3b. We recall that our model allows an arena to contain a maximum number of n items, some of which might not be active (i.e. with $\pi_i^t = 0$). When an item loses public attention and does not receive any for several epochs, we consider it as extinguished and substitute it with a new one with a starting popularity equal to 0. Hence, when the rhythm of collective attention accelerates, the number of items receiving attention in a given time-window increases. One closing observation should be made concerning the concentration of collective attention on different items. At every epoch of our simulations, we can evaluate the Gini index of how popularity is distributed among different attention matters. Being the Gini index, an index of inequality, the higher its value, the higher the concentration of popularity on few items. Figure 18.3c shows how the average instant Gini index changes with α. Higher trendiness boost amplifies the difference between successful and unsuccessful attention matters, creating a situation in which, at each iteration, most of the available attention is captured by a minority of over-visible items.

Regimes characterized by few items capturing the majority of the attention but incapable of sustaining it for a long time might not be able to raise structured and constructive debate.

In the previous work, we analyzed this kind of accelerated rhythms under the name of *junk news bubbles* (Castaldo et al. 2022) and discussed the potential harm for the public debate coming

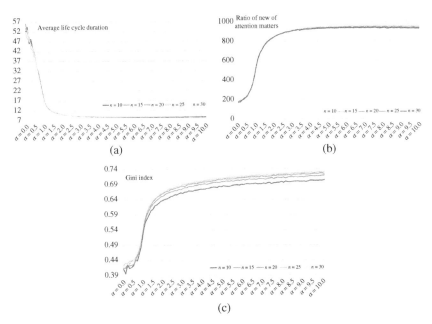

Figure 18.3 (a) Mean life cycle. (b) Ratio of new items getting attention in 10 000 epoches. (c) Gini index of instant distribution of attention.

from this syncopated rhythm of attention that is at the same time increasingly dispersed and increasingly concentrated.

Here, we would primarily stress that recommendation systems, given their enormous impact in diffusing content, should definitely be included in models considering online content diffusion as, otherwise, those models could be mis-interpreted.

18.4 Conclusion

In the last several years, the control systems community has been drawn toward working on social dynamics. Most attention has been devoted to mathematical models to describe decision-making and the evolution of opinions or beliefs: this focus is testified by surveys (Proskurnikov and Tempo 2018; Jia et al. 2015) and is confirmed by the contributions to this book, which include a chapter on social diffusion (Zino and Cao 2023). Some of these models have taken into account the concurrent evolution of interpersonal appraisals or individual susceptibilities (Mei et al. 2017; Amelkin et al. 2017; Bizyaeva et al. 2020) or the role of discrete actions such as voting and deliberations (Varma and Morărescu 2017; Ceragioli and Frasca 2018; Mei et al. 2019). In several cases, this research has nicely combined theoretical investigation and data analysis (Friedkin et al. 2016; Fontan and Altafini 2021). Despite the growing scope of this research, the community should be aware that the field of potential questions in the social and economic sciences is even broader, including marketing, innovation dynamics, and information diffusion in social media.

In this chapter, we have argued about the importance of attention dynamics in social media, by focusing on the important case study of YouTube. We have provided a review of the relevant literature, emphasizing key facts that shape this dynamics, like the boundedness of collective attention and the role that novelty and popularity play in the diffusion of contents online. We have highlighted how recommendation systems function as gateways of information (and thus, conversely, of attention) and we have described a simple dynamical model of collective attention. We hope that this introduction can foster the interest of the control system community into these problems and contribute to expand the research in cyber-social systems.

Acknowledgments

This research has been partly supported by CNRS through the 80 PRIME MITI project "Disorders of Online Media" (DOOM). This work is also supported by the European Union – Horizon 2020 Program under the scheme "INFRAIA-01-2018-2019 – Integrating Activities for Advanced Communities," Grant Agreement n.871042, "SoBigData++: European Integrated Infrastructure for Social Mining and Big Data Analytics" (www.sobigdata.eu).

References

Meta's investor earnings report for 3Q (2021). https://investor.fb.com/home/default.aspx (accessed 20 January 2022).

Twitter's investors report for 3Q (2021). https://s22.q4cdn.com/826641620/files/doc_financials/2021/q3/Final-Q3\stquote21-Shareholder-letter.pdf (accessed 20 January 2022).

Amelkin, V., Bullo, F., and Singh, A.K. (2017). Polar opinion dynamics in social networks. *IEEE Transactions on Automatic Control* 62 (11): 5650–5665.

Anderson, A., Maystre, L., Anderson, I. et al. (2020). Algorithmic effects on the diversity of consumption on spotify. *Proceedings of the Web Conference 2020*, 2155–2165.

Bakshy, E., Messing, S., and Adamic, L.A. (2015). Exposure to ideologically diverse news and opinion on Facebook. *Science* 348 (6239): 1130–1132.

Barabási, A.-L. (2005). The origin of bursts and heavy tails in human dynamics. *Nature* 435 (7039): 207–211.

Bass, F.M. (1969). A new product growth for model consumer durables. *Management Science* 15 (5): 215–227.

Bettencourt, L.M.A., Cintrón-Arias, A., Kaiser, D.I., and Castillo-Chavez, C. (2006). The power of a good idea: quantitative modeling of the spread of ideas from epidemiological models. *Physica A: Statistical Mechanics and its Applications* 364: 513–536.

Bild, D.R., Liu, Y., Dick, R.P. et al. (2015). Aggregate characterization of user behavior in Twitter and analysis of the retweet graph. *ACM Transactions on Internet Technology* 15 (1): 1–24.

Bizyaeva, A., Franci, A., and Leonard, N.E. (2020). A general model of opinion dynamics with tunable sensitivity. *arXiv preprint arXiv:2009.04332*.

Castaldo, M., Venturini, T., Frasca, P., and Gargiulo, F. (2022). Junk news bubbles modelling the rise and fall of attention in online arenas. *New Media & Society* 24 (9): 2027–2045.

Ceragioli, F. and Frasca, P. (2018). Consensus and disagreement: the role of quantized behaviors in opinion dynamics. *SIAM Journal on Control and Optimization* 56 (2): 1058–1080.

Ceragioli, F., Frasca, P., Piccoli, B., and Rossi, F. (2021a). Generalized solutions to opinion dynamics models with discontinuities. In: *Crowd Dynamics, Modeling and Simulation in Science, Engineering and Technology*, vol. 3 (ed. N. Bellomo and L. Gibelli), 11–47. Cham: Birkhäuser.

Ceragioli, F., Frasca, P., and Rossi, W.S. (2021b). Modeling limited attention in opinion dynamics by topological interactions. In: *Network Games, Control and Optimization* (ed. S. Lasaulce, P. Mertikopoulos, and A. Orda), 272–281. Cham: Springer International Publishing. ISBN 978-3-030-87473-5.

Cha, M., Kwak, H., Rodriguez, P. et al. (2007). I tube, you tube, everybody tubes: analyzing the world's largest user generated content video system. *Proceedings of the 7th ACM SIGCOMM Conference on Internet Measurement*, IMC '07, 1–14. New York, NY, USA: Association for Computing Machinery.

Cha, M., Kwak, H., Rodriguez, P. et al. (2009). Analyzing the video popularity characteristics of large-scale user generated content systems. *IEEE/ACM Transactions on Networking* 17 (5): 1357–1370.

Cheng, H.-T., Koc, L., Harmsen, J. et al. (2016). Wide & deep learning for recommender systems. *CoRR*, abs/1606.07792.

Chica, M. and Rand, W. (2017). Building agent-based decision support systems for word-of-mouth programs: a freemium application. *Journal of Marketing Research* 54 (5): 752–767.

Citton, Y. (2014). *Pour une écologie de l'attention*. Paris: Seuil.

Coleman, J.S., Katz, E., and Menzel, H. (1957). The diffusion of an innovation among physicians. *Sociometry* 20: 253–270.

Colleoni, E., Rozza, A., and Arvidsson, A. (2014). Echo chamber or public sphere? Predicting political orientation and measuring political homophily in Twitter using big data. *Journal of communication* 64 (2): 317–332.

Covington, P., Adams, J., and Sargin, E. (2016). Deep neural networks for YouTube recommendations. *Proceedings of the 10th ACM Conference on Recommender Systems*, RecSys '16, 191–198.

Crane, R. and Sornette, D. (2008). Robust dynamic classes revealed by measuring the response function of a social system. *Proceedings of the National Academy of Sciences of the United States of America* 105 (41): 15649–15653.

Crogan, P. and Kinsley, S. (2012). Paying attention: toward a critique of the attention economy. *Culture Machine* 13: 1–29.

Daley, D.J. and Kendall, D.G. (1964). Epidemics and rumours. *Nature* 204; 1118.

Deschâtres, F. and Sornette, D. (2005). Dynamics of book sales: endogenous versus exogenous shocks in complex networks. *Physical Review E* 72: 016112.

Dewes, C., Wichmann, A., and Feldmann, A. (2003). An analysis of internet chat systems. *Proceedings of the 3rd ACM SIGCOMM Conference on Internet Measurement*, IMC '03, 51–64. New York, NY, USA: Association for Computing Machinery.

Dezsö, Z., Almaas, E., Lukács, A. et al. (2006). Dynamics of information access on the web. *Physical Review E* 73: 066132.

Downs, A. (1972). Up and down with ecology: the "issue-attention cycle". *The Public Interest* 28: 38–50.

Dubois, E. and Blank, G. (2018). The echo chamber is overstated: the moderating effect of political interest and diverse media. *Information, Communication & Society* 21 (5): 729–745.

Figueiredo, F., Benevenuto, F., and Almeida, J.M. (2011). The tube over time: characterizing popularity growth of YouTube videos. *Proceedings of the 4th ACM International Conference on Web Search and Data Mining*, WSDM '11, 745–754. New York, NY, USA.

Fontan, A. and Altafini, C. (2021). A signed network perspective on the government formation process in parliamentary democracies. *Scientific Reports* 11 (1): 1–17.

Friedkin, N.E., Proskurnikov, A.V., Tempo, R., and Parsegov, S.E. (2016). Network science on belief system dynamics under logic constraints. *Science* 354 (6310): 321–326.

Garrett, R.K. (2009). Echo chambers online?: Politically motivated selective exposure among Internet news users. *Journal of Computer-Mediated Communication* 14 (2): 265–285.

Ge, Y., Zhao, S., Zhou, H. et al. (2020). Understanding echo chambers in e-commerce recommender systems. In *Proceedings of the 43rd International ACM SIGIR Conference on Research and Development in Information Retrieval*, 2261–2270.

Geschke, D., Lorenz, J., and Holtz, P. (2019). The triple-filter bubble: using agent-based modelling to test a meta-theoretical framework for the emergence of filter bubbles and echo chambers. *British Journal of Social Psychology* 58 (1): 129–149.

Gillespie, T. (2016). #trendingistrending: when algorithms become culture. In: *Algorithmic Cultures: Essays on Meaning, Performance and New Technologies*. London and New York: Routledge.

Goldhaber, M.H. (1997). The attention economy and the Net. *First Monday* 2 (4). https://doi.org/10.5210/fm.v2i4.519.

Google YouTube Terms of Service (2022). Advertiser-friendly content guidelines. https://support.google.com/youtube/answer/6162278?hl=en (accessed 20 January 2022).

Gupta, P., Goel, A., Lin, J. et al. (2013). WTF: the who to follow service at Twitter. *Proceedings of the 22nd International Conference on World Wide Web*, WWW '13, 505–514. New York, NY, USA: Association for Computing Machinery.

Harder, U. and Paczuski, M. (2006). Correlated dynamics in human printing behavior. *Physica A: Statistical Mechanics and Its Applications* 361 (1): 329–336.

Hartonen, T. and Alava, M.J. (2013). How important tasks are performed: peer review. *Scientific Reports* 3: 1679/1–5.

Hawkes, A.G. and Oakes, D. (1974). A cluster process representation of a self-exciting process. *Journal of Applied Probability* 11 (3): 493–503.

Hilgartner, S. and Bosk, C.L. (1988). The rise and fall of social problems: a public arenas model. *American Journal of Sociology* 94 (1): 53–78.

Jia, P., MirTabatabaei, A., Friedkin, N.E., and Bullo, F. (2015). Opinion dynamics and the evolution of social power in influence networks. *SIAM Review* 57 (3): 367–397.

Jiang, R., Chiappa, S., Lattimore, T. et al. (2019). Degenerate feedback loops in recommender systems. *Proceedings of the 2019 AAAI/ACM Conference on AI, Ethics, and Society*, 383–390.

Jin, F., Dougherty, E., Saraf, P. et al. (2013). Epidemiological modeling of news and rumors on Twitter. *Proceedings of the 7th Workshop on Social Network Mining and Analysis*, SNAKDD '13, New York, NY, USA: Association for Computing Machinery.

Jo, H.-H., Pan, R.K., and Kaski, K. (2012). Time-varying priority queuing models for human dynamics. *Physical Review E* 85: 066101. https://doi.org/10.1103/PhysRevE.85.066101.

Johansen, A. (2001). Response time of internauts. *Physica A: Statistical Mechanics and its Applications* 296 (3): 539–546.

Johansen, A. (2004). Probing human response times. *Physica A: Statistical Mechanics and its Applications* 338 (1–2): 286–291.

Johansen, A. and Sornette, D. (2000). Download relaxation dynamics on the WWW following newspaper publication of URL. *Physica A: Statistical Mechanics and its Applications* 276 (1): 338–345.

Kämpf, M., Tismer, S., Kantelhardt, J.W., and Muchnik, L. (2012). Fluctuations in Wikipedia access-rate and edit-event data. *Physica A: Statistical Mechanics and its Applications* 391 (23): 6101–6111.

Karsai, M., Kivelä, M., Pan, R.K. et al. (2011). Small but slow world: how network topology and burstiness slow down spreading. *Physical Review E* 83: 025102.

Karsai, M., Jo, H.-H., and Kaski, K. (2018). *Bursty Human Dynamics, SpringerBriefs in Complexity*. Springer.

Kiesling, E., Günther, M., Stummer, C., and Wakolbinger, L.M. (2012). Agent-based simulation of innovation diffusion: a review. *Central European Journal of Operations Research* 20 (2): 183–230.

Latour, B. (2002). Gabriel tarde and the end of the social. In: *The Social in Question. New Bearings in the History and the Social Sciences* (ed. P. Joyce), 16. London: Routledge.

Latour, B., Jensen, P., Venturini, T. et al. (2012). 'The whole is always smaller than its parts': a digital test of Gabriel Tardes' monads. *The British Journal of Sociology* 63 (4): 590–615.

Lorenz-Spreen, P., Mønsted, B.M., Hövel, P., and Lehmann, S. (2019). Accelerating dynamics of collective attention. *Nature Communications* 10 (1): 1759.

Lu, Y., Zhang, P., Cao, Y. et al. (2014). On the frequency distribution of retweets. *Procedia Computer Science* 31: 747–753. 2nd International Conference on Information Technology and Quantitative Management, ITQM 2014.

Marois, R. and Ivanoff, J. (2005). Capacity limits of information processing in the brain. *Trends in Cognitive Sciences* 9 (6): 296–305.

McCombs, M. (2005). A look at agenda-setting: past, present and future. *Journalism Studies* 6 (4): 543–557. https://doi.org/10.1080/14616700500250438.

McCombs, M.E. and Shaw, D.L. (1972). The agenda-setting function of mass media. *Public Opinion Quarterly* 36 (2): 176.

Mei, W., Friedkin, N.E., Lewis, K., and Bullo, F. (2017). Dynamic models of appraisal networks explaining collective learning. *IEEE Transactions on Automatic Control* 63 (9): 2898–2912.

Mei, W., Cisneros-Velarde, P., Chen, G. et al. (2019). Dynamic social balance and convergent appraisals via homophily and influence mechanisms. *Automatica* 110: 108580.

Meta (2022a). Instagram community guidelines, a. https://help.instagram.com/477434105621119/?helpref=uf_share (accessed 20 January 2022).

Meta (n.d.a). Community standards - objectionable content. https://help.instagram.com/477434105621119 (accessed 20 January 2022).

Mitra, S., Agrawal, M., Yadav, A. et al. (2009). Characterizing web-based video sharing workloads. *Proceedings of the 18th International Conference on World Wide Web*, WWW '09, 1191–1192. New York, NY, USA: Association for Computing Machinery.

Mryglod, O., Holovatch, Y., and Mryglod, I. (2012). Editorial process in scientific journals: analysis and modeling. *Scientometrics* 91 (1): 101–112.

Naumov, M., Mudigere, D., Shi, H.-J.M. et al. (2019). Deep learning recommendation model for personalization and recommendation systems. *CoRR*, abs/1906.00091.

Nguyen, T.T., Hui, P.-M., Harper, F.M. et al. (2014). Exploring the filter bubble: the effect of using recommender systems on content diversity. *WWW '14: Proceedings of the 23rd International Conference on World Wide Web*, April 2014, 677–686.

Oliveira, J.G. and Vazquez, A. (2009). Impact of interactions on human dynamics. *Physica A: Statistical Mechanics and its Applications* 388 (2–3): 187–192. January

Pariser, E. (2011). *The Filter Bubble: What the Internet Is Hiding from You*. Penguin UK.

Proskurnikov, A.V. and Tempo, R. (2018). A tutorial on modeling and analysis of dynamic social networks. Part II. *Annual Reviews in Control* 45: 166–190.

Qiu, X., Oliveira, D.F.M., Shirazi, A.S. et al. (2017). Limited individual attention and online virality of low-quality information. *Nature Human Behaviour* 1: Article number: 0132.

Rand, W. and Rust, R.T. (2011). Agent-based modeling in marketing: guidelines for rigor. *International Journal of Research in Marketing* 28 (3): 181–193.

Rand, W., Herrmann, J., Schein, B., and Vodopivec, N. (2015). An agent-based model of urgent diffusion in social media. *Journal of Artificial Societies and Social Simulation* 18 (2): 1.

Richier, C., Altman, E., Elazouzi, R. et al. (2014). Bio-inspired models for characterizing YouTube viewcout. *2014 IEEE/ACM International Conference on Advances in Social Networks Analysis and Mining (ASONAM 2014)*, 297–305.

Rocha, L.E.C., Liljeros, F., and Holme, P. (2011). Simulated epidemics in an empirical spatiotemporal network of 50,185 sexual contacts. *PLoS Computational Biology* 7 (3): 1–9. https://doi.org/10.1371/journal.pcbi.1001109.

Rossi, W.S. and Frasca, P. (2020). Opinion dynamics with topological gossiping: asynchronous updates under limited attention. *IEEE Control Systems Letters* 4 (3): 566–571.

Rossi, W.S., Polderman, J.W., and Frasca, P. (2021). The closed loop between opinion formation and personalised recommendations. *IEEE Transactions on Control of Network Systems* 9 (3): 1092–1103.

Roth, C., Mazieres, A., and Menezes, T. (2020). Tubes & bubbles. Topological confinement of YouTube recommendations. *PLoS ONE* 15 (4): e0231703. https://doi.org/10.1371/journal.pone.0231703.

Samad, T. (2023). Human-in-the-loop control and cyber–physical–human systems: applications and categorization. In: *Cyber–Physical–Human Systems: Fundamentals and Applications* (ed. A.M. Annaswamy, P.P. Khargonekar, F. Lamnabhi-Lagarrigue, and S.K Spurgeon). Hoboken, NJ: John Wiley & Sons, Inc.

Sears, D.O. and Freedman, J.L. (1967). Selective exposure to information: a critical review. *Public Opinion Quarterly* 31 (2): 194–213.

Silverman, C. (2016). This analysis shows how viral fake election news stories outperformed real news on facebook. https://www.buzzfeednews.com/article/craigsilverman/viral-fake-election-news-outperformed-real-news-on-facebook (accessed 20 January 2021).

Simon, H.A. (1971). Designing organizations for an information rich world. In: *Computers, Communications, and the Public Interest* (ed. M. Greenberger), 37–72. The Johns Hopkins Press: Baltimore, MD: Johns Hopkins Press.

Solsman, J.E. (2018). YouTube's AI is the puppet master over most of what you watch. https://www.cnet.com/news/youtube-ces-2018-neal-mohan/ (accessed 20 January 2022).

Sunstein, C. (2001). *Republic.Com*, Princeton University Press.

Tarde, G. (1890). *Les lois de l'imitation*. Paris: Félix Alcan.

Tarde, G. (1893). *Monadologie et sociologie*. Paris: Les empêcheurs de penser en rond. ISBN 1554423996.

Terranova, T. (2012). Attention, economy and the brain. *Culture Machine* 13: 1–19.

Toole, J.L., Cha, M., and González, M.C. (2012). Modeling the adoption of innovations in the presence of geographic and media influences. *PLoS ONE* 7 (1): e29528.

Varma, V.S. and Morărescu, I.-C. (2017). Modeling stochastic dynamics of agents with multi-leveled opinions and binary actions. *2017 IEEE 56th Annual Conference on Decision and Control (CDC)*, 1064–1069. IEEE.

Vázquez, A., Oliveira, J.G., Dezsö, Z. et al. (2006). Modeling bursts and heavy tails in human dynamics. *Physical Review E* 73 (3): 036127.

Vazquez, A., Rácz, B., Lukács, A., and Barabási, A.-L. (2007). Impact of non-poissonian activity patterns on spreading processes. *Physical Review Letters* 98: 158702.

Venturini, T. and Latour, B. (2010). The social fabric: digital traces and quali-quantitative methods. *Proceedings of Future En Seine 2009*. Paris: Editions Future en Seine.

Weng, L., Flammini, A., Vespignani, A., and Menczer, F. (2012). Competition among memes in a world with limited attention. *Scientific Reports* 2 (1): 335.

Wojcicki, S. (2021). My mid-year update to the YouTube community. https://www.theverge.com/2021/4/9/22375702/google-updates-youtube-ad-targeting-hate-speech (accessed 20 January 2022).

Xiong, F., Liu, Y., Zhang, Z.-j. et al. (2012). An information diffusion model based on retweeting mechanism for online social media. *Physics Letters A* 376 (30): 2103–2108.

Zhao, Z., Hong, L., Wei, L. et al. (2019). Recommending what video to watch next: a multitask ranking system. *Proceedings of the 13th ACM Conference on Recommender Systems*, RecSys '19, 43–51. New York, NY, USA: Association for Computing Machinery.

Zhou, R., Khemmarat, S., and Gao, L. (2010). The impact of YouTube recommendation system on video views. *Proceedings of the 10th ACM SIGCOMM Conference on Internet Measurement*, IMC '10, 404–410. New York, NY, USA: Association for Computing Machinery.

Zino, L. and Cao, M. (2023). Social diffusion dynamics in cyber–physical–human systems. In: *Cyber–Physical–Human Systems: Fundamentals and Applications* (ed. A.M. Annaswamy, P.P. Khargonekar, F. Lamnabhi-Lagarrigue, and S.K Spurgeon). Hoboken, NJ: John Wiley & Sons, Inc.

Zuboff, S. (2019). *The Age of Surveillance Capitalism*. New York: Hachette.

19

Cyber–Physical–Social Systems for Smart City

Gang Xiong[1,2], Noreen Anwar[3], Peijun Ye[3], Xiaoyu Chen[3,4], Hongxia Zhao[3], Yisheng Lv[3], Fenghua Zhu[1,2], Hongxin Zhang[5], Xu Zhou[6], and Ryan W. Liu[7]

[1] *The Beijing Engineering Research Center of Intelligent Systems and Technology, Institute of Automation, Chinese Academy of Sciences, Beijing, China*
[2] *The Guangdong Engineering Research Center of 3D Printing and Intelligent Manufacturing, The Cloud Computing Center, Chinese Academy of Sciences, Beijing, China*
[3] *The State Key Laboratory for Management and Control of Complex Systems, Institute of Automation, Chinese Academy of Sciences, Beijing, China*
[4] *School of Artificial Intelligence, University of Chinese Academy of Sciences, Beijing, China*
[5] *The State Key Laboratory of CAD & CG, Zhejiang University, Hangzhou, China*
[6] *Computer Network Information Center, Chinese Academy of Sciences, Beijing, China*
[7] *Department of Navigation Engineering, School of Navigation, School of Computer Science and Artificial Intelligence, Wuhan University of Technology, Wuhan, China*

19.1 Introduction

An emerging trend in modern technology and an increase in the world's population present an unprecedented challenge that shifts the responsibility to cities. Our society faces significant challenges, and therefore, to alleviate the arising problems in economic, demographic, social, and environmental concerns, technology plays a vital role in developing smart cities. It leverages information and communication technologies to tackle the problems and elevates the levels of service, sustainability, inhabitants' well-being, and economic development. The concept of Smart Cities is to embed intelligence into everyday products or services to increase the efficiency with which some rudimentary but critical functions that are performed. The application of information and digital technology to deliver intelligent and smart innovative solutions to the demanding requirements of an urban ecosystem, including infrastructure, governance, transportation, health-care, and security, is usually the core focus of the Smart Cities. The development of intelligent systems has already begun in this direction globally; governments are working to make things "Smart," mostly in parallel and in partnership. Great projects to better the lives of citizens are hidden under the wave of new ideas.

The history of human development has demonstrated that productive activities in corresponding eras should be synchronized with social, scientific, and technological changes and be conducted and managed in a smart framework. The cyber–physical–social systems (CPSSs) concept was defined in 2010 for the first time (Kopackova and Libalova 2017). Cyberspace, physical space, and social space are integrated. CPSS highlights the in-depth interplay between the physical, virtual, and social worlds. CPSS integrates the social systems and the physical systems via smart human–machine cyber-space interaction and ensures that such complex systems are managed

Cyber–Physical–Human Systems: Fundamentals and Applications, First Edition.
Edited by Anuradha M. Annaswamy, Pramod P. Khargonekar, Françoise Lamnabhi-Lagarrigue, and Sarah K. Spurgeon.

and controlled (Kyriazopoulou 2015; Nowakowski and Mrówczyńska 2016). The need to strike a balance between social development and economic progress in an era of rapid urbanization is the primary driver of global interest in smart cities. A lot of research has been done on the physical problems of intelligent cities and how new technology might facilitate them.

The improvement of energy consumption, health care, traffic, education, and services involves the development of a strategy that integrates all these areas into a well-articulated systemic and global vision (Guzdial 2009). However, many scholars regard the city as a communal entity that reflects one homogenous body with one voice; social advancements and worldwide challenges such as globalization, urbanization, environment, and climate change are developing new technologies, products, and services aging. These stumbling blocks make for fantastic growing conditions. The CPSS is a new research topic in which the physical, cyber, and social worlds (as they relate to human factors) are integrated based on the interactions between these worlds in real-time. Smart city technology has the potentiality to significantly improve cities' effectiveness and efficiency, which is critical given the predicted rapid rise of urban populations over the next few decades. A smart city is one that reacts to today's challenges while also improving the quality of life for its citizens. It also assures that the city satisfies the economic, social, and environmental needs of future generations. In a nutshell, it is a good location to live, with the highest possible quality of life and efficient resource utilization. The global smart cities market will expand from US\$ 622 billion to US\$ 1 trillion in 2019 and US\$ 3.48 trillion by 2026, at a CAGR of 21.28%. The fundamental driving force of smart city development is urbanization. There has also been a rise in global life expectancy. There are currently 617.1 million individuals over the age of 65 in the world, with that figure expected to climb to 1.6 billion between 2035 and 2050.

Around two-thirds of cities have already been involved in smart city technology, and many others are considering it. Increased federal financing and robust partnerships between local governments and private-sector tech companies will help solidify the reality of smart cities.

A smart city requires a cyber–physical infrastructure with new technology platforms and particular portability, security, privacy, and big data policies. Cloud computing is one of the most prominent models for storing data, owing to its originality in terms of real-time analysis application responses and stringent security assumptions. It is critical to understand that the usefulness of smart city technology extends beyond data gathering and sharing; it is incorporated in a closed-loop network that includes a sensor, communicator, decision-maker, and actuator, all of which serve to keep the individual in the loop. As a result, a balance between physical and cyber components is essential, as these components solve critical issues of protection, safety, privacy, and energy management required by wireless data and the actuation system (Zhiyong et al. 2019; Allwinkle and Cruickshank 2011). As a result, when creating and implementing CPSS, designers must include human traits to empower users, robots, and gadgets to communicate.

It is a huge problem for this paradigm to integrate heterogeneous distributed devices into the software's collaborative environment because of the nature of the program. For instance, cameras and other sensing devices embedded in electrical lines or heating, ventilating, air-conditioning (HVAC) modules cannot connect to mobile phones, desktops, PCs, or servers. This interaction can be paralleled with the Internet's emergence in the 1980s and mid-1990s. The integration of information, audio, video, and novel applications into a constrained space has started, and the programming stage has been created for different industries and companies. New platforms with mobility and secure preparation of massive volumes of data are projected to emerge within the cyber–physical system (CPS), where the Internet is rapidly evolving. We know that IoT is one part of a broader network. However, in a smart city, the Internet is simply the structure of a transport packet from one PC to another and transfers data from different sensor machines to servers, and alternatively, orders data flow into the actuator system to perform a real-time function, for example controlling traffic headlights or HVAC components, the transmission of basic medical resources, or emergence. Smart technology in the realm of smart city normally needs an innovative way to fit all users and be conveniently accessible.

According to the CPSS, artificial intelligence and human societies will undergo a significant transformation. With automation as a piece of technological knowledge, the cyber, physical, and social environments are all seamlessly blended into the present intelligent era of information and communication technology. One of the most important innovations in automation information is the combination of information from the representation and machine learning approaches (Neirotti et al. 2014; Finger and Razaghi 2017). It is the information representations that contain the knowledge and convert it to a qualitative description and data structure format. The rapid expansion of CPSS innovations has resulted in the development of new mass applications, which have an impact on such advancements in the field of smart cities (Hayat 2016; Koop and van Leeuwen 2017; Finger and Razaghi 2017; Baig et al. 2017).

This section digs deep into the detailed definition and working of smart cities with CPSS. It also explains the importance and innovation of smart cities with the use of CPSS.

19.2 Social Community and Smart Cities

CPSSs are complicated physical systems that interact with many distributed computing parts to monitor, control, and manage them. Now, an important field of research and innovation is established, including robots, smart cities, buildings, transportation solutions, medical implants, aircraft, and many more. A fundamental necessity for establishing a cyber-capable workforce is to recognize how human cyber skills in realistic cyber operating settings may be challenged, assessed, and rapidly developed. Nowadays, our societies face major challenges in terms of climate change, energy efficiency, renewable energy, the control of diseases, increased traffic congestion, etc.

With smart cities, the goal should be to boost economic development, while also keeping an eye on the environment and making technological advancements. To benefit citizens, progress necessitates the establishment of a partner environment. A smart city's components include the following: intelligent infrastructure, intelligent buildings, intelligent transportation, intelligent energy, intelligent healthcare, intelligent technology, intelligent governance, intelligent education, and intelligent residents. Smart cities are characterized by sustainability, urbanization, quality of life (QoL), and intelligence. Climate change, energy, pollution, and social difficulties are all important factors in determining the viability of a smart city's long-term sustainability. The emotional and financial well-being of residents can be used to determine the overall QoL. The urbanization features of the intelligent city involve several factors and indicators including technology, infrastructure, governance, and economy. The intelligence of a smart city is conceived as an aspiration to improve the city's economic, social, and environmental norms and its residents. Different widely cited components of city intelligence include intelligent economics, intelligent people, smart governance, smart transportation, and intelligent lifestyle.

19.2.1 Smart Infrastructure

Through the development of smart infrastructures, technology can play a significant role in relieving those developing challenges. The aim behind intelligent infrastructures is to integrate intelligence into everyday items or services to make certain rudimentary but critical functions more efficient. The trend toward the development of smart systems has already begun. Smart city is one of the fields of application for CPSSs. The intelligent city is a municipality that uses ICTs to improve operational efficiency, share information with the public, and improve the quality of government services as well as the protection of the public (Kopackova and Libalova 2017; Kyriazopoulou 2015; Nowakowski and Mrówczyńska 2016). As a complex system, an intelligent city comprises many smart subsystems. CPSS is also the means through which artificial intelligence and machine-learning applications of the future generation can be deployed and are an increasing source of large data. CPSS is deeply embedded in physical processes of smart cities and is being used in very different applications. Regardless of the application domain, a CPSS possesses three fundamental characteristics:

(1) CPSS is very closely linked to its environment, which is any behavior change in its environs that results in changes in the behavior of the CPSS and vice versa.
(2) There are numerous heterogeneous components that make up CPSS, each with varying degrees of capability. Observation sensors that are deeply embedded into physical processes have limited capability, while the organizations that handle them are significantly more capable. Variability in CPSS workflows could lead to processing, communication, and storage capacity issues.
(3) To provide the typically coordinated service, CPSS frequently requires communication channels between its components, either inherent in or outside its physical processes (Guzdial 2009). This differs from ordinary stand-alone embedded systems. To make each of the CPS elements practical, many QoS concerns must be addressed: management of the cyber and physical processes, security, energy efficiency, interoperability, and sustainability. The security and safety of each CPS must be considered from the beginning of the development process.

Cities have evolved into cyber–physical information systems, posing new challenges for urban experience development, use of live laboratory techniques, as well as participation by the public. Reference (Zhiyong et al. 2019) tried to make product-system service design more widespread. They investigate the ramifications of a new cyber–physical public design paradigm.

It is necessary to observe smart cities as a system of CPSSs, which are the result of the integration of multiple technologies that work together to give seamless services to end-users, to enable the technology for such a setting. A large amount of data perceived from the environment and/or produced by citizens themselves can be gathered, stored, and processed using such technologies, which in turn foster social interactions.

A smart city with CPSS is a closed-loop system that collects, processes, decides, controls, and optimizes data. The effect could be disastrous, as in the instance of a "smart parking" APP that alerts cars to available spaces. As a result, several drivers converge on the few slots, increasing traffic congestion. New ways are required to deal with CPSS and citizens' increasing demands for reliable, secure, and cheap services. IoT technologies are the backbone of cyber–physical systems and are widely used in smart cities. Innovative and holistic approaches are required to evaluate both the component resources and their interrelationships in smart cities.

Smart infrastructure and the building of smart cities manage rainfall, sanitary, and sewerage networks (Allwinkle and Cruickshank 2011; Neirotti et al. 2014; Finger and Razaghi 2017; Hayat 2016; Koop and van Leeuwen 2017). These networks also reduce energy consumption without sacrificing comfort and safety by using sensors of heating, ventilating, air-conditioning, and lighting (HVAC&L) (Allwinkle and Cruickshank 2011; Neirotti et al. 2014; Finger and Razaghi 2017; Hayat 2016; Koop and van Leeuwen 2017; Finger and Razaghi 2017; Baig et al. 2017; Mariani and Giacaglia 2018; Bai and Huang 2012). Car-to-car communication and navigation systems, as well as vehicle-to-fixed location communication and navigation systems, are examples of smart transportation, e.g. car-to-infrastructure (Kobayashi et al. 2017; Chen et al. 2017; Melo et al. 2017). In addition, ITS encompasses the integrated transportation systems of rail, water, and air (Gunes et al. n.d.). A solution for location-based challenges and an advanced driver assistance system (ADAS) for safety concerns are among the goals of Intelligent Transportation (Qu et al. 2010; Nunes et al. 2015; Xiong et al. 2015; Wang et al. 2017).

19.2.2 Smart Energy

Smart energy integrates the dispersed renewable energy sources, distributes it efficiently, and optimizes power use. The power system inspector is the information from essential components of the power system and determines the control signal, and then returns them for optimized power grid operations to the basic components of the power system (Orumwense and Abo-Al-Ez 2019; Cao et al. 2020). CPSS is a large and heterogeneous interconnected transmission and distribution system with a significant load that has a risk of being targeted by a cyber-attack. The backbone of an intelligent energy system is the intelligent grid or smart grid (Yohanandhan et al. 2020). The smart grid effectively incorporates the actions and behaviors of all linked users, such as (i) consumers, (ii) generators, and (iii) combined consumer and generator users, into a formal definition. The use of ICT plays a crucial role within an intelligent grid in (i) supporting energy consumption management, (ii) supplying renewable energy sources generating power generation, and (iii) facilitating the location-dependent, point-of-sale transactional services for plug-in hybrid electric vehicle (PEVs); and (iv) strengthening customer interactions.

In Smart Grid CPSS systems, resource management robots are used to connect, operate, and operate next-generation power and electricity grids. It provides better connections and operations to generators and distributors, as well as for prosumers and provides reliability and protection of electricity supply (Xue and Yu 2017; Cheng et al. 2018). As economic factors have an important effect on power consumption, societal sensors are mounted to record huge social signals and smart sensors to gather electricity data based on the theoretical framework for CPSS (Chenxi et al. 2020).

19.2.3 Smart Transportation

There are a variety of technologies used in smart city transportation, which is referred to as "Intelligent Transportation Systems" or simply "ITS." A wide range of ITS applications have helped to improve transportation for people and products for more than three decades, not just within cities but also outside of them.

There are two major components to smart mobility in a smart community: intelligent traffic control management and self-driving vehicles. Traffic management's goal is a system in which all cars and traffic signs and control bases may share data and make appropriate decisions in a safe and optimal environment. When it comes to electric automobiles, we should pay close attention. Autonomous or driverless vehicles will play a significant role in future transportation because humans will be largely absent from the process of decision-making. It will be possible to make better decisions using a cloud-based system that allows all smart transportation system components to communicate all necessary data. Figure 19.1 shows a structure for intelligent traffic control with autonomous cars that are linked to one another. It is also connected with the smart grid.

Traffic management facilities such as traffic signs and traffic lights, as well as autos, can all be seen connected via the cloud in this diagram. By sending and receiving data, they may build a robust database. Lots of specifics are being researched in the field of intelligent traffic control for self-driving vehicles in this regard. Some examples include image recognition, traffic flow estimation, and the identification of various optimization issues.

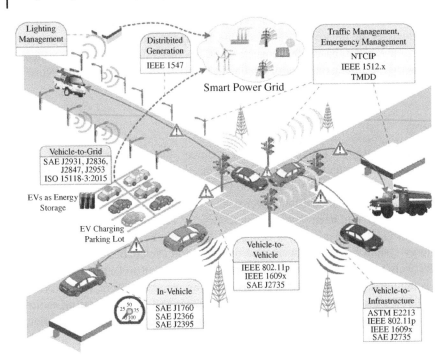

Figure 19.1 Smart transportation scheme (Turner and Uludag 2015).

Smart traffic signs and smart roads are also crucial criteria for smart transportation. In a smart city, all traffic signs and road infrastructure can make choices using data from vehicles and cameras covering all roads and intersections. The data from the start, destination, and traffic flow will be collected in real-time, and the appropriate decisions will be sent to all parts of a smart transportation system. Figure 19.2 shows an example of a smart traffic control management system.

Path planning analysis is a tough issue in the notion of smart traffic control (Amini et al. 2017). Smart traffic signs and autonomous cars make path planning in broad environments a significant data problem. To solve the multiagent path planning problem, M. J. S. B and Nambiar (Jasna et al. 2018) recommended using game theory. They compared it to the usual A* technique in their research. Path planning using stochastic obstacle avoidance is proposed in (Younis and Moayeri 2016). They are interested in single-agent path planning.

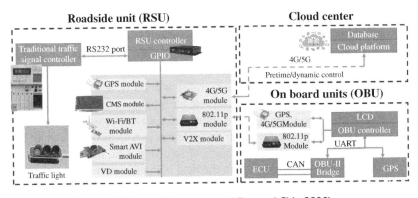

Figure 19.2 Smart traffic control management (Lee and Chiu 2020).

This area may have various issues to consider. There is also a real-time path planning challenge in a smart community that considers all smart city agents' interactions. The issue of security will be a hot topic in the coming years. Cyber-attacks are a worry because IoT is used in smart cities. Securing smart communities may be one of the future challenges.

19.2.4 Smart Healthcare

The development of medical big data analytics, together with media cloud computing, offered the potential to process, store, and manage media content relevant to health care for sophisticated decision-making (Younis and Moayeri 2016; Solanas 2014). However, given the existence of a large number of intelligent sensors, devices, media content, and intelligent communication information, it is a challenging task to devise and develop an intelligent health-monitoring framework combined with stakeholders in intelligent cities (e.g. smart homes, smart hospitals, patients, and medical professionals) (Patsakis et al. 2014; Olshansky 2016). Different data-driven approaches to machine learning and artificial intelligence (AI) understand healthcare needs and the action of health services for essential patients (Poongodi et al. 2021). There are still many initiatives and gaps to assist intelligent cities. Here, the authors proposed a patient monitoring and ambulance tracking system to diagnose health problems quickly. Figure 19.3 has shown the impact of the suggested approach on intelligent healthcare services in smart cities (Da et al. n.d.). Intelligent healthcare includes developing sensors, intelligent hospitals, and intelligent emergency response.

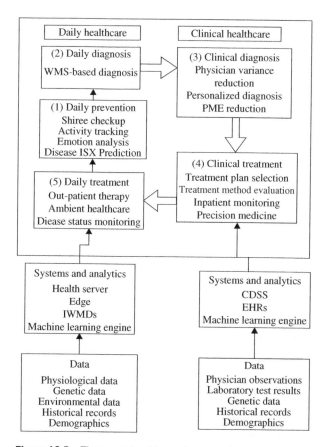

Figure 19.3 The smart healthcare framework.

Various methods such as ICT, cloud computing, smartphone Apps, and advanced data analytics are employed to conduct smart hospitals.

This section discussed the goals and drivers of CPSS for smart cities. It also includes pointers to recent reports from academies, governments, professional societies, etc. to provide a fuller appreciation of smart cities.

19.3 CPSS Concepts, Tools, and Techniques

19.3.1 CPSS Concepts

A smart, sustainable city is an inventive city using information communication and technologies, ICTs, and other means of improved quality of life, the efficiency of urban operations and services, and competitiveness while guaranteeing that it meets the economic, social, and environmental demands of present and coming generations. Developing more inclusive information and communications technology (ICT) standards is critical to developing a more inclusive approach to developing smart cities. Making cities more intelligent helps to improve city services and improves the quality of life of inhabitants. ICT is critical to advancing toward smarter city settings. The creation and integration of smart city apps could be supported by smart city software platforms. Nevertheless, before these platforms can be extensively used, the ICT community must address major scientific and technological difficulties.

There are numerous ways to define "smart cities." Changing the adjective "smart" to something like "intelligent" or "digital" might lead to a slew of conceptual variations. The term "smart city" is a generalization that is applied in a variety of contexts. Creating a smart city doesn't have a single blueprint or a single definition that applies to everyone. Then comes descriptions of government organizations, business associations, technology leaders, and educational institutions.

(1) It is a city that is poised to make a seamless transition from its current economic model to one based on self-determination, independence, and consciousness among its residents (Giffinger et al. 2007).

(2) A low-carbon economy and innovation are supported by using ICT in smart cities. ICT-enabled smart cities save money, improve service delivery, and quality of life, and reduce ecological maps (Giffinger et al. 2007).

(3) Using infrastructure coordination, smart sustainable city (SSC) aims to discover Hitachi's vision for environmental and lifestyle safety and convenience around the world. Built with the help of SSCs, infrastructure includes two levels: a lifestyle infrastructure that supports residents' daily activities, and an information technology infrastructure that connects them (Yoshikawa et al. 2012).

(4) Similarly, "intelligent" cities (clusters, regions) are multilayer territorial innovation systems that bring together knowledge-intensive activities, institutions for learning and invention, and digital places for communication and interaction. The key quality of intelligence is invention and tackling new issues (Komnions 2010) (Figure 19.4).

The purpose of "smart" city development is to provide quality and innovative services to the public, businesses, and visitors, while creating a secure, pleasant, and inclusive urban environment. To do this, a smart city must properly integrate three levels.

(1) Human capabilities and knowledge-intensive activities are included in the physical layer.

(2) Institutional layer contains adequate institutional procedures for fostering social cohesion in the pursuit of new knowledge and innovative approaches to problems. In more detail,

Figure 19.4 The Concept of "Smart" Cities.

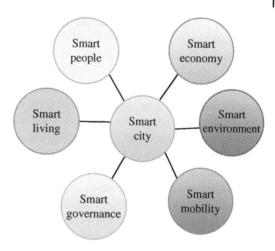

institutions and processes for information dissemination, knowledge transfer, and cooperative new product development are involved.

(3) Infrastructural digital layer that supports both individual and collective action by incorporating a variety of ICT infrastructure, tools, applications, and content.

Cities that have been deemed "smart" or "intelligent" have been defined using highly subjective concepts. Sensors and other interactive media in "smart" cities help disseminate information and foster citizen participation in civic life. However, it appears that "intelligent" cities rely more on collective/collaborative intelligence, innovation systems, and web-based collaborative spaces than other cities. Integrating the urban environment's physical, institutional, and digital components is critical in both cases.

19.3.2 CPSS Tools

In human-centered smart cities, collaboration between different sectors is essential. It is not just about finding new ways to collaborate and cooperate with city citizens in this form of collaboration. In addition, it requires the creation of whole innovation ecosystems that are accessible to all stakeholders. Human-centered smart cities are growing in terms of innovation and service creation as their ecosystems grow. Smarter cities can contribute to optimizing the use of resources and infrastructure toward greater sustainability. One option is to combine creatively the huge volumes of data created from multiple urban sources such as sensor networks, transmission systems, user devices, and social networks to create integrated services and applications, improve urban services and make better use of urban resources. However, using all these data sources properly and effectively is a challenge.

It is helpful to think of cities as interconnected ecosystems where many distinct (and sometimes conflicting) interests must come together to ensure a healthy environment and a high standard of living for all. A city's most pressing issues can be solved with the help of new technology, which makes it easier for people from different regions of the city to collaborate. "Smart City" is a concept that has gained a lot of attention in recent years from legislators and business leaders as well as residents. No single definition exists for what constitutes a smart city, but it may be summed up as any community that makes use of modern information and communication technology to enhance the quality of life of its citizens while simultaneously promoting long-term prosperity.

Smart city plans have traditionally concentrated on "top-down" approaches, even though all smart city proposals aim to improve the daily lives of citizens. The idea of a "smart city" has gained traction in political debates that take residents' needs into account but are ultimately aimed at creating and enforcing rules at the institutional level. As clients, testers, or users, citizens are often overlooked as creative and innovative thinkers and makers.

Several urban services, such as transportation control, air quality, waste management, health-care, public safety, water, and energy, have been put forward for development. Most of these solutions, however, focus on a particular area, tackle a specific problem, and have been designed from scratch with little software reuse. They do not interact and result in duplication of work, incompatible solutions, and unoptimized use of resources. Integrating all these domains into a cohesive solution requires fundamental software infrastructure and services. A novel, compre-hensive software platform might provide these basic functions, facilitating the development of advanced smart cities applications. An integrated middleware environment that aids software developers in creating, executing, deploying, and maintaining smart city applications.

The enabling technologies are used in state-of-the-art software platforms for smart cities as a tool, and there are four main categories: Internet of things (IoT) (Atzori et al. 2010) applied to control sensors and actuators responsible for retrieving information from the city; Big data (Mayer-Schönberger and Cukier 2013) to support storage and processing of the data collected from the city; Cloud computing (Armbrust et al. 2010), to provide elasticity to the services and data storage; and cyber–physical systems (White et al. 2010) to enable the interaction of systems with the city environment. This architecture includes components for the implementation of the smart cities software platform based on the most frequent enabling technologies, needs, and problems investigated in this research:

(1) Intelligent cities need a great number of historical and real-time data to work correctly.
(2) Self-sufficient and intelligent cars are increasingly incorporated into mobile applications. Consequently, cars may communicate with one other and towns. This can alleviate congestion, prevent traffic accidents, and enhance navigation.
(3) Video monitoring can help to enforce traffic, promote public safety, construct smart lighting systems, and function as a crime sensor.
(4) Cities can identify and respond to polluters by air quality sensors, pinpoint green areas with low pollution levels and offer people notifications about air quality status.
(5) Intelligent cities can preserve their data for three primary places.
(6) For analytical purposes, smart cities require vast amounts of data. Cloud data systems in their data centers use solid-state drives, delete redundant data, and encrypt data transmission. In general, cloud-based systems offer more flexible payment choices than on-site data centers.
(7) Edge computing allows you to process data near the source. Border computing can be less expensive than transmitting data to remote storage sites and subsequently to the relevant municipal authorities. In some cities, edge-computing features such as artificial intelligence (AI) traffic control are already under development. Traffic management AI uses smart automation to identify traffic congestion and accidents and to respond more quickly to various conditions.
(8) The advantages of cloud and edge computing are combined with hybrid data storage solutions. Hybrid data storage allows cities to make some new decisions on conditions and on rich storage arrays based on real-time notifications.

19.3.3 CPSS Techniques

19.3.3.1 IoT in Smart Cities

The rapid advancement of wireless technology has resulted in a significant shift in the way people go about their daily lives. They are utilizing sophisticated equipment based on the most recent technology for their everyday needs at home. Residents of modern cities worldwide can take advantage of this profitable opportunity. Increased Internet usage by citizens would result in greater Internet penetration, and IoTs would play a critical role in this regard.

Since the idea of a smart city was originally established, IoT technology has been a significant element in intelligent urban development. As technology progresses and more countries become the connection of the next generation, IoT technology continues to rise and has a greater impact on our way of living. Taking use of IoTs, on the other hand, is only part of the tale. Combining IoTs with AI in "Smart Machines" is required to imitate intelligent behavior and arrive at an accurate and reliable conclusion without the need for human involvement. Combining artificial intelligence and IoT information systems has now become a necessary precondition for achieving information system success. It is critical for the success of an information system to identify the factors that influence it. The IoT can be interpreted as a relaxing combination of three elements; it is an interaction between people and individuals through the Internet; it is an interaction between people and things via the Internet. It is an interaction between things using the Internet.

Why Is IoT Important for Smart Cities? According to the estimates from the software-defined network (SDN) Improve IoTs safety study, by 2025, there will be about 75.44 billion IoT linked devices, as shown in Figure 19.5. The Internet of Things could become one of the smartest collective and collaborative systems in history, with over 7.33 billion mobile users by 2023 and more than 1.105 million wearable devices users by 2022 (Hayajneh et al. 2020).

The cities must comprehend the benefits and opportunities of the IoT for intelligent cities with the scope for so much promise and opportunity across a wide range of industries, including urban transportation, security, sustainability, maintenance, healthcare, and management.

Sophisticated connectivity is one of the key building elements of smart urban development in the next generation. Citizens and governments will be linked in ways we have never seen before. IoT will provide smart cities with enormous prospects and benefits, but this level of interconnection will bring its own set of issues.

Smart City IoT Solutions IoT is an infrastructure including physical items, modern automobiles, buildings, and even crucial electrical components that are interconnected on an online basis so that they can acquire and exchange data among themselves. The IoTs, that is "things," have

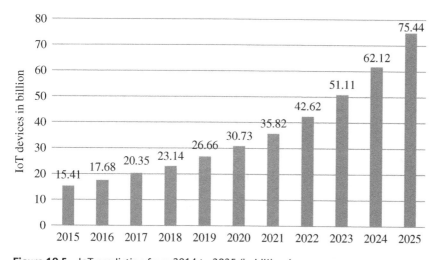

Figure 19.5 IoT prediction from 2014 to 2025 (in billions).

the priority and the ability to arrange and communicate without human interaction with others (Cristea et al. 2013). Each person has over six Internet-connected devices (Evans 2011). IoT aspires to make the Internet more ubiquitous and immersive. This includes home appliances, monitoring, surveillance cameras, sensors, displays, actuators, and vehicles. The IoT will improve the creation of apps that exploit large amounts of data created by objects to provide new services to businesses, people, and the government.

IoT Related Technologies in Smart Cities The IoT and its related technologies use the Internet to consolidate disparate devices. In this context, all gadgets available should also be connected to the Internet (Hammi et al. 2018) to ease accessibility. Sensors, actuators, cloud, data science, communication technology, artificial intelligence, and machine learning may grow to expand utilization in different locations for data analysis and collection. If devices are integrated into IoT, then an analysis system can be used to determine how well they work together. Some ongoing projects involve the surveillance of bikers, automobiles, and public parking lots, and sensor services for collecting specific data that can be used. Many IoT applications in the service domain use an IoT infrastructure to help with air and noise pollution, vehicle mobility, and surveillance systems (Figure 19.6).

(1) Intelligent Traffic and Parking Management, Public and Private Urban Transportation, and Logistics:
    One of the most important data sources in a typical smart city is traffic management data, from which people and government agencies alike stand to gain greatly by properly managing and analyzing the data (Neyestani et al. 2015). Residents will be able to plan their arrival time at a place using traffic management data (Hammi et al. 2018), too. Different cars' arrival and departure could be tracked for numerous parking lots located throughout the city by making use of smart parking (Neyestani et al. 2015). As a result, the quantity of cars in each region should be considered when constructing the layout. Tracking of vehicles or locations is a critical study topic in vehicular ad-hoc networks (VANETs). In general, retrieving multimodal data is a tough undertaking, but with the development of sensor networks, it has become possible. Several frameworks for vehicular tracking have been presented, such as multimodal 3D learning and multimodal data fusion (Zhang et al. 2018) (Figure 19.7).

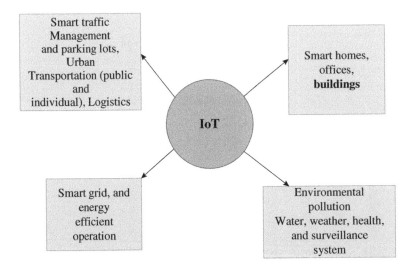

Figure 19.6 The main application of IoT for Smart City.

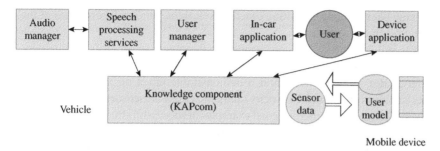

Figure 19.7 In-car positioning architecture.

Additionally, new parking lots should be constructed wherever there is a higher-than-expected availability of automobiles (Neyestani et al. 2015).

Video surveillance is not new. Smart surveillance cameras have been available for a long time, and they are widely utilized to enforce traffic laws worldwide. Law enforcement uses automatic number plate recognition (ANPR) or automatic license plate recognition (ALPR) cameras to locate stolen cars, control traffic, collect tolls, and discourage crime.

(2) Smart Homes

The use of advanced IoT technologies to equip homes and buildings may help reduce the consumption of resources associated with buildings (such as electricity and water) while also increasing the happiness of the people who live there (El-Baz and Bourgeois 2015). Data gathered by sensors might be used to maintain smart houses under constant surveillance and management (Hammi et al. 2018). Smart houses have the distinct advantage of being suitable, as connected devices can control more services and free up the citizen to conduct other tasks.

(3) Smart Grid

To develop an automated and distributed advanced energy delivery network, the smart grid makes use of modern technology such as intelligent and autonomous controllers, advanced data management software, and two-way communications between power utilities and consumers. If IoT technology is applied to the power network, it will have a significant impact on the cost-effectiveness of electricity generation, consumption, transmission, and distribution (Neyestani et al. 2015).

(4) System for Monitoring and Detecting Pollution in the Environment, Water, Weather, and other Variables

We cannot call a city unintelligent if its residents are unhappy living there. To avoid this problem, a smart city should monitor pollution levels and share that information with its residents, especially those who are ill or need medical attention. The environmental and noise data realization was provided in a separate module (Rathore et al. 2016). Water and weather systems can use different sensors, such as temperature, humidity, wind, rain, speed, and pressure, to offer useful information to smart cities (Neyestani et al. 2015). Furthermore, from the people's perspective, security and safety are the most important aspects of a smart city. The smart city should be constantly monitored to achieve this goal. However, data analysis and criminal detection improve smart city security (Rathore et al. 2016).

IoT's potential is infinite. Our urban hubs may be transformed into intelligent, sustainable, and effective environments with wide-ranging implementation, careful deployment and management, IoT, urban data platforms, Big Data, and artificial intelligence. The secret to the success of every industry, from healthcare to manufacturing to transit to education, is the common use of information. By collecting data and implementing practical solutions, our intelligent cities of the next generation are smarter than ever.

Connected streetlights and lighting systems are popular to increase a smart city's productivity. The fundamental advantages of smart lighting include reducing energy and maintenance costs, increasing public safety, safer transportation, and a demonstrable effect on the environment. Intelligent streetlights can also function as electricity chargers, emission sensors, and broadband wireless connection points. There are numerous good examples of successful streetlights, but some remarkable short success stories are here. Copenhagen, Denmark –the smart lighting solution in Copenhagen has substantially reduced energy expenses by almost 60%. Bristol, UK – local authorities have replaced the traditional streetlights in cities with a new system in Bristol. This has saved more than £1 million per year. Barcelona, Spain – Barcelona boasts a network of intelligent street lighting providing mobile broadband connectivity. They can potentially be used in the future for more IoT capability.

Smart security cameras are growing increasingly advanced and may be used to forecast future crimes. They may also be able to recognize and track pedestrians. These types of extensive surveillance are unpopular since many citizens express privacy concerns and question their data usage.

The IoT network combines several techniques. These technologies do not effectively determine an IoT element's condition. As a result, managerial costs and complexity rise. The suggested method uses artificial neural networks to forecast an element's state. The rate of service, loss, and specific lost load are utilized to determine an element's status. The artificial neural networks ANN-IoT model combines probabilistic and multilayered perceptron NN. The multilayered perceptron forecasts the state, whereas the probabilistic network analyses the sample for a class – the multilayered perceptron trains on available statistical data. The multilayered perceptron output connects to the probabilistic NN input (Estimote 2018). The rise of IoT devices raises security concerns. Thus, securing IoT devices requires a secure mechanism. The proposed technique starts with a testbed overview. The IoT testbed is built using machine learning, imitated edge devices with Arduino Uno, Arduino boards with Wi-Fi, and temperature sensors (Gimbal 2018; Das et al. 2018).

A cloud-based machine learning method is proposed to lessen customer reluctance to disclose sensitive data and build a scheme to solve privacy difficulties. The system suggests doing some initial calculations on the client and sending the results (partial data) to the server. The server will receive encrypted data (encryption should be holomorphic), which can be processed without converting to plaintext (Gharaibeh et al. 2017; Anagnostopoulos et al. 2017). The system proposes to achieve privacy by making data conversion irreversible at the server. Because the user handles the first processing, this technique improves performance.

Fog computing requires advanced cyber-attack detection schemes. It acts as a service provider for distributed nodes in IoT fog computing. Its distributed environment requires authorization and interference handling. Cloud computing has also transformed the business society sphere significantly. It currently does not support fog-to-things computing. It is used in offline cyber-security breach detection activities from the cloud to smart things. Fog computing with a distributed design minimizes the storage memory and calculating control of security operations from IoT devices while also reducing Cloud latency issues. Due to their dispersion and resource restrictions, IoT Fog nodes are the most efficient phase of attack detection. Deep learning (DL) is used to detect attacks on the Fog level. It necessitates data distribution and model handling. Currently, the Fog network uses stochastic gradient descent (SGD). However, a distributed network that requires parallel computations is not an option. The IoT collects massive amounts of data, which are challenging to manage for the SGD application (Handte et al. 2016). Menouar et al. propose a smart personal health adviser (SPHA) to assess users' general health (Menouar et al. 2017). It is possible to examine physiological states by collecting data from many sources, such as structured data (user's medical record), text data (doctors' diagnoses), and video and picture (medical device) data. Psychological indicators, such as speech and facial expressions, are retrieved using temporal Bayesian fusion (Kotenko et al. 2015). Cañedo et al. propose a deep learning strategy for dependable diabetic therapy (DLRT) that distinguishes phony or genuine glucose levels (Cañedo and Skjellum 2016). The OneTouch meter measures glucose in the patient's finger blood and sends it to the insulin pump. A sensor must be implanted inside the patient to monitor blood glucose levels, and the insulin pump does its job. The glucose reading is then sent to a mobile device (smartphone or laptop) for uploading to a server.

19.3.3.2 Big Data in Smart Cities

Choosing a big-data analysis platform for a smart city is difficult due to the variety of data and requirements. When choosing a big data platform, factors including the underlying data processing architecture, data storage, data analytics support, and powerful visualizations should be considered. As a smart city generates so much data, it is best to handle it locally using cloudlets, edge, and fog computing. Data generating is feasible in practically every human activity, promising new insights into our reality. This data availability shows how Big Data may help optimize resource utilization while making informed decisions. AI can analyze Big Data by training computers to think like humans.

Taxonomy and Big-Data Characteristics in a Smart-City Medical, energy, traffic, and environmental monitoring are just a few of the applications that generate data continuously in smart cities. The rate of data creation for different sensors varies, making data processing difficult (Xiong et al. 2015). For example, GPS sensors generate data in seconds, but temperature sensors generate data hourly. The quality of data generated by intelligent cities is critical due to the variety of sources. To ensure data quality, the data source should be dependable, and the sampling frequency should be increased.

The structure of taxonomical five dimensions is separated into five vertical computing facilities: computer infrastructure, storage infrastructure, data variation, data analytics, and data visualization.

Computer Infrastructure This refers to the various processing systems that are generally utilized for massive data volumes from intelligent cities. The computer infrastructure can process the data in real/near real-time or batch mode depending on the data requirements. Hadoop is popular for batch processing, for example, whereas Spark is utilized for real-time processing.

Infrastructure Storage The data collected from smart cities, from multimedia to text, are very varied. Many sensor data are structured by nature and so various types of databases are necessary in addition to the regular relationship structure. Therefore, the second vertical storage infrastructure is used to determine the type of storage needed depending on the type of large data. In addition to SQL-based storage systems such as Oracle, MySql, among others, smart cities also require NoSQL (MongoDB, Aerospike, HBase, Cassandra, etc.) and NewSQL systems based on HStore, VoltDB, etc.

Data Variety The unstructured data are generated by smart cities. Time series data are collections of values or occurrences collected throughout time. Streaming data are those that arrive continuously, such as sensor data or Internet traffic. Sequence data are ordered elements or events recorded with or without a sense of time (Cristea et al. 2013), for example, smart-retail data and human DNA. As a result of this, data from social networks, the Internet, and human body area networks are

Table 19.1 Mapping between data types and analytical technique.

Data type	Analytical technique
Time series	Hidden Markov models, Markov random fields, dynamic time warping (DTW), etc.
Streaming	K-means, distributed SVD, linear SVM, Kernel SVM, parallel tree learning, etc.
Sequence	Hidden Markov models, sequence alignment algorithms such as BLAST (basic local alignment search tool)
Graph	Collaborative filtering, page rank, singular value decomposition (SVD)
Spatial	Spatial autoregressive model (SAR), Markov random field (MRF) based Bayesian classifier algorithms
Multimedia	Motion vector analysis, Various types of statistical models, association rule mining-based algorithms

ideally suited to be described as graph data structures. Spatial data include information from remote sensing, geographic information systems, and medical imaging. The last type is multimedia data. Each data type has its own properties and is evaluated using different data mining techniques. Table 19.1 shows the relationship between data categories and data mining methodologies.

Data Analytics For producing predictions, finding patterns, locating hidden information, or making decisions, machine-learning algorithms are utilized (Evans 2011). Analytical techniques must be adopted based on the requirements. Surveillance algorithms are used for classification, prediction/regression, and source signal separation. In big unlabeled datasets, typical supervised techniques cannot be applied. Semi-supervised learning techniques use structural commonalities between labeled and unlabeled data to generalize functional mapping across large datasets efficiently. The goal of reinforcement learning is to maximize a reward function.

Data Visualization One of the primary benefits of big data in a smart city infrastructure is that they help human stakeholders appreciate the value of data by visualizing it. Spatial visualization schemes allow data objects to be mapped to specific points on a coordinate system, simplifying complex data sets. Line charts, bar charts, scatter plots, among others are examples. The abstract visualization technique summarizes large-scale data before visualizing them (Hammi et al. 2018) – heterogeneous data aggregation, histogram binning, etc. Third, interactive visualization allows for real-time visualization and user involvement, for example Microsoft pivot table and tableau.

Bigdata Technology Platform for "Smart Cities" This section provides a quick introduction to big data analytical systems that collect data, perform required analytics using appropriate data mining techniques, and visualize data. Choosing a big-data analysis platform for a smart city is difficult due to the wide variety of data and requirements. When choosing a big-data platform, factors including the fundamental data processing infrastructure, data storage, data analytics support, and powerful visualizations should be considered. Given the massive amounts of data generated by smart cities, it is always best to handle and analyze data closer to the source using cloudlets, edge, and fog computing.

Table 19.2 lists the smart-city big-data analytics solutions. In real-time data analytics systems like SAP-Hana, applications that require ongoing monitoring may be easily monitored.

Table 19.2 Smart city big data analytics solutions.

Name	Key features	Disadvantages
Hadoop (Hopkins and Hawkin 2018)	* Handles wide variety of structured, unstructured, and semistructured data * Uses cluster of commodity hardware * Economical * Easily scalable * Faster processing due to parallelism * Fault tolerant due to replication	* Not suitable for small datasets * Has high I'O overhead * Lacks encryption at storage and network level
Cloudera Data Hub (Diaconita 2018)	* All the advantages of Hadoop * Enterprise grade platform ideally suited to smart cities	Does not have its own hardware and software systems * Dependent on third parties for privacy and security issues
SAP-Hana (Vera-Baquero 2018)	* Handles text and unstructured data * In-memory platform mainly catering to different types of transactional needs and hence fast * Low latency * Inbuilt support for R language providing power visualizations	* Extremely expensive * Very limited flexibility due to strict hardware restrictions * Leis powerful when compared to Hadoop-based systems
MongoDB (Santos 2018)	* The query-response time is short * Supports the concept of horizontal scalability * Particularly suitable for conducting real time analysis	* Provides very limited support for shared extension and complex queries * Limited support for event types of EPCIS standard
CiDAP (Cheng 2015)	* Handles wide variety of structured, unstructured, and semistructured data * Suited for IoT platforms by supporting the publish-subscribe paradigm * Highly scalable in terms of storage and processing * Support for powerful real-time analytics * Support huge data volumes (up to 50 TB) * Strong data compression scheme and hence suitable for smart industries	* Closed system, i.e. not interoperable with other vendors * Very limited security * There is no support for anomaly detection
Infobright (Sleazak 2010)	* Very efficient due to the use of data skipping technology and columnar design	* High cost * Table modifications consume a lot of time * All queries cannot be answered optimally using the Info bright optimizer
Map (Hopkins and Hawkin 2018; Gohar 2018)	* All the advantages of Hadoop * Better system recovery features * Provides enhanced security	* High overall system complexity
IoTDSF (Gouveia et al. 2016)	* Capability of cross-platform data access * Supports the concept of horizontal scalability * Supports multitenant management	* Requires support from numerous database adapters * More evaluation is needed based upon performance metric

Massive analytic systems are utilized for large data sets. Massive analytics are possible using Hadoop and Cloudera (Neyestani et al. 2015; Hammi et al. 2018). These systems are typically off-line; thus, a speedy reaction is not necessary. In some cases, data can be less than a cluster's memory. Therefore, memory-level analytics may be used, which is ideal for real-time analytics. MongoDB is an example (Zhang et al. 2018). The platforms listed in Table 19.2 are indicative of the various analytical systems addressed previously.

This section focuses on the CPSS concepts, tools, and techniques for smart cities from the technical perspective. The main objective of this section is to explain the integration of engineering and social-behavioral sciences.

19.4 Recent Research Advances

A smart city's objective consists of increasing the quality of life by collecting data from interconnected sensors, devices, and people. Perpetual urban challenges, for instance, security, waste management, transportation, and traffic, are dealt with through the use of data to increase efficiency, but to do so, all data must be stored in a place where all parties – private and government – can readily access and use it. Environment cloud will assist break down intergovernmental silos that prevent different departments from communicating and understanding other departments' data-based priorities – an issue that has been considered a major barrier to smart city adoption. The continuing perpetuation of the IoTs has caused demonstrable security problems, and safety is also a major part of the new product.

Technological breakthroughs contribute to making intelligent cities a reality in the building industry through CPS. CPS links information and the physical world using IoTs.

19.4.1 Recent Research Advances of CASIA

The integrated building information models (BIM) with the IoT and CPS sensor technology are becoming increasingly popular in the built-in environment. Future developments could see digital twins being used, which could create new CPS opportunities employing surveillance, simulation, and optimization technologies. However, researchers often fail to take full account of the safety consequences. To date, BIM data and cybersecurity principles cannot be assimilated widely, and security has thus far been disregarded. Reference (Kopackova and Libalova 2017) examines empirical research on IoT applications in the built environment and analyses real-world IoT applications to better construction, human lives, and cybersecurity. This study analyses state-of-the-art research on digital twins in the built environment, IoT, BIM, urban cities, and cyber security to answer these issues. The outcomes of the study confirmed the relevance of using IoT and BIM digital twins. In addition, eight reference zones throughout Europe were recognized for their contributions to the progress of IoT science. Therefore, this article analyses the usage of digital twins in CPS to come up with proposals to expand BIM requirements to promote compliance with IoT, enhance cyber safety, and combine digital twins and city standards into smart cities of the future.

The indivisible CPS parts are a typical complex structure of CPSS, which can be controlled and managed by traditional theories and techniques. The artificial societies, computational experiments, and parallel execution (ACP) approach is introduced (Kyriazopoulou 2015), using data-driven models for the social system. A similar description of the physical–social system (PSS), i.e. the cyber systems, is applied. Computer experiments are used for validation of the control scheme. Parallel execution will ultimately take place in the CPSS step by step. In this work, the mode of travel like by car, by bike, on foot can be determined by

$$P = \frac{\exp\left(a_k D_{ij} + \beta_k \log\left(C_j\right) + \gamma_k\right)}{1 - \exp\left(a_k D_{ij} + \beta_k \log\left(C_j\right) + \gamma_k\right)}$$

where P indicates the probability that an agent takes the kth type of activity; a_k, β_k are the coefficient corresponding to the k-class activities. D_{ij} denotes the distance between site i and j; C_j represents the area of the jth site.

Intelligent city services, intelligent apps, and intelligent devices are the tools and artifacts that challenge and sometimes even disturb conventions, standards, and behavioral rites, and so lead to different behavioral shifts on the individual level, the group, and society at large. In this regard, the smart city can be seen as a laboratory of its sort to examine the intricate interplay between humans and computers from a multidimensional perspective. This study (Nowakowski and Mrówczyńska 2016), examines the present smart city monitoring technologies to determine their primary limitations and sources of often justified controversy. It is believed that the utility of mesh technology should be studied to avoid these limitations in the first place. Second, it is also stated that citizens must not only bring their people back into the smart city debate but also highlight ways by which they might participate in the co-design of intelligent city solutions and in urban decision-making. To overcome all these imperatives, smart cities are designed as an intelligent service system and a wireless integrated enhanced mesh (WIMTE) intelligent city surveillance system is therefore developed (Figure 19.8).

Many studies have been done on CPSS due to its flexibility, reliability, and security. Due to its contributions, CPSS has a favorable impact on the city and municipal affairs, and smart citizens should be utilized to construct the smart city community. The study in reference (Zhiyong et al. 2019) aims to use the CPSS infrastructure to report local problems and use the teams waiting to solve them more efficiently. Using CPSS architecture, the system's stakeholders, cyber, physical, and social aspects, are explored in levels. This allows changing job assignment algorithms directly at the cyber layer. The proposed system is simulated using a greedy approach for the job assignment.

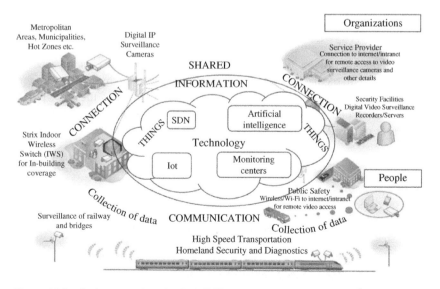

Figure 19.8 An integrated model for WIMTE: smart cities as smart service systems (adapted from: Strix Systems) (Kashef et al. 2021).

In reference (Allwinkle and Cruickshank 2011), a network of edge computers for autonomous vehicle control has been built. Each car has a cyber-network agent programmer on the edge communicating with other vehicles. The advantage of connected vehicle control over the standalone car is sophisticated global optimization control. For all vehicles in the society, the network edge computer estimated "social welfare." This chapter outlines an edge computer's control architecture. Also, discussed are some experimental outcomes for controlling autonomous vehicles. The minimizing formula used by this chapter is the sum of waiting time for all vehicles under maximized social welfare (MSW) that is

$$\sum_{i=1}^{n_1} D_{V1_i} + \sum_{j=1}^{n2} D_{V2_j} = \text{Delay total}$$

n_1 and n_2 are the number of waiting vehicles.

With the rapid advancement of artificial intelligence, humanity is entering the era of intelligent connectivity. As a result, the demand for sensors is surging. Traditional sensors require external power to operate, and the energy consumption is considerable for widely distributed and intermittent equipment, making it difficult to design green and healthy applications. However, self-powered sensors using triboelectric nanogenerators (TENG) can harvest energy from the environment and store it (Neirotti et al. 2014). These self-powered sensors are essential for smart cities, smart homes, smart transportation, environmental monitoring, and biomedicine. This study covers recent research advances on self-powered sensors based on TENG in the IoT, robotics, human–computer interaction, and intelligent medical sectors. TENG's working mechanism and the basic principles of self-powered sensors are shown in Figure 19.9.

TENG is Maxwell's displacement current in the equation below, which is the current generated by the changing electric field plus the polarization change rate of the medium. P_s is the term in the current displacement vector D to derive the output power of TENG:

$$\nabla \times H = J' + \frac{\partial D'}{\partial t'}$$

$$J_D = \frac{\partial D}{\partial t} = \varepsilon \frac{\partial E}{\partial t} = \frac{\partial P_s}{\partial t}$$

Currently, the rapidly growing notion is smart city development with self-reliant capabilities to handle societal concerns. Any firm can grow by upgrading existing systems and expanding services to fulfill all human requirements. Immersive technologies including virtual reality, mixed reality, and human–computer interaction have multiplied. These technologies are used in industry, education, entertainment, media, IT, and real estate. Mixed reality and human–computer interaction are evolving faster. In (Finger and Razaghi 2017) the study outlines the conceptual framework for future application development in the above areas. People and computers, notably digital technologies, connect readily. With the integration of new technologies, these encounters

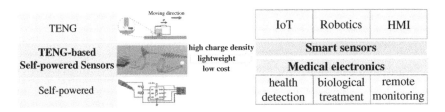

Figure 19.9 Schematic illustration of TENG-based self-powered sensors (Jiang et al. 2021).

have become more natural. The model uses computer-linked hand motions and displays the results on a head-mounted display.

19.4.2 Recent Research in European Union

In today's fast-paced society, smart cities are essential because they are a way to improve the world. The standard of living for residents of cities rises as cities grow, so new ways must be found to combine these two trends. According to the United Nations, more than 55% of the world's population lives in urban areas (United Nations 2015).

Urbanization is posing new challenges to European cities. The use of clean energy and modern buildings are a few of the issues that must be addressed to reduce the environmental impact of urbanization. Clean energy for all Europeans is a plan from the EU to help cities become more environmentally friendly, sustainable, and livable. Smart cities are a good way to do this. The European Commission's "European Innovation Partnership on Smart Cities and Communities" project shows that "smart cities" are becoming more important in the EU. European Union citizens' lives are affected by new energy policies meant to make their lives better. Smart cities are a good example of how this works. There are now new business models in big cities that use technology to improve people's lives. Technology-based business models are important because they improve the quality of life for residents, cut down on greenhouse gas emissions, and create jobs in the IT industry, which encourages new ideas. Social change is also made when experts from different fields work together to build a new kind of city, the "smart city," in this case. For this study, the researchers thought there was a lot of support in the European Union for "smart city" projects that used information technology and artificial intelligence. For people in the European Union, law, economics, and technology all work together to create synergy effects. Qualitative research methods are used to look at the research hypothesis, and case studies are also used to look at how it works in the real world (e.g. Austria, Finland, Romania). For the case studies, two very different countries in the European Union are looked at because of a new trend toward being more eco-friendly (EC 2018).

For this ambitious policy goal, energy efficiency must improve. These developments are aided by IoT, AI, and consumer awareness. We must improve building efficiency and consumer awareness. The energy performance of buildings directive (EPBD) establishes and evaluates such concepts and developments. It follows the EU's new energy policy (EC 2018). Buildings consume 40% of energy and emit 36% of CO_2 in European cities. A zero-energy building solution must be found by 2020 (EC "Buildings." 2018). The policy goal is zero energy and carbon new buildings.

These policy changes should be accompanied by new technologies such as the IoTs and AI (SET-Plan). This strategy will help the EU achieve a low-carbon economy by 2050. The European Commission published a roadmap for a low-carbon economy by 2050 in 2011. The documents also discuss "smart cities" as defined by the EU's Smart Cities and Communities Initiative. This initiative's goal is to improve the quality of life for EU citizens by collaborating with all stakeholders (European Commission 2015). The EU's 20–20–20 strategy calls for the development of energy-efficient cities, and ICT and AI can help by opening new business opportunities for citizens and residents alike. Using AI to improve smart cities makes sense, but only if citizens are willing to use it. Smart cities could create new markets for products like smart homes and smart cars. The European Union's smart cities initiative has created new markets and products. New communication technologies and networks must be used to ensure that energy systems in a smart city are safe from cyberattacks (O'Dwyer et al. 2019). If you have a system that controls or

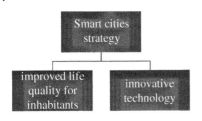

Figure 19.10 Pillars for improving smart cities.

collects data, you will soon face new high-level processes from the people who run things for you. It will also be part of new security policies for smart cities to think about how to make them safe (Lacinak and Ristvej 2017). Figure 19.10 shows the two main ways to improve smart cities.

People in the EU start thinking about how to better their lives. They are looking for the best resources to help them implement green energy policies; therefore, smart cities become an example of this. Eurostat created the EU Urban Audit to keep an eye on issues like sustainability. National statistics workers help create this tool. It is based on data from Eurostat yearbooks. This Eurostat tool is useful for assessing the smart city landscape across Europe. This means we no longer must worry about a lack of data or a single method.

A "smart city" is a place where people can live comfortably while conserving resources and utilizing cutting-edge technology. Vienna, the Austrian capital, achieves this (StadtWien 2019). From now until 2050, Vienna aims to save energy, money, and resources. As shown in Figure 19.11, Vienna's smart city has many unique features.

Life quality is a long-term goal in developing smart cities, as shown in the Vienna case study. Technology can help achieve this goal by assisting in the implementation of new sustainable energy policies. Given the importance of technologies in implementing the concept of smart cities, new business models based on information technology and artificial intelligence may emerge. Life quality was cited as the main goal of smart cities by 55% of survey respondents. Technology was important for 30% of those who polled, while energy conservation was important for 15%. The city of Vienna was chosen for this case study because it is known for its urban development (BBVA API_Market 2021).

In Graz, Austria, energy efficiency is part of improving the quality of life (Smart City Graz 2021). This concept is implemented in "Smart City Project Graz" (Hammerl 2014).

Smart cities are gaining popularity in Romania. Cluj-Napoca has made impressive progress in this direction. It has created a low-emission public transportation system using cutting-edge technology like electric buses. It also tries to reduce noise pollution and provide residents with smart parking. Figure 19.12 depicts its main features.

Smart communities, smart environment, smart living, and smart government are all implemented in Oradea, Romania, to support the digitalization of local government (Oradea 2021).

Energy efficiency is one of the main goals of the EU's energy policy. Sadly, most customers don't know about or aren't interested in this goal. Only 10% of the people who took the survey thought that energy efficiency was important for sustainable development. Only 7% of the people who took the survey thought that smart cars could help make new energy policies happen. In other EU countries (such as Austria and Romania), people live differently (e.g. Austria and Romania).

Figure 19.11 Attributes of the smart city Vienna.

Figure 19.12 Development of smart cities in Romania: Cluj-Napoca. Source: Adapted from Smart City Romania Association (2019).

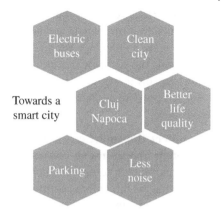

The European Commission's circular economy indicator is an important one that could show how people think about energy policies. Figure 19.12 shows that the value of this indicator varies a lot from country to country in the EU.

Application programming interfaces APIs for smart cities can be developed to increase the city's digital connectivity for residents. In Finland, smart city applications are already available. In Finland, the goal is to have connected smart city applications. This means that citizens in smart cities will have similar access to data (Data Business Finland 2021). The cities, such as Helsinki, Tampere, Espoo, and Turku, benefit from technology (European Data Portal 2021).

Technology has an impact on every aspect of society. This has considerable social significance and benefits.

19.4.3 Future Research Challenges and Visions

The smart city concept began as an aspiring approach to city development; several technological signs of progress, such as artificial intelligence (AI) and the Internet of Things (IoT), have encouraged a realistic and practical evaluation of the intelligent city framework. Today's cities face major issues such as expanding population, physical and social infrastructure shortages, environmental and regulatory demands, dwindling revenue bases, and budgets with higher expenditures. They need to find innovative and intelligent ways of dealing with the complexity of urban life, issues ranging from pollution, overpopulation, and urban sprawl to inadequate housing, high unemployment, management of resources, environmental protection, and rising crime rates. The smart city projects will develop scientific and technology research encompassing the several layers of networked services and applications in the face of the scientific and technological challenges and propose new answers to the future Internet and smart cities problems.

By 2050, the world population will be over 9 billion, with 80% living in cities. Cities now house just over half of the world's 7 billion inhabitants. While cities only account for 2% of the Earth's landmass, they consume 80% of its natural resources. The old resource delivery systems rely on city expansion and excessive physical and social resources consumption. Local action is vital in a low carbon future, even as urbanization increases global emissions. Lacking legally mandated international climate action and little state leadership, cities are currently creating methods to address climate change consequences through mitigation and adaptation. Cities are our best hope for overcoming the chronic twin challenges of "Global Warming," "Food Security," and ushering in a new era of "Green Economy" in many ways.

The major intellectual challenge society faces is accepting that as we develop new digital technologies. Therefore, we should also study their application, implementation, and social impact. The challenge is to develop truly smart cities that will improve the quality of life for all habitats intelligently. As new forms of data and advice are implemented using crowdsourcing, informed citizens begin to make a difference. Mobile and other applications generate new forms of preference elicitation while the economy moves online, and material cash is disappearing. These profound shifts require equally powerful science, which Future ICT will utilize.

For this purpose, we may identify five important scientific difficulties that relate to the six goals listed above.

To connect the smart city's infrastructure to its operational functioning and planning through management, control, and optimization: ICT is rapidly becoming integrated into the smart city in terms of materials and infrastructure, while wireless solutions proliferate in ways that are difficult to comprehend. Simultaneously, advances in computation and data analytics are utilizing these same technologies to influence the planning of such cities. There needs to be a concerted effort to demonstrate how such advances may be incorporated so that cities can truly become smart in terms of how their planners and citizens use such technologies to improve the quality of life. This form of multimodality poses a significant obstacle.

To Investigate the Concept of the City as an Innovation Laboratory: ICT is being developed to boost the efficiency of energy systems, to improve the delivery of city-based services ranging from utilities to retail, and to improve communications and transportation. The potential of creating real-time models of cities using frequently observed data is now a distinct possibility, and smart cities should evolve cognitive functions in the form of laboratories – which enable their monitoring and creation. A city's competitive edge is critical to such intelligence.

To develop technologies that promotes equity, fairness, and a higher standard of living in cities: Efficiency and equity must coexist. New technologies tend to polarize and divide on a variety of levels, and we need to investigate how future and emerging technologies might be used to better urban and transportation planning, as well as economic and community development. While the smart city has the potential to eliminate the digital gap, it will also create new divides, which we must foresee and plan for.

To create portfolios of urban simulations that serve as a guide for future designs: As the real-time city and its sensing capabilities advance toward delivering knowledge about longer-term changes, the building of urban simulation models will gain a new sense of urgency. Aggregate models will be replaced by disaggregate models, and our study will investigate a variety of different types of models, building on and expanding the complexity sciences. We believe it is critical to developing numerous models of the same circumstance because a pluralistic approach is critical to improving our knowledge of this complexity.

To create technologies that facilitates broad participation: The new information and communication technologies are mostly network-based and enable substantial interactions across a variety of areas and scales. Coordination and integration processes using state-of-the-art data systems and distributed computing must include mechanisms for citizens to contribute and combine their personal knowledge with that of specialists building these technologies. Concerns about privacy as well as security are critical components of this dilemma.

To maintain and improve urban population mobility: New ICT has the potential to significantly improve mobility on a variety of levels, leisure, and social activities, and so enabling citizens to increase their satisfaction with life (Table 19.3).

This section of the chapter highlights the future research challenges. The objective of presenting future challenges and vision is to provide new researchers in this field guidance of information technology services.

Table 19.3 Research challenges and major requirements.

Feature	Applications	Benefits	Research challenges	Major requirements
Security (Kopackova and Libalova 2017; Kyriazopoulou 2015)	ITS, e-healthcare, smart schools, logistics	Secure attack-free operating environment for service deployment	(1) The absence of standardized security solutions without compromising data integrity (2) Secure deployment and integration of device and network cloud-based services (3) Efficient early identification of dangers both inner and outsider	Identification of network vulnerabilities as weak entry points for various attacks
Trust (Kyriazopoulou 2015)	ITS, e-healthcare	Ensures user confidence that the services requested are free of vulnerabilities	(1) Efficient decentralized system of confidence management (2) Intelligent trust evaluation and compromised IoT network during service outage	Decentralized trust paradigm that avoids a single point of network failure
Privacy (Kyriazopoulou 2015)	ITS, e-healthcare	Provides the data protection and network user privacy	Ensure anonymity of IoT network users for the use of certain services	Robust encryption and cryptographic technologies network model
Risk management (Nowakowski and Mrówczyńska 2016)	ITS, indoor e-healthcare	Ensures security through the identification of unclear IoT network events and threats	(1) Low-cost and efficient risk management systems capable of successfully identifying newly discovered assaults (2) Rapid and dynamic risk decision-making processes to mitigate recognized hazards	(1) Threat modeling to discover network threats (2) Identify threat actors and asset-based threat modeling
Interoperability (Guzdial 2009)	ITS, smart home, personal e-health ecosystem	Provides a platform for communicating two IoT devices from different domains	Integration of locked-in vendor devices	Generic, centralized, flexible and open reference models for integration and communication devices (e.g, IP, CoAP)

(Continued)

Table 19.3 (Continued)

Feature	Applications	Benefits	Research challenges	Major requirements
Low power and low-cost communication	ITS, smart meters, e-healthcare	Provides a wide range of applications with low-cost communication in IoT-based intelligent cities	How to extend the battery life of Internet of Things devices?	Microelectronics and wireless communication improvements to enable low-cost communication and increased battery life
Big data (Zhiyong et al. 2019)	Smart meters, ITS, e-healthcare	Improves IoT network performance by processing useful information identified by verified sources (e.g. traffic data analysis can reduce the processing of traffic congestion)	(1) The lack of proper instruments to handle information created enormously (2) Privacy and security protection of users (3) Efficient central data collection and information	(1) Centralized centers for large data processing (2) Public awareness of the safe use of IoT resources
Connectivity	ITS, waste management, e-healthcare, smart industry	Ensures IoT devices can connect from different domains	How can connectivity be ensured in a large range of IoT devices without a communication and mobility network?	(1) Efficient use of spectrum for communicating IoT devices (2) Smart use of all possible media (e.g. Wi-Fi, 3G, LTE, WiMAX) (3) Development of gossip-based methods to connect IoT devices without a communication network

19.5   Conclusions

"Smart cities" is the latest notion for developing future cities. The key to integrating a sustainable future with continuing economic growth and employment creation is anticipated in smart cities. There are several definitions of a smart city, including sustainable, sustainable, intelligent, and green. However, the unifying denominator seems to be access to data and clever tools to link knowledge and drive change. Mart Cities' strategic utilization of innovative and high-tech, ICT-based solutions to connect inhabitants and city technology on a shared platform is the main thing compared to "Eco Cities" and "Sustainable Cities."

The intelligent city concept is intimately linked to digital technologies. Several aspects of the usage of digital technology in intelligent cities are of concern. One is the lack of adaptation to the possibilities of new technologies in city structures. Another is the lack of broad city restructuring before implementing smart city technologies. Applying IoT and other modern digital technology solutions to ancient, nonupdated urban structures limits the ease of implementing intelligent city projects. In smart cities, non-Internet technology coexists with Internet technology. Many people still do not have access to digital technology. A good smart city solution takes the real environment of the city into account and does not prevent the use of old digital and nondigital technology.

Through the prism of the city, we will examine the fundamental problems of social evolution. This will entail developing strategies and methodologies for dealing with evolving systems that are becoming more complex because of technological advancement and increased prosperity while also analyzing and anticipating these issues using the same technologies increasing that complexity. Additionally, this includes the creation of statistical mechanics for cognitive systems. We will advance the science and art of urban simulation, which we feel is firmly grounded in a sound understanding of how the city functions in space and time as an economic entity and social artifact. This will entail incorporating our models into new theories of the contemporary city, founded on new economic geography, urban economics, agent-based conceptualizations of social and economic systems, and novel methods of mobility and communication. We must create novel approaches for combining spatial and related databases and will continue to advance data mining techniques for very big data sets in the terabyte range. These will necessitate significant advancements in neural networks, machine learning, and evolutionary computation.

We will produce new ideas for helping cities to reach their full potential through increased intelligence. Smart cities serve as incubators for ever-smarter ideas, and we will illustrate this paradigm in situ with many exemplar cities. Smart cities are competing cities, and we will examine how cities across the spectrum of smartness can adapt to new initiatives and strengthen their competitive edge. We believe that great cities and smart city systems must represent this competitive spirit in an interactive evolutionary environment, ensuring that no city falls behind or advances too quickly. We will create new web-based interactive environments to assist a broader spectrum of citizen activists and groups in comprehending and designing the city and society they have an interest and stake.

Smart cities are also equitable cities. We will create an infrastructure accessible to a wider variety of interests and groups with varying degrees of skill and action, enabling everyone to participate. Our emphasis on efficiency while maintaining a sense of equality is key to this approach. The web-based interactive platforms that we believe are necessary for the citizen science that we believe should be standard in the smart city will allow for the advancement and balance of fairness and competitiveness. We believe that many of the ways we will develop will be founded on concepts about how groups compete and collaborate. The infrastructure, expertise, and data that will distinguish the smart city will make equity easy to construct, and such cities' quality of life will increase.

Acknowledgments

This work was supported in part by National Natural Science Foundation of China under Grants U1909204, 62076237, U19B2029, and U1811463, Chinese Guangdong's S&T project (2019B1515120030, 2020B0909050001), Youth Innovation Promotion Association Chinese Academy of Sciences under Grant 2021130, China Academy of Railway Sciences Corporation Limited Project: RITS2021KF03.

References

Allwinkle, S. and Cruickshank, P. (2011). Creating smart-er cities: an overview. *Journal of Urban Technology* 18: 1–16.

Amini, S., Gerostathopoulos, I., and Prehofer, C. (2017). Big data analytics architecture for real-time traffic control. In: *2017 5th IEEE International Conference on Models & Technologies for Intelligent Transportation Systems*, 710–715. Tum Llcm.

Anagnostopoulos, T., Zaslavsky, A., Kolomvatsos, K. et al. (2017). Challenges and opportunities of waste management in IoT-enabled smart cities: a survey. *IEEE Transactions on Sustainable Computing* 2: 275–289.

Armbrust, M., Fox, A., Griffith, R. et al. (2010). A view of cloud computing. *Communications of the ACM* 53 (4): 50–58.

Atzori, L., Iera, A., and Morabito, G. (2010). The internet of things: a survey. *Computer Networks* 54 (15): 2787–2805.

Bai, Z. and Huang, X. (2012). Design and implementation of a cyber-physical system for building smart living spaces. *International Journal of Distributed Sensor Networks* 2012: 764186-1–764186-9.

Baig, Z.A., Szewczyk, P., Valli, C. et al. (2017). Future challenges for smart cities: cyber-security and digital forensics. *Digital Investigation* 22: 3–13.

BBVA API_Market (2021). http://www.bbvaapimarket.com/en/api-world/how-apis-are-powering-smart-cities/ (accessed on 28 January 2021).

Cañedo, J. and Skjellum, A. (2016). Using machine learning to secure IoT systems. In: *14th IEEE Annual Conference on Privacy, Security and Trust (PST2016)*, 219–222. Auckland, New Zealand: IEEE.

Cao, Y., Li, Y., Liu, X., and Rehtanz, C. (2020). *Cyber-Physical Energy and Power Systems*. Singapore: Springer.

Chen, Y., Ardila-Gomez, A., and Frame, G. (2017). Achieving energy savings by intelligent transportation systems investments in the context of smart cities. *Transportation Research Part D: Transport and Environment* 54: 381–396.

Cheng, B., Longo, S., Cirillo, F., Bauer, M. et al. (2015). Building a big data platform for smart-cities: experience and lessons from santander. In: *2015 IEEE International Congress on Big Data*, 592–599. New York, NY, USA: IEEE.

Cheng, L., Yu, T., Zhang, X., and Yang, B. (2018). Parallel cyber physical social systems based smart energy robotic dispatcher and knowledge automation: concepts, architectures, and challenges. *IEEE Intelligent Systems* 34 (2): 54–64.

Chenxi, H., Yuan, H., Zhang, J.J. et al. (2020). Mid-long term electricity consumption forecasting analysis based on cyber-physical-social system architecture. In: *2020 IEEE 16th International Conference on Automation Science and Engineering (CASE)*, 564–569. IEEE.

Cristea, V., Dobre, C., and Pop, F. (2013). Context-Aware Environments for the Internet of Things.

Da, T., Wang, T., Wang, T. et al. A systematic review and meta-analysis of user acceptance of consumer-oriented health information technologies. *Computers in Human Behavior* 104: 106147.

Das, A., Dash, P., and Mishra, B.K. (2018). An innovation model for smart traffic management system using internet of things (IoTs). In: *Cognitive Computing for Big Data Systems Over IoT: Frameworks, Tools and Applications*, 355–370. Cham, Springer International Publishing.

Data Business Finland (2021). http://www.databusiness.fi/content/uploads/201710/20171109_HarmonisedSmartCityAPIs_WEB.pdf (accessed 27 January 2021).

Diaconita, V. (2018). Hadoop oriented smart cities architecture. *Sensors* 18 (4): 1–20.

EC (2018). Energy and smart cities. https://ec.europa.eu/energy/en/topics/technology-and-innovation/energy-and-smart-cities (accessed 10 December 2018).

EC "Buildings." (2018). https://ec.europa.eu/energy/en/topics/energy-efficiency/buildings (accessed on 10 December 2018).

El-Baz, D. and Bourgeois, J. (2015). Smart cities in Europe and the alma logistics project. *ZTE Communications* 13 (4): 10–15.

Estimote (2018). The Physical World. Software-defined. https://estimote.com (accessed 31 May 2018).

European Commission (2015). Towards an integrated strategic energy technology (SET) plan: accelerating the European energy system transformation. In: *Communication from the Commission C*, 6317 final. Brussels, Belgium: European Commission.

European Data Portal (2021). Open Smart City APIs, Portalul European De Date (europeandataportal.eu). http://www.europeandataportal.eu/ro/news/open-smart-city-apis (accessed 27 January 2021).

Evans, D. (2011). The Internet of Things: How the Next Evolution of the Internet Is Changing Everything.

Finger, M. and Razaghi, M. (2017). Conceptualizing *smart cities*. *Informatik-Spektrum* 40: 6–13.

Gharaibeh, A., Salahuddin, M.A., Hussini, S.J. et al. (2017). Al-fuqaha, smart cities: a survey on data management, security, and enabling technologies. *IEEE Communication Surveys and Tutorials* 19: 2456–2501.

Giffinger, R., Fertner, C., Kramar, H. et al. (2007). *Smart Cities: Ranking of European Medium Sized*. Centre of Regional Science, Vienna University of Technology.

Gimbal (2018). Relevance reimagined. https://gimbal.com (accessed 24 September 2018).

Gohar, M. (2018). SMART TSS: defining transportation system behavior using big data analytics in smart cities. *Sustainable Cities and Society* 41: 114–119.

Gouveia, J.P., Seixa, J., and Giannakidis, G. (2016). Smart city energy planning: integrating data and tools. In: *Proceedings of the 25th International Conference Companion on World Wide Web*, 345–350. New York, NY, USA: Geneva, Association for Computing Machinery.

Gunes, V., Peter, S., Givargis, T., and Vahid, F. A survey on concepts, applications, and challenges in cyber physical systems. *KSII Transactions on Internet and Information Systems* 8 (12).

Guzdial, M. (2009). Education teaching computing to everyone. *Communications of the ACM* 52 (5): 31–33.

Hammerl, B. (2014). Smart City Labs als Möglichkeitsraum für technologische und soziale Innovationen zur Steigerung der Lebens qualität in Städten. *Proceedings of the REAL CORP 2014 Tagungsband*, Vienna, Austria (21–23 May 2014). www.corp.at (accessed on 14 January 2021).

Hammi, B., Khatoun, R., Zeadally, S., and Khoukhi, A.F.L. (2018). IoT technologies for smart cities. *IET Networks* 7 (1): 1–13.

Handte, M., Foell, S., Wagner, S. et al. (2016). An internet of-things enabled connected navigation system for urban bus riders. *IEEE Internet of Things Journal* 3: 735–744.

Hayajneh, A., Zakirul, A.,.M., Bhuiyan, A., and McAndrew, I. (2020). Improving Internet of Things (IoT) security with software-defined networking (SDN). *Computers* 9 (1): 8.

Hayat, P. (2016). Smart cities: a global perspective. *India Q* 72: 177–191.

Hopkins, J. and Hawkin, P. (2018). Big data analytics and IoT in logistics: a case study. *The International Journal of Logistics Management* 29 (2): 575–591.

Jasna S. B., Supriya P., and Nambiar, T.N.P. (2018). Application of Game Theory in Path Planning of Multiple Robots. *2017 IEEE International Conference on Intelligent Computing, Instrumentation and Control Technologies (ICICICT)*.

Jiang, M., Yi, L., Zhu, Z., and Jia, W. (2021). Advances in smart sensing and medical electronics by self-powered sensors based on triboelectric nanogenerators. *Micromachines* 12 (6): 698.

Kashef, M., Visvizi, A., and Troisi, O. (2021). Smart city as a smart service system: human-computer interaction and smart city surveillance systems. *Computers in Human Behavior* 124: 106923.

Kobayashi, A.R.K., Kniess, C.T., Serra, F.A.R. et al. (2017). Smart sustainable cities: bibliometric study and patent information. *International Journal of Innovation and Sustainable Development*. 5.

Komnions, N. (2010). The architecture of intelligent cities. In: *Conference Proceedings Intelligent Environments 06*, 53–61. Institution of Engineering and Technology.

Koop, S.H. and van Leeuwen, C.J. (2017). The challenges of water, waste and climate change in cities. *Environment, Development and Sustainability* 19: 385–418.

Kopackova, H. and Libalova, P. (2017). Smart city concept as socio-technical system. In: *2017 International Conference on Information and Digital Technologies (IDT)*, 198–205. Zilina, Slovakia: IEEE.

Kotenko, I., Saenko, I., Skorik, F., and Bushuev, S. (2015). Neural network approach to forecast the state of the internet of things elements. In: *XVIII IEEE International Conference on Soft Computing and Measurements (SCM2015)*, 133–135. St. Petersburg, Russia: IEEE.

Kyriazopoulou, C. (2015). Smart city technologies and architectures: a literature review. In: *2015 International Conference on Smart Cities and Green ICT Systems (SMARTGREENS)*, 1–12. Lisbon, Portugal: IEEE.

Lacinak, M. and Ristvej, J. (2017). Smart city, safety and security. *Procedia Engineering* 192: 522–527.

Lee, W.-H. and Chiu, C.-Y. (2020). Design and implementation of a smart traffic signal control system for smart city applications. *Sensors* 20 (2): 508.

Mariani, E. and Giacaglia, M.E. (2018). Guidelines for electronic systems designed for aiding the visually impaired people in metro networks. In: *Advances in Human Aspects of Transportation* (ed. N.A. Stanton), 1010–1021. Basel, Switzerland: Springer International Publishing.

Mayer-Schönberger, V. and Cukier, K. (2013). *Big Data: A Revolution that Will Transform How We Live, Work, and Think*. Houghton Mifflin Harcourt.

Melo, S., Macedo, J., and Baptista, P. (2017). Guiding cities to pursue a smart mobility paradigm: an example from vehicle routing guidance and its traffic and operational effects. *Research in Transportation Economics* 65: 24–33.

Menouar, H., Guvenc, I., Akkaya, K. et al. (2017). Uav-enabled intelligent transportation systems for the smart city: applications and challenges. *IEEE Communications Magazine* 55: 28.

Neirotti, P., De Marco, A., Cagliano, A.C. et al. (2014). Current trends in Smart City initiatives: some stylised facts. *Cities* 38: 25–36.

Neyestani, N., Damavandi, M.Y., and Catalão, M.S.-k.J.P.S. (2015). Modeling the PEV traffic pattern in an urban environment with parking lots and charging stations. In: *PowerTech, 2015 IEEE Eindhoven, Eindhoven*, 1–6. IEEE.

Neyestani, N., Damavandi, M.Y., Shafie-Khah, M. et al. (2015). Allocation of plug-in vehicles' parking lots in distribution systems considering network-constrained objectives. *IEEE Transactions on Power Systems* 30: 2643–2656.

Nowakowski, K., P. and Mrówczyńska, B. (2016). How to improve WEEE management? Novel approach in mobile collection with application of artificial intelligence. *Waste Management* 50: 222–233.

Nunes, D.S., Zhang, P., and Silva, J.S. (2015). A survey on human in the loop applications towards an internet of all. *IEEE Communication Surveys and Tutorials* 17 (2): 944–965.

O'Dwyer, E., Pan, I., Acha, S., and Shah, N. (2019). Smart energy systems for sustainable smart cities: current developments, trends and future directions. *Applied Energy* 237: 581–597.

Olshansky, S.J. (2016). The future of smart health. *Computer* 49 (11): 14–21.

Oradea (2021). Oradea. O Strategie Pentru un-Oras, "Inteligent: Oradea Smart City," (Translation: Oradea—A Strategy for a Smart City). http://www.oradea.ro/stiri-oradea/o-strategie-pentru-un-oras-inteligent-oradea-smart-city (accessed 29 December 2020).

Orumwense, E.F. and Abo-Al-Ez, K. (2019). A systematic review to aligning research paths: energy cyber-physical systems. *Cellular Logistics* 6 (1) https://doi.org/10.1080/23311916.2019.1700738.

Patsakis, R., Venanzio, P., Bellavista, A., and Solanas, M.B. (2014). Personalized medical services using smart cities infrastructures. In: *IEEE International Symposium on Medical Measurements and Applications (MeMeA)*, 1–5. Lisboa: IEEE.

Poongodi, M., Sharma, A., Hamdi, M. et al. (2021). Smart healthcare in smart cities: wireless patient monitoring system using IoT. *The Journal of Supercomputing* 1–26.

Qu, F., Wang, F.Y., and Yang, L. (2010). Intelligent transportation spaces: vehicles, traffic, communications, and beyond. *IEEE Communications Magazine* 48 (11): 136–142.

Rathore, M.M., Ahmad, A., Paul, A., and Rho, S. (2016). Urban planning and building smart cities based on the internet of things using big data analytics. *Computer Networks*.

Santos, J. (2018). City of things: enabling resource provisioning in smart cities. *IEEE Communications Magazine* 56 (7): 177–183.

Sleazak, D. (2010). Infobright analytic database engine using rough sets and granular computing. In: *2010 IEEE International Conference on Granular Computing, GrC 2010, San Jose, California, USA*, 432–437.

Smart City Graz (2021). www.smartcitygraz.at (accessed on 30 January 2021).

Smart City Romania Association (2019). http://romaniansmartcity.ro/2017/07/26cluj-napoca-un-viitor-smart-city/ (accessed 18 January 2019).

Solanas (2014). Smart health: a context aware health paradigm within smart cities. *IEEE Communications Magazine* 52 (8): 74–81.

StadtWien (2019). https://smartcity.wien.gv.at/site/buergerinnen (accessed 18 January 2019).

Turner, S. and Uludag, S. (2015). Towards smart cities: interaction and synergy of the smart grid and intelligent transportation systems. Smart Grid: Networking, Data Management and Business Models.

United Nations (2015). The 2030 agenda for sustainable development. https://sustainabledevelopment.un.org/hlpf/2018 (accessed 31 December 2018).

Vera-Baquero, A. (2018). Big data analysis of process performance: a case study of smart cities. *Big Data in Engineering Applications* 44: 41–63.

Wang, F.Y., Zheng, N.N., Cao, D. et al. (2017). Parallel driving in CPSS: a unified approach for transport automation and vehicle intelligence. *IEEE/CAA Journal of Automatica Sinica* 4 (4): 577–587.

White, J., Clarke, S., Groba, C. et al. (2010). R&D challenges and solutions for mobile cyber-physical applications and supporting internet services. *Journal of Internet Services and Applications* 1 (1): 45–56.

Xiong, G., Zhu, F., Liu, X. et al. (2015). Cyber physical social system in intelligent transportation. *IEEE/CAA Journal of Automatica Sinica* 2 (3): 320–333.

Xue, Y. and Yu, X. (2017). Beyond smart grid cyber physical social system in energy future [point of view]. *Proceedings of the IEEE Instrument, Electra, Electronics Engineering* 105 (12): 2290–2292.

Yohanandhan, R.V., Elavarasan, R.M., Manoharan, P., and Mihet-Popa, L. (2020). Cyber-physical power system (CPPS): a review on modeling, simulation, and analysis with cyber security applications. *IEEE Access* 8: 151019–151064.

Yoshikawa, Y., Sato, A., Hirasawa, S. et al. (2012). Hitachi's vision of the smart city. *Hitachi Review* 61 (3): 111–118.

Younis, O. and Moayeri, N. (2016). Cyber-physical systems: a framework for dynamic traffic light control at road intersections. In: *2016 IEEE Wireless Communications and Networking Conference (WCNC) – Doha, Qatar*, 1–6. Doha, Qatar: IEEE, vol. 4, no. 6.

Zhang, Y., Song, B., Du, X., and Guizani, M. (2018). Vehicle tracking using surveillance with multimodal data fusion. *IEEE Transactions on Intelligent Transportation Systems* 19: 2353–2361.

Zhiyong, F., Chao, C., Wang, H., and Wang, Y. (2019). Toward the participatory human-centered community an exploration of cyber-physical public design for urban experience. *IET Cyber-Physical Systems: Theory & Applications* 4 (3): 209–213.

Part VI

Concluding Remarks

20

Conclusion and Perspectives

Anuradha M. Annaswamy[1], Pramod P. Khargonekar[2], Françoise Lamnabhi-Lagarrigue[3], and Sarah K. Spurgeon[4]

[1] Department of Mechanical Engineering, Massachusetts Institute of Technology, Cambridge, MA, USA
[2] Department of Electrical Engineering and Computer Science, University of California, Irvine, CA, USA
[3] CNRS, CentraleSupelec, University of Paris-Saclay, Gif-sur-Yvette, France
[4] Department of Electronic and Electrical Engineering, University College London, London, UK

20.1 Benefits to Humankind: Synthesis of the Chapters and their Open Directions

Cyber–physical–human systems (CPHS) have at their heart seamless integration of the cyber, the physical, and the human (Lamnabhi-Lagarrigue et al. 2017; Netto and Spurgeon 2017). This paradigm produces an important shift in the balance across the three elements which in turn creates both challenges and opportunities. Until relatively recently, technological solutions have focused primarily on delivering engineered products and processes to support humans who are in some sense consumers of what is delivered. The focus of the control community has frequently been on decisions informed by physical sensors and delivered by actuators which are almost entirely physically based. The cyber dimension has also been largely focused on mechanisms for information exchange (including human–computer interfaces) as well as computation. The human perspective has of course been a valuable lens through which we evaluate such systems, but the human has not been viewed as an integral and primary element of the system through design and development to implementation. The CPHS paradigm integrates human aspects into the overall framework with individuals, or even societies, naturally embedded within systems, within controllers and/or within multiagent loops, and have an important part to play as actors within systems throughout the design cycle rather than in a priori evaluation tasks. This change in focus ensures that greater attention is paid to the human benefits, interests, and flourishing, which is essential to create systems appropriate to tackle the challenges of twenty-first century society such as climate change and energy management.

The integration of humans within systems creates many benefits. Safety is an important example of such benefits. In an industrial environment when working collaboratively with robots, however well trained the work force, individuals may experience cognitive fatigue which may have clear implications for workplace safety where humans are engaging in mutual cooperation with unconstrained robotic systems. A management framework such as air traffic control could perhaps be greatly enhanced by considering not only the physical dynamics of multiple air vehicles but also the influence on the system of cognitive fatigue both in pilots and other decision-making authorities. In terms of safety considerations in general, the CPHS paradigm is essential to developing

Cyber–Physical–Human Systems: Fundamentals and Applications, First Edition.
Edited by Anuradha M. Annaswamy, Pramod P. Khargonekar, Françoise Lamnabhi-Lagarrigue, and Sarah K. Spurgeon.

safe solutions to city transport solutions which may involve integration of autonomous vehicles, vehicles driven by humans as well as pedestrians. Appropriate integration of humans in these scenarios is essential to ensuring safety.

In terms of sustainability, the potential benefits are enormous. Humans are embedded in a variety of environments including buildings, cities, and larger infrastructures and have a huge potential to suitably engage with these infrastructures and enable an efficient energy use. The creation of new CPHS also creates exciting opportunities to live and work in new ways that may not only be sustainable but also safe. For example, robots working with humans in an undersea environment may not only reduce environmental impacts but also create safer workspaces in open seas and environmentally friendly ways in undersea environments.

The opportunities afforded by CPHS for healthcare and well-being are considerable. From the need to serve an aging population to the desire to optimally manage critical care facilities, CPHS have a part to play across the entire spectrum in this sector. As the availability of sensors, data, software, and automation is becoming increasingly pervasive, and given that humans are integral to the overall infrastructure, participating in various roles ranging from patients to healthcare professionals, there are several rich opportunities for the CPHS paradigm and framework. A prime example is the realization of personalized medicine which will enhance both outcomes and patient satisfaction and ensure optimal access and equal treatment to all even with scarce resources, varying needs, and disparate constraints.

The potential benefits of CPHS in supporting well-being via integration between the physical and mental state of individuals are very attractive and compelling. Neurological disorders impact affected individuals not just in terms of the impacts of the disorder itself but also in terms of the impact on quality of life. Indeed, there may well be coupling between these elements. One example is the a reduction of spasticity and pain by recovering mobility via wearable robotics and neuro-prostheses thereby leading to an improved satisfaction with daily life which in can increase participation in wider activities creating a virtuous circle of improved health and well-being.

Whatever the application domain, appropriate integration of the human dimension within the CPHS paradigm is thus necessary for continuing societal evolution that embraces many of the opportunities afforded by advances in technology. In all of these sectors, human trust that they will be safe and comfortable within any CPHS is a necessary ingredient for greater human flourishing.

Significant challenges remain in realizing the visions articulated above. Indeed, these challenges inform much of the material in this book as well as future directions for research. The first of these challenges is the highly uncertain nature of individuals and groups of individuals (societies). The areas of fatigue, aging, neurological disorders, and other well-being impacts, which are certainly time varying phenomena and dependent on the specific individual characteristics, have already been noted in the above as has the potential impacts on individuals and societies. Much research is happening in this area as described in Part 2 *Fundamental Concepts and Methods,* including behavioral modeling, the study of social diffusion, the development of various types of inclusive models of humans including phenomena such as collaboration and the integration of complexity across the physical and human space. We recommend that utmost care be exercised in reliance on models to capture human benefits and interests. History of technology is replete with examples of harms to subgroups of humans from advanced technologies where due consideration of essential value of each human being was not properly considered. Clearly, a lot more research along this direction is necessary. Similar challenges remain in all other application domains such as transport, robotics, healthcare, and general sociotechnical systems. In all cases, validation platforms including living laboratory environments where real-time and at scale studies can be carried out across all sectors need to be developed in order for CPHS to realize their full benefit to society.

20.2 Selected Areas for Current and Future Development in CPHS

Earlier chapters in this book capture much of the mainstream research in CPHS. The goal of this section is to discuss a few ideas that complement previous chapters and have the potential to become fruitful directions for future research.

20.2.1 Driver Modeling for the Design of Advanced Driver Assistance Systems

Advanced driver assistance systems have been the subject of much research in recent decades at the disciplinary interface between control engineering and human factors. In essence, it is a question of defining driver models and system design strategies based on knowledge from psychology, behavioral neuroscience, or ergonomics.

The issue of driver assistance systems has come a long way from the introduction of Electronic Stability Program (ESP) and Anti-lock Braking System (ABS) in vehicles to the most recent developments in autonomous vehicles. From a CPHS perspective, engineering and human factors issues should be approached under the scope of vehicle and human dynamics modeling. The postulate is that the better the prediction of the driver's behavior, the better the interaction. This could be ensured by a "gray box" approach, with parameters that are as meaningful and interpretable as possible with respect to the sensorimotor and cognitive processes involved in driving (Mars and Chevrel 2017). The implementation of an approach based on driver models can be considered at different levels of the control of the driving activity. To date, this approach has been applied to steering control modeling, for instance, with application cases like haptic shared control of the steering wheel (Abbink et al. 2012; Mars et al. 2014) and estimation of the driver's distraction state (Ameyoe et al. 2015). It consisted in synthetizing a control law based on a driver model (Saleh et al. 2013) or in ensuring coadaptation between human and automation via the identification of a driver model parameters (Zhao et al. 2020). Some attempt to extend the approach to the modeling of the motorcyclist behavior has also been proposed (Loiseau et al. 2020).

The development of autonomous vehicles is a natural progression of this topic and can be viewed as an ultimate goal and poses myriad challenges. One of them is related to the change of the status of the driver, who is removed from the operational control loop and in some sense transitions to an outer-loop and plays supervisory and tactical roles (Petit et al. 2021). With these new roles of the human, new modeling approaches need to be developed for the emerging CPHS, even if the general problem remains the same: how to make the vehicle's behavior safe, compatible, and understandable for humans. This constitutes an important avenue of future research within CPHS.

20.2.2 Cognitive Cyber–Physical Systems and CPHS

A major goal of cybernetics (Wiener 1948, 1961) was to understand "control and communication in animal and machine." It is one of the major foundations for the field of modern control theory and a central component of cyber–physical systems. As we have seen in previous chapters, the CPHS field is concerned about the various modes of interaction between CPS and humans. We believe that the success of CPHS will be measured, in significant part, by its contributions to the future development of cyber–physical systems that enhance humans as individuals, communities, and societies. This naturally leads to very important issues of ethics which will be covered in Section 20.3.

The field of cognitive science deals with scientific understanding of the nature of human cognition. Ulric Neisser, one of the founders of the field of cognitive psychology, gave one of the most used definitions of cognition (Neisser 1967): "The term 'cognition' refers to all processes by which the sensory input is transformed, reduced, elaborated, stored, recovered, and used." Major cognitive functions include – perception, attention, memory, reasoning, problem-solving, and knowledge representation.

As we think about CPHS, the above definition of cognition can motivate future conceptions of CPHS. We define cognitive CPS (Khargonekar 2019; Mortlock et al. 2022) as cyber–physical systems that have cognitive capacities or functions such as perception, memory, reasoning, problem solving. Indeed, from this point of view, we suggest that the cognitive CPS concept allows us to combine CPS with artificial intelligence (including machine learning) and cognitive neuroscience. Indeed, the fields of artificial intelligence and machine learning have a wealth of approaches to mimicking cognitive functions. These include connectionist approaches such as artificial neural networks as well as logic-and reasoning-based approaches.

These cognitive functions can be *designed* into cyber–physical systems through specific architectures and algorithms. For example, it is easy to imagine a CPS with a memory functionality with some algorithms for storing, updating, and retrieval which will depend on the goal of the CPS (Muthirayan and Khargonekar 2022). In a similar vein, the use of deep learning for computer vision can enable visual perception to be designed into CPS. Reinforcement learning approaches integrated into CPS can provide problem-solving capabilities to cognitive CPS. It is exciting to think about future possibilities in cognitive CPS as our understanding and capabilities to design cognitive functions with sensing, communications, networking, control, and computation improves.

An alternative approach to incorporating cognitive capabilities in CPS is to allow CPS to *evolve* to have these cognitive functions. One can begin a cognitive CPS with some baseline functionality and build in capacity for further development through online learning. Such an approach may well be most suited and even necessary for fully autonomous CPS. It is also possible to speculate about specific evolutionary approaches that are inspired by biological and sociological mechanisms of evolution and development. These mechanisms can be seen at the level of genomes, brains, minds, and cultures.

It is quite likely that both the intentional design as well as evolutionary development paths will be explored and combined. We also envision collections of cognitive CPS.

While there is much interest in artificial general intelligence, there is strong evidence that the most productive directions combine the complementary capabilities of humans and machines. We posit that cognitive CPS as envisioned above can work with humans for much benefit to humans. It is possible to imagine cognitive CPS that communicate effectively with humans. They can also enhance trust between CPS and humans which will be essential for their long-term acceptance. Cognitive CPS can be designed to be such that they can be controlled by humans and not disobey them. Indeed, cognitive CPS can be designed to observe the famous Asimov's laws of robotics.

20.2.3 Emotion–Cognition Interactions

There is currently tremendous investment in developing intelligent personal assistants in the forms of robots, IoT devices, and online chatbots, mobile devices. Apple's Siri, Microsoft's Cortana, Microsoft's XiaoIce, Google Assistant, Facebook M, Amazon's Alexa … are becoming an integral part of our lives. More precisely, a race is underway to design virtual agents with the ability to display human-like emotions, to respond appropriately to human emotional expressions, to understanding their needs, to anticipate them, to establish an emotional connection, to offer support and assistance. Their benefits for behavioral health care could be very important. However, due to the market pressure, many of these current devices already in the market, do not guarantee the nonexistence of serious adverse psychological effects, see for instance (Hudlicka 2017):

– "Our emotions, perhaps even more so than our thoughts, are likely the most personal and private aspects of our lives. The development and use of applications that sense, infer, monitor, or aim to model our emotions therefore presents considerable and as yet unexplored ethical challenges…"

– "Virtual agents acting as coaches in serious therapeutic games may be designed to induce affection, so that they are viewed as empathic, that their message is trusted, and they can be more persuasive."

– "By inducing attachment and trust, virtual affective agents have the potential to mimic aspects of human relationships and humans can thus, at least theoretically, enter into relationships with agents."

Who can measure the consequences of such manipulations of emotions? What about the risk of virtual relationships replacing actual human relationships?

In parallel with this invasion on the market of uncontrolled and immature designs, research is developing slowly, too far upstream. As mentioned by (Shum et al. 2018), "the daunting challenge for such systems is that they must work well in many open domain scenarios. It is also critical to develop failsafe mechanisms to ensure that these systems do not disadvantage and harm anyone, physically or mentally." Symbolic models of cognition–emotion interactions are relevant for research in psychopathology (genetic predispositions, trauma, chronic, or acute stress) and the mechanisms of psychotherapeutic action, as well as their applicability to the development of technologies for behavioral health. Emotions (affective states) and models are for instance introduced in (Scherer 2009): "The event and its consequences are appraised with a set of criteria on multiple levels of processing. The result of the appraisal will generally have a motivational effect, often changing or modifying the motivational state before the occurrence of the event." The current approaches mainly use AI and Deep Learning for designing these virtual agents. The CPHS community has a great role to play in the successful design of these control devices, for instance regarding, see in particular (Hudlicka 2017):

– Sensors design. Current systems of emotion detection are still very rudimentary, recognizing only a few very characteristic emotion from facial expressions, gestures, posture, voice, and messages.

– Modeling "the multiple modalities of emotions, and their interactions. The understanding of these interactions is limited, and most computation-friendly affective theories focus on the cognitive modality. This limits the ability of contemporary emotion models to accurately model the mechanisms of psychopathology and therapeutic action.

– Model validation. The development of validation methods and criteria poses a significant challenge for research models aiming to emulate biological mechanisms mediating affective processing and cognition-emotion interactions."

These approaches need to involve collaborations between emotion researchers in psychology and neuroscience, systems and control researchers, and computational modelers.

20.3 Ethical and Social Concerns: Few Directions

The ongoing developments offered by CPHS technologies are impacting human behavior and society in many ways, and we expect the impact to accelerate. While these technologies have the potential to bring about benefits to humankind, they also give rise to profound ethical and social concerns (Lamnabhi-Lagarrigue and Samad 2023). As such, the design, development, and application of these technologies require careful reflection, debate, and deliberation.

The CPHS control community needs to create awareness of these ethical issues, from the design to the implementation and to be engaged in the ethical issues related to the technical developments.

Open questions are for instance: how the CPHS control community can foresee and establish an understanding of the range of ethical concerns and how can we promote responsible research and innovation? What could be technical actions related to designing control strategies in order to achieve ethical requirements as well as ethical constraints that must be taken into account in the innovation process and design? How to ensure safety and to understand uncertainty in the context of human decision-making. What human states and models can be chosen in a given application domain? What would be the best (data-driven) robust to uncertainty control approaches for mitigating this uncertainty that learn and adapt? How to focus on education and public awareness at all levels, including engineers and scientists, toward the development of ethically aware and responsible products and services?

20.3.1 Frameworks for Ethics

As we think about ethics in the context of CPHS, we can leverage the extensive literature on the well-developed field of ethics in engineering; see, for example, (Martin and Schinzinger 2004), Professional societies such as the National Society of Professional Engineers, the Association for Computing Machinery, the Institute of Electrical and Electronics Engineering, etc., have codes of ethics for their members. These codes offer guidance on ethical issues that will arise in research, development, and deployment of CPHS. In addition, the rapid progress in artificial intelligence and machine learning is giving rise to a large body of literature on ethical issues in AI and ML. A relatively comprehensive summary of ethics in AI can be found in the recent paper by (Fjeld et al. 2020). Recently, IEEE has launched a new initiative entitled Ethically Aligned Design (IEEE 2019). CPHS ethics can be informed by these considerations as well.

CPHS can be rightly regarded as an emerging technology. From this perspective, the relatively recent literature on ethics of emerging technologies is quite relevant. Specifically, the papers by Moor and Brey present very interesting perspectives that can shed light on current and future ethical issues in CPHS. The recent study (Khargonekar and Sampath 2020) builds on these prior works and presents a framework for research, development, and deployment of CPHS. They frame their analysis along two dimensions: (i) stage of development of CPHS and (ii) locus of decision-making. In this two-dimensional framework, they analyze how ethical issues in CPHS can be examined. See also an earlier study by (Thekkilakattil and Dodig-Crnkovic 2015) on ethics of CPS.

While the abovementioned studies offer some foundations for consideration of ethical issues in CPHS, it is quite likely that CPHS will lead to entirely new ethical issues as this research field develops in the coming years. It is expected that artificial intelligence and machine learning technologies will be integrated into future CPHS. In this case, all the issues of ethics in AI and ML will take on new forms as these technologies are integrated into the physical world and human lives. For example, self-driving cars which integrate advanced ML and AI techniques into automobiles have already led to numerous ethical concerns. It is possible to envision new types of data collection, intrusive surveillance, biological and behavioral interventions, and socioeconomic-political control enabled by newer forms of CPHS. These developments will require the research communities to ensure human autonomy, agency, and benefits as novel CPHS technologies across their full life cycle: visioning, research, development, innovation, and large-scale commercialization.

20.3.2 Technical Approaches

In all application domains and all levels of human interaction, key challenges from a control theory perspective include (i) ensuring robustness, security, stability, predictability, learning, and adaptation of the designs, and (ii) understanding safety and uncertainty in the context of human decision-making.

Exploring these notions from a theoretical and dynamic point of view, and characterizations thereof, will be essential in deploying theoretic solutions into real-world applications. These notions, and guarantees obtained via control theory more generally, will involve the development of robust to uncertainty human models likely including data-driven approaches for mitigating this uncertainty. This needed research directions have been introduced in some chapters of this first CPHS book. They open avenues of challenging multidisciplinary research for the young generation.

20.4 Afterword

The Editors of this first CPHS book are very grateful to all the authors for their outstanding chapters where several topics of this emerging discipline at the intersection between engineering and social-behavioral sciences have been nicely described together with some of their future developments. A key objective was to represent the intersection which is central to CPHS across CPS and social-behavioral science.

This book has sought to be more than a collection of research articles and has strived to achieve an appropriate level of consistency and uniformity through making connections between chapters. We hope that this book will become a key resource for the research community, university professors, graduate students, industry experts, and government agencies, who seek to familiarize themselves with the current state of knowledge in CPHS.

The next CPHS book, planned in 2024, will be inspired by the works of the Systems and Control research community, as foreseen in the results of the IEEE CSS Workshop on Control for Societal Challenges, dedicated to develop a scientific roadmap for the future of the Systems and Control discipline, Control for Societal-Scale Challenges: Roadmap 2030, under the leadership of Anuradha M. Annaswamy, Karl H. Johansson, and George J. Pappas (https://ieeecss.org/control-societal-scale-challenges-roadmap-2030). The roadmap features CPHS among six broad methodological challenges that the community needs to address. This next CPHS book is also envisioned to summarize the contributions that will emerge during the IFAC workshops, CPHS2022 and CPHS2024. CPHS2022 (https://www.cphs2022.org) and its successor will focus in particular on modeling, design, analysis, control, verification, and certification of CPHS, including theoretical, algorithmic, computational, and experimental aspects, with emphasis on modeling, analysis, and control of integrated CPHS, the multitude of humans and autonomous elements, social and societal aspects of CPHS, interface transparency between humans and CPS, and dynamic interaction and teaming of humans and autonomy.

References

Abbink, D.A., Mulder, M., and Boer, E.R. (2012). Haptic shared control: smoothly shifting control authority? *Cognition, Technology & Work* 14: 19–28. https://doi.org/10.1007/s10111-011-0192-5.

Ameyoe, A., Chevrel, P., Le Carpentier, E. et al. (2015). Identification of a linear parameter varying driver model for the detection of distraction. *IFAC-PapersOnLine* 48 (26): 37–42. https://doi.org/10.1016/j.ifacol.2015.11.110.

Fjeld, J., Achten, N., Hilligoss, H., et al. (2020). Principled artificial intelligence: mapping consensus in ethical and rights-based approaches to principles for AI, Berkman Klein Center Research Publication No. 2020-1.

Hudlicka, E. (2017). Computational modeling of cognition–emotion interactions: theoretical and practical relevance for behavioral healthcare. In: *Emotions and Affect in Human Factors and Human-Computer Interaction* (ed. M.P. Jeon), 383–436. Springer.

IEEE (2019). Ethically Aligned Design, IEEE Press. https://standards.ieee.org/news/2017/ead_v2.html.

Khargonekar, P.P. (2019). Cognitive cyber-physical systems: vision for the next CPS frontier. Presentation presented at the ESWEEK 2019, New York. https://cpb-us-e2.wpmucdn.com/faculty.sites.uci.edu/dist/8/644/files/2019/10/PPK_ESWeek_4.pdf

Khargonekar, P.P. and Sampath, M. (2020). *A framework for ethics in cyber-physical-human systems,* IFAC World Congress 2020. *IFAC-PapersOnLine* 53 (2): 17008–17015. https://faculty.sites.uci.edu/khargonekar/files/2019/11/Ethics_CPHS.pdf.

Lamnabhi-Lagarrigue, F. and Samad, T. (2023). Social, orgsanizational, and individual impacts of automation. In: *Handbook of Automation* (ed. S.Y. Nof). Springer (in press).

Lamnabhi-Lagarrigue, F., Annaswamy, A., Engell, S. et al. (2017). Systems and control for the future of humanity, research agenda: current and future roles, impact and grand challenges. *Annual Reviews in Control* 43: 1–64. Part 4.7: Cyber-Physical and Human Systems, contributed by Ruzena Bajcsy and Mariana Netto.

Loiseau, P., Boultifat, C.N.E., Chevrel, P. et al. (2020). Rider model identification: neural networks and quasi-LPV models. *IET Intelligent Transport Systems* 14 (10): 1259–1264. https://doi.org/10.1049/iet-its.2020.0088.

Mars, F. and Chevrel, P. (2017). Modelling human control of steering for the design of advanced driver assistance systems. *Annual Reviews in Control* 44: 292–302. https://doi.org/10.1016/j.arcontrol.2017.09.011.

Mars, F., Deroo, M., and Hoc, J.M. (2014). Analysis of human-machine cooperation when driving with different degrees of haptic shared control. *IEEE Transactions on Haptics* 7 (3): 324–333. https://doi.org/10.1109/TOH.2013.2295095.

Martin, M.W. and Schinzinger, R. (2004). *Ethics in Engineering*, 4e. McGraw-Hill Education.

Mortlock, T., Muthirayan, D., Yu, S.-Y. et al. (2022). Graph learning for cognitive digital twins in manufacturing systems. *IEEE Transactions on Emerging Topics in Computing* 10 (1): 34–45.

Muthirayan, D. and Khargonekar, P.P. (2022). Memory augmented neural network adaptive controllers: performance and stability. *IEEE Transactions on Automatic Control* 68 (2): 825–838.

Neisser, U. (1967). *Cognitive Psychology*: Classic Edition, 2014, (First Edition 1967). New York: Psychology Press.

Netto, M. and Spurgeon, S. (ed.) (2017). Special section on CPHS. *Annual Reviews in Control* 44 (2017): 249–374. (Special Section Eds.)

Petit, J., Charron, C., and Mars, F. (2021). Risk assessment by a passenger of an autonomous vehicle among pedestrians: relationship between subjective and physiological measures. *Frontiers in Neuroergonomics* 2: 682119. https://doi.org/10.3389/fnrgo.2021.682119.

Saleh, L., Chevrel, P., Claveau, F. et al. (2013). Shared steering control between a driver and an automation: stability in presence the of driver behavior uncertainty. *IEEE Transactions on Intelligent Transportation Systems* 14 (2): 974–983. https://doi.org/10.1109/TITS.2013.2248363.

Scherer, K.R. (2009). Emotions are emergent processes: they require a dynamic computational architecture. *Philosophical Transactions of the Royal Society B* 364 (1535): 3459–3474.

Shum, H.Y., He, X.D., and Li, D. (2018). From Eliza to XiaoIce: challenges and opportunities with social chatbots. *Frontiers of Information Technology and Electronic Engineering* 19: 10–26. https://doi.org/10.1631/FITEE.1700826.

Thekkilakattil, A. and Dodig-Crnkovic, G. (2015). Ethics aspects of embedded and cyber-physical systems. *IEEE 39th Annual Computer Software and Applications Conference (COMPSAC)*, 39–44, Taichung, Taiwan (1–5 July 2015).

Wiener, N. (1948). *Cybernetics or Control and Communication in the Animal and the Machine*. Paris: Hermann & Cie.

Wiener, N. (1961). *Cybernetics, or Control and Communication in the Animal and the Machine*, 2e. Cambridge, Massachusetts: The MIT Press.

Zhao, Y., Chevrel, P., Claveau, F., and Mars, F. (2020). Towards a driver model to clarify cooperation between drivers and haptic guidance systems. *2020 IEEE International Conference on Systems, Man, and Cybernetics (SMC)*, 1731–1737, Toronto, ON, Canada (11–14 October 2020).

Index

Cyber–Physical–Human Systems: Fundamentals and Applications, First Edition.
Edited by Anuradha M. Annaswamy, Pramod P. Khargonekar, Françoise Lamnabhi-Lagarrigue, and Sarah K. Spurgeon.
© 2023 The Institute of Electrical and Electronics Engineers, Inc. Published 2023 by John Wiley & Sons, Inc.

.

Printed and bound by CPI Group (UK) Ltd, Croydon, CR0 4YY

27/10/2024

14580679-0005